IUTAM-ISIMM SYMPOSIUM ON MATHEMATICAL MODELING AND PHYSICAL INSTANCES OF GRANULAR FLOWS

To learn more about AIP Conference Proceedings, including the
Conference Proceedings Series, please visit the webpage
http://proceedings.aip.org/proceedings

IUTAM-ISIMM SYMPOSIUM ON MATHEMATICAL MODELING AND PHYSICAL INSTANCES OF GRANULAR FLOWS

Reggio Calabria, Italy   14 – 18 September 2009

EDITORS

Joe D. Goddard
University of California, San Diego
La Jolla, California

James T. Jenkins
Cornell University
Ithaca, New York

Pasquale Giovine
Mediterranean University of Reggio Calabria
Reggio Calabria, Italy

SPONSORING ORGANIZATIONS
International Union of Theoretical and Applied Mathematics
The Italian Institute of Higher Mathematics
U. S. National Science Foundation
Regional Council of Reggio Calabria
Province of Reggio Calabria
Mediterranean University of Reggio Calabria
International Society for the Interaction of Mechanics and Mathematics

Melville, New York, 2010
AIP CONFERENCE PROCEEDINGS ■ VOLUME 1227

Editors:

Prof. Joe D. Goddard
Department of Mechanical and Aerospance Engineering
University of California, San Diego
9500 Gilman Drive
La Jolla, CA 92093-0411

E-mail: jgoddard@ucsd.edu

Prof. James T. Jenkins
School of Civil and Environmental Engineering
Cornell University
Ithaca, NY 14853

E-mail: jtj2@cornell.edu

Prof. Pasquale Giovine
Dipartimento di Meccanica e Materiali
Università Mediterranea di Reggio Calabria
Via Graziella, 1
Località Feo di Vito
I-89125 Reggio di Calabria
Italy

E-mail: giovine@unirc.it

Authorization to photocopy items for internal or personal use, beyond the free copying permitted under the 1978 U.S. Copyright Law (see statement below), is granted by the American Institute of Physics for users registered with the Copyright Clearance Center (CCC) Transactional Reporting Service, provided that the base fee of $30.00 per copy is paid directly to CCC, 222 Rosewood Drive, Danvers, MA 01923. For those organizations that have been granted a photocopy license by CCC, a separate system of payment has been arranged. The fee code for users of the Transactional Reporting Services is: ISBN/978-0-7354-0772-5/10/$30.00

© 2010 American Institute of Physics

Permission is granted to quote from the AIP Conference Proceedings with the customary acknowledgment of the source. Republication of an article or portions thereof (e.g., extensive excerpts, figures, tables, etc.) in original form or in translation, as well as other types of reuse (e.g., in course packs) require formal permission from AIP and may be subject to fees. As a courtesy, the author of the original proceedings article should be informed of any request for republication/reuse. Permission may be obtained online using Rightslink. Locate the article online at http://proceedings.aip.org, then simply click on the Rightslink icon/"Permission for Reuse" link found in the article abstract. You may also address requests to: AIP Office of Rights and Permissions, Suite 1NO1, 2 Huntington Quadrangle, Melville, NY 11747-4502; Fax: 516-576-2450; Tel.: 516-576-2268; E-mail: rights@aip.org.

L.C. Catalog Card No. 2010901600
ISBN 978-0-7354-0772-5
ISSN 0094-243X
Printed in the United States of America

CONTENTS

Preface ... ix
List of Participants ... xi
Committees and Sponsors ... xiii

VISCO-PLASTICITY AND MICROMECHANICS OF RATE-DEPENDENT FLOW

The Plastic Potential, Double-Slip, Double-Spin and Viscoplasticity ... 3
D. Harris

Nonequilibrium Liquid Theory for Sheared Granular Liquids ... 19
H. Hayakawa, S.-H. Chong, and M. Otsuki

Steady, Inclined Flow of a Mixture of Grains and Fluid over a Rigid Base ... 31
J. T. Jenkins and D. Berzi

Microstructure and Particle-Phase Stress in a Dense Suspension ... 41
J. T. Jenkins and L. La Ragione

From Powders to Collapsing Soil/Living Quicksand: Discrete Modeling and Experiment ... 50
D. Kadau

The Hard-Particle Model for a Dense Granular Flow Down an Inclined Plane ... 58
V. Kumaran

On the Hyperbolicity of a Two-Fluid Model for Debris Flows ... 72
C. Mineo and M. Torrisi

Rheology of Confined Granular Flows ... 79
P. Richard, A. Valance, J.-F. Métayer, J. Crassous, M. Louge, and R. Delannay

Temporal Dynamics in Density Relaxation ... 89
A. D. Rosato, V. Ratnaswamy, D. J. Horntrop, O. Dybenko, and L. Kondic

Solid-Fluid Transition and the Formation of Ripples in Vertically Oscillated Granular Layers ... 100
O. Sano

Internal Erosion in Gas-Flow Weak Conditions ... 115
H. Steeb

A New Approach Based on Langevin Type Equation for Driven Granular Gas under Gravity ... 135
J. Wakou and M. Isobe

ELASTO-PLASTICITY AND MICROMECHANICS OF RATE-INDEPENDENT FLOW

Non-Polar Continuum Modeling, Experiment, Numerical Simulation, Statistical Mechanics and Statistical Physics

Response of a Cohesionless Packing to a Point Load ... 151
E. H. B. Amar, D. Clamond, N. Fraysse, and J. Rajchenbach

kGamma Distributions in Granular Packs ... 157
T. Aste, G. W. Delaney, and T. Di Matteo

Stress Transmission and Incipient Yield Flow in Dense Granular Materials ... 167
R. Blumenfeld

Incremental Response of a Model Granular Material by Stress Probing with DEM Simulations ... 183
F. Froiio and J.-N. Roux

On the Coarse-Graining of Grains ... 198
I. Goldhirsch

Macroscopic Stress from Dynamic, Rotating Granular Media 208
    S. Luding

Simple Interaction Model for Partially Wet Granular Materials 214
    N. Mitarai and H. Nakanishi

Shear Induced Diffusion in Dense Granular Flows 221
    A. V. Orpe, C. H. Rycroft, and A. A. Kudrolli

Equilibrium of Granular Clusters: Influence of Boundary Curvature and Contact Properties 230
    R. Pignatelli

Force Transmission in Cohesive Granular Media 240
    F. Radjai, V. Topin, V. Richefeu, C. Voivret, J.-Y. Delenne, E. Azéma, and S. El Youssoufi

How Granular Materials Deform in Quasistatic Conditions 260
    J.-N. Roux and G. Combe

Theory of Random Packings 271
    C. Song, P. Wang, and H. A. Makse

A Plasticity Model with Microstructure Evolution for Quasistatic Granular Flows 280
    J. Sun and S. Sundaresan

Polar Continuum Modeling: Experiment and Numerical Simulation

Modelling Limit States within the Framework of Hypoplasticity 290
    E. Bauer

**Homogenisation of Discrete Media towards Micropolar Continua:
A Computational Approach** 306
    W. Ehlers

Remarks on Constitutive Laws for Dry Granular Materials 314
    P. Giovine

Granular Hypoplasticity with Cosserat Effects 323
    J. D. Goddard

The Stress in a Slowly Sheared Granular Column 333
    V. Mehandia and P. R. Nott

SEGREGATION

Particle Size Segregation in Granular Avalanches: A Brief Review of Recent Progress 343
    J. M. N. T. Gray

Size Segregation in Dry Granular Flows of Binary Mixtures 363
    M. Larcher and J. T. Jenkins

The Gray-Thornton Model of Granular Segregation 371
    M. Shearer, L. B. H. May, N. Giffen, and K. E. Daniels

**Preliminary Investigations on the Rheology and Boundary Stresses Associated with Granular
Mixtures** 379
    B. Yohannes and K. M. Hill

WAVES

**Micro-Macro Transition and Linear Wave Propagation in Three-Component Compacted Granular
Materials** 391
    B. Albers

Three-Dimensional Lattice-Based Dispersion Relations for Granular Materials 405
    N. P. Kruyt

The Waveguide Theory for Booming Sand Dunes 416
    N. M. Vriend and M. L. Hunt

Wave Propagation in Strongly Nonlinear Two-Mass Chains 425
    S.-Y. Wang, E. B. Herbold, and V. F. Nesterenko

ABSTRACTS

An Advanced Numerical Method to Describe Order Dynamics in Nematics 437
    A. Amoddeo, G. Lombardo, and R. Barberi

Mixing Equilibrium in Two-Density Fluidized Beds by DEM 438
    A. Di Renzo and F. P. Di Maio

Plane Waves in Porous Media 439
    L. Fiorino

Granular Jets and Hydraulic Jumps on an Inclined Plane 440
    C. G. Johnson and J. M. N. T. Gray

Coupled ODEs Model for the Dynamics of Dunes 441
    H. Nishimori, A. Katsuki, H. Sakamoto, and H. Niiya

Saltating Particles in a Turbulent Boundary Layer 442
    A. Valance, A. O. El Moctar, P. Dupont, I. Cantat, and J. T. Jenkins

Author Index 443

Preface

Joint IUTAM-ISIMM Symposium on Mathematical Modeling and Physical Instances of Granular Flows
Reggio Calabria, Italy, 14-18 September 2009

The following is a collection of papers presented at the above symposium, which was conducted under the auspices of the International Union of Theoretical and Applied Mechanics (IUTAM) and the International Society for the Interaction of Mechanics and Mathematics (ISIMM). A list of participants and committee members is given below.

Representing the main themes of the symposium, papers are grouped topically in the Table of Contents as follows:

Visco-plasticity and micromechanics of rate-dependent flow

Elasto-plasticity and micromechanics of rate-independent flow
> Non-polar continuum modeling: experiment, numerical simulation, statistical mechanics and statistical physics
>
> Polar continuum modeling: experiment and numerical simulation

Segregation

Waves

Abstracts of other papers

The last category includes papers that are being submitted for publication elsewhere or that were deemed tangentially relevant to the principal themes of the symposium. The full-length papers range from research summaries to survey-type articles. This body of work should constitute a valuable resource to research students and researchers in the fields of granular mechanics and physics, geomechanics and continuum mechanics.

The Symposium was made possible by financial support from the International Union of Theoretical and Applied Mechanics, the Gruppo Nazionale di Fisica Matematica of the Istituto Nazionale di Alta Matematica, the U.S. National Science Foundation, the Presidenza del Consiglio Regionale della Calabria, the Provincia di Reggio Calabria, the Università "Mediterranea", the Faculty of Engineering and the Department of Mechanics and Materials of Reggio Calabria.

This volume is dedicated to the memory of Professor Ioannis Vardoulakis, member of the Scientific Committee, and a major force in civil engineering and mechanics. He will be particularly remembered for his scholarship, collegiality, and visionary applications of structured continuum models to granular mechanics and geomechanics. His untimely passing deprived us of what promised to be an insightful contribution to the Symposium, prior to which he had expressed the wish "to present some thoughts that puzzle me these days concerning the affinity of Cosserat continuum theory and the behavior of granular matter."

The Editors

J.D. Goddard
J.T. Jenkins
P. Giovine

Participants

Prof. Andreas Acrivos
Levich Institute
City College of New York
U.S.A.

Dr. Antonino Amoddeo
Dip. Meccanica e Materiali
U. Mediterranea Reggio Calabria
Italy

Dr. Diego Berzi
DIIAR - Sez. Ing. Idraulica
Politecnico di Milano
Italy

Dr. Vito Cimmelli
Dept. Math. & Computer Sci.
U. Basilicata, Potenza
Italy

Dr. Lucia Fiorino
Dip. Meccanica e Materiali,
Universit di Reggio Calabria,
Italy

Prof. Joe Goddard
Dept. Mech. & Aerospace Eng.
UC San Diego
U.S.A.

Dr. David Harris
School of Mathematics
University of Manchester
U.K.

Prof. Kolumban Hutter
c/o Lab. Hydraulics, Hydrology
& Glaciolology, ETH Zürich
Switzerland

Dr. Bettina Albers
Fach. Grundbau & Bodenmech.
Technische Universität Berlin
Germany

Dr. Tomaso Aste
Res. School of Physical Sci.
& Eng., A. N. U.
Australia

Dr. Rafi Blumenfeld
Earth Sci. & Eng., Imperial
College & Cavendish Lab.,
Cambridge U.
U.K.

Dr. Alberto Di Renzo
Dip. Ing. Chimica e Materiali
Università della Calabria
Italy

Dr. Francesco Froiio
Tribol. et Dynam. des Systèmes
Ecole Centrale de Lyon
France

Prof. Isaac Goldhirsch
School of Mech. Eng.
Tel Aviv University
Israel

Dr. Hisao Hayakawa
Yukawa Inst. Theo. Physics
Kyoto University
Japan

Dr. Salvatore Iacono
Department of Mathematics
University of Messina
Italy

Mr. Sotirios Alevizos
(correspond.) Lab. Geomaterials
N.T.U. Athens
Greece

Prof. Erich Bauer
Institut für Baumechanik
Technical University of Graz,
Austria

Prof. Gianfranco Capriz
Dipartimento di Matematica
Università di Pisa
Italy

Prof. Wolfgang Ehlers
Institut für Mechanik
Universität Stuttgart
Germany

Prof. Pasquale Giovine
Dip. Meccanica e Materiali
U. Mediterranea Reggio Calabria
Italy

Prof. Nico Gray
School of Mathematics
University of Manchester
U.K.

Dr. Kimberly Hill
Department of Civil Engineering
University of Minnesota
U.S.A.

Dr Masaharu Isobe
Graduate School of Engineering
Nagoya Institute of Technology
Japan

Prof. James Jenkins
Dept. Theo. & Appl. Mech.
Cornell University
U.S.A.

Mr. Christopher Johnson
School of Mathematics
University of Manchester
U.K.

Dr. Dirk Kadau
Institute for Building Materials
ETH Zürich
Switzerland

Dr. D.V. Khakhar
Dept. Chemical Engineering
I.I.T. Bombay
India

Dr. Niels Kruyt
Dept. Mechanical Engineering
Universiteit Twente
The Netherlands

Prof. V. Kumaran
Dept. Chemical Engineering
Indian Institute of Science
India

Dr. Luigi LaRagione
Dip. Ingegneria Civile e
Ambientale
Politecnico di Bari, Italy

Dr. Michele Larcher
Dip, Ingegneria Civile e
Ambientale
Università di Trento, Italy

Dr. Stefan Luding
Dept. of Mechanical Engineering
Universiteit Twente
The Netherlands

Dr. Hernan Makse
Physics Department
City College of New York
U.S.A.

Dr. Carmen Mineo
Dept. Math. & Computer Sci.
Università di Catania
Italy

Dr. Namiko Mitarai
Niels Bohr Institute,
University of Copenhagen
Denmark

Prof. Vitali Nesterenko
Dept. Mech. & Aerospace Eng.
UC San Diego
U.S.A.

Prof. Hiraku Nishimori
Dept. Maths. & Life Sci.
Hiroshima University
Japan

Prof. Prabhu Nott
Dept. Chemical Eng.
Indian Institute of Science
India

Dr. Ashish Orpe
Chemical Engineering Division
National Chemical Laboratory,
India

Ing. Rossella Pignatelli
Dip. Ingegneria Strutturale
Polytecnico Milano
Italy

Dr. Thorsten Pöschel
Eng. Adv. Materials
U. Erlangen-Nürnberg
Germany

Dr. Farhang Radjai
Lab. Mécanique et Genie Civil
CNRS, U. Montpellier 2
France

Dr. Jean Rajchenbach
Lab. Phys. Mat. Condenseé
U. Nice, Sophia Antipolis
France

Mr. Patrick Richard
Institut de Physique de Rennes
Université de Rennes
France

Dr. Pierre Rognon
School of Civil Engineering
The University of Sydney
Australia

Dr. Anthony Rosato
Dept. Mech. and Indust. Eng.
New Jersey Inst. Technology
U.S.A.

Dr. Jean-Noël Roux
Unité de Recherches Navier
Université Paris Est
France

Prof. Osamu Sano
Department of Applied Physics
Tokyo U. Agriculture & Tech.
Japan

Prof. Michael Shearer
Department of Mathematics
North Carolina State University,
U.S.A.

Prof. Holger Steeb
Institut für Mechanik
Ruhr-Universität Bochum,
Germany

Dr. Jin Sun
Dept. Chemical Engineering
Princeton University
U.S.A.

Prof. Sankaran Sundaresan
Dept. Chemical Engineering
Princeton University
U.S.A.

Dr. Alexandre Valance
Insitut de Physique de Rennes
Université de Rennes I
France

Dr. Nathalie Vriend
Ctr. Mathematical Sciences
University of Cambridge
U.K.

Prof. Krzysztof Wilmanski
Depart. Building Engineering
U. Zielona Gora
Poland

Scientific Committee

G. Capriz (Italy)   R. Connelly (U.S.A.)   J. D. Goddard (U.S.A., Chairman)
J. T. Jenkins (U.S.A.)   O. Sano (Japan)   I. Vardoulakis (Greece)
K. Wilmanski (Poland)   C. Cercignani (Italy, IUTAM representative)

Editorial Subcommittee

G. Capriz (Associate)   P. Giovine (Editor)   J. D. Goddard (Editor)
J. T. Jenkins (Editor)   O. Sano (Associate)   K. Wilmanski (Associate)

Local Organizing Committee

Antonino Amoddeo   Michele Buonsanti   Pasquale Giovine (Presidente)
Nicola Moraci   Francesco Oliveri   Adolfo Santini

Sponsors

International Union of Theoretical and Applied Mechanics

Università degli Studi
Mediterranea di Reggio Calabria

Dipartimento di
Meccanica e Materiali

Gruppo Nazionale di
Fisica Matematica, INDAM

Provincia di Reggio Calabria

U.S. National
Science Foundation

Presidenza del Consiglio
Regionale della Calabria

VISCO-PLASTICITY AND MICROMECHANICS
OF RATE-DEPENDENT FLOW

The Plastic Potential, Double-slip, Double-spin and Viscoplasticity

David Harris

School of Mathematics, University of Manchester, UK

Abstract. In this paper we describe two classical models for rate-independent behaviour of granular materials, namely the plastic potential and the double shearing model, emphasising their ill-posedness. We then describe a model, called the double-slip and double-spin model which generalises the plastic potential model and is closely related to the double shearing model. This new model eliminates the causes of the ill-posedness in the classical models and provides a suitable basis for the analysis of the deformation and flow of granular materials in the rate-independent regime.

There has been considerable recent interest in the intermediate regime between densely-packed, rate-independent, quasi-static flow and the rate-dependent dilute gaseous regime. In this intermediate regime the material also exhibits a degree of rate-dependence. The natural extension of a rate-independent plasticity model to incorporate rate-dependent material behaviour is by way of viscoplasticity. The archetypal example here is the Bingham material which generalises a von Mises type plasticity model and introduces a viscosity parameter into the model. We propose an extension of the double-slip and double-spin model to incorporate viscosity, thereby extending the range of the model to incorporate rate-dependent behaviour. The new model is then applied to a simplified problem of pipe flow.

Keywords: rate-dependence, rate-independence, plasticity, plastic potential, double-shearing, double-slip and double-spin, pipe flow
PACS: 46.35+z,47.57.Gc,62.20fq

INTRODUCTION

It is a common-place that granular materials may exhibit mechanical behaviour analogous to all three thermodynamic phases, i.e. they may behave in a solid-like, liquid-like or gaseous-like manner in response to their environment. In this paper we shall initially consider equations which are intended to govern solid-like behaviour of granular materials and then use these equations as a basis for extending the model to incorporate liquid-like behaviour. For slow loading under quasi-static conditions an appropriate class of model to use is the elastic-plastic or rigid-plastic model in which the material response is rate-independent. An example of such a model is the plastic potential model and in this case the constitutive relations are characterised by two scalar valued functions of the stress tensor.

Although much effort has been put into the development of such models, see for example [1], [2], particularly in the civil and geotechnical engineering community, they are not without their problems. Since their major use is for quasi-static loading the inertia terms are omitted from the equations of motion, thereby preventing the true evolution of the system from being calculated. When the inertia terms are incorporated, in the case of the plastic potential model in which the yield function is distinct from the plastic potential (a non-associated flow rule) the model is ill-posed for the Cauchy initial value problem, see [3] and [4], in the sense that the solution is discontinuous with regard to the initial data.

An alternative type of rigid-plastic model is the physically or kinematically-based flow rules based upon the concept of double sliding or shearing for planar flows. In the original formulation due to de Josselin de Jong, [5], the double sliding free rotating model, the kinematic hypothesis assumed that the deformation comprises the superposition of two simultaneous shears and two simultaneous spins. One of the spins was obtained from the anti-symmetric part of the velocity gradient tensor (the spin or half the vorticity) while the other was due to a primitive kinematic quantity called a free rotation. This latter quantity was both indeterminate and indeterminable. The resulting theory consisted of partial differential inequalities. In order to render the model determinate, Spencer, [6], proposed the double shearing model in which the primitive free rotation quantity was replaced by a derived quantity associated with the stress tensor, namely the time rate of change of the angle that the greater principal stress direction makes with the positive x-axis. This successfully reduced the model to one consisting of partial differential equations, but also reduced the degree of freedom that the material may exhibit. Although it is not emphasised in its name, the double shearing model is also based on a double simultaneous rotation, namely the rate of rotation of the principal stress axes and the spin. The double shearing model is also ill-posed, see [4].

In both cases, the ill-posedness is due to a loss of hyperbolicity caused by the way the terms appearing in the equations interact. Plasticity models have an unusual mathematical structure in that the planar steady-state equations may be either hyperbolic, parabolic or elliptic depending upon the properties of the yield and plastic potential functions. It is argued very strongly here that plasticity models are best used with a yield function that, for steady-state planar flows, gives rise to spatial hyperbolicity. For quasi-static flows of the plastic potential model, the contributions from the stress equilibrium equations and the flow rule uncouple and give rise to distinct velocity and stress characteristics. For an associated flow rule (identical yield and plastic potential functions) the stress and velocity characteristics coincide whereas for a non-associated flow rule they are distinct. For time dependent flows the characteristics remain unaltered from the steady-state case and so, for rigid-plastic flows, the model is parabolic and a proper formulation requires boundary conditions.

In the case of the double shearing model the cause of the ill-posedness is the presence of the stress-rate in the equations. The system is elliptic.

One way to eliminate these causes of ill-posedness is to (a) use an alternative spin to that of the principal axes of stress and (b) to ensure the coincidence of the stress and velocity characteristics. Such a method has been used in [7]. In the planar case, terms which are formally identical to terms appearing in the double shearing model (the "non-coaxial" terms) are employed to ensure that, if they exist, then the stress and velocity characteristics coincide. However, instead of the in-plane spin of the principal axes of stress, the model is embedded into a Cosserat continuum and the intrinsic spin vector is used instead. This enables the existence of the stress and velocity characteristics to be assured.

In both the kinematically-based or plastic potential based models, the yield function is a scalar function of the stress components and the yield condition an inequality satisfied by the values of this function which may be assumed to take only non-positive values. A simple method of introducing rate-dependent material behaviour is to allow the material to go "beyond" yield, i.e. to allow the yield function to take on positive values while also introducing a new material parameter, the viscosity, into the constitutive relation. In this way, rate-dependent effects and fluid-like characteristics may be introduced into the model, which is called a rigid/viscoplastic model.

In this paper, we incorporate two types of dissipation into the model, namely, in addition to the rate-independent Coulomb solid friction we incorporate a rate-dependent viscosity. Thus, we generalise a Bingham fluid in two ways, firstly we incorporate a pressure-dependent yield function (instead of using the pressure independent von Mises yield function) and secondly, an extra term is incorporated, a "non-coaxial" term, into the constitutive relation. Rate-dependence and dynamic plasticity, including a discussion of the Bingham fluid, is dealt with in the book by Cristescu [8].

The plan of the paper is as follows, firstly, we describe two classical models for rate-independent plasticity and present a closely related model which is well-posed. Secondly, we give a derivation of an extension to this model which incorporates viscosity. The intended application of the new model is to either single or multi-phase flows in which the solids fraction is high (dense flows), but either the presence of a fluid phase means that the mixture exhibits rate-dependent behaviour, or for a single phase granular flow, the material is "just" beyond yield, by which we mean that the material is still densely, but not tightly, packed and the nature of the contacts between grains contains a proportion of non-impulsive, finite duration, sliding and rolling contacts; the remaining contacts comprising impulsive, almost instantaneous collisions. Intuitively, the assumption is that there exists a sufficiently high proportion of contacts which are still in yield in order to be able to use the yield point as a scalar measure of the "distance" of the total stress state from the yield response. Thirdly, we consider an analogue of a well-known Bingham-type problem, and analyse a particular type of granular system undergoing pipe-flow. Finally we consider boundary conditions for the system which may, or may not, be suitable for this problem.

MATHEMATICAL FORMULATION - CONTINUUM MECHANICS

The role of continuum mechanics in the mechanics of granular materials has proven difficult to evaluate. The reason is evident. The discrete nature of real granular materials together with a natural length scale large in comparison with those of physical situations in which continuum mechanics has provided a successful theoretical framework brings into question the appropriateness of a model based upon continuous field quantities. The rapid expansion of computing power has also been a powerful stimulus to the development and exploitation of discrete models.

For small scale systems comprising a small number of grains, a discrete description is plainly required. However, for a sufficiently large number of grains the methods of continuum mechanics not only become feasible but for sufficiently large systems, also necessary, as the number of calculations required to solve a problem outstrips the available computing power.

However, the problems with continuum modelling have been exacerbated by the fact that while a standard continuum appears to suffice for the description of many of the phenomena exhibited by granular materials, no classical continuum model has been universally accepted as an adequate basis for their description. Since much effort has been put into such models one may conjecture that a standard continuum is, in fact, insufficient, for otherwise it is likely to already have been found. One possible way out of this problem is to use an enhanced continuum and indeed, here, we adopt a Cosserat continuum. However, Cosserat effects appear to be somewhat rather elusive to verify experimentally and consequently, we adopt here the following tentative hypothesis. The ill-posedness of the classical models is due to the inadequate nature of the classical continuum which requires enriching with quantities associated with rotation. However, we put forward the hypothesis that in many situations the magnitudes of the quantities missing from the standard continuum are small and not important for the overall material response, that, in fact, the crucial role for the Cosserat quantities is to stabilise the material response. This hypothesis is not intended to hold universally, in situations where there is significant net grain rotation such as in the interior of a shear band the Cosserat quantities will be important and need to be taken into account. In this paper we emphasise the role of the quantities appearing in a standard continuum.

An Eulerian description of the motion is adopted and we use the following notation. Let **x** denote the position vector of a spatial point relative to an origin O. Let **v** denote the velocity at place **x** and time t. The symmetric part of velocity gradient tensor, denoted by **d**, is called the deformation-rate tensor and the anti-symmetric part of the velocity gradient tensor, denoted by **s**, is called the spin tensor. The components of **s** are a half times the vorticity vector components.

The first Cosserat quantity that we introduce is the intrinsic spin vector, denoted by Ω, at place **x** and time t. The anti-symmetric tensor dual to the intrinsic spin vector is denoted by Ω_a. The intuitive physical interpretation of the intrinsic spin is that it represents a spatial average (over a suitably defined representative volume element) of the angular velocities of the grains. Neither the spin nor the intrinsic spin are suitable quantities to appear in a constitutive relation since they are not frame indifferent. However, their difference is frame indifferent and we define the *relative spin* or *relative vorticity* to be the spin (i.e. half the vorticity) relative to the intrinsic spin. It is denoted by ω and defined by

$$\omega = \mathbf{s}^T - \Omega_a. \tag{1}$$

where the superscript ^{T} denotes transpose. Thus ω is an anti-symmetric tensor. It is objective (it is defined as above to emphasise that physically, it is the difference between two spins and writing it in this way makes its objectivity evident) and hence is a kinematic tensor suitable to appear in constitutive equations. We assert that the relative spin is a crucial kinematic quantity in the theoretical description of granular materials.

The Cauchy stress tensor is denoted by σ and in a theory incorporating the intrinsic spin it is not necessarily a symmetric tensor. The second Cosserat quantity that we introduce is the couple stress tensor, denoted by μ. Physically, this represents the effect of a couple exerted across a surface in the same way that the stress tensor represents the effect of a force exerted across the surface. The bulk density, denoted by ρ, differs from and is less than the grain density. The final Cosserat quantity that we introduce is the moment of inertia tensor, denoted by **I**, which physically represents the resistance of the material to rotation.

Governing equations

We shall take the governing equations for a Cosserat continuum to comprise three balance laws. Firstly, there is the translational equations of motion

$$\rho \dot{\mathbf{v}} = \nabla \cdot \sigma + \rho \mathbf{F}, \tag{2}$$

where the superposed dot denotes the material derivative, **F** denotes the body force, which are formally identical to the corresponding equations for a standard material, except that now, the Cauchy stress tensor may not be symmetrical. Secondly, the rotational equation of motion

$$\rho I \dot{\Omega} = \nabla \cdot \mu + \sigma_\times + \rho \mathbf{G}, \tag{3}$$

where $\sigma_\times = \varepsilon : \sigma$, ε denotes the permutation tensor and **G** denotes the body couple. In the absence of Cosserat quantities, i.e. intrinsic spin and couple stress, this equation reduces to the standard continuum expression for the symmetry of the Cauchy stress. Finally, in common with a standard continuum, there is the continuity equation for the bulk density

$$\dot{\rho} + \rho \nabla \cdot \mathbf{v} = 0. \tag{4}$$

Note that in the rotational equation of motion, we have taken the moment of inertia tensor **I** to be constant. In more sophisticated treatments the moment of inertia tensor **I** is allowed to vary and to satisfy a conservation law or, more generally, an evolution law. We will not need this extra generality, and make the simplest assumption that **I** is constant.

Classical versus Cosserat continua

A major disadvantage of using a Cosserat continuum as compared to a standard continuum is the increase in the number of field quantities, the complexity of the resulting equations and the increase in the number of parameters which may be difficult to interpret physically and even more difficult to measure and quantify. Given the elusive nature of determining Cosserat effects experimentally, there is good reason to believe that couple stress and intrinsic spin may often be of negligible magnitude. There are exceptions to this, Cosserat quantities are likely to be significant at boundaries and in localisation (shear banding). In the remaining parts of the field it seems reasonable to suppose that one may neglect the rotational part of motion and in such regions flows may be considered to be such that Ω=constant. Furthermore, in the natural reference frame in which the problem is formulated we may take $\Omega=\mathbf{0}$, i.e. the intrinsic spin is neglected. In such cases, the couple stress may then also be neglected, $\mu = \mathbf{0}$ and the Cauchy stress tensor assumed to be symmetric

$$\sigma^T = \sigma. \tag{5}$$

In this paper we emphasise the idea that, while requiring Cosserat concepts for a complete specification of the model, a class of flows in which only the quantities defined in a standard continuum are non-zero may be of importance. We shall refer to such flows as standard flows. Thus, henceforth in this paper, we assume that the intrinsic spin Ω and couple stress μ are both identically zero and that the Cauchy stress σ is symmetric, i.e. that equ. (5) holds. This enables us to simplify the exposition and omit equations between quantities which are identically zero.

Two scalar functions

In the classical plasticity models that we wish to consider, we shall suppose that the constitutive properties of the material are determined by two scalar quantities, namely the yield function, denoted by $f = f(\sigma)$ and the plastic potential function, denoted by $g = g(\sigma)$. The condition of requiring the existence of two such scalar functions can be dispensed with in plasticity models, but this assumption greatly simplifies the calculation (or indeed measurement) of the material moduli. The yield condition restricts σ to states such that $f(\sigma) \leq 0$. The strict inequality sign corresponds to a rigid material response and the equality sign corresponds to a state of yield - and in this case the material may deform and flow non-rigidly.

It is convenient to define second order tensor quantities

$$\mathbf{f} = (f_{ij}) = \frac{\partial f}{\partial \sigma}, \qquad \mathbf{g} = (g_{ij}) = \frac{\partial g}{\partial \sigma}. \tag{6}$$

In fact **f** and **g** effectively determine the properties of the model. For an isotropic material f and g may depend upon only the stress invariants and it is common to assume an isotropic response when using plasticity models. For an anisotropic material f and g may depend upon the individual components of the stress tensor. It is well known that, even for an initially isotropic material, anisotropy develops during the course of the deformation.

Plastic Potential Model

The well-known plastic potential model is used extensively in civil and geotechnical engineering. As stated in the introduction, a crucial early contribution is due to Drucker and Prager, [1], and the basic model has been the basis of much modelling and generalisation, for example the Critical State Soil mechanics model, [2]. It comprises, in addition to the yield condition, the so-called flow rule

$$\mathbf{d} = \dot{\lambda}\mathbf{g}, \tag{7}$$

where $\dot{\lambda}$ denotes the loading index. When used as a pre-failure model, hardening is incorporated into the model via work-, strain or density-hardening which enables $\dot{\lambda}$ to be calculated by following the history of the deformation. For a perfectly plastic material λ is an arbitrary scalar multiplier, the calculation of the value of which is to be determined as part of the solution to the problem. However, in the case of planar deformations λ may be eliminated algebraically and the following two equations obtained,

(a) The dilatancy equation
$$d_{11} + d_{22} = \sin v \left[(d_{11} - d_{22}) \cos 2\psi + 2d_{12} \sin 2\psi \right], \tag{8}$$

where v denotes the angle of dilatancy and is determined in terms of the plastic potential and ψ denotes the angle that the algebraically greater principal stress direction makes with x-axis.

(b) The equation for the coaxiality of **d** and σ
$$(d_{11} - d_{22}) \sin 2\psi - 2d_{12} \cos 2\psi = 0. \tag{9}$$

If **f** satisfies the condition
$$f_{12}^2 \geq f_{11} f_{22}. \tag{10}$$

then for quasi-static planar steady-state flows the equations governing the model are "spatially" hyperbolic. For a non-associated flow rule (i.e. f and g are distinct) there are four distinct characteristic directions (for an associated flow rule, i.e. $f = g$ there are two repeated characteristic directions). Since the contributions from the stress and velocity fields uncouple we may refer to the stress and velocity characteristic directions, so in the former case the stress and velocity characteristics are distinct, while for the latter they are coincident.

It is useful to note that in the case of the Mohr-Coulomb yield condition the stress characteristic direction coincides with the direction in which the maximum resistance to shearing is most likely to be attained - we may call this the yield direction. Similarly, the velocity characteristic directions coincide with what may be termed the slip directions, i.e. the material slips on itself in these directions. So for the non-associated flow rule the yield and slip directions are distinct, i.e. the material "yields in one direction and slips in another". This does not appear to be physically realistic and this physical defect is mirrored mathematically by the ill-posedness of the governing equations. If now the inertia terms are retained in the equations of motion it turns out that the characteristic directions are unaltered, so the system is degenerate and may be said to be parabolic "in time". At a given point in space the four distinct characteristic directions divide the plane into eight segments. With regard to these segments, the system behaves in a manner analogous to the forward and backwards heat equation and so the Cauchy problem appears to be inherently ill-posed. It does not seem possible to regularise this with boundary conditions.

Double shearing model

The planar equations governing the double-shearing model for incompressible materials were proposed in Spencer [6] and are

(a) the incompressibility condition
$$d_{11} + d_{22} = 0, \tag{11}$$

(b) the equation governing the non-coaxiality of **d** and σ
$$(d_{11} - d_{22}) \sin 2\psi - 2d_{12} \cos 2\psi = 2 \sin \phi \left(\dot{\psi} + s_{12} \right). \tag{12}$$

where ϕ denotes the angle of internal friction and the superposed dot denotes the material derivative. The double-shearing model was extended to encompass dilatant materials by Mehrabadi and Cowin, [9], and the governing equations are

(a) dilatancy equation
$$d_{11} + d_{22} = \frac{\sin v^*}{\cos(\phi - v^*)} \left[(d_{11} - d_{22}) \cos 2\psi + 2d_{12} \sin 2\psi \right], \tag{13}$$

where v^* is a dilatancy parameter which is related to the plastic potential dilatancy parameter v by
$$\tan v^* = \frac{\sin v \cos \phi}{1 - \sin v \sin \phi}. \tag{14}$$

(b) the equation governing the non-coaxiality of **d** and σ

$$(d_{11} - d_{22})\sin 2\psi - 2d_{12}\cos 2\psi = 2\frac{\sin(\phi - v^*)}{\cos v^*}(\dot{\psi} + s_{12}), \tag{15}$$

The double-shearing model was constructed with the specific intention that the stress and velocity characteristic directions coincide in the case where the yield function is that for the Mohr-Coulomb yield condition. This was done by ensuring that the yield and slip directions as defined above are identical. The construction is successful from this point of view. Unfortunately, the yield and slip directions do not coincide with the characteristic directions for the model because there are no real characteristic directions - the model is spatially elliptic.

The model is based upon the idea that the flow is a simultaneous superposition of two (possibly dilatant) shearing motions. However, expressing this idea in equation form leads to the inclusion of the spin (or vorticity) in the constitutive equations, i.e. it leads to equations which are not frame indifferent. To overcome this problem, a second spin is required and the difference between the two spins taken, since then, any two observers in relative motion will measure the difference between the spins, and this is an objective quantity. However, in a standard continuum the only other source for incorporating a second spin is the Cauchy stress tensor and consequently the spin of the principal axes of stress was incorporated into the model. It should be noted that although the standard way of writing the double shearing model is as a sum of terms $\dot{\psi} + s_{12}$, this is easily written as the difference $\dot{\psi} - s_{21}$. Now, unfortunately, the presence of the material derivative of ψ has the unintended side effect on the characteristic equation of ensuring that its roots are complex - so the model has no real characteristics. It may be called spatially elliptic. It should be emphasised that the yield and slip directions do exist and are indeed coincident, but they are no longer characteristic directions.

More detailed expositions of the double-shearing model are given in the original papers by Spencer [6] and Mehrabadi and Cowin [9]. An excellent review of the model, its properties and solutions known at the time of publication in 1982 is presented in Spencer [10].

Double slip and double spin model

In order to resolve these issues of ill-posedness, Harris and Grekova, [7] proposed a model in which a non-coaxial term of the double-shearing type is added to the plastic potential flow rule. The non-coaxial term is chosen in such a way as to ensure that the yield and slip directions are coincident for planar flows. The stress-rate term in the double shearing model is replaced by the intrinsic spin and thus a description of the model requires a Cosserat rather than a standard continuum. Thus, a Cosserat continuum is introduced into the description of the model in order to ensure frame indifferent governing equations and this is irrespective of the magnitude of the Cosserat terms.

It should be noted that the term $\dot{\psi}$ is inherently only applicable in two dimensions (there is no three-dimensional analogue), explaining why the model was developed for planar flows and why the Mohr-Coulomb yield condition was adopted (since then yield is independent of the intermediate principal stress, the flow is, essentially, two dimensional, although not necessarily planar). It also explains why a three-dimensional version of the double shearing model was not forthcoming until seventeen years after the planar version, in [10]. In contrast, the intrinsic spin Ω, being a vector, is indeed three-dimensional. The issue then arises as to the form that a three-dimensional non-coaxial term should take in order that it reduces to the double-shearing type non-coaxial term for planar deformations. It turns out that a term with the same form as the rotational terms in the co-rotational (Jaumann) derivative has the correct properties and is of the form $\omega \cdot \sigma - \sigma \cdot \omega$, where ω is the relative spin. The three dimensional double-slip and double-spin rigid-plastic non-coaxial flow rule, [11] is

$$\mathbf{d} = \dot{\lambda}\mathbf{g} + \gamma(\omega \cdot \sigma - \sigma \cdot \omega), \tag{16}$$

where γ denotes a non-coaxiality parameter which is chosen to ensure that, for planar flows, the stress and velocity characteristics coincide. γ is determined by f and g and the current state of stress. Now equ. (16) is applicable to both Mohr-Coulomb materials (i.e. yield is independent of the intermediate principal stress) and to materials in which yield is dependent on all three principal stresses.

Equ. (16) is the form that the constitutive equation takes in the case of standard flows, i.e. equ. (5) holds and for such flows equ. (16), together with a yield condition for the symmetric stress tensor σ, equ. (2), (5) and (4) form a complete theory which is, in effect, a classical model embedded in a Cosserat continuum. The author is proposing that for many problems in the flow of granular materials this model is sufficient for analysing the flow of granular materials.

It should, perhaps, be pointed out that for a general flow (i.e. one in which Cosserat quantities are not negligible) the stress tensor will not, in general, be symmetric and for such flows, in the right-hand side of equ. (16), σ is replaced by σ^s, where the superscript s denotes the symmetric part, i.e. equ. (16) governs the *symmetric* part of the flow. A second constitutive equation is then required which governs the rotational part of the motion. This extra equation is, in general, a relation between the couple stress, the intrinsic spin gradient, the non-symmetric part of the Cauchy stress tensor and the relative spin tensor. A complete theory is then provided by equ. (2), (3), (4), equ. (16) involving σ^s and the constitutive equation governing the rotational part of the motion. Since we are neglecting Cosserat effects in this paper we omit the equation governing rotational motion. It will be considered in a future publication.

In component form, for planar deformations

$$d_{11} = \dot{\lambda} g_{11} + 2\gamma \omega_{12} \sigma_{12}, \tag{17}$$

$$d_{22} = \dot{\lambda} g_{22} - 2\gamma \omega_{12} \sigma_{12}, \tag{18}$$

$$d_{12} = \dot{\lambda} g_{12} - \gamma \omega_{12} (\sigma_{11} - \sigma_{22}), \tag{19}$$

$$d_{33} = \dot{\lambda} g_{33} = 0. \tag{20}$$

Note that, in general, the equation $g_{33} = 0$ determines principal stress σ_3. This equation is absent for a Mohr-Coulomb material.

For isotropic Mohr-Coulomb materials, γ is given by, see [11],

$$\gamma = \frac{1}{2q_\sigma} \frac{\sin \phi - \sin \nu}{1 - \sin \phi \sin \nu}, \tag{21}$$

where q_σ denotes the maximum in-plane shear stress,

$$q_\sigma = \frac{1}{2} \left[(\sigma_{11} - \sigma_{22})^2 + (\sigma_{12} + \sigma_{21})^2 \right]^{1/2}, \tag{22}$$

For a material in which yield depends upon all three principal stresses, q_σ is replaced by the second invariant of the deviatoric stress tensor.

The planar double-slip and double-spin equations may also be written in a form analogous to the double shearing model, namely

(a) dilatancy equation

$$d_{11} + d_{22} = \sin \nu \left[(d_{11} - d_{22}) \cos 2\psi + 2 d_{12} \sin 2\psi \right], \tag{23}$$

(b) the equation governing the non-coaxiality of **d** and σ

$$(d_{11} - d_{22}) \sin 2\psi - 2 d_{12} \cos 2\psi = 2 \frac{\sin (\phi - \nu)}{1 - \sin \phi \sin \nu} (\Omega - s_{21}), \tag{24}$$

where

$$\tan 2\psi = \frac{\sigma_{12} + \sigma_{21}}{\sigma_{11} - \sigma_{22}}, \tag{25}$$

where ψ denotes the angle that the greater principal direction of the symmetric part of the stress makes with the x-axis. We note that the planar double slip and double spin dilatancy equation, (23), is identical to the plastic potential dilatancy equation, (8), while the planar double slip and double spin non-coaxiality equation, (24), is similar in form to the double-shearing non-coaxiality equation (15).

For the standard flows considered in this paper, $\sigma_{12} = \sigma_{21}$. For a general flow (i.e. one in which Cosserat effects are not negligible) equ. (24) becomes the equation governing the non-coaxiality of **d** and the symmetric part of the stress σ^s. For such general flows, the constitutive equation governing the rotational part of the motion provides the extra equation required to close the system and obtain a complete theory. Again, this equation will be presented in a future publication devoted to Cosserat effects in granular materials.

VISCOPLASTICITY

A continuum model for granular materials does not directly model individual grains nor their interactions with their nearest neighbours, nevertheless, the structure of the equations and the material quantities and parameters appearing in them must reflect some averaged aspects of the micro-structure. In the case of rate-independent quasi-static plasticity models we envisage the state of the granular material to be that of high packing fraction, with each grain in contact with several neighbours. Contacts between grains are of finite duration, and so the interactions are "non-impulsive" and deformation and flow occur by the grains slipping and rolling against each other. With this informal and intuitive picture in mind, we now begin to relax these assumptions. A granular system is highly sensitive to its environment, particularly when non-cohesive. External boundaries and gravity play a crucial role in the state of the system. If the degree of confinement is such as to allow the packing fraction to reduce while simultaneously there is an increase in the energy input into the system from its environment (either through the boundaries or due to gravity or some other body force), a proportion of the grains may develop velocity fluctuations in regions of reduced packing fraction giving rise to almost instantaneous collisions between grains. This allows the possibility of extra mobility of the grains to occur giving rise to diffusive motion. In this way an extra pressure and and viscous effects come into play. In the terrestrial environment granular materials always comprise a mixture of solid grains and fluid, either gas, liquid or a mixture of both. One source of energy input into the solid system may be due to fluid motion from an external source. Boundary conditions may also be a cause of energy input and increase the kinetic energy of the solid grains. In this section we wish to model this transitional motion. In a representative volume element, a sufficiently high proportion of the grains remain sufficiently closely packed to retain solid characteristics, in particular solid Coulomb friction, but there is also assumed to be a proportion of grains less closely packed so that their velocity fluctuations and their ability to diffuse through the material may become significant.

In this section we generalise an argument due to Spencer [12], in which he developed a visco-plastic version of the incompressible double shearing model. His purpose was to regularise the double shearing equations, i.e. to incorporate a dissipative mechanism that would circumvent the problem of ill-posedenss. Here, we carry out a similar construction for the three dimensional double-slip and double spin model. But our purpose is very different. Our basic rate-independent model is already well-posed and the purpose of introducing viscosity into the model is to widen the scope of the model and incorporate additional physical effects into the model. The first step is to split the right hand side of equ. (16) into two terms,

$$\mathbf{d}_1 = \dot{\lambda}\mathbf{g}, \qquad \mathbf{d}_2 = \gamma(\omega\cdot\sigma - \sigma\cdot\omega). \tag{26}$$

then the total plastic deformation-rate $\mathbf{d}$ is

$$\mathbf{d} = \mathbf{d}_1 + \mathbf{d}_2, \tag{27}$$

thus, we separate the part of the deformation-rate obtainable from a plastic potential from the non-coaxial part. Bearing in mind the above intuitive picture of the system, we now suppose that the material to be in a state just "beyond" yield, i.e. either there is an additional inter-granular stress due to collisions between a proportion of the grains or there is a contribution to the stress from the fluid component and we interpret this as enabling the existence of stress states σ such that

$$f(\sigma) \geq 0. \tag{28}$$

where f denotes the yield function in (6). Now suppose that σ_0 corresponds to a state of stress which lies on the yield surface, but is otherwise unspecified, i.e.

$$f(\sigma_0) = 0 \tag{29}$$

and which also satisfies the plastic potential

$$g(\sigma_0) = 0. \tag{30}$$

Now, the expression for $\mathbf{d}_2$ is a determinate relationship between the stress, deformation-rate and relative spin and thus is a complete constitutive equation. However, the expression for $\mathbf{d}_1$ contains an unknown multiplier $\dot{\lambda}$ and so is an incomplete constitutive relation. For the purpose of determining the quantity $\dot{\lambda}$ we shall assume a general constitutive relation of the form

$$\sigma - \sigma_0 = F(\mathbf{d}_1). \tag{31}$$

where σ denotes the current stress, σ_0 a stress point on the yield surface and F denotes an arbitrary function of its argument $\mathbf{d}_1$. Choose σ_0 by making stationary the viscous rate of dissipation with respect to σ_0. Given that, in stress space, σ is a fixed but arbitrary state of stress outside the yield surface and σ_0 is a variable stress state on the yield surface, the stationary point corresponds to a minimum value of the work-rate. Thus we find stationary points of the work-rate $\dot{W}$

$$\dot{W} = \boldsymbol{\sigma} : \mathbf{d}. \tag{32}$$

Now, for a symmetric state of stress and for isotropic materials it may easily (relative to principal axes of stress) be checked that $\boldsymbol{\sigma} : \mathbf{d}_2 = 0$. However, this relationship fails for anisotropic materials. Assuming isotropy, it follows that

$$\dot{W} = \boldsymbol{\sigma} : \mathbf{d}_1 \tag{33}$$

and eliminating $\boldsymbol{\sigma}$ using equ.(31) and imposing the constraint equ. (30)

$$\dot{W} = (\boldsymbol{\sigma}_0 + F(\mathbf{d}_1)) : \mathbf{d}_1 - \alpha g(\boldsymbol{\sigma}_0) \tag{34}$$

where α denotes a Lagrange multiplier. Differentiating equ.(34) with respect to $\boldsymbol{\sigma}_0$ gives, in component form,

$$\frac{\partial \dot{W}}{\partial \sigma_{ij}^0} = d_{ij}^1 - \alpha \frac{\partial g}{\partial \sigma_{ij}^0} \tag{35}$$

where σ_{ij}^0, d_{ij}^1 denote the components of $\boldsymbol{\sigma}_0$, $\mathbf{d}_1$, respectively. At a stationary value of $\dot{W}$

$$\frac{\partial \dot{W}}{\partial \sigma_{ij}^0} = 0 \tag{36}$$

and hence

$$d_{ij}^1 = \alpha \frac{\partial g}{\partial \sigma_{ij}^0}, \tag{37}$$

which is consistent with equ. (16). Thus, we shall suppose that in a flow "beyond" yield (16) we may take as our constitutive equation

$$\mathbf{d} = \alpha \mathbf{g} + \gamma (\boldsymbol{\omega} \cdot \boldsymbol{\sigma} - \boldsymbol{\sigma} \cdot \boldsymbol{\omega}), \tag{38}$$

We now introduce a viscosity parameter into equ (38) via the Lagrange multiplier α in the following way. Suppose that α is dependent upon $\boldsymbol{\sigma}$, $\boldsymbol{\sigma}_0$ via some scalar function h of their components. In fact we shall suppose that

$$\alpha = \frac{1}{2\eta} [h(\boldsymbol{\sigma}) - h(\boldsymbol{\sigma}_0)], \tag{39}$$

where η is a viscosity coefficient, as a measure of the "distance" of the current stress point from the yield surface. The rate of working may now be written as

$$\dot{W} = \boldsymbol{\sigma} : \mathbf{d}_1 = \alpha \boldsymbol{\sigma} : \mathbf{g} \tag{40}$$

Eliminating α using (39) and also incorporating the constraint (29)

$$\dot{W} = \frac{\boldsymbol{\sigma} : \mathbf{g}}{2\eta} [h(\boldsymbol{\sigma}) - h(\boldsymbol{\sigma}_0)] - \beta f(\boldsymbol{\sigma}_0) \tag{41}$$

where β is another Lagrange multiplier. Again, for a stationary point of $\dot{W}$ with respect to $\boldsymbol{\sigma}_0$, we obtain

$$\frac{\partial \dot{W}}{\partial \sigma_{ij}^0} = -\frac{\boldsymbol{\sigma} : \mathbf{g}}{2\eta} \frac{\partial h}{\partial \sigma_{ij}^0} - \beta \frac{\partial f}{\partial \sigma_{ij}^0} = 0 \tag{42}$$

Evidently, a stationary point is obtained by choosing

$$h = f \tag{43}$$

together with

$$\beta = -\frac{\boldsymbol{\sigma} : \mathbf{g}}{2\eta}, \tag{44}$$

i.e. we use the yield function as the measure of the "distance" from yield. The expression for the work-rate equ. (41) then becomes

$$\dot{W} = \frac{\sigma:\mathbf{g}}{2\eta}[f(\sigma) - f(\sigma_0)] + \frac{\sigma:\mathbf{g}}{2\eta} f(\sigma_0) \tag{45}$$

$$= \frac{\sigma:\mathbf{g}}{2\eta} f(\sigma). \tag{46}$$

Comparing this with equ.(40), we see that

$$\alpha = \frac{f(\sigma)}{2\eta} \tag{47}$$

Hence, from equ.(38) the required viscoplastic constitutive equation is

$$\mathbf{d} = \begin{cases} \frac{f(\sigma)}{2\eta} \mathbf{g} + \gamma(\omega \cdot \sigma - \sigma \cdot \omega) & \text{if } f(\sigma) > 0 \\ 0 & \text{if } f(\sigma) \le 0 \end{cases} \tag{48}$$

Note that in the quasistatic limit, both $f(\sigma) \to 0$ and $\eta \to 0$ and so their ratio is indeterminate and hence the viscoplastic model is consistent with the rate-independent plasticity model and reduces to equ.(16) in the quasistatic limit, when λ is interpreted as in indeterminate parameter.

PIPE FLOW

As an example of the application of the viscoplastic constitutive equation we consider pipe flow and consider the conditions under which the simple structure of the classical solution for a Bingham material carries over into the constitutive equation considered here. In general, the Bingham-type solution will necessarily fail since its success relies on a separation of the axial and transverse stresses. In the model presented here, the yield function contains both types of stress component, thereby preventing their separation. As a way of enabling a Bingham-type simple solution, we consider a particular physical system and adopt an assumption which will re-instate the separation of the stress variables. Consider a heavily laden mixture of liquid (say water) and grains which together completely fill the volume occupied by the interior of the pipe. We shall assume that the flow has reached a steady state in which the the volume fraction of solids is uniform in both space and time, i.e. the density of the mixture is constant. The system is considered as a single phase mixture in which the total stress τ is made up of the pore pressure p and the intergranular stress σ (often called the *effective* stress),

$$\tau = -p\mathbf{I} + \sigma \tag{49}$$

where $\mathbf{I}$ denotes the identity tensor. In order to retain the structure of the Bingham-type solution we make the assumption that the yield and plastic potential functions are functions of the effective stress (and *not* the total stress). This then allows, on the one hand, the liquid pressure gradient down the pipe to transport the mixture, but on the other hand the pore pressure in the liquid does not contribute to the yield of the solids phase. It should also be noted that the "extra" shear stress due to viscoplastic effects *does* contribute to the yield of the solids phase, i.e. its physical origin is an intergranular stress and not a fluid stress.

Consider a long pipe with a circular cross-section of radius R. We shall consider the flow between two stations distant L apart in the central part of the pipe, sufficiently far from the ends so that end effects may be neglected. Take an origin O at the centre of the circular cross-section at the left-hand station. It is convenient to use cylindrical polar coordinates with z the axial distance from O, r the radial distance from O and θ the circumferential angle. We shall consider a flow from left to right which has had time to settle down to a steady state and we shall assume the flow to be that of telescopic shear

$$v_r = v_\theta = 0, \qquad v_z = v(r) \tag{50}$$

then the components of the rate of deformation and spin tensors in cylindrical coordinates

$$d_{rr} = \frac{\partial v_r}{\partial r}, \qquad d_{\theta\theta} = \frac{v_r}{r} + \frac{1}{r}\frac{\partial v_\theta}{\partial \theta}, \qquad d_{zz} = \frac{\partial v_z}{\partial z}, \tag{51}$$

$$d_{r\theta} = \frac{1}{2}\left(\frac{1}{r}\frac{\partial v_r}{\partial \theta} - \frac{v_\theta}{r} + \frac{\partial v_\theta}{\partial r}\right), \qquad d_{rz} = \frac{1}{2}\left(\frac{\partial v_r}{\partial z} + \frac{\partial v_z}{\partial r}\right), \tag{52}$$

$$d_{\theta z} = \frac{1}{2}\left(\frac{\partial v_\theta}{\partial z} + \frac{1}{r}\frac{\partial v_z}{\partial \theta}\right) \tag{53}$$

and

$$s_{r\theta} = \frac{1}{2}\left(\frac{1}{r}\frac{\partial v_r}{\partial \theta} - \frac{v_\theta}{r} - \frac{\partial v_\theta}{\partial r}\right), \quad s_{rz} = \frac{1}{2}\left(\frac{\partial v_r}{\partial z} - \frac{\partial v_z}{\partial r}\right). \tag{54}$$

$$s_{\theta z} = \frac{1}{2}\left(\frac{\partial v_\theta}{\partial z} - \frac{1}{r}\frac{\partial v_z}{\partial \theta}\right) \tag{55}$$

become,

$$d_{rr} = d_{\theta\theta} = d_{zz} = d_{\theta z} = d_{r\theta} = 0, \quad d_{rz} = \frac{1}{2}v'(r) \tag{56}$$

and

$$s_{\theta z} = s_{r\theta} = 0, \quad s_{rz} = -\frac{1}{2}v'(r). \tag{57}$$

where the prime denotes differentiation with respect to r. We also assume that $v' \leq 0$, i.e., that the velocity is a monotonically decreasing function of r and that $v'(0) = 0$, i.e. the velocity profile is smooth at the axis of the pipe.

The stress tensor for the rigid/viscoplastic double-slip and double-spin granular material is assumed to take the form

$$\tau_{ij} = -p\delta_{ij} + \sigma_{ij} + 2\eta d_{ij}. \tag{58}$$

Consider a simple generalisation of the standard Bingham solution for pipe flow such that

- the pressure $p > 0$ is a function of z only,
- the normal stresses are no longer all equal, but their difference is a function of r only,
- the shear stress is a function of r only.

In particular, we shall assume the total stress tensor to be symmetric and to take the form

$$\tau_{rr} = \tau_{\theta\theta} = -p(z), \quad \tau_{zz} = -p(z) + \sigma(r), \tag{59}$$

$$\tau_{\theta z} = \tau_{r\theta} = 0, \quad \tau_{rz} = -\tau(r) + \eta v'(r). \tag{60}$$

The equations of motion in cylindrical polar coordinates,

$$\rho \dot{v}_r = \frac{\partial \tau_{rr}}{\partial r} + \frac{1}{r}\frac{\partial \tau_{r\theta}}{\partial \theta} + \frac{\partial \tau_{zr}}{\partial z} + \frac{\tau_{rr} - \tau_{\theta\theta}}{r} + \rho F_r \tag{61}$$

$$\rho \dot{v}_\theta = \frac{\partial \tau_{r\theta}}{\partial r} + \frac{1}{r}\frac{\partial \tau_{\theta\theta}}{\partial \theta} + \frac{\partial \tau_{z\theta}}{\partial z} + \frac{2\tau_{r\theta}}{r} + \rho F_\theta \tag{62}$$

$$\rho \dot{v}_z = \frac{\partial \tau_{rz}}{\partial r} + \frac{1}{r}\frac{\partial \tau_{z\theta}}{\partial \theta} + \frac{\partial \tau_{zz}}{\partial z} + \frac{\tau_{rz}}{r} + \rho F_z. \tag{63}$$

reduce, for a steady state flow in the absence of body forces, to

$$\frac{\partial \tau_{rz}}{\partial r} + \frac{\partial \tau_{zz}}{\partial z} + \frac{\tau_{rz}}{r} = 0. \tag{64}$$

From equations (59), (60) and (64),

$$-\frac{dp}{dz} = -\frac{d\tau_{rz}}{dr} - \frac{\tau_{rz}}{r}. \tag{65}$$

The left-hand side is independent of r, the right-hand side independent of z and so each side is equal to a constant A,

$$\frac{dp}{dz} = A, \tag{66}$$

$$\frac{d\tau_{rz}}{dr} + \frac{\tau_{rz}}{r} = A, \tag{67}$$

Let $p(0) = p_0$, $p(L) = p_L$ with $p_L < p_0$ then equ. (66) gives

$$p(z) = \frac{p_L - p_0}{L} z + p_0, \tag{68}$$

and equ.(67) becomes,

$$\frac{d\tau_{rz}}{dr} + \frac{\tau_{rz}}{r} = \frac{p_L - p_0}{L}. \tag{69}$$

a linear first order ode with solution

$$\tau_{rz} = \frac{p_L - p_0}{2L} r + \frac{B}{r}, \tag{70}$$

where B is an arbitrary constant of integration. In order for the stress to be bounded on the axis of the pipe, we take $B=0$, and so τ_{rz} is linear in r,

$$\tau_{rz} = \frac{p_L - p_0}{2L} r. \tag{71}$$

The solution for the shear stress is thus identical to that for a classical Bingham fluid. Now, the Coulomb yield function for the effective stress reads,

$$f(\sigma_{rr}, \sigma_{zz}, \tau_{rz}) = \frac{1}{2} \left[(\sigma_{rr} - \sigma_{zz})^2 + 4\tau_{rz}^2 \right]^{1/2} + \frac{1}{2} (\sigma_{rr} + \sigma_{zz}) \sin\phi - c\cos\phi \geq 0 \tag{72}$$

where c denotes the cohesion. Thus,

$$f(\sigma, \tau_{rz}) = \frac{1}{2} \left(\sigma^2 + 4\tau_{rz}^2 \right)^{1/2} + \frac{1}{2} \sigma \sin\phi - c\cos\phi \geq 0 \tag{73}$$

and the corresponding plastic potential is

$$g(\sigma_{rr}, \sigma_{zz}, \tau_{rz}) = \frac{1}{2} \left[(\sigma_{rr} - \sigma_{zz})^2 + 4\tau_{rz}^2 \right]^{1/2} + \frac{1}{2} (\sigma_{rr} + \sigma_{zz}) \sin\nu + k \geq 0 \tag{74}$$

where k is a constant, i.e.

$$g(\sigma, \tau_{rz}) = \frac{1}{2} \left(\sigma^2 + 4\tau_{rz}^2 \right)^{1/2} + \frac{1}{2} \sigma \sin\nu + k \geq 0 \tag{75}$$

Now,

$$g_{rr} = \frac{1}{2} (\sigma_{rr} - \sigma_{zz}) \left[(\sigma_{rr} - \sigma_{zz})^2 + 4\tau_{rz}^2 \right]^{-1/2} + \frac{1}{2} \sin\nu \tag{76}$$

$$g_{zz} = -\frac{1}{2} (\sigma_{rr} - \sigma_{zz}) \left[(\sigma_{rr} - \sigma_{zz})^2 + 4\tau_{rz}^2 \right]^{-1/2} + \frac{1}{2} \sin\nu \tag{77}$$

$$g_{rz} = 2\tau_{rz} \left[(\sigma_{rr} - \sigma_{zz})^2 + 4\tau_{rz}^2 \right]^{-1/2} \tag{78}$$

i.e.

$$g_{rr} = -\frac{1}{2} \sigma \left(\sigma^2 + 4\tau_{rz}^2 \right)^{-1/2} + \frac{1}{2} \sin\nu \tag{79}$$

$$g_{zz} = \frac{1}{2} \sigma \left(\sigma^2 + 4\tau_{rz}^2 \right)^{-1/2} + \frac{1}{2} \sin\nu \tag{80}$$

$$g_{rz} = 2\tau_{rz} \left(\sigma^2 + 4\tau_{rz}^2 \right)^{-1/2} \tag{81}$$

The visco-plastic constitutive equation (48) becomes

$$d_{rr} = \frac{1}{2\eta} f(\sigma, \tau_{rz}) g_{rr} + 2\gamma \omega_{rz} \tau_{rz} = 0 \tag{82}$$

$$d_{zz} = \frac{1}{2\eta} f(\sigma, \tau_{rz}) g_{zz} - 2\gamma \omega_{rz} \tau_{rz} = 0 \tag{83}$$

$$d_{rz} = \frac{1}{2\eta} f(\sigma, \tau_{rz}) g_{rz} - \gamma \omega_{rz} (\sigma_{rr} - \sigma_{zz}) \qquad (84)$$

and for the first two equations to be consistent it is necessary that $\sin v = 0$. This is obvious physically given the assumed velocity field and the fact that the flow is in a steady state. Thus, from equ. (21)

$$d_{rr} = -d_{zz} = \frac{1}{2\eta} f(\sigma, \tau_{rz}) g_{rr} + \frac{2\sin\phi}{(\sigma^2 + 4\tau_{rz}^2)^{1/2}} \omega_{rz} \tau_{rz} = 0 \qquad (85)$$

$$d_{rz} = \frac{1}{2\eta} f(\sigma, \tau_{rz}) g_{rz} - \frac{\sin\phi}{(\sigma^2 + 4\tau_{rz}^2)^{1/2}} \omega_{rz} (\sigma_{rr} - \sigma_{zz}) \qquad (86)$$

The equations thus reduce to

$$f(\sigma, \tau_{rz}) \sigma = 2\eta \sin\phi v'(r) \tau_{rz}. \qquad (87)$$

$$f(\sigma, \tau_{rz}) \tau_{rz} = \frac{1}{2} \eta \left[(\sigma^2 + 4\tau_{rz}^2)^{1/2} - \sigma \sin\phi \right] v'(r) \qquad (88)$$

Thus

$$\frac{f(\sigma, \sigma_{rz})}{\eta v'(r)} = \frac{2\sin\phi \tau_{rz}}{\sigma} = \frac{(\sigma^2 + 4\tau_{rz}^2)^{1/2} - \sigma \sin\phi}{2\tau_{rz}} \qquad (89)$$

The second equation gives,

$$\sin\phi (\sigma^2 + 4\tau_{rz}^2) = (\sigma^2 + 4\tau_{rz}^2)^{1/2} \sigma. \qquad (90)$$

Thus, the constitutive equation requires that $\sigma > 0$ (unless τ_{rz} is also zero), i.e. that τ_{zz} is the algebraically greater principal stress. Equ. (90) may be solved for $\sigma > 0$ to obtain

$$\sigma(r) = -2\tau_{rz} \tan\phi \qquad (91)$$

since $\tau_{rz} < 0$ from equ. (71). Substituting into the inequality (73) gives the condition on τ_{rz} for yield to be exceeded, namely,

$$|\tau_{rz}| \geq \frac{c \cos^2 \phi}{1 + \sin^2 \phi}. \qquad (92)$$

Compared with a classical Bingham fluid, the magnitude of the shear stress required to exceed yield is reduced, since in a granular system with internal friction, the unequal principal stresses contribute to yield. The yield condition becomes satisfied at $r = r_0$, where

$$r_0 = \frac{2cL \cos^2 \phi}{(p_0 - p_L)(1 + \sin^2 \phi)}. \qquad (93)$$

and so, for a cohesionless system, the yield condition is everywhere exceeded, except on $r = 0$. Thus, the mixture moves as a rigid plug with velocity $v = v_0$, say, for $r < r_0$ and we see that the effect of the angle of internal friction is to reduce the radius of the central rigid-plug, again, this effect is due to the principal stresses being unequal and contributing to yield. From the first equation in (89), for $r > r_0$ the velocity is determined from

$$v'(r) = \frac{f(\sigma, \sigma_{rz})}{4\eta \sin\phi} \frac{\sigma}{\tau_{rz}} \qquad (94)$$

i.e.

$$v'(r) = -\frac{1}{2\eta \cos^2 \phi} \left[(1 + \sin^2 \phi) |\tau_{rz}| - c \cos^2 \phi \right] \qquad (95)$$

Hence, we obtain, using equ. (71)

$$v(r) = C - \frac{1}{2\eta \cos^2 \phi} \left[(1 + \sin^2 \phi) \frac{p_0 - p_L}{4L} r^2 - cr \cos^2 \phi \right] \qquad (96)$$

where C denotes an arbitrary constant of integration. It is of interest to note that this equation generalises the solution for a Bingham material, to which it reduces when $\phi = 0$. Finally, from the second of equ. (60)

$$\tau = \eta v'(r) - \tau_{rz} = -\frac{1}{2\cos^2 \phi} \left[(1 + \sin^2 \phi) \frac{p_0 - p_L}{2L} r - c \cos^2 \phi \right] - \frac{p_L - p_0}{2L} r. \qquad (97)$$

The new analytic solution presented here indicates, and gives some insight into, the effect of the presence of the angle of internal friction on the classical solution for a Bingham material.

BOUNDARY CONDITIONS

There are a number of alternative boundary conditions that may be considered, see [8] for a general discussion.

1. No slip condition
Suppose that the mixture sticks at the wall, i.e. let $v = 0$ when $r = R$ then

$$v(r) = \frac{1}{2\eta \cos^2 \phi} \left[(1 + \sin^2 \phi) \frac{p_0 - p_L}{4L} (R^2 - r^2) - c(R - r) \cos^2 \phi \right], \tag{98}$$

Although the liquid phase sticks to to the wall, there is no reason to suppose that the solids phase will. Consequently, regarding the mixture as sticking to the wall is probably not physically realistic. The velocity v_0 of the rigid plug is obtained by substituting $r = r_0$ in equ. (98).

2. Slip at the wall
If the mixture slips at the wall then suppose that as $r \to R$ the velocity is such that $v \to v_w$, say. Then

$$v(r) = v_w + \frac{1}{2\eta \cos^2 \phi} \left[(1 + \sin^2 \phi) \frac{p_0 - p_L}{4L} (R^2 - r^2) - c(R - r) \cos^2 \phi \right] \tag{99}$$

3. Wall Friction
If the material slips at the wall, we consider the effect of the pipe on the mixture, i.e. the nature of the shearing stress exerted by the wall on the mixture. Firstly, we note that if the pipe wall is smooth then the mixture may move as a rigid body at constant speed and with $f(\sigma) < 0$ everywhere, i.e. the mixture is nowhere in yield. So, we shall suppose that the pipe is rough. Since the mixture is in global equilibrium, a balance of forces on the whole body between the two stations $z = 0$ and $z = L$ gives

$$\pi R^2 (p_0 - p_L) = 2\pi R L \tau_w \tag{100}$$

where τ_w denotes the frictional force exerted by the pipe on the mixture. Thus,

$$\tau_w = \frac{R(p_0 - p_L)}{2L}. \tag{101}$$

The nature of this frictional force will now be considered.

a. Liquid viscosity
Suppose that the influence of the liquid causes a boundary layer in which viscous effects dominate the pipe/mixture boundary. Let the boundary layer be confined to a "small" thickness h in which the material behaves like the liquid. Let γ_l denote the viscosity of the liquid then

$$\tau_{zr} = \gamma_l \frac{\partial v_z}{\partial r} \tag{102}$$

gives

$$\tau_w = \gamma_l \frac{v_w}{h} \tag{103}$$

so

$$v_w = \frac{R(p_0 - p_L) h}{2L \gamma_l} \tag{104}$$

b. Mixture viscosity
Let γ_m denote a pipe frictional parameter such that at the wall we postulate a friction law of the form, see [8],

$$\tau_w = \mu_m (-\tau + 2\eta d_{rz}) \tag{105}$$

From the second equation in (59), we obtain at the wall

$$\tau_w = \mu_m (p_0 - p_L) R \tag{106}$$

c. Solid Coulomb friction
Finally suppose there is a coefficient of friction μ_p between the pipe and the mixture. The stress condition at the wall is then

$$\tau_w = \mu_p \sigma_w \tag{107}$$

where σ_w denotes the normal stress at the wall. Then $\sigma_w = \tau_{rr}(R) = -p_z$. Thus, the frictional force at the wall is a function of z and so this boundary condition is not a good choice in view of our stated intention of separating axial and radial stress effects. It is probably also not physically appropriate to adopt a boundary condition suitable for dry Coulomb friction in a mixture saturated with liquid.

Finally, we note that we have neglected Cosserat effects at the wall. However, Cosserat effects are likely to become important in a narrow annulus in the vicinity of the pipe boundary. The kinematic Cosserat boundary condition requires relating the velocity to the intrinsic spin. A boundary condition for the couple stress is also required. Finally, we note that the presence of Cosserat effects does not alter the boundary conditions for the Cauchy stress which involves the traction across the boundary, since the formula for the traction is unaffected by the non-symmetry of the Cauchy stress. Since we have neglected Cosserat effects we do not pursue this point any further.

CONCLUSIONS

In this paper we have presented and summarised a well-posed rate-independent plasticity model, called the double-slip and double-spin model. This model generalises the classical plastic potential model using concepts from the double shearing model and from the theory of a Cosserat continuum. The double-slip and double-spin model eliminates two kinds of ill-posedness, one suffered by the plastic potential model, the other by the double-shearing model. The double-slip and double-spin model is thus a well-founded model for use under both quasi-static and dynamic conditions.

However, it is usual, when inertial rate effects are introduced into the formulation (and here we mean both translational and rotational inertia), to also include rate dependency into the constitutive equation. It is also the case that while civil and geotechnical engineering deal with quasi-static problems in which rate effects are often neglected (usually using the plastic potential model, which we regard as an "incomplete" constitutive law) current research interests in other disciplines, for example chemical engineering and physics, deal with rate dependent effects and, in particular, the transitional regime between rate-independent and granular gas flow, has recently received much attention. These reasons have provided motivation for extending the double-slip and double-spin model to incorporate rate-dependent, i.e. viscous and diffusive, behaviour. The resulting visco-plastic model presented here incorporates both viscosity and solid friction.

The paper also considers a boundary value problem, namely that of a simplified pipe flow. This is one of the standard examples in the flow of a Bingham fluid. The intention is to retain the crucial simplifying characteristic of the Bingham solution, namely that there is a separation of axial and radial effects. By considering a specific physical system, namely a very heavily laden flow of grains immersed in an incompressible liquid with the interior of the pipe completely filled with solids and liquid, we are able to retain this structure for the solution. This is achieved by a device often used in civil and geotechnical engineering, namely decomposing the stress into a total and effective stress and assuming that the liquid axial pressure gradient causes the transport of material whereas the effective stresses are only radial and that the yield condition is a function of the effective and not the total stress. Using this device, we are able to present an analytic solution similar to that obtained for a Bingham fluid. However, if this separation does not occur, i.e. if a z-dependent stress component appears in the yield function then no simplification occurs and the solution will be much more complicated.

Finally, we have briefly considered boundary conditions which may be appropriate for the physical system considered here.

REFERENCES

1. D.C. Drucker and W. Prager, *Q. Appl. Math.*, **10**, 157–65 (1952).
2. A. Schofield and P. Wroth, *Critical State Soil Mechanics*, McGraw-Hill, London (1968).
3. D.G. Schaeffer, "Mathematical issues in the continuum formulation of slow granular flow" in *Two phase waves in fluidized beds, sedimentation and granular flows*, D.D. Joseph and D.G. Schaeffer (eds.), University of Minnesota 1990.
4. D. Harris, *Acta Mechanica*, **146**, 199–225 (2001).
5. G. de Josselin de Jong, Statics and kinematics of the failable zone of a granular material, Doctoral Thesis, pp.119, Delft: Uitgeverij Waltmann 1959.
6. A.J.M. Spencer, *J. Mech. Phys. Solids*, **12**, 337–51 (1964).
7. D. Harris and E.F. Grekova, *Journal of Engineering Mathematics*, **52**, 107–35 (2005).
8. N.D. Cristescu, *Dynamic Plasticity*, World Scientific Publishing, Singapore (2007).
9. M.M. Mehrabadi and S.C. Cowin, *J. Mech. Phys. Solids*, **26**, 269–84 (1978).

10. A.J.M. Spencer, "Deformation of ideal granular materials" in *The Rodney Hill 60th. Anniversary Volume*, H.G. Hopkins and M.J. Sewell (eds.), Oxford: Pergamon Press, 607-52 1981.
11. D. Harris, "Double slip and spin: dilatant shear in a reduced Coserrat model", in *Modern Trends in Geomechanics*, W. Wu. and H.-S. Yu (eds.), Springer Proceedings in Physics, 329–46 (2006).
12. A.J.M. Spencer, "A viscoplastic double-shearing theory", in *Trends in Applications of Mathematics to Mechanics* Y. Wang and K. Hutter (eds) Shaker Verlag, Aachen, 509-18 2005.

Nonequilibrium liquid theory for sheared granular liquids

Hisao Hayakawa*, Song-Ho Chong† and Michio Otsuki**

*Yukawa Institute for Theoretical Physics, Kyoto University, Kyoto 606-8502, Japan
†Institute for Molecular Science, Okazaki 444-8585, Japan
**Department of Physics and Mathematics, Aoyama Gakuin University, Sagamihara 229-8558, Japan

Abstract. A nonequilibrium liquid theory for uniformly sheared granular liquids is developed starting from SLLOD equations. We derive a generalized Green-Kubo formula and demonstrate that it yields the nonequilibrium steady-state average which is essentially independent of the choice of the initial condition. It is also shown that the fluctuating hydrodynamics can be derived from Mori-type equations for density and current-density fluctuations if one considers a weak-shear and small-dissipation limit along with the Markovian approximation.

Keywords: Sheared Granular Liquids, generalized Green-Kubo formula, Liquid Theory
PACS: 05.40.-a,05.70.Ln, 45.70.-n, 61.20.Lc

INTRODUCTION

Assemblies of dense granular materials behave unlike usual materials. One may think that a liquid phase is absent for granular assemblies since they lack attractive interaction. Indeed, the gas kinetic theory such as the Boltzmann-Enskog theory has historically been employed in describing dense granular flows, where correlation effects appear only through the contact value of the radial distribution function.[1, 2, 3, 4, 5, 6] This approach is powerful and semi-quantitatively accurate even for considerably dense systems. In these days, however, we have recognized the relevancy of the concept of "granular liquids" because correlation effects in granular flows, such as long-time correlations and long-range correlations, turned out to be relevant as in molecular liquids.[7, 8, 9, 10, 11, 12, 13]

There have been some developments in the granular liquid theory starting from microscopic basic equations such as the Liouville equation. The Liouville equation of granular fluids was first discussed by Schofield and Oppenheim[14] long time ago, and Brey et al.[15] further developed such a formulation. Recently, Dufty et al.[16, 17] discussed in detail the Green-Kubo formula of freely cooling granular gases starting from the Liouville equation. The Green-Kubo formula for granular fluids has been discussed in various contexts[18, 19, 20, 21, 22], and these studies suggest that some correction terms to the conventional Green-Kubo formula are necessary for granular fluids.

There is an advantage in using the liquid theory. It is known that liquid theories such as the mode-coupling theory (MCT) for supercooled liquids are commonly used to describe the glass transition of molecular liquids or colloidal assemblies.[23, 24] Liu and Nagel[25] proposed that the jamming transition is a fundamental transition in glassy and granular materials. Since then many aspects of similarities between the conventional glass transition and the jamming transition have been investigated [26], where some researchers have used granular materials to study dynamical heterogeneity in glassy materials. [27, 28, 29, 30, 31] Along this line it appears natural to apply a liquid theory to dense granular assemblies.

Nevertheless, the jamming transition under a plane shear exhibits some distinct aspects from conventional glass transitions, e.g., the jamming transition depends strongly on details of microscopic interactions between particles. In fact, one cannot use a naive MCT [32] in describing sheared jamming transitions. Instead, the jamming transition of frictionless granular particles is believed to be a continuous transition at a critical density above which elastic moduli and the yield stress become nonzero, and there are scaling laws in the vicinity of the critical point as observed in conventional critical phenomena. [33, 34, 35, 36, 37, 38]

Quite recently, Chong and Kim[39] have reformulated the liquid theory of sheared dense molecular liquids with the Gaussian thermostat. Later, Chong et al.[40] extended their formulation to soft granular liquids under a plane shear, and found the existence of the generalized Green-Kubo formula and integral fluctuation theorem. Chong et al.[41] have further developed an MCT for uniformly sheared granular liquids. This generalized Green-Kubo formula applies not only to linear regime but also to nonequilibrium states arbitrarily far from equilibrium. Thus, one can avoid the use of hydrodynamic equations including Burnett and super-Burnett terms which occasionally exhibit a divergent behavior.

The aim of this paper is to outline our formulation of the liquid theory for dense sheared granular materials. In the next section, we summarize the microscopic starting equations, such as SLLOD equations, Liouville equations, and nonequilibrium distribution function. In section III, we present the transient time correlation function formalism and discuss steady state properties. We then derive the generalized Green-Kubo formula. It is demonstrated that the generalized Green-Kubo formula yields the nonequilibrium steady-state average which is essentially independent of the specific choice of the initial condition. In section IV, we outline the liquid theory beyond the generalized Green-Kubo formula, and discuss its connection to the fluctuating hydrodynamics. The paper is summarized in section V.

MICROSCOPIC STARTING EQUATIONS

In this section, we derive exact microscopic equations and relations which serve a basis in constructing a nonequilibrium liquid theory for uniformly sheared frictionless granular particles.

SLLOD equations of motion

Let us consider a system of N smooth granular particles of mass m in a volume V under a stationary shear characterized by the shear-rate tensor κ. We assume that each granular particle is a soft-sphere, and the contact force acts only on the normal direction. Under a homogeneous shear, the velocity profile is given by $\kappa \cdot \mathbf{r}$ at position $\mathbf{r}$. The Newtonian equations of motion describing such a homogeneously sheared system are given by the SLLOD equations [42]

$$\dot{\mathbf{r}}_i = \frac{\mathbf{p}_i}{m} + \kappa \cdot \mathbf{r}_i, \qquad (1)$$

where $\mathbf{r}_i$ refers to the position of the ith particle, $\dot{\mathbf{r}}_i = d\mathbf{r}_i/dt$, and

$$\dot{\mathbf{p}}_i = \mathbf{F}_i^{(\mathrm{el})} + \mathbf{F}_i^{(\mathrm{vis})} - \kappa \cdot \mathbf{p}_i. \qquad (2)$$

Here Eq.(1) is the definition of the peculiar momentum $\mathbf{p}_i$ satisfying $\sum_i \mathbf{p}_i = 0$, $\mathbf{F}_i^{(\mathrm{el})} = \sum_{k \neq i} \mathbf{F}_{ik}^{(\mathrm{el})}$ is the conservative force exerted on the ith particle by other particles with

$$\mathbf{F}_{ik}^{(\mathrm{el})} = -\frac{\partial u(r_{ik})}{\partial \mathbf{r}_{ik}} = \Theta(\sigma - r_{ik}) f(d - r_{ik}) \hat{\mathbf{r}}_{ik}, \qquad (3)$$

where σ is the diameter of each grain, $u(r_{ik})$ is the pairwise potential, $\mathbf{r}_{ik} = \mathbf{r}_i - \mathbf{r}_k$, $r_{ik} = |\mathbf{r}_{ik}|$, $\hat{\mathbf{r}}_{ik} = \mathbf{r}_{ik}/r_{ik}$, and $\Theta(x)$ is the Heviside function satisfying $\Theta(x) = 1$ for $x > 0$ and $\Theta(x) = 0$ for $x < 0$. The actual elastic repulsive force $f(x)$ is proportional to $x^{3/2}$ for three dimensional systems, but we sometimes use a simpler form $f(x) \propto x$. Similarly, the viscous dissipative force $\mathbf{F}_i^{(\mathrm{vis})}$ is represented by a sum of two-body contact forces as $\mathbf{F}_i^{(\mathrm{vis})} = \sum_{j \neq i} \mathbf{F}_{ij}^{(\mathrm{vis})}$ with

$$\mathbf{F}_{ij}^{(\mathrm{vis})} = -\hat{\mathbf{r}}_{ij} \mathscr{F}(r_{ij}) (\mathbf{g}_{ij} \cdot \hat{\mathbf{r}}_{ij}) \equiv -\hat{\mathbf{r}}_{ij} \Theta(d - r_{ij}) \gamma(\sigma - r_{ij}) (\mathbf{g}_{ij} \cdot \hat{\mathbf{r}}_{ij}). \qquad (4)$$

In Eq. (4) we have introduced $\mathbf{g}_{ij} \equiv \mathbf{v}_i - \mathbf{v}_j = (\mathbf{p}_i - \mathbf{p}_j)/m + \kappa \cdot (\mathbf{r}_i - \mathbf{r}_j)$ with the velocity of ith particle $\mathbf{v}_i \equiv d\mathbf{r}_i/dt$. The viscous function $\gamma(x)$ is proportional to $\sqrt{x}$ for three dimensional systems, but we sometimes use a simpler model $\gamma(x) = const.$

It should be noted that SLLOD equations (1) and (2) reduce to the Newtonian equation of motion

$$m\ddot{\mathbf{r}}_i = \mathbf{F}_i^{(\mathrm{el})} + \mathbf{F}_i^{(\mathrm{vis})}, \qquad (5)$$

if one eliminates the peculiar momentum $\mathbf{p}_i$. This means that the SLLOD equations are equivalent to the Newtonian equation of motion under the Lees-Edwards boundary condition. [42] We also note that frictionless sheared granular particles under a constant pressure boundary behave as those under the Lees-Edwards boundary condition in the vicinity the jamming transition.[43] On the other hand, it is hard to extract physical essences from actual frictional granular assemblies under a physical boundary condition. Thus, a set of SLLOD equations is a natural starting point in constructing a liquid theory of granular particles. We also notice that a granular flow on an inclined slope can be approximately described by a uniformly sheared flow except for the boundary layers.[44]

We stress that an energy sink term is necessary even for a system of a sheared molecular liquid. Indeed, the system heats up without the energy sink. In simulations, one usually introduces a thermostat, and an experimental apparatus for the real system plays a role of the thermostat. We have already confirmed that behaviors of uniformly sheared granular liquids with small inelasticity are almost the same as those for a model of a molecular liquid with a velocity rescaling thermostat.[11, 12]

We have assumed that the interaction between granular particles is described by a soft-core model, which differs from most of the conventional treatment of the granular gas kinetic theory and the MCT for sheared granular liquids where a hard-core model is adopted.[32] The soft-core model has several advantages: (i) the model is more realistic than the hard-core model, (ii) one can apply this formulation to very dense systems even in the vicinity of the jamming transition, and (iii) one can avoid the use of the pseudo-Liouvillian.

The Liouville equation

For nonequilibrium systems described by the SLLOD equations, the Liouville equation is commonly used. [42] The time evolution of phase variables whose time dependence comes solely from that of the phase space point $\Gamma = (\mathbf{r}^N, \mathbf{p}^N)$ is determined by

$$\frac{d}{dt}A(\Gamma) = \dot{\Gamma} \cdot \frac{\partial}{\partial \Gamma} A(\Gamma) \equiv i\mathscr{L} A(\Gamma). \tag{6}$$

The operator $i\mathscr{L}$ is referred to as the Liouvillian. The formal solution to this equation can be written as

$$A(\Gamma, t) = \exp(i\mathscr{L}t) A(\Gamma). \tag{7}$$

On the other hand, the Liouville equation for the nonequilibrium phase-space distribution function $\rho(\Gamma, t)$ is given by

$$\frac{\partial \rho(\Gamma, t)}{\partial t} = -\left[\dot{\Gamma} \cdot \frac{\partial}{\partial \Gamma} + \Lambda(\Gamma)\right]\rho(\Gamma, t) \equiv -i\mathscr{L}^\dagger \rho(\Gamma, t), \tag{8}$$

where the phase space contraction factor $\Lambda(\Gamma)$ is defined by

$$\Lambda(\Gamma) \equiv \frac{\partial}{\partial \Gamma} \cdot \dot{\Gamma} = \sum_i \left(\frac{\partial}{\partial \mathbf{r}_i} \cdot \dot{\mathbf{r}}_i + \frac{\partial}{\partial \mathbf{p}_i} \cdot \dot{\mathbf{p}}_i\right). \tag{9}$$

For our model (1) - (4) one easily obtains its explicit form:

$$\Lambda(\Gamma) = -\frac{1}{m}\sum_{i,j} \Theta(d - r_{ij})\gamma(d - r_{ij}) < 0. \tag{10}$$

The formal solution to the Liouville equation (8) reads

$$\rho(\Gamma, t) = \exp(-i\mathscr{L}^\dagger t)\rho(\Gamma, 0). \tag{11}$$

From Eqs. (6) and (8) we readily obtain the relation

$$i\mathscr{L}^\dagger(\Gamma) = i\mathscr{L}(\Gamma) + \Lambda(\Gamma). \tag{12}$$

One can show that the following adjoint relations hold [42]:

$$\int d\Gamma [i\mathscr{L} A(\Gamma)] B(\Gamma) = -\int d\Gamma A(\Gamma) [i\mathscr{L}^\dagger B(\Gamma)], \tag{13}$$

$$\int d\Gamma [e^{i\mathscr{L}t} A(\Gamma)] B(\Gamma) = \int d\Gamma A(\Gamma) e^{-i\mathscr{L}^\dagger t} B(\Gamma). \tag{14}$$

If the phase-space contraction factor $\Lambda(\Gamma)$ is identically zero, then $i\mathscr{L}^\dagger = i\mathscr{L}$ holds, and the Liouvillian becomes self-adjoint or Hermitian. In general, this is not the case for nonequilibrium dissipative systems.

Nonequilibrium distribution function

Let us consider an equilibrium system to which a constant shear rate $\dot{\gamma}$ satisfying $\kappa_{\alpha\beta} = \dot{\gamma}\delta_{\alpha x}\delta_{\beta y}$ is applied at time $t = 0$, and thereafter the system evolves according to the SLLOD equations (1) and (2). The Liouvillian is given by $i\mathscr{L} = i\mathscr{L}^{(\text{el})}$ for $t < 0$ and

$$i\mathscr{L} = i\mathscr{L}^{(\text{el})} + i\mathscr{L}_{\dot{\gamma}} + i\mathscr{L}^{(\text{vis})} \tag{15}$$

for $t > 0$, where the unperturbed adiabatic or the elastic part ($i\mathscr{L}^{(\text{el})}$), the shear part ($i\mathscr{L}_{\dot{\gamma}}$), and the viscous part ($i\mathscr{L}^{(\text{vis})}$) are respectively given by

$$i\mathscr{L}^{(\text{el})} = \sum_i \left[\frac{\mathbf{p}_i}{m} \cdot \frac{\partial}{\partial \mathbf{r}_i} + \mathbf{F}_i^{(\text{el})} \cdot \frac{\partial}{\partial \mathbf{p}_i} \right], \tag{16}$$

$$i\mathscr{L}_{\dot{\gamma}} = \sum_i \left[(\kappa \cdot \mathbf{r}_i) \cdot \frac{\partial}{\partial \mathbf{r}_i} - (\kappa \cdot \mathbf{p}_i) \cdot \frac{\partial}{\partial \mathbf{p}_i} \right], \tag{17}$$

$$i\mathscr{L}^{(\text{vis})} = \sum_i \mathbf{F}_i^{(\text{vis})} \cdot \frac{\partial}{\partial \mathbf{p}_i}. \tag{18}$$

Here, we assume that the initial distribution is given by the canonical one

$$\rho(\Gamma, 0) = \rho_{\text{eq}}(\Gamma) \equiv \frac{e^{-\beta H(\Gamma)}}{Z(\beta)}; \quad Z(\beta) \equiv \int d\Gamma e^{-\beta H(\Gamma)}, \tag{19}$$

where $\beta \equiv 1/T$ is the inverse temperature in the initial state, and H is the total Hamiltonian defined by

$$H = \sum_i \left\{ \frac{\mathbf{p}_i^2}{2m} + \frac{1}{2} \sum_{k \neq i} u(r_{ik}) \right\}. \tag{20}$$

It should be noted that the effect of the shear appears in H through $\mathbf{p}_i^2/2m = m(\mathbf{v}_i - \kappa \cdot \mathbf{r}_i)^2/2$. We also note that there holds a trivial relationship

$$i\mathscr{L}^{(\text{el})} \rho_{\text{eq}} = 0. \tag{21}$$

It might appear that our formulation depends strongly on our choice of the initial condition, Eq. (19). However, we will argue below that nonequilibrium steady-state properties therefrom are insensitive to such a choice.

From Eq. (8) the time evolution of the distribution function can be written as

$$\rho(\Gamma, t) = e^{-i\mathscr{L}^\dagger t} \rho_{\text{eq}}(\Gamma). \tag{22}$$

With the identity

$$e^{-i\mathscr{L}^\dagger t} = 1 + \int_0^t ds\, e^{-i\mathscr{L}^\dagger s} (-i\mathscr{L}^\dagger), \tag{23}$$

Eq. (22) can be expressed as

$$\rho(\Gamma, t) = \rho_{\text{eq}}(\Gamma) + \int_0^t ds\, e^{-i\mathscr{L}^\dagger s} (-i\mathscr{L}^\dagger) \rho_{\text{eq}}(\Gamma). \tag{24}$$

From Eqs. (10), (12), (15)-(18), and (21) we get

$$i\mathscr{L}^\dagger \rho_{\text{eq}}(\Gamma) = i\mathscr{L}_{\dot{\gamma}} \rho_{\text{eq}}(\Gamma) + i\mathscr{L}^{(\text{vis})} \rho_{\text{eq}}(\Gamma) + \Lambda(\Gamma) \rho_{\text{eq}}(\Gamma). \tag{25}$$

The first term on the right hand side in this expression is given by

$$i\mathscr{L}_{\dot{\gamma}} \rho_{\text{eq}}(\Gamma) = \beta \sum_i \left[(\kappa \cdot \mathbf{r}_i) \cdot \mathbf{F}_i^{(\text{el})} + (\kappa \cdot \mathbf{p}_i) \cdot \frac{\mathbf{p}_i}{m} \right] \rho_{\text{eq}}(\Gamma)$$

$$= \kappa : \sigma^{(\text{el})} \rho_{\text{eq}}(\Gamma) = \dot{\gamma} \sigma_{xy}^{(\text{el})} \rho_{\text{eq}}(\Gamma) \tag{26}$$

where $\sigma^{(\text{el})}$ denotes the elastic stress tensor whose element is given by

$$\sigma^{(\text{el})}_{\alpha\beta} = \sum_i \left[\frac{p_i^\alpha p_i^\beta}{m} + r_i^\alpha F_i^{(\text{el})\beta} \right] \quad (\alpha, \beta = x, y, z). \tag{27}$$

In the final equality of Eq. (26) we have used the specific form $\kappa_{\alpha\beta} = \dot{\gamma}\delta_{\alpha x}\delta_{\beta y}$. On the other hand, from Eq. (18), $i\mathscr{L}^{(\text{vis})}\rho_{\text{eq}}(\Gamma)$ is given by

$$\begin{aligned} i\mathscr{L}^{(\text{vis})}\rho_{\text{eq}}(\Gamma) &= -\beta\rho_{\text{eq}}(\Gamma)\sum_i \mathbf{F}_i^{(\text{vis})} \cdot \frac{\mathbf{p}_i}{m} \\ &= \beta\left[\sum_i (\kappa\cdot\mathbf{r}_i)\cdot\mathbf{F}_i^{(\text{vis})}\right]\rho_{\text{eq}}(\Gamma) - \frac{\beta}{2}\sum_{i,k}\mathbf{F}_{ik}^{(\text{vis})}\cdot\mathbf{g}_{ik}\rho_{\text{eq}}(\Gamma). \end{aligned} \tag{28}$$

It is convenient to introduce Rayleigh's dissipation function $\mathscr{R}$ as

$$\mathscr{R} \equiv -\frac{1}{4}\sum_{i,k}\mathbf{g}_{ik}\cdot\mathbf{F}_{ik}^{(\text{vis})} = \frac{1}{4}\sum_{i,k}\Theta(d-r_{ik})\gamma(d-r_{ik})(\mathbf{g}_{ik}\cdot\hat{\mathbf{r}}_{ik})^2. \tag{29}$$

$\rho(\Gamma,t)$ can then be written as

$$\rho(\Gamma,t) = \rho_{\text{eq}}(\Gamma) + \int_0^t ds\, e^{-i\mathscr{L}^\dagger s}[\rho_{\text{eq}}(\Gamma)\Omega(\Gamma)], \tag{30}$$

where $\Omega(\Gamma)$ is the nonequilibrium work function defined by

$$\Omega(\Gamma) \equiv -\beta\dot{\gamma}\sigma_{xy}(\Gamma) - 2\beta\mathscr{R}(\Gamma) - \Lambda(\Gamma) \tag{31}$$

in terms of the total stress tensor

$$\sigma_{\alpha\beta} = \sum_i \left[\frac{p_i^\alpha p_i^\beta}{m} + r_i^\alpha (F_i^{(\text{el})\beta} + F_i^{(\text{vis})\beta}) \right] \quad (\alpha, \beta = x, y, z). \tag{32}$$

One can show that the equilibrium average of the nonequilibrium work function is zero:

$$\langle \Omega(\Gamma) \rangle_c = 0. \tag{33}$$

Hereafter, we shall reserve the notation $\langle \cdots \rangle_c$ for representing the averaging over the initial canonical distribution function $\rho(\Gamma, 0) = \rho_{\text{eq}}(\Gamma)$:

$$\langle \cdots \rangle_c \equiv \int d\Gamma \rho_{\text{eq}}(\Gamma) \cdots. \tag{34}$$

GENERALIZED GREEN-KUBO FORMULA

Transient time-correlation function formalism and generalized Green-Kubo formula

In contrast to the case of equilibrium quantities, the nonequilibrium ensemble average $\langle A(t) \rangle_c$ of a phase variable $A(t)$ depends explicitly on the time t past since the start of the shearing. Using the nonequilibrium phase-space distribution function $\rho(\Gamma,t)$, $\langle A(t) \rangle_c$ can be expressed as

$$\langle A(t) \rangle_c = \int d\Gamma \rho(\Gamma,0) A(t) = \int d\Gamma \rho(\Gamma,t) A(0), \tag{35}$$

where the second equality follows from Eq. (14).

Substituting Eq. (30) into Eq. (35) and then using Eq. (13), one obtains

$$\langle A(t) \rangle_c = \langle A(0) \rangle_c + \int_0^t ds \langle A(s)\Omega(0) \rangle_c. \tag{36}$$

The expression (36) relates the nonequilibrium value of a phase variable $A(t)$ at time t to the integral of the *transient* time-correlation function. Indeed the integrand in Eq. (36), $\langle A(s)\Omega(0)\rangle_c$, is the correlation between the nonequilibrium work function in the initial state and A at time s after the shearing force is turned on. It should be remembered, however, that the dynamics inside the brackets $\langle\cdots\rangle_c$ is governed by the granular SLLOD equations, and only averages like $\langle A(0)\rangle_c$ coincide with equilibrium quantities.

The system is in a nonequilibrium steady state if the ensemble averages of all phase variables become time-independent. Let us notice that the long-time limit of Eq. (36) approaches a constant, and hence, the integral is convergent if the system displays *mixing*. [42] This feature can be demonstrated by taking a time derivative of Eq. (36):

$$\frac{d}{dt}\langle A(t)\rangle_c = \langle A(t)\Omega(0)\rangle_c. \quad (37)$$

If the system displays mixing [42], all long-time correlations between phase variables vanish. With the aid of Eq.(33) we obtain $\lim_{t\to\infty}(d/dt)\langle A(t)\rangle_c = 0$ and

$$\lim_{t\to\infty}\langle A(t)\rangle_c = \langle A\rangle_{ss}. \quad (38)$$

Here, the steady-state average, denoted by $\langle\cdots\rangle_{ss}$ hereafter, is obtained from the long-time limit of Eq. (36):

$$\langle A\rangle_{ss} = \langle A(0)\rangle_c + \int_0^\infty ds \langle A(s)\Omega(0)\rangle_c. \quad (39)$$

Because of Eq.(33), Eq.(39) can be rewritten as

$$\langle A\rangle_{ss} = \langle A(0)\rangle_c + \int_0^\infty ds \langle \Delta A(s)\Omega(0)\rangle_c, \quad (40)$$

where $\Delta A(s) \equiv A(s) - A(s\to\infty)$.

Equation (39) is the generalized Green-Kubo formula which relates the steady-state average to the time-correlation function describing transient dynamics evolving from an initial equilibrium towards a final steady state. One can easily show that Eq. (39) reduces to the conventional Green-Kubo formula if the external force is weak and the dissipative force is neglected, i.e., for small $\dot\gamma$ and $\gamma(x) = 0$. For example, by setting $A = \sigma_{xy}$ in Eq. (39), one obtains for the steady-state shear stress defined via $\sigma_{ss} \equiv -\langle\sigma_{xy}\rangle_{ss}/V$

$$\sigma_{ss} = -\frac{\langle\sigma_{xy}(0)\rangle_c}{V} - \frac{1}{V}\int_0^\infty ds\,\langle\sigma_{xy}(s)\Omega(0)\rangle_c. \quad (41)$$

When $\gamma(x) = 0$, there hold $\langle\sigma_{xy}(0)\rangle_c = 0$ and $\Omega = -\beta\dot\gamma\sigma_{xy}$ [see Eq. (31)], and Eq. (41) formally reduces to

$$\sigma_{ss} = \frac{\beta\dot\gamma}{V}\int_0^\infty ds\,\langle\sigma_{xy}(s)\sigma_{xy}(0)\rangle_c. \quad (42)$$

For small $\dot\gamma$, one can replace the Liouvillian governing the dynamics of $\sigma_{xy}(s)$ in the integrand by that for a quiescent equilibrium state, and hence, Eq. (42) is the conventional Green-Kubo formula for the viscosity η defined via $\eta \equiv \sigma_{ss}/\dot\gamma$. It should be noted that the conventional derivation of Green-Kubo formula requires a convergent factor $e^{-\varepsilon t}$ for the integrand with taking the limit $\varepsilon \to 0$. Similarly, a small dissipation is also necessary for Eq.(41) to obtain a convergent result. Namely, the dissipation plays a role of the convergent factor. Otherwise, the system is heated up and cannot reach a steady state.

Therefore, with the aid of Eq.(41), the viscosity η satisfies

$$\eta = -\frac{\langle\sigma_{xy}(0)\rangle_c}{\dot\gamma V} - \frac{1}{\dot\gamma V}\int_0^\infty ds\,\langle\sigma_{xy}(s)\Omega(0)\rangle_c \quad (43)$$

in general situations. We should stress that Eq. (41) or Eq.(43) is the full-order expression, and applies to nonequilibrium states arbitrarily far from equilibrium. Thus, we do not have to worry about Burnett or super-Burnett terms which occasionally exhibit an unstable behavior. In other words, the viscosity η or the steady shear stress σ_{ss} involves effects of nonlinear rheology, and is free from the magnitude of the deviation from a reference state.

On the initial condition

In this subsection, let us demonstrate that $\langle A \rangle_{ss}$ is independent of the choice of the initial condition such as the initial temperature and initial distribution.[40] This result is highly nontrivial, because Eq. (39) appears to depend on the choice of the initial canonical distribution.

From Eqs. (11) and (19) one obtains the Kawasaki representation [42]

$$\rho(\Gamma,t) = \exp\left[-\int_0^t ds\, \Lambda(-s)\right] \frac{e^{-\beta H(-t)}}{Z(\beta)} = \rho_{eq}(\Gamma) \exp\left[\int_0^t ds\, \Omega(-s)\right]. \tag{44}$$

In the second equality we have introduced the nonequilibrium work function $\Omega(t) = e^{i\mathcal{L}t}\Omega(\Gamma)$ at time t.

Since

$$\frac{\partial}{\partial \beta}\left\{\frac{e^{-\beta H(-t)}}{Z(\beta)}\right\} = [\langle H \rangle_c - H(-t)] \frac{e^{-\beta H(-t)}}{Z(\beta)}, \tag{45}$$

one obtains from Eqs. (44) and the definition of the average for $t \to \infty$

$$\begin{aligned}
\frac{\partial}{\partial \beta} \langle A(t) \rangle_c &= \int d\Gamma A(0)[\langle H \rangle_c - H(-t)]\rho(\Gamma,t) \\
&= \langle A(t) \rangle_c \langle H \rangle_c - \langle A(t) H \rangle_c \to 0,
\end{aligned} \tag{46}$$

i.e., $\langle A \rangle_{ss} = \lim_{t\to\infty} \langle A(t) \rangle_c$ is independent of the inverse temperature β of the initial equilibrium state. Thus, $\langle A \rangle_{ss}$ is uniquely specified by the "thermodynamic" parameters $(N,V,\dot{\gamma})$ characterizing the nonequilibrium steady state.

One can prove a stronger statement that the average of any variable in the steady state is invariant if the initial condition can be expanded in an orthogonal polynomial of the kinetic energy associated with the Gaussian function or the exponential function such as the Laguerre bi-polynomial and the Hermite polynomial. Let us demonstrate that the average under the initial condition expanded in the Sonine polynomial, which is related to the Laguerre bi-polynomial, is the same as the one under the canonical initial condition. We assume the following initial condition

$$\rho_{in}(\Gamma) \equiv \frac{e^{-\beta H}}{Z(\beta)}\left\{1 + \sum_{l=1}^{\infty} a_l S_{3/2}^{(l)}\left(\beta \frac{p^2}{2m}\right)\right\}, \qquad p^2 \equiv \sum_i \mathbf{p}_i^2, \tag{47}$$

where a_l and $S_{3/2}^{(l)}(x)$ are respectively the expansion parameter and the Sonine polynomial which satisfies the orthogonal condition

$$\int_0^\infty dx\, e^{-x} x^m S_m^{(p)}(x) S_m^{(q)}(x) = \frac{\Gamma(m+p+1)}{p!} \delta_{p,q} \tag{48}$$

with the Gamma function $\Gamma(x)$. It should be noted that the Sonine expansion around the Gaussian has widely been used for the description of freely cooling granular gases[2], but Eq.(47) is more general than the case of freely cooling cases. Indeed, any function of the kinetic energy can be expanded in an orthogonal polynomial. Thus, the only assumption adopted here is that the initial condition can be represented by a product of the canonical distribution and a function of the kinetic energy.

Let us denote the average under $\rho_{in}(\Gamma)$ as

$$\langle A(t) \rangle_{in} \equiv \int d\Gamma \rho_{in}(\Gamma) A(t). \tag{49}$$

We also introduce the difference between the average under the initial condition (47) and the one under the canonical initial condition (19):

$$\delta A(t) \equiv \langle A(t) \rangle_{in} - \langle A(t) \rangle_c. \tag{50}$$

One immediately obtains

$$\begin{aligned}
\delta A(t) &= \int d\Gamma A(\Gamma)(\rho_{in}(\Gamma(t)) - \rho_c(\Gamma(t))) \\
&= \int d\Gamma A(\Gamma,t)(\rho_{in}(\Gamma) - \rho_c(\Gamma)),
\end{aligned} \tag{51}$$

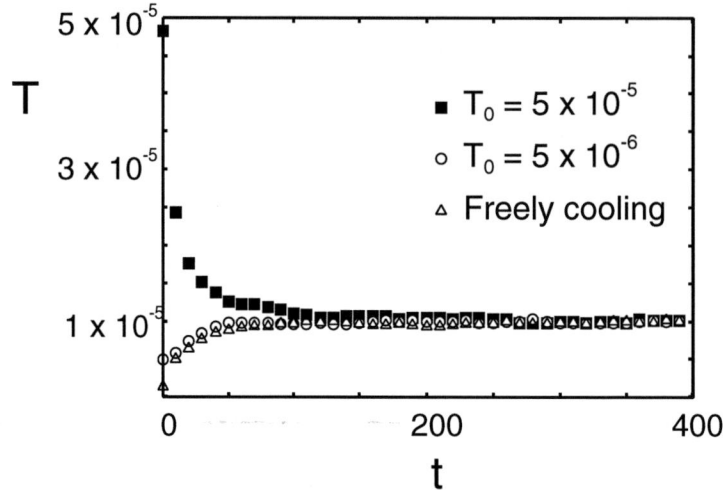

FIGURE 1. The time evolution of the granular temperature starting from $T_0 = 5 \times 10^6$, $T_0 = 5 \times 10^5$, and the homogeneous cooling state. See the text in details.

where we have used Eq.(14). Substituting (47) into (51) we obtain

$$\begin{aligned}\delta A(t) &= \sum_{l=1}^{\infty} a_l \int d\Gamma \frac{e^{-\beta H}}{Z(\beta)} A(\Gamma,t) S_{3/2}^{(l)}\left(\beta \frac{p^2}{2m}\right) \\ &= \sum_{l=1}^{\infty} a_l \left\langle A(\Gamma,t) S_{3/2}^{(l)}\left(\beta \frac{p^2}{2m}\right)\right\rangle_c \to \sum_{l=1}^{\infty} a_l \langle A(\Gamma,t)\rangle_c \left\langle S_{3/2}^{(l)}\left(\beta \frac{p^2}{2m}\right)\right\rangle_c\end{aligned} \quad (52)$$

in the limit $t \to \infty$ with the aid of the mixing property. If we use $S_{3/2}^{(0)}(x) = 1$ and Eq. (48), we get the relation

$$\left\langle S_{3/2}^{(l)}\left(\beta \frac{p^2}{2m}\right)\right\rangle_c = 0. \quad (53)$$

Thus, we obtain
$$\lim_{t \to \infty} \delta A(t) = 0, \quad (54)$$

which is the end of proof. Thus, the steady state starting from Eq.(47) is equivalent to the one from the canonical distribution (19).

To demonstrate the irrelevancy of the choice of a specific initial condition, we shall present computer-simulation results for two-dimensional soft granular particles. The simulations have been done for the canonical initial condition and for the homogeneous cooling state. The system consists of polydisperse 4000 grains of diameters $0.7\sigma_0$, $0.8\sigma_0$, $0.9\sigma_0$ and σ_0, and the number of grains of each diameter is 1000. The total area fraction is $\phi = 0.5$. We have adopted the linear spring model to represent the elastic repulsion during a contact. All variables are non-dimensionalized by the maximum diameter of grains σ_0, its mass m and the spring constant k. The dissipation appears through the linear viscous damping with its coefficient $\eta_{damp} = 1.0$. (m, k, and $\eta_{damp} = 1.0$ are common to all the grains.) The applied shear rate is $\dot{\gamma} = 0.0005$. The freely cooling initial condition has been prepared by performing a granular simulation in the absence of shear up to $t = 200$ starting from the canonical distribution at $T_0 = 5 \times 10^{-5}$.

Figure 1 shows the result of the granular temperature defined by the kinetic energy. It is easily seen that all the results starting from different initial conditions converge to a unique steady kinetic temperature in the long time limit.

Figure 2 displays the radial distribution function $g(r)$ for the largest grains in a steady state. We find that the radial distribution function $g(r)$ in the steady state is independent of the initial condition. Thus, our numerical results support our theoretical prediction that steady-state properties are independent of the choice of a specific initial condition.

Thus, the generalized Green-Kubo formula (41) yields the same steady-state average irrespective of the initial condition. We should note that most of nonequilibrium generalizations of the Green-Kubo formula assume the

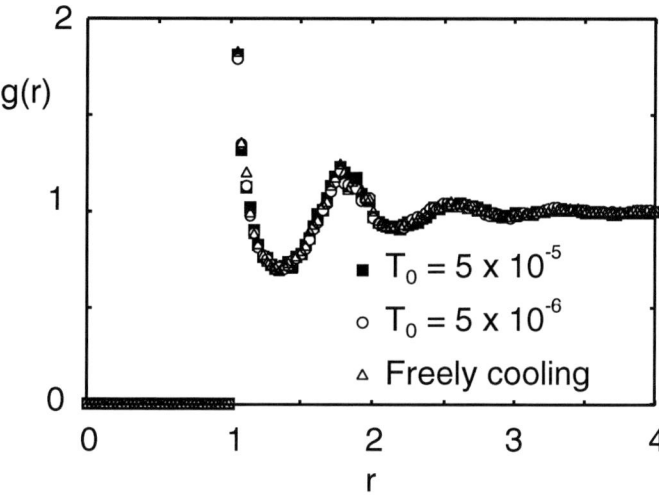

FIGURE 2. The radial distribution function in the steady state. The legend is common with that in Fig.1.

existence of a nonequilibrium steady distribution function and strongly depends on its steady distribution ρ_{ss}. This standard method has several difficulties such as (i) the determination of ρ_{ss} is difficult, and (ii) the distribution might not approach a steady value, because there are in general no compatible solutions of both $i\mathscr{L}A_{ss}(\Gamma) = 0$ and $i\mathscr{L}^\dagger \rho_{ss}(\Gamma) = 0$. On the other hand, our formulation is free from such difficulties, and we can calculate, e.g., the steady shear stress under the canonical initial condition and the obtained result is independent of the adopted initial condition.

OUTLINE OF THE LIQUID THEORY BEYOND GREEN-KUBO FORMULA

In this section, we briefly explain how to use the generalized Green-Kubo formula to describe sheared granular liquids. Because of the limitation of the length of this paper, we shall skip details of the derivation, and the interested reader is referred to ref.[41].

Because the energy is not a conserved quantity which quickly relaxes to a steady value in the uniform shear, the relevant hydrodynamic variables are the density fluctuations

$$n_{\mathbf{q}}(t) \equiv \sum_i e^{i\mathbf{q}\cdot\mathbf{r}_i(t)} - N\delta_{\mathbf{q},\mathbf{0}}, \tag{55}$$

and the current density fluctuations $j_{\mathbf{q}}^\lambda$ defined by

$$j_{\mathbf{q}}^\lambda = \sum_i \frac{p_i^\lambda}{m} e^{i\mathbf{q}\cdot\mathbf{r}_i}. \tag{56}$$

We introduce the projection operator $\mathscr{P}$ onto these variables:

$$\mathscr{P}X \equiv \sum_{\mathbf{k}} \langle X n_{\mathbf{k}}^* \rangle \frac{1}{NS_k} n_{\mathbf{k}} + \sum_{\mathbf{k}} \langle X j_{\mathbf{k}}^{\mu*} \rangle \frac{1}{Nv_T^2} j_{\mathbf{k}}^\mu, \tag{57}$$

where $S_k \equiv \langle n_{\mathbf{k}}(t) n_{\mathbf{k}}(t)^* \rangle / N$ and $v_T \equiv \sqrt{T/m}$. The complementary projection operator shall be defined by $\mathscr{Q} \equiv I - \mathscr{P}$.

Let us introduce

$$R_{\mathbf{q}}^\lambda(t) \equiv e^{i\mathscr{Q}\mathscr{L}\mathscr{Q}t} R_{\mathbf{q}}^\lambda \tag{58}$$

with

$$R_{\mathbf{q}}^\lambda \equiv \mathscr{Q}i\mathscr{L}j_{\mathbf{q}}^\lambda = i\mathscr{L}j_{\mathbf{q}}^\lambda - iq_\lambda \frac{v_T^2}{S_q} n_{\mathbf{q}} - iB_{\mathbf{q}}^\lambda n_{\mathbf{q}} + A_{\mathbf{q}}^{\lambda\mu} j_{\mathbf{q}}^\mu, \tag{59}$$

where $i\mathscr{L} \equiv i\mathscr{L}^{(\text{el})} + i\mathscr{L}^{(\text{vis})}$, and

$$A_{\mathbf{q}}^{\lambda\mu} = \frac{n}{m}\int d\mathbf{r}\,(1-e^{i\mathbf{q}\cdot\mathbf{r}})\,\hat{r}^{\lambda}\hat{r}^{\mu}\,\mathscr{F}(r)g(r), \tag{60}$$

$$iB_{\mathbf{q}}^{\lambda} = -\frac{\dot{\gamma}}{mS_q}\left\{n\int d\mathbf{r}\,\hat{r}^{\lambda}\hat{x}\hat{y}\,r\mathscr{F}(r)g(r)e^{i\mathbf{q}\cdot\mathbf{r}} + n^2\int d\mathbf{r}\int d\mathbf{r}'\,\hat{r}^{\lambda}\hat{x}\hat{y}\,r\mathscr{F}(r)g^{(3)}(\mathbf{r},\mathbf{r}')e^{i\mathbf{q}\cdot\mathbf{r}'}\right\}, \tag{61}$$

in terms of the pair and triple correlation functions

$$ng(r) = \frac{1}{N}\sum_{\substack{i,j\\i\ne j}}\langle\delta(\mathbf{r}-\mathbf{r}_{ij})\rangle, \tag{62}$$

$$n^2 g^{(3)}(\mathbf{r},\mathbf{r}') \equiv \frac{1}{N}\sum_{i\ne j\ne l}\langle\delta(\mathbf{r}-\mathbf{r}_{ij})\delta(\mathbf{r}'-\mathbf{r}_{il})\rangle. \tag{63}$$

One finds the following continuity equations for the sheared system relating the partial time derivative of $n_{\mathbf{q}}(t) = e^{i\mathscr{L}t}n_{\mathbf{q}}$ to $j_{\mathbf{q}}^{\lambda}(t) = e^{i\mathscr{L}t}j_{\mathbf{q}}^{\lambda}$:

$$\left[\frac{\partial}{\partial t} - \mathbf{q}\cdot\kappa\cdot\frac{\partial}{\partial \mathbf{q}}\right]n_{\mathbf{q}}(t) = i\mathbf{q}\cdot\mathbf{j}_{\mathbf{q}}(t), \tag{64}$$

and

$$\left[\frac{\partial}{\partial t} - \mathbf{q}\cdot\kappa\cdot\frac{\partial}{\partial \mathbf{q}}\right]j_{\mathbf{q}}^{\lambda}(t) = iq_{\lambda}\frac{v_T^2}{S_q}n_{\mathbf{q}}(t) + iB_{\mathbf{q}}^{\lambda}n_{\mathbf{q}}(t) - A_{\mathbf{q}}^{\lambda\mu}j_{\mathbf{q}}^{\mu}(t) + R_{\mathbf{q}}^{\lambda}(t) - \int_0^t ds\,M_{\mathbf{q}}^{\lambda\mu}(s)e^{i\mathscr{L}(t-s)}j_{\mathbf{q}(s)}^{\mu}$$
$$+ \int_0^t ds\,iL_{\mathbf{q}}^{\lambda}(s)e^{i\mathscr{L}(t-s)}n_{\mathbf{q}(s)} - \int_0^t ds\,N_{\mathbf{q}}^{\lambda\mu}(s)e^{i\mathscr{L}(t-s)}j_{\mathbf{q}(s)}^{\mu}, \tag{65}$$

where we have introduced the following memory kernels

$$M_{\mathbf{q}}^{\lambda\mu}(t) \equiv \frac{1}{Nv_T^2}\langle R_{\mathbf{q}}^{\lambda}(t)R_{\mathbf{q}(t)}^{\mu*}\rangle, \tag{66}$$

$$L_{\mathbf{q}}^{\lambda}(t) \equiv -i\frac{1}{NS_{q(t)}}\langle R_{\mathbf{q}}^{\lambda}(t)\mathscr{Q}[n_{\mathbf{q}(t)}^{*}\Omega(0)]\rangle, \tag{67}$$

$$N_{\mathbf{q}}^{\lambda\mu}(t) \equiv -\frac{1}{Nv_T^2}\langle R_{\mathbf{q}}^{\lambda}(t)\mathscr{Q}[j_{\mathbf{q}(t)}^{\mu*}\Omega(0)]\rangle. \tag{68}$$

The equations (64)-(68) are the exact equations for uniformly sheared granular liquids governed by Eqs. (1)-(4).

To obtain a closure to these equations, one has to introduce some approximations such as the mode-coupling approximation. The details of such approximations will be reported elsewhere.[41] Instead, here, let us briefly explain what equations can be obtained under the Markovian approximations in the case of a weak shear and an elastic limit.

It is straightforward to show that Eq.(65) reduces to

$$\left[\frac{\partial}{\partial t} - \mathbf{q}\cdot\kappa\cdot\frac{\partial}{\partial \mathbf{q}}\right]j_{\mathbf{q}}^{\lambda}(t) \approx iq_{\lambda}\frac{v_T^2}{S_q}n_{\mathbf{q}}(t) - \int_0^t ds\,M_{\mathbf{q}}^{\lambda\mu}(s)e^{i\mathscr{L}(t-s)}j_{\mathbf{q}(s)}^{\mu} + R_{\mathbf{q}}^{\lambda}(t) \tag{69}$$

under the weak shear and the elastic limit. Thus, the effect of shear appears only through the convective deformation of the wave number in the elastic and the unsheared limit. This equation still includes the non-Markovian memory kernel.

For many situations in a liquid state far from the jamming transition, one can ignore memory effects. When we adopt the Markovian approximation with the assumption that the system is isotropic, it is known that the memory kernel can be approximately given by its hydrodynamic limit[45]

$$M_{\mathbf{q}}^{\lambda\mu}(t) \approx \hat{q}_{\lambda}\hat{q}_{\mu}q^2 v_1\delta(t) + (\delta_{\lambda\mu} - \hat{q}_{\lambda}\hat{q}_{\mu})q^2 v_2\delta(t), \tag{70}$$

where $\hat{q}_\lambda \equiv q_\lambda/q$, ν_1 and ν_2 are the bulk kinetic viscosity and the kinetic viscosity, respectively. Equation (69) then reduces to

$$\left[\frac{\partial}{\partial t} - \mathbf{q} \cdot \kappa \cdot \frac{\partial}{\partial \mathbf{q}}\right] j_\mathbf{q}^\lambda(t) \approx iq_\lambda \frac{v_T^2}{S_q} n_\mathbf{q}(t) - \nu_1 q_\lambda (\mathbf{q} \cdot \mathbf{j}_\mathbf{q}(t)) - \nu_2 q^2 (\delta_{\lambda\mu} - \hat{q}_\lambda \hat{q}_\mu) j_\mathbf{q}^\lambda(t) + R_\mathbf{q}^\lambda(t), \qquad (71)$$

where $R_\mathbf{q}^\lambda(t)$ satisfies the fluctuation-dissipation relation

$$\langle R_\mathbf{q}^\lambda(t) R_\mathbf{k}^\mu(0) \rangle = \delta(t)\delta(\mathbf{q}+\mathbf{k}) N v_T^2 \{\hat{q}_\lambda \hat{q}_\mu q^2 \nu_1 + (\delta_{\lambda\mu} - \hat{q}_\lambda \hat{q}_\mu) q^2 \nu_2\}. \qquad (72)$$

Equation (71) combined with Eqs. (64) and (72) is the equation of fluctuating hydrodynamics.

Let us notice that both the equal-time long-range correlation function[12] and the long-time tails of autocorrelation functions[11, 13] can be discussed within the framework of the fluctuating hydrodynamics. Our liquid theory presented here, therefore, provides not only a microscopic basis of the fluctuating hydrodynamics, but also a basis of both the long-time tails and the long-range correlations for sheared granular liquids.

DISCUSSION AND CONCLUSION

Discussion

Now, let us compare our formulation with previous formulations of the Green-Kubo formula for granular fluids.[16, 18, 19, 20, 21, 22] We first stress that our formulation is unique in that it applies to a nonequilibrium steady state of uniformly sheared granular liquids, while the previous ones deal with the Green-Kubo formula for granular gases in a freely cooling state. We also note that Green-Kubo formula in the previous studies is the linear response theory to a nonequilibrium "steady state" (which includes a homogeneous cooling state), but our method is a nonlinear response theory to the initial canonical state. In practice, the application of the linear response theory to freely cooling granular gases has some technical problems; (i) one cannot take the $t \to \infty$ limit of the integral of the time-correlation function because the granular particles quickly lose their kinetic energy, (ii) thus the behavior of the steady state strongly depends on the cut-off time of the integration, and (iii) the determination of a nonequilibrium steady distribution is difficult. Let us also notice that that freely cooling states cannot be realized in experiments. On the other hand, our method which is the nonlinear response theory to the initial canonical distribution has several advantages such as (i) the steady state under a uniform shear can be approximately realized in many situations, and (ii) the steady state is almost independent of the choice of the initial condition, although our method cannot be generalized to a response theory to a reference state.

Nevertheless, it is remarkable that the Green-Kubo formula for freely cooling granular gases has a similar structure to our generalized Green-Kubo formula. Indeed, Eq. (139) in ref.[16] is the essentially same as Eq.(39) where their expression corresponds to the integral $\eta \propto \int_0^\infty dt \langle \sigma_{xy}(0) [e^{-i\mathcal{L}^\dagger t} \Omega(\Gamma)] \rangle$ in our context.

Conclusion

This paper summarizes our recent studies on the granular liquid theory under the uniform shear. We demonstrated that there exists the generalized Green-Kubo formula in sheared granular liquids. We also showed that it yields the nonequilibrium steady-state average which is essentially independent of a specific choice of the initial condition. It should be noted that our formulation does not rely on the presence of the steady distribution function which is hard to obtain, but any averaged quantity in the steady state can be obtained under the simple initial condition. In the previous section, we outlined how we can apply our formulation to characterize the behaviors of granular liquids. One of the most important conclusions in this paper is that the granular liquid theory starting from the Liouville equation reduces to the fluctuating hydrodynamics if one considers a weak-shear and elastic limit along with the Markovian approximation. Thus, this paper provides a microscopic support of the fluctuating hydrodynamics which is known to give accurate results of the time correlations and the spatial correlations.

We have skipped detailed derivation of Mori-type generalized Langevin equation and MCT. This will be discussed in another paper.[41] We also note that we can derive the integral fluctuation theorem along the same line, though there is no microscopic time reversal symmetry. [40] The formulation of granular liquid theory, thus, gives an interesting subject in pure nonequilibrium statistical mechanics.[46]

ACKNOWLEDGMENTS

This work was partially supported by Ministry of Education, Culture, Science and Technology (MEXT), Japan (Nos. 20740245, 21015016, 21540384 and 21540388), and by the Global COE program " The Next Generation of Physics, Spun from Universality and Emergence" from MEXT Japan. The author also thanks the Yukawa International Program for Quark-Hadron Sciences at Yukawa Institute for Theoretical Physics, Kyoto University. The numerical calculation was carried out on Altix3700 BX2 at YITP in Kyoto University.

REFERENCES

1. I. Goldhirsch, Annu. Rev. Fluid Mech. **35**, 267 (2003).
2. N. V. Brilliantov and T. Pöschel, *Kinetic Theory of Granular Gases*(Oxford University Press, Oxford, 2004).
3. J. T. Jenkins and M. W. Richman, Phys. Fluids **28**, 3485 (1985).
4. V. Garzo and J. W. Dufty, Phys. Rev. E **59**, 5895 (1998).
5. J. F. Lutsko, Phys. Rev. E **72**, 021306 (2005).
6. K. Saitoh and H. Hayakawa, Phys. Rev. E **75**, 021302 (2007).
7. V. Kumaran, Phys. Rev. Lett. **96**, 258002 (2006).
8. A. V. Orpe and A. Kudrolli, Phys. Rev. Lett. **98**, 238001 (2007).
9. C. H. Rycroft, A. V. Orpe and A. Kudrolli, Phys. Rev. E **80**, 031305 (2009).
10. V. Kumaran, Phys. Rev. E **79**, 011301 (2009): ibid 011302.
11. M. Otsuki and H. Hayakawa, J. Stat. Mech.: Theory Exp. (2009) L08003.
12. M. Otsuki and H. Hayakawa, Phys. Rev. E **79**, 021502 (2009).
13. M. Otsuki and H. Hayakawa, submitted to EuroPhys. J. E (arXiv:0907.4462).
14. J. Schofield and I. Oppenheim, Physica A **181**, 89 (1992), ibid **187**, 210 (1992), ibid **204**, 555 (1994).
15. J. J. Brey, J. W. Dufty and A. Santos, J. Stat. Phys. **87**, 1051 (1997).
16. J. W. Dufty, A. Baskaran, and J. J. Brey, Phys. Rev. E **77**, 031310 (2008).
17. J. W. Dufty, A. Baskaran, and J. J. Brey, Phys. Rev. E **77**, 031311 (2008).
18. I. Goldhirsch and T. C. van Noije, Phys. Rev. E **61**, 3241 (2000).
19. J. W. Dufty and J. J. Brey, J. Stat. Phys. **109**, 433 (2002).
20. J. W. Dufty, J. J. Brey, and J. Lutsko, Phys. Rev. E **65**, 051303 (2002).
21. J. Lutsko, J. J. Brey, and J. W. Dufty, Phys. Rev. E **65**, 051304 (2002).
22. J. J. Brey, M. J. Ruiz-Montero, P. Maynar, and M. I. Garcia de Soria, J. Phys.: Condens. Matter **17**, S2489 (2005).
23. S. P. Das, Rev. Mod. Phys. **76**, 785 (2004).
24. W. Götze, Complex Dynamics of Glass-Forming Liquids (Oxford Univ. Press, Oxford 2008).
25. A. J. Liu and S. R. Nagel, Nature **396**, 21 (1998).
26. M. Miguel and M. Rubi, Jamming, Yielding and Irreversible Deformation in Condensed Matter (Springer-Verlag, Berlin, 2006).
27. O. Dauchot, G. Marty, and G. Biroli, Phys. Rev. Lett. **95**, 265701 (2005).
28. A. R. Abate and D. J. Durian, Phys. Rev. E. **74**, 031308 (2006).
29. A. R. Abate and D. J. Durian, Phys. Rev. E. **76**, 021306 (2007).
30. F. Lechenault, O. Dauchot, G. Biroli and J. P. Bouchaud, Euro. Rev. Lett. **83**, 46003 (2008).
31. K. Watanabe and H. Tanaka, Phys. Rev. Lett. **100**, 158002 (2008).
32. H. Hayakawa and M. Otsuki, Prog. Theor. Phys. **119**, 381 (2008).
33. C. S. O'Hern, S. A. Langer, A. J. Liu and S. R. Nagel, Phys. Rev. Lett. **88**, 075507 (2002).
34. C. S. O'Hern, L. E. Silbert, A. J. Liu and S. R. Nagel, Phys. Rev. E **68**, 011306 (2003).
35. P. Olsson and S. Teitel, Phys. Rev. Lett. **99**, 178001 (2007).
36. T. Hatano, J. Phys. Soc. Jpn. **77**, 123002 (2008).
37. M. Otsuki and H. Hayakawa, Prog. Theor. Phys. **121**, 647 (2009).
38. M. Otsuki and H. Hayakawa, Phys. Rev. E **80**, 011308 (2009).
39. S.-H. Chong and B. Kim, Phys. Rev. E **79**, 021203 (2009).
40. S.-H. Chong, M. Otsuki and H. Hayakawa, arXiv:09061930.
41. S.-H. Chong, H. Hayakawa and M. Otsuki, in preparation.
42. D. J. Evans and G. P. Morriss, Statistical Mechanics of Nonequilibrium Liquids (Cambridge University Press, Cambridge, 2008) 2nd Edition.
43. T. Hatano, M. Otsuki and S. Sasa, J. Phys. Soc. Jpn. **76**, 023001 (2007).
44. N. Mitarai and H. Nakanishi, Phys. Rev. Lett. **94**, 128001 (2005).
45. U. Balucani and M. Zoppi, Dynamics of the Liquid State (Oxford University Press, Oxford 1994).
46. S.-H. Chong, M. Otsuki and H. Hayakawa, submitted to Prog. Theor. Phys. Suppl.

Steady, Inclined Flow of a Mixture of Grains and Fluid over a Rigid Base

James T. Jenkins[a] and Diego Berzi[b]

[a] *Department of Theoretical and Applied Mechanics*
Cornell University, Ithaca, New York 14853 USA
[b] *Department of Environmental, Hydraulic, Infrastructures and Surveying Engineering*
Politecnico di Milano, Milan, Italy

Abstract. We simplify a recent theory for steady, gravity-driven flow of a highly concentrated granular-fluid mixture, with the assumption that the flow is shallow, and we solve, in an approximate way, for the variation of height in a steady, non-uniform, inclined flow over a rigid bed in absence of sidewalls. We then analyze the influence of the approximations on the shape of the wave.

Keywords: Debris flow, granular-fluid mixture
PACS: 45.70._n, 47.55._t

INTRODUCTION

Recently, Berzi & Jenkins [1,2,3] proposed a simple theory based on a linear rheology for the particle interactions, turbulent shearing of the fluid, buoyancy, and drag. They provided a complete analytical description of the steady, uniform flow of a granular-fluid mixture over an erodible bed contained between frictional sidewalls. In order to obtain such analytical solution, they assumed a constant concentration in the particle-fluid region and the similarity of the particle and fluid velocity profiles. The predictions of this description compared favourably with the measurements in experiments on steady, uniform granular-fluid flows by Armanini et al. [4] and Larcher et al. [5]. As seen in the experiments, the particle and fluid velocity distributions, the flow depths, and the free surface inclination were completely determined by the particle and fluid volume fluxes. The prediction of the depth of the particles is a unique feature of the theory; it is a consequence of the incorporation of the sidewall forces.

Here, we first simplify the analytical relations between the particle and fluid volume fluxes, the particle and fluid depths and the angle of inclination of the free surface obtained by Berzi and Jenkins [1,2,3] in a steady, uniform flow and then we solve, in an approximate, depth-averaged way, the motion of a steady, non-uniform granular-fluid wave over a rigid bed in absence of sidewalls. Such a granular-fluid flow has been experimentally investigated by Davies [6]. We will show that the theory can reproduce the main qualitative features of these flows. In particular, the often observed bulbous shape of those waves naturally results as a consequence of the non-uniformity in the flow depths, without making use of an artificially imposed non-uniformity in the particle concentration [7].

THEORETICAL FRAMEWORK

We let ρ denote the fluid mass density, c the particle concentration, g the gravitational acceleration, W the channel width, σ the particle specific mass, d the particle diameter, η the fluid viscosity, U the fluid velocity, and u the particle velocity. The Reynolds number $R \equiv \rho d (gd)^{1/2}/\eta$ characterizes the fall velocity of the particles. In what follows, we phrase the momentum balances and constitutive relations in terms of dimensionless variables, with lengths made dimensionless by d, velocities by $(gd)^{1/2}$, and stresses by $\rho \sigma g d$.

Steady, Uniform Flow

We take $z = 0$ to be the top of the grains, $z = h$ to be the position of either the erodible or the rigid bed, and H to be the height of the water above a bed of inclination ϕ. The inclination of the free surface is θ; it coincides with ϕ in a steady, uniform flow. The degree of saturation, $\xi = H/h$, is greater than unity in the over-saturated case and less than unity in the under-saturated. Sketches of over- and under-saturated flows are depicted in Fig.1, together with a generic velocity profile for the particles.

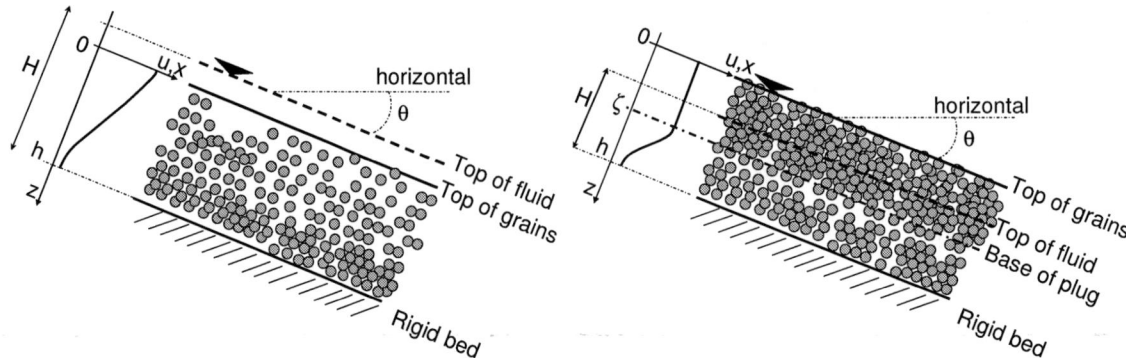

FIGURE 1. Sketch of steady, (a) over- and (b) under-saturated, uniform flows over rigid beds. Also shown are typical velocity profiles for the particles in the two configurations.

We assume that it is possible to apply the rheology proposed by the French group GDR MiDi [8]. This rheology provides the particle stress ratio $\mu \equiv s/p$ and the concentration c as unique functions of the inertial parameter $I \equiv |\dot\gamma|/(p/c)^{1/2}$, where s is the particle shear stress, p the particle pressure and $\dot\gamma$ is the strain rate. In this case, $|\dot\gamma| = -u'$; where here and in what follows, a prime indicates a derivative with respect to z.

We consider highly concentrated flows, in which the functions are approximately linear [9]:

$$\mu = \breve{\mu} + \chi I \tag{1}$$

and

$$c = \hat{c} - bI, \tag{2}$$

where $\breve{\mu}$ and $\hat{c}$ are the minimum stress ratio and the maximum concentration, respectively, and χ and b are material coefficients. The quantities $\breve{\mu}$ and $\hat{c}$ characterize the bed, at which $I = 0$; $\breve{\mu}$ is the tangent of the angle of repose and $\hat{c}$ is the concentration at dense, random packing. We expect (1) and (2) be valid up to a minimum concentration $\check{c}$ for which the flow can still be considered dense. Upon combining (1) and (2), we determine the stress ratio $\hat{\mu} = \breve{\mu} + \chi(\hat{c} - \check{c})/b$ that corresponds to the minimum concentration $\check{c}$. We will show in the following how this largest value of the stress ratio limits the applicability of the present theory.

The balances of fluid momentum transverse and parallel to the flow, in the region in which both phases are present, are

$$P' = \cos\theta/\sigma \tag{3}$$

and

$$S' = (1-c)\sin\theta/\sigma - cC(U-u)/\sigma, \tag{4}$$

respectively, where P is the fluid pressure, S the fluid shear stress, and C is the dimensionless drag,

$$C \equiv (3|U-u|/10 + 18.3/R)/(1-c)^{3.1}, \tag{5}$$

derived by Dallavalle [10], with the concentration dependence suggested by Richardson & Zaki [11]. When an upper clear fluid layer is present, the distribution of the fluid shear stress can be obtained from (4) with $c = 0$.

The balances of particle momentum transverse and parallel to the flow are

$$p' = (1 - 1/\sigma) c \cos\theta \tag{6}$$

and

$$s' = c \sin\theta + cC(U - u)/\sigma \tag{7}$$

respectively, where p is the particle effective pressure (total pressure minus the pore pressure), and s is the particle shear stress. The balances for the particles when an upper dry layer is present can be obtained from (6) and (7) by letting σ become infinite.

Here, in the mixture, we ignore the turbulent shear stress in the fluid relative to gravity and drag and neglect the friction of the sidewalls. In the clear fluid layer, we assume that the turbulent mixing length is proportional to the thickness of the layer:

$$S = -(1 - \hat{c}) k^2 (H - h)^2 |U'| U', \tag{8}$$

where $k = 0.20$, half the value of Karman's constant. We also assume that the concentration is approximately constant and at its maximum value, $c = \hat{c}$.

With these assumptions, and considering the surface at $z = 0$ as free of particle stress, it is possible to obtain the particle stress ratio μ as a function of z from the momentum balances (3), (4), (6) and (7):

$$\mu = \frac{\sigma z + (1 - \hat{c})[z - h(1 - \alpha)]/\hat{c}}{\sigma z - z + h(1 - \alpha)} \tan\theta + \frac{\sigma S^*}{[\sigma z - z + h(1 - \alpha)] \hat{c} \cos\theta} \tag{9}$$

(for details of this derivation, see Berzi and Jenkins [2]), where $\alpha = H/h$ in an under-saturated flow and unity otherwise; and, when the flow is oversaturated, $S^* = \sin\theta (H - h)/\sigma$ is the fluid shear stress at the top of the particles. Then, using the linear rheology, $\mu = \breve{\mu} + \chi I$, with $I = \sigma u'/\{[(\sigma - 1)z + h(1 - \alpha)] \hat{c} \cos\theta\}$, in (9), we can obtain an equation for u' that we may integrate for the particle velocity.

Over-saturated Flows

When the flow is oversaturated, the aforementioned equation for u' may be integrated once and the result averaged through the depth of the particles to provide an expression for the average particle velocity u_A:

$$u_A = u_b + \frac{2}{15} \frac{h^{1/2}}{\chi} \frac{1}{\hat{c}(\sigma - 1)} \left(\frac{\sigma - 1}{\sigma} \cos\theta\right)^{1/2} \left[(5H - 2h) \tan\theta + 3\hat{c}(\sigma - 1) h (\tan\theta - \breve{\mu})\right]. \tag{10}$$

If we neglect the velocity slip at the base, u_b, and assume that the inclination is small enough to replace $\cos\theta$ by unity, then

$$u_A = \frac{2}{15} \frac{h^{1/2}}{\chi} \frac{1}{\hat{c} \sigma^{1/2} (\sigma - 1)^{1/2}} \left[(5H - 2h) \tan\theta + 3\hat{c}(\sigma - 1) h (\tan\theta - \breve{\mu})\right]. \tag{11}$$

Similarly, upon integrating in succession (4) and (8) in the absence of particles and averaging the result through the depth of the clear fluid, we obtain the average fluid velocity U_{cm} there:

$$U_{cm} = U_0 + 2(\sin\theta)^{1/2} (H - h)^{1/2} /(5k), \tag{12}$$

where the subscript 0 denotes a value at $z=0$. If, in the mixture, we neglect the difference between the average fluid and particle velocities, the average fluid velocity U_m in the mixture equals u_A and

$$U_0 = u_0 = u_A + \frac{4}{15}\frac{1}{\hat{c}\sigma^{1/2}(\sigma-1)^{1/2}}\frac{h^{1/2}}{\chi}\left[(5H-4h)\tan\theta + \hat{c}(\sigma-1)h(\tan\theta - \breve{\mu})\right]. \qquad (13)$$

With these, the average fluid velocity U_A through the depth of the flow is

$$U_A = [u_A h + U_{cm}(H-h)]/H. \qquad (14)$$

Under-saturated Flows

Given the definition of the inertial parameter and the constitutive relations, we may calculate the location, ζ, of the base of the plug; given that, in the plug, the shear rate and, therefore, the inertial parameter I are both zero, the linear rheology $\mu = \breve{\mu} + \chi I$ implies that ζ is the value of z at which $\mu = \breve{\mu}$ in (9):

$$\zeta = \frac{\tan\theta - \hat{c}(\tan\theta - \breve{\mu})}{\tan\theta + (\sigma-1)\hat{c}(\tan\theta - \breve{\mu})}(h-H). \qquad (15)$$

Again, upon integrating for the particle velocity in the region of shearing and averaging the result over the depth of the region of shear:

$$u_m = \frac{2}{3}\frac{1}{\hat{c}(\sigma-1)^2\sigma^{1/2}}\frac{1}{\chi}\Bigg[\Big\{(\sigma h - H)^{3/2}\left[\tan\theta + \hat{c}(\sigma-1)(\tan\theta - \breve{\mu})\right] - 3(h-H)(\sigma h - H)^{1/2}\sigma\tan\theta$$

$$-\frac{1}{(\sigma-1)}\frac{1}{(h-\zeta)}\Big\{\frac{2}{5}\Big[(\sigma h - H)^{5/2} - ((\sigma-1)\zeta + h - H)^{5/2}\Big]\left[\tan\theta + \hat{c}(\sigma-1)(\tan\theta - \breve{\mu})\right] \qquad (16)$$

$$-2(h-H)\Big[(\sigma h - H)^{3/2} - ((\sigma-1)\zeta + h - H)^{3/2}\Big]\sigma\tan\theta\Big\}\Bigg],$$

for $h-H \leq z \leq h$.

In the upper, dry layer of an under-saturated flow, the ratio of shear stress to pressure is, with the assumption of constant concentration,

$$\mu = \tan\theta, \qquad (17)$$

The stress ratio μ must exceed its minimum value $\breve{\mu}$ to have a non-zero inertial parameter. If $\tan\theta < \breve{\mu}$, then, for $0 \leq z \leq h-H$, the shear rate is zero; if, instead, $\tan\theta > \breve{\mu}$, then the dry layer is subject to shearing. In this, we can employ the linear rheology (1) in (17), the definition of the inertial parameter and the pressure distribution $p = \hat{c}z\cos\theta$ to obtain

$$u' = -(\cos\theta)^{1/2}(\tan\theta - \breve{\mu})z^{1/2}/\chi. \qquad (18)$$

Upon integrating between z and ζ, with $z \leq \zeta$, and using the boundary condition $u(\zeta) = u_p$, we obtain the distribution of the particle velocity in the dry shear layer; upon integrating this distribution between 0 and ζ we then obtain the mean value of the particle velocity in the dry shear layer:

$$u_{dm} = \frac{2}{5} \frac{(\cos\theta)^{1/2}}{\chi} \zeta^{3/2} (\tan\theta - \breve{\mu}) + u_p. \tag{19}$$

where

$$u_p = \frac{2}{3} \frac{1}{\hat{c}(\sigma-1)^2 \sigma^{1/2}} \frac{1}{\chi} \Big\{ \Big[(\sigma h - H)^{3/2} - ((\sigma-1)\zeta + h - H)^{3/2} \Big] \big[\tan\theta + \hat{c}(\sigma-1)(\tan\theta - \breve{\mu}) \big]$$

$$-3(h-H)\Big[(\sigma h - H)^{1/2} - ((\sigma-1)\zeta + h - H)^{1/2}\Big]\sigma\tan\theta \Big\}. \tag{20}$$

So, in this case, the average particle velocity is

$$u_A = \begin{cases} u_m(1-\zeta/h) + u_{dm}\zeta/h, & \text{if } \tan\theta \geq \breve{\mu} \\ u_m(1-\zeta/h) + u_p \zeta/h, & \text{if } \tan\theta < \breve{\mu} \end{cases} \tag{21}$$

and the average fluid velocity is, simply,

$$U_A = u_m(1-\zeta/h) + u_p(\zeta - h + H)/h. \tag{22}$$

In our application of the expressions for the average particle and fluid velocities in steady uniform flow to the more complicated problem of steady, non-uniform flow, we regard the velocities as being specified and investigate the possibility of solving analytically for the inclination as a function of the velocities and depths.

Steady, Non-uniform, Flow

As do Berzi and Jenkins [3], we consider the propagation of a steady wave of a granular-fluid mixture over a rigid bed in absence of sidewalls. This particular flow configuration has been experimentally investigated by Davies [6] and seems to possess some features in common with natural debris flows.

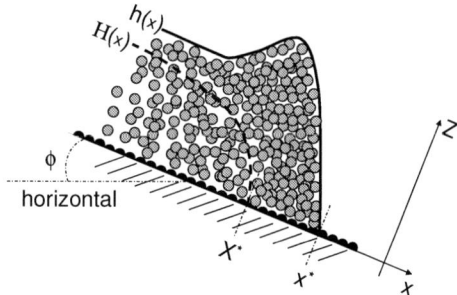

FIGURE 2. Frame of reference used for the analysis of a steady, non-uniform flow over a rigid bed.

A sketch of the flow configuration with the frame of reference is depicted in Fig. 2. In contrast to Fig. 1, the origin of the Z-axis is at the rigid bed of inclination ϕ and the coordinate Z increases towards the free surface. The origin of the x-axis is taken to be somewhere upslope. Both the particle and the fluid depth are functions of x, with x* and X* indicating the position of the particle and fluid snout, respectively. At each value of x, the ratio H/h can be greater than, equal to, or less than unity (over-, fully and under-saturated flow); also, either h or H can vanish (clear fluid or dry granular flow).

Wave Body

We first focus on the governing equations for the part of the wave in which both particles and fluid are present, $x<\min(x^*,X^*)$, and consider the wave front to the next sub-section.

The longitudinal momentum balance for the fluid is

$$\frac{1}{\sigma}\frac{dU}{dt} = \frac{1}{\sigma}\sin\phi - \frac{\partial P}{\partial x}, \qquad (23)$$

if $h \leq Z \leq \beta h$ (the clear fluid layer); and

$$\frac{1}{\sigma}(1-c)\frac{dU}{dt} = (1-c)\frac{1}{\sigma}\sin\phi - \frac{c}{\sigma}C(U-u) + \frac{\partial S}{\partial Z} - \frac{\partial\left[(1-c)P\right]}{\partial x} - P\frac{\partial c}{\partial x}, \qquad (24)$$

if $0 \leq Z \leq \alpha h$ (the mixture layer), where, when $H/h \leq 1$, $\alpha = H/h$ and $\beta = 1$ and, when $H/h > 1$, $\beta = H/h$ and $\alpha = 1$. The term involving the derivative of concentration along the flow accounts incorporates the buoyancy of the particles. The corresponding longitudinal momentum balance for the particles is

$$c\frac{du}{dt} = c\sin\phi + \frac{\partial s}{\partial Z} - \frac{\partial p}{\partial x}, \qquad (25)$$

if $\alpha h \leq Z \leq h$ (the dry layer); and

$$c\frac{du}{dt} = c\sin\phi + \frac{1}{\sigma}cC(U-u) + \frac{\partial s}{\partial Z} - \frac{\partial(p+cP)}{\partial x} + P\frac{\partial c}{\partial x}, \qquad (26)$$

if $0 \leq Z \leq \alpha h$ (the mixture layer), where $p+cP$ is the total particle pressure. When the inertial terms and the non-uniformity in the x-direction vanish, and $\phi = \theta$, (24) and (26) reduce to (4) and (7).

Berzi and Jenkins [3] depth-average the longitudinal momentum balances (23)-(26), with concentration constant and equal to $\hat{c}$, and obtain

$$\left[\beta - 1 + \alpha(1-\hat{c})\right]h\frac{dU_A}{dt} = \frac{1}{\sigma}\left[\beta - 1 + \alpha(1-\hat{c})\right]h\sin\phi - \frac{\hat{c}}{\sigma}\int_0^{\alpha h}C(U-u)dZ - (1-\hat{c})\int_0^{\alpha h}\frac{\partial P}{\partial x}dZ - \int_h^{\beta h}\frac{\partial P}{\partial x}dZ \qquad (28)$$

and

$$\hat{c}h\frac{du_A}{dt} = \hat{c}h\sin\phi + \frac{\hat{c}}{\sigma}\int_0^{\alpha h}C(U-u)dZ + \int_0^h\frac{\partial s}{\partial Z}dZ - \int_0^h\frac{\partial p}{\partial x}dZ - \hat{c}\int_0^{\alpha h}\frac{\partial P}{\partial x}dZ, \qquad (29)$$

respectively, where,

$$\left[\beta - 1 + \alpha(1-\hat{c})\right]h\frac{dU_A}{dt} \equiv \int_h^{\beta h}\frac{dU}{dt}dZ + (1-\hat{c})\int_0^{\alpha h}\frac{dU}{dt}dZ \qquad (30)$$

and

$$\frac{du_A}{dt} \equiv \frac{1}{h}\int_0^h\frac{du}{dt}dZ. \qquad (31)$$

They make the fundamental assumption that the resistances due to the internal shear stresses and the drag force in (23) through (26) can be approximated by the expressions valid when the flow is steady and uniform, i.e. when they balance only the component of the particle weight in the x-direction:

$$\frac{\hat{c}}{\sigma}\int_0^{\alpha h} C(U-u)dZ = \frac{1}{\sigma}\left[\beta - 1 + \alpha(1-\hat{c})\right]h\sin\Theta \qquad (32)$$

and

$$-\frac{\hat{c}}{\sigma}\int_0^{\alpha h} C(U-u)dZ - \int_0^h \frac{\partial s}{\partial Z}dZ = \hat{c}h\sin\theta. \qquad (33)$$

The angles Θ and θ coincide with the inclination ϕ of the bed in a steady, uniform flow. However, in the steady, non-uniform flow considered here, these three angles are different: the inclination ϕ of the rigid, bumpy bed is known, while Θ and θ can be evaluated from the solutions for the steady, uniform flows using the local values of the non-uniform heights. That is, we specify u_A and U_A. Then, the specification of u_A, h and H results in a value of θ that, in general, is different from ϕ. Similarly, the specification of U_A results in an inclination Θ different from both ϕ and θ. The angles agree if the depths of the fluid and particles become equal to their uniform values.

Berzi and Jenkins [3] employ Leibniz's rule to write the integral of derivatives as derivatives of integrals and assume that the distributions of pressure in the fluid and particles are hydrostatic. They focus on steady, non-uniform flows characterized fluid and particle velocities that are equal and independent of x [7]. Then, (28) and (29) reduce to:

$$\frac{dH}{dx} = \tan\phi - \tan\Theta \qquad (34)$$

and

$$\left(1 - \frac{\alpha}{\sigma}\right)\frac{dh}{dx} + \frac{\alpha}{\sigma}\frac{d(\beta h)}{dx} = \tan\phi - \frac{\sin\theta}{\cos\phi}. \qquad (35)$$

Equations (34) and (35) can be further simplified if we assume that the angles ϕ, θ, and Θ are small:

$$\frac{dH}{dx} = \tan\phi - \tan\Theta \qquad (36)$$

and

$$\left(1 - \frac{\alpha}{\sigma}\right)\frac{dh}{dx} + \frac{\alpha}{\sigma}\frac{d(\beta h)}{dx} = \tan\phi - \tan\theta. \qquad (37)$$

Solutions of the two ordinary differential equations (36) and (37) provide the evolution of the depths $h(x)$ and $H(x)$, when both particles and fluid are present, $x < \min(x^*, X^*)$, once the value of the angle of inclination of the rigid bumpy bed, ϕ, the common value of the average velocities, $u_A = U_A$, and two boundary conditions are specified. The two boundary conditions are the values of h and H at $x = \min(x^*, X^*)$. In the following sub-section we show how we set these boundary conditions.

Wave Front

The natural boundary conditions at the snouts are the vanishing of the particle and fluid depths, i.e. $h(x^*) = 0$ and $H(X^*) = 0$. In general, x^* is different from X^*, so that the front of the wave could be either a clear fluid ($x^* < X^*$) or dry ($x^* > X^*$). Davies' [6] careful qualitative description of his experiments refers to steady waves with a dry front. Consequently, we limit our analysis to this case.

If the front is dry, then, for $X^* \leq x \leq x^*$, $H = 0$; while h evolves according to:

$$\frac{dh}{dx} = \tan\phi - \tan\theta. \qquad (38)$$

In this simple case, we can obtain $\tan\theta$ as an explicit function of u_A and h from the solution for a steady, uniform, dry flow. Using (11) and (13),

$$\tan\theta = \breve{\mu} + \chi u_A / \left[(2h/5)h^{1/2}\right]. \qquad (39)$$

Equation (38) implies that, for a dry front, dh/dx must be negative at $x = x^*$, where $h = 0$. Given that the stress ratio is constant and equal to $\tan\theta$ and that we can have a dense flow when $\breve{\mu} \leq \mu \leq \hat{\mu}$, this condition constrains the angle of inclination of the rigid bed ($\breve{\mu} \leq \tan\phi \leq \hat{\mu}$). It implies that it is not possible to have a steady wave with a dry granular front, if the angle of inclination of the rigid bed is less than the angle of repose of the granular material. Equation (38) with (39) can be easily integrated for $X^* \leq x \leq x^*$. The value of $h = h^*$ at $x = X^*$ can then be used as a boundary condition, together with $H(X^*) = 0$, to solve (36) and (37). We use $\tan\Theta = \breve{\mu} + \chi U_A h^{*-1/2}$ for $H(X^*) = 0$ in (36). The control parameters in the problem are the inclination ϕ of the rigid bumpy bed, the common value of the average velocities $u_A = U_A$ (necessary, together with H and h, to evaluate $\tan\theta$ and $\tan\Theta$ at every step of the integration) and the positions x^* and X^* of the snouts.

Results

Here, we present the results for the steady, non-uniform flow of a mixture of water and the plastic cylinders for the particle properties $\sigma = 1.54$, $\chi = 0.5$, $\breve{\mu} = 0.41$ and $\hat{c} = 0.69$. Our interest is in testing several approximations that can be made in determination of $\tan\theta$ in terms of u_A, h, and H from (11) or (21) and $\tan\Theta$ in terms of U_A, h, and H from (14) or (22). These determinations are employed to predict flow in a moving bed device such as that employed by Davis [6].

Davies' apparatus was a rectangular flume 50 mm in width and 2 m in length, with its ends closed to retain both the fluid and the granular material. The channel bed was a corrugated nylon belt, driven by a system of rollers, and moving at a controlled, constant velocity. In the experiments, the belt velocity, the inclination of the channel, and the volumes of fluid and particles in the granular-fluid mixture were the control parameters. Fixing the belt velocity was equivalent to fixing the average velocities $u_A = U_A$ of the flow in the laboratory frame of reference (Fig. 2). In the apparatus, the mean velocity of the fluid and particles with respect to the sidewalls was zero. In this case, the frictional force exerted by the sidewalls on the particles is, on average, small.

We employ a fourth-order Runge-Kutta method to solve the three differential equations (33), (34) and (35). In doing this, we adopt values of the control parameters, in particular, the average velocity and bed inclination large enough to have a particle depth greater than 10 diameters for the most of the flow. At each step in the Runga-Kutta method, we must evaluate the equilibrium angles Θ and θ for the fluid and the particles.

To do this for known values of h and H at a given x, we evaluate ζ and we find the slope, corresponding to $\tan\Theta$ or $\tan\theta$, that satisfies the condition on the average velocity for the fluid or the particles, respectively. Here, we do this in three different ways: exactly, using a numerical scheme; approximately, ignoring $\tan\theta - \breve{\mu}$ relative to $\tan\theta$ in ζ alone; and approximately, ignoring $\tan\theta - \breve{\mu}$ relative to $\tan\theta$ in ζ, u_A and U_A. The last two determinations are valid only if $\tan\theta$ is close to the minimum slope for having a flow; but, given that they can be done analytically, it is worth analyzing their influence on the results.

In Fig. 3, we first show the influence of the neglect of the fluid shear stress in the mixture, by comparing the numerical exact solution of the present theory with the solutions obtained using the theory of Berzi and Jenkins [3]. The latter solutions had the fluid shear stress in the mixture, S, in (4), determined by either the small-scale or large-scale turbulence (mixing length equal to one tenth of a particle diameter or proportional to H, respectively), and thus allowed for a difference between fluid and particle velocity in the mixture layer; they also took into account the presence of slip at the bed. The results were obtained for a steady wave of PVC cylinders and water on a rigid bed at an inclination of $\theta = 23°$, with $u_A = U_A = 10$, $x^* = 1000$ and $X^* = 600$. The neglect of the slip velocity at the base, whose influence can be seen in the dry front, seems a reasonable approximation; while the fluid shear stress has a substantial influence on the shape of the tail, but not on the peak.

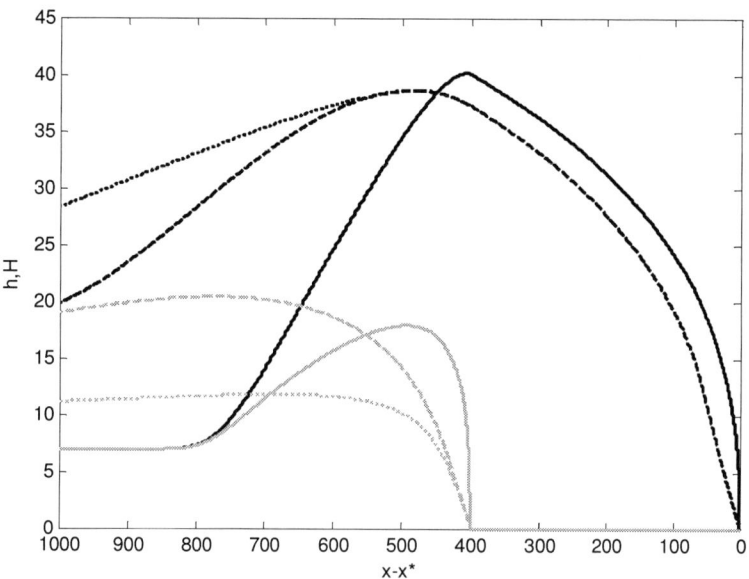

FIGURE 3. Profiles of particle height h (black lines) and fluid height H (gray lines) when the fluid shear stress is neglected (solid lines), determined by large-scale fluid turbulence (dashed lines), and determined by small-scale fluid turbulence (dotted lines).

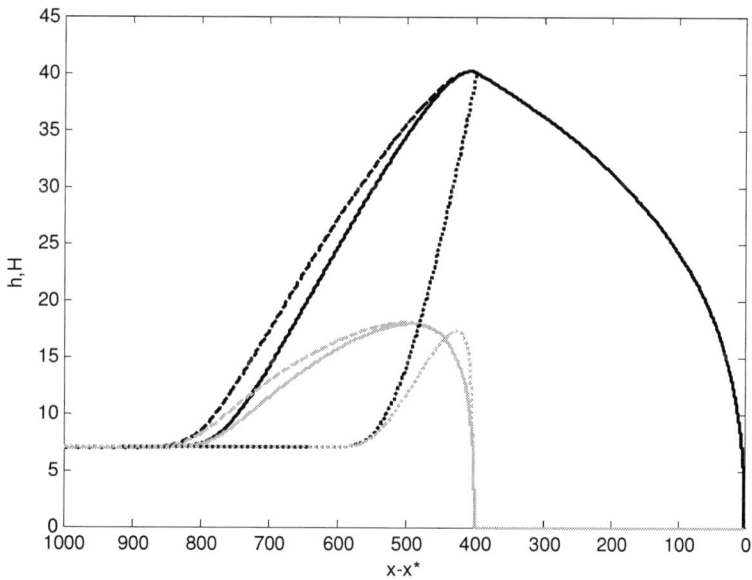

FIGURE 4. Profiles of particle height h (black lines) and fluid height H (gray lines) obtained exactly (solid lines), ignoring $\tan\theta - \bar{\mu}$ relative to $\tan\theta$ in ζ alone (dashed lines), and ignoring $\tan\theta - \bar{\mu}$ relative to $\tan\theta$ in ζ, u_A and U_A (dotted lines).

In Fig. 4, we show the differences in the computed heights of the particles and the fluid for the three different determinations of $\tan\theta$ and $\tan\Theta$. It shows that $\tan\theta - \bar{\mu}$ can be neglected relative to $\tan\theta$ in ζ, without substantially affecting the shape of the wave, but that it should be kept in the expressions for u_A and U_A. As already mentioned, the neglect of $\tan\theta - \bar{\mu}$ relative to $\tan\theta$ in ζ permits explicit analytical expression to be obtained for $\tan\theta$ and $\tan\Theta$. These can easily be implemented in mathematical models for the determination of flow configurations more complicated than those considered in the present work.

CONCLUSIONS

We employed analytical relations between particle and fluid average velocities, particle and fluid depths, and angle of inclination of the free surface to solve for the steady motion of a long granular-fluid wave over a rigid bed. We tested the sensitivity of the predicted profiles to the methods employed for determining the internal resistance of the flow and the importance of the turbulent shear stress and the method of computing it.

ACKNOWLEDGEMENT

We are grateful to Professor E. Larcan for his continued interest in this work. J. T. Jenkins acknowledges financial support from the Region of Lombardia and the hospitality of the Section of Hydraulics of the D.I.I.A.R. Department at the Politecnico di Milano.

REFERENCES

1. D. Berzi and J. T. Jenkins, *J. Fluid Mech.* **608**, 393–410 (2008).
2. D. Berzi and J. T. Jenkins, *Phys. Rev. E* **78**, 011304 (2008).
3. D. Berzi and J. T. Jenkins, *J. Fluid Mech.* **641**, 359–387 (2009).
4. A. Armanini, A., H. Capart, L. Fraccarollo and M. Larcher, *J. Fluid Mech.* **532**, 269–319 (2005).
5. M. Larcher, L. Fraccarollo L., A. Armanini. and H. Capart, *J. Hydr. Res.* **45**, 1–13 (2007).
6. T. R. H. Davies, *New Zealand J. Hyd.* **29**, 18–46. (1990).
7. O. Hungr, *Earth Surf. Proc. Lndfrms.* **25**, 483–495 (2000).
8. G. D. R. Midi, *Euro. Phys. J. E* **14**, 341–365 (2004).
9. F. da Cruz, S. Emam, M. Prochnow, J._N. Roux and F. Chevoir, *Phys. Rev. E* **72**, 021309 (2005).
10. J. Dallavalle, *Micromeritics.* Pitman (1943).
11. J. F. Richardson and W. N. Zaki, *Trans. Inst. Chem. Engrs.* **32**, 35–53 (1954).
12. J. T. Jenkins and E. Askari, *J. Fluid Mech.* **223**, 497–508 (1991).
13. O. Pouliquen, *Phys. Fluids* **11**, 1956–1958 (1999).
14. T. Takahashi, *Debris Flow,* IAHR Monograph, Balkema, Rotterdam ,1991.
15. R. Iverson, *Rev. Geophys.* **35**, 245–296 (1997).

Microstructure and Particle-Phase Stress in a Dense Suspension

James T. Jenkins[a] and Luigi La Ragione[b]

[a]*Department of Theoretical and Applied Mechanics, Cornell University, Ithaca, NY 14853*
[b]*Dipartimento di Ingegneria Civile e Ambientale Politecnico di Bari, 70125 Bari, Italy*

Abstract. We calculate the microstructure and particle phase stress in the slow flow of a dense monolayer of identical spheres subjected to pure straining in a viscous liquid. The spheres are assumed to interact through lubrication forces and a short-range repulsion. This repulsive force breaks the symmetry of the approach and departure of a typical pair and is responsible for the asymmetry in the particle concentration about axes at 45^0 to the principal axes of straining. This asymmetry in particle concentration is responsible for a particle pressure and influence the shear stress associated with the particle phase.

Keywords: Concentrated suspensions, microstructure, particle-phase stress
PACS: 66. 20 Ej, 83.80.Hj, 83.10.Gr

MODEL

We consider a concentrated monolayer of identical spheres of radius a that are subjected to pure straining in a viscous liquid. We consider pure straining, in order to minimize the complication of particle rotations; however, we note that without particle rotations, moment equilibrium is not satisfied. The situation of interest is indicated in Fig. 1. Particles are assumed to interact through lubrication forces and a short-range repulsion.

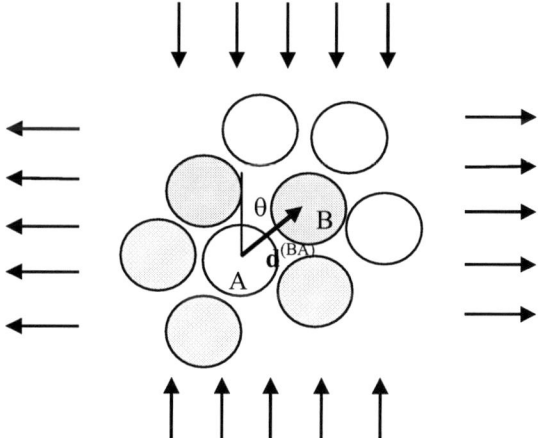

FIGURE 1. The pair A-B with the neighbors of sphere A shaded.

We focus attention on a typical pair of neighboring spheres, with the vector $\mathbf{d}^{(BA)}$ from the center of A to that of B making an angle of θ with the principal axis of compression, and determine the relative velocity of their centers by requiring that the sum of forces on each be zero, assuming that their neighbors move with the average velocity appropriate to their position [1,2]. The presence of the repulsive force breaks the symmetry of the approach and departure of the pair. Knowledge of how the components of the relative velocity parallel and perpendicular to their line of centers vary with its orientation with respect to the principal axes of straining permits the calculation of the variation of the particle pair distribution with orientation to the flow. Because of the asymmetry in approach and departure, this pair distribution is asymmetric with respect to axes intermediate to the principal axes of straining. The asymmetry in pair distribution is associated with the existence of a particle pressure and influences the magnitude of the particle shear stress.

FORCE EQUILIBRIUM

We write the equilibrium of forces for particle A interacting with its neighbors:

$$F_i^{(BA)} + \sum_{n \neq B}^{N^{(A)}} F_i^{(nA)} = 0, \tag{1}$$

where $\mathbf{F}^{(BA)}$ is, for example, the force exerted by B on A and $N^{(A)}$ is the number of neighbors of A. In what follows, $s^{(BA)}$ is the separation of the edges of spheres A and B, a hat denotes a unit vector, $\mathbf{v}^{(BA)}$ is the velocity of the center of B relative to A, a dot indicates a time derivative, $\hat{t}^{(BA)} = (\cos\theta, -\sin\theta)$ is the unit vector perpendicular to $\mathbf{d}^{(BA)}$, $\mathbf{D}$ is the symmetric part of the spatial gradients of the average velocity, and ν is the area fraction occupied by the sphere in the monolayer.

We treat the interaction between A and B exactly, as the sum of a lubrication force transmitted through a liquid with viscosity μ [3] and a short-range force characterized by a parameter F_0 [4]:

$$F_i^{(BA)} = 6\pi a \mu K_{ij}^{(BA)} v_j^{(BA)} - \frac{F_0}{s^{(BA)}} \hat{d}_i^{(BA)}, \tag{2}$$

where

$$v_i^{(BA)} = \dot{s}^{(BA)} \hat{d}_i^{(BA)} + 2a\dot{\theta}^{(BA)} \hat{t}_i^{(BA)} \tag{3}$$

and

$$K_{ij}^{(BA)} = \frac{1}{4} \frac{a}{s^{(BA)}} \hat{d}_i^{(BA)} \hat{d}_j^{(BA)} + \frac{1}{6} \ln\left(\frac{a}{s^{(BA)}}\right)\left(\delta_{ij} - \hat{d}_i^{(BA)} \hat{d}_j^{(BA)}\right). \tag{4}$$

We treat the interaction between sphere A and sphere n, where $n \neq B$, in a mean-field manner and employ the average separation $\bar{s}$ between their edges:

$$F_i^{(nA)} = 12\pi a^2 \mu K_{ij}^{(nA)} D_{jk} \hat{d}_k^{(nA)} - \frac{F_0}{\bar{s}} \hat{d}_i^{(nA)}, \tag{5}$$

and

$$K_{ij}^{(nA)} = \frac{1}{4} \frac{a}{\bar{s}} \hat{d}_i^{(nA)} \hat{d}_j^{(nA)} + \frac{1}{6} \ln\left(\frac{a}{\bar{s}}\right)\left(\delta_{ij} - \hat{d}_i^{(nA)} \hat{d}_j^{(nA)}\right). \tag{6}$$

For the average separation of the edges of the spheres, we employ the value of the mean free path from the kinetic theory for circular disks [5], $\bar{s} = \sqrt{\pi} a / (8\nu g_0)$, where the radial distribution function, g_0, is that appropriate for dense flows [6], $g_0 = 7.26(0.82 - 0.69)/(0.82 - \nu)$. The use of the average separation for the neighbors of A other than B is rather approximate, but the force balance is solved for the relative velocity for every orientation of the pair A-B. This provides a means for both testing and improving upon the approximation.

The components of the force balance (1) parallel and perpendicular to the line of centers of A-B are

$$0 = \frac{3\pi}{2}\frac{a^2\mu}{s^{(BA)}}\dot{s}^{(BA)} - \frac{F_0}{s^{(BA)}} + 3\pi\frac{a^3\mu}{s}\hat{d}_i^{(BA)}J_{ijk}D_{jk} - \frac{F_0}{s}\hat{d}_i^{(BA)}Y_i \quad (7)$$

and

$$0 = \ln\left(\frac{a}{s^{(BA)}}\right)\dot{\theta}^{(BA)} + \frac{3}{2}\frac{a}{s}\hat{t}_i^{(BA)}J_{ijk}D_{jk}, \quad (8)$$

respectively. In these,

$$J_{ijk} \equiv \overline{\sum_{n\neq B}^{N^{(A)}}\hat{d}_i^{(nA)}\hat{d}_j^{(nA)}\hat{d}_k^{(nA)}} = \alpha_1\hat{d}_i^{(BA)}\hat{d}_j^{(BA)}\hat{d}_k^{(BA)} + \alpha_2\left(\delta_{ij}\hat{d}_k^{(BA)} + \delta_{ik}\hat{d}_j^{(BA)} + \delta_{jk}\hat{d}_i^{(BA)}\right) \quad (9)$$

and

$$Y_i \equiv \overline{\sum_{n\neq B}^{N^{(A)}}\hat{d}_i^{(nA)}} = \xi\hat{d}_i^{(BA)}, \quad (10)$$

where the over-bar indicates a conditional average over orientations, given the presence of sphere B,

$$\overline{\Psi_{i..k}} = 2\int_0^\pi I(\theta')\hat{d}_i..\hat{d}_k d\theta', \quad (11)$$

in which $I(\theta)\,d\theta$ is the average number of neighbors, other than B, within the element of angle $d\theta$. To calculate the coefficients α_1, α_2, and ξ in the continuum representations of **J** and **Y**, we employ, as a first approximation, a distribution, with sphere B located at $\theta' = 0$, that is insensitive to the orientation of the pair A-B with respect to the principal axes of straining:

$$I(\theta') = \begin{cases} 0, & \text{if } 0 < \theta' < \pi/3 \\ 3(k-2)/2\pi, & \text{if } \pi/3 < \theta' < \pi/2 \\ k/2\pi, & \text{if } \pi/2 < \theta' < \pi \end{cases} \quad (12)$$

in which k is the average number of neighbors per particle - the coordination number. Given the presence of particle B, this distribution function ensures that there are k/2 neighbors uniformly distributed above and below the diameter that is perpendicular to the line of centers. With this, the coefficients are

$$\alpha_1 = -\frac{2k-6}{3\pi}, \quad \alpha_2 = -\frac{(9\sqrt{3}-16)k + 6(8-3\sqrt{3})}{24\pi}, \quad \text{and} \quad \xi = -\frac{(3\sqrt{3}-4)k + 6(2-\sqrt{3})}{2\pi}. \quad (13)$$

Eqs. (7) and (8) may be written in terms of these coefficients as

$$3\pi a\mu \frac{a}{s^{(BA)}} \dot{s}^{(BA)} = 2F_0 \left(\frac{1}{s^{(BA)}} + \xi \frac{1}{\bar{s}} \right) - 6\pi a^2 \mu \frac{a}{s}(\alpha_1 + 2\alpha_2)\hat{d}_i^{(BA)} D_{ij} \hat{d}_j^{(BA)} \qquad (14)$$

and

$$\ln\left(\frac{a}{s^{(BA)}}\right)\dot{\theta}^{(BA)} = -3\frac{a}{s}\alpha_2 \hat{t}_i^{(BA)} D_{ij} \hat{d}_j^{(BA)} . \qquad (15)$$

These equations describe the radial and circumferential motion of sphere B relative to sphere A. We next characterize the influence of the repulsive force on this motion.

In pure straining,

$$D_{ij} = \begin{bmatrix} \dot{\gamma} & 0 \\ 0 & -\dot{\gamma} \end{bmatrix}; \qquad (16)$$

so,

$$\hat{t}_i^{(BA)} D_{ij} \hat{d}_j^{(BA)} = \dot{\gamma}\sin 2\theta \quad \text{and} \quad \hat{d}_i^{(BA)} D_{ij} \hat{d}_j^{(BA)} = -\dot{\gamma}\cos 2\theta \qquad (17)$$

We make lengths dimensionless by a, introduce $F = 2F_0/(3\pi\mu\dot{\gamma}a^3)$, and write $d\gamma = \dot{\gamma}\,dt$. Then, Eqs. (14) and (15) become

$$\frac{1}{s}\frac{ds}{d\gamma} = F\left(\frac{1}{s} + \frac{\xi}{\bar{s}}\right) + \frac{2}{s}(\alpha_1 + 2\alpha_2)\cos 2\theta \qquad (18)$$

and

$$\ln\left(\frac{1}{s}\right)\frac{d\theta}{d\gamma} = -3\frac{1}{s}\alpha_2 \sin 2\theta . \qquad (19)$$

Upon eliminating the parameter γ between Eqs. (18) and (19), we obtain

$$\frac{ds}{d\theta} = -\frac{1}{3}\frac{F\bar{s} + [\xi F + 2(\alpha_1 + 2\alpha_2)\cos 2\theta]s}{\alpha_2 \sin 2\theta}\ln\left(\frac{1}{s}\right). \qquad (20)$$

We integrate this equation numerically for $\nu = 0.69$ and $k = 5.5$ to determine the separation between the edges of the spheres as a function of the orientation of a pair, both in the presence and absence of a repulsive force. The constant of integration is determined by the requirement that the average of s be $\bar{s}$. In the Appendix, we indicate how an approximation to this equation may be integrated analytically. In Fig. 2, we show the plots of dimensionless separation versus circumferential angle for $F = 0$ and $F = 10^{-4}$. For the latter, there is a small asymmetry in the neighborhood of $\theta = \pi/4$.

MICROSTRUCTURE

We next determine the effect that the small asymmetry in particle separation has on the distribution $A(\theta)$ of particles around the circumference of a typical particle, where $A(\theta)$ is defined so that $A(\theta) d\theta$ is the average number of particles in the increment of angle $d\theta$.

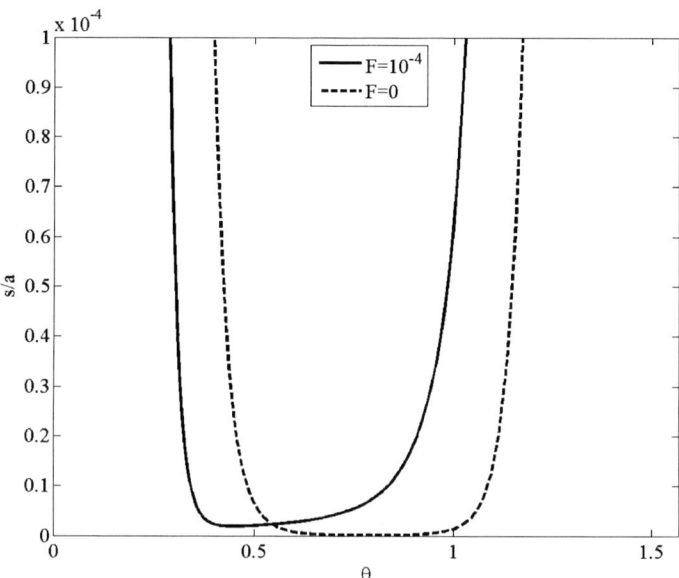

FIGURE 2. Dimensionless edge separation versus circumferential angle for $\nu = 0.69$ and $k = 5.5$ with and without a dimensionless repulsive force of 10^{-4}.

When the distribution is isotropic,

$$A = \frac{k}{2\pi}. \qquad (21)$$

However, when the circumferential velocity varies with position along the circumference, anisotropy develops in order to maintain a constant flux of particles. Then

$$A(\theta)\dot{\theta}(\theta) = \text{constant}, \qquad (22)$$

where the dot indicates a time derivative, subject to

$$\int_0^{2\pi} A(\theta) d\theta = k. \qquad (23)$$

We use the relations between s and θ obtained in the numerical solutions of Eq. (20) to determine $\dot{\theta}$, given by Eq. (19), as a function of θ. In Fig. 3, we plot the resulting dimensionless circumferential velocity as a function of orientation in the presence and absence of a repulsive force. There is a significant asymmetry in the circumferential velocity about $\theta = \pi / 4$.

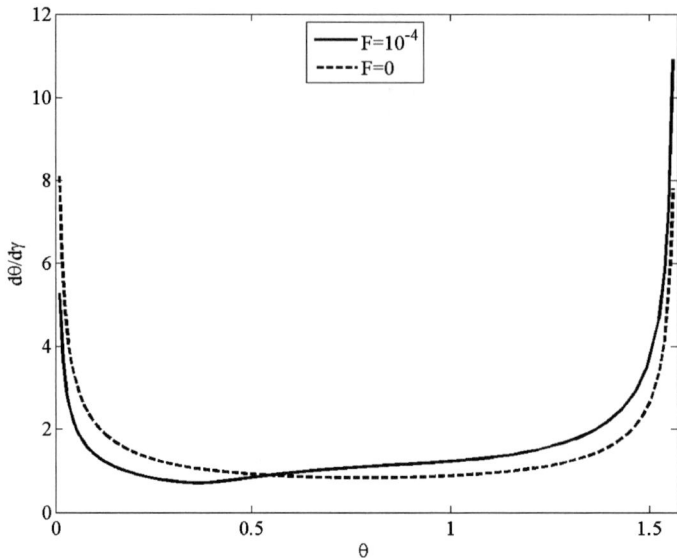

FIGURE 3. Dimensionless circumferential velocity versus circumferential angle for ν = 0.69 and k = 5.5 with and without a dimensionless repulsive force of 10^{-4}.

Finally, we substitute the numerical values of the circumferential velocity in Eq. (22), with the constant determined by Eq. (23), and calculate the distribution of particles around the circumference of a typical particle. The result is shown in Fig. 4. The asymmetry in this distribution has an important influence on the stress associated with the particle phase.

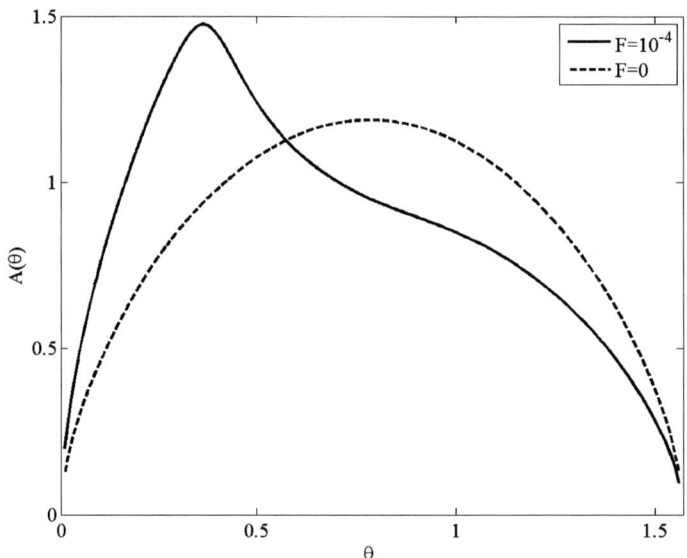

FIGURE 4. Circumferential distribution of particles versus circumferential angle for ν = 0.69 and k = 5.5 with and without a dimensionless repulsive force of 10^{-4}.

STRESS

The particle-phase stress **T** may be written as an integral over the first moment of distribution of inter-particle forces weighted by the number n of particles per unit area (cf., [7]):

$$T_{ij} = na \int_0^{2\pi} A F_i \hat{d}_j d\theta = na \int_0^{2\pi} A \left(6\pi a \mu K_{ik} v_k - F_0 \frac{a}{s} \hat{d}_i \right) \hat{d}_j d\theta, \tag{24}$$

where, **K** and **v** are given by (4) and (3), respectively. Employing these in Eq. (24), we obtain

$$T_{ij} = \frac{3\pi}{2} na^3 \mu \int_0^{2\pi} A \left\{ \left[\frac{2F_0}{3\pi a^3 \mu} \left(\frac{a}{s} + \xi \frac{a}{s} \right) - \frac{a}{s} (\alpha_1 + 2\alpha_2) \hat{d}_m D_{mn} \hat{d}_n \right] \hat{d}_i \hat{d}_j \right.$$

$$\left. - 2 \frac{a}{s} \alpha_2 \left(\hat{t}_m D_{mn} \hat{d}_n \right) \hat{t}_i \hat{d}_j \right\} d\theta - na \int_0^{2\pi} A F_0 \frac{a}{s} \hat{d}_i \hat{d}_j d\theta, \tag{25}$$

in which we note that the terms proportional to 1/s cancel.

We replace the number density by the area fraction using $n = \nu / (\pi a^2)$ and introduce the dimensionless stress, $\mathbf{t} = 2\mathbf{T}/(3 a \dot{\gamma} \mu)$. Then,

$$t_{ij} = \nu \int_0^{2\pi} A \frac{1}{s} \left\{ \left[F\xi + (\alpha_1 + 2\alpha_2) \cos 2\theta \right] \hat{d}_i \hat{d}_j - 2\alpha_2 \sin 2\theta \, \hat{t}_i \hat{d}_j \right\} d\theta. \tag{26}$$

The dimensionless pressure, p, and shear stress, S, are given as combinations of components of the dimensionless stress:

$$p \equiv -\frac{1}{2}(t_{xx} + t_{yy}) = -2\nu \int_0^{\pi/2} A \frac{1}{s} \left[F\xi + (\alpha_1 + 2\alpha_2) \cos 2\theta \right] d\theta \tag{27}$$

and

$$S \equiv \frac{1}{2}(t_{xx} - t_{yy}) = -2\nu \int_0^{\pi/2} A \frac{1}{s} \left\{ \left[F\xi + (\alpha_1 + 2\alpha_2) \cos 2\theta \right] \cos 2\theta + 2\alpha_2 \sin^2 2\theta \right\} d\theta. \tag{28}$$

For $\nu = 0.69$ and $k = 5.5$, we carry out numerical integrations for the pressure and shear stress using the particle distributions shown in Fig. 4 and give the results in Table 1. As a consequence of the asymmetry in the circumferential distribution of particles about $\theta = \pi/4$, there is a non-zero pressure associated with the particle phase and the magnitude of the particle shear stress is influenced.

TABLE 1. Particle pressure and shear stress with and without a repulsive force

Component	$F = 10^{-4}$	$F = 0$
p	14.14	0.02
S	53.26	49.37

It is interesting to see what part of the particle pressure and the particle shear stress are directly associated with the repulsive force and what parts are due to the hydrodynamics [4]. Of course, there is an indirect influence of the repulsive force to the hydrodynamics through the circumferential anisotropy. The cause of this anisotropy is that of the edge separation directly caused by the repulsive force. We separate the stress into hydrodynamic and repulsive contributions and denote them by the superscripts H and R, respectively:

$$t_{ij}^H = \nu \int_0^{2\pi} A \frac{1}{s} \left[(\alpha_1 + 2\alpha_2) \cos 2\theta \, \hat{d}_i \hat{d}_j - 2\alpha_2 \sin 2\theta \, \hat{t}_i \hat{d}_j \right] d\theta \tag{29}$$

and

$$t_{ij}^R = \nu \int_0^{2\pi} A \xi F \frac{1}{s} \hat{d}_i \hat{d}_j d\theta. \tag{30}$$

The numerical values of the separate contributions to the pressure and shear stress are given in Table 2. The direct contributions from the repulsive force are negligible.

TABLE 2.

Contribution	$F = 10^{-4}$
p^H	14.13
p^R	0.01
S^H	53.26
S^R	0.00

CONCLUSIONS

A rough implementation of particle equilibrium for particles assumed to interact as pairs through points of near contact produces expressions for the normal and tangential components of their relative velocities as functions of their orientation with respect to the flow. From these velocities, the separation of neighboring particles and their circumferential distribution can be determined as functions of orientation. The existence of asymmetries in the circumferential distribution of particles influences the shear stress and produces a pressure associated with viscous interactions between particle pairs at points of near contact. We hope that the extension of the methods employed here to more complicated flows will help in the interpretation of existing numerical simulations [4], physical experiments [8], and phenomenological theories [9].

ACKNOWLEDGMENTS

The authors are grateful for support from Strategic Plan-119, Regione Puglia (Italy) and Gruppo Nazionale della Fisica Matematica.

REFERENCES

1. J. Jenkins, D. Johnson, L. La Ragione and H. Makse, *J. Mech. Phys. Solids* **53**, 197-225 (2005).
2. L. La Ragione and J. T. Jenkins, *Proc. Roy. Soc. Lond.* A **463**, 735-758 (2007).
3. D. J. Jeffrey and Y. Onishi, *J. Fluid Mech.* **139**, 261-290 (1984).
4. A. Singh and P. R. Nott, *J. Fluid Mech.* **412**, 279-301 (2000).
5. J. T. Jenkins and M. W. Richman, *Phys. Fluids* **28**, 3485-3494 (1985).
6. S. Torquato, *Phys. Rev. E* **51**, 3170-3184 (1995).
7. J. F Brady and J. F. Morris, *J. Fluid Mech.* **348**, 103-139 (1997).
8. A. Singh and P. R. Nott, *J. Fluid Mech.* **490**, 293-320 (2003).
9. J. D. Goddard, *J. Fluid Mech* **568**, 1-17 (2006).

APPENDIX

The differential equation that relates the separation and angle is

$$\frac{ds}{d\theta} = -\frac{1}{3}\frac{F\bar{s} + [\xi F + 2(\alpha_1 + 2\alpha_2)\cos 2\theta]s}{\alpha_2 \sin 2\theta}\ln\left(\frac{1}{s}\right). \tag{A1}$$

In this, we take $s = s^0 + Fs^1$ and assume that $F \ll 1$. Then, at lowest order

$$\frac{d}{d\theta}\ln s^0 = \frac{2(\alpha_1 + 2\alpha_2)}{3\alpha_2}\frac{\cos 2\theta}{\sin 2\theta}\ln s^0, \tag{A2}$$

which has the solution

$$s^0 = \exp(C\sin 2\theta)^{\frac{\alpha_1 + 2\alpha_2}{3\alpha_2}}, \tag{A3}$$

where C is a constant to be determined by the condition on the average of the full solution. At next order,

$$\frac{ds^1}{d\theta} - \frac{2}{3}\frac{(\alpha_1 + 2\alpha_2)}{\alpha_2}\frac{\cos 2\theta}{\sin 2\theta}\left(\ln s^0 + \frac{1}{s^0}\right)s^1 = \frac{2}{3}\frac{1}{\alpha_2}\frac{1}{\sin 2\theta}(\bar{s} + \xi s^0)\ln s^0, \tag{A4}$$

which may be solve by quadrature using an integrating factor.

From Powders to Collapsing Soil/Living Quicksand: Discrete Modeling and Experiment

Dirk Kadau

Institute of Building Materials, ETH Zürich, Switzerland

Abstract. The discrete element method constitutes a general class of modeling techniques to simulate the microscopic behavior (i.e., at the particle scale) of granular/soil materials. We present a variant of the contact dynamics method, originally developed to model compact and dry systems with lasting contacts. This variant accounts for the cohesive nature of fine powders and soils. The attractive force plays an important role in the stabilization of large voids, leading to highly porous systems as e.g. in fine cohesive powders. Using this model we can explain how large pores are stabilized, and which contact laws are crucial. The model is applied for compaction of fine powders and compared to experiments, both showing a power law behavior in the high stress regime. We use a modification of the model to investigate the "quicksand" behavior of a collapsing soil material. Our contact dynamics model as a microscopic description is adjusted to capture the essential physical processes underlying the dynamics of generation and collapse of the system. Our physical model is validated with real data obtained from in situ measurements performed with a specific type of natural quicksand at the shore of drying lagoons. Cyanobacteria form an impermeable crust, giving the impression of stable ground. After breaking the crust a person rapidly sinks to the bottom of the field. We measured the shear strength of the material before and after perturbation and found a drastic change. We show that the shear strength behavior of our collapsing soil model is consistent with the behavior of this quicksand, for both the unperturbed and the collapsed phases of the material. We also investigate how deep the object can be pushed in and how well the intruder is captured by the material after it collapsed above the intruder. During the penetration process we measured the relation between the driving force and the resulting velocity of the intruder. We also investigated the influence of different strength of viscous drag acting on the grains.

Keywords: Granular matter, Contact dynamics, Simulations, Discrete element method, Quicksand, Collapsing soil
PACS: 47.57.-s,45.70.Mg,83.80.Hj

INTRODUCTION

Since beginning of human civilization the handling of granular materials is of great importance. Nowadays, in many engineering problems their mechanical behavior plays an important role. In process engineering, e.g., the storage and transport of granular matter is a central issue [1], in civil engineering major issues involve soil stability or landslides [2]. For the modeling of the mechanical behavior very often continuum models [3, 4, 5] are applied. Contrarily, particle based models, so called discrete element methods (DEM), consider the microscopic properties of particles in contact to model the mesoscopic/macroscopic mechanical behavior of particle collectives [6, 7]. In most cases "soft particle molecular dynamics" (often simply called "DEM") is used [8, 9, 10, 11, 12, 6]. Here we will use the contact dynamics approach [13, 14], designed for modeling in the limit of perfect rigidity. After briefly introducing the main concepts of the method, we present an extension of the method including cohesive forces as well as rolling friction. We first apply the method to dry powder in a shear tester, where we study the influence of particle elasticity by comparing contact dynamics to "soft particle molecular dynamics". In the following we show results for the pore stabilization of cohesive powders and their compaction behavior which is validated by experimental results. Finally, we show that a modified version could be applied to the mechanical behavior of a collapsing soil/"living quicksand". In this case our model results can be directly compared to field experiments which are also briefly presented in this paper.

CONTACT DYNAMICS METHOD

The discrete element method denotes a general class of modeling techniques to simulate the microscopic behavior (i.e. at the particle scale) of granular/soil materials. In most cases "soft particle molecular dynamics" is used, where the contact forces are calculated according to their local causes, i.e. the local deformations leading e.g. to an elastic restoring force. Nowadays this method is well established resulting in a huge number of publications, e.g. [15, 16, 8, 9,

10, 11, 12, 6]. Note that nowadays still most simulations model non cohesive particles, although models for cohesive particles are also well established in the meantime [17, 18, 19, 6].

Here, we use an alternate method, a variant of the contact dynamics method, as described in the following. The method was originally developed to model compact and dry systems with lasting contacts [14, 13]. The technique is conceptually based on the exact implementation of non-smooth contact laws, which means that the steric volume exclusion for perfectly rigid particles and the Coulomb friction law are strictly implemented. Compared to "soft particle molecular dynamics" in contact dynamics the time step can be chosen much larger as it is not necessary to solve the detailed elastic behavior at each contact. However, for a contact network the solution for the contact forces considering the non-smooth contact laws can be only determined iteratively. This involves a pseudo-elastic behavior when the number of iterations is not chosen large enough [20, 21]. To assure that in the simulations rigid particles are modeled a high number of iterations is needed, leading to a scaling of the computation time with the square of the number of particles. Still the method is advantageous for the often considered case of slow deformations and rigid particles [22]. Note that for these cases commonly in "soft particle molecular dynamics" softer particles than the real ones are modeled. Furthermore, if one is willing to allow for this pseudo-elastic behavior in contact dynamics simulations, which is analogous to the elastic behavior in "soft particle molecular dynamics", the computation time for both methods show a linear scaling. A more detailed comparison of the performance of the two method can be found in [22].

The absence of cohesion between particles can only be justified in dry systems on scales where the cohesive force is weak compared to the gravitational force on the particle, i.e. for dry sand and coarser materials, which can lead to densities close to that of random dense packings. However, an attractive force plays an important role in the stabilization of large voids [23], leading to highly porous systems as e.g. in fine cohesive powders, in particular when going to very small grain diameters. In the nanometer range of particle sizes, the cohesive force becomes the dominant force, so that particles stick together upon first contact. In the meanwhile also for contact dynamics a few simple models for cohesive particles are established [24, 25, 26, 23]. Here we consider the bonding between two particles in terms of a cohesion model with a constant attractive force F_c acting within a finite range d_c, so that for the opening of a contact a finite energy barrier $F_c d_c$ must be overcome. In addition, we implement rolling friction between two particles in contact, so that large pores can be stable in the system [23, 27, 28, 29, 30].

Summarizing, we use the following contact laws for the simulations presented in this paper. Perfect volume exclusion (Signorini condition) is assumed in normal direction, where a cohesive force is added as described above. For tangential direction Coulomb friction is applied (Coulomb law). Additionally, for the contact torques rolling friction is applied as a threshold law similar to the Coulomb law. A detailed description of this model can be found in [23]. For the simulation of the collapsing soil/living quicksand the formation and breakage of cohesive bonds has to be considered additionally. This will be described in more detail later in this paper.

Applications

Influence of particle elasticity in shear-testers

An important aspect of this study is the link between the microscopic particle properties (in particular contact properties) to the macroscopic behavior for non-cohesive powders. We compared results for perfectly rigid particles, which could be achieved by using contact dynamics, to results for soft elastic particles using "soft particle molecular dynamics". For the steady state flow in shear testers we could show that the particle elasticity plays a minor role [31]. The behavior is dominated by the collective dynamics rather than the properties of the individual contacts. For the investigated case also the distribution of forces are similar, i.e. the force statistics of both methods are comparable [31].

Pore stabilization in cohesive powders

A central question again has been the influence of the implemented contact models on the macroscopic properties. In very fine powders as e.g. nano-powders a very high pore volume of more than 90% can be observed [32]. We found in our simulations that very large pores are not mechanically stable and therefore cannot be obtained in stable configurations when using a cohesive force alone. Additionally, a friction force enhanced by the attractive force as well as rolling friction and torsion friction (3 dimensional simulations) are crucial [23, 28]. For the compaction behavior of these powders we found a power law in the simulations for high stresses when plotting the excess porosity depending on stress [23, 27, 29, 30]. To get the excess porosity one has to subtract the minimal porosity, which theoretically is obtained at infinitely high stress. Whereas the minimal porosity cannot be reached in experiments due to the breakup of the grains, in the simulations it can be easily determined by switching off the cohesion. The behavior predicted by the simulations could be confirmed for highly cohesive powders in experiments using a biaxial shear box [27, 29, 30]. By this, for relatively high compaction stresses an alternative description to the law of Kawakita [33], which had been widely used in powder technology, could be suggested [29, 30].

COLLAPSING SOIL/LIVING QUICKSAND

Despite the ubiquitous appearance of quicksand in adventure books and movies, its origin and physico-chemical behavior still represent controversial scientific issues in the fields of soil and fluid mechanics [34]. It has been argued repeatedly [35] that, because the density of sludge is typically larger than that of water, a person cannot fully submerge, and therefore cannot be really "swallowed" by any quicksand.

The fluidization of a soil due to an increase in ground water pressure, which in fact is often responsible for catastrophic failures at construction sites, is called by engineers the "quick condition" [36, 37]. Another source of fluidization can be vibrations either from an engine [38] or through an earthquake [39]. While this "liquefaction" phenomenon can essentially happen with any soil [40], it is known that samples taken from natural quicksand usually show quite specific, but anomalous rheology depending on the peculiarities of the material composition and structure [41]. Also in dry quicksand [42, 43, 44] a fragile/metastable structure leads to interesting material behavior, although the microscopic material properties are obviously quite different. Nevertheless, wet and dry quicksand are more similar than one would think. In this section we show results of real and *in situ* quicksand measurements. Based on these results we develop and numerically solve a modified version of the previously presented simulation model as a simple physical model for this quicksand/collapsing soil. The model is capable of incorporating at the microscopic level the essential structural and dynamical features of the material when subjected to different types of perturbations. Here, we focus on two aspects of the problem, namely the shear resistance property of the material and the penetration of an intruder. which will later be used to compare to simulations.

Experiments

The *in situ* measurements are performed in a natural reserve called Lençois Maranhenses located in the North-East of Brazil [45, 46]. We found that at the shore of many drying lakes in this place, it is common to find a special type of quicksand consisting of an impermeable crust lying above a metastable suspension of grains (fig. 1a). Provided one does not exert on the surface a pressure higher than $p_c = 10 - 20$ kPa, it is possible to step on it and the surface will elastically deform in a very similar way to what happens when one walks on a waterbed. These deformations visibly extend over a couple of meters. If the pressure p_c is exceeded, the surface cracks and a person/object starts sinking in [45]. We observe the separation of the excess pore-water from a repacked and wet sand soil [47] with pronounced shear thinning behavior [41]. Because the collapse of this quicksand is irreversible, we had to study its rheology and strength *in situ*.

By placing light plates on the surface we could walk on the quicksand without visually modifying it and make various measurements before and after the collapse (fig. 1a). The most striking result concerns the shear strength τ measured with a vane rheometer [49, 45, 48] are shown in Fig. 1b. The vane consists of four thin blades, which cut out a cylindrical area of the matarial when rotated. The measured shear stress is mostly that of the highly sheared region at the surface of the cylinder. By this assumption, additionally knowing the geometry, the shear stress can be calculated from the applied torque on the vane. With the field vane rheometer used here both the threshold stress and the steady state stress can be measured for quasistatic deformation, i.e. low shear rates [49]. We measured the maximal shear strength. A more detailed description of the measurements presented here can be found in [45, 48], a general description or critical discussion of interpretating the results gained by the relative complex geometry of the vane is presented e.g. in [49, 50, 51, 52]. As the blades of the vane are very thin, only a small region of the original structure will be destroyed when pushing in the rheometer which ensures that the unperturbed shear strength could be measured [48]. Before destroying the metastable structure, τ is essentially constant up to the bottom of the basin where it rapidly

 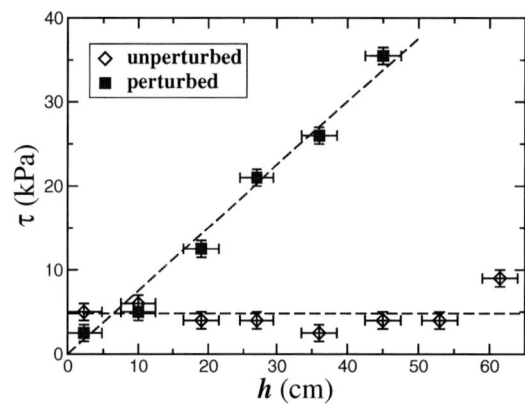

FIGURE 1. a) Example of a quicksand field in the Lençois Maranhenses. White plates (left in the figure) protect the soil from collapsing when walking on them. After the collapse excess pore water comes to the surface (perturbed spot on the right). b) Experimental measurements of the shear strength as a function of depth before (open diamonds) and after (filled squares) the collapse of the quicksand. The least squares fit to data of a linear function $\tau = ah$ gives $a = 0.75 \pm 0.03$ kPa/cm after the collapse. The shear strength of the unperturbed quicksand (before the collapse) follows an approximately constant behavior $\tau \approx 5$ kPa until reaching the bottom, which was at about 60 cm below the surface for this case (Data taken from [48]).

increases. After the system collapsed and the water came out (see perturbed spot in fig. 1a), τ linearly increases with depth h:

$$\tau(h) = ah, \quad (1)$$

with $a = 0.75 \pm 0.03$ kPa/cm. Our measurements indicate that the quicksand is essentially a suspension with depth independent shear strength. After the collapse, it becomes a soil dominated by the Mohr-Coulomb friction criterion for its shear strength. The material undergoes a cross-over from a yield stress material, i.e. a more fluid-like behavior to a Coulomb material, i.e. more solid-like behavior after the collapse.

The physico-chemical and biological analysis of this collapsing soil show the presence of a huge amount of *cyanobacteria* as well as *diatomacea* of various types [45]. During the drying of the lakes, these organisms are responsible for the formation of the elastic and impermeable crust which prevents further water from evaporating out of the metastable suspension of grains below in the bubble. As reported in previous studies [53, 54], we confirmed that the cohesion force between grains inside the suspension is mediated by the *cyanobacteria* present in the system.

Simulations

In the case of collapsing soils/"living quicksand" our cohesion model has to be modified. One also has to take into account the time necessary for bonds to appear, i.e. during relatively fast processes new bonds will not be formed, whereas for longterm processes bonds are allowed to form at a particle contact. Finally, gravity also cannot be neglected in the model since the particle diameter is usually well above the micron-size. For simplicity, however, the surrounding pore water is not explicitly considered but only taken into account as a buoyant medium, reducing the effective gravity acting onto the grains.

The comparison between the real data and our contact dynamics model can only be made qualitatively due to obvious difficulties in obtaining reasonably precise experimental measurement of any *in situ* microscopic parameters of the quicksand. The simulation has been specially designed to mimic the likely processes of formation and (local) collapse during the shear strength measurements reported before. In this way, the unperturbed quicksand is modeled by the ballistic deposition of particles driven by gravity. During this relatively slow process, bonds are allowed to form when particles stick to each other. As previously discussed, we justify the cohesive bonds in this case as being mediated by the bacteria living in the suspension. The particles are sequentially deposited on top of the already existing structure at random horizontal positions with a fixed time interval between each grain to allow for relaxation of the structure. In principle, an extremely large time between two depositions should be used to represent the slow drying process of the lakes taking place over a couple of months. The procedure adopted here is to start generating a structure using a

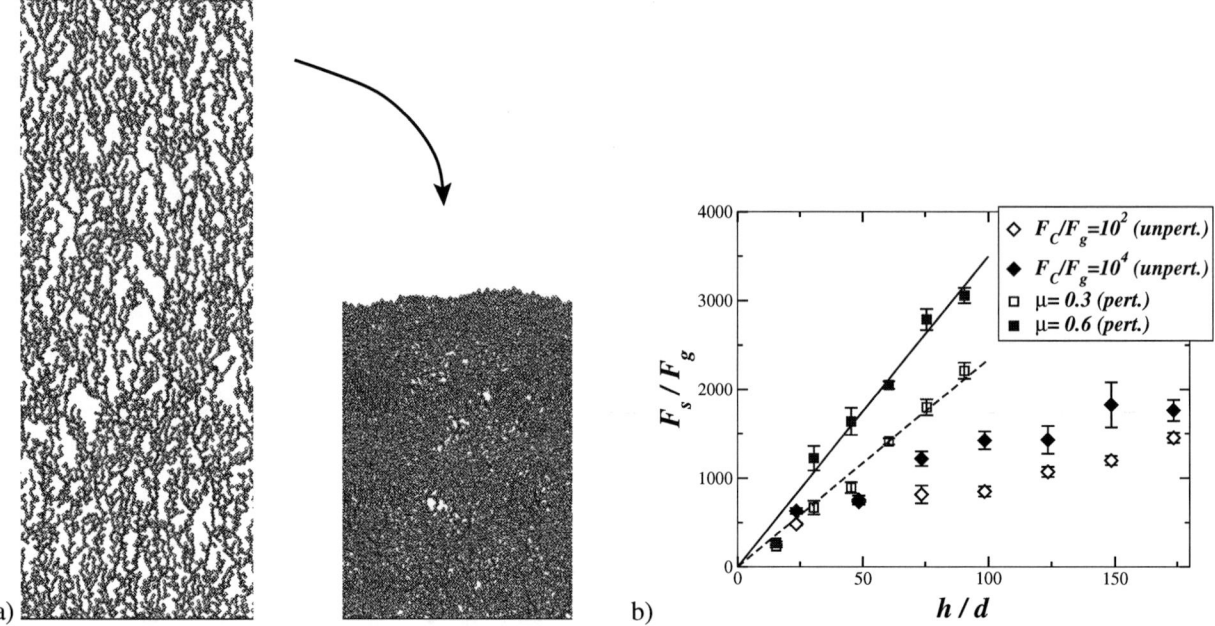

FIGURE 2. a) Typical realizations of the granular system prepared for computational simulations of shear strength. Left we show the (unperturbed) granular network of cohesive disks generated by ballistic deposition and contact dynamics, right the (perturbed) compact structure obtained after the collapse of the system due to a piston-like pushing force applied at the top. b) Dependence of the simulated shear strength (here, the shear force F_s normalized by the gravity F_g acting on each grain, see also text) on the depth for unperturbed and collapsed systems. The friction coefficient for both unperturbed cases with different cohesive forces F_c is $\mu = 0.3$. Compared to the collapsed system, both for $F_c/F_g = 10^4$ (diamonds), the unperturbed grain structure shows only a weaker dependence on the depth (circles). This is consistent with the behavior of the real quicksand (see Fig. 1b). Moreover, an increase by a factor of 100 in the cohesion force does not change substantially the behavior of the shear strength. Also shown are the results of simulations with the unperturbed system for two different values of the friction coefficient. Expectedly, a higher friction coefficient leads to higher shear strength (Data taken from [48]).

relatively short time interval, then generating new configurations when gradually increasing the time interval. We only stop when the packing generated with the highest time interval has, within a given tolerance, the same volume fraction as the previous one. We show in Fig. 2a (left) that this procedure results in a highly porous, and therefore tenuously connected, network of grains. Note that using a shorter time interval for relaxation of the structures would lead to more compact structures, as in this case large parts of the structure would collapse at once leading to a "piston-like" compaction with strong effects of inertia. This is mostly avoided when using sufficiently large time intervals between each deposition, where only small parts of the structure settle at once, as described in the previous.

Shear Strength

With the purpose to simulate the shear strength experiments performed after the collapse of the quicksand, we implement in our model a piston-like pushing force acting on the particles at the top. In this situation, the collapse of the granular network that starts from the upper region of the system is analogous to a gradual increase in the effective piston weight, leading to an acceleration of the compaction mechanism. During this process, bonds will break without reforming again up to the point in which most of the bonds are broken (see Fig. 2a, right). We calculate the shear strength of the unperturbed and perturbed structure as follows. At a given depth, we apply a constant force in the horizontal direction to a randomly chosen particle and observe how far it can move in this direction. By changing the force using nested intervals, one can calculate within a given numerical tolerance the minimum force necessary to move this particle at a distance that is sufficiently large to sample the disordered porous geometry (e.g., approximately 20 particle diameters away). This procedure is then repeated for different particles at the same depth to produce an average shear strength value. We preferred this kind of measurement against the traditional ones as the procedure used here involves only the smallest possible regions of the structures, thus best suited to being able to measure unperturbed

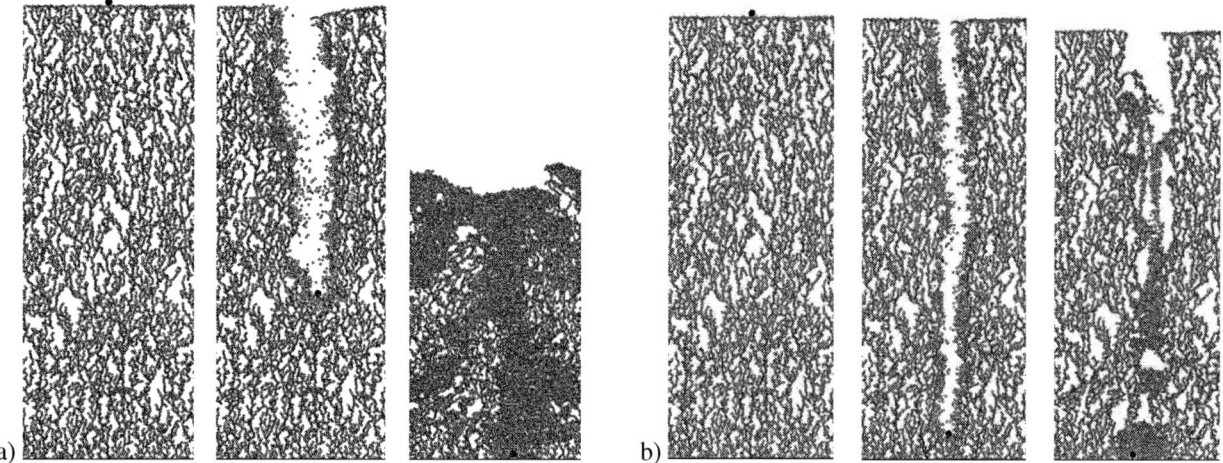

FIGURE 3. a) Snapshots from computational simulations showing a typical realization of the penetration process of an intruder into a very loose cohesive packing. The (unperturbed) collapsing suspension/soil is shown left. As shown in the middle the movement of the intruder is responsible for the partial destruction of the granular structure along its trajectory. At the end (right), the intruder rests under a compact mass of (perturbed) material. b) Snapshots for simulations with strong viscous drag acting on the grains. Using the viscous drag coefficient of water for the typical grain sizes of the experiment would not lead to significant change in the behavior. Therefore, we show here simulations using a 15 times higher viscous drag coefficient, leading to a significantly smaller region of collapse. Note that in the right snapshot still a few grains are settling.

material properties and depth dependence within the simulated systems of restricted size. For a direct simulation of the boundary value problem of the vane experiments much larger systems would be needed. This could be helpful also for further interpretation of the experimental results obtained by the vane rheometer (see also previous section).

The results in Fig. 2b show the variation with depth of the shear strength averaged over 10 different particles for both unperturbed and perturbed systems. In agreement with the real quicksand experiments, the model results obtained with the unperturbed system show only a minor dependence of the shear strength on the depth. Additionally, we observe that a significant increase in the cohesion force intensity (e.g., by a factor of one hundred) does not lead to a substantially different behavior of the shear strength. This is a consequence of two competing effects. With decreasing the cohesion force the shear resistance decreases when hypothetically keeping the same density. For our fragile structures, however, this decrease additionally leads to an increasing density, and a larger shear strength is necessary to move a particle in a more compact medium (i.e., a high density packing). As also shown in Fig. 2b, the shear strength increases linearly with depth for the case of the collapsed material. Such a behavior is again consistent with our experimental findings of the real quicksand: a similar cross-over from a yield stress material to soil dominated by the Mohr-Coulomb friction criterion is observed. Expectedly, as we observe in the same plot, an increase in the friction coefficient generally increases the average shear strength. A more detailed discussion on a microscopic basis can be found in ref. [48].

Penetration tests

We also carry out simulations of an object (intruder) being pushed inside and subsequently removed from the fragile cohesive granular structures. A large disk of low density (half of the grain density) is then pushed with constant force into the granular structure. Only if the force exceeds a certain threshold it penetrates the medium. The sequence of snapshots shown in Fig. 3a corresponds to different stages of the penetration process into a typical realization of the loose packing. The initial (unperturbed) network structure (Fig. 3a, left), as described earlier, is a result of the aggregation of cohesive disks by means of a gravity driven process using contact dynamics. Once the intruder breaks through, the highly porous material collapses under the action of gravity. We observe the creation of a channel (Fig. 3a, middle) which finally collapses over the descending intruder. At the end, the partial destruction of the network due to the penetration process generates a large amount of compact (perturbed) material over the intruder. As shown in Fig. 3a (right), the larger disk is buried under the debris of particles. This results clearly suggests that objects lighter than water can be effectively captured in a quicksand with collapsing soil properties. Furthermore, if we assume that cohesive bonds can be restored by some particular physico-chemical or biological mechanism, the force needed to remove the intruder disk could be significantly higher than the originally penetration force. A more quantitative confirmation for this interesting behavior can be found in ref. [48]. We also investigated the penetration behavior and its dependence when varying parameters such as the density of the material and the size of the intruder, as well as the dynamical behavior during the penetration process [55]. The influence of the surrounding fluid is estimated by adding a viscous drag exerted by the fluid to the particles. Using a viscous drag coefficient of water for grains of average diameter corresponding to the experimental ones does not change the behavior significantly, i.e. the penetration depth and the region of perturbed material do not change. However, when choosing a 15 times higher value for the drag coefficient, the behavior changes significantly. In this case the intruder only destroys the fragile network within and very close to the channel as illustrated in fig. 3b. Additionally the collapse time (i.e. the closing of the channel) takes much longer.

SUMMARY/CONCLUSION

We presented an extension to contact dynamics for cohesive particles and showed that it can be successfully applied for cohesive powders. On this basis, we have developed a microscopic model that is capable to simulate the most important features of a real collapsing suspension/soil, namely a natural quicksand that can be found in the North-East of Brazil. Shed by bacteria in a highly unstable granular skeleton, this "living" quicksand can catastrophically collapse and, during this rapid segregation, irreversibly bury objects lighter than water. Our simulations yield that one could need a force three times one's weight to get out of such morass. Moreover, our model results are also compatible with the shear strength behavior of the real material in both unperturbed and collapsed phases.

ACKNOWLEDGMENTS

The author thanks the organizers of MPGF09 for the organization of the symposium, H. J. Herrmann and J. S. Andrade Jr. for their support in general, as well as the organizers of MPGF09 and the ETH Zurich for financial support.

REFERENCES

1. J. Schwedes, *Granular Matter* **5**, 1–43 (2003).
2. K. Hutter, "Important aspects in the formation of solid-fluid debris-flow models," in *these proceedings*, 2010.
3. G. Capriz, "A continuum theory for granular flows," in *these proceedings*, 2010.
4. P. Giovine, "Constitutive laws for granular materials," in *these proceedings*, 2010.
5. J. D. Goddard, "Generalized granular hypoplasticity," in *these proceedings*, 2010.
6. S. Luding, "Structure and stress anisotropy in granulates with friction and adhesion," in *these proceedings*, 2010.
7. I. Goldhirsch, "On the coarse graining of grains," in *these proceedings*, 2010.
8. S. Luding, *European Journal of Environmental and Civil Engineering* **12**, 785–826 (2008).
9. F. Froio, and J. N. Roux, "Incremental response of model granular materials: results of DEM simulations," in *these proceedings*, 2010.
10. K. Hill, B. Yohannes, L. Hsu, and W. Dietrich, "Rheology of dense sheared granular mixtures: boundary stresses," in *these proceedings*, 2009.
11. A. Rosato, V. Ratnaswamy, D. J. Horntrop, and L. Kondic, "Density relaxation of granular matter via Monte Carlo and Discrete Element Simulations," in *these proceedings*, 2010.
12. J. N. Roux, "How granular materials deform in quasistatic conditions," in *these proceedings*, 2010.
13. J. J. Moreau, *Eur. J. Mech. A-Solid* **13**, 93–114 (1994).
14. M. Jean, and J. J. Moreau, "Unilaterality and dry friction in the dynamics of rigid body collections," in *Contact Mechanics International Symposium*, Presses Polytechniques et Universitaires Romandes, Lausanne, 1992, pp. 31–48.
15. R. G.-R. R., H. Herrmann, and S. McNamara, "Powder and Grains," A.A. Balkema Publ. Leiden, 2005, vol. 1 & 2.
16. M. Nakagawa, and S. Luding, editors, *Powders and Grains 2009*, vol. 1145, AIP Conf. Proc., 2009.
17. S. Luding, "Molecular Dynamics Simulations of Granular Materials," in *The Physics of Granular Media*, edited by H. Hinrichsen, and D. Wolf, Wiley-VCH, Weinheim, 2004, pp. 299–324.
18. N. Mitarai, and F. Nori, *Adv. Phys.* **55**, 1–45 (2006).
19. N. Mitarai, "Simple model for wet granular materials with liquid clusters," in *these proceedings*, 2010.
20. T. Unger, L. Brendel, D. E. Wolf, and J. Kertész, *Phys. Rev. E* **65**, 061305 (2002).
21. T. Unger, and J. Kertesz, "The Contact Dynamics Method for Granular Media," in *Modelling of Complex Systems*, American Inst. of Physics, Melville, New York, 2003, pp. 116–38.

22. L. Brendel, T. Unger, and D. Wolf, "Contact Dynamics for Beginners," in *The Physics of Granular Media*, edited by H. Hinrichsen, and D. Wolf, Wiley-VCH, Weinheim, 2004, pp. 325–40.
23. D. Kadau, G. Bartels, L. Brendel, and D. E. Wolf, *Phase Transit.* **76**, 315–31 (2003).
24. A. Taboada, N. Estrada, and F. Radjai, *Phys. Rev. Lett.* **97**, 098302 (2006).
25. V. Richefeu, M. E. Youssoufi, and F. Radjaï, *Phys. Rev. E* **73**, 051304 (2006).
26. F. Radjai, N. Estrada, V. Richefeu, and A. Taboada, "Force transmission in cohesive granular media," in *these proceedings*, 2010.
27. D. Kadau, L. Brendel, G. Bartels, D. Wolf, M. Morgeneyer, and J. Schwedes, *Chemical Engineering Transactions* **3**, 979–84 (2003).
28. G. Bartels, T. Unger, D. Kadau, D. Wolf, and J. Kertész, *Granular Matter* **7**, 139–43 (2005).
29. L. Brendel, D. Kadau, D. Wolf, M. Morgeneyer, and J. Schwedes, *AIDIC Conference Series* **6**, 55–65 (2003).
30. M. Morgeneyer, M. Röck, J. Schwedes, L. Brendel, D. Kadau, D. Wolf, and L.-O. Heim, *Schriftenreihe Mechanische Verfahrenstechnik: Behavior of Granular Media* **9**, 107–36 (2006).
31. D. Kadau, D. Schwesig, J. Theuerkauf, and D. Wolf, *Granular Matter* **8**, 35–40 (2006).
32. T. Schwager, D. E. Wolf, and T. Pöschel, "Fractal substructure of a nanopowder," in *these proceedings*, 2010.
33. K. Kawakita, and K. H. Lüdde, *Powder Technology* **4**, 61–8 (1969).
34. H. Freundlich, and F. Juliusburger, *T. Faraday Soc.* **31**, 769–73 (1935).
35. E. R. Smith, *Ohio J. Sci.* **46**, 327–8 (1946).
36. R. F. Craig, *Soil Mechanics*, E & FN Spon, New York, 1997.
37. U. El Shamy, and M. Zeghal, *Engrg. Mech.* pp. 413–26 (2005).
38. D. A. Huerta, V. Sosa, M. C. Vargas, and J. C. Ruiz-Suarez, *Phys. Rev. E* **72**, 031307 (2005).
39. K. Kruszelnicki, *New Scientist* **152**, 26–9 (1996).
40. T. W. Lambe, and R. V. Whitman, *Soil Mechanics*, Wiley, New York, 1969.
41. A. Khaldoun, E. Eiser, G. Wegdam, and D. Bonn, *Nature* **437**, 635 (2005).
42. D. Lohse, R. Rauhe, R. Bergmann, and D. van der Meer, *Nature* **432**, 689–90 (2004).
43. J. Royer, E. Corwin, A. Flior, M.-L. Cordero, M. Rivers, P. Eng, and J. H.M., *Nature Physics* **1**, 164–7 (2005).
44. G. Caballero, R. Bergmann, , D. van der Meer, A. Prosperetti, and D. Lohse, *Phys. Rev. Lett.* **99**, 018001 (2007).
45. D. Kadau, H. Herrmann, J. Andrade Jr., A. Araújo, L. Bezerra, and L. Maia, *Brief Communications to Granular Matter* **11**, 67–71 (2009).
46. J. A. D. Kadau, H.J. Herrmann, "Mechanical behaviour of "living quicksand": simulation and experiment," in [16], pp. 981–4.
47. W. R. Parker, *Nature* (1966).
48. D. Kadau, H. Herrmann, and J. Andrade Jr., *Eur. Phys. J. E* **30**, 275–81 (2009).
49. C. R. I. Clayton, M. C. Matthews, and N. E. Simons, *Site Investigation*, Blackwell Science, Oxford, 1995.
50. J. D. Sherwood, *Journal of Non-Newtonian Fluid Mechanics* **154**, 109–19 (2008).
51. S. Savarmand, M. Heniche, V. Bechard, F. Bertrand, and P. J. Carreau, *Journal of Rheology* **51**, 161–77 (2007).
52. H. A. Barnes, and Q. D. Nguyen, *Journal of Non-Newtonian Fluid Mechanics* **98**, 1–14 (2001).
53. A. Danin, *J. Arid Env.* **167**, 409 (1978).
54. A. Danin, *J. Arid Env.* **21** (1991).
55. D. Kadau, H. Herrmann, and J. Andrade Jr., Simulations of the penetration process in fragile granular structures (2009), in preparation.

The hard-particle model for a dense granular flow down an inclined plane

V. Kumaran

Department of Chemical Engineering, Indian Institute of Science, Bangalore 560 012, India.

Abstract. Perfectly hard particles are those which experience an infinite repulsive force when they overlap, and no force when they do not overlap. In the hard-particle model, the only static state is the isostatic state where the forces between particles are statically determinate. In the flowing state, the interactions between particles are instantaneous because the time of contact approaches zero in the limit of infinite particle stiffness. Here, we discuss the development of a hard particle model for a realistic granular flow down an inclined plane, and examine its utility for predicting the salient features both qualitatively and quantitatively. We first discuss Discrete Element simulations, that even very dense flows of sand or glass beads with volume fraction between 0.5 and 0.58 are in the rapid flow regime, due to the very high particle stiffness.

An important length scale in the shear flow of inelastic particles is the 'conduction length' $\delta = (d/(1-e^2)^{1/2})$, where d is the particle diameter and e is the coefficient of restitution. When the macroscopic scale h (height of the flowing layer) is larger than the conduction length, the rates of shear production and inelastic dissipation are nearly equal in the bulk of the flow, while the rate of conduction of energy is $O((\delta/h)^2)$ smaller than the rate of dissipation of energy. Energy conduction is important in boundary layers of thickness δ at the top and bottom. The flow in the boundary layer at the top and bottom is examined using asymptotic analysis. We derive an exact relationship showing that the a boundary layer solution exists only if the volume fraction in the bulk decreases as the angle of inclination is increased. In the opposite case, where the volume fraction increases as the angle of inclination is increased, there is no boundary layer solution.

The boundary layer theory also provides us with a way of understanding the cessation of flow when at a given angle of inclination when the height of the layer is decreased below a value h_{stop}, which is a function of the angle of inclination. There is dissipation of energy due to particle collisions in the flow as well as due to particle collisions with the base, and the fraction of energy dissipation in the base increases as the thickness decreases. When the shear production in the flow cannot compensate for the additional energy drawn out of the flow due to the wall collisions, the temperature decreases to zero and the flow stops. Scaling relations can be derived for h_{stop} as a function of angle of inclination.

Keywords: Dense granular flow, hard particle
PACS: 45.50.-j,45.70.Mg,51.10.+y

INTRODUCTION

A complete fluid-dynamical description of a granular flow, similar to the Navier-Stokes equations for a simple fluid, does not exist at present. Rheological models have been proposed for granular flows, but we have not yet made the progression to solving the equations analytically in well-specified limits in order to obtain a deeper physical understanding of these flows. Such a progression can be made only if analytical techniques, such as boundary layer analysis, can be applied to granular flows. In the present paper, we show how the boundary layer analysis can be carried out for a specific well-studied class of granular flows, which is the flow down an inclined plane.

The flow of a granular material down an inclined plane has been studied extensively using computer simulations[1, 2, 3, 4]. These incorporate sophisticated particle interactions for relatively large systems with heights of up to hundreds of particle diameters, and provide detailed information about the density, mean velocity and granular temperature (mean square fluctuating velocity) profiles. One of the most remarkable features observed in simulations of the granular flow down an inclined plane[1] is that the volume fraction of the particles is a constant in the bulk of the flow, the granular temperature and all stress components are linear functions of height, and the mean velocity increases as the square root of the height from the bottom of the layer. The volume fraction is found to be, within numerical accuracy, independent of position in the flow (apart from two thin layers at the top and bottom of thickness about 3-5 particle diameters where the volume fraction varies with position), independent of the total height of the material, and dependent only on the angle of inclination of the inclined plane. In addition, it is found that as the height of the flowing layer is decreased, there is a minimum height below which the flow stops at a fixed angle of inclination.

There have been several studies of this flow, both phenomenological and kinetic theory based[5, 6, 10, 12, 11, 13, 14, 15, 16, 17, 18, 19, 20, 21, 22] but a clear explanation for several of these features is still lacking. In the early constitutive relations of Savage[5], the stress is divided into a frictional 'Mohr-Coulomb' stress and a part that depends on the rate of deformation. Both of these components are written using tensor expansions in the rate of deformation tensor and the volume fraction gradient, and are simplified for simple parallel flows. A similar approach was used by Johnson and Jackson[6], where the stress is divided into two parts, a frictional stress and a kinetic stress.

$$\sigma = \sigma_c + \sigma_f \qquad (1)$$

where σ_c and σ_f are the collisional and frictional parts of the stress. For the frictional part, the ratio of the shear and normal stresses is considered to be a constant (yield criterion), while a simple volume-fraction-dependent form is used for the normal stress. In contrast to the general expression of Savage[5], the kinetic stress is written in a manner similar to Lun et al[7], where the constitutive relation for the stress contains an isotropic pressure and a viscous part. An additional energy balance equation is written to obtain the 'granular temperature', which is required for evaluating the stress and the viscosity. The dependences on the viscometric coefficients on the volume fraction are modeled using equations obtained from the Chapman-Enskog procedure for dense gases.

This simple addition of the frictional and kinetic parts implies that the material elements in the frictional and kinetic regimes provide parallel resistance for the flow. If the resistances are in series, then the strain rates have to be added up, while the stress across all the elements remains the same. In the earlier theories, no microscopic justification is provided for adding up the two stress elements.

A different form of the constitutive relation has been used recently by Pouliquen and Forterre [8] and Jop et al[9], where a 'friction law' is proposed for the relation between the shear and normal stresses. Here, the ratio of the normal and shear stresses is written in terms of the 'inertia paramter' I, which is the ratio of the strain rate and the square root of the pressure. Here again, the ratio of the shear and normal stresses can be considered as the sum of a frictional part (independent of volume fraction and strain rate) and a second term which depends on the 'inertia parameter' $I = (\dot\gamma d/\sqrt{p/\rho})$, where $\dot\gamma$ is the strain rate, d is the particle diameter, p is the normal stress and ρ is the density. The 'kinetic' part is proportional to $(I_0/I - 1)^{-1}$, where I_0 is a constant. Clearly, this decreases to zero as the strain rate goes to zero, but with a more complicated form than the earlier expressions[5, 6]. For this, we can expand the ratio of the normal and shear stresses as,

$$\frac{\sigma}{p} = \mu_s + S' \qquad (2)$$

where μ_s is one component of the static (yield) part, and S' is written as,

$$\frac{1}{S'} = \frac{1}{\mu'}\left(1 + \frac{I_0}{I}\right) \qquad (3)$$

The above equation can be recast as,

$$\frac{1}{S'} = \frac{1}{\mu'} + \frac{1}{S_k} \qquad (4)$$

Clearly, this is a series combination of a frictional element $(1/\mu')$ and a second 'kinetic' component which depends on the strain rate through the inertia parameter I

$$S_k = (\sigma_k/p) = \frac{I_0}{\mu'}\frac{\dot\gamma d}{\sqrt{p/\rho}} \qquad (5)$$

If this is expressed in terms of the strain rate, we obtain a constitutive relation for the kinetic shear stress of the form,

$$\sigma_k = \frac{I_0}{\mu'}\sqrt{p\rho}(\dot\gamma d) \qquad (6)$$

This is similar to the equation for the viscous shear stress in kinetic theory, except that $\sqrt{p/\rho}$ is substituted for the square root of the temperature. However, the microscopic reason for this specific arrangement of series and parallel resistances to flow is not clear at present. Moreover, what is added up is the series and parallel resistances for the ratio of shear and normal stresses in the case of Pouliquen[8], and not the stresses themselves. Jop et al[9] just used a simple extension of the stress equation, where the stress tensor is written as the sum of the pressure and shear stress, and the shear stress is a scalar function times the rate of deformation tensor, in their analysis. The flow is assumed to be incompressible in order to determine the pressure. This form of the stress tensor implies that the principal axes of the shear stress and rate of deformation tensor are parallel. In simulations[1, 4], there is always a significant second normal stress difference.

The frictional stress equation, which is a relation between the shear and normal stresses, cannot be considered a 'constitutive equation'. A constitutive equation should relate the surface stresses of a volume element to the local deformation of that element. Both the pressure and the shear stress should be determined in terms of the local rate of deformation tensor, the volume fraction and any other 'state variables' used to describe the flowing state of the material. The yield condition in frictional constitutive relations is obtained from the frictional stress, which is the minimum stress required for the initiation of flow. The microscopic origin of the frictional stress is the friction between grains in the static arrangement, which has to be overcome for the initiation of flow. While the frictional stress does account for the initiation of flow at a finite angle of inclination, it cannot explain the cessation of flow at a finite angle when the angle of inclination is decreased. Since the grains are in the flowing state prior to flow cessation when the angle of inclination is decreased, the condition for the cessation of flow should be obtainable from a kinetic stress alone. This is one of the major deficiencies in frictional-kinetic models which we attempt to overcome in the present analysis using a kinetic description alone.

It has been realised, for some time now, that equations based on kinetic theory for smooth inelastic particles cannot be used to explain these dense flows. In particular, if the Newton's law for viscosity, with the viscosity coefficient obtained by the Chapman-Enskog procedure for dense gases, is used as the constitutive relation, the model prediction is that the volume fraction increases as the angle of inclination increases[19]. This is clearly unphysical, since it implies that when we take a static granular layer and decrease the angle of inclination, it starts to flow. It is necessary to include the Burnett terms in the constitutive relation, and carry out the calculation for the rough particle model, in order to obtain expansion of the layer as the angle of inclination is increased.

It is generally assumed that these flows form a different state in which correlations play a dominant role in determining the rheology. Ertas and Halsey[11] postulated regions of correlated motion of length scale large compared to the particle diameter, where the angular velocity of neighbouring particles are anti-correlated. However, there has been no evidence of this from simulations so far. Jenkins[16, 17] postulated that there are long range correlations which result in an energy dissipation rate which is smaller than that calculated from kinetic theory. However, both simulations[23] and experiments[24] have not shown any long range correlations either in space or time. In fact, there is evidence[24, 25, 26] that the velocity autocorrelations decay faster in sheared inelastic flows than in the shear flow of elastic particles. Similarly, the mean angular velocity in the bulk of the flow does not show a significant departure from one-half of the local vorticity[27], indicating the micropolar effects[13] may not be important.

Here, we summarise a series of studies which seek to examine whether the hard-particle model is relevant for a dense granular flows down an inclined plane. Of specific interest is whether the model is 'realistic', that is, whether the time period of particle interactions in real flows is small compared to the time between successive interactions in the flow. Another issue is whether a hydrodynamic theory, based on the hard-particle model, can qualitatively and quantitatively reproduce all the salient features in the flow down an inclined plane. The specific features of interest here are the cessation of flow when the angle of inclination reduces below a minimum value (angle of repose), the constant volume fraction in the bulk of the flow excluding boundary layers of thickness 3-5 particle diameters at the top and bottom, the existence of a boundary layer length scale where the properties are different from the bulk of the flow, the minimum height at which the flow stops.

MULTI-BODY CONTACTS OR INSTANTANEOUS COLLISIONS?

Two limiting situations are in competition in the dense flow of granular materials. The first is the high volume fraction limit, where the volume fraction approaches the close packing volume fraction. In this case, it is expected that the system will be in the multi-body contact regime, where an average particle is in simultaneous contact with its neighbours. The number of contacts will increase as the volume fraction is increased when the particle stiffness is kept a constant. The second limit is that of high particle stiffness (spring constant). Since the time of an interaction is inversely proportional to the square root of the spring constant[1], the time period of an interaction decreases as the stiffness increases. For perfectly hard (infinite stiffness) particles, the period of an interaction is zero. As the particles become stiffer and stiffer, the period of an interaction will decrease at constant volume fraction. In practical granular flows, the relative approach to these two limits (high volume fraction and high spring stiffness) will determine whether the system is in the instantaneous collision regime or the multi-body contact regime.

Reddy and Kumaran[4, 27] studied the effect of increasing spring stiffness in a dense flow of spherical particles down an inclined plane. The Discrete Element simulations was used, where particle interactions are modeled using a linear spring-dashpot model. The rough-particle model was used where the relative velocity tangential to surfaces at contact is reversed in an interaction, and the damping constant in the normal direction was adjusted so that the normal coefficient of restitution is 0.9. For these interaction parameters, the flow ceases when the angle of inclination decreases below 21°, and the steady flow becomes unstable when the angle of inclination increases above 25°. The simplest measure of the particle interaction regime measured by the authors is the co-ordination number, which is the number of particles in simultaneous contact with the test particle. The co-ordination number was determined in the bulk of the flow, where the volume fraction is a constant, as a function of the scaled spring constant $(k_n/(mg/d))$ in the range $(k_n/(mg/d)) = 10^5 - 10^8$, where m and d are the particle mass and diameter, and g is the acceleration due to gravity. It is found that the average co-ordination number decreases as the spring constant increases in this range. The co-ordination number has a value between 2-4 when the spring constant $(k_n/(mg/d))$ is less than 10^5, whereas the co-ordination number is less than 1 for all angles above 21° when $(k_n/(mg/d))$ is higher than 10^8. Therefore, for real granular flows, it is necessary to have an estimate of the particle stiffness in order to determine whether the particle interaction is binary or multi-body.

An approximate way to determine the particle stiffness is to use dimensional analysis. For a linear spring-dashpot model, the spring stiffness had dimensions of $k_n \sim MT^{-2}$, where M and T are the mass and time dimensions. For a homogeneous spherical particle made of a linear elastic material, the spring stiffness will be a function of the Young's modulus E (with dimensions $ML^{-1}T^{-2}$), the dimensionless Poisson ratio and the particle mass and diameter. From dimensional analysis, it can be inferred that the spring stiffness has to scale as (Ed). However, there has been disagreement in literature about whether the value of E to be used is that for the sand or glass material itself, or whether the modulus of a loose assembly of sand should be used. Some authors[1, 4] have based their estimates of the spring stiffness on the Young's modulus of the sand particle itself. The Young's modulus for real materials, such as sand and glass beads, is in the range $10^{11} N/m^2$. On this basis, the spring constant can be estimated as $10^8 N/m$ for particles of diameter 1 mm, and $10^7 N/m$ for particles of diameter 0.1 mm. In contrast, Campbell[28, 29] has determined the spring stiffness based on the speed of sound through a loose assembly of sand grains, which is about 100 m/s[30]. Using this, the Young's modulus can be estimated as $2.5 \times 10^7 N/m^2$ if we use a value of $2500 kg/m^3$ for the density of sand, and the spring stiffness is in the range $2.5 \times 10^3 - 2.5 \times 10^4 N/m$ for sand particles with diameter in the range 0.1-1 mm.

Experimental results have more recently become available which directly measure the stiffness of contacts between particles[31, 32]. In these experiments, two individual grains of sand were mounted on pins of diameter 2 mm and pressed against each other. The normal displacement and the normal force were simultaneously measured, and the particle contact stiffness was inferred from these. The radius of curvature of the surfaces at contact were in the range 0.05 to 8.2 mm. There are several surprising results from these experiments. One of these is that for several different types of sand, the force law at small forces is linear. This because in the initial stages, the deformation is due to the compression of asperities on the surface, instead of the compression of two smooth surfaces in which the contact area increases with time. The spring constant for the linear contact law at low forces is in the range $0.2 - 2 \times 10^6 N/m$ for a large class of materials. This spring constant is an order of magnitude smaller than the estimates of Reddy and Kumaran[4] for particles of $100 \mu m$ diameter, but is about two to three orders of magnitude larger than that obtained using the speed of sound through loose sand. In their calculations, Reddy and Kumaran[27] used the linear spring constant of $k_n = 10^6 N/m$ as a reference value based on the experiments of Cole & Peters[31, 32]. For particles with mass density of $2500 kg/m^3$, this results in a dimensionless spring stiffness $(k_n/(mg/d))$ in the range 10^7 to 10^9 for particles with diameter in the range $100 \mu m$ to $1mm$.

The contact model more appropriate for smooth particles is the Hertzian contact model, where the force of interaction is proportional to the $k'_n \delta^{3/2}$, where δ is the overlap distance. This model takes into account the increase in the area of contact of the particles as the overlap increases. Cole and Peters[31, 32] found that the contact between smooth particles is well described by the Hertzian contact model. The spring constant in the Hertzian contact regime is well predicted by the Mindlin-Deresiewicz theory[33], and that the k'_n scales as $Ed^{1/2}$. The numerical results are about 20% lower than the value that would be obtained if the elasticity modulus of the material is inserted in the Mindlin-Deresiewicz relation. The elasticity modulus for materials such as sand and glass is $E \sim 10^{11} N/m^2$, and if we assume a mass density of $2500 kg/m^3$, the non-dimensional spring stiffness $(k'_n/(mg/d^{3/2}))$ varies in the range $10^9 - 10^{10}$ for particles with diameter in the range $100 \mu m$ to $1mm$.

The effect of a change in the spring stiffness on the average co-ordination number (number of particles that are in simultaneous contact with a test particle) was examined by Reddy and Kumaran[4] for the flow down an inclined plane of spherical particles, using the Discrete Element (DEM) simulation technique. Both the linear and the Hertzian contact laws were studied for scaled spring stiffnesses in the range 10^4 to 10^8; it is difficult to obtain results for scaled spring stiffness greater than about 10^8 due to computational limitations. For this inclined plane flow, the flow stops when the angle of inclination is below 21°, while the flow becomes unstable when the angle of inclination is above 25°, and so there is only a small range of angles of inclination where the flow is stable. The simulations were carried out for spring stiffness in the range 10^4 to 10^8; higher stiffness could not be examined due to computational limitations. For all angles of inclination, it was found that the co-ordination number decreases monotonically as the angle of inclination is increased, and it does not seem to asymptotically reach a constant value. The co-ordination number is in the range 2-4 at the lowest spring stiffness of 10^4, but it decreases below 1 when the scaled spring stiffness was increased to 10^8. The only exception is the lowest angle of inclination of 21°, where the co-ordination number is between 1 and 2 at the highest spring stiffness of 10^8; however, even in this case, the co-ordination number shows a monotonic decrease with spring stiffness.

As reported in the first simulations of Silbert et al[1] of the flow down an inclined plane, the stresses follow Bagnold's law (stress proportional to the square of the strain rate). This is a dimensional necessity in the regime where particles interact via binary collisions, because there are no time scales apart from the inverse of the strain rate. A more surprising result is that the numerical value of the stress in the flow down an inclined plane changes little as the spring constant is increased, and the system transitions from the multi-body contact regime to the binary collision regime. Reddy and Kumaran[4] analysed this by examining the 'force ratio', which is the average ratio of magnitudes of the second-largest and the largest forces on a particle which is in simultaneous contact with two or more particles. This force ratio was found to be small (less than about 0.2) even when the co-ordination number is larger than 1. This indicates that there is only one large force acting on a particle even when it is in the multi-body contact regime, and this may explain why the Bagnold coefficients do not change much as the spring constant decreases and the system goes from the binary contact regime to the multi-body contact regime.

DYNAMICS IN THE BULK OF THE FLOW:

The 'Bagnold' coefficients, which are the ratios of the different components of the stress and the square of the strain rate, were calculated using both Discrete Element (DE) simulations and Event Driven (ED) simulations in Reddy and Kumaran[4]. It should be noted the ED simulations are performed at constant strain rate, but the strain rate does vary with position in the DE simulations. Collisions are considered to be instantaneous in the ED simulations. It is found that the theoretical predictions of the Bagnold coefficients are in quantitative agreement with both DE and ED simulations provided the pair distribution function obtained from the simulations is inserted into the theory. However, it is found that the pair distribution function in a sheared granular flow is significantly larger than that in an equilibrium fluid of elastic particles. If this pair distribution function is incorporated in the theory, then it is possible to get agreement to within about 25% for the Bagnold coefficients.

Another interesting feature of the flow down an inclined plane is the near constant volume fraction in the bulk of the flow. The lack of variation in the bulk of the flow volume fraction was explained on the basis of the hard-particle model[34]. In that analysis, the 'conduction length $\delta = (d/(1 - e_n)^{1/2})$' was identified at the distance over which there is a balance between the rate of conduction of energy, and the rate of dissipation of energy due to inelastic collisions. Here, d is the particle diameter and e_n is the normal coefficient of restitution. When h is large compared to the conduction length, the rate of conduction of energy can be neglected in the leading approximation, and the rate of production of energy due to mean shear is balanced by the rate of dissipation due to inelastic collisions. It has been shown[34, 18, 19, 20] that as a consequence of the balance between the rates of production and dissipation, the volume fraction is a constant in the leading approximation. The correction to the volume fraction[34] is $O(\delta/h)^2$ smaller than the leading order volume fraction.

The dynamics of a homogeneously sheared granular material composed of hard inelastic particles was analysed using event driven (ED) simulations[21, 22]. The ED simulations were carried out for volume fractions greater than about 0.4, and both smooth particles (tangential relative velocity of surfaces of contact unchanged in a collision) and rough particles (tangential relative velocity reversed in a collision) were considered. At such high volume fractions, ED simulations encounter numerical errors due to inelastic collapse (an infinite number of collisions in finite time). Inelastic collapse is observed only for particles with constant coefficient of restitution. If a more realistic velocity dependent coefficient of restitution (in which the coefficient of restitution tends to 1 at low velocities) is used, there is no inelastic collapse. Inelastic collapse manifests as particle overlaps in the simulations, when the time between collisions becomes comparable to the numerical precision. In the simulations, we checked for particle overlaps, and only used data for which there was no particle overlap for the entire duration of the simulation. No special technique was used to avoid collapse, in order to avoid any biasing of simulation results.

The important conclusions of this study was as follows:

1. The relative arrangement of particles was analysed using the icosahedral order parameter Q_6 in three dimensions, and the planar order parameter q_6 in the plane perpendicular to the gradient direction. It was found that for shear flows of sufficiently large size, the system continues to be in the random state, with Q_6 and q_6 close to zero, even for volume fractions between $\phi = 0.5 - 0.6$; in contrast, for a system of elastic particles in the absence of shear, the system orders (crystallises) at $\phi = 0.49$[35, 36]. The result that there is no crystalline ordering in the sheared state is an important one, because it indicates that an isotropic pair distribution function can be used in theories for a sheared granular flow. The pair distribution function in a sheared granular flow has also been analysed[37], and it is found that there is no signature of crystallisation in the sheared state.

2. Another important quantity measured in the simulations was the variation of the collision frequency with volume fraction and coefficient of restitution. The strain rate in the flow is related to the 'granular temperature' (mean square of the particle fluctuating velocity) by the energy balance equation, in contrast to an equilibrium system where the temperature is set by the temperature of the surroundings. All time scales can be non-dimensionalised by the inverse of the strain rate, and the variation of all dynamical quantities (collision frequency, stress, strain rate) were scaled by suitable powers of the strain rate. An analysis of the variation of the collision frequency (scaled by the strain rate) revealed some interesting results. By plotting the inverse of the collision frequency as a function of volume fraction, it was found that the collision frequency at constant strain rate diverges at a volume fraction ϕ_{ad} (volume fraction for arrested dynamics) which is lower than the random close packing volume fraction 0.64 in the absence of shear. The volume fraction ϕ_{ad} decreases as the coefficient of restitution is decreased from $e_n = 1$; ϕ_{ad} has a minimum of about 0.585 for coefficient of restitution e_n in the range $0.6 - 0.8$ for rough particles, and is slightly larger for smooth particles.

3. In addition, the dissipation rate and all components of the stress diverge proportional to the collision frequency in the close packing limit. In contrast, the momentum transport per collision, which depends on the particle fluctuating velocities (granular temperature) does not diverge. This indicates that the increase in the rate of momentum transport due to an increase in the collision frequency is responsible for the divergence of the stress.

4. The qualitative behaviour of the increase in the stress and dissipation rate are well captured by results derived from kinetic theory, but the quantitative agreement is lacking even if the collision frequency obtained from simulations is used to calculate the pair correlation function used in the theory. This was in agreement with previous calculations[18, 19], where it was found that the qualitative variation of the stresses with volume fraction were well predicted by kinetic theories based on the Enskog approximation, but there were quantitative differences between theory and simulations.

5. In order to ascertain the reason for this discrepancy, the distribution of relative velocities between colliding particles in shear flows of inelastic spheres is analysed in the volume fraction range $0.4 - 0.64$[22]. It has been shown earlier[38] that the relative velocity distribution for colliding particles is different from the product of the single particle distributions. Since momentum transfer and energy dissipation are primarily collisional in this regime, these are determined by the pre-collisional relative velocities between particles. In kinetic theories, the molecular chaos approximation is used, where the two-particle velocity distribution is the product of the single-particle velocity distributions. The single particle velocity distributions were found to be well approximated by Gaussian distributions even for very dense flows with coefficients of restitution as low as 0.6. It was found that the pre-collisional relative velocities normal to the surfaces of contact showed the largest violation of the molecular chaos approximation. The distribution of pre-collisional normal relative velocities (along the line joining the centers of the particles) is an exponential distribution for particles with low normal coefficient of restitution in the range $0.6 - 0.7$, in contrast to the Gaussian distribution for the normal relative velocity in an elastic fluid in the absence of shear.

6. The distribution of relative tangential velocities of the particles were closer to Gaussian distributions even at coefficients of restitution as low as 0.6, and they did not show a significant violation of the molecular chaos approximation.

7. Empirical relations are formulated for the relative velocity distribution, in order to calculate the collisional contributions to the pressure, shear stress and the energy dissipation rate in a shear flow. These theoretical predictions were found to be in quantitative agreement with both the ED simulations for the homogeneous shear flow of inelastic particles, as well as the flow down an inclined plane.

It should be noted that the theoretical calculation only includes the velocity distribution between a pair of colliding particles, whereas there is no such limitation in the simulations. If the dominant contribution to the stress were due to long range correlations, such as stress chains or dense clusters, there would not be agreement, because there is no biasing towards correlated regions in the theoretical calculation. So the agreement indicates that the only correlation effect that is important is the modification of the two-particle relative velocity distribution, and long-range correlations do not seem to affect the stresses.

8. An issue of importance is whether the angle of repose (angle at which the flow stops when the inclination is decreased) can be predicted by formulations based on the hard particle model. The hard-particle analysis[21, 22] specifically found that this angle can be predicted by the hard-particle model, as follows. Since all time scales are scaled by the strain rate in the event-driven simulations, the Bagnold coefficients are numerically equal to the stresses. As the volume fraction increases, the stresses diverge at the volume fraction for arrested dynamics ϕ_{ad}. However, the ratio of the shear and normal stress tend to a finite value as the stresses diverge. Since the ratio of shear and normal stresses in the flow down an inclined plane is equal to $\tan(\theta)$, the ratio of stresses in the limit $\phi \to \phi_{ad}$ is the tangent of the angle of repose (angle at which stress diverges at constant strain rate, or strain rate goes to zero at constant stress). This angle was determined using the stresses calculated from the relative velocity distribution, and was found to be in numerical agreement with simulation results.

BOUNDARY LAYER SOLUTIONS:

The flow in the 'conduction boundary layers' of thickness comparable to the conduction length at the bottom and top was analysed[20]. Here, an exact condition (within the assumptions of the hard-particle model) was obtained showing that a boundary layer solution is possible if and only if the volume fraction decreases as the angle of inclination increases. There is no boundary layer solution which matches the solution in the bulk for parameter regions where the volume fraction increases as the angle of inclination increases. We briefly review this calculation in order to clarify the assumptions involved.

The granular material is composed of hard sphere particles of diameter d flowing down a plane inclined at an angle θ to the horizontal. A Cartesian coordinate system is used, where the velocity and velocity gradient are in the x and y directions respectively. The mass of a particle is set equal to 1 for simplicity. The shear and normal stress balances, and the constant ratio of the shear and normal stresses, are

$$\begin{aligned}(d\sigma_{xy}/dy) &= -\rho g \sin(\theta), \\ (d\sigma_{yy}/dy) &= \rho g \cos(\theta), \\ (\sigma_{xy}/\sigma_{yy}) &= -\tan(\theta).\end{aligned} \quad (7)$$

The energy equation at steady state is

$$\frac{d}{dy} K \frac{dT}{dy} + \mu \dot{\gamma}^2 - D = 0, \quad (8)$$

where K is the thermal conductivity, D is the rate of dissipation of energy, T is the 'granular temperature', μ is the viscosity and $\dot{\gamma}$ is the strain rate.

The expressions for the shear and normal stresses are

$$\begin{aligned}\sigma_{xy} &= \mu \dot{\gamma}, \\ \sigma_{yy} &= -p\end{aligned} \quad (9)$$

where p is the pressure, μ is the viscosity.

It is convenient to express the viscometric coefficients and the dissipation coefficients as a product of two functions, one of which is a dimensionless function of volume fraction, and the other is a product of suitably chosen powers of the granular temperature and particle diameter, the latter having the same dimensions as the viscometric function under consideration. (Note that the granular temperature has dimensions of the square of the velocity, since the mass is set equal to 1.) From dimensional analysis, it can be inferred that $p = p_\phi (T/d^3)$ $K = K_\phi (T^{1/2}/d^2)$, $\mu = \mu_\phi (T^{1/2}/d^2)$, and $D = D_\phi \varepsilon^2 (T^{3/2}/d^4)$. where the variables with subscript ϕ are dimensionless functions of the volume fraction, and $\varepsilon = (1-e_n)^{1/2}$ is the small parameter used in the expansion[18, 19] to determine the constitutive relations, where e_n is the normal coefficient of restitution. The parameter ε^2 is written separately in the expression for the rate of dissipation of energy in order to ensure that the rate of dissipation goes to zero in the limit of elastic collisions. The strain rate can be expressed in terms of the temperature using equations 9 for the stresses, and 7 for the ratio of the stresses,

$$\tan(\theta) = \frac{\mu_\phi(\dot{\gamma}d/T^{1/2})}{p_\phi T}$$
$$= \frac{\mu_\phi G}{p_\phi} \quad (10)$$

Using equation 10, $G = (\dot{\gamma}d/T^{1/2})$ can be calculated as a function of the volume fraction. It should be noted that the function G is a constant in the bulk, though the strain rate $\dot{\gamma}$ and the temperature T do vary with height.

In the bulk of the flow, there is a balance between the rate of production of energy due to shear and the rate of dissipation due to inelastic collisions. A consequence of this is that the volume fraction is a constant, and the temperature is a linear function of height[34, 19]. However, within distances comparable to the conduction length from the boundaries, the rate of conduction is not negligible, and it is necessary to rescale the y coordinate by the conduction length $\delta = (d/\varepsilon)$. The 'inner' coordinate near the bottom surface is defined as $y^\dagger = (y/\delta)$. Using this inner coordinate, the y momentum equation and the energy equation can be rewritten as

$$\frac{d(p_\phi T)}{dy^\dagger} = -\frac{6\phi g \delta \cos(\theta)}{\pi} \quad (11)$$

$$\frac{d}{dy^\dagger} K_\phi T^{1/2} \frac{dT}{dy^\dagger} = -T^{3/2}\left(\frac{\mu_\phi G(\phi, \tan(\theta))^2}{\varepsilon^2} - D_\phi\right) \quad (12)$$

where $\varepsilon = (1 - e_n)^{1/2}$, and p_ϕ, K_ϕ, μ_ϕ and D_ϕ are functions of the volume fraction for the rough particle model[19]. In the momentum conservation equation 11, the left side of the equation scales as (gh), since the temperature at the bottom is proportional to gh, while the right side scales as $g\delta$. Therefore, in an asymptotic expansion in the parameter (δ/h), the right side of equation 11 can be neglected in comparison to the left side, and the leading order solution for the temperature is,

$$T = \frac{6\phi_o g(h-y)\cos(\theta)}{\pi p_\phi} \quad (13)$$

where ϕ_o is the volume fraction in the bulk, and the constant of integration on the right side of equation 13 is determined from the requirement that the upward pressure has to balance the weight of material above per unit area.

The spatial evolution equation for the volume fraction is obtained by inserting the expressions for the temperature 13 into the energy conservation equation 12

$$\frac{d}{dy^\dagger} K_\phi T^{1/2} \frac{dT}{dy^\dagger} = -T^{3/2}\left(\frac{\mu_\phi G(\phi, \tan(\theta))^2}{\varepsilon^2} - D_\phi\right) \quad (14)$$

Using equation 13 to express T in terms of ϕ, we obtain,

$$\frac{d}{dy^\dagger}\left(-\frac{p'_\phi K_\phi}{(p_\phi)^{5/2}} \frac{d\phi}{dy^\dagger}\right) = -\frac{1}{(p_\phi)^{3/2}}\left(\frac{\mu_\phi G(\phi, \tan(\theta))^2}{\varepsilon^2} - D_\phi\right) \quad (15)$$

where primes denote derivatives with respect to ϕ. The above equation can be simplified to obtain a second order differential equation for ϕ,

$$\frac{d^2\phi}{dy^{\dagger 2}} + E(\phi)\left(\frac{d\phi}{dy^\dagger}\right)^2 = F(\phi) \quad (16)$$

where

$$E(\phi) = \frac{(p_\phi)^{5/2}}{(p_\phi)' K_\phi} \left(\frac{(p_\phi)' K_\phi}{(p_\phi 2)^{5/2}}\right)'$$

$$F(\phi) = \frac{p_\phi}{(p_\phi)' K_\phi}\left(\frac{\mu_\phi G(\phi, \tan(\theta))^2}{\varepsilon^2} - D_\phi\right) \quad (17)$$

where primes denote derivatives with respect to ϕ. Equation 16 is a non-linear equation, which has to be solved for a specific model for the pressure, viscosity and thermal conductivity, in order to obtain the volume fraction as a function of height in the boundary layer. It is also a second order differential equation in $y^\dagger$, which requires two boundary conditions. One of these is the matching condition in the limit $y^\dagger \to \infty$,

$$\phi \to \phi_o \text{ for } y^\dagger \to \infty \tag{18}$$

The second condition, which is the condition on the temperature or temperature gradient at the bottom, is not necessary for proving the existence of solutions, and so we do not discuss this in detail. The second condition is the condition for either the temperature or the flux at the base. The Jenkins and Richman[39] conditions at the bottom surface balance the heat flux towards the base J with the rate of dissipation of energy D. Since the heat flux, $-K(dT/dz)$, where K is $T^{1/2}$ times a function of volume fraction, and D is proportional to $T^{3/2}$, the Jenkins and Richman conditions have the form,

$$\frac{dT}{dy} = \beta T \tag{19}$$

where β is a positive function of the volume fraction at the base when the base is dissipative. If there is an energising base which supplies energy to the flow, the function β is negative, and the temperature decreases with an increase in height from the surface. It is also possible to consider other boundary conditions, such as a constant temperature or a constant heat flux at the surface. However, before imposing boundary conditions, we first whether solutions exist for equation 16 in the limit $y^\dagger \to \infty$.

The equation 16 is most conveniently expressed in terms of the departure from the volume fraction ϕ_o in the bulk,

$$\frac{d^2(\phi - \phi_o)}{dy^{\dagger 2}} + E(\phi) \left(\frac{d(\phi - \phi_o)}{dy^\dagger} \right)^2 = F(\phi) \tag{20}$$

For $\phi \to \phi_o$, the above equation can be linearised in the difference $(\phi - \phi_o)$ about $\phi = \phi_o$. It should be noted that $F(\phi_o) = 0$ in the bulk of the flow where there is a balance between the rate of production due to shear and the rate of dissipation due to inelastic collisions, and therefore $F(\phi)$ can be approximated as $(dF(\phi)/d\phi)|_{\phi=\phi_o} (\phi - \phi_o)$. The resulting linear equation is

$$\frac{d^2(\phi - \phi_o)}{dy^{\dagger 2}} = H(\phi_o)(\phi - \phi_o) \tag{21}$$

where

$$H(\phi_o) = \left. \frac{dF(\phi)}{d\phi} \right|_{\phi=\phi_o} \tag{22}$$

The solution of this equation, consistent with the requirement that the perturbation to the volume fraction should be finite for $y^\dagger \to \infty$, is

$$\phi = \phi_o + C \exp(-\sqrt{H(\phi_o)} y^\dagger) \tag{23}$$

where C is a constant to be determined from the boundary conditions. Clearly, exponentially decaying solutions which satisfy the matching condition for $y^\dagger \to \infty$ exist only if $H(\phi_o)$ is positive. If $H(\phi_o)$ is negative, there are no solutions that satisfy the boundary condition for $y^\dagger \to \infty$. Thus, a boundary layer solution for the energy field requires that $H(\phi)$ is positive at $\phi = \phi_o$.

The function $H(\phi_o)$ has been calculated as a function of the coefficient of restitution for different volume fractions for the hard-particle model[20]. For the smooth-particle model, the coefficient $H(\phi_o)$ is always negative, indicating that there are no boundary layer solutions. For the rough particle model, $H(\phi_o)$ is negative when the coefficient of restitution is very close to 1, indicating that a boundary layer solution cannot be obtained in the limit of elastic interparticle collisions. However, $H(\phi_o)$ changes sign and assumes positive values as the coefficient of restitution decreases. This implies that an exponentially decreasing boundary layer solution exists when the coefficient of restitution is less than a maximum value, and a boundary layer solution does not exist when the coefficient of restitution increases beyond this value.

Thus, the above analysis indicates that a boundary layer solution does not exist under all conditions, but only for specific parameter values. In particular, there is a critical dependence on the variation of the parameter $(\frac{\mu_\phi G(\phi, \tan(\theta))^2}{\varepsilon^2} - D_\phi)$ with ϕ about $\phi = \phi_o$, and it is necessary to evaluate this parameter in order to assess whether a boundary layer solution is possible or not. This is in contrast to viscous boundary layer solutions in high Reynolds number fluids flows, for example, which always exist. However, it is still necessary to solve a non-linear equation, equation 20, in order to obtain the actual volume fraction and temperature profiles in the boundary layer. There are difficulties with existence and uniqueness due to the non-linear nature of the equation. There is an additional complication, which is the rapid variation of the pair distribution function with volume fraction as the limit of close packing is approached.

We now relate the existence of a boundary layer solution to the variation of the volume fraction with angle of inclination in the bulk. In the study of Kumaran[19], the volume fraction was related to the angle of inclination using the ratio of stresses using equation 10. For later convenience, we refer to the right side of equation 10 as $P(\phi, G)$.

$$P(\phi, G) = \frac{\mu_\phi G}{p_\phi} \qquad (24)$$

The parameter G, which provides the relation between the strain rate and the temperature was obtained from the balance between the rates of production and dissipation of energy. It is convenient to work with the function $Q(\phi, G)$, which is defined as the difference between the (ratio of the rates of production and dissipation) and 1, which has to be zero in the bulk for $h \gg \delta$,

$$Q(\phi, G) = \left(\frac{\mu_\phi G^2}{D_\phi \varepsilon^2} - 1 \right) = 0 \qquad (25)$$

Equation 25 is used to obtain G in terms of the volume fraction, and inserted into equation 24, to obtain the angle of inclination as a function of volume fraction. Note that G tends to a finite value in the limit of close packing $\phi \to \phi_c$, since both μ_ϕ and D_ϕ diverge proportional to χ. Therefore, $P(\phi, G)$ also tends to a finite value in this limit. In Kumaran[19], a flow was considered to be physically realistic if the volume fraction decreases as the angle of inclination increases. This requires that the derivative of $P(\phi, G)$ with respect to ϕ at constant $Q(\phi)$ has to be negative. Therefore, the volume fraction decreases as the angle of inclination increases only for

$$\left. \frac{\partial P(\phi, G)}{\partial \phi} \right|_{Q(\phi, G)=0} < 0 \qquad (26)$$

In the present boundary layer analysis, we showed that a boundary layer solution exists only for

$$\begin{aligned} H(\phi) &= \left. \frac{p_\phi D_\phi}{p'_\phi K_\phi} \frac{d}{d\phi} \left(\frac{\mu_\phi G(\phi, \tan(\theta)^2)}{D_\phi \varepsilon^2} - 1 \right) \right|_\phi \\ &> 0 \end{aligned} \qquad (27)$$

Note that $H(\phi)$ can be written in terms of the derivative of the function $Q(\phi, G)$ with respect to ϕ,

$$H(\phi) = \left. \frac{p_\phi D_\phi}{p'_\phi K_\phi} \frac{dQ}{d\phi} \right|_{P(G,\phi)=\tan(\theta)} \qquad (28)$$

It can be easily shown that the factors in front of $(dQ/d\phi)|_{P(G,\phi)=\tan(\theta)}$ are positive, if the pressure and its derivative with respect to volume fraction are positive, as required for stability. Therefore, $H(\phi)$ is positive when $(dQ/d\phi)|_{P(G,\phi)=\tan(\theta)}$ is positive, and a boundary layer solution exists only when

$$\left. \frac{dQ}{d\phi} \right|_{P(G,\phi)=\tan(\theta)} > 0 \qquad (29)$$

The equations 26 and 29 are compared, using the identity

$$\left. \frac{dP}{dQ} \right|_\phi \left. \frac{\partial Q}{\partial \phi} \right|_P = - \left. \frac{dP}{d\phi} \right|_Q \qquad (30)$$

It can easily be verified that the derivative $(dP/dQ)|_\phi$, calculated by first substituting for G in terms of Q from equation 25, is always positive. Therefore, equation 30 indicates that conditions 26 and 29 are either satisfied simultaneously, or are violated simultaneously. This provides the principal result that a boundary layer solution at the bottom boundary for the flow down an inclined plane exists if and only if the volume fraction in the bulk decreases as the angle of inclination is increased.

MINIMUM HEIGHT h_{stop} AS A FUNCTION OF ANGLE OF INCLINATION:

A salient feature of the granular flow down an inclined plane is the h_{stop} vs. angle of inclination curve[1, 40]. In experiments, after a granular layer has flowed down an incline, there is a residual layer of thickness h_{stop} remaining on the inclined plane, and the thickness of this layer decreases as the angle of inclination increases. It is found that h_{stop} is a decreasing function of the angle of inclination, and the volumetric flow rate is better correlated to the ratio (h/h_{stop}), rather than (h/d), where h is the height of the flow and d is the particle diameter.

In order to examine the reasons for the cessation of flow in thin layers, the boundary layer analysis was extended to thin layers, whose thickness is comparable to the boundary layer thickness[20]. An energy balance argument can be used to explain the physical reason for the minimum height below which flow stops. The flow is driven by a balance between the source of energy due to the mean shear and the dissipation due to inelastic collisions. There is dissipation of energy due to particle collisions in the flow as well as due to particle collisions with the base in this case. Since inter-particle collisions take place throughout the flow whereas wall collisions take place only at the boundary, the dissipation due to wall collisions form a larger fraction of the total dissipation as the height of the flowing layer is decreased. When the shear production in the flow cannot compensate for the additional energy drawn out of the flow due to the wall collisions, the temperature decreases to zero and the flow stops. Thus, this analysis predicts that h_{stop} is non-zero only when the base is dissipative. If there is a transfer of energy from the base to the flowing material, then h_{stop} should be equal to zero. Further, for a dissipative base, the analysis provides scaling results for the dependence of h_{stop} on the angle of inclination for a slightly dissipative base as well as for a highly dissipative base. These scaling results arise as follows.

For a dense flow, a uniform asymptotic approximation is obtained by using the pair distribution function, χ, as the independent variable instead of the volume fraction ϕ. Since the pair distribution function diverges in the close packing limit, χ^{-1} can be used as a small parameter in an asymptotic expansion. The pressure, viscosity, thermal conductivity and the dissipation rate diverge proportional to χ near close packing, and so we can approximate these as,

$$\begin{aligned} p &= (p_c T \chi / d^3) \\ K &= (K_c T^{1/2} \chi / d^2) \\ \mu &= (\mu_c T^{1/2} \chi / d^2) \\ D &= (D_c \varepsilon^2 T^{3/2} \chi / d^4) \end{aligned} \tag{31}$$

From equation 10, the function G can be approximated as,

$$G = \frac{\tan(\theta) p_c}{\mu_c} \tag{32}$$

It is not sufficient to truncate the right side of equation 14 at just the leading term in an expansion in χ, because this term is zero in the leading approximation for $\phi = \phi_o$.

$$-T^{3/2} D_\phi \left(\frac{\mu_\phi G(\phi, \tan(\theta))^2}{D_\phi \varepsilon^2} - 1 \right) \tag{33}$$

The ratio $(\mu_\phi G(\phi, \tan(\theta))^2 / D_\phi \varepsilon^2)$ tends to a constant in for $\chi \gg 1$ in the close packing limit, since μ_ϕ and D_ϕ are proportional to χ in this limit, while G is a constant. Therefore, equation 33 has to have the form,

$$T^{3/2} D_\phi \left(R'_c + \frac{R_c}{\chi} - 1 \right) \tag{34}$$

correct to $O(\chi^{-1})$, where R'_c and R_c are constants. Also, $(\mu_\phi G(\phi, \tan(\theta))^2 / D_\phi \varepsilon^2)$ is identically equal to 1 for the outer solution $\phi = \phi_o$, or $\chi = \chi_o$, where χ_o is the pair distribution function in the outer region. Therefore, $(R'_c - 1) = -(R_c / \chi_o)$ in equation 34, and the right side of equation 14 is

$$T^{3/2} D_c \chi R_c \left(\frac{1}{\chi} - \frac{1}{\chi_o} \right) \tag{35}$$

It can be shown[20] that R_c is positive when $H(\phi_o)$ in equation 27 is positive.

In the dense limit, the temperature 13 can be rewritten as,

$$T = \frac{6\phi_o gY \cos(\theta)}{\pi p_c \chi} \tag{36}$$

In the above equation, $Y = (h - y)$ is the distance from the top surface, and we have approximated $\phi = \phi_o$ throughout the layer, because the variation in the volume fraction is $O(\delta/h)^2$ smaller than the volume fraction. With this, the energy balance equation 12 becomes,

$$\frac{d}{dY}\left(\frac{K_c}{d^2}\chi^{1/2}Y^{1/2}\frac{d}{dY}\left(\frac{Y}{\chi}\right)\right) = -\frac{R_c D_c \varepsilon^2}{d^4}\frac{Y^{3/2}}{\chi^{1/2}}\left(\frac{1}{\chi} - \frac{1}{\chi_o}\right) \tag{37}$$

We can obtain an equation for χ using the substitution $\chi = (\chi_o Y / hv^2)$,

$$\frac{d^2 v}{dY^2} + \frac{1}{Y}\frac{dv}{dY} + \frac{R_c D_c \varepsilon^2}{2K_c \chi_o d^2}\left(v - \frac{hv^3}{Y}\right) = 0 \tag{38}$$

Using a scaled co-ordinate, $Y^* = (Y/h)$, we obtain a non-linear differential equation for v,

$$\frac{d^2 v}{dY^{*2}} + \frac{1}{Y^*}\frac{dv}{dY^*} + Ah^2\left(v - \frac{v^3}{Y^*}\right) = 0 \tag{39}$$

where

$$A = (R_c D_c \varepsilon^2 / 2K_c \chi_o d^2). \tag{40}$$

The zero flux condition at the top surface $Y = 0$, and the boundary condition 19 at the bottom surface $Y = h$ can be expressed in terms of the variable v,

$$\begin{aligned}\frac{dv}{dY^*} &= 0 \text{ at } Y^* = 0 \\ \frac{dv}{dY^*} &= \frac{\beta hv}{2} \text{ at } Y^* = 1\end{aligned} \tag{41}$$

Equation 39 can be solved, subject to boundary conditions 41, to obtain the χ and T as a function of height. In the solution[20], it is found that as the height h is decreased, the temperature decreases to zero everywhere at a finite height h if the base is dissipative ($\beta < 0$). Since the strain rate is proportional to $T^{1/2}$, the strain rate is also identically zero at this height. This height is identified as h_{stop}, the minimum height required for flow. The dependence of h_{stop} on the angle of inclination has to be determined numerically. However, physical arguments can be used in the limits of slightly and highly dissipative bases. If the base is highly dissipative ($\beta \gg 1$), it is equivalent to having a zero temperature boundary condition in equation 19. By dimensional analysis, (Ah^2) has to be a constant when the flow stops, because A and h are the only dimensional parameters. Therefore, $(h_{stop}/d) \propto (Ad^2)^{-1/2}$. In the limit $\beta \ll 1$ (base with little dissipation), one would expect h_{stop} to be proportional to β, and a function of A. For dimensional consistency, h_{stop} has to be proportional to (βA^{-1}). Therefore, $(h_{stop}/d) \propto (\beta/(Ad))$.

In order to obtain a relation between h_{stop} and the angle of inclination, it is necessary to relate A to the angle of inclination. Form equation 40, we find that $A \propto \chi_o^{-1}$, where χ_o is the pair distribution function in the bulk. If $\tan(\theta_c)$ is the angle for the cessation of flow, then $\chi_o \to \infty$ at $\tan(\theta) = \tan(\theta_c)$, because the volume fraction is equal to the random close packing volume fraction, and the pair distribution function diverges. For $\tan(\theta) > \tan(\theta_c)$, we would expect $(1/\chi_o) \sim (\tan(\theta) - \tan(\theta_c))$ in the linear approximation. Explicitly expressions for the dependence of χ_o on $(\tan(\theta) - \tan(\theta_c))$ for the smooth and rough particle models have been derived[20], but we restrict attention to a scaling arguments here.

Using $(1/\chi_o) \sim (\tan(\theta) - \tan(\theta_c))$, we find that $A \sim (\tan(\theta) - \tan(\theta_c))$. Therefore, the dependence of (h_{stop}/d) on $(\tan(\theta) - \tan(\theta_c))$ are as follows. If the rate of dissipation of energy at the base is low, then the scaling

$$\frac{h_{stop}}{d} \propto (\tan(\theta) - \tan(\theta_c))^{-1/2} \tag{42}$$

is obtained. In contrast, for a highly dissipative base, we obtain the scaling

$$\frac{h_{stop}}{d} \propto (\tan(\theta) - \tan(\theta_c))^{-1} \tag{43}$$

CONCLUSIONS:

The studies summarised here show that the hard particle model is both realistic and useful for describing the flow down an inclined plane, and all salient features of the flow can be captured using the hard-particle model. Some important conclusions are,

1. The binary collision approximation is valid for the granular flow of materials such as sand and glass, with heights up to about 40 particle diameters, down an inclined plane. This was found to be true for all angles of inclination, except for the lowest angle at which there is initiation of flow.
2. Though the binary collision approximation is valid, the molecular chaos approximation is not. There are significant correlation effects, due to which the pre-collisional two-particle distribution function is very different from the single-particle distribution function.
3. When the effect of correlations is incorporated in a kinetic formulation, it is possible to predict the cessation of flow at a finite angle of inclination (angle of repose).
4. For thick flowing layers, the rate of conduction of energy is small compared to the rates of production and dissipation. The rate of conduction of energy is significant only in 'conduction boundary layers' of thickness $\delta = (d/(1-e_n)^{1/2})$ at the top and bottom, where d is the particle diameter and e_n is the coefficient of restitution.
5. In the bulk, there is a balance between the rates of production and dissipation. A direct consequence of this is that the volume fraction is a constant in the leading approximation in (δ/h) where δ is the boundary layer thickness and h is the height of the flow. All components of the stress tensor are proportional to the square of the strain rate.
6. An analysis of the boundary layer equations indicate that a boundary layer solution exists if and only if the volume fraction decreases with an increase in the angle of inclination. In the opposite case where the volume fraction increases with an increase in the angle of inclination, no steady flow is possible.
7. The minimum height h_{stop}, at which the flow stops, depends on the nature of dissipation at the base. For a dissipative base, as the thickness of the layer is decreased, there is a minimum height at which the rate of production of energy due to shear in the bulk is not sufficient to compensate for the rate of dissipation in the base. This results in the cessation of flow at $h = h_{stop}$.

ACKNOWLEDGMENTS

The author would like to thank the Department of Science and Technology, Government of India for financial support.

REFERENCES

1. L. E. Silbert, D. Ertas, G. S. Grest, T. C. Halsey, D. Levine and S. J. Plimpton, Phys. Rev. E, **64**, 51302, (2001).
2. L. E. Silbert, G. S. Grest, S. J. Plimpton & D. Levine, *Phys. Fluids* **14**, 2637-2646, (2002).
3. N. Mitarai & H. Nakanishi, *Phys. Rev. Lett.* **94**, 128001, 2005.
4. K. A. Reddy & V. Kumaran, *Phys. Rev. E*, **76**, 061305, (2007).
5. S. B. Savage, *J. Fluid Mech.*, **92**, 53-96, (1979).
6. P. C. Johnson and J. R. Jackson, *J. Fluid Mech.*, **176**, 67-94, (1987).
7. C. K. K. Lun, S. B. Savage, D. J. Jeffrey, & N. Chepurniy, *J. Fluid Mech.*, **140**, 223-256, (1984).
8. O. Pouliquen & Y. Forterre, *J. Fluid Mech.* **453**, 133âĂŞ-151 (2002).
9. P. Jop, Y. Forterre and O. Pouliquen, *Nature* **441**, 727-730, (2006).
10. M.-Y. Louge, and S. C. Keast, *Phys. Fluids*, **13**, 1213-1233, (2001).
11. D. Ertas, and T. C. Halsey, *Europhys. Lett.* **60**, 931-934 (2002).
12. L. Bocquet, J. Errami, and T. C. Lubensky, *Phys. Rev. Lett.*, **89**, 184301 (2002).
13. N. Mitarai, H. Hayakawa and H. Nakanishi, *Phys. Rev. Lett.*, **88**, 174301, (2002).
14. M.-Y. Louge, *Phys. Rev. E*, **67**, 061303-061313 (2003).
15. GDR MiDi *Euro. Phys. J. E*, **14**, 341-365, (2004).
16. J. T. Jenkins, *Phys. Fluids*, **18**, 103307-103315 (2006).
17. J. T. Jenkins, *Granular Matter*, **10**, 47-52 (2007).
18. V. Kumaran, *J. Fluid Mech.*, **506**, 1-43, (2004).
19. V. Kumaran, *J. Fluid Mech.*, **561**, 1-42, (2006).
20. V. Kumaran, *J. Fluid Mech.*, **599**, 120-168, (2008).
21. V. Kumaran, *J. Fluid Mech.*, **632**, 109-144, (2009).

22. V. Kumaran, *J. Fluid Mech.*, **632**, 145-198, (2009).
23. O. Baran, D. Ertas, T. C. Halsey, G. S. Grest, and J. B. Lechman, *Phys. Rev. E*, **74**, 051302-051311 (2006).
24. A. V. Orpe, V. Kumaran, K. A. Reddy, and A. Kudrolli, *Europhys. Lett.*, **84**, 64003, (2008).
25. H. Hayakawa and M. Otsuki, *Prog. Theor. Phys. Supplement*, **178**, 49-55, 2009.
26. M. Otsuki and H. Hayakawa, *J. Stat. Mech. - Theor. & Exp.* L08003, (2009).
27. K. A. Reddy & V. Kumaran, *J. Fluid Mech.*, submitted.
28. C. S. Campbell, *J. Fluid Mech.*, **539**, 273-297, (2005).
29. C. S. Campbell, *Powder Tech.*, **162**, 208-229, (2006).
30. R. J. Bathurst, & L. Rothenburg, *Trans. ASME: J. Appl. Mech.*, **55**, 17-23, (1988).
31. D. M. Cole, and J. F. Peters, *Granular matter*, **9**, 309-321,(2007).
32. D. M. Cole, and J. F. Peters, *Granular matter*, **10**, 171-185, (2008).
33. R. D. Mindlin, and H. Deresiewicz, *ASME J. Appl. Mech.* **20**, 327-344, (1953).
34. V. Kumaran, *Europhys. Lett.*, **73**, 1-7, (2006).
35. V. S. Kumar and V. Kumaran, *J. Chem. Phys.*, 123, 114501-114513, (2005).
36. V. S. Kumar and V. Kumaran, *J. Chem. Phys.*, **124**, 204508 (2006).
37. M. Otsuki and H. Hayakawa, *Phys. Rev. E*, **80**, 11308 (2009)
38. N. Mitarai, and H. Nakanashi, *Phys. Rev. E*, 031305, (2007).
39. J. T. Jenkins and M. Richman, *Arch. Rat. Mech. Anal.*, **87**, 355-377, (1985).
40. O. Pouliquen, *Phys. Fluids*, **11**, 542-548, (1999).

On the hyperbolicity of a two-fluid model for debris flows

C. Mineo and M. Torrisi

Department of Mathematics and Computer Science
viale A. Doria, 6, 95125 Catania, Italy
mineo@dmi.unict.it
torrisi@dmi.unict.it

Abstract. We consider the system of partial differential equations associated with the mathematical model for debris flows proposed by E.B. Pitman and L. Le (*Phil. Trans. R. Soc. A*, **363**, 1573-1601, 2005) and analyze the problem of the hyperbolicity of the model.

Keywords: Debris flows, mixtures, hyperbolic systems, avalanches
PACS: 45.70.-n, 45.70.Ht

INTRODUCTION

In recent decades, there has been an increasing frequency of occurrence of catastrophic events, especially in mountain areas, interfering with the man-made structures of the area itself. This has generated great interest, both in research and in land planning, to the phenomena of intense transport by geophysical mass flows, generally identified with the terms debris flows, avalanches or landslides. These flows are composed of mixture of soil and rock with a significant quantity of interstitial fluid. They often originate in sediments, present in the form of natural deposits, mobilized by alluvial events resulting in the formation of floods of sediments that are propagated with elevated speed. Similar phenomena can be consequent to an eruption, where slow effusions of lava, flows of red-hot ash and rocks race along the surface of the mountain at speeds of up to hundreds of meters per second. These flows can melt snow, creating a muddy mixture of ash, rock and water.

Geophysical mass flows, can synthetically be described as gravitational motions of sediments, since the dominant force for the solid component is the gravity. They are also affected by the actions associated with the granular interactions and, in particular, inertial effects due to collisions among the granules and the deformation of the mixture. Therefore, the traditional practice of supposing that solid and liquid inextricably unite to form a single-phase material should be avoided, and the distinct properties and interactions of the flows of fluid and solid constituents should be emphasized.

In this short paper we consider a two-component mixture of a solid phase and a fluid phase that, in general could describe the so-called debris flows. The model which we take in consideration in the present paper is due to E.B. Pitman and L. Le [1] and begins from the fluid mechanical description of fluidized beds shown from Anderson and Jackson in [2]. The plan of the paper is the following. In section 2, the Pitman and Le two-fluid model [1] is briefly described. In section 3, after rewriting their system in matrix form, the reality of roots of characteristic equation is discussed. The conclusion are given in section 4.

THE TWO-FLUID MODEL

Following [1], we consider a system of equations in three dimensions, describing a thin layer of a mixture of incompressible solid, frictional granular material and interstitial incompressible fluid, each of constant specific density ρ^s and ρ^f, respectively. In the following, we neglect any erosion of the base and assume that the components of the mixture are moving over a variable terrain. At any point of the terrain, we consider a Cartesian coordinate system xyz, defined so that the plane xy is tangent to the basal surface and z axis is the normal direction. Moreover, it is assumed that the mixture is moving without a preferred direction in the xy-plane.

Mass and momentum equations [2] for the two constituent phases are written as

$$\partial_t(\rho^s\varphi) + \nabla \cdot (\rho^s\varphi\mathbf{v}) = 0, \tag{1}$$

$$\partial_t(\rho^f(1-\varphi)) + \nabla \cdot (\rho^f(1-\varphi)\mathbf{u}) = 0, \tag{2}$$

$$\rho^s\varphi(\partial_t\mathbf{v} + (\mathbf{v}\cdot\nabla)\mathbf{v}) = -\nabla \cdot \mathbf{T}^s - \varphi\nabla \cdot \mathbf{T}^f + \mathbf{f} + \rho^s\varphi\mathbf{g}, \tag{3}$$

and

$$\rho^f(1-\varphi)(\partial_t\mathbf{u} + (\mathbf{u}\cdot\nabla)\mathbf{u}) = -(1-\varphi)\nabla \cdot \mathbf{T}^f - \mathbf{f} + \rho^f(1-\varphi)\mathbf{g}, \tag{4}$$

where φ is the solid volume fraction, $\mathbf{v}$ and $\mathbf{u}$ are the solid and fluid velocities, $\mathbf{T}^s$ and $\mathbf{T}^f$ the solid and fluid stress tensors, and $\mathbf{g}$ the gravity acceleration vector. Moreover, $\mathbf{f}$ represents all non-buoyant interaction forces of the fluid on the particle; these are characterized using a simple drag interaction:

$$\mathbf{f} = (1-\varphi)\beta(\mathbf{u}-\mathbf{v}), \tag{5}$$

with the phenomenological function β [14] given by

$$\beta = \frac{(\rho^s - \rho^f)\varphi g}{v_T(1-\varphi)^m}, \tag{6}$$

where v_T is a characteristic velocity, g is the magnitude of the gravitational force and m is related to the Reynold number of the flow.

For a more direct comparison of theory with the experiments, we exclude the viscosity, so the only fluid stress considered is a pressure. Moreover, a Coulomb friction law, expressing collinearity between shear stresses and normal stresses is used:

$$T^{sxz} = -sgn(v)v^b T^{szz}, \tag{7}$$

where $v^b = \tan\phi_{bed}$ is the friction coefficient and ϕ_{bed} is the angle of friction for bed slip. It is worth stressing that in recent works, it has been shown that the Coulomb friction coefficient of the particle phase at the bed can be considered, at least for particular values of the concentration of particles, to depend on a certain inertial parameter, see e.g [3, 4, 5] and references therein.

Finally, an earth pressure relation is employed for the solid phase [7, 8, 9, 10, 13]; this assumes that the basal shear stress is proportional to the normal stress,

$$T^{sxx} = K_{act/pas}T^{szz}, \tag{8}$$

where, in x direction, $K_{act/pas} := \alpha_{xx}$, is the earth pressure coefficient, determined by the equation:

$$K_{act/pas} = 2\frac{1 \mp [1 - \cos^2(\phi_{int})[1 + \tan^2(\phi_{bed})]]^{1/2}}{\cos^2(\phi_{int})} - 1, \tag{9}$$

in which $\phi = \phi_{int}$ denotes the internal friction angle, the sign $\mp$ depends on whether the sediment mass is locally extending ("active" coefficient $\partial v/\partial_x > 0$, with - sign) or compressing ("passive" coefficient $\partial v/\partial_x < 0$, with + sign) as it deforms and moves downslope.

In agreement with the celebrated paper of S.B. Savage and K. Hutter [7], the earth pressure coefficient can be obtained with reference to the Mohr stress circle and Coulomb failure (for its derivation see also Apendix C in [6]). In the special case where $\phi_{bed} = 0$, equation (9) reduces to the classical Rankine [11, 12] definitions commonly used in soil mechanics. An exceptional case for the behavior of the earth coefficient can occurs if the bed has maximum roughness. In this case, $\phi_{bed} = \phi_{int}$ and (9) gives a single-valued expression:

$$K_{act/pas} = \frac{1 + \sin^2(\phi_{int})}{1 - \sin^2(\phi_{int})}, \tag{10}$$

This indicates that slab of Coulomb material moves with zero velocity divergence. Finally, when the sediment mixture is static, it is assumed that $K_{act/pas} = 1$ ($T^{sxx} = T^{szz}$). In this case, the stress state is statically indeterminate and stresses are insufficient for full mobilization of frictional forces. Figure 1.[8] shows a graph of some values of the earth pressure coefficient obtained from equation (9).

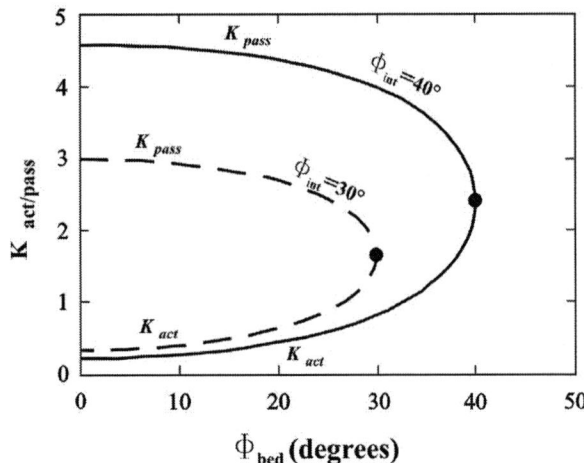

FIGURE 1. Graph of the active and passive earth pressure coefficients obtained from equation (9) as functions of ϕ_{bed} for two typical values of ϕ_{int}. In the graph, the bullet indicates where the coefficient has the unique value obtained from equation (10) for $\phi_{bed} = \phi_{int}$.

The balance equations, through a lengthy process of derivation, are averaged over flow depth and are *suitably* [1] scaled by using the shallow flows assumption $\varepsilon = H/L \ll 1$, where H is a characteristic thickness and L is the characteristic flow length. The upper surface of the flow is stress-free for both constituents; then, material surface conditions at the upper free surface and tangent motion conditions at the base are imposed for both phases as kinematic boundary conditions. Finally, we assume the bottom surface $z_b = z_b(x,y)$, which describes the bottom topography, to be flat, that is, $z_b = 0$.

Then, in the one dimensional case and removing the limiting cases $h = 0$, $\varphi = 0$, $\varphi = 1$, we write below, omitting the lengthy procedure of their derivation, the resulting Pitman-Le model equations:

$$\begin{cases} h_t + [\varphi v + (1-\varphi)u]h_x + h(v-u)\varphi_x + h\varphi v_x + h(1-\varphi)u_x = 0, \\ \varphi_t + \frac{1}{h}[\varphi(v-u)(1-\varphi)]h_x + [v - \varphi(v-u)]\varphi_x + \varphi(1-\varphi)v_x - \varphi(1-\varphi)u_x = 0, \\ v_t + (a+b)h_x + \frac{ah}{2\varphi}\varphi_x + vv_x = (1-\gamma)\frac{(1-\varphi)^{1-m}}{v_T}(u-v) - sgn(v)v^b(1-\gamma)g, \\ u_t + ch_x - \frac{ch}{2(1-\varphi)}\varphi_x + uu_x = -\left(\frac{1-\gamma}{\gamma}\right)\frac{\varphi(1-\varphi)^{-m}}{v_T}(u-v), \end{cases} \quad (11)$$

where

$$\gamma = \frac{\rho^f}{\rho^s} < 1, \quad a = \varepsilon(1-\gamma)\alpha_{xx}g, \quad b = \varepsilon\gamma g, \quad c = \varepsilon g. \quad (12)$$

and not only have (5), (6), (7) and (8) been taken into account, but also that

$$T^{szz} = (1-\gamma)\varphi g\frac{h}{2}, \quad (13)$$

and

$$T^{fzz} = g\frac{h}{2}. \quad (14)$$

ON THE WAVE PROPAGATION VELOCITIES

The system (11) in matrix form becomes

$$\partial_t \mathbf{U} + \mathbf{A}\partial_x \mathbf{U} = \mathbf{F}, \tag{15}$$

where

$$\mathbf{U} = \begin{bmatrix} h \\ \varphi \\ v \\ u \end{bmatrix} \tag{16}$$

$$\mathbf{A} = \begin{bmatrix} \varphi v + (1-\varphi)u & h(v-u) & h\varphi & h(1-\varphi) \\ \frac{\varphi(1-\varphi)(v-u)}{h} & v - \varphi(v-u) & \varphi(1-\varphi) & -\varphi(1-\varphi) \\ a+b & \frac{ah}{2\varphi} & v & 0 \\ c & -\frac{ch}{2-2\varphi} & 0 & u \end{bmatrix} \tag{17}$$

$$\mathbf{F} = \begin{bmatrix} 0 \\ 0 \\ (1-\gamma)\frac{(1-\varphi)^{1-m}}{v_T}(u-v) - sgn(v)v^b(1-\gamma)g \\ -\left(\frac{1-\gamma}{\gamma}\right)\frac{\varphi(1-\varphi)^{-m}}{v_T}(u-v) \end{bmatrix} \tag{18}$$

and $\partial_t = \partial/\partial_t, \partial_x = \partial/\partial_x$.

Following well known techniques, the characteristic equation

$$P(\lambda) := det(\mathbf{A} - \lambda \mathbf{I}) = 0 \tag{19}$$

of the system is

$$(\lambda - v)^2(\lambda - u)^2 - \frac{h}{2}[a(1+\varphi) + 2\varphi b](\lambda - u)^2 - \frac{ch}{2}(2-\varphi)(\lambda - v)^2 + \frac{ch^2}{2}(a+b\varphi) = 0. \tag{20}$$

The problem of hyperbolicity of the system is related to the reality of all of the roots of this equation.

In [1], the case of the parameter $\varepsilon = 0$, has been considered and it has been observed that in this case, taking (12) into account, equation (20) reduces to:

$$(\lambda - v)^2(\lambda - u)^2 = 0 \tag{21}$$

That is, there are two pairs of double real roots.

Another special case is the case $u = v$. In this case, the characteristic polynomial reduces to:

$$(\lambda - v)^4 - \frac{h}{2}[a(1+\varphi) + 2\varphi b - c(2-\varphi)](\lambda - v)^2 + \frac{ch^2}{2}(a+b\varphi) = 0, \tag{22}$$

which can be considered a quartic equation in the unknown $\lambda - v$. The coefficient of $(\lambda - v)^2$ is always negative while the discriminant can be shown to be positive, at least for physically relevant values of γ and α_{xx}, in these cases, the quartic equation admits two real positive roots for λ^2; this results in four real roots for the characteristic polynomial.

Even if both previous cases do not have physical meanings, they allow:

- to affirm that for ε very small the system is still hyperbolic,

- to compute, starting from the case $u = v$, the roots for $u \neq v$ by using asymptotic methods, see [1].

In view of further applications we observe that the constant vector

$$\mathbf{U_0} := \begin{bmatrix} h_0 \\ \varphi_0 \\ 0 \\ 0 \end{bmatrix}, \tag{23}$$

with $h_0 > 0$ and $0 < \varphi_0 < 1$, gives a two parameters family of solutions for the system (15). Here we show as the system can be hyperbolic around a constant state $\mathbf{U_0}$.

After having put $\mathbf{U} = \mathbf{U_0}$, the characteristic equation becomes a quartic equation in λ:

$$P_0(\lambda) := \lambda^4 - \frac{h_0}{2}[a(1+\varphi_0) + 2\varphi_0 b - c(2-\varphi_0)]\lambda^2 + \frac{ch_0^2}{2}(a+b\varphi_0) = 0, \tag{24}$$

whose roots are written as

$$\lambda_0^i = \pm \frac{1}{2}\sqrt{\mathscr{A} \pm h_0\sqrt{\Delta_0}} \qquad i = 1,2,3,4 \tag{25}$$

where

$$\mathscr{A} = h_0[(a-2c) + \varphi_0(a+2b+c)], \tag{26}$$

and

$$\Delta_0 = (a-2c)^2 + \varphi_0^2(a+2b+c)^2 + 2\varphi_0(a-2c)(a+2b+c) - 8c(a+b\varphi_0). \tag{27}$$

The eigenvalues λ_0^i $i = 1,2,3,4$ will be real if and only if both $\mathscr{A}$ and Δ_0 are non negative. In fact, due to Descartes rule, it follows that the values of

$$\lambda^2 = \frac{1}{4}\left(\mathscr{A} \pm h_0\sqrt{\Delta_0}\right) \tag{28}$$

are not only real but also non-negative. Then all the eigenvalues λ_0^i are real.

From (26) and (27), taking (12) into account, it is a simple matter to ascertain that in order for $\mathscr{A}$ and Δ_0 to both be non-negative it is necessary and sufficient that both:

$$(1+\varphi_0)(1-\gamma)\alpha_{xx} - 2 + \varphi_0(2\gamma+1) \geq 0 \tag{29}$$

and

$$(1+\varphi_0)^2(1-\gamma)^2\alpha_{xx}^2 + 2(1-\gamma)(2\gamma\varphi_0 - \varphi_0 - 6)\alpha_{xx} + \varphi_0(4\varphi_0\gamma(\gamma+1) + \varphi_0 - 8\gamma) \geq 0 \tag{30}$$

So, for values of α_{xx} and γ satisfying the previous conditions, we can say that the system is hyperbolic, at least in a suitable neighborhood of $\mathbf{U_0}$. For instance, for $\varphi_0 = 0,6$ from (29) and from (30) it is possible to ascertain that in order the eigenvalues be real, we must require that

$$\alpha_{xx} \geq \frac{\gamma(1,9\gamma - 6,2) + 4,2 + 2\sqrt{\gamma(-0,4\gamma^2 + 4\gamma - 6,9) + 3,2}}{\gamma(2,6\gamma - 5,1) + 2,6}, \tag{31}$$

at least for values of γ of relevant physical interest (e.g. $\gamma = 0,4$).

By assuming $\lambda, h, \varphi, u, v$ depending from a small parameter η and analytic with respect to it, we can write:

$$\begin{cases} \lambda^i = \lambda_0^i + \eta\lambda_1^i + \eta^2\lambda_2^i ....., \\ h = h_0 + \eta h_1 + \eta^2 h_2......, \\ \varphi = \varphi_0 + \eta\varphi_1 + \eta^2\varphi_2......, \\ u = \eta u_1 + \eta^2 u_2......, \\ v = \eta v_1 + \eta^2 v_2....... \end{cases} \tag{32}$$

By truncating the previous expansions at the first power of η and after having substituted them in the $P(\lambda)$, we obtain, by neglecting all powers of η greater than one, an approximate characteristic polynomial:

$$P_1(\lambda^i) := P_0(\lambda_0^i) + \eta \left\{ 2[2\lambda_1^i - (u_1+v_1)]\lambda_0^{i3} + \frac{1}{2}[(\varphi_0 h_1 + \varphi_1 h_0)(c-a-2b) - h_1(a+2c)]\lambda_0^{i2} + \right.$$
$$\left. + h_0\{(u_1 - \lambda_1^i)[a(1+\varphi_0) + 2b\varphi_0] + (v_1 - \lambda_1^i)c(2-\varphi_0)\}\lambda_0^i + ch_0\left[\frac{b}{2}(h_0\varphi_1 + 2h_1\varphi_0) + ah_1\right] \right\} = 0. \quad (33)$$

Then by requiring:

$$\begin{cases} P_0(\lambda_0^i) = 0 \\ 2[2\lambda_1^i - (u_1+v_1)]\lambda_0^{i3} + \frac{1}{2}[(\varphi_0 h_1 + \varphi_1 h_0)(c-a-2b) - h_1(a+2c)]\lambda_0^{i2} + \\ + h_0\{(u_1 - \lambda_1^i)[a(1+\varphi_0) + 2b\varphi_0] + (v_1 - \lambda_1^i)c(2-\varphi_0)\}\lambda_0^i + ch_0\left[\frac{b}{2}(h_0\varphi_1 + 2h_1\varphi_0) + ah_1\right] = 0, \end{cases} \quad (34)$$

it follows, trivially that

$$\lambda_0^i = \pm \frac{1}{2}\sqrt{\mathscr{A} \pm h_0\sqrt{\Delta_0}}, \qquad i = 1,2,3,4 \quad (35)$$

and

$$\lambda_1^i = \frac{-2(v_1+u_1)\lambda_0^{i3} + \frac{1}{2}[(\varphi_0 h_1 + \varphi_1 h_0)(c-a-2b) - h_1(a+2c)]\lambda_0^{i2}}{\lambda_0^i[4\lambda_0^{i2} - ah(1+\varphi_0) - ch(2-\varphi_0) - 2hb\varphi_0]} +$$
$$+ \frac{\{cv_1 h_0(2-\varphi_0) + u_1 h_0[a(1+\varphi_0) + 2b\varphi_0]\}\lambda_0^i + ch_0\left[\frac{b}{2}(h_0\varphi_1 + 2h_1\varphi_0) + ah_1\right]}{\lambda_0^i[4\lambda_0^{i2} - ah(1+\varphi_0) - ch(2-\varphi_0) - 2hb\varphi_0]}. \qquad i = 1,2,3,4. \quad (36)$$

So,

$$\lambda^i \approx \pm \frac{1}{2}\sqrt{\mathscr{A} \pm h_0\sqrt{\Delta_0}} + \eta \left\{ \frac{-2(v_1+u_1)\lambda_0^{i3} + \frac{1}{2}[(\varphi_0 h_1 + \varphi_1 h_0)(c-a-2b) - h_1(a+2c)]\lambda_0^{i2}}{\lambda_0^i[4\lambda_0^2 - ah(1+\varphi_0) - ch(2-\varphi_0) - 2hb\varphi_0]} + \right.$$
$$\left. + \frac{\{cv_1 h_0(2-\varphi_0) + u_1 h_0[a(1+\varphi_0) + 2b\varphi_0]\}\lambda_0^i + ch_0\left[\frac{b}{2}(h_0\varphi_1 + 2h_1\varphi_0) + ah_1\right]}{\lambda_0^i[4\lambda_0^2 - ah(1+\varphi_0) - ch(2-\varphi_0) - 2hb\varphi_0]} \right\}, \quad i = 1,2,3,4. \quad (37)$$

CONCLUSIONS

In this short paper, we recall briefly the two-fluid model describing the motion of debris flows [1] over a surface $z_b = z_b(x,y)$. After having put the governing equations in the (1+1)-dimensional case in the matrix form, we derive the characteristic equation for the eigenvalues. This equation is a complete equation of fourth degree. However, to derive the roots explicitly is a non-trivial task, except for some special cases. The knowledge that in these cases the roots are real, allows us to appeal to perturbation arguments to find, in general, their approximate values. In this spirit, we have found the relations between the ratio $\rho_f/\rho_s = \gamma$ and the earth pressure coefficient α_{xx} that allows to consider the two-fluid model to be hyperbolic, at least around a two parameter family of constant solutions $\mathbf{U_0}$.

ACKNOWLEDGMENTS

C.M. and M.T. acknowledge the financial support from P.R.A. (ex 60%) of University of Catania and from G.N.F.M. of INdAM.

REFERENCES

1. E. B. Pitman and L. Le, *Phil. Trans. R. Soc. A*, **363**, 1573-1601 (2005).
2. T. B. Anderson and R. Jackson, *Ind. Eng. Chem. Fundam*, **6**, 527-539 (1967).

3. GDR MiDi, *Eur. Phys. J. E*, **14**, 341-365 (2004).
4. F. da Cruz, S. Emam, M. Prochnow, J. N. Roux, and F. Chevoir, *Phys. Rev. E*, **72**, 021309 (2005).
5. D.Berzi and J.T.Jenkins, *J.Fluid Mech.*, **608**, 393-410, (2008).
6. R. M. Iverson, *Rev. Geophys*, **35**, 245-296 (1997).
7. S. B. Savage, and K. Hutter *J. Fluid Mech.*, **199**, 177-215 (1989).
8. R. M. Iverson, and R. P. Denlinger *Journal of Geophysical Research*, **106**, 537-552 (2001).
9. R. M. Iverson, and R. P. Denlinger *Journal of Geophysical Research*, **106**, 553-566 (2001).
10. J. M. N. T. Gray, M. Wieland and K. Hutter, *Proc. R. Soc. London, Ser. A*, **455**, 1841-1874 (1999).
11. W. J. M. Rankine, *Philos. Trans. R. Soc. London*, **147**, 9–27 (1857).
12. T. W. Lambe, and R. V. Whitman, *Soil Mechanics*, John Wiley, New York, 1979, pp. 553
13. K. Terzaghi, "The shearing resistance of saturated soils and the angle between planes of shear," in *First International Conference on Soil Mechanics*, 1936, pp. 54-56.
14. J. F. Richardson, and W. N. Zaki, *Sedimentation and fluidization: part I. Trans*, **32**, 35-53 (1954).

Rheology of confined granular flows

Patrick Richard*, Alexandre Valance*, Jean-François Métayer*, Jérôme Crassous*, Michel Louge† and Renaud Delannay*

*Université Rennes 1, Institut de Physique de Rennes, UMR CNRS 6251, 263 av. Général Leclerc, 35042 Rennes cedex FRANCE
†Sibley School of Mechanical and Aerospace Engineering, Cornell University, Ithaca, NY 14853, USA

Abstract. The properties of confined granular flows on a heap are studied through numerical simulations and experiments. We address how such system can be simulated with period boundaries in the flow direction. The packing fraction and velocity profiles are found to be described by one length scale. The dependence of the kinematic properties on the number of grains and on micromechanical parameters (coefficient of restitution and coefficient of friction) is described. Our results show that the friction at the sidewalls gradually decreases and that this decrease can be explained by the intermittent motion of the grains in the quasistatic part of the flow.

Keywords: granular flows, surface flows, confinement, friction weakening
PACS: 45.70.-n, 64.70.P-

INTRODUCTION

Surface flows of granular materials are frequently observed in industrial processes and in nature. In such type of flows, the moving particles appear to be limited to a surface layer with a "frozen" bulk region below. Interestingly, the gas, liquid and solid behaviors of granular systems are present at the same time in such flows. Moreover the influence of sidewalls is important [1, 2]. It is thus the ideal type of flow to test existing theories or to inspire new ones which aim is to describe and predict the whole behavior of flowing granular matter. Most studies concerning these flows have been conducted in two different configurations. The first one is a rotating drum [3, 5, 4, 6], a cylinder, partly filled with granular material, rotating around its axis at a controlled speed. Stationary states can be studied in such kind of configuration but it does not permit fully developed flows (i.e. flows that are invariant in the main flow direction). The second configuration consists in pouring grains at a constant flow rate between two parallel sidewalls with a gap W [7, 1]. In this configuration, a static heap slowly forms by trapping grains at its top. After a transient, the growth of the heap stops and the flowing layer at its surface reaches a steady state. Studying this configuration, Taberlet et al. [1] reported the "side-wall stabilized heap" (SSH) regime, where frictional sidewalls decrease the shear stress in the flow. Using a balance of momentum for the mobilized layer, one can derive [1] an approximate linear scaling law linking the free surface angle θ, the height of the layer, h, and the width of the channel, W:

$$\tan\theta = \mu_i + \mu_w \frac{h}{W}, \tag{1}$$

where μ_i and μ_w are effective friction coefficients. All the details of this derivation can be found in [1]. Thus, confining walls play a major role in the momentum balance if the second term on the right in this equation dominates over the first. This equation also points out that the relevant parameter to evaluate the effect of the wall friction on the flow is h/W and not W/d, where d is the diameter of the grains.

It has been shown that the "frozen" bulk layer below the flow is not purely static but exhibit slow creeping motion [8, 9]. In the following we will refer to this region as "quasistatic" layer.

Let us consider now a layer of initesimal thickness dy, parallel to the free surface and located at the depth y. A simple force balance yielded the ratio S/N of the mean shear and normal stresses on planes parallel to the free surface leads to an equivalent of :

$$S(y)/N(y) = \tan\theta - \mu_w y/W, \tag{2}$$

where y is the downward coordinate normal to the surface (see Figure 1). Assuming that the coefficient μ_w does not depend on y, this equation leads to a paradox: the effective friction S/N reverses sign at finite depth.

In this paper we report numerical and experimental results on the rheology of confined granular flows between two

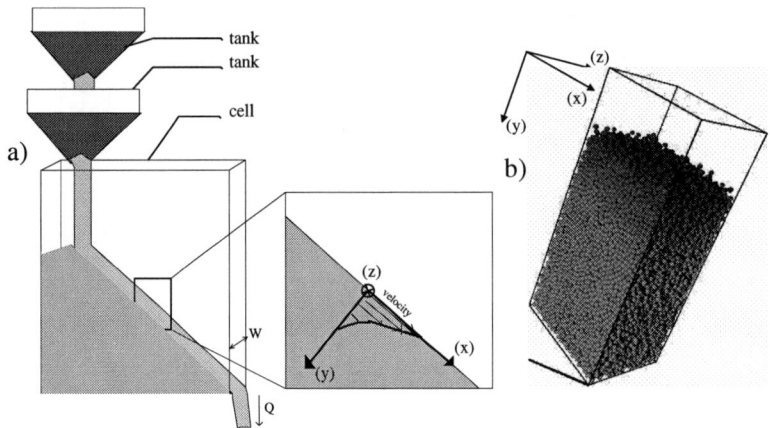

FIGURE 1. (color online) Sketch of the apparatus, closed bottom and left, and bound by two plane, parallel, frictional sidewalls. The outut flow rate is Q (a) and a typical 3D snapshot for chute flow: $N_g = 24,000$ grains, with $W/d = 20$ (d is the grain diameter), angle of inclination $\theta = 35°$, coefficient of restitution $e = 0.88$, and friction coefficient $\mu = 0.50$ (b). The color of the grains corresponds to the velocity in the stream wise direction : v_x. It varies from yellow (light gray) ($v_x = 0$) to red (dark gray) (maximum value of the velocity).

vertical sidewalls. We will show that it is possible to simulate such kind of flows using periodic boundary conditions. Then, we will report that the effective sidewall friction coefficient depends on the depth. In doing so, we will resolve the above mentioned paradox.

The outline of this paper is the following. The next section is devoted to the description of our experimental set-up and of the numerical methods. The third section deals with the study of the velocity and packing fraction of these flows. Then, we will report the evolution of the sidewall friction with the depth. The conclusion and some perspectives will be presented in the last section.

EXPERIMENTAL AND NUMERICAL METHODS

Experimental set-up

Our experimental set-up (Figigure 1) is similar to the one used by Taberlet et al. [1]. It consists of two 1200 mm × 1200 mm parallel and vertical glass plates separated by a width W. Glass-beads of diameter $d = (500 \pm 100)\,\mu$m are continuously poured between these plates through a "double-hopper". The lower hopper is continuously filled with particles by the upper one. The aperture of the lower hopper precisely controls the input flow rate Q, defined as the mass of material entering the channel per unit of time and per unit of width. In such a set-up, Q and W are the only control parameters. When the system reaches a steady-sate, the input flow rate is equal to the output flow rate which is measured using an electronic weighting of material falling out of the channel. All our experiments are conducted in a temperature and humidity-controlled room (20° C and 50% of humidity). The electrostatic effects are minimized by passing the grains through a metal sieve connected to the ground prior to all experiments. The beads are painted in black to limit light reflection. A Photron APX RS camera of 1024×1024 resolution for 50 to 30,000 Hz frame rate tracks rapid grains, a Nikon D200 reflex photo camera at 12 images/minute is used for slow grains. This range of frequency is large enough to track grains whose velocity is between 3.10^{-4} m/s and 10 m/s (fast camera) and between 1.10^{-6} m/s and 3.10^{-2} m/s (photo camera). The bead size is about 30 pixels. A tracking code deduces the velocity of a given grain from the difference between its successive positions, assuming that the tracked grains motion is less that one radius between two images. To estimate the precision of our measures, and the smallest velocity we can detect, the position of grains in an immobile packing have been tracked. The displacements measured are artifacts due to spot light variations inducing fluctuations in the gray level of each pixel. The articfacts detected are lower than $1/200$ mm. The minimum average velocity that can be detected is then $(f/200)$ mm.s^{-1} where f (in s^{-1}) is the frequency of the camera. The angle of inclination of the flow, θ, is also measured by image processing.

The packing fraction is deduced from the absorption of γ-rays by the granular material. The channel is placed between the source of γ-rays (Cs137) and a scintillator which measures the intensity of the beam. This intensity follows the Beer-Lambert law [10]: $I(y) = I_0 \exp(-\alpha W \nu(y))$, where I_0 is the intensity measured when the channel is empty, α the absorption constant which depends on the material ($\alpha = 0.188$ cm^{-1} for glass), W the channel width and ν the packing fraction averaged in the direction z perpendicular to the sidewalls. More details on this methods can be found in [11, 12].

Numerical methods

Computer simulations have turned out to be a powerful tool to investigate the physics of granular flow, especially valuable as they reveal flow details, such as stresses, which experiments cannot provide. A very popular simulation scheme is an adaptation of the classical Molecular Dynamics technique [13]. It consists of integrating Newton's equations of motion for a system of "soft" grains starting from a given initial configuration. This requires giving an explicit expression for the forces that act between grains. Let us consider two overlapping spheres i and j. The overlap δ leads to a normal force $F_n = k_n \delta - \gamma_n \dot{\delta}$, where k_n is a spring constant, γ_n a viscous damping. The damping is used to obtain an inelastic collision. The tangential force used is that of the well-known regularized Coulomb law $F_t = min(\mu F_n, k_t u_t)$ where μ is friction coefficient, k_t a tangential spring stiffness and u_t is the tangential displacement of the contact. The latter is set to zero at the initiation of a contact and its rate of change du_t/dt is equal to the sliding velocity. Note that the rigid body motion around the contact is taken into account to ensure that the tangential displacement is always in the local tangent plane of the contact. For most of the results presented here, the following values of the parameters are used: particle diameter $d = 0.5$ mm (with a small uniform polydispersity $\pm 20\%$ to prevent crystalization), density $\rho = 2500$ kg.m^{-3}, $k_n = 5.6075 \times 10^6$ mg/d, where m is the mean mass of a grain, $\gamma_n = 130.89\, m\sqrt{g/d}$, $k_t = 2k_n/7$ and (unless specified) $\mu = 0.5$. These values lead to a restitution coefficient $e = 0.88$. When studying the effect of the coefficient of restitution, k_n is kept constant and γ_n is tune to obtain the chosen value of e. Impacts against the sidewalls are treated as collisions with a sphere of infinite mass and radius, which mimics a large flat surface. Note that the value of friction coefficient and the value of the coefficient of restitution are the same for the wall-grain and the grain-grain interactions.

The use of periodic boundary conditions for flows on a rough bottom and without sidewalls is straightforward: the angle of the free surface is equal to the angle of inclination of the rough bottom. This is no more the case for confined flows on a heap, since the angle of the free surface depends on the input flow rate. Two kind of simulations can be performed:

- A "full system" simulation where the whole system is simulated from the input of the grains to the output. A constant number flow rate is imposed and, as in experiments, above a critical value [1], grains get trapped underneath the flowing layer, and the SSH grows slowly [14] until its free surface reaches a steady angle.
- A periodic boundary conditions simulation, the angle of inclination is set to a given value. The flow rate Q may evolve towards a constant value.

These two simulations are therefore fundamentally different. The "full system simulation" is the closest to the experiments : the flow rate is imposed and the angle of inclination evolves towards a steady value. In the case of the PBC simulation, the angle of inclination is imposed and the flow rate evolves towards a steady value. Interestingly both types of simulation give the same $\tan\theta$ .vs. Q curve [15] and all the flow properties are found to be very close [16]. Surprisingly, in the periodic system, grains below the flowing layer remain nearly immobile despite the steep inclination. Thus, the confining walls permit the establishment of a stable equilibrium on the erodible heap at an angle far exceeding what is observed without them. Nonetheless, above an angle θ_{max}, grains accelerate, and, below θ_{min}, they come to rest. These critical inclinations of the flowing layer are functions of simulation parameters. Remarkably, they also depend on the relative channel width (W/d), even though the grain diameter d does not appear explicitly in Eq. (1). An important point is that θ_{max} also depends on the number of grains used in the simulation. This point will be addressed latter. In both systems, Q is measured by counting the number of grains flowing through a surface perpendicular to the confining walls per unit time. Note that, in periodic simulations, we observed that flow properties became insensitive to the domain length L along the stream wise direction when $L > 10d$. Therefore, all the periodic simulations were performed using $L = 25d$.

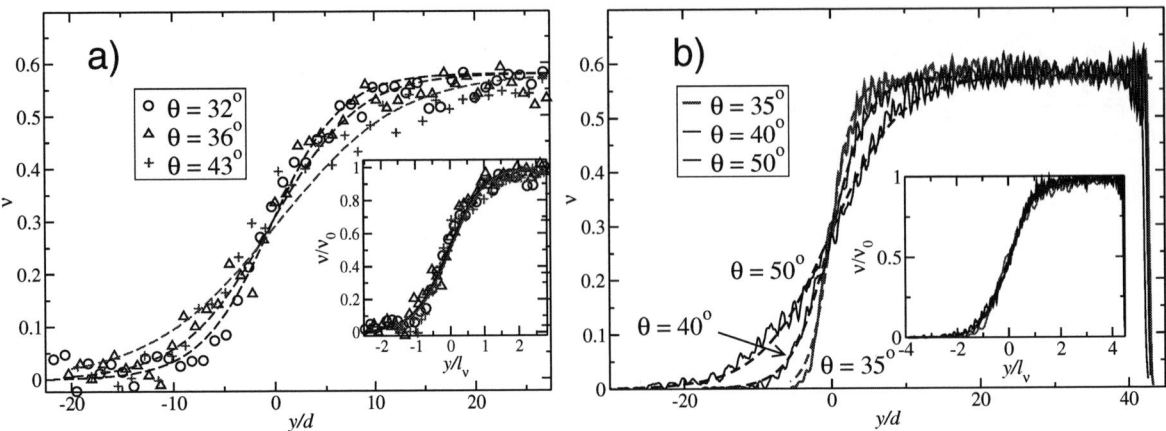

FIGURE 2. (color online) Profiles of the packing fraction ν for different inclinations shown versus depth for experiments $W/d = 18$ (a) and numerical simulations with $W/d = 10$ (b). The two insets show how ν/ν_0 versus y/l_ν collapse on a master curve.

PACKING FRACTION AND STREAM-WISE VELOCITY

Packing fraction and velocity profiles

Figure 2 shows profiles for the packing fraction ν along the downward direction y perpendicular to the free surface. Two distinct regions can be clearly identified: a quasi-static region with a packing fraction approaching 0.58, toped by flowing layer where the volume fraction decreases drastically as one approaches the free surface. The packing fraction profiles can be very well fitted by an exponential function of the following form:

$$\nu(y) = \frac{\nu_0}{2}\left[1 + \tanh(y/l_\nu)\right] = \frac{\nu_0}{1 + \exp(-2y/l_\nu)}, \qquad (3)$$

where ν_0 and l_ν are fitting parameters corresponding respectively to the packing fraction in the quasi-static region and to the characteristic length scale over which the packing fraction varies. Note that the origin of the y axis has been chosen such that $\nu(y=0) = \nu_0/2$. The two parameters ν_0 and l_ν completely characterize the packing fraction profiles. If the y coordinate is made dimensionless using the length scale l_ν instead of the grain diameter d, all the curves collapse on a master curve which is nothing but a tangent hyperbolic function. We find that the packing fraction in the quasi-static region is independent of the inclination angle and equal to $\nu_0 = 0.58$, while the length scale l_ν increases linearly with the tangent of the inclination angle (Figure 3):

$$\frac{l_\nu}{W} = \eta\left[\tan\theta - \tan\theta_0\right]. \qquad (4)$$

The fit coefficients are $\theta_0 \approx 20°$ (20° experimentally and 23° numerically) and $\eta \approx 1.20$ (1.20 experimentally and 1.22 numerically). Equation (4) is thus equivalent to the simple relation between the flow inclination and the relative height h/W of the flowing layer:

$$\tan\theta = \tan\theta_0 + \mu_w(h/W), \qquad (5)$$

where $\tan\theta_0$ and μ_w are interpreted respectively as an internal friction angle of the granular material and an effective friction coefficient at the side walls that accounts for rolling or sliding contacts [1].

The vertical velocity profiles are reported in Figure 4 for experiments (a) and numerical simulations (b). Velocity profiles in the flowing layer share the same length scale l_ν and all the curves collapse on a single curve $v_x/(2l_\nu\sqrt{g/d})$ .vs. y/l_ν. Some differences are observed for small angles of inclination corresponding to flows that are close to jamming. The shear rate is nearly constant in the interval $-1 < y/l_\nu < 1$. Beneath it, $v_x/(2l_\nu\sqrt{g/d})$ smoothly transitions into a decaying exponential. Deeper, as experiments in the quasistatic region will show, creeping velocities drop on a shorter scale [8, 9, 17]. Other works report that the shear rate is no more constant in the case of large channels for rotating drum [6] and confined chute flows [2].

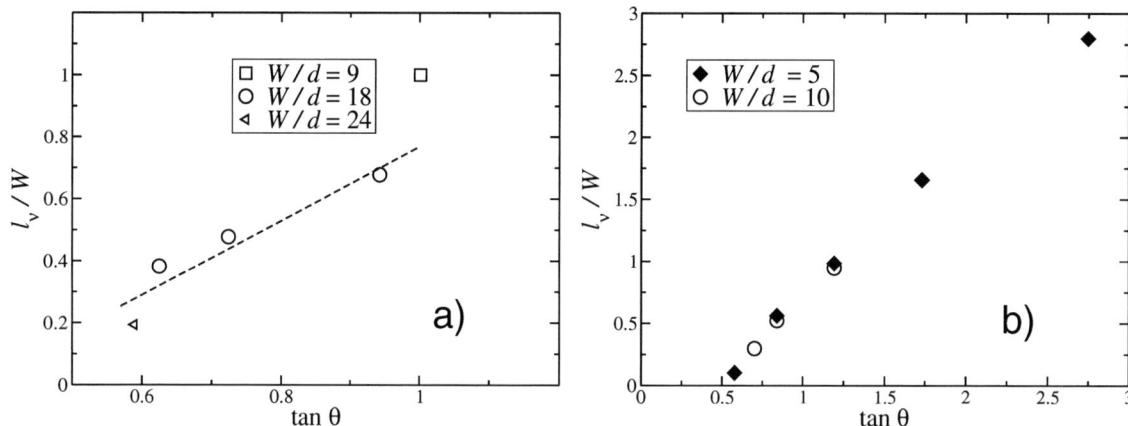

FIGURE 3. Variations of the scale l_v with $\tan\theta$ for experiments (a) and numerical simulations (b).

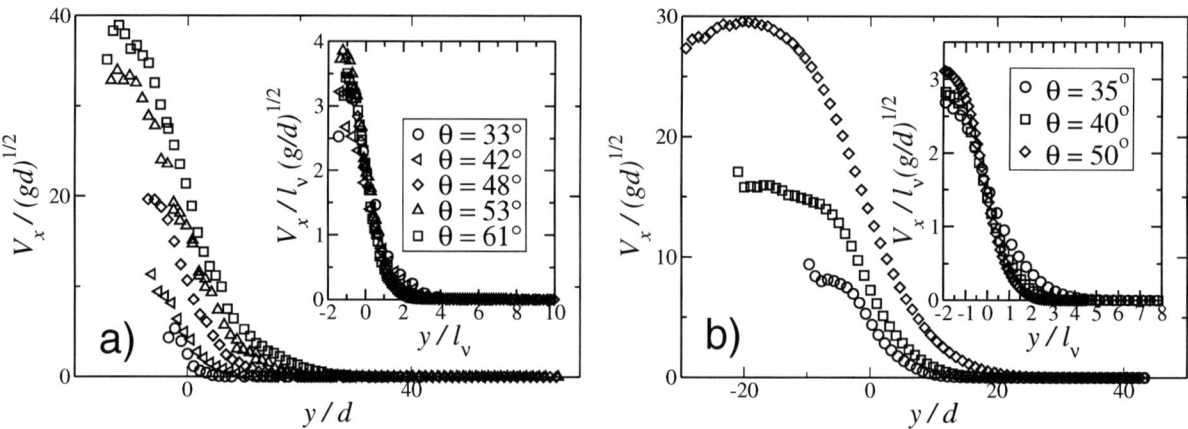

FIGURE 4. Profiles of stream wise velocity $v_x/\sqrt{gd}$ for experiments with $W/d = 9$ (a) and numerical simulations with $W/d = 10$ (b). The insets shows how profiles collapse in the flowing layer if distance and velocity are made dimensionless, respectively, with l_v and the characteristic speed $2l_v(g/d)^{1/2}$.

Interestingly, the packing fraction v and the stream wise velocity v_x share the same characteristic length l_v. An important point is that both velocity and packing fraction are significantly different to what is observed for flows on a bumpy bottom [18]. In this latter case the packing fraction is found to be constant within the flow which is not the case for confined flows. Note that this difference may be a consequence of the steeper angles of inclination needed to observe confined granular flows. The differences on velocity profiles are even more dramatic. For flows on bumpy bottom the velocity profiles are concave whereas, for flows on a heap, the profiles are convex. This dramatic difference stresses the importance of the boundaries (rough and bumpy .vs. erodible "bottom") and of the confinement in granular flows.

Effect of the number of grains

The numerical results reported in Figures 2, 3 and 4 have been obtained using 12,000-grains simulations with $W/d = 10$. This number of grains is large enough to form a quasi-static layer atop which the flow occurs. As reported in previous section, in the PBC simulations, the input parameters are the gap between the sidewalls and the angle of inclination. Another parameter is the number of grains used for the simulation : N_g. Let us assume that this is number is large enough to form the quasistatic layer. Do the properties of the flow depend on this number, i.e. on the height of the quasistatic layer? To address this question, we report on Figure 5 the variations of packing fraction,

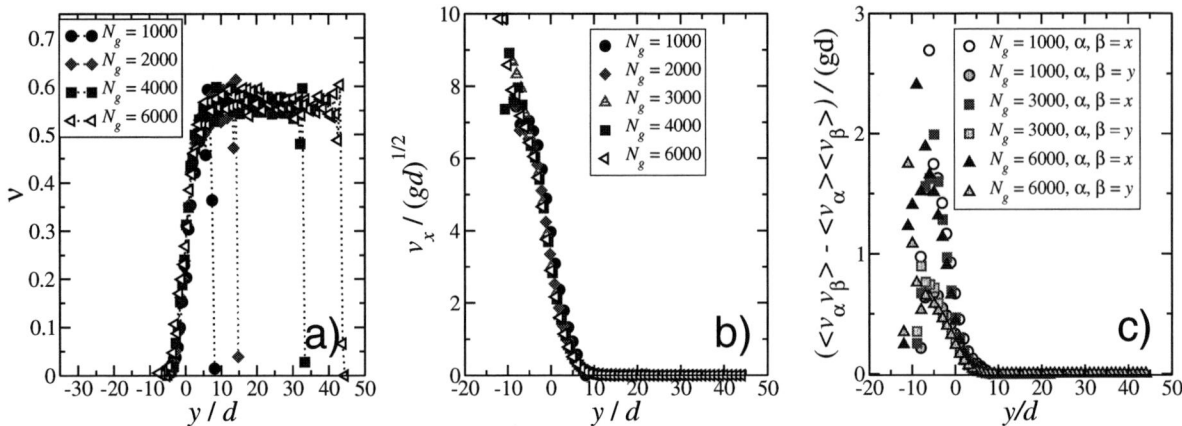

FIGURE 5. (color online) Effect of the number of grains used in the PBC simulation with $W/d = 5$ and $\theta = 40°$ on (a) the packing fraction profiles, (b) the stream-wise velocity profile and (c) the velocity fluctuation profiles (c). In the latter, open symbols : $\alpha = x, \beta = x$, full symbols $\alpha = y, \beta = y$.

stream-wise velocity and velocity fluctuations profiles for a large range of N_g (from $N_g = 1,000$ to $N_g = 6,000$), for $W/d = 5$ and $\theta = 40°$. All the curves collapse, except that the quasistatic layer is more or less developed depending on N_g. This demonstrates that the properties of the flowing layer depend very weakly on N_g if this number of grains is large enough to create a basal quasistatic layer. This means that the interactions between the flowing and the quasistatic layers are limited to a very short distance. As reported in [7, 8, 9] the velocity profile in the quasistatic region decreases exponentially. Thus, it is tempting to speculate that this short distance corresponds to the characteristic length of this decay which is of the order of magnitude of the grain size.

The effect of the number of grains on the properties of the quasistatic layer will be discussed in another paper. Note that although the reported results are obtained for a specific configuration ($W/d = 5$ and $\theta = 40°$) the conclusions reported here are generic to all the values of θ and W/d we have used.

Effect of micromechanical parameters

In this subsection, we report the influence of two micromechanical parameters: the coefficient of restitution and the friction coefficient. These two coefficients have dramatic effects on the collision between two grains but do they affect significantly the macroscopic properties of the flow ?

Effect of the coefficient of restitution. Figure 6a and Figure 6b report respectively the velocity profiles in the direction of the flow and the packing fraction profiles for different values of the coefficient of restitution. We explore a large range of coefficients of restitution from $e = 0$ to $e = 0.88$. The Figures show that variation in the coefficient of restitution has little effect on the flow behavior. Among the differences, we can observe that:

- the packing fraction of the quasistatic layer is slightly higher for $e = 0$.
- the thickness of the gaseous layer (where the grains describe ballistic trajectories) increases with e.

These weak differences are consistent with what has been observed for flows on a bumpy bottom [18] where the effect of e is also found to be very little. The weak effect of the coefficient of restitution can be understood in terms of non-local dissipation. As reported by Rajchenbach [19], in granular flows, a nearly infinite number of collisions occur in a finite time. Thus the kinetic energy of a grain colliding the flow is attenuated very rapidly. Consequently, the grain-grain coefficient of restitution e has a very little effect since the collisions are virtually almost completely inelastic. This is why the observed rheology appears almost independent of the binary restitution coefficient e of the grains.

Effect of the coefficient of friction. Another microscopic parameter we have explored is the friction coefficient. We tune this coefficient from 0.1 to 0.5. As mentioned in the method section, the friction coefficient is the same for

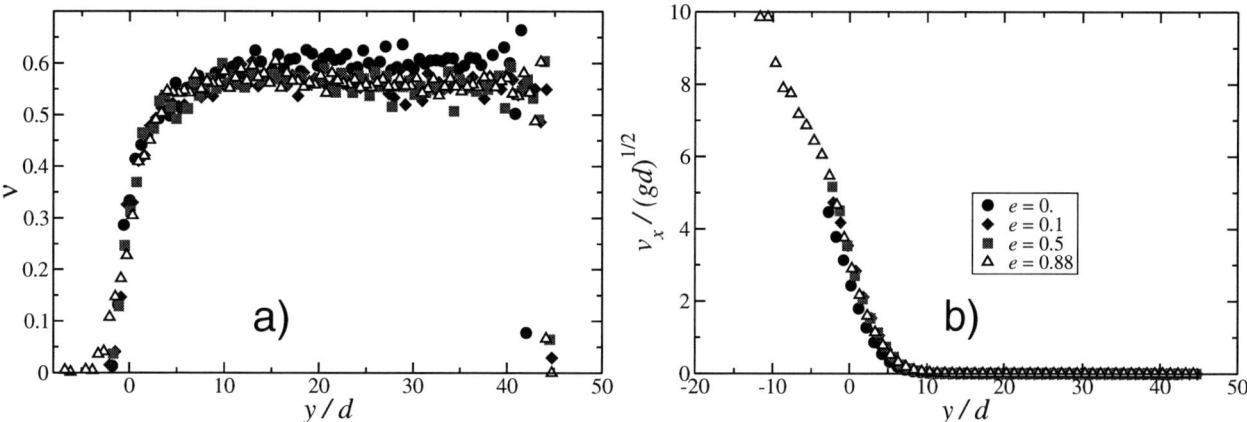

FIGURE 6. (color online) Vertical profiles of the packing fraction (a) and of the velocity (b) for different coefficients of restitution. The gap between sidewalls is $W/d = 5$, the friction coefficient $\mu = 0.5$ and the angle of the flow is 40°.

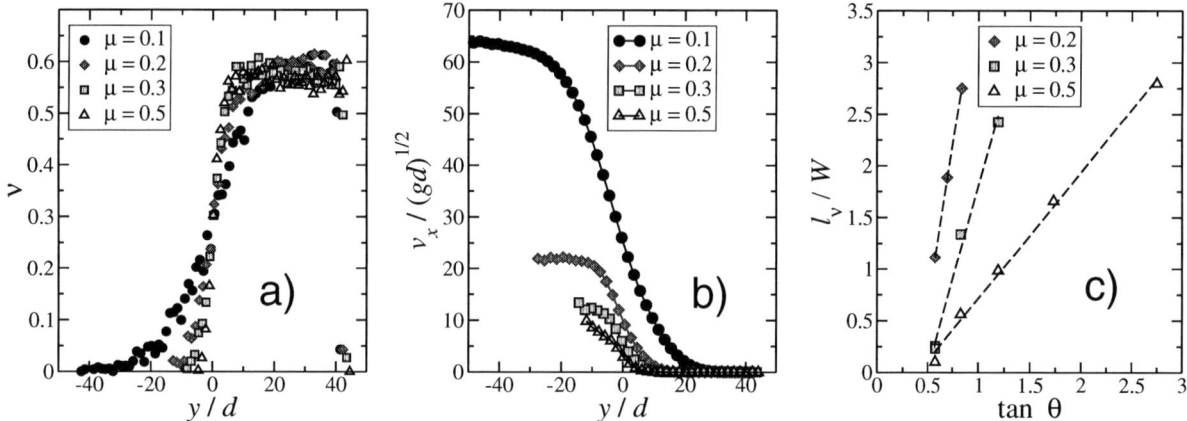

FIGURE 7. (color online) Vertical profiles of the packing fraction (a) and of the velocity (b) for different friction coefficients. The evolution of the characteristic length l_v with the packing fraction with the angle of inclination is reported in (c).

the grain-grain and for the grain-wall interactions. Contrary to the coefficient of restitution, this parameter has an important effect on the flow properties (see Figure 7). Decreasing the friction coefficient increases the characteristic length l_v. As expected decreasing this coefficient also increases the velocity and the shear rate. Note that these results are somewhat different from what is observed for granular flows on a bumpy bottom [18]. In such a system, the friction coefficient has little effect on the packing fraction. This confirms the importance of dissipation by friction in confined granular flows. The next step of this study is to tune independently these two friction coefficients to point out precisely their relative importance on the flow properties.

FRICTION WEAKENING

Unlike earlier studies [1, 2], which treated μ_w as a constant, our simulations show that the resultant sidewall friction $\mu_\tau = |\vec{\tau_w}|/|\sigma_{zz}^w|$, which we compute as the magnitude ratio of the surface force $\vec{\tau_w} = \sigma_{zx}^w \vec{e_x} + \sigma_{zy}^w \vec{e_y}$, and normal stress σ_{zz}^w on sidewalls, weakens with depth, as shown in Figure 8 for different inclinations. Let us recall that the co-ordinates used are defined in Figure 1. Close to the free surface, μ_τ is close to the grain-grain friction coefficient and varies little. At greater depths μ_τ decreases sharply. Here again, μ_τ .vs. y/l_v collapses on a master curve except for angles corresponding to flows close to the jamming transition (see insets of Figure 8).

A full characterization of sidewall friction requires the knowledge of its direction. To address this point we record in

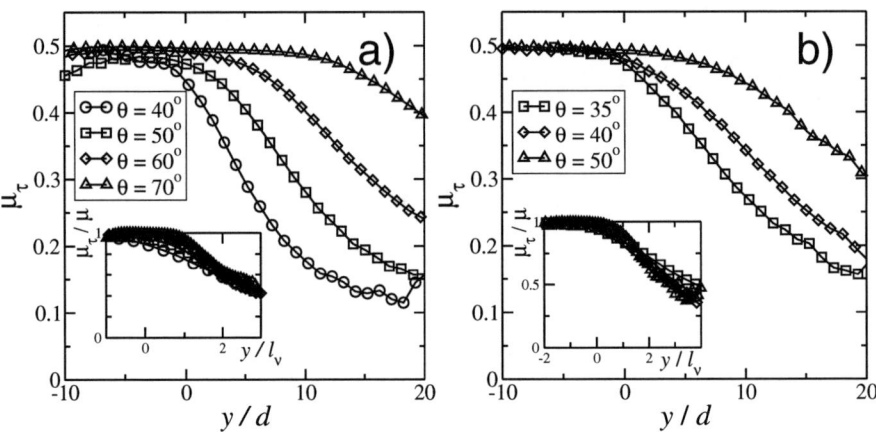

FIGURE 8. Variation of the resultant sidewall friction μ with depth for several θ for $W/d = 5$ and $W/d = 10$. The inset reveals a master curve of μ_τ/μ .vs. $y = l_v$ which becomes linear for $y/l_v > 2$ (dashed line).

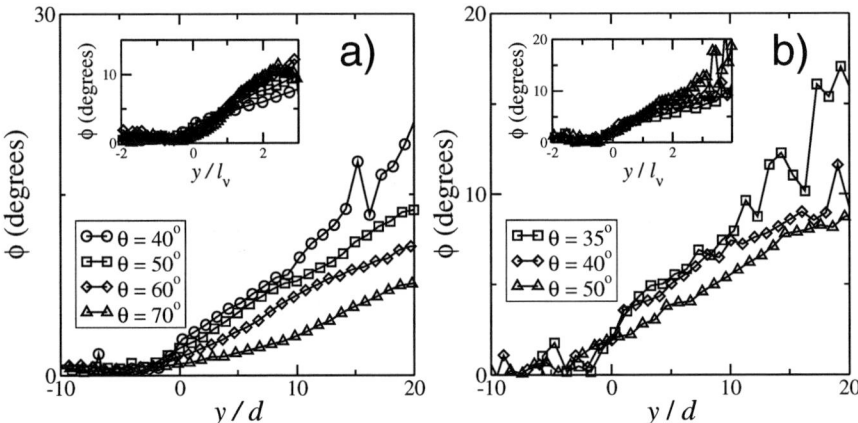

FIGURE 9. Variation of the angle of the resultant sidewall friction ϕ with depth for several θ and for $W = 5d$ (a) and $W = 10d$ (b). The inset reveals a master curve of $\tan\phi$ .vs. y/l_v.

our simulations the direction ϕ of the wall friction $\vec{\tau_w} = -|\vec{\tau_w}|(\cos\phi\,\vec{e_x} + \sin\phi\,\vec{e_y})$, which varies with y/l_v. An angle ϕ equals to zero means that the direction of the friction is the opposite of the main direction of the flow. In the flowing layer, Figure 9 shows that friction is pointed against the flows. It rotates progressively with depth and is more and more pointed against gravity.

In order to understand the decrease of the effective wall friction coefficient we studied the trajectories of the grains. Figure 10 reports the trajectory of a grain. This trajectory (the successive positions of the particle tracked in each image) has been obtained experimentally for a grain in contact with a sidewall and that belongs to the quasistatic layer where grains exhibit creeping. We clearly observe fluctuating motion that reveals a caging motion: the grains seems to be trapped in a finite area before escaping and being trapped again. The evolution of y with time (Figure 10, inset) confirms the presence of quick rearrangements between two long periods of trapping. Thus, the motion of the grains in the quasistatic layer is intermittent. This explains the very small values of stream wise average velocity measured in the quasistatic layer (from $1\,\text{mm.s}^{-1}$ to less than $1\,\text{nm.s}^{-1}$) [9]. Similarly to what is observed during granular compaction [20] a large motion is not a motion from one cage to another one, but a cage deformation. The quick rearrangements observed become less and less frequent as one goes deeper in the flow. While trapped there, grains describe an oscillatory motion with zero mean displacement, thus contributing negligibly to the mean resultant wall friction force $\vec{\tau_w}$. As trapping duration grows with depth the resultant wall friction μ_τ weakens as shown in Figure 8.

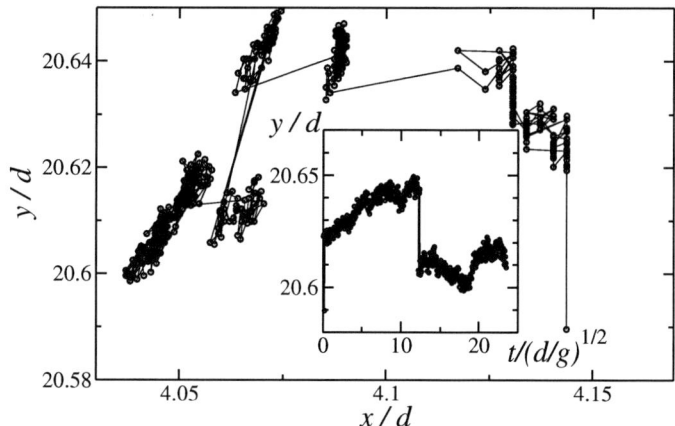

FIGURE 10. Trajectory (obtained experimentally) of a grain initially located at $y/l_v \approx 3$. Inset : depth of the grain .vs. dimensionless time.

CONCLUSION

In this paper we show that we are able to simulate granular flows on a heap using periodic boundary conditions in the flow direction. The effect of the number of grains in the periodic cell has been addressed and we demonstrated that the properties of the flowing layer do not depend on this number if it is large enough to form a quasi-static layer. The effect of the binary coefficient of restitution is found to be very weak whereas the effect of the friction coefficient is found to be dramatic. We observe strong differences between flows on a heap and flows on a rough bottom. Our results demonstrate that intermittent motion weakens the resultant wall friction on a length scale l_v. The existence of such depth is necessary and sufficient to yield $S/N = \tan\theta - \mu_w h/W$, relation which, as Taberlet [1] observed, gives internal friction in terms of inclination and apparent depth h of the flowing layer. In doing so, we resolved the chief paradox of their analysis, which had predicted negative friction at large depths. We also identified the origin of μ_w. These results open new questions on granular flows. What is the effect of the friction coefficient on the characteristic length l_v? Which friction coefficient dominates ? The wall-grain or the grain-grain ? The case of bumpy sidewalls should also be explored since it can correspond to natural or industrial situations.

ACKNOWLEDGMENTS

This work was supported by ANR grant (ANR-05-BLAN-0273), NSF travel grant INT-0233212 and CNRS (PICS France-USA). The authors acknowledge stimulating discussions with James Thomas Jenkins and Diego Berzi. We also thank Paul Sanchez for his preliminary work on the friction weakening.

REFERENCES

1. N. Taberlet, P. Richard, A. Valance, W. Losert, J-M. Pasini, J. T. Jenkins and R. Delannay, *Phys. Rev. Lett.* **91**, 264301(2003).
2. P. Jop, Y. Forterre and O. Pouliquen, *Journal of Fluid Mechanics* **541**, 167-192 (2005).
3. J. Rajchenbach, *Phys. Rev. Lett.* **65**, 2221 (1990)
4. N. Taberlet, P. Richard,1 and E. J. Hinch, *Phys. Rev. E* **73**, 050301 (2006).
5. A. Orpe and D.V. Khakhar, *Phys. Rev. E* **64**, 031302 (2001)
6. G. Félix, V. Falk, and U. DŠOrtona, *Eur. Phys. J. E* **22**, 25-31 (2007)
7. P.-A. Lemieux and D. J. Durian, *Phys. Rev. Lett.* **85**, 4273–4276 (2000).
8. T. S. Komatsu, S. Inagaki, N. Nakagawa, and S. Nasuno , *Phys. Rev. Lett.* **86**, 1757 (2001).
9. J. Crassous, J-F. Metayer, P. Richard and C. Laroche, *J. Stat. Mech.* P03009 (2008).
10. P. Yates, *Chemical calculations*, Nelson Thornes, Cheltenham, UK (2002) ISBN 0-7487-7075-5.
11. P. Philippe and D. Bideau, *Europhys. Lett.* **60** (5) 677-683 (2002).
12. Ph. Ribière, P. Richard, P. Philippe, D. Bideau and R. Delannay, *Eur. Phys. J. E* **22** 249–253 (2007)
13. B. Smit and D. frenkel, *Understanding Molecular Simulation. From Algorithms to Applications*, Academic Press (2001).

14. N. Taberlet, P. Richard, E. Henry and R. Delannay, *Europhys. Lett.* **68** 515-521 (2004).
15. N. Taberlet, P. Richard and Renaud Delannay, *Computers & Mathematics with Applications*, **55**, 230 - 234, (2008).
16. N. Taberlet, PhD thesis, university of Rennes I (2005).
17. P. Richard, A. Valance, J.F Métayer, P. Sanchez, J. Crassous, M. Louge and R. Delannay, *Phys. Rev. Lett.* **101**, 248002 (2008).
18. L. E. Silbert, D. Ertas, Gary S. Grest, Thomas C. Halsey, Dov Levine, and Steven J. Plimpton, *Phys. Rev. E*, **64**, 051302 (2001).
19. J. Rajchenbach, *Phys. Rev. Lett.* **90**, 144302 (2003).
20. P. Ribière, P. Richard, R. Delannay, D. Bideau, M. Toiya, and W. Losert, *Phys. Rev. Lett.*, 268001 (2005).

LIST OF SYMBOLS

N_g	number of grains
d	grain diameter
W	gap between sidewalls
h	height of the flowing layer
θ	angle of the flow
S	shear stress
N	normal stress
y	depth
μ_i	effective internal friction coefficient
μ_w	effective friction coefficient at the sidewalls
Q	output flow rate
e	grain-grain coefficient of restitution
μ	grain-grain friction coefficient
ν	packing fraction
ν_0	packing fraction in the quasi-static region
l_ν	characteristic length over which the packing fraction varies
v_x	velocity in the main direction of the flow
f	frequency of the camera
I	intensity of the gamma-ray beam after if passes through the medium
I_0	intensity of the gamma-ray beam without medium
α	Absorption coefficient
F_n	normal force
F_t	tangential force
k_n	normal spring constant
γ_n	normal viscous damping
δ	normal overlap between grain
k_t	tangential spring constant
u_t	tangential displacement of the contact
g	strength of the gravitational field
m	mean mass of a grain
ρ	grain density
θ_{min}	Angle below which the granular medium does not flow
θ_{max}	Angle above which the granular medium accelerates
η	fit coefficient for equation 4
θ_0	fit coefficient for equation 4
L	size of the periodic cell in the x direction
μ_τ	resultant sidewall friction coefficient
σ_{ij}^w	stress at the walls
τ_w	shear stress at the walls
ϕ	angle of the resultant sidewall friction
t	time

Temporal Dynamics in Density Relaxation

A. D. Rosato[a], V. Ratnaswamy[a], D. J. Horntrop[b], O. Dybenko[a], L. Kondic[b],

[a]*Granular Science Laboratory, Department of Mechanical and Industrial Engineering*
[b]*Department of Mathematical Sciences and Center for Applied Mathematics and Statistics*
New Jersey Institute of Technology
Newark, NJ 07102, USA

Abstract. The density relaxation phenomenon is modeled using both Monte Carlo and discrete element simulations to investigate the effects of regular taps applied to a vessel having a planar floor filled with monodisperse spheres. Results suggest the existence of a critical tap intensity which produces a maximum bulk solids fraction. We find that the mechanism responsible for the relaxation phenomenon is an evolving quasi-ordered packing structure propagating upwards from the plane floor.

Keywords: density relaxation, tapping, discrete element simulations, packing density
PACS: 45.70.-n; 05.65.+b; 02.70.Ns; 81.05.Rm

1. INTRODUCTION

The development of predictive models capable of capturing the quasi-static to rapid regimes that may possibly occur within a single granular flow will necessarily require a clear insight into the mechanisms responsible for variations in density within the material. Within this context, it is important to understand how bulk solids undergo an increase in density when subjected to time-dependent external forces – a phenomenon known as 'density relaxation'. This process is essentially the opposite of vibration-induced fluidization, where the material exhibits behavior akin to that of a dense gas.

The results reported in this paper provide significant new insight on density relaxation in granular materials exposed to discrete taps imposed by the motion of a flat plane floor. In the spirit of studies reported in the literature on this phenomenon, a rather special material consisting of monodisperse spheres was selected, thereby ensuring that the effects of particle property differences (such as segregation, e.g., [1-7]) are eliminated. Our investigations, consisting of discrete element simulation modeling with inelastic, monodisperse, frictional spheres, revealed a clear picture of the dynamical mechanism responsible for density relaxation, namely, the upward progression of organized layers induced by the plane floor as the taps evolve.

The remainder of this paper is organized as follows. In Section 2, we give a brief survey of the extensive literature on the packing of spherical materials. This comprises a discussion of experimental findings in addition to an overview of the wide ranging theoretical and computational results appearing in the literature. Section 3 contains a summary of the granular dynamics simulation method and model that is used in our computational investigation. Select results from our simulations presented in Section 4 demonstrate the importance of the nature and strength of the applied tap on the development and propagation of organization in a granular system with a perfectly planar floor. The summary and conclusions are given in Section 5.

2. SELECT OVERVIEW OF THE LITERATURE

From a historical perspective, studies on density relaxation have their basis in the extensive literature (see, for example [8-10]) on the packing of particles [11-25] where the concern was often in characterizing loose and dense random structures. As early as 1611, Kepler looked into the hexagonal geometry of the snowflake, while fifty years later, Hooke explored the packing of circles and spheres. In 1727, Hales carried out experiments by pressing dried peas into a container to form rather regular polyhedra that he assumed were regular dodecahedra. Analogous investigations by Comte de Buffon in 1753 resulted in this experiment being known as the 'peas of Buffon'. In 1899, Slichter [26] was the first to report on analytical expressions for the porosity in beds of uniform spheres. Many years later in 1939, Hales experiments were repeated by Marvin using lead shot, in which a pressed close-packed configuration formed regular dodecahedron (each face being a rhombus). However, a pressed randomly poured assembly revealed that the predominant structure consisted of irregular 14-faced polyhedra and no rhombic dodecahedra. The question of how many rigid spheres could be packed around another sphere of the same size was conjectured by Newton to be 12, while Scottish astronomer Gregory argued in 1694 that 13 could fit. (For a historical overview, see [27]). Almost two centuries later, in 1883, Barlow [28] found a hexagonal close packing in which each sphere touched twelve others. Shortly thereafter in 1887, Thompson [29] found that it was possible to fill Euclidean space without voids with truncated octahedrons – a polyhedron containing fourteen faces, eight of which are hexagons and six are squares. Two hundred sixty-two years after Gregory's conjecture, Leech [30] proved that while there was enough space for 13.397 spheres, there were no arrangements that made this possible. Rogers [21] pursued this idea and in 1958, he proved that if there was a regular arrangement of uniform spheres more dense than hexagonal close packing, its solid fraction could be no larger than $\sqrt{18}\,[\cos^{-1}(1/3) - (\pi/3)] \approx 0.7797$.

The concept of 'random loose' and 'random dense' was introduced by Oman and Watson [31] in 1944 to describe two limiting cases of random packings of uniform spheres. In 1960 this notion was quantified by Scott [32], who carried out experiments in cylindrical containers using 3 mm steel ball bearings. Loose random packing of solids fraction $v_{rl} \approx 0.59$ were obtained by simply pouring the ball bearings into the cylinder, while dense random packings $v_{rd} \approx 0.63$ were created by vibrating the containers. Improved experiments by Scott and Kilgour in 1969 [33] led to what is accepted today as the solids fraction of a dense random packing of uniform spheres, i.e., $v_{rd} = 0.6366 \pm 0.0005$. In 1990 Onoda and Liniger [34] carried out experiments on spheres settling in viscous liquids to stable configurations resulting in an extrapolated random loose density of 0.555 ± 0.005. An investigation coupling statistical mechanics theoretical approach with two-dimensional dissipative molecular dynamics simulations has been reported [35] in which a new lower bound, called 'random very loose packing' of disks was identified. An explanation of why an equivalent lower bound has not been found in physical experiments with disks is given based on an entropy argument. However, the question as to whether or not analogous results for sphere packings may exist was not addressed. Two years earlier, Dong et al. [36] reported on the settling of uniform spherical particles in a liquid to examine the role of various inter-particle forces on the final density, indicating that the random loose solids fractions can vary from 0.554 to 0.64. It is worth noting here that there have also been a large number of numerical approaches aimed at understanding this topic. (See, for example [12, 20, 22-24, 37-44]. Recent theoretical reports on random close packings include a local, stochastic paradigm of Clusel et al. [45], a statistical mechanics model of Song et al. [46] predicting an upper density bound of 0.634 for the random close packing of spheres, and estimates of the glass transition for a hard sphere liquid based on a replica approach [47, 48].

Early studies on density relaxation (a term that was coined in the mid-1990's) were primarily concerned with how optimal packings could be produced through the use of discrete taps or continuous vibrations [49-51]. In an early paper, Stewart observed in experiments that the maximum improvement of the solids fraction occurred when the bed of particles was subjected to high frequency vertical vibration and low displacement amplitude. Similar results were obtained by Evans. Experiments by Ayer and Soppet suggested that the maximum packing density of uniform steel shot may be attained by applying continuous vibrations to the containment vessel with accelerations $6 \leq \Gamma \leq 12$. Data on the solids fraction for different container-to-sphere diameter ratios D/d yielded a limiting solids fraction which was then fit to the following expression: $v_{max} = 0.635 - 0.216 \cdot e^{-0.313 D/d}$. In experiments on dry sand, D'Appolonia et al. [52] found that the highest solids fractions were obtained when the cylindrical container was subjected to vertical oscillations (frequency ω, amplitude a) having an acceleration $\Gamma \equiv a\omega^2/g \approx 2$. An increase of acceleration $\Gamma > 2$ led to the elimination of stable configuration and an ensuing decrease of solids fraction. Later Dobry and Whitman [53] reported that the most rapid densification of sand in a cylinder took place at accelerations Γ in the range from 0.9 to 1.1, while the highest bulk density was obtained when $1.1 \leq \Gamma \leq 1.3$. At higher accelerations between 1.3 and 2.0, the density either stabilized or continued to increase.

In 1957, Macrae [54] studied the effect of taps applied to a cylindrical vessel filled with powders of coal and zirconium. He observed that a sufficiently large increase of impact velocity (or in other words, the energy absorbed by the bed) lead to the destruction of existing particle assemblies. He hypothesized that this was caused by a reduction the surface friction, thereby weakening the force network maintaining the system in a 'metastable' equilibrium configuration. That is, the system's internal force network responsible for maintaining its structural integrity was disrupted. Evolution of the system to states having greater bulk densities could be achieved by applying taps of impact velocity smaller than a critical value. In later experimental work, Takahashi and Suzuki [51] proposed an empirical expression for tapped density evolution in terms of first-order rate kinetics as a function of the reciprocal of the tap number $\eta = 1/N$, i.e., $dV^*/d\eta = -kV^*$, where $V^* \equiv (V_0 - V)/V_0 - V_\infty$ is the fractional volume compaction, and V_0 and V_∞ is the initial (poured) and final fractional volume, respectively. The compaction parameter V^* was viewed as the integration of a continuous distribution of microscopic holes within the packing, with each having a characteristic filling rate constant k, so that $V^* = \int_0^\infty f(k)\exp(-k/N)dk$. Subsequently, they found that their experimental compaction data could be well fit using Weibull's distribution function $f(k) = b \cdot m \cdot k^{m-1} \exp(-bk^m)$. When $k = 1$, their model reduced to Kawakita's compaction equation for ceramic powders $V^* = \int_0^\infty b\exp(-bk)\exp(-k/N)dk$.

The effect of detached, vertical sinusoidal taps (frequency ω, amplitude a) applied to a tall, narrow cylindrical vessel filled with 2mm mono-disperse, soda-lime glass spheres was experimentally studied by Knight et al. [55] using a noninvasive, capacitive technique to measure the evolution of solids fraction from its initial value V_o. They first attempted to match their data to the Kohlraush-William-Watts (KWW) law given by $v(t) = v_\infty - \Delta v_\infty \exp\left[-(t/\tau)^\beta\right]$; although the resulting fit was reasonable for the initial phase of the experiment, there were significant deviations for a large number of taps. Consequently, in order to cope with the logarithmically long times to reach the steady state, they selected a four-parameter phenomenological model of the form $v(z) = v_\infty - \frac{v_\infty - v_0}{1 + B\ln(1+z/\tau)}$, where z is the number of taps, v_∞ is a steady-state density (dependent on the acceleration history), τ is a relaxation time, and B is an undetermined constant that depends on $\Gamma \equiv a\omega^2/g$. Subsequent papers by Nowak et al. [56] reported on reversible and irreversible branches of the solids fraction evolution as a function of vibration intensity Γ.

Nicolas et al. [57] carried out interesting experimental studies on the behavior of cyclically sheared spherical glass beads in a box, in which the sidewall rotated in parallel clockwise/counterclockwise motions through an angle θ where 0° corresponded to vertical walls. They observed a characteristic stroboscopic increase of solids fraction from an initial value 0.61 up to 0.66 for a fixed θ = 2.7⁰. An increase of the angular amplitude lead to greater compaction (for instance, 0.68 for angle amplitude θ = 5.4⁰ and 0.693 for θ = 10.7⁰). They also reported on a reversible-irreversible volume fraction evolution depending on variable shear angle, somewhat analogous to the results of Nowak et al.'s experiments. The influence of shear and `crystallization by shear' was also observed in subsequent experiments [58-60], while crystallization by horizontal shaking has been reported [61].

A resurgence of interest (partly triggered by experiments [55, 56, 62, 63]) in uncovering particle-level mechanisms responsible for observed macroscopic behavior has resulted in theoretical studies involving free volume arguments [64], parking lot paradigms [65-68], and stroboscopic decay approaches [69, 70]. Computational investigations involving stochastic and deterministic simulations [71-78] have also been carried out to directly address the coupling between the detailed particle dynamics and bulk behavior. In this regard, we note the single particle and collective dynamics [72], jump and push filling mechanisms identified in [79], and the appearance of local structural order evidenced by a second peak in the radial distribution function [12]. The formation of an ordered structure was also discussed in [80], where it was shown that there is a narrow range of densities where disordered and crystalline structures are expected to coexist. In addition, crystallization of cohesive and noncohesive particles was explored experimentally [81, 82], where it was found that vibrational annealing can be successfully used to form almost perfect structures. Important progress related to density relaxation has also been made in characterizing glass and jamming transitions in granular materials [83].

3. DISCRETE ELEMENT MODEL

The discrete element method involves the numerical integration of the equations of motion governing the evolution of particles that interact via idealized, dissipative contact forces. A detailed review of this method and its application can be found in [84, 85]. For the investigations reported in this paper, we used inelastic, monodisperse, spherical particles that obey the soft-sphere models developed by Walton and Braun [86], in which interaction forces are functions of an allowed overlap between pairs of particles. After initial contact, this overlap is interpreted as the deformation that generates the repulsive force between particles. Energy loss in the normal direction (i.e., along center line of contacting spheres) is produced by linear loading (K_1) and unloading springs (K_2), which results in a constant restitution coefficient $e = \sqrt{K_1/K_2}$. We remark that the latter relation is found by solving the differential equations of motion for two colliding spheres over the loading and unloading phases to determine their relative approach and separation velocities. This model has been shown to reproduce (in experiments and finite element computations) the nearly linear loading behavior for spherical surfaces that experiences plastic deformation of the order of one percent of a particle diameter [87]. In the tangential direction, a Mindlin-Deresiewicz like model [88, 89] is used in which tangential stiffness is a function that decreases with tangential displacement until full sliding occurs at the friction limit μ. Although contacts that experience rotational with tangential sliding are not a feature of this model, particle rotation is possible through the transmission of tangential impulses. Explicit integration of the equations of motion for the system of spheres is done using a Verlet leap-frog algorithm, in which the time step $\Delta t = \pi e/p \sqrt{m/2K_1}$ (where m is the mass of a particle) is determined by dividing the duration of the loading period into p steps (typically $p \approx 40$). For the simulation studies described in the next section, we chose acrylic spheres (mass density $\rho = 1200$ kg/m^3) with $e = 0.9$ and $\mu = 0.10$, in close correspondence with Louge's experimental measurements [90] on acrylic spheres (i.e., $\mu = 0.096 \pm 0.006$, $e = 0.934 \pm 0.009$).

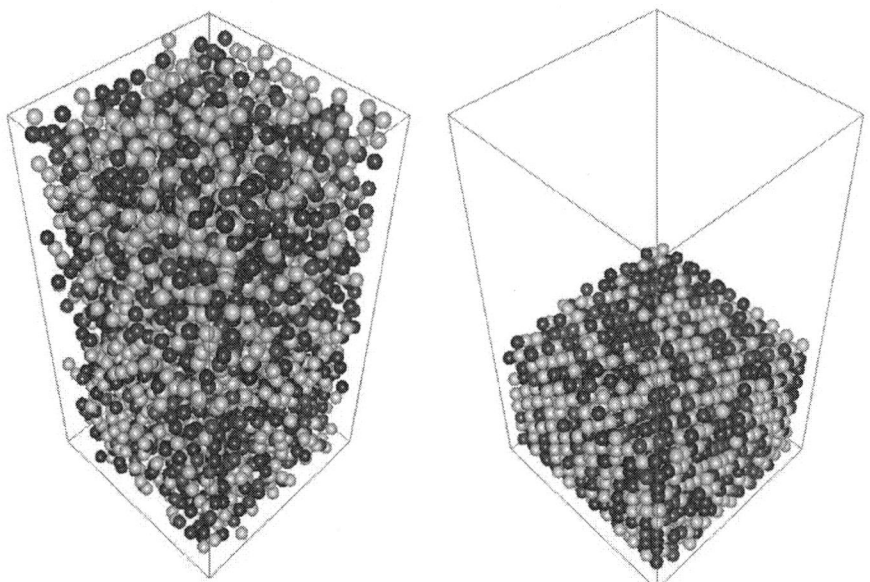

FIGURE 1. Snapshot of the initial configuration of particles randomly placed in the computational box (left) and the system after the particles have settled under gravity (right).

The computational region is a laterally periodic box (length x width = $12d$ x $12d$) into which particles of diameter d randomly placed at $t = 0$ within the domain settled under gravity to a fill depth of approximately $22d$. We emphasize that the floor is a flat plane having the same restitution and friction coefficients as the particles. For the friction coefficient that was chosen, the settling process produced bulk solids fractions[1] corresponding to a poured loose random packing. Fig. 1 shows snapshots of the initial random configuration and the poured or settled packing. However, by tuning the value of the friction coefficient μ, it is possible to generate a wide range of solids fractions.

---

[1] In order to eliminate the effect of the free surface and to achieve consistency across all realizations, bulk solids fractions were computed by slicing off or removing the top twenty percent of the assembly.

This behavior is demonstrated in the simulation results of Fig. 2, which were obtained by averaging twenty independent realizations. Large friction coefficients produce less dense systems after pouring due to the formation of bridges and arches within the structure [91-93].

3.1 Evolution of the Bulk Solids Fraction

Significant variations in the solids fraction evolution were observed for poured configurations that were statistically indistinguishable with regard to their initial bulk solids fractions and distributions of nearest neighbors and free volume. This finding is in accord with intuition in that the details of the microstructure may have a strong effect on the path the system takes as it advances to a dense structure. We demonstrate our finding with an ensemble of twenty poured realizations with bulk solids fractions v_o lying in the range 0.6088 ± 0.0035. For each realization,

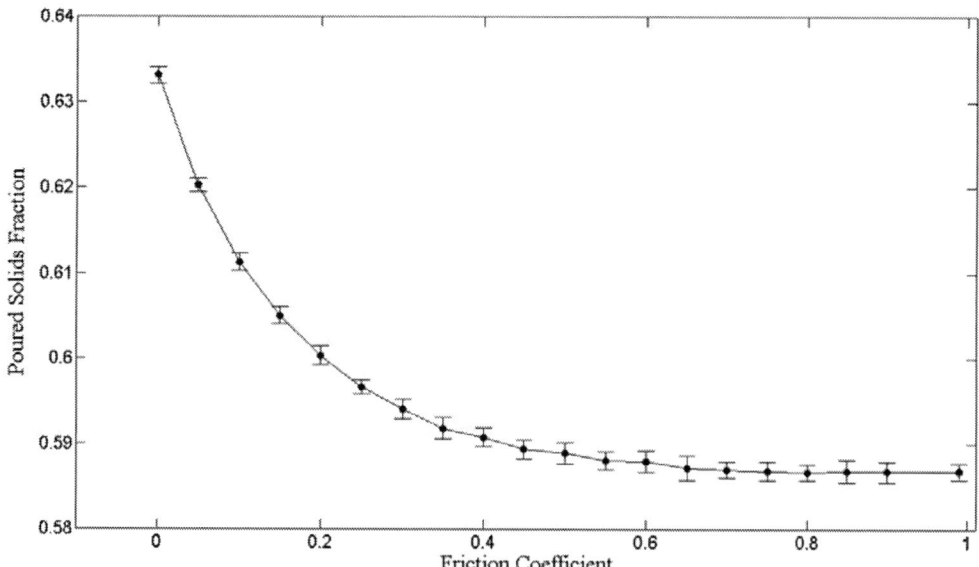

FIGURE 2. The solids fraction resulting from gravitational settling the particles depends on the value of the friction coefficient. Each point on the graph represents an average taken over 20 discrete element realizations. The vertical lines represent the deviation from the average. The solid line connects the data points in order to show the trend.

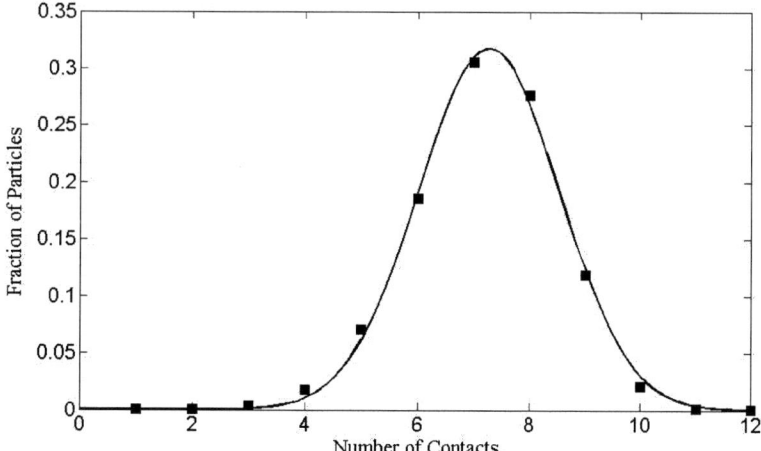

FIGURE 3. The points are ensemble-averaged coordination numbers for 20 poured realizations. The solid line is a fit ($R^2 = 0.9975$) to $f(x) = \dfrac{1}{\beta\sqrt{\pi}} \exp\left[-\left(\dfrac{x-\alpha}{\beta}\right)^2\right]$, where $\alpha = 7.271$, $\beta = 1.256$.

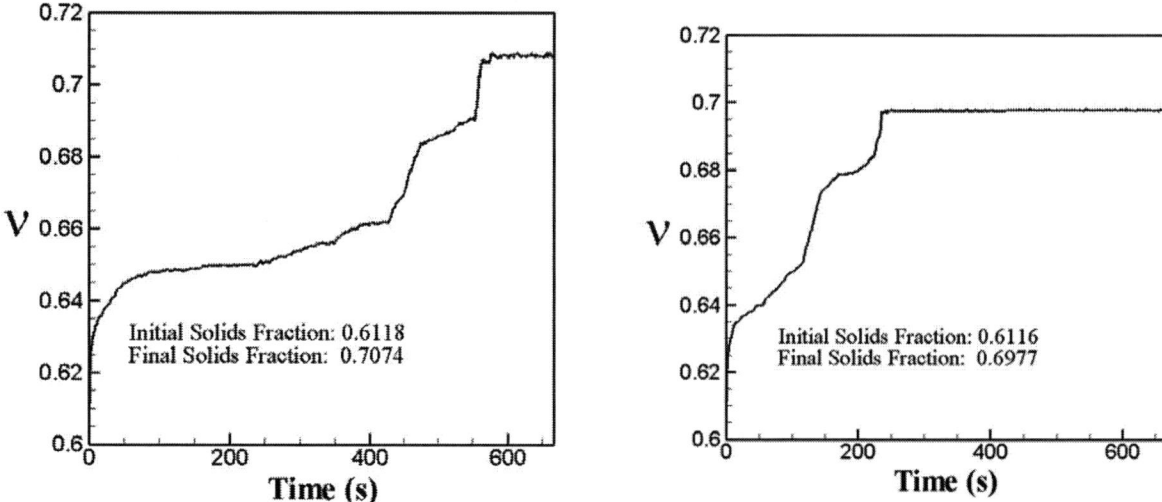

FIGURE 4. Evolution of the bulk solids fraction of two realizations, in which v_o is the initial solids fraction

the coordination number distribution was computed by simply counting the number of spheres touching[2] each particle. Although not shown here, each of these individual distributions was statistically indistinguishable from the coordination number distribution computed using the ensemble average of the distributions (Fig. 3). The Voronoi diagrams were also computed for each realization. From these diagrams, we computed the distribution of free volume, defined as the difference between the volumes of a Voronoi polyhedron and its encapsulated sphere. Although not included in this paper, we also observed very little visual difference between the free volume distributions of the ensemble of 20 realizations, which were quite similar (both qualitatively and quantitatively) to the ensemble-averaged distribution. Thus, there appears to be very little statistical difference between the members of the ensemble in the context of bulk density, and the distributions of coordination number and free volume. However, as can be seen in Fig. 4, there is a rather striking difference between the solids fraction evolutions. Taps were applied to each realization for the same time duration under identical vibration parameters, i.e., $a/d = 0.4413$ and $f = 7.5$ Hz, corresponding to $\Gamma = 2$. The ensemble-averaged trajectory (for which $\langle v \rangle|_{t \sim 680} = 0.6997 \pm 0.0058$) is given in Fig.5. In some of the trajectories, rather dramatic jumps or steep gradients in the solids fraction (see Fig. 4) occurred, which at first glance appeared to be spurious. However, further consideration of these trajectories suggested that the jumps were caused by collective reorganizations of the microstructure that took place over small time durations relative to the time of tapping. The later study demonstrates the sensitive dependence of the trajectories on the details of the microstructure, and the importance of using ensemble averages in these simulations rather than only single realizations despite the deterministic dynamics. It has been reported in experiments [94] that although the final or stationary state is affected by tap parameters, this state does not depend on the initial bulk density. In our simulations, we have only considered the effect of the initial state on the evolution of the system. We have not completed sufficient case studies of extended tap duration to examine the statistics of the final state, although our preliminary results indicate agreement with the experimental findings.

4. EFFECT OF TAP AMPLITUDE ON THE SOLIDS FRACTION EVOLUTION

In this section, the effect of the amplitude of the tap at a fixed frequency on the evolution of the solids fraction is discussed. The initial (poured) assembly of spheres is tapped by applying harmonic displacement oscillations to the flat floor consisting of a half-sine wave (amplitude a and period τ), followed by a relaxation time t_r of sufficient duration to ensure that upon collapse a quiescent state of minimal kinetic energy is attained. In order to demonstrate the sensitivity of the evolution on the tap amplitude, we selected a fixed frequency $f = 7.5$ Hz and two displacement

---

[2] Touching spheres are those who distance between centers lies within $1.05d$.

amplitudes (i.e., a/d = 0.1103 and 0.4413) corresponding to dimensionless accelerations $\Gamma = 0.5$ and $\Gamma = 2.0$, respectively. The evolution of microstructure was quantified by the vertical distribution of the number of particle

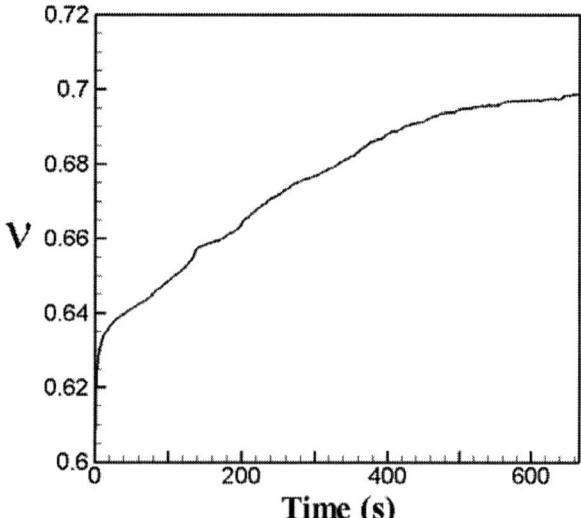

FIGURE 5. Evolution of the ensemble-averaged bulk solids fraction over 20 discrete element simulations in which f = 7.5 Hz, a/d = 0.4413, Γ =2. The initial value is $\langle v_o \rangle = 0.6088$.

centers (normalized by the total number of particles in the system) $\bar{n}$ over the duration of 898 taps for an ensemble of 20 realizations. Fig. 6 shows $\bar{n}$ for a/d = 0.4413 at t = 0s and at three subsequent times, t = 67s, t = 267s and t = 534s, as a function of the normalized distance from the floor y/d. The poured assembly (t = 0s) exhibits the characteristic local ordering within a few layers adjacent to the plane floor and a rather flat distribution beyond this. As the number of taps increases, there is growth in the number of peaks in the distribution that signals the evolution of structural ordering starting from the floor. Upon completion of the 898 taps, the ensemble-average bulk solids fraction $\langle v \rangle \approx 0.7$ However, for the case with the smaller amplitude (a/d = 0.1101), we observed no significant change in the distribution of centers as it remained flat (except for a few layers near the wall) and only a minimal increase in the bulks solids fraction upon completion of thousands of taps. This is shown in Fig. 7, which compares the evolution of the ensemble averaged bulk solids fraction at the two tap amplitudes. We hypothesize that the sensitivity of the densification process on tap amplitude at a fixed frequency may be related to the ability of the tap to dilate the assembly a sufficient amount to allow particles to slowly rearrange in such a way as to reduce the potential energy of the system. Indeed, one can imagine two extremes that engender no microstructure development or appreciable increase in bulk density: (i) an aggressive, energetic tap that greatly expands the system so that any existing microstructure is destroyed upon contraction and, (ii) a low energy tap that merely transmits a stress wave through the contact network, but does not dislodge or disturb the structure to any appreciable degree. We are currently investigating this conjecture.

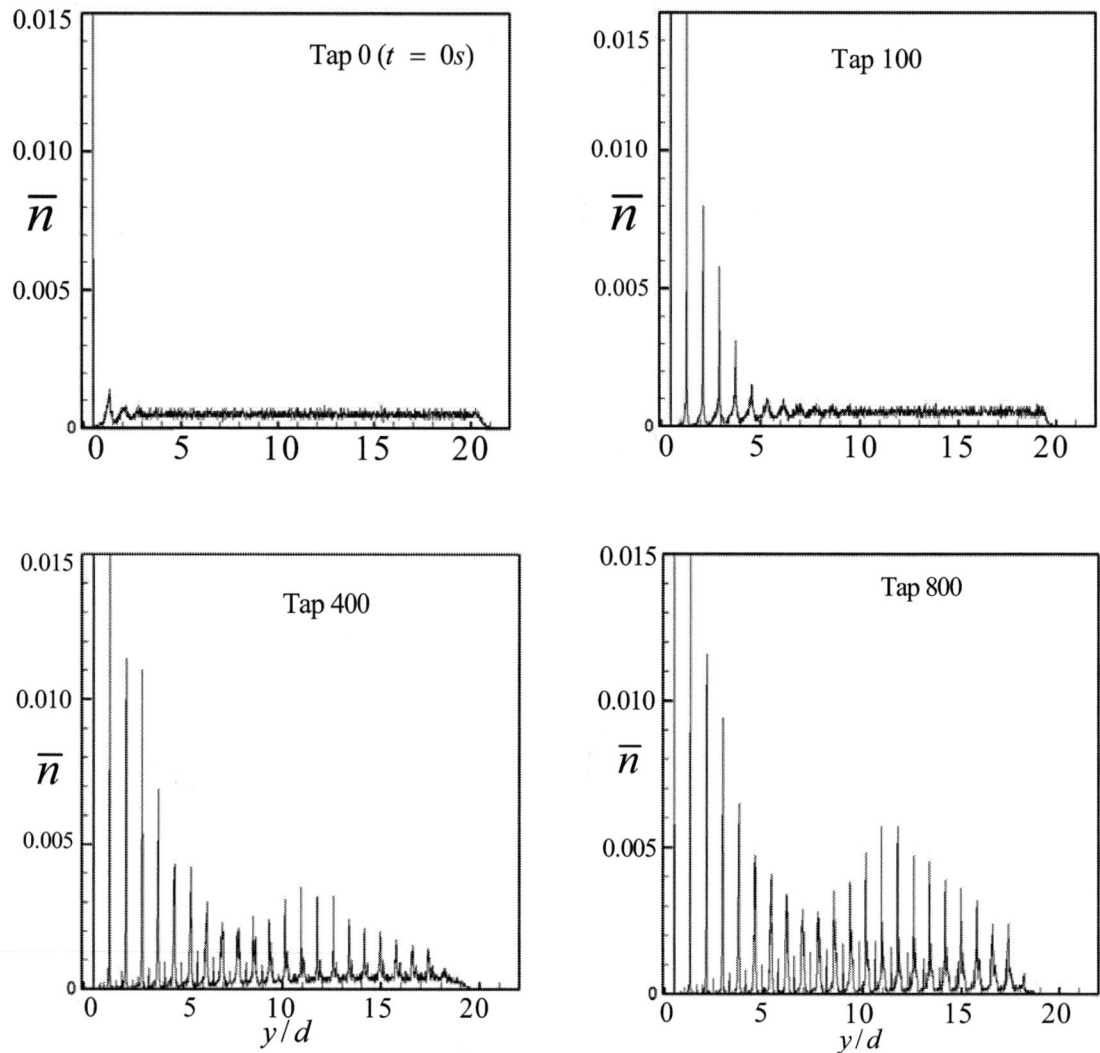

FIGURE 6. Ensemble-averaged (over 20 realizations) distributions of the particle centers measured from the floor ($y/d = 0$) to the top surface for $a/d = 0.4413$, $f = 7.5$ Hz ($\Gamma \approx 2.0$) at $t = 0$s, $t = 67.3$s (tap 100), $t = 267.3$s (tap 400) and $t = 533.9$s (tap 800). The corresponding solids fractions are $\langle v \rangle = 0.6088, 0.6430, 0.6743, 0.6955$.

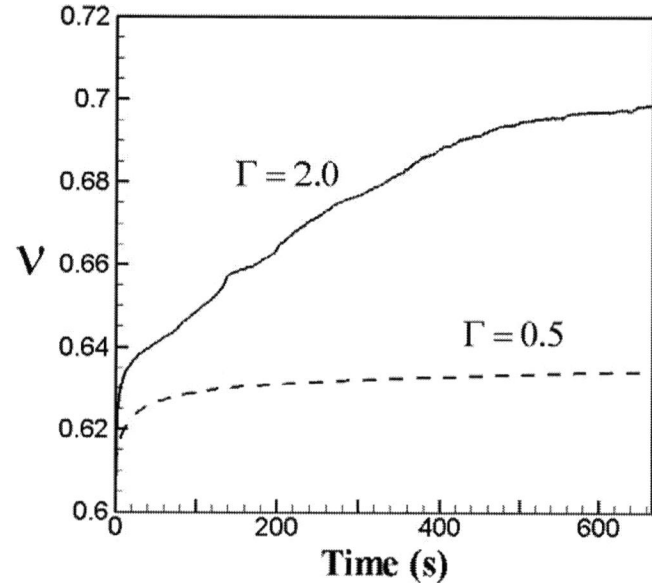

FIGURE 7. Evolution of the ensemble-averaged bulk solids fraction (over 20 realizations) for $\Gamma = 0.5$ (f = 7.5 Hz, a/d = 0.1103) and $\Gamma = 2$ (f = 7.5 Hz, a/d = 0.4413)

5. SUMMARY

In this paper, a concise review of the literature on density relaxation of monodisperse spheres is given. This was followed by a summary of our recent discrete element simulation results on density relaxation for a system consisting of a laterally periodic box filled with uniform spheres that are subjected to tapping through the time-dependent motion of a planar floor. The special material consisting of monodisperse spheres that eliminated any effects of particle property differences was chosen in order to understand the extent of the system's evolution towards a perfect, dense crystalline state. At a fixed tap frequency, a rather significant effect of the tap amplitude on the densification process was observed. The simulation results showed for the proper tap amplitude, there was a progression of organized layers from the plane floor well beyond the region immediately adjacent to the flat boundary. We speculate that the sensitivity of the densification process on the amplitude may be related to the ability of the tap to dilate the assembly a sufficient amount to allow particles to slowly rearrange so as to reduce the potential energy of the system. In future work, we will study the dilation behavior of the system over a time scale associated with the tap period, as well as carry out a complete investigation of the dependence of the relaxation process on material parameters.

6. ACKNOWLEDGMENTS

The authors thank D. Blackmore, P. Singh and J. T. Jenkins (Cornell University) for discussions of this work, and T. Duong who wrote the code to generate the Voronoi constructions. DJH acknowledges partial support from the grant NSF-DMS-0406633 and LK from the grants NSF-DMS-0605857 and NSF-DMS-0835611. Partial computational support was provided by NSF-DMS-0420590. In addition, a portion of this research was done using resources of the Open Science Grid, which is supported by the National Science Foundation and the U.S. Department of Energy's Office of Science.

REFERENCES

1. A. Dziugys, B. Peters, H. Hunsinger, et al., *Gran. Matt.* **9**, 387 (2007).
2. K. Liffman, K. Muniandy, M. Rhodes, et al., *Gran. Matt.* **3**, 205 (2001).
3. T. Poschel and H. J. Herrmann, *Europhys. Lett.* **29**, 123 (1995).
4. A. D. Rosato, D. L. Blackmore, N. Zhang, et al., *Chem. Eng. Sci.* **57**, 265 (2002).
5. Q. Shi, G. Sun, M. Hou, et al., *Phys. Rev. E* **75** (2007).
6. J. Sun, F. Battaglia, and S. Subramaniam, *Phys. Rev. E* **74** (2006).
7. J. C. Williams, *Powder Technol.* **15**, 245 (1976).
8. T. Aste and D. Weaire, *The Pursuit of Perfect Packing* (Taylor and Francis, New York, 2008).
9. J. H. Conway and N. J. A. Sloane, *Sphere Packings, Lattices and Groups* (Springer-Verlag, New York, 1991).
10. D. J. Cumberland and R. J. Crawford, *The Packing of Particles* (Elsevier, 1987).
11. C. R. A. Abreu, F. W. Tavares, and M. Castier, *Powder Technol.* **134**, 167 (2003).
12. D. J. Adams and A. J. Matheson, *J. Chem. Phys.* **56**, 1989 (1972).
13. R. F. Benenati and C. B. Brosilow, *AIChE Journal* **8**, 359 (1962).
14. C. H. Bennett, *J. Appl. Phys.* **6**, 2727 (1972).
15. J. D. Bernal, in *Symposium on Liquids: Structure, Properties, Solids Interactions*, edited by T. J. Hughel (Elsevier 1965), p. 25.
16. J. D. Bernal and J. Mason, *Nature* **188**, 910 (1960).
17. J. L. Finney, *Proc. Roy. Soc. Lond. A* **319**, 479 (1970).
18. K. Gotoh, W. S. Jodrey, and E. M. Tory, *Powder Technol.* **20**, 233 (1978).
19. W. A. Gray, *The Packing of Solid Particles* (Chapman and Hall, Ltd., London, 1968).
20. M. J. Powell, *Powder Technol.* **25**, 45 (1980).
21. C. A. Rogers, *Proc. Lond. Math. Soc.* **8**, 609 (1958).
22. S. Torquato, T. M. Truskett, and P. G. Debenedetti, *Phys. Rev. Lett.* **84**, 2064 (2000).
23. E. M. Tory, B. H. Church, M. K. Tan, et al., *Can. J. Chem. Eng.* **51**, 484 (1973).
24. W. M. Visscher and M. Bolsterli, *Nature* **239**, 504 (1972).
25. K. Z. Y. Yen and T. K. Chaki, *J. Appl. Phys.* **71**, 3164 (1992).
26. C. S. Slichter, (US Geol. Survey, 19th Annual Report, Part II, 1899), p. 295.
27. T. C. Hales, *Discrete and Computational Geometry* **36**, 5 (2006).
28. W. Barlow, *Nature* **29**, 186 (1883).
29. W. Thompson, *Phil. Mag.* **24**, 503 (1887).
30. J. Leech, *Math. Gaz.* **40**, 22 (1956).
31. A. O. Oman and K. M. Watson, *Natl. Patrol. News* **36**, 795 (1944).
32. G. D. Scott, *Nature* **188**, 908 (1960).
33. G. D. Scott and D. M. Kilgour, *Brit. J. Appl. Phys.* **2**, 863 (1969).
34. G. Y. Onoda and E. G. Liniger, *Phys. Rev. Lett.* **64**, 2727 (1990).
35. M. P. Ciamarra and A. Coniglio, *Phys. Rev. Lett.* **101**, 128001 (2008).
36. K. J. Dong, R. Y. Yang, R. P. Zou, et al., *Phys. Rev. Lett.* **96**, 145505 (2006).
37. H. A. Makse, D. L. Johnson, and L. M. Schwartz, *Phys. Rev. Lett.* **84**, 4160 (2000).
38. H. G. Matuttis, S. Luding, and H. J. Herrmann, *Powder Technol.* **109**, 278 (2000).
39. G. E. Mueller, *Powder Technol.* **159**, 105 (2005).
40. C. O'Hern, S. A. Langer, A. J. Liu, et al., *Phys. Rev. Letter* **88**, 075507 (2002).
41. T. G. Owe Berg, R. L. McDonald, and R. J. Trainor, *Powder Technol.* **3**, 183 (1969/1970).
42. S. Remond, J. L. Gallias, and A. Mizrahi, *Gran. Matt.* **10**, 329 (2008).
43. W. Soppe, *Powder Technol.* **62**, 189 (1990).
44. J. Tobochnik and P. M. Chapin, *J. Chem. Phys.* **88**, 5824 (1988).
45. M. Clusel, E. I. Corwin, A. O. N. Siemens, et al., *Nature* **460**, 611 (2009).
46. C. Song, P. Wang, and H. Makse, *Nature* **453**, 629 (2008).
47. G. Parisi and F. Zamponi, *J. Stat. Mech.* **2009**, p03026 (2009).
48. G. Parisi and F. Zamponi, *J. Chem. Phys.* **123** (2005).
49. J. E. Ayer and F. E. Soppet, *J. Am. Ceram. Soc.* **48**, 180 (1965).
50. P. E. Evans and R. S. Millman, *Powder Met.* **7**, 50 (1964).
51. M. Takahashi and S. Suzuki, *Ceramic Bulletin* **65**, 1587 (1986).

52. D. D'Appolonia and E. D'Appolonia, in *3rd Asian Regional Conference on Soil Mechanics and Foundation Engineering* (Jerusalem Academic Press, 1967), p. 1266.
53. R. Dobry and R. V. Whitman, (American Society of Testing Materials, Philadelphia, 1973), p. 156.
54. J. C. Macrae, W. A. Gray, and P. C. Finlason, *Nature* **179**, 1365 (1957).
55. J. B. Knight, C. G. Fandrich, C. N. Lau, et al., *Phys. Rev. E* **51**, 3957 (1995).
56. E. R. Nowak, J. B. Knight, M. L. Povinelli, et al., *Powder Technol.* **94**, 79 (1997).
57. M. Nicolas, P. Duru, and O. Pouliquen, *Europhys. Lett.* **60**, 677 (2002).
58. K. E. Daniels and R. Behringer, *Phys. Rev. Lett.* **94**, 168001 (2005).
59. N. W. Mueggenburg, *Phys. Rev. E* **71**, 031301 (2005).
60. J.-C. Tsai and J. Gollub, *Phys. Rev. E* **70**, 031303 (2004).
61. O. Pouliquen, M. Nicolas, and P. D. Weidman, *Phys. Rev. Lett.* **79**, 3640 (1997).
62. E. R. Nowak, A. Grushin, A. C. B. Barnum, et al., *Phys. Rev. E* **63**, 020301(R) (2001).
63. E. R. Nowak, J. B. Knight, E. B. Naim, et al., *Phys. Rev. E* **57**, 1971 (1998).
64. T. Boutreux and P. G. DeGennes, *Physica A* **244**, 59 (1997).
65. K. L. Gavrilov, *Phys. Rev. E* **58**, 2107 (1998).
66. A. J. Kolan, E. R. Nowak, and A. V. Tkachenko, *Phys. Rev. E* **59**, 3094 (1999).
67. J. Talbot, G. Tarjus, and P. Viot, *Eur. Phys. J. E* **5**, 445 (2001).
68. M. Wackenhut and H. J. Herrmann, *Phys. Rev. E* **68**, 041303 (2003).
69. S. J. Linz, *Phys. Rev. E* **54**, 2925 (1996).
70. S. Linz and A. Dohle, *Phys. Rev. E* **60**, 5737 (1999).
71. D. Arsenovic, S. B. Vrhovac, Z. M. Jaksic, et al., *Phys. Rev. E* **74**, 061302 (2006).
72. G. C. Barker and A. Mehta, *Phys. Rev. A* **45**, 3435 (1992).
73. G. Liu and K. E. Thompson, *Powder Technol.* **113**, 185 (2000).
74. L. F. Liu, Z. P.Zhang, and A. B. Yu, *Physica A* **268**, 433 (1999).
75. M. Nicodemi and A. Coniglio, *Phys. Rev. Lett.* **82**, 916 (1999).
76. M. Nicodemi, A. Coniglio, and H. J. Herrmann, *Phys. Rev. E* **59**, 6830 (1999).
77. S. Remond and J. L. Galias, *Physica A* **369**, 545 (2006).
78. L. A. Pugnaloni, M. Mizrahi, C. M. Carlevaro, et al., *Phys. Rev. E* **78**, 051305 (2008).
79. X. Z. An, R. Y. Yang, K. J. Dong, et al., *Phys. Rev. Lett.* **95**, 205502 (2005).
80. A. V. Anikeenko, N. N. Medvedev, and T. Aste, *Phys. Rev. E* **77**, 031101 (2008).
81. O. Carvente and J. C. Ruiz-Suarez, *Phys. Rev. Lett.* **95**, 018001 (2006).
82. O. Carvente and J. C. Ruiz-Suarez, *Phys. Rev. E* **78** (2008).
83. M. Nakagawa and S. Luding, in *AIP Conference Proceedings* (AIP, Golden, CO, 2009), Vol. 1145.
84. H. P. Zhu, Z. Y. Zhou, R. Y. Yang, et al., *Chem. Eng. Sci.* **62**, 3378 (2007).
85. H. P. Zhu, Z. Y. Zhou, R. Y. Yang, et al., *Chem. Eng. Sci.* **63**, 5728 (2008).
86. O. R. Walton and R. L. Braun, *J. Rheol.* **30**, 949 (1986).
87. O. R. Walton, in *Particulate Two-Phase Flow*, edited by M. C. Roco (Butterworths, Boston, 1992), p. 884.
88. R. D. Mindlin and H. Deresiewicz, *J. Appl. Mech.* **20**, 327 (1953).
89. O. R. Walton, *Mechanics of Materials* **16**, 239 (1993).
90. M. Y. Louge, in *http://www.mae.cornell.edu/microgravity/impact-table.html* (Cornell University, Ithaca, NY, 1999).
91. L. A. Pugnaloni, G. C. Barker, and A. Mehta, *Advances in Complex Systems* **4**, 289 (200).
92. L. A. Pugnaloni and G. C. Barker, *Physica A* **337**, 428 (2004).
93. G. C. Barker, in *The Chemical Physics of Food*, edited by P. Belton (Blackwell Publishing, LtD, Oxford, 2007), p. 135.
94. P. Ribiere, P. Richard, P. Philippe, et al., *Eur. Physics J. E* **22**, 249 (2007).

Solid-fluid transition and the formation of ripples in vertically oscillated granular layers

Osamu Sano

Department of Applied Physics, Tokyo University of Agriculture and Technology, Koganei, Tokyo 184-8588

Abstract. A vertically oscillated granular layer in a narrowly-spaced container under atmospheric pressure was observed. Above a certain critical forcing (frequency and amplitude) depending on the layer height, the layer exhibits parametrically excited undulations and ripples. A density wave propagating along the layer was elucidated by subtracting successive images taken by the high-speed video camera. A new model taking into account of the refraction of density wave due to the vertical distribution of the propagation velocity is proposed to explain the wavelength of the ripple. The formation of the ripples reflects the fluid-like part of the layer, and not the total thickness of the layer, in determining the dispersion relation, which is in accordance with our experimental findings.

Keywords: granular layer, vertical vibration, dilatancy, density wave, ripple, wavelength
PACS: 45.70.Qj, 47.54.-r, 62.20.mq, 64.70.D-

INTRODUCTION

Granular materials can bridge the gap between particulate systems and continuum media, between microscopic properties (i.e. particle size, layer number, ...) and macroscopic properties (i.e. layer height, wavelength, ...) and between solid-like behavior and fluid-like behavior. One of the phenomena that attracts attention on the vibrating granular materials is the formation of regular planar patterns like square cells, triangle/hexagonal cells, stripes[1, 2, 3, 4, 5, 6, 7, 8], or quasi-crystal patterns[9]. Their vertical structures correspond to either the parametrically excited ripples similar to those observed on the surface of fluids[10, 11, 12, 13, 14, 15] or undulations, i.e. the wavy deformation of a layer with its width almost unchanged[16, 17, 18]. In our previous papers[19, 20], we have experimentally shown the occurrence of undulation and classified its eigen-modes. Our findings are as follows: Above a certain critical forcing, the layer shows parametrically excited undulation that is characterized by eigen-modes. Each particle exhibits a small relative displacement, but its orbit remains within almost vertical line segment. This deformation is ascribed to the dilatancy and buckling of the layer at the impingement of the layer, if the latter has a layer number greater than about three in agreement with previous papers[3, 21, 22, 23], so that the squeezing of the middle layer is achieved. The undulation retains a solid-like character, whose eigen-modes are explained in terms of a theory of elasticity[24]. Further increase of the forcing yields stationary ripples, reminiscent of the parametric instability in vibrated fluid layers called the Faraday instability. The dispersion relation is insensitive to the material constant of the constituent grains, and a fluid-like behavior characterized by macroscopic variables like the layer height and kinematic viscosity is recognized. Analogy to Faraday waves in fluid motion, however, is only partially fulfilled; we should note that the occurrence of the void space between the layer bottom and the bottom wall of the container, the extremely large ratio of wave height to wavelength, ... are not observed in the fluid motion. The reported dispersion relations are of the form $\lambda/h = a + b/(f\sqrt{h/g})^\alpha$, where α varies between -2 and -1[3, 6, 11, 12, 14, 15, 25, 26]. Here λ is the wavelength, h is the layer height, g is the acceleration of gravity, f is the frequency of external forcing, and a, b are some constants. One of the reasons of this disagreement will be the choice of h, because it is not necessarily the total layer thickness h that is relevant to the ripple formation. In the case of a thicker layer under weaker external forcing, the lower part of the layer will transfer the upward momentum from the bottom wall, but this region is rather immobilized, and only the upper part is relevant to the formation of ripples.

The nonuniform distribution of particles introduces difficulty in the modeling, so that the most of the theories for vibrating granular material assume freely evolving granular gasses or an assembly of particles in almost uniform densest packing, both of which assume periodic boundary or infinite extension[2]. The experimental system, however, has a finite volume, so that it includes the particle motion near the boundary whose properties are not the same as those in the bulk, and yet whose contour shape is sometimes decisive in characterizing the whole system. One of the typical properties is the wavelength, whose saturation with the increase of the layer thickness has already been pointed out. In particular, Clément et al.[12, 13] proposed an empirical relation for the dispersive regime in the form $\lambda/\sqrt{N_h} = \lambda^*(d) + g^*/f^2$ (where $N_h = 2h/\sqrt{3}d$, d is the grain diameter, and λ^*, g^* are constants), or $(2\pi f)^2/gk = A + Bkh$ (where k is the wave number and A, B are constants) reflecting both shallow water and deep water characteristics in the fluid motion. They pointed out the deviation of their experimental data from the above-mentioned relations for thicker layers in addition to the so far admitted large frequency off-set, and attributed this disagreement to a "complex interplay" between the global deformation of the granular layer (i.e., arching and peak amplitude) and the internal dynamics of pressure/density waves due to vertical impact. But they did not explicitly show how the surface deformation is generated. The wavelength saturation was also investigated by Melo et al. in 3D case[3], and the necessity of using different scaling before and after the transition of a "grain mobility" $\tilde{v}_{gm} \equiv 2\pi af/\sqrt{gd} = \Gamma/(2\pi f\sqrt{d/g})$ has been mentioned by Umbanhowar et al.[25]. The latter presented the data that a wavelength scaling is governed by the layer height h for $\tilde{v}_{gm} > 2.5$ (with larger Γ or smaller layer number N), but fails for $\tilde{v}_{gm} < 2.5$. Their grain mobility transition is closely related to solid-fluid transition (see also [27]), but detailed dependence on Γ, h, N, etc. was not shown, so that a convincing explanation on the wavelength selection is yet to be given.

As far as the ripple regime is concerned, we need to know the effective layer height h^* that reflects the part of the layer in fluid states, which, however, may depend on the forcing condition as well as the material properties. This requires understanding of the formation mechanism taking into account of the individual particle motion and the collective motion of the particles. The collision of particles and the clustering of particles in 1D system have been reported[22], in which a clustered state that depends on N, Γ, ε (ε being a restitution coefficient) and a complicated transition regime was identified. Furthermore, one-dimensional shock wave propagation in vertically oscillated thick granular materials has been examined, and the layer compaction and the propagation of high temperature front were shown[28, 29]. These results, however, do not lead to the surface wave formation because of the system's dimensionality. In the 2D system, density wave propagation[13], shock waves past a wedge[30], and those emitted from the kinks of the ripples[31, 32], have been reported in the numerical simulation and experiment, but they are not connected to the physical mechanism on the 2D ripple formation. The occurrence of surface waves was analyzed by Goddard and Didwania[33] on the basis of hydrodynamic stability theory. The latter approach, however, is the one assuming the whole granular layer in a fluid state.

We have recently shown the possibility of the surface deformation of a granular layer due to the propagation of a density wave, which is generated by the horizontal dilatation of the material at the impingement on the container wall[34]. If the compression exceeds a certain critical magnitude, the layer will undergo a local buckling, which will develop into a fluid-like wave in a *thinner* granular layer, where all the constituent particles are mobilized. On the other hand, if the compression is not as large as the one mentioned above, or a granular layer is *thicker*, the density wave can propagate non-uniformly because of nonuniform distribution of particle number density and bulk modulus. It will propagate faster along the layer in the lower condensed region, whereas it will propagate slower near the upper free surface where particle number density and bulk modulus is smaller. Then the wave front will deflect upward, and the surface deformation, or ripples, can be generated. In our model, it is not necessarily the total layer thickness but the upper mobilized part of the layer, that is relevant to the ripple formation. The vertical variation of the density wave velocity, and hence the amount of refraction is crucial for wavelength selection. In this paper, experimental data showing the above-mentioned density wave propagation in vertically vibrated thin granular layers and a continuum model to explain the wavelength of the ripples in vertically vibrating granular layers are given.

EXPERIMENT

We performed an experiment on the pattern formation of a granular layer, confined in a narrowly-spaced rectangular container of horizontal dimensions $L(=81\text{-}146 \text{ mm})$ and $W(=3\text{-}10\text{mm})$, that was vibrated vertically $z = a\sin(2\pi ft + \phi)$. Here z is the vertical displacement, a and f are, respectively, the amplitude and frequency of the externally applied vibration, and ϕ is the phase. All experiments were performed under atmospheric pressure, and the cross-sectional view of the patterns was observed by a high-speed video camera. The particles we tested were aluminum spheres of $d = 1.1$mm, alumina spheres of $d = 1.1$mm and $d = 0.55$mm, and lead spheres of $d = 1.1$ mm. Under the present conditions, the effect of the air viscosity was negligible, and the observed patterns were quasi-twodimensional. We show typical patterns in Figs. 1(a) and 1(b). Figure 1(a) is an example of "undulation", which is characterized by arch-like deformation of the layer with an integer or half-integer number of waves. The ridges and troughs alternate with a period $2/f$, and the feet of the arches are almost always in contact with the bottom wall. The layer thickness is almost unchanged (Fig.2(a)), and the motion of the constituent particles remains within a short vertical line segment

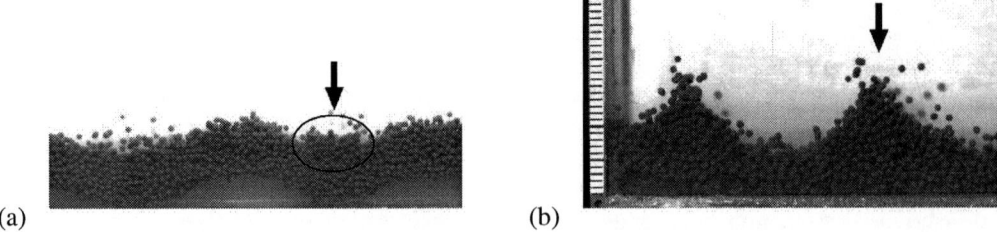

FIGURE 1. (a) Undulation of a layer of lead beads (d =1.1 mm, h =10.3 mm, f =30.0 Hz, a =2.10 mm), and (b) Ripples on a layer of lead beads (d =1.1 mm, h =5.9 mm, f =27.5 Hz, a =2.21 mm).

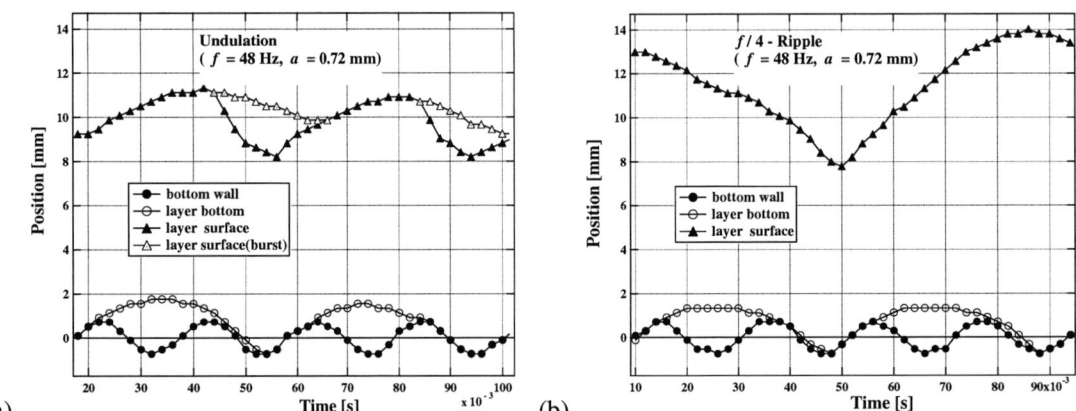

FIGURE 2. Time trajectory of the layer of lead spheres (d =1.1mm, h =8.7mm) for (a) undulation and (b) $f/4$-ripple, respectively. Measurement is made at an arch-foot position of undulations, whereas it is at a peak-valley position of ripples, as shown by arrows in Figs.1(a) and 1(b). The upper and lower boundary positions of the burst are shown by open and solid triangles, respectively, in Fig.2(a).

(Fig.3(a)). In this sense, the behavior is *solid-like*. Closer look at the contour shape of the undulation, however, reveals compacted regions near the feet of the arches, above which an assembly of loosely contacting particles is superposed (denoted here as "burst" and shown by the circle in Fig.1(a)). The latter regions sometimes move upward, but they look irrelevant to the force chains of the main solid-like regions below. On the other hand, Fig.1(b) is a snapshot of the "ripple" pattern, in which crests and valleys alternate with a period $2/f$, $4/f$ (Fig.2(b)) or in general $2n/f$ (n: integer). The motion of the constituent particles apparently shows that the behavior is similar to those observed in vibrating *fluid* layer (Fig.3(b)). However, an extraordinarily large ratio of wave amplitude to wavelength, and the presence of the almost cusped crests as well as a clear space between the layer bottom and the container should be remarked.

Figure 4 shows the time variation of the undulation (generated under the same condition as in Fig.1(a)) observed by a high speed video camera with a sampling time 1/1000 second, in which two successive images are subtracted to show the relative displacement of the particles. In the first figure Fig.4(a), for example, the positions of the particles at frame 132 is subtracted from frame 133, the latter being $\Delta t = 1$ ms later than the former. The part of the particles that collides with the bottom wall (i.e., the region near the feet of undulation) is compressed and solidified, so that the latter region is described darker and smoother. White lines in the figures give a rough sketch of the boundaries of these darker (immobilized) regions. To do this we chose, for example, a small rectangular region (d) [in Fig.4(b)] along the layer, and measured the image density level as shown in Fig.4(e). Although the latter level changes gradually from higher to lower region, we can identify the boundary of mobilized-immobilized regions within an accuracy of a few particle diameter. A sequence of frames shows the propagation of the compressed regions, or density wave, toward the ridges of the undulation along the layer (fronts of a density wave are shown by). The velocity of this density wave is of order 10 m/s, which implies an extremely small effective Young's modulus, i.e., 10^{-4} times smaller than that of an ordinary elastic material.

A similar density wave was observed in the ripple. In Fig.5(a), the crest located near the center of this picture is lowered until the layer becomes almost flat (Fig.5(b)), during which the layer is in contact to the bottom wall and is

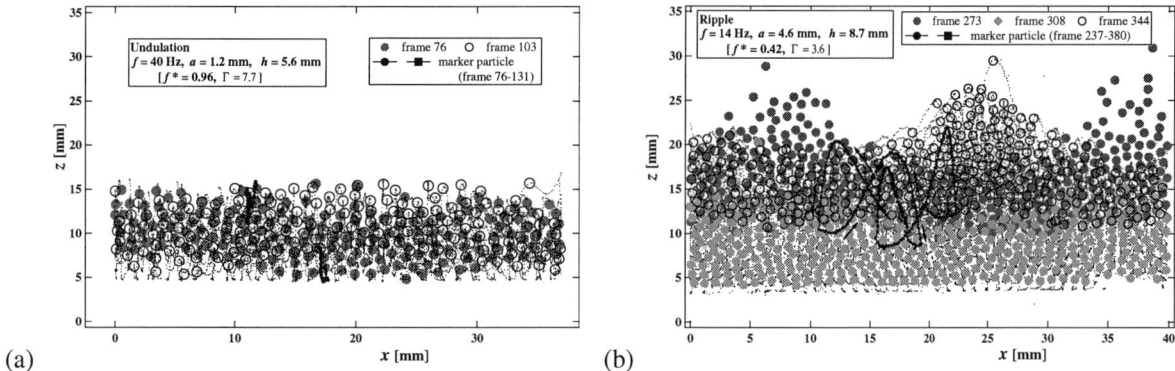

FIGURE 3. Trajectory of a particle in the layer in (a) undulation ($f = 40$Hz, $a = 1.2$mm, $h = 5.6$mm; $f^* \equiv f\sqrt{h/g} = 0.96$, $\Gamma = 7.7$) and (b) ripple ($f = 14$Hz, $a = 4.6$mm, $h = 8.7$mm; $f^* = 0.42$, $\Gamma = 3.6$) relative to the container wall.

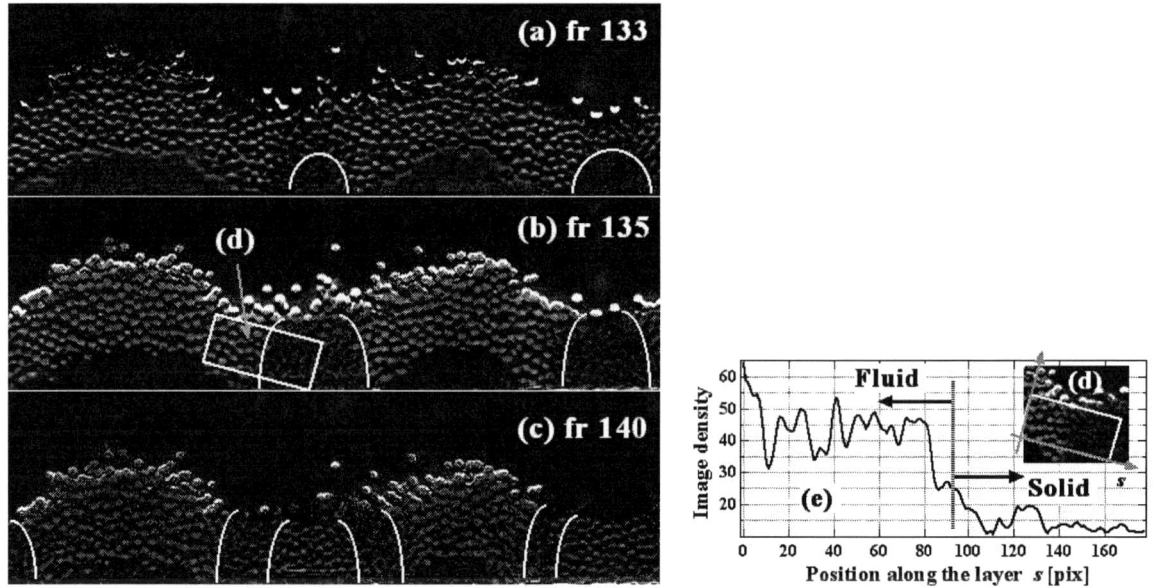

FIGURE 4. Subtraction of successive images with time difference $\Delta t = 1$ ms of the undulation mode in the layer of lead beads ($d = 1.1$ mm, $f = 30.0$ Hz, $a = 2.10$ mm): (a) frame 133, (b) frame 135, and (c) frame 140. Figure 4(e) is an example to show how the boundary between the solid and fluid states in the region shown by (d) [in Fig.4(b)] is determined from the image density variations. Fronts of a density wave are given by white lines in (a)-(c).

compressed. The origin of the time $t = 0$ is chosen when the container is at its highest position (at about frame 33 in the present case), around which the contour of the free surface exhibits a sharp peak (as shown by dotted lines in Fig.5(a)). The layer detaches at the bottom wall and moves upward almost as a rigid body (Fig.5(c)). Some local regions that are compressed and solidified appear (Fig.5(d)). The compressed regions develop both horizontally and upward (Figs.5(e)-(g)). Similar compressed regions starting from different positions collide, which lift the surface regions that are previously valleys and form the new crests, or ripples (Fig.5(h)). In the latter figure, the free surface contour at frame 176 ($t = 1.966T$) is shown by the dotted lines.

Figure 6 shows the vertical distribution of the number density at a peak-valley position. The ordinate is the vertical position measured from the bottom of the layer, which is normalized by the average layer thickness h_0 (=10.3 mm in the present case). On the other hand, the abscissa is the number of particles averaged in the test section of $5d$(horizontal) $\times 2d$(vertical). The data a($t/T = 0.258$) corresponds to the layer that undergoes highest compression, whereas the data e($t/T = 1.908$) corresponds to the layer that is at its maximum expansion. The surface wave develops for t/T larger

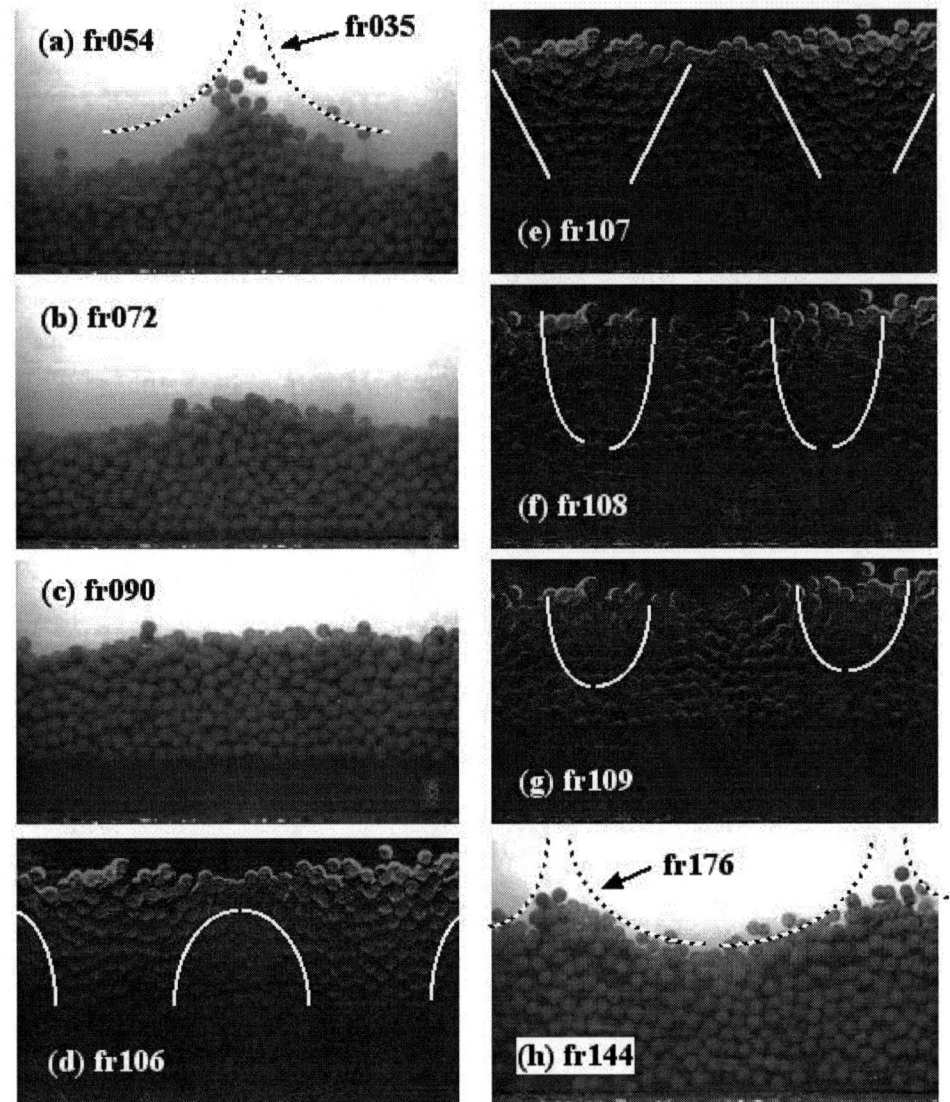

FIGURE 5. Time sequence of the $f/4$ ripple mode in the layer of lead beads (d =1.1 mm, f =27.5 Hz, a =2.07 mm) between $t=0$ (frame 33) and $t=2T$ (frame 178): (a) $t/T = 0.261$, (b) $t/T = 0.509$, (c) $t/T = 0.756$, and (h) $t/T = 1.499T$, where $T = 1/f$. The layer is almost flat and moves upward for $t/T = 0.3 - 0.7$. Subtraction of successive images with time difference $\Delta t = 1$ ms at (d) $t/T = 0.976$, (e) $t/T = 0.990$, (f) $t/T = 1.004$, and (g) $t/T = 1.018$. Sketch of the propagation of a density wave is given by white lines that are determined similarly to Fig.4.

than about 1 (from c to e). Note that the region with larger velocity gradient is limited to a few diameter's thickness from the free surface, and that the lower part has almost constant number density. We also recognize the negative density gradient near the lower boundary of the layer at time $t/T = 1.908$, which corresponds to downward expansion of this part as was observed experimentally.

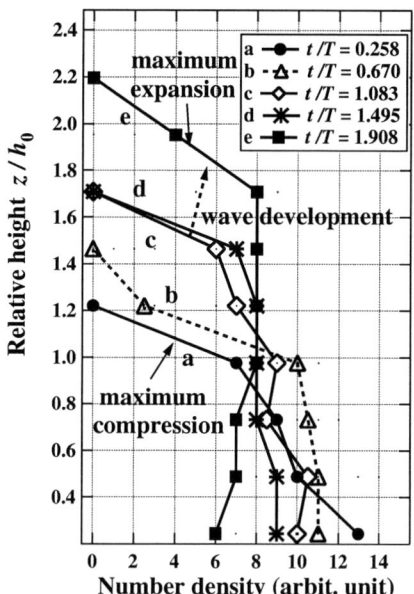

FIGURE 6. Vertical distribution of number density. Density gradient near the surface is almost constant and is limited within a region of a few diameter thickness, whereas number density is almost constant in the lower part.

THEORETICAL MODEL

A. Dilatancy induced buckling and the formation of undulations

In an earlier paper[24], we reported on dilatancy induced buckling and the formation of undulations of various wave numbers. If the layer number is at least three and the impingement of the layer is larger than a certain critical value but is not so large as to blow off the upper most layer of particles, the layer is highly squeezed at the collision with the bottom wall and will dilate in the horizontal direction. The presence of the side walls, however, restricts the horizontal dimension, so that buckling will occur and the layer will undulate. This is similar to the process that takes place in an elastic plate (having effective Young's modulus E^* and geometric moment of inertia I) which is compressed on both sides by a force F.

The contour shape of the lower boundary of the granular layer is given by

$$z = \frac{2k}{\alpha}[1 - cn(\alpha s)], \tag{1}$$

$$x = \frac{2}{\alpha}\left[E\left(am(\alpha s), k\right) - \frac{1}{2}\alpha s\right], \tag{2}$$

where s is the length along the layer, $\alpha = \sqrt{F/(E^*I)}$, $cn(\alpha s)$ is Jacobi's elliptic function, $E(\phi, k)$ is the elliptic integral of the second kind, $\phi = am(\alpha s)$ is the amplitude function, and the constant k is determined by the boundary conditions that reflect the amplitude of the wave. Details for the expressions (1) and (2) can be found in [24].

We show typical wave forms described by eqs.(1) and (2) in Fig.7. The forces are applied horizontally to the layer at its ends, so that the number of undulation is either integer (Fig.7(a)) or half integer (Fig.7(b)), but the number itself depends on the magnitude of compression and hence the externally applied frequency, amplitude and the phase.

B. Density wave propagation and the free surface deformation

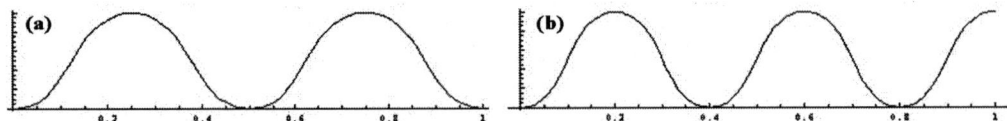

FIGURE 7. Example of undulation mode: (a) Symmetric mode with respect to the midpoint and wave number 2 (S_2), and (b) antisymmetric mode with respect to the midpoint and wave number 2.5 (A_3).

We shall consider a density wave propagation in an otherwise uniform semi-infinite two-dimensional elastic layer of thickness h, to which an impulsive pressure p_0 is applied at the end of the layer $x = 0$[34]. Here we have taken the Cartesian coordinate system with the x and z axes in the horizontal and vertical directions, respectively (Fig.8). We denote the displacements in the x and z directions by u and w, respectively, and the pressure by p. The governing equations for $\mathbf{u} = (u, w)$ and p are

$$p = -K^* \left(\frac{\partial u}{\partial x} + \frac{\partial w}{\partial z} \right), \tag{3}$$

$$\rho \frac{\partial^2 u}{\partial t^2} = K^* \left(\frac{\partial^2 u}{\partial x^2} + \frac{\partial^2 u}{\partial z^2} \right), \tag{4}$$

$$\rho \frac{\partial^2 w}{\partial t^2} = K^* \left(\frac{\partial^2 w}{\partial x^2} + \frac{\partial^2 w}{\partial z^2} \right), \tag{5}$$

where ρ and K^* are the effective density and the effective bulk modulus, respectively. Initially the layer is at rest:

$$\mathbf{u} = \mathbf{0} \quad \text{for} \quad t \leq 0, \tag{6}$$

and the boundary condition at the end $x = 0$ is

$$-K^* \frac{\partial u}{\partial x} = p_0. \tag{7}$$

The boundary condition at the bottom for $t > 0$ is

$$w = 0 \quad \text{at} \quad z = 0. \tag{8}$$

On the free surface denoted by $z = h + \zeta(x,t), (\zeta \ll h)$, we apply the kinematic condition

$$\frac{\partial w}{\partial t} = \frac{\partial \zeta}{\partial t} + \frac{\partial u}{\partial t} \frac{\partial \zeta}{\partial x}, \tag{9}$$

and the dynamical condition

$$\frac{\partial u}{\partial x} + \frac{\partial w}{\partial z} = 0. \tag{10}$$

The solution valid to the first order of amplitude is obtained in the form

$$u = \sum_n \cos\left(\frac{\pi n z}{2h}\right) f_n(x,t), \tag{11}$$

$$w = \sum_n \sin\left(\frac{\pi n z}{2h}\right) g_n(x,t), \tag{12}$$

where

$$\mathbf{u} = \mathbf{0} \quad \text{for} \quad t \leq x/v, \tag{13}$$

and

$$u = \sum_{m=0}^{\infty} \frac{(-1)^m 4 v p_0}{\pi K^*(2m+1)} \cos(q_{2m+1} z) \int_\xi^t J_0\left(q_{2m+1} v \sqrt{\tau^2 - \xi^2}\right) d\tau \tag{14}$$

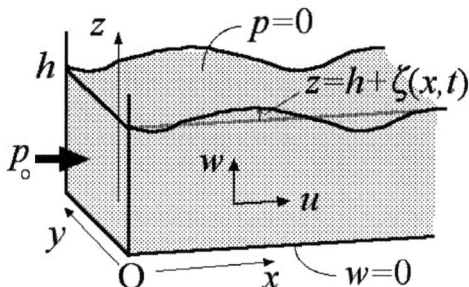

FIGURE 8. Definition sketch of the granular layer and the propagation of the surface wave.

for $t > x/v$. Here $\xi = x/v, q_n = \pi n/(2h), v = \sqrt{K^*/\rho}$ and J_0 is the zeroth order Bessel function of the first kind. The lowest mode of the displacement is

$$u_1 = \frac{4vp_0}{\pi K^*}\cos(q_1 z)U(\xi,t), \qquad (15)$$

where

$$U(\xi,t) = \int_\xi^t J_0(q_1 v\sqrt{\tau^2 - \xi^2})d\tau. \qquad (16)$$

Figure 9(a) shows the lowest mode displacement u_1 at $z = 0$ and its time dependence, where only the main part U with $qv = 1$ is given. The displacement due to an impulsive pressure p_0 at $x = 0$ at time $t = 0$ propagates in the positive direction of x, in which the alternate appearances of the compressed and rarefied regions with the increase of time are recognized. The stress distribution s_1 due to u_1 is proportional to w_1:

$$w_1 = \frac{8hp_0}{\pi^2 K^*}\sin(q_1 z)W(\xi,t), \qquad (17)$$

where

$$W(\xi,t) = q_1 v\xi \int_\xi^t \frac{1}{\sqrt{\tau^2 - \xi^2}}J_1(q_1 v\sqrt{\tau^2 - \xi^2})d\tau - 1, \qquad (18)$$

and J_1 is the Bessel function of the first kind of order one. Figure 9(b) shows the time sequence of the stress distribution s_1, where the main part W with $qv = 1$ is given. At the free surface $z \approx h$, the stress becomes maximum, whose peak appears near the propagation front of the displacement, followed by the second peak, third peak, etc., with decreasing magnitudes. Note that the magnitude of the stress at the wave front is enhanced as it proceeds, which may induce local buckling that leads to a *burst-like* or *spike-like* contour shape of the free surface (see Fig.10). Even if the magnitude of the stress is not as strong as the one that induces local buckling, collision of density waves of opposite directions that are generated at the feet of arches of undulation may exert highly concentrated stresses at the ridges of the layer. Consequently local buckling and the spike-like deformations of the free surface will also be expected, which can trigger the ripples.

C. Density wave refraction and the formation of ripples

Taking into account of the density wave propagation along the granular layer under vertical vibration, we consider a new model that can explain the ripple formation in the vertically nonuniform distribution of the wave propagation. It is well-known that the general formula of the wave refraction under the refractive index $n(z)$ is given by

$$n(z)\sin\varphi = C, \qquad (19)$$

where

$$\sin\varphi = \frac{dx}{\sqrt{dx^2 + dz^2}} = \frac{1}{\sqrt{1 + z'^2}}, \qquad (20)$$

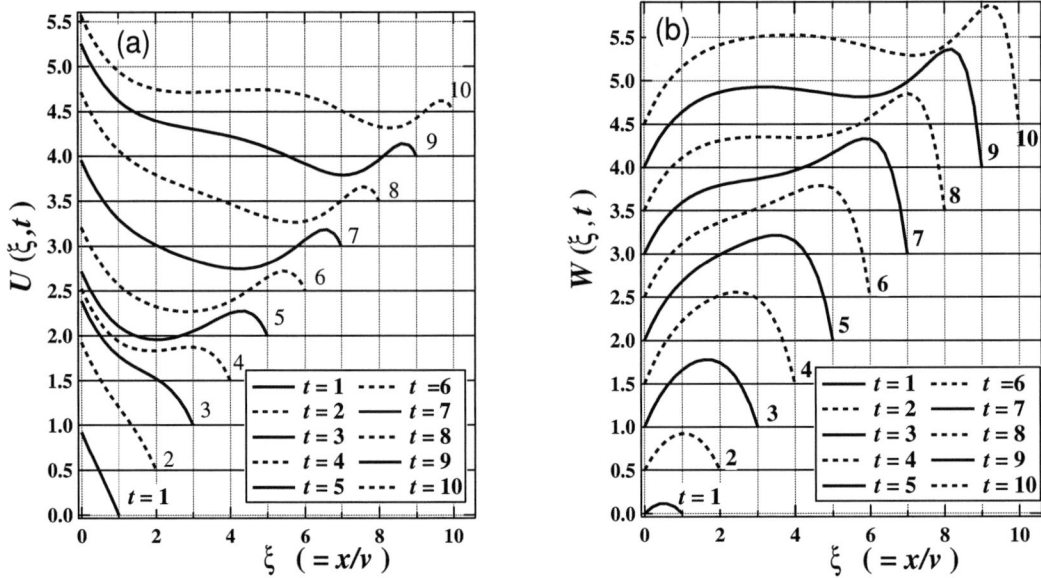

FIGURE 9. (a) The lowest mode displacement u_1 at $z = 0$ (arbitrary unit), and (b) stress distribution and its propagation on the surface of the layer (arbitrary unit).

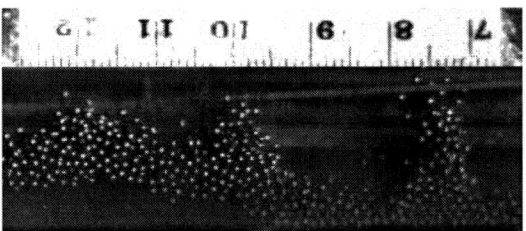

FIGURE 10. Spikes on the undulation.

$$n = \frac{v_0}{v(z)}, \qquad (21)$$

v_0 and v are the reference and the local wave velocities, respectively, and C is a constant to be given by the initial condition. Figure 11 is the definition sketch of the wave propagation in a material that has stepwisely different refractive index. By combining eqs.(19)-(21), we find that the equation that determines the propagation of the density

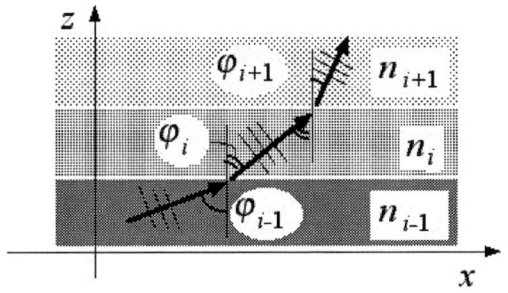

FIGURE 11. Definition sketch of the wave deflection.

wave in a material with refractive index that continuously varies with depth is

$$\frac{dz}{dx} = \sqrt{\left(\frac{v_0}{v}\right)^2 - 1}. \qquad (22)$$

Here we have chosen $C = 1$ at some reference position (x_0, z_0) in a region with uniform distribution of particles, where the density wave propagates horizontally.

Obviously $v(z)$ should approach a reference speed v_0 in a homogeneous region far down from the surface, whereas it tends to 0 as the free surface is approached. This can be anticipated from Fig.6, and the relation

$$v = \sqrt{\frac{K^*}{\rho}} = \sqrt{\frac{dp}{d\rho}} = \sqrt{\frac{dp/dz}{d\rho/dz}}. \qquad (23)$$

In general, dp/dz is a small constant in a granular layer, so that $d\rho/dz$ will be decisive. Thus in our case, v is smaller near the upper surface, whereas it is large and almost constant in the lower region, which leads us to consider the following models:

(i) model 1

$$v(z) = \frac{1}{2}v_0\left\{1 - \tanh[\alpha(z - h_0)]\right\}, \qquad (24)$$

where h_0 is the average height, and α is a constant that characterizes the effective "thickness" of the free surface region, and

(ii) model 2

$$v(z) = \begin{cases} v_0 & (z \leq h_{min}) \\ v_0\left(1 - \frac{z - h_{min}}{h_{max} - h_{min}}\right) & (h_{min} \leq z \leq h_{max}) \end{cases}, \qquad (25)$$

where h_{min} is the height of the layer below which the material is uniform, and h_{max} is the height of the peak.

In the first model, we obtain a solution:

$$z = h_0 - \frac{1}{2\alpha}\log\left|2\alpha^2(x - x_0)^2 - \frac{1}{2}\right|, \qquad (26)$$

by solving the differential equation (22) using (24). Typical wave form is shown in Fig.12(a), where the amplitude becomes minimum at $x = x_0$ and infinity at $x = x_0 \pm 1/(2\alpha)$. Physically the positions $x = x_0 \pm 1/(2\alpha)$ are regarded as the peak positions, and the behavior between these points is repeated, so that the wavelength is given by

$$\lambda = \frac{1}{\alpha}. \qquad (27)$$

As α decreases, which means that the fluid-like region near the upper surface becomes broader, the wavelength becomes longer. Reliable data for α is needed in order to validate this model.

In the second model we have

$$(x - x_0)^2 + (z - h_{max})^2 = (h_{max} - h_{min})^2, \qquad (28)$$

and

$$\lambda = 2(h_{max} - h_{min}), \qquad (29)$$

by solving the differential equation (22) using (25). A typical wave form is shown in Fig.12(b).

In order to compare our second model with the experimental data, we further approximate the collision of a granular layer as follows: We assume that the particles are in nearly closest packing with vertical center-to-center separation $\delta (> d)$, but that the particles arranged in one vertical line are in contact to the neighboring particles. When the lowest particle collides with the bottom wall, it will move upward by the amount $\Delta z \equiv \delta - d = \beta d$ during a certain collision time Δt, where β will presumably be proportional to the externally applied amplitude a, which is discussed in the last section.

The motion of this particle will be opposed by the force from the neighboring particles, which is approximated by the restoring force with a certain elastic spring constant k^*. Here we have neglected the dissipation due to the frictional force, because of the simplicity and because of our observation that the energy dissipation is much more important in a downward phase of the layer where the particles flows down the slope of the crest (see Figs.5(a) and 5(b)) than they

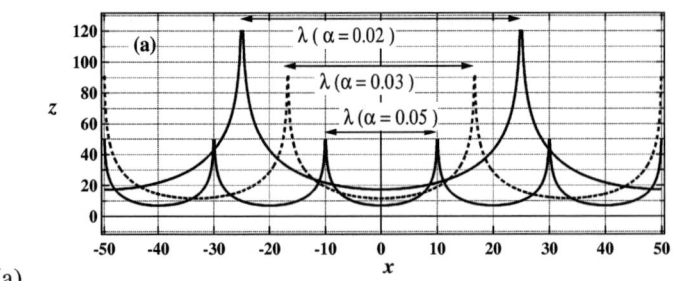

FIGURE 12. Surface deformation: (a) model 1, and (b) model 2.

are in an upward phase (see Figs.5(c)-5(g)). In the latter, particles are in contact to each other to form a force network, and their arrangement is only slightly varied to give a vertical compression. Thus the motion of the lowest particle is described by

$$z_1 = \frac{d}{2} + \frac{F_0}{k^*}(1 - \cos \omega t), \tag{30}$$

where $\omega = \sqrt{k^*/m}$, m is the mass of the particle, and F_0 is the force applied to the particle from the container. The lowest particle will then collide with the upper next one after a time t_0, which is approximately equal to

$$t_0 \approx \sqrt{\frac{2m\beta d}{F_0}}, \tag{31}$$

when the collision time is sufficiently short. After successive collisions, the upper-most particle will be pushed up with the velocity

$$v_N = \frac{F_0}{k^*}\omega \sin \omega t_0 \approx \sqrt{\frac{2F_0\beta d}{m}}. \tag{32}$$

In this approximation, the maximum height of the column of particle h_{max} is

$$h_{max} = (N-1)\delta + d + \frac{F_0 \beta d}{mg}, \tag{33}$$

whereas the minimum height h_{min} is Nd, from which the wave height $h_{max} - h_{min}$ is calculated. Consequently the wavelength λ is estimated to be

$$\lambda = 2(N-1)\beta d \sqrt{1 + \frac{F_0}{(N-1)mg}},$$

$$\approx 2h\beta \sqrt{1 + \frac{F_0 d}{mgh}} \quad \text{for} \quad N \gg 1, \tag{34}$$

where we have approximated $h \approx (N-1)d$. The collision time Δt is estimated to be $c_0 T (= c_0/f)$, where c_0 is a small constant whose value may be different depending on whether the ripple has a period $f/2$ or $f/4$, during which

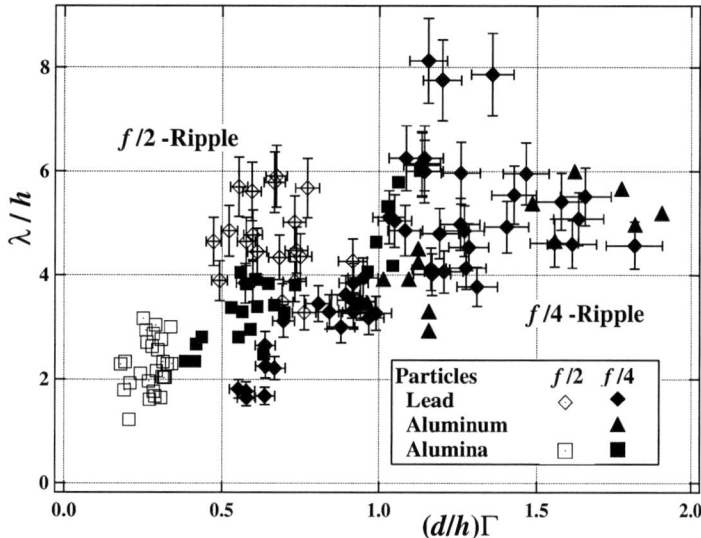

FIGURE 13. Wavelength selection in a layer with $h/d > 4$: Open squares and diamonds for $f/2$-ripple ($\Gamma = 2.45\text{-}6.90$), and closed squares, diamonds and triangles for $f/4$-ripple ($\Gamma = 4.67\text{-}12.7$).

the lowest particle acquires an additional momentum $\Delta p \approx 2\pi f a m$, so that the force applied to this particle will be estimated to be $F_0 \approx 2\pi f^2 a m/c_0$. Thus our final expression is given by

$$\frac{\lambda}{h} \approx 2\beta\left(1+\frac{2\pi a f^2 d}{g h c_0}\right) = 2\beta\left(1+c_1\frac{d}{h}\Gamma\right), \qquad (35)$$

where $\Gamma = (2\pi f)^2 a/g$ is the non-dimensional acceleration and c_1 is a constant. Putting aside the two unknowns β and c_1, which can be determined from detailed experimental data, we show the part of the relation between λ/h and $\Gamma d/h$ in Fig.13. Figure 13 shows the data for $f/2$-ripple with $\Gamma=2.45\text{-}6.90$ in open squares and diamonds, and the data for $f/4$-ripple with $\Gamma=4.67\text{-}12.7$ in closed squares, diamonds and triangles. Although there are some scattering of data around the linear dependence, favorable agreement is recognized irrespective of the difference of materials, particle sizes and layer height.

CONCLUSIONS AND DISCUSSION

An experimental data showing the propagation of density wave in a vertically vibrated thin granular layer and a continuum model to explain the enhancement of the stress near the free surface as the density wave proceeds are given. Furthermore a new model is proposed to explain the wavelength of the ripples in a vertically vibrating granular layer. In a *thiner* granular layer, all the constituent particles are mobilized above a certain critical external forcing, so that the fluid-like wave will be realized. In a *thicker* granular layers, however, the lower part of the particles is immobilized, so that only the upper part of the layer will be relevant to the ripple formation. In order to estimate the wavelength in this case, we take into account of the spatial distribution of the propagation speed of the density wave. In the lower condensed region where particles are almost uniformly distributed, the density wave will propagate faster and uniformly along the layer. On the other hand, the particle number density and bulk modulus will decrease toward the free surface, which results in the decrease of density wave velocity. Then the wave front will deflect upward, and the surface deformation will be generated for large enough forcing. In our model, it is not the total layer thickness that determines the wavelength. Rather the vertical variation of the density wave velocity, and hence the amount of refraction is crucial for wavelength selection, i.e. smaller propagation velocity gradient in the vertical direction exhibits larger wavelength, and vice verse.

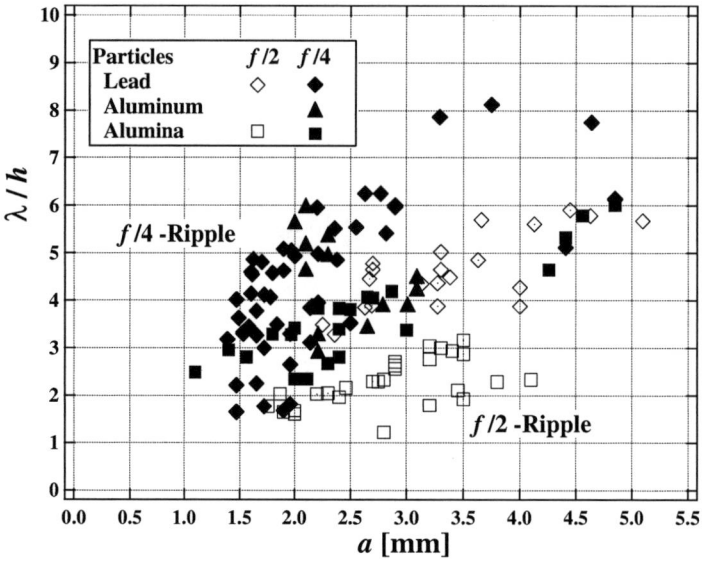

FIGURE 14. Wavelength vs. amplitude in a layer with $h/d > 4$: Open squares and diamonds for $f/2$-ripple ($\Gamma = 2.45$-6.90), and closed squares, diamonds and triangles for $f/4$-ripple ($\Gamma = 4.67$-12.7).

Our expression (35) is similar to the results shown by Clément and Labous[13]. The latter implicitly gives a roughly linear increase of the saturation wavelength λ_{sat} with the number of layers N_h, i.e.

$$\frac{\lambda_{\text{sat}}}{d} = c_2 + c_3 N_h, \tag{36}$$

where c_2 and c_3 are constants to fit the data. The expression (36) can be rewritten as

$$\frac{\lambda}{h} = \tilde{c}_3 + \tilde{c}_2 \frac{d}{h} \tag{37}$$

where $\tilde{c}_3$ and $\tilde{c}_2$ are re-scaled constants of c_3 and c_2. The expression (37) is the same form as eq.(35) except the factor Γ.

The proportionality of the wavelength λ to external forcing amplitude a has been remarked in the previous papers (e.g. [3, 12, 9]), which is accounted for by our relation (35) on the assumption that β is proportional to a. We show the $\lambda/h - a$ relation in Fig.14 for $f/2$-ripple and $f/4$-ripple. In spite of the scattering of data, the proportionality is generally admitted. The difference of the slope for different materials will be mainly due to the magnitude of the friction. Indeed we have measured the angle of repose in our preliminary experiment, which gives $30.6°$, $30.9°$ and $33.1°$ for lead, aluminum and alumina spheres of almost the same sizes. Smaller values of β seem to reflect larger angle of repose, or larger friction. The same remarks will hold for Fig.13, where the data points for alumina will shift downward by the modification of β, a tendency that improves the linear relation between λ/h and $(d/h)\Gamma$. At this moment the effect of friction and hence the dissipation is not quantitatively taken into account, nor the direct measurement of the density wave velocity across the layer is made. It is our future investigation to refine our wavelength selection mechanism taking into account of the dissipation effect.

In the case of much thicker layer, we observed separated region of fluid-like and solid-like ones. Figures 15 and 16 show the particle configurations in a test section of 5 mm width and above the bottom of the layer (shown by rectangles), in ascending and descending phases, respectively, of the $f/4$ ripple. The test particles were lead spheres of $d = 1.1$ mm with average layer height $h = 11.9$ mm, which were vibrated at a frequency $f = 27.5$ Hz and an amplitude $a = 1.96$ mm. Particles in the lower boundary are in contact to the bottom wall between $t = 0$ and 10 ms, but they detach the bottom wall of the container afterward. In the ascending phase, particles in the lower layers (about 5 layers in the present case) look almost packed and behave solid-like, so that their relative motions are hardly recognized whether or not the layer is in contact to the bottom wall or not. On the other hand, the particles near the upper boundary look dilated and loosely compacted. In the descending phase, lower part of the layer gradually moves downward, whereas

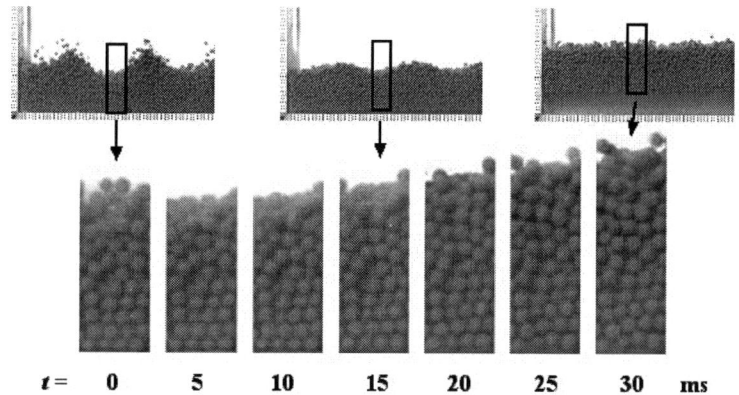

FIGURE 15. Particle configuration in ascending phase. Lower boundary of the layer is chosen as a baselines in the lower figures. The layer is in contact to the bottom wall between $t = 0$ and 10 ms, but levitates afterward.

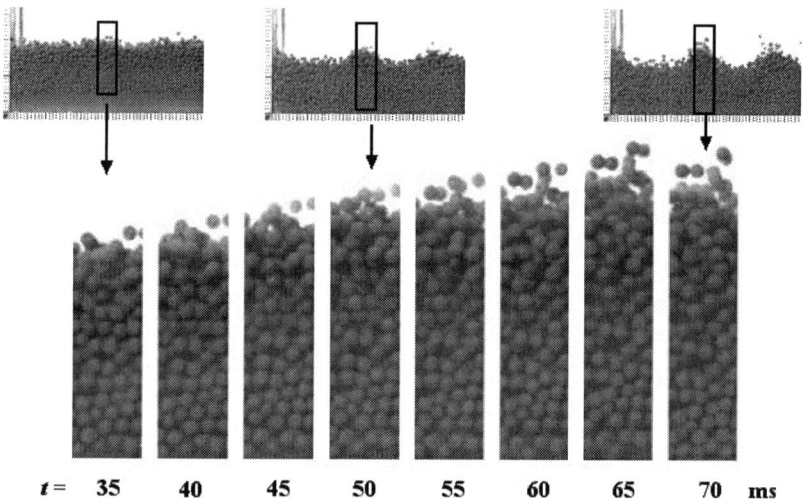

FIGURE 16. Particle configuration in descending phase. Lower boundary of the layer is chosen as a baselines in the lower figures. The layer collides with the bottom wall at $t \approx 70$ ms.

the upper part remains almost at its highest position (i.e. peak formation) so that the spacing between upper and lower layers increases. The separation of the whole layer into the lower solid-like region and upper fluid-like region will be the 2D version of the 1D clustering described by Luding et al.[22]. Our model assumes a granular material that may change from solid-like to fluid-like properties in a single layer, so that it cannot directly be extended to the layer that is disconnected by the space. It will require further knowledge on the mechanism of clustering, which is also left for our future investigation.

ACKNOWLEDGMENTS

This work was partially supported by a Grant-in-Aid for Scientific Research (C) from the Ministry of Education, Science and Culture, Japan.

REFERENCES

1. M. Faraday, *Phil. Trans. R. Soc. London* **52**, 299–340 (1831).
2. H. M.Jaeger and S.R.Nagel, *Science* **255**, 1523–1531 (1992); H. M.Jaeger, S.R.Nagel and R.P.Behringer, *Rev. Mod. Phys.* **68**, 1259–1273 (1996); I. S. Aranson and L. S. Tsimring, *Rev. Mod. Phys.* **78** (2006) 641–692.
3. F. Melo, P.Umbanhowar and H.L.Swinney, *Phys. Rev. Lett.* **72**, 172–175 (1994).
4. F. Melo, P.Umbanhowar and H.L.Swinney, *Phys. Rev. Lett.* **75**, 3838–3841 (1995).
5. P. Umbanhowar, F.Melo and H.L.Swinney, *Nature* **382**, 793–796 (1996).
6. T. H.Metcalf, J.B.Knight and H.Jaeger, *Physica* **A236**, 202–210 (1997).
7. C. Bizon, M.D.Shattuck, J.B.Swift, W.D.McCormick and H.L.Swinney,*Phys. Rev. Lett.* **80**, 57–60 (1998).
8. C. Bizon, M.D.Shattuck and J.B.Swift, *Phys. Rev. E* **60**, 7210–7216 (1999).
9. O. Sano, A.Ugawa and K.Suzuki, *Forma* **14**, 321–329 (1999).
10. S. Luding, E.Clément, J.Rajchenbach and J.Duran, *Europhys. Lett.* **36**, 247–252 (1996).
11. K. M.Aoki and T.Akiyama, *Phys. Rev. Lett.* **77**, 4166–4169 (1996).
12. E. Clément, L.Vanel, J.Rajchenbach and J.Duran, *Phys. Rev. E* **53**, 2972–2975 (1996).
13. E. Clément and L.Labous, *Phys. Rev. E* **62**, 8314–8323 (2000).
14. A. Ugawa, K. Suzuki, and O.Sano, *Proc. 4th Intern. Conf. Micromech. Granular Media, Sendai*, 537-540 (2001).
15. A. Ugawa and O.Sano, *J. Phys. Soc. Jpn.* **71**, 2815–2819 (2002).
16. S. Douady, S.Fauve and C.Laroche, *Europhys. Lett.* **8**, 621–627 (1989).
17. Y. Lan and A.D.Rosato, *Phys. Fluids* **9**, 3615–3624 (1997).
18. A. Goldshtein, M.Shapiro, L.Moldavsky and M.Fichman, *J. Fluid Mech.* **287**, 349–382 (1995).
19. A. Ugawa, and O.Sano, *J. Phys. Soc. Jpn.* **72**, 1390–1395 (2003).
20. K. Kanai, A.Ugawa, and O.Sano, *J. Phys. Soc. Jpn.* **74**, 1457–1463 (2005).
21. S. MacNamara and W. R. Young, *Phys. Fluids A*, **5**, 34–45 (1993).
22. S. Luding, E.Clément, A. Blumen, J.Rajchenbach and J.Duran, *Phys. Rev. E***49**, 1634–1646 (1994); S. Luding, H. J. Herrmann and A. Blumen, *Phys. Rev. E*, **50**, 3100–3108(1994).
23. B. Bernu, F. Delyon and R. Mazighi, *Phys. Rev. E*, **50**, 4551–4559 (1994).
24. O. Sano, *Phys. Rev.* **E72**, 051302, 1–7 (2005).
25. P. Umbanhowar, and H.L.Swinney, *Physica A* **274**, 344–362 (2000).
26. J. Bougie, J. Kreft, J. B. Swift, and H. L. Swinney, *Phys. Rev. E* **71**, 021301 (2005).
27. N. Mujica and F. Melo, *Phys. Rev. Lett.*, **80**, 5121–5124 (1998).
28. J. Bougie, S. -J. Moon, J.B.Swift, and H. L. Swinney, *Phys. Rev. E* **66**, 051301 (2002).
29. K. Huang, G.Miao, P.Zhsng, Y.Yun and R.Wei, *Phys. Rev. E***73**, 041302 (2006).
30. E. C. Rericha, C. Bizon, M.D. Shattuck, and H. L. Swinney, *Phys. Rev. Lett.* **80**, 014302 (2002).
31. S. -J. Moon, M.D.Shattuck, C.Bizon, D. I.Goldman, J.B.Swift and H.L.Swinney, *Phys. Rev. E* **65**, 011301,1–10 (2002).
32. S. -J. Moon, D. I. Goldman, J. B. Swift and H. L. Swinney, *Phys. Rev. Lett.***91**, 134301(2003).
33. J. D. Goddard, and A. K. Didwania, *Mech. Materials* **41**, 637–651 (2009).
34. O. Sano and A.Takei, *AIP Conf. Proc.* **1145**, 729–732 (2009).

Internal erosion in gas-flow weak conditions

Holger Steeb

Ruhr-University Bochum, Institute of Mechanics, Universitätsstr. 150, D-44 780 Bochum, Germany

Abstract. We present a continuum-based formulation of volumetrical erosion phenomena in gas-flow formations. The investigated model is based on a so-called hybrid three-phase mixture formulation, i.e. a superimposed continuum consisting of a solid matrix, the compressible pore gas and the fluidized/eroded particles. The theory establishes a non-linear instationary diffusion equation with respect to the pore pressure in combination with a set of mass and constitutive equations. We discuss analytical and numerical solutions of the set of equations under quasi-static and dynamic conditions and compare the results with various well-established sub-models.

Keywords: Mixture theory, internal erosion, gas-flow, outbursts
PACS: 47.11.St;47.56.+r;92.40.Gc

MOTIVATION

Several authors of publications in various geo- and hydromechanical disciplines have pointed out that internal erosion processes in polydisperse granulates, in a first temporal phase also called suffusion, plays a significant role in the design of protective filters and in the stability of dam constructions [1, 2, 3]. Furthermore, internal erosion poses a particular threat to embankment dams or dykes since these kinds of processes usually occur undetected within the structure of the embankment or in the subsoil. According to an investigation of Foester et al., cf. [4, 5], approximatively 46 % of all dams showing dysfunctions have problems with internal erosion. Therefore, internal erosion mechanisms appear to be the main cause of dam instabilities. As the internal erosion process is a non-visible phenomenon, this often leads to a destruction of these engineering constructions with catastrophic consequences or, at least, to a failure of the serviceability. In 2001, suffusion and subsequently, piping caused a reservoir failure of the Malana gravity dam [6]. Due to the failure of the Malana gravity dam the entire contents of the reservoir of about 220,000 cubic meters of water escaped within 20 minutes which has caused settlements of 5 to 6 meters in an area measuring about 500 square metres partly below the dam body without destruction. In part, the failure of the levees in New Orleans, US, through hurricane Katrina (August 29, 2005), was also caused by internal erosion and piping. Nicholson stated in an ASCE testimony [7], that a series of sinkholes, together with *boils* near the inboard toe of the embankment, were found at the London avenue canal, north, across from the breach. These observations give rise to the conclusion that soil movement, i.e. internal erosion and piping occured. A further prominent example of a catastropic failure where internal erosion mechanism were at least a participant is the Teton dam accident, cf. [8]. Besides internal erosion mechanism in cohesionless soils, internal erosion plays a significant role in cohesive soils, e.g. reservoir rocks or various types of silts. In petroleum engineering, e.g. in the North-sea sandstone reservoir, the prediction of the so-called sand production rate of the reservoir sandstone is important to run the petroleum well economically, cf. [9, 10, 11, 12, 13].

All the previously described erosion and solid-liquid phase transition models refer to viscous (incompressible) fluid flow in cohesive soils like reservoir rocks or cohesionless sand-gravel mixtures, or, in other words: mixtures with material incompressible constituents $\rho^{\alpha R} = \rho_0^{\alpha R} = $ const. In contrast to erosion processes introduced by a pore liquid, sudden outbursts in rock or coal seams are caused by the flow process of a compressible pore gas. For an overview we draw attention to the review article of Shepherd et al. [14]. In order to understand and predict such catastrophic failures of gas-flow through coal, often combined with the loss of numerous miners or expensive equipment, cf. [15], it is an important task to develop sophisticated models which take the special aspects of erosive gas-flow through porous media into account, cf. [16, 17, 18]. Furthermore, the volumetric sand production during the operation of (compressible) hydrocarbon production in reservoir sandstones is another phenomenon which needs to be considered in that class. All these applications fall in the field of solid-liquid phase transformations within compressible fluid-saturated porous media. Generally speaking, several constituents are involved during the phase transition process. On the one hand, the mixture consists of the solid skeleton, e.g. the sandstone or coal matrix φ^s. The effective density of the skeleton keeps constant during the process, which results in a material incompressible constituent. On the other hand, the pore gas φ^g has to be regarded as a compressible constituent. In the present approach, we omit a wet gas phase and, therefore, another incompressible fluid phase. Taking such a further incompressible liquid constituent into account, one obtains a partially-saturated mixture. Additionally, a further fluid phase takes the eroded particles φ^a – *the fines* – into account and models the solid-liquid phase transformation process. Modeling aspects will be discussed later.

We begin by introducing the main modelling assumptions. The present model is based on the thermodynamically consistent Theory of Porous Media (TPM) which extends the classical Mixture Theory (MT) by the concept of volume fractions, cf. [19, 20, 21, 22, 23, 24]. As a consequence, the whole microstructure is represented on the macro-scale by the scalar fields of volume fractions n^α with $n^\alpha := dv^\alpha/dv$. The volume of the Representative Elementary Volume (REV) of the mixture is introduced as dv, while the partial volume occupied by the constituent φ^α is described by dv^α. Thus, we can introduce the *saturation condition* $\sum_\alpha n^\alpha = 1$, cf. comments in Hutter et al. [25]. Next, we have to distinguish the *partial* density $\rho^\alpha := dm^\alpha/dv$, i.e. the mass of the constituent φ^α within the REV with respect to the volume of the mixture, from the so-called *effective* or *true* density $\rho^{\alpha R} := dm^\alpha/dv^\alpha$. Obviously, we obtain the mixture's density ρ at the REV as the sum of the individual partial densities, i.e. $\rho = \sum_\alpha \rho^\alpha$. Furthermore, the partial density is related to the effective densities by $\rho^\alpha = n^\alpha \rho^{\alpha R}$. Often, the sum of the volume fractions of all fluid constituents φ^β is denoted as the porosity $\phi := \sum_\beta n^\beta$. Quantities with respect to the reference configuration $\mathscr{B}_0$ of the skeleton, are abbreviated with the index $(\bullet)_0$.

As we are mainly interested in both the flow and the erosion process, we consider a rigid solid skeleton $\rho^{sR} = \text{const}$ and $\mathbf{v}_s = \mathbf{0}$. Thus the problem is driven by a set of partial balances of mass for the solid and the gas constituent and the partial balance of momentum of the gas phase, respectively. The theory establishes a non-linear instationary diffusion equation with respect to the pore pressure in combination with an evolution equation of the porosity and set of constitutive equations.

Kinematical Assumptions

In order to develop the model, we begin by introducing the governing volume fractions of the three-phase mixture model with the constituents φ^α and $\alpha \in \{s, g, a\}$. First, we introduce the porosity $\phi(\mathbf{x},t)$ and the concentration field $c(\mathbf{x},t)$, respectively

$$\begin{aligned}\phi &= \frac{dv^g + dv^a}{dv} = n^g + n^a \quad \text{or} \quad n^a = c\phi, \\ c &= \frac{dv^a}{dv^g + dv^a} = \frac{n^a}{n^a + n^g} \quad \text{or} \quad n^g = \phi(1-c).\end{aligned} \quad (1)$$

As we are mainly interested in the kinematical behavior of the internal, i.e. volumetrical, erosion process, the velocity of the stationary solid skeleton is assumed to be zero

$$\mathbf{x}'_s =: \mathbf{v}_s = \mathbf{0}. \quad (2)$$

Note that equation (2) restricts the investigation to hydraulical effects. Nevertheless, this is the main part in the internal erosion process. Fluxes are only defined for the compressible gas phase φ^g and the fluidized particles φ^a. Volume and mass discharges for the single constituents $(\mathbf{q}_\beta)$ and the total mixture $(\mathbf{q})$ are introduced as

$$\begin{aligned}\mathbf{q}_g &= \phi(1-c)\mathbf{v}_g, \\ \mathbf{q}_a &= c\phi\mathbf{v}_a, \\ \mathbf{q} &= \mathbf{q}_g + \mathbf{q}_a = \phi([1-c]\mathbf{v}_g + c\mathbf{v}_a).\end{aligned} \quad (3)$$

Partial and Total Balance Equations

According to the material incompressibility condition of the solid constituent and the introduced kinematical relations, we obtain a set of volume balances for the constituents φ^s and φ^a. The partial mass balance of the gas phase $\varphi^{\mathfrak{g}}$ is a storage equation, i.e. a conservation law

$$\partial_t(\phi) = \hat{n}^{\mathfrak{a}},$$
$$\partial_t(c\phi) + \operatorname{div}\mathbf{q}_{\mathfrak{a}} = \hat{n}^{\mathfrak{a}}, \quad (4)$$
$$\partial_t\left([\phi - \phi c]\rho^{\mathfrak{g}R}\right) + \operatorname{div}\left(\rho^{\mathfrak{g}R}\mathbf{q}_{\mathfrak{g}}\right) = 0,$$

with $\hat{n}^{\mathfrak{a}} = -\hat{n}^{\mathfrak{s}}$ and $\hat{n}^{\mathfrak{g}} = 0$. Other interesting Mixture Theory-based models for erosion phenomena (sand production, piping) taking a volume or mass production term into account have been proposed by [9, 26, 27]. The storage equation of the mixture is obtained by weighting equations (4)$_1$ and (4)$_2$ with the effective density of the gas-phase $\rho^{\mathfrak{g}R}$ and summing up the resulting partial mass balance equations. Thus, we finally obtain the mixture's mass balance

$$\phi(1-c)[\rho^{\mathfrak{g}R}]'_{\mathfrak{g}} + \rho^{\mathfrak{g}R}\operatorname{div}\mathbf{q} = \phi(1-c)\partial_t(\rho^{\mathfrak{g}R}) + \operatorname{grad}\rho^{\mathfrak{g}R}\cdot\mathbf{q}_{\mathfrak{g}} + \rho^{\mathfrak{g}R}\operatorname{div}\mathbf{q} = 0. \quad (5)$$

In equation (5), we have introduced the material time derivative $[\bullet]'_\alpha := \partial_t[\bullet] + \operatorname{grad}[\bullet]\cdot\mathbf{v}_\alpha$. It is obvious that in the case of material incompressible constituents $\rho^{\mathfrak{g}R} = \operatorname{const}$, the mass balance of the mixture (5) is reduced to the standard continuity equation of the total fluid discharge, i.e. $\operatorname{div}\mathbf{q} = 0$.

Note that the developed kinematical erosion model is not closed with the given set of mass- and volume balance equations. To solve the closure problem for adiabatic processes, we have to take the partial balances of momentum of $\varphi^{\mathfrak{a}}$ and $\varphi^{\mathfrak{g}}$ into account and, additionally, we have to formulate several constitutive equations for the specific volume production $\hat{n}^{\mathfrak{a}}$, the compressible gas density $\rho^{\mathfrak{g}R}$ and two of the three discharges, e. g. $\{\mathbf{q}^{\mathfrak{a}},\mathbf{q}^{\mathfrak{g}}\}$ (the third one, i.e. ($\mathbf{q}$) is calculated from the kinematic restriction (3)$_3$). Note that this leads to a complete, but complex model as various cross effects combined with material parameters, like the famous Yuster effect of multiphase flow in porous media [28, 29, 30], have to be taken into account. As the aim of the present study is to formulate a simpler model of erosion in gas-flow weak conditions, we concentrate on the main physical aspects taking further simplifications of the flow process into account.

First, we assume that the concentration of the eroded fines $c(\mathbf{x},t)$ is in the low range, described by $0 < c \ll 1$. Thus, we obtain a simplification of the inherent discharges

$$\mathbf{q} \approx \mathbf{q}_{\mathfrak{g}} \approx \phi\mathbf{v}_{\mathfrak{g}}, \quad (6)$$
$$\mathbf{q}_{\mathfrak{a}} \approx \mathbf{0}.$$

With this assumption in hand, we need to develop only one constitutive equation for the total discharge $\mathbf{q}(\mathbf{x},t)$. Furthermore, the set of PDEs is reduced to an evolution equation for the porosity and a storage equation for the mixture, cf. equation (4).[1]

$$\partial_t(\phi) = \hat{n}^{\mathfrak{a}},$$
$$\partial_t(\phi\rho^{\mathfrak{g}R}) + \operatorname{div}(\rho^{\mathfrak{g}R}\mathbf{q}) = 0. \quad (7)$$

Constitutive Equations

Next, we formulate the remaining constitutive equations for the (barotropic) gas phase, i.e. $p(\rho^{\mathfrak{g}R})$ and the filter velocity $\mathbf{q}$. Taking the compressibility of the pore gas into account, the effective density of the pore gas depends on the pore pressure

$$p \propto \left[\rho^{\mathfrak{g}R}\right]^{1/\gamma}. \quad (8)$$

Further on, we assume that the pore gas is driven by the polytropic constitutive law for *adiabatic* decompression, cf. Sommerfeld [32, § 7]. Thus, it is assumed that the system is isolated as far as energy is concerned. Now, equation (8)

---

[1] It has to be stated, that no sink and source terms appear in the mixture's balance equation. This is in disagreement with the work of Paterson et al. [31, 17], Valliappan and Wohua [18] and others. In the work of the previously mentioned authors, a source term in the mixture's mass balance (7)$_2$ is introduced to take the desorption of gas into the micro-pores of coal into account. This double-porosity effect of coal could also be modeled in the context of the classical mixture theory by an additional mass exchange term between the coal $\varphi^{\mathfrak{s}}$ and the gas phase $\varphi^{\mathfrak{g}}$. In the present approach, this phenomenon is neglected with respect to simplicity.

can be rewritten with respect to an equilibrium state

$$\frac{\rho^{gR}}{\rho_0^{gR}} = \left[\frac{p}{p_0}\right]^{1/\gamma}, \qquad (9)$$

with the equilibrium values p_0 and ρ_0^{gR}. The quantities p_0 and ρ_0^{gR} can be related e.g. to the (given) athmospheric pressure and the corresponding effective gas density, respectively. The polytropic exponent $\gamma = c_p/c_v$ describes the ratio of specific heat of the pore gas under constant pressure and volume.[2] Alternatively, we investigate the *isothermal* case. Thus, the polytropic exponent is reduced to the identity $\gamma = 1$, and Boyle-Mariotte's law is obtained

$$\rho^{gR} = \frac{p + p_0}{\bar{R}^g \Theta} = \frac{p}{\bar{R}^g \Theta} + \rho_0^{gR}, \qquad (11)$$

where $\bar{R}^g$ is the specific gas constant and Θ is the absolute temperature.[3] The direct momentum exchange [22] between the viscous pore gas and the rigid solid skeleton is obtained from the entropy principle in the form

$$\hat{\mathbf{p}}^g = \hat{\mathbf{p}}_{eq}^g + \hat{\mathbf{p}}_{neq}^g = p\,\mathrm{grad}\,\phi - \frac{\phi\mu^{gR}}{k^s}\mathbf{q} - \frac{\phi\rho^{gR}B}{k^s}|\mathbf{q}|\mathbf{q}. \qquad (12)$$

We split the exchange of momentum into the equilibrium part $\hat{\mathbf{p}}_{eq}^g = p\,\mathrm{grad}\,\phi$ and a non-linear non-equilibrium part $\hat{\mathbf{p}}_{neq}^g$ depending on the total discharge $\mathbf{q}$. If non-linear cross effects are taken into account, it is worth to mention, that so-called nozzling terms $\hat{\mathbf{p}}_{neq}^g(\mathrm{grad}\,\phi)$, similar to the terms proposed in [26], can be included additionally. This issue will be investigated in future in more detail and compared to experimental observations. Combining equation (12) with the partial balance of momentum of the gas phase gives Forchheimers extension of Darcy's law

$$\rho^{gR}\left[\frac{\mathbf{q}}{\phi}\right]_g' + \mathrm{grad}\,p = -\frac{\mu^{gR}}{k^s}\mathbf{q} - \frac{\rho^{gR}B}{k^s}|\mathbf{q}|\mathbf{q}, \qquad (13)$$

with the inertia term

$$\rho^{gR}\left[\frac{\mathbf{q}}{\phi}\right]_g' = \rho^{gR}\left[\frac{1}{\phi}\partial_t(\mathbf{q}) - \frac{\mathbf{q}}{\phi}\partial_t(\phi) + \frac{1}{\phi^2}\mathrm{grad}\,\mathbf{q}\cdot\mathbf{q} - \left(\frac{\mathbf{q}}{\phi^3}\otimes\mathrm{grad}\,\phi\right)\cdot\mathbf{q}\right] \qquad (14)$$

and the form drag coefficient $B\,[\,\mathrm{m}\,]$. Furthermore, a constitutive equation for the erosion process itself, i.e. for the volume exchange $\hat{n}^a$ [1/s] or the density exchange $\hat{\rho}^a = \hat{n}^a\rho^{aR}$, cf. equation (4), is described in generic form as

$$\hat{n}^a = \xi(c,\phi)|\mathbf{q}|, \qquad \text{or} \qquad \hat{\rho}^a = \xi(c,\phi)\rho^{sR}|\mathbf{q}|. \qquad (15)$$

According to theoretical results previously obtained by a classical evaluation process of the entropy principle [27], the volume production $\hat{n}^a$ is assumed to be proportional to the intensity of the total fluid discharge $\mathbf{q}$, i.e. proportional to $|\mathbf{q}|$. Originally, this constitutive equation was proposed by H.A. Einstein [33, 34, 35] and later on it was applied in

---

[2] Following [32], the polytropic exponent γ is calculated, e.g. for bi-atomic molecules (O_2, with $n \equiv 5$) as

$$\frac{c_p}{c_v} = 1 + \frac{2}{n} = 1.4, \qquad (10)$$

while $n = 5$ is the number of degrees of freedom for the $n = 3$ independent translations and $n = 2$ rotations, respectively.

[3] The partial time derivative of the effective pore gas density is given as

$$\partial_t(\rho^{gR}) = \frac{\rho^{gR}}{\gamma p_0}\left[\frac{p}{p_0}\right]^{1/\gamma - 1}\partial_t(p),$$

while the material time derivative of the pore pressure is prescribed in the case of small concentrations $c \ll 1$ as

$$(p)_g' = \partial_t(p) + (1-c)\phi\,\mathrm{grad}\,p\cdot\mathbf{q}_g \approx \partial_t(p) + \frac{1}{\phi}\mathrm{grad}\,p\cdot\mathbf{q}.$$

filtration kinetics by Sakthivadivel and Irmay [34]. Thus, we may call it the Einstein-Sakthivadivel law. We want to remark that the Einstein-Sakthivadivel law is thermodynamically-consistent with respect to the entropy principle, cf. the framework for incompressible erosive mixtures in the work of Steeb and Diebels [27].

To be more specific, it is assumed that $\xi(c,\phi)$ is a non-linear function with respect to the homogenized behavior of the microstructure of the overall mixture, i.e. the constitutive equations depends on the microscopic field quantity ϕ

$$\xi = \frac{1}{H}\left[\exp(1 - \frac{\phi}{\phi_{cr}}) - 1\right]. \tag{16}$$

Equation (16) reflects the physically-feasible assumption, that the erosion process is more active in intact regions with moderate porosities $\phi \approx \phi_0$ as it is in highly eroded regimes with $\phi \mapsto \phi_{cr}$. The saturation-type behavior of the phase-transition process is governed by the so-called critical porosity ϕ_{cr}, with $\phi_0 \leq \phi_{cr} \leq 1.0$. The empirical erosion modulus $\Lambda = 1/H$ [1/m] is chosen as

$$H = \frac{1}{\Lambda}, \tag{17}$$

which is in agreement with the original work of Einstein [33] and Sakthivadivel and Irmay [34]. It is obvious that the erosion modulus has to be calibrated experimentally, e.g. in cavity tests. Nevertheless, it has to be remarked that such results are still not available for gas-flow formations.

Without going into further technical details, we want to mention that various modifications of the Einstein-Sakthivadivel law can be motivated with respect to first experimental results with incompressible constituents. This especially concerns the function ξ, which could be an even more sophisticated function, e.g. a function depending on the whole set of thermodynamical process variables.

In erosion processes, the microstructure of the porous medium (i.e. porosity) is changed with respect to the solid-liquid phase transformation. Thus, the characteristic size of the micro-pores (e.g. the diameter of the cavities) increases within the process, as solid material from the side wall of the pores is transported away. From the macroscopic point of view, this is felt as a change in the (effective) intrinsic permeability k^s [m^2]. As the remaining material parameter of this microstructure phenomenom in a continuum-mechanical setting is just the scalar-valued porosity, it is obvious that the permeability is modeled as a (non-linear) function given by

$$k^s = k_0^s f(\phi). \tag{18}$$

In the present model, we propose a one-parameter power law ansatz with respect to the intrinsic permeability k^s, cf. Ehlers and Eipper [36]

$$k^s = k_0^s \left[\frac{\phi}{\phi_0}\right]^m. \tag{19}$$

Note that k_0^s is the initial undisturbed intrinsic permeability which can be determined in classical non-erosive permeability tests. The second material parameter, i.e. the power m, has to be determined in erosion experiments measuring the effective permeability in time. Applying pressure boundary conditions or an hydraulic gradient in the experimental setup, the discharge of the total amount fluid in unit time Q [m/s], could be measured, e.g. by installing a flowmeter.

Isothermal Gas Flow Through Porous Media

Before we start analyzing the phase transition phenomena, we would like to discuss the non-linear diffusion phenomenon of compressible gas flow through a rigid porous medium. In this restricted situation, the porosity $\phi = \phi_0$ is constant and equation (7)$_1$ can be removed from the set of governing equations.
Furthermore, we restrict ourselves to *isothermal* conditions, thus the polytropic exponent is $\gamma = 1$.

Boundary Value Problem

The boundary value problem for isothermal gas flow through a rigid porous medium ($\phi = \phi_0$ const) is written as

$$\partial_t(\rho^{gR}) + \frac{1}{\phi}\text{div}(\rho^{gR}\mathbf{q}) = 0, \qquad \forall\, \mathbf{x} \text{ in } \mathscr{B} \times t, \tag{20}$$

with the balance of momentum of the gas phase and the constitutive equations for the momentum production, cf. equation (13), and a rate form of the polytropic equation of the pore pressure p, cf. equation (9)

$$\frac{1}{\phi}\rho^{gR}\left[\partial_t(\mathbf{q}) + \frac{1}{\phi}\operatorname{grad}\mathbf{q}\cdot\mathbf{q}\right] + \operatorname{grad} p = -\frac{\mu^{gR}}{k^s}\mathbf{q} - \frac{\rho^{gR} B}{k^s}|\mathbf{q}|\mathbf{q},$$

$$\partial_t\left[\frac{p}{\rho^{gR}}\right] + \frac{1}{\phi}\operatorname{grad}\left[\frac{p}{\rho^{gR}}\right]\cdot\mathbf{q} = 0. \tag{21}$$

The boundary conditions are given as

$$p(\mathbf{x},t) = \bar{p}, \quad \forall\,\mathbf{x}\text{ on }\Gamma_D \times t \quad \text{and} \quad \mathbf{q}(\mathbf{x},t)\cdot\mathbf{n} = \bar{q}, \quad \forall\,\mathbf{x}\text{ on }\Gamma_N \times t, \tag{22}$$

and the initial conditions are given as

$$p(\mathbf{x},t_0) = p_0 \quad \forall\,\mathbf{x}\text{ at }\mathscr{B} \times t_0. \tag{23}$$

Weak Form

Furthermore, the weak form of the boundary value problem is obtained in the following form

$$\int_{\mathscr{B}}[\phi\,\partial_t(\rho^{gR})]\delta p\,dv - \int_{\mathscr{B}}[\rho^{gR}\mathbf{q}]\cdot\operatorname{grad}\delta p\,dv = -\int_{\Gamma_N}\rho^{gR}\bar{q}\delta p\,da \quad \forall\,\delta p \in \mathscr{V}, \tag{24}$$

and the Galerkin finite element problem reads: *Find $p \in \mathscr{U}_h$ such that*

$$\int_{\mathscr{B}}[\phi\,\partial_t(\rho^{gR}_h)]\delta p_h\,dv - \int_{\mathscr{B}}[\rho^{gR}_h\mathbf{q}_h]\cdot\operatorname{grad}\delta p_h\,dv = -\int_{\Gamma_N}\rho^{gR}_h\bar{q}_h\delta p_h\,da \quad \forall\,\delta p_h \in \mathscr{V}_h. \tag{25}$$

If we assume a linear constitutive equation for the non-equilibrium part of the momentum production (Darcy regime), quasi-static conditions (zero inertia forces) and the special case of isothermal conditions (polytropic exponent $\gamma = 1$), we are able to insert the constitutive equations for the effective density of the gas $\rho^{gR} = (\rho^{gR}_0/p_0)p$ and Darcy's law $\mathbf{q} = -(k^s/\mu^{gR})\operatorname{grad} p$, (cf. (13)) into (7)$_2$ and we obtain[4]

$$\partial_t(p) = \frac{k^s}{\phi\,\mu^{gR}}(\operatorname{grad} p\cdot\operatorname{grad} p + \operatorname{div}\operatorname{grad} p) \tag{26}$$

or

$$\partial_t(p) = \frac{k^s}{2\phi\,\mu^{gR}}\Delta p^2. \tag{27}$$

---

[4] Note that we can simplify

$$\operatorname{div}\operatorname{grad}(p^2) =: \Delta p^2 = \operatorname{div}(2p\operatorname{grad} p),$$
$$= 2(\operatorname{grad} p\cdot\operatorname{grad} p + p\operatorname{div}\operatorname{grad} p),$$
$$= 2(\operatorname{grad} p\cdot\operatorname{grad} p + p\Delta p),$$

leading to

$$\frac{1}{2}\Delta p^2 = \operatorname{grad} p\cdot\operatorname{grad} p + p\Delta p.$$

Boundary Value Problem

Furthermore, the governing equations of the IBVP can be written in a dimensionless form introducing the following quantities:

$$\mathbf{q} = \frac{k^{\mathfrak{s}} \tilde{p}}{\mu^{\mathfrak{g}R} \tilde{l}} \mathbf{v}^\star, \qquad p = \tilde{p}\, p^\star, \qquad \rho^{\mathfrak{g}R} = \tilde{\rho}^{\mathfrak{g}R} \rho^\star, \tag{28}$$

$$\operatorname{grad}[\bullet] = \frac{1}{\tilde{l}} \operatorname{grad}^\star[\bullet], \qquad \operatorname{div}[\bullet] = \frac{1}{\tilde{l}} \operatorname{div}^\star[\bullet], \qquad t = \frac{\phi \tilde{l}^2 \mu^{\mathfrak{g}R}}{k^{\mathfrak{s}} \tilde{p}} t^\star.$$

Note that $\tilde{l}$, $\tilde{p}$ and $\tilde{\rho}^{\mathfrak{g}R}$ are typical length, pressure and density scales. Inserting the set of quantities (28) into the governing equations (20) and (21), we finally obtain the set of dimensionless equations describing the initial boundary value problem in the domain $\mathscr{B}^\star \times t^\star$

$$\begin{aligned}
\partial_{t^\star}(\rho^\star) + \operatorname{div}^\star(\rho^\star \mathbf{v}^\star) &= 0, \\
\mathscr{C}_1 \rho^\star \left[\partial_{t^\star}(\mathbf{v}^\star) + \operatorname{grad}^\star \mathbf{v}^\star \cdot \mathbf{v}^\star \right] + \operatorname{grad}^\star p^\star &= -\mathbf{v}^\star - \mathscr{C}_2 \rho^\star |\mathbf{v}^\star| \mathbf{v}^\star, \\
\partial_{t^\star} \left[\frac{p^\star}{\rho^\star} \right] + \operatorname{grad}^\star \left[\frac{p^\star}{\rho^\star} \right] \cdot \mathbf{v}^\star &= 0,
\end{aligned} \tag{29}$$

with

$$\mathscr{C}_1 = \frac{(k^{\mathfrak{s}})^2 \tilde{p} \tilde{\rho}^{\mathfrak{g}R}}{\phi^2 \tilde{l}^2 (\mu^{\mathfrak{g}R})^2} \qquad \text{and} \qquad \mathscr{C}_2 = \frac{k^{\mathfrak{s}} \tilde{\rho}^{\mathfrak{g}R} B \tilde{p}}{(\mu^{\mathfrak{g}R})^2 \tilde{l}}. \tag{30}$$

The boundary conditions are given as

$$p^\star(\mathbf{s},t^\star) = \bar{p}^\star, \quad \forall\, \mathbf{s} \text{ on } \Gamma_D^\star \times t^\star \quad \text{and} \quad \mathbf{v}^\star(\mathbf{s},t^\star) \cdot \mathbf{n} = \bar{v}^\star, \quad \forall\, \mathbf{s} \text{ on } \Gamma_N^\star \times t^\star, \tag{31}$$

and the initial conditions at $t_0^\star$

$$p^\star(\mathbf{s}, t_0^\star) = p_0^\star \qquad \forall\, \mathbf{s} \text{ at } \mathscr{B}^\star \times t_0^\star. \tag{32}$$

The dimensionless position vector is introduced as $\mathbf{s}$. The dimensionless numbers $\mathscr{C}_1$ and $\mathscr{C}_2$ represent the importance of inertia forces or the non-linear Forchheimer drag forces, cf. [37]. The importance of the Forchheimer contribution to the drag forces can be analyzed in more detail. Neglecting inertia forces, equation (29)$_2$ can be written as

$$-\operatorname{grad}^\star p^\star = (1 + \mathscr{C}_2 \rho^\star |\mathbf{v}^\star|)\, \mathbf{v}^\star. \tag{33}$$

Introducing the unit vectors

$$\mathbf{e}_{[\mathbf{v}^\star]} = \frac{\mathbf{v}^\star}{|\mathbf{v}^\star|} \qquad \text{and} \qquad \mathbf{e}_{[\operatorname{grad}^\star p^\star]} = \frac{\operatorname{grad}^\star p^\star}{|\operatorname{grad}^\star p^\star|}, \tag{34}$$

and incorporating the directions of the pressure gradient and the velocity field

$$-\mathbf{e}_{[\operatorname{grad}^\star p^\star]} = \mathbf{e}_{[\mathbf{v}^\star]}, \tag{35}$$

we are able to rewrite (33) as

$$\left\{ |\operatorname{grad}^\star p^\star| \right\} \mathbf{e}_{[\mathbf{v}^\star]} = \left\{ (1 + \mathscr{C}_2 \rho^\star |\mathbf{v}^\star|) |\mathbf{v}^\star| \right\} \mathbf{e}_{[\mathbf{v}^\star]}. \tag{36}$$

Now, the terms in the brackets $\{\circ\}$ can be compared and we obtain a quadratic equation in the norm of the velocity $|\mathbf{v}^\star|$ which can be solved

$$\mathscr{C}_2 \rho^\star |\mathbf{v}^\star|_{1,2} = -\frac{1}{2} \pm \frac{1}{2} \left[1 + 4\mathscr{C}_2 \rho^\star |\operatorname{grad}^\star p^\star| \right]^{1/2}. \tag{37}$$

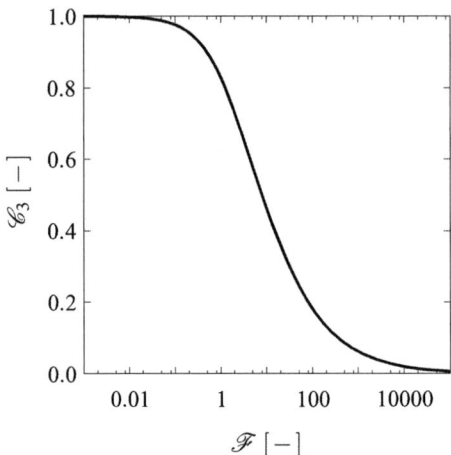

FIGURE 1. Influence of Forchheimer term on velocity. The dimensionless number $\mathscr{F}$ is introduced as the so-called Forchheimer number $\mathscr{F} := 4\mathscr{C}_2 \rho^\star |\operatorname{grad}^\star p^\star|$.

Note that only the positive root is of physical interest, as it satisfies the conditions $|\mathbf{v}^\star| \equiv \mathbf{0}$ for $|\operatorname{grad}^\star p^\star| \equiv \mathbf{0}$. Finally, equation (37) can be inserted in equation (33) leading to

$$\mathbf{v}^\star = -2\left[1+\sqrt{1+4\mathscr{C}_2 \rho^\star |\operatorname{grad}^\star p^\star|}\right]^{-1} \operatorname{grad}^\star p^\star =: -\mathscr{C}_3 \operatorname{grad}^\star p^\star. \tag{38}$$

The influence of the quadratic Forchheimer contribution is shown in Figure (1). It can be observed that the Forchheimer term significantly *decreases* the velocity of the pore fluid for values of $\mathscr{F} \gg 1$. Note that the dimensionless Forchheimer number can also be expressed in physical quantities, cf. the comments in [38].

Weak Form

To discuss the basic physical behavior of the non-linear problem, we start with the most simple investigation corresponding to a quasi-static formulation of the initial boundary value problem given by equations (29) and (31), (32). Inertia forces are neglected in (29)$_2$. Furthermore, we assume a linear behavior of Darcy-type. Thus, equation (29)$_2$ can be simplified to

$$\operatorname{grad}^\star p^\star = -\mathbf{v}^\star. \tag{39}$$

The weak form of the boundary value problem is given by

$$\int_{\mathscr{B}^\star} [\partial_{t^\star}(\rho^\star)]\delta p \, dv - \int_{\mathscr{B}^\star} [\rho^\star \mathbf{v}^\star] \cdot \operatorname{grad}\delta p \, dv = -\int_{\Gamma_N^\star} \rho^\star \bar{v}^\star \delta p \, da \quad \forall\, \delta p \in \mathscr{V}. \tag{40}$$

and the Galerkin finite element problem reads: *Find $p^\star \in \mathscr{U}_h$ such that*

$$\int_{\mathscr{B}^\star} [\partial_{t^\star}(\rho^\star)_h]\delta p_h \, dv - \int_{\mathscr{B}^\star} [\rho_h^\star \mathbf{v}_h^\star] \cdot \operatorname{grad}\delta p_h \, dv = -\int_{\Gamma_N^\star} \rho_h^\star \bar{v}_h^\star \delta p_h \, da \quad \forall\, \delta p_h \in \mathscr{V}_h. \tag{41}$$

Assuming similar conditions as denoted in the previous section, we finally obtain one non-linear diffusion equation in the dimensionless pore pressure $p^\star$

$$\partial_{t^\star}(p^\star) = \frac{1}{2}\Delta^\star\left([p^\star]^2\right). \tag{42}$$

In equation (42) we have introduced the (dimensionless) Laplace operator $\Delta^\star$. In weak form it is given as

$$\int_{\mathscr{B}^\star} [\partial_{t^\star}(p^\star)]\delta p \, dv + \int_{\mathscr{B}^\star} [p^\star \operatorname{grad} p^\star] \cdot \operatorname{grad}\delta p \, dv = -\int_{\Gamma_N^\star} p^\star \bar{v}^\star \delta p \, da \quad \forall\, \delta p \in \mathscr{V}, \tag{43}$$

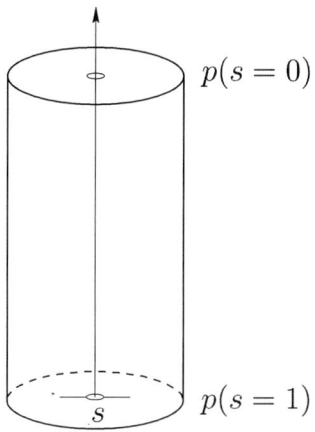

FIGURE 2. One-dimensional isothermal gas flow through porous media

and the Galerkin finite element problem reads: *Find $p^\star \in \mathcal{U}_h$ such that*

$$\int_{\mathcal{B}^\star} [\partial_{t^\star}(p_h^\star)] \delta p_h \, dv + \int_{\mathcal{B}^\star} [p_h^\star \operatorname{grad} p_h^\star] \cdot \operatorname{grad} \delta p_h \, dv = - \int_{\Gamma_N^\star} p_h^\star \bar{v}_h^\star \delta p_h \, da \qquad \forall \, \delta p_h \in \mathcal{V}_h. \tag{44}$$

One-Dimensional Flow - Analytical solutions

Considering an one-dimensional physical setup of *steady gas flow* through a porous column, de Ville [37] derived an analytical solution for the pressure and the velocity distribution. The gas flow is driven by prescribed boundary conditions on the bottom $p(s=0) = 1$ and the top of the specimen $p(s=0) = 0$. The cross section of the column is described as $A = 1$. The geometrical and boundary conditions, respectively, are sketched in Figure 2. The one-dimensional problem can be derived from equations (29), and reads

$$\begin{aligned}
\mathcal{C}_1 \rho^\star v^\star \frac{dv^\star}{ds} + \frac{dp^\star}{ds} &= -v - \mathcal{C}_2 \rho^\star (v^\star)^2, \\
\frac{d(\rho^\star v^\star)}{ds} &= 0, \\
\frac{d(p^\star/\rho^\star)}{ds} &= 0.
\end{aligned} \tag{45}$$

For $\mathcal{C}_1 = 0$ and $\mathcal{C}_2 = 0$, the steady-state solution of the dimensionless pressure field $p^\star$ and the dimensionless velocity field $v^\star$ reads, cf. [37],

$$\begin{aligned}
v^\star(s) &= \frac{1}{2(1-s)^{1/2}}, \\
p^\star(s) &= (1-s)^{1/2},
\end{aligned} \tag{46}$$

or for the adiabatic case

$$\begin{aligned}
v^\star(s) &= \frac{\gamma}{1+\gamma} \left[\frac{1 - \alpha^{\frac{1+\gamma}{\gamma}}}{(1 - (1-\alpha^{\frac{\gamma+1}{\gamma}})s)^{\frac{1}{1+\gamma}}} \right], \\
p^\star(s) &= \left[1 - (1-\alpha^{\frac{1+\gamma}{\gamma}})s \right]^{\frac{\gamma}{\gamma+1}}.
\end{aligned} \tag{47}$$

The spatially-distributed solutions are sketched for variations of the boundary term α, cf. Figures (3), and for variations in the polytropic exponent γ, cf. Figure (4). Especially the non-linear character of the pore pressure and velocity distribution in space, cf. Figure (3), is remarkable and could be used as a benchmark in first numerical implementations in order to control the accuracy of finite element solutions. Notice that a dominant Forchheimer term, cf. equation $(45)_1$, is responsible for a solution with sharp gradients near the drained boundary $p^\star(1) = 0$, cf. Figure (6). Therefore we use problem-adapted finite element meshes, cf. Figure (6), right.

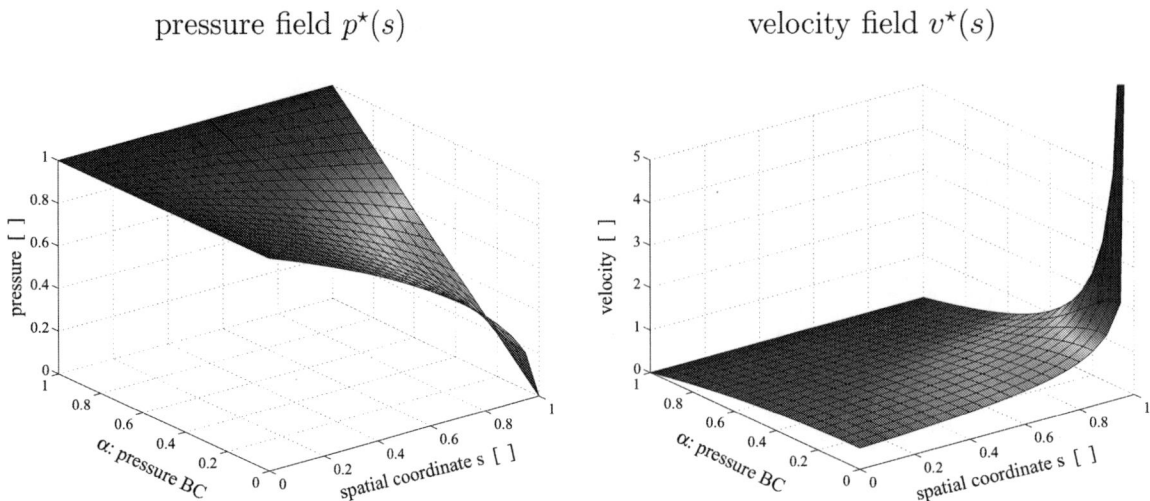

FIGURE 3. Analytical solution of dimensionless pressure field $p^\star(s)$ and velocity field $v^\star(s)$ for isothermal conditions, i.e. $\gamma = 0$.

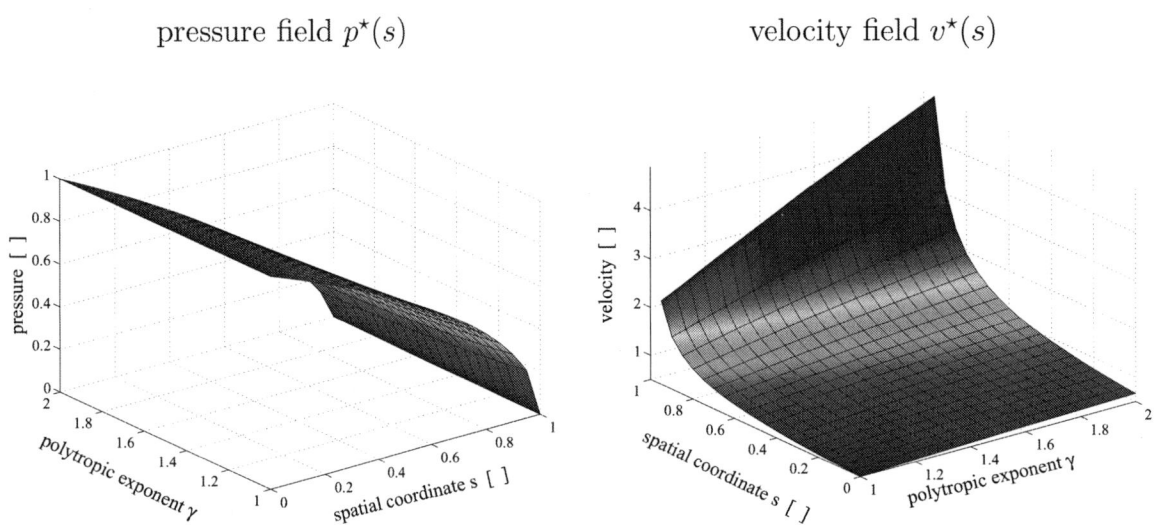

FIGURE 4. Analytical solution of dimensionless pressure field $p^\star(s)$ and velocity field $v^\star(s)$ for adiabatic conditions, i.e. $\gamma = [1:2]$. The boundary condition $p(s=1) \equiv \alpha$ is set to $\alpha = 0$.

One-Dimensional Flow - Numerical Solutions

Next, we study the behavior in the time domain. Thus, no analytical solutions can be derived. Therefore, we calculate the results by non-linear and transient Galerkin finite element techniques as described before. The governing finite element problem of the dimensionless equation is given in equation (20). For the linear Darcy-regime, we obtain the dimensionless pressure and velocity distribution $p^\star(s,t^\star)$ and $v^\star(s,t^\star)$ within the time domain $t^\star = [0:1]$. The results are shown in Figure (5). Again, the sharp gradients in the solution of the dimensionless velocity and pore pressure gradient for $t^\star \to 0$ is remarked and could be used in numerical benchmark tests.

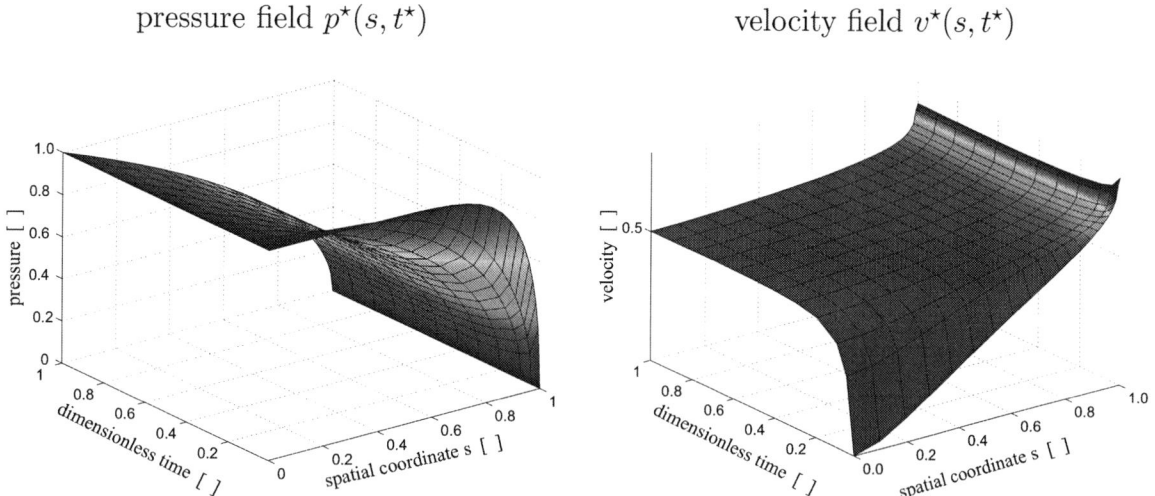

FIGURE 5. Numerical solution of dimensionless pressure field $p^\star(s)$ and velocity field $v^\star(s)$ for isothermal conditions, i.e. $\gamma = 1$. The boundary condition $p(s=1) \equiv \alpha$ is set to $\alpha = 0$. The results at $t^\star = 1$ correspond to the *steady state* solution. Linear Q1-elements are applied in combination with a Backward-Euler-scheme.

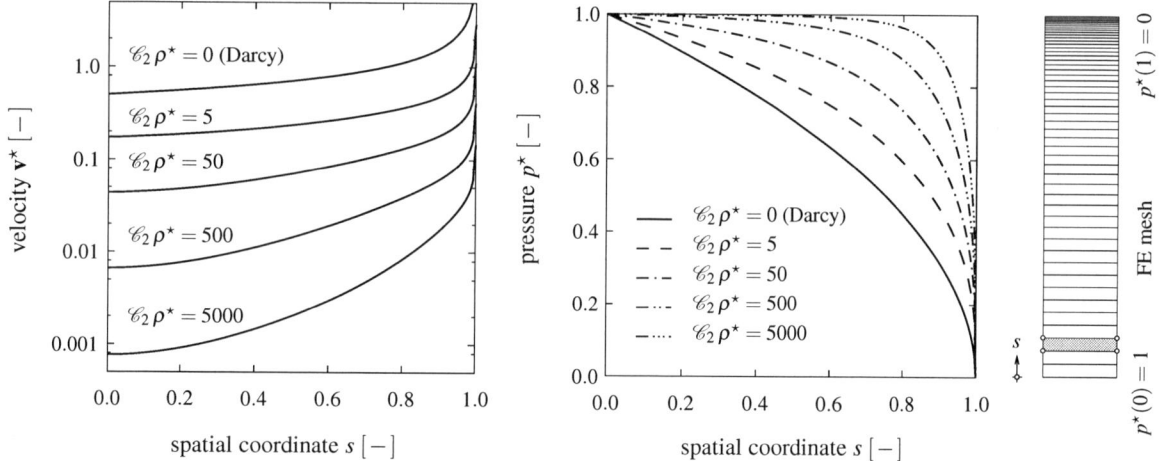

FIGURE 6. Numerical solution of the one-dimensional column experiment. Steady-state distribution of velocity $\mathbf{v}^\star(s)$ and pressure field $p^\star(s)$, respectively.

Wave Propagation of Gas Through Porous Media

Let us consider the set of dimensionless equations (29). These equations can be linearized around an equilibrium state. Finally, this is leading to

$$\begin{aligned}
\mathscr{C}_1 \rho_0^\star \partial_{t^\star}(\mathbf{v}^\star) + \mathrm{grad}^\star p^\star + \mathbf{v}^\star &= \mathbf{0}, \\
\partial_{t^\star}(\rho^\star) + \mathrm{div}(\rho_0^\star \mathbf{v}^\star) &= 0, \\
\partial_{t^\star}(p^\star) - \frac{p_0^\star}{\rho_0^\star}\partial_{t^\star}(\rho^\star) &= 0.
\end{aligned} \qquad (48)$$

The rate form of the constitutive equation of the pore gas, (cf. equation (48)$_3$), can be inserted into the mass balance (48)$_2$. Thus, we get

$$\text{div}^\star(\rho_0^\star \mathbf{v}^\star) = -\frac{\rho_0^\star}{p_0^\star}\partial_{t^\star}(p^\star),$$
$$\text{div}^\star(\rho_0^\star [\partial_{t^\star}\mathbf{v}^\star]) = -\frac{\rho_0^\star}{p_0^\star}\partial_{[tt]^\star}(p^\star). \quad (49)$$

Note that we have defined the abbreviation $\partial_{[tt]^\star}(p^\star) := \partial^2(p^\star)/\partial[t^\star]^2$. Furthermore, we build the dimensionless Laplace operator $\Delta^\star(p^\star := \text{div}^\star \text{grad}^\star(p^\star))$. Building the divergency of the balance of momentum of the pore gas, (cf. (48)$_1$), and inserting equations (49) into it, will give us the dimensionless wave equation of compressible gas flow though a rigid porous skeleton

$$\partial_{[tt]^\star}(p^\star) + \frac{1}{\rho_0^\star \mathscr{C}_1}\partial_{t^\star}(p^\star) - \frac{p_0^\star}{\rho_0^\star \mathscr{C}_1}\Delta^\star(p^\star) = 0. \quad (50)$$

Next, we choose the standard harmonic ansatz for the pressure field $p^\star$

$$p^\star = P^\star \exp(i[kx - \omega t]). \quad (51)$$

Inserting (51) into the wave equation (50) leads to a quadratic equation in the complex wave number k

$$\left[-\omega^2 + \frac{p_0^\star}{\rho_0^\star \mathscr{C}_1}k^2 - i\frac{1}{\rho_0^\star \mathscr{C}_1}\omega\right]P^\star = 0, \quad (52)$$

or

$$\mathbf{A} \cdot \mathbf{U} = \mathbf{0}. \quad (53)$$

with $\mathbf{U} = [P^\star]$ which is fullfilled for $\mathbf{A} \equiv \mathbf{0}$. Finally, we obtain the quadratic expression of the wave number [5]

$$k^2 = \left[\frac{1}{v_p^2}\right]\omega^2 + i\left[\frac{1}{p_o^\star}\right]\omega. \quad (54)$$

We have introduced the characteristic velocity $v_p^2 := p_0^\star/(\rho_0^\star \mathscr{C}_1)$. Finally, we could calculate the low and high frequency limits of the speed of propagation and the attenuation of monochromatic waves

$$\lim_{\omega \to 0}\left(\frac{\omega}{\text{Re}(k)}\right)^2 = 0, \qquad \lim_{\omega \to \infty}\left(\frac{\omega}{\text{Re}(k)}\right)^2 = v_p^2,$$
$$\lim_{\omega \to 0}\text{Im}(k) = 0, \qquad \lim_{\omega \to \infty}\text{Im}(k) = \frac{v_p}{2p_0^\star}. \quad (55)$$

The quadratic solution k^2 of equation (54) can be compared to the solutions of the Biot equations [39] describing wave propagation phenomena in compressible porous media. If we reduce Biot's equations to the case of a rigid skeleton, the resulting phase velocities and attenuation characteristics of the compressional wave, cf. equations (55), can be compared. Thus, the compressional wave (55) is identical to the so-called 2nd P-wave, Biot-wave or slow wave, cf. dispersion relations in Figures (7) and (8).

---

[5] Taking into account the physical dimensions of the problem, the wave equation of gas flow through porous medium can be written as

$$\partial_{tt}(p) + \frac{\mu^{\mathfrak{g}R}\phi}{k^s \rho_0^{\mathfrak{g}R}}\partial_t(p) - \frac{p_0}{\rho_0^{\mathfrak{g}R}}\Delta(p) = 0,$$

leading to a quadratic equation in the wave number

$$k^2 = \left[\frac{\rho^{\mathfrak{g}R}}{p_0}\right]\omega^2 + i\left[\frac{\mu^{\mathfrak{g}R}\phi}{p_0 k^s}\right]\omega.$$

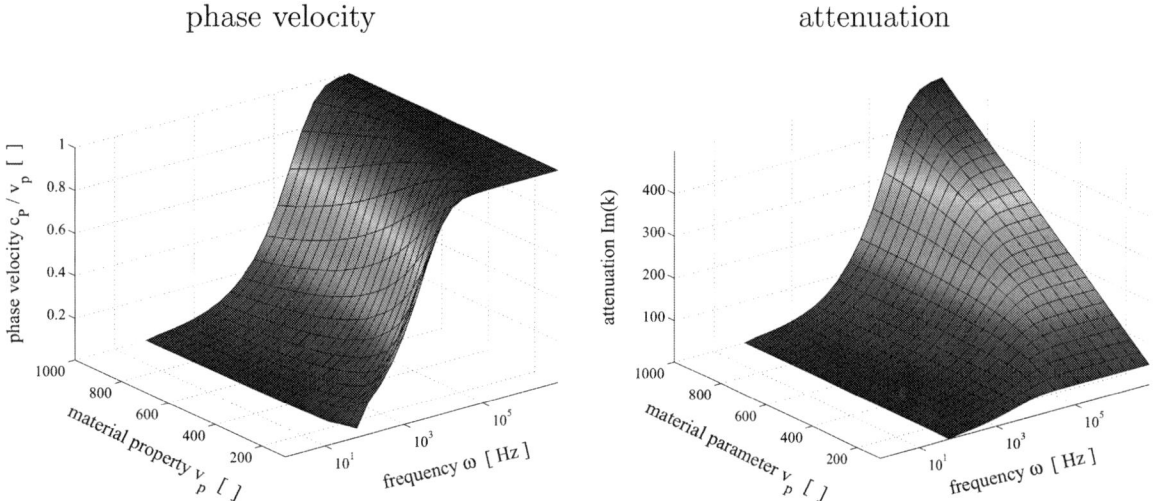

FIGURE 7. Speed of propagation and attenuation of monochromatic waves. The first dimensionless parameter describing propagating waves v_p is variied in the range $v_p = [100 : 1000]$, while the second quantity $p_0^\star = 1$ is held constant.

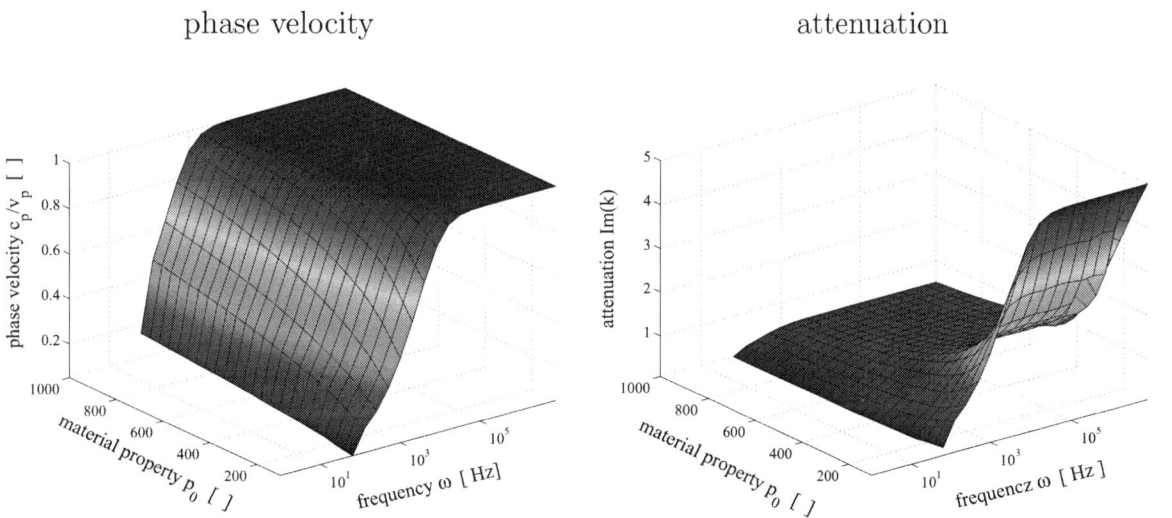

FIGURE 8. Speed of propagation and attenuation of monochromatic waves. The second dimensionless parameter describing attenuation of waves $p_0^\star$ is varied in the range $p_0^\star = [100 : 1000]$, while the second quantity $v_p = 1000$ holds constant.

Outbursts in Coal: Numerical Modeling

Let us now come back to the complete non-linear erosion model. As outbursts are characterized through a phase transition process from a solid to a gaseous state, as denoted in the introduction of the present section, the governing equations have to take an evolving porosity and an evolving permeability into account. According to the previously made assumptions, cf. comments in previous sections and equations (7), the governing mass balance equation of the mixture in dimensionless form describing outbursts in coal is given by

$$\partial_{t^\star}(\phi^\star \rho^\star) + \mathrm{div}^\star(\rho^\star \mathbf{v}^\star) = 0, \tag{56}$$

with the porosity related to a typical porosity scale which is described by the initial porosity ϕ_0 and the intrinsic permeability related to the permeability of the initial state k_0^s

$$\begin{aligned} \phi &= \phi_0 \phi^\star, & \xi &= \frac{\xi^\star}{\tilde{l}}, & k^s &= k_0^s \left[\frac{\phi}{\phi_0}\right]^m = k_0^s [\phi^\star]^m, \\ \mathbf{q} &= \frac{k_0^s \tilde{p}}{\mu^{gR} \tilde{l}} \mathbf{v}^\star, & p &= \tilde{p} p^\star, & \rho^{gR} &= \tilde{\rho}^{gR} \rho^\star, \\ \text{grad}[\bullet] &= \frac{1}{\tilde{l}} \text{grad}^\star[\bullet], & \text{div}[\bullet] &= \frac{1}{\tilde{l}} \text{div}^\star[\bullet], & t &= \frac{\phi_0 \tilde{l}^2 \mu^{gR}}{k_0^s \tilde{p}} t^\star. \end{aligned} \quad (57)$$

The set of constitutive equations, (cf. the mass transfer equations (15) and (16), the quasi-static form of the balance of momentum of the pore gas (13) and (29) and the ideal gas law (9) and (29)$_3$), is given as

$$\begin{aligned} \partial_{t^\star}(\phi^\star) &= \xi^\star |\mathbf{v}^\star|, \\ f(\phi^\star) \, \text{grad}^\star p^\star = [\phi^\star]^m \, \text{grad}^\star p^\star &= -\mathbf{v}^\star - \mathscr{C}_2 \rho^\star |\mathbf{v}^\star| \mathbf{v}^\star, \\ \partial_{t^\star}\left[\frac{p^\star}{\rho^\star}\right] &= 0. \end{aligned} \quad (58)$$

Note that the evolution of the porosity $\phi^\star$ described by the porosity function $f(\phi^\star) = [\phi^\star]^m$ corresponds to the power-law ansatz of [36], cf. (19). Taking into account the constitutive equation of Einstein-Sakthivadivel-type [33], we obtain

$$\xi^\star(\phi^\star) = \Lambda^\star \left[\exp(\tilde{\phi}_{cr} - \phi^\star) - 1\right], \quad (59)$$

with $\Lambda^\star = \Lambda/\tilde{l}$. Again, we can reformulate (58)$_2$ and obtain a non-linear equation for the velocity $\mathbf{v}^\star$ depending on the pressure gradient and the porosity field, respectively

$$\mathbf{v}^\star = -2(\phi^\star)^m \left[1 + \sqrt{1 + 4[\phi^\star]^m \mathscr{C}_2 \rho^\star |\text{grad}^\star p^\star|}\right]^{-1} \text{grad}^\star p^\star =: -\mathscr{C}_4 \, \text{grad}^\star p^\star. \quad (60)$$

The intensity of the velocity field is given by

$$|\mathbf{v}^\star| = \left|\frac{1}{2\mathscr{C}_2 \rho^\star}\left[-1 + \sqrt{1 + 4[\phi^\star]^m \mathscr{C}_2 \rho^\star |\text{grad}^\star p^\star|}\right]\right|. \quad (61)$$

Equation (60) is interesting, as it contains two opposite effects decreasing or increasing the characteristic velocity $\mathbf{v}^\star$. As outlined in Figure (1), the quadratic Forchheimer contribution *decreases* the velocity as the intensity of the viscous drag force $\hat{\mathbf{p}}_{neq}^g$ is increasing with increasing form drag coefficient B or, equivalently, with increasing dimensionless Forchheimer number $\mathscr{F}$. Contrary to the Forchheimer effect, the solid-liquid phase transition process, i.e. the removing of mass from the solid skeleton, *increases* the velocity. Note that the phase transition process is captured in (60) by the term $(\phi^\star)^m$. Keeping the pressure gradient constant, these effects are outlined in Figure (9) from a quantitative point of view. In (9)$_1$, the Darcy regime ($\mathscr{F} = 0$) is shown. Note that the velocity of the pore gas is significantly increased during the erosion process (increasing $\phi^\star$). The exponent m captures this porosity-dependent permeability evolution of the intrinsic quantity $k^s(\phi^\star)$, cf. equation (57). In Figure (9)$_2$, we keep the exponent m constant but take into account the Forchheimer contribution. Comparing Figure (9)$_1$ and Figure (9)$_2$, we observe that the Forchheimer term slows down the increase of velocity.

The weak form of the boundary value problem is given by

$$\int_{\mathscr{B}^\star} \left[p^\star \xi^\star |\mathbf{v}^\star| + \phi^\star \partial_{t^\star}(p^\star)\right] \delta p \, dv - \int_{\mathscr{B}^\star} [p^\star \mathbf{v}^\star] \cdot \text{grad}\, \delta p \, dv = -\int_{\Gamma_N^\star} p^\star \bar{v}^\star \delta p \, da \quad \forall\, \delta p \in \mathscr{V}, \quad (62)$$

and the Galerkin finite element problem reads: *Find $p^\star \in \mathscr{U}_h$ such that*

$$\int_{\mathscr{B}^\star} \left[p_h^\star \xi^\star |\mathbf{v}_h^\star| + \phi^\star \partial_{t^\star}(p_h^\star)\right] \delta p_h \, dv - \int_{\mathscr{B}^\star} [p_h^\star \mathbf{v}_h^\star] \cdot \text{grad}\, \delta p_h \, dv = -\int_{\Gamma_N^\star} p_h^\star \bar{v}_h^\star \delta p_h \, da \quad \forall\, \delta p_h \in \mathscr{V}_h. \quad (63)$$

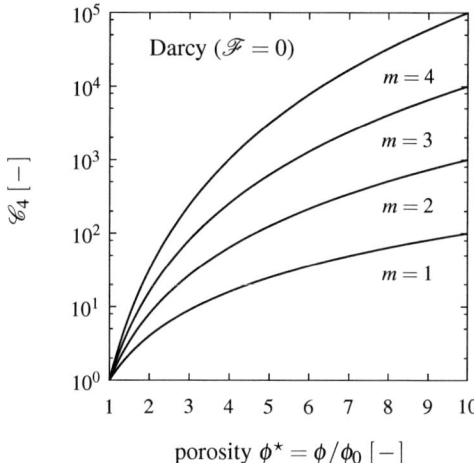
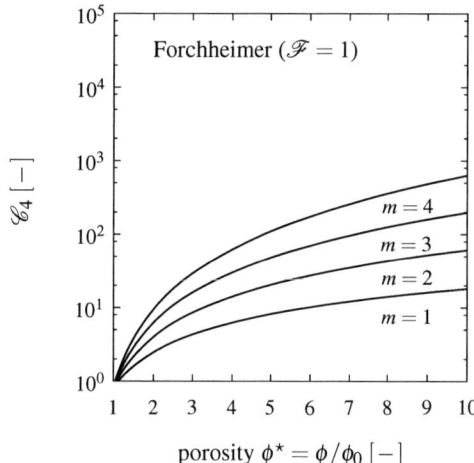

FIGURE 9. Influence of porosity evolution $\phi^\star$, exponent m and non-linear Forchheimer contribution $\mathscr{F}$ on the velocity $\mathbf{v}^\star$.

Note that the proposed finite element problem has to be solved in combination with an appropriate time-stepping scheme and a local Newton-Raphson scheme for the non-linear porosity-evolution equation $(58)_1$. The overall problem is highly non-linear. Therefore, it has to be linearized accurately and solved on the global level with a further Newton-Raphson scheme. With respect to the numerical stability, we propose an alternative coupled formulation in the primary variables $[p_h^\star, \phi_h^\star]^T$. With respect to *notational simplicity*, we do not take into account the second order Forchheimer contribution, i.e. $\mathscr{C}_2 \equiv 0$. The first part of the set of equations in weak form is, again, given by the weak form of the balance of mass of the mixture, cf. (62). Note that we are able to plug in the Darcy-type equation $(58)_2$, the evolution of the porosity $(58)_1$ with (59) and the ideal gas law $(58)_3$. The second equation is prescribed by the balance of volume of the solid skeleton, i.e. the evolution equation of the porosity $(58)_2$.

$$
\begin{aligned}
\text{(I.)} \quad & \int_{\mathscr{B}^\star} \left[\phi^\star \, \partial_{t^\star}(p^\star) \right] \delta p \, \mathrm{d}v \\
& + \int_{\mathscr{B}^\star} \left[f(\phi^\star) \, p^\star \, \mathrm{grad}^\star \, p^\star \right] \cdot \mathrm{grad}^\star \, \delta p^\star \, \mathrm{d}v \\
& + \int_{\mathscr{B}^\star} \left[f(\phi^\star) \, \xi^\star(\phi^\star) \, p^\star \, | \mathrm{grad}^\star \, p^\star | \right] \delta p \, \mathrm{d}v \quad = \quad -\int_{\Gamma_N^\star} p^\star \bar{v}^\star \, \delta p \, \mathrm{d}a, \\
\text{(II.)} \quad & \int_{\mathscr{B}^\star} \left[\partial_{t^\star}(\phi^\star) - \xi^\star(\phi^\star) \, f(\phi^\star) \, | \mathrm{grad}^\star(\phi^\star) | \right] \delta \phi \, \mathrm{d}v \quad = \quad 0.
\end{aligned}
\tag{64}
$$

If we will further discuss the behaviour of the set of equations (64), it is usefull to introduce a ratio α which expresses the character of the equation depending on the functions f and $\xi^\star$. Note that the relation α of the porosity-dependent material functions determines the character of a instationary non-linear convection-diffusion equation. Nevertheless, we observe that with *increasing* erosion ($\phi^\star \to \tilde{\phi}_{cr}$) the convective part is *decreasing*, cf. Figure (10)

$$
\alpha(\phi^\star) := \frac{f(\phi^\star)}{|\xi^\star(\phi^\star) f(\phi^\star)|} = \frac{1}{\xi^\star(\phi^\star)}.
\tag{65}
$$

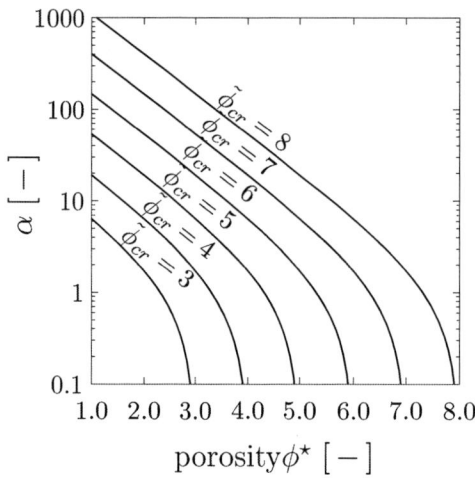

FIGURE 10. Influence of the relation $\alpha(\phi^\star)$ which describes the character of the non-linear PDE, cf. [40]. As α is decreasing during the erosion process, it is guaranteed that the quasi-static character of $(64)_1$ is elliptic.

One-dimensional CO_2 Flow Through a Homogeneous Coal Seam

In the first numerical example, we investigate a homogeneous coal seam saturated with a pore gas with finite length $\tilde{l} = 1$, cf. Figure (11). The coal seam is embedded into impermeable and rigid rock layers; thus, we have a typical one-dimensional erosive flow problem. Note that analogous numerical investigations have only been made with finite difference schemes based on a pure pressure formulation, cf. the numerical investigations in [17]. With respect to the applied Dirchlet boundary conditions of pressure-type at the inner domain of the coal seam ($p(s = 1) = p(1) = 1$) and at the mine face ($p(s = 0) = p(0) = 0$), we observe gas flow from the seam to the mine opening. With respect to the dimensionless form of the governing equations, we are able to study the influence of the porosity function $f(\phi^\star)$ and the equation $\xi^\star(\phi^\star)$, responsible for the mass transfer from the coal skeleton to the gas phase. The discrete problem is based on the mixed Galerkin finite element formulation in weak form (64). The finite element mesh is based on 100 Q1P1 elements, i.e. quadrilaterals with linear shape functions for the pressure $p_h^\star$ and the porosity $\phi_h^\star$. The total number of degrees of freedom (DOFs) is 404. An implicit backward Euler scheme for the time discretization with a dimensionless time step length of $\Delta t^\star = 1 \times 10-4$ [-] is applied. According to (56), the following model parameters are chosen for the example shown in (12): $\phi_0 = 1$, $m = 9$, $\tilde{\phi}_{cr} = 3$ and $\Lambda^\star = 0.4$ in combination with the material parameters in Table 1. A steep pressure gradient is developed combined with a *boundary recession*, i. e. the athmospheric pressure $p(0)$, is moving inwards the specimen. Note again, that the boundary recession could be explained by the highly non-linear character of the erosion problem in gas-flow weak conditions. In the completely eroded, i. e. weakened domain of the specimen assigned with $\hat{n} = 0$ and $\phi^\star = \phi_{cr}^\star$, the material has a much larger porosity ($\phi_{cr}^\star = 3\phi_0$) as in the original domain. Furthermore, the hydraulic conductivity, i. e. the permeability of the eroded domain, has increased drastically with respect to the chosen porosity function $f(\phi^\star)$. With respect to the chosen material parameters, the increase in permeability is more then 4 order of magnitues $f(\phi_{max}^\star) = 3^m = 19683$. It could be observed in the spatial distribution of the pore pressure, cf. $p(s,t)$ in Figure (12), that a steep pressure gradient is developed combined with a *boundary recession*, i.e. the athmospheric pressure $p(0)$, is moving inwards the coal seam. Note again, that the boundary recession could be explained by the highly non-linear character of the erosion problem in gas-flow weak conditions. In the completely eroded, i.e. weakened domain of the specimen assigned with $\hat{n}^a = 0$ and $\phi^\star = \phi_{cr}^\star$, the material has a much larger porosity ($\phi_{cr}^\star = 3\phi_0$) as in the original domain. Furthermore, cf. $\hat{n}^a(s,t)$ in Figure (12), we observe that the distributed erosion front is also moving inwards with the character of a non-linear disperse wave. In areas of high pressure gradients, internal erosion is most active.

TABLE 1. Typical material parameters of coal saturated with carbon dioxide (CO_2), neglecting adsorption and double-porosity effects, cf. comments in [31, 17]. The pressure in the coal seam p_i is about 8.0 [MPa] = 80 [bar], corresponding to a reservoir depth of 350 [m] below surface. The pressure at the mine face p_0 is athmospheric.

Material parameter	Symbol	Value	Unit
dynamical viscosity	μ^{gR}	1×10^5	Pa s
intrinsic permeability	k^s	10	mD
initial porosity	ϕ_0	0.1	–
CO_2 pressure (coal seam)	p_i i.e. $p(1)$	8.0	MPa
CO_2 pressure (mine face)	p_0 i.e. $p(0)$	0.1	MPa
CO_2 density (coal seam)	ρ_0^{gR}	155.1	Kg/m^3
CO_2 density (mine face)	ρ_0^{gR}	1.94	Kg/m^3

DISCUSSION AND CONCLUSION

In the present paper we investigate non-linear flow conditions and internal erosion of a compressible pore gas in a porous constituent. We present various (dimensionless) sub-models which describe Darcy- and Forchheimer flow and wave propagation phenomena. The models are formulated in dimensionless weak form suitable for stable and reliable finite element implementation. Characteristic non-linear effects of the sub-models are highlighted. Furthermore, we discuss the aspects of internal erosion or solid-liquid phase transformation. Non-linear physical effects like the decrease in seapage velocity govern by the Forchheimer effect and the opposite phenomenon, the increase in seapage govern by the erosion process are discussed. The numerical results are highlighting the main aspects of the erosion process in gas-flow weak conditions, especially the character of the formation and evolution of a boundary layer-type solution (boundary recession). The results are in agreement with the simplified one-dimensional solution presented in the work of Paterson [31] with respect to outbursts in coal.

As the models are formulated within the framework the phenomenological continuum Mixture Theory, or to be more specific the Theory of Porous Media, the discussed models can be straightforwardly extended towards more

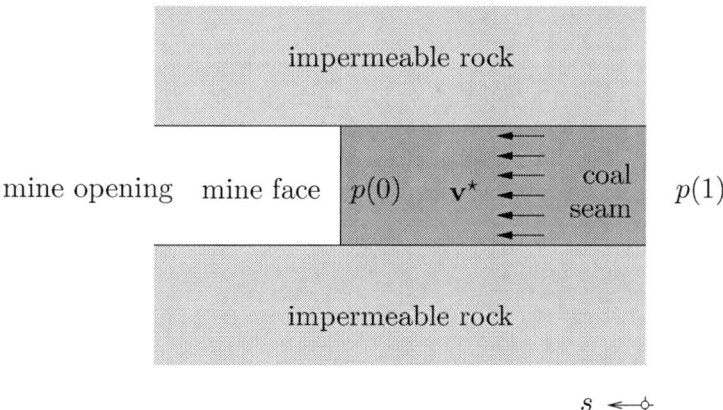

FIGURE 11. One-dimensional boundary value problem of outbursts in coal mines.

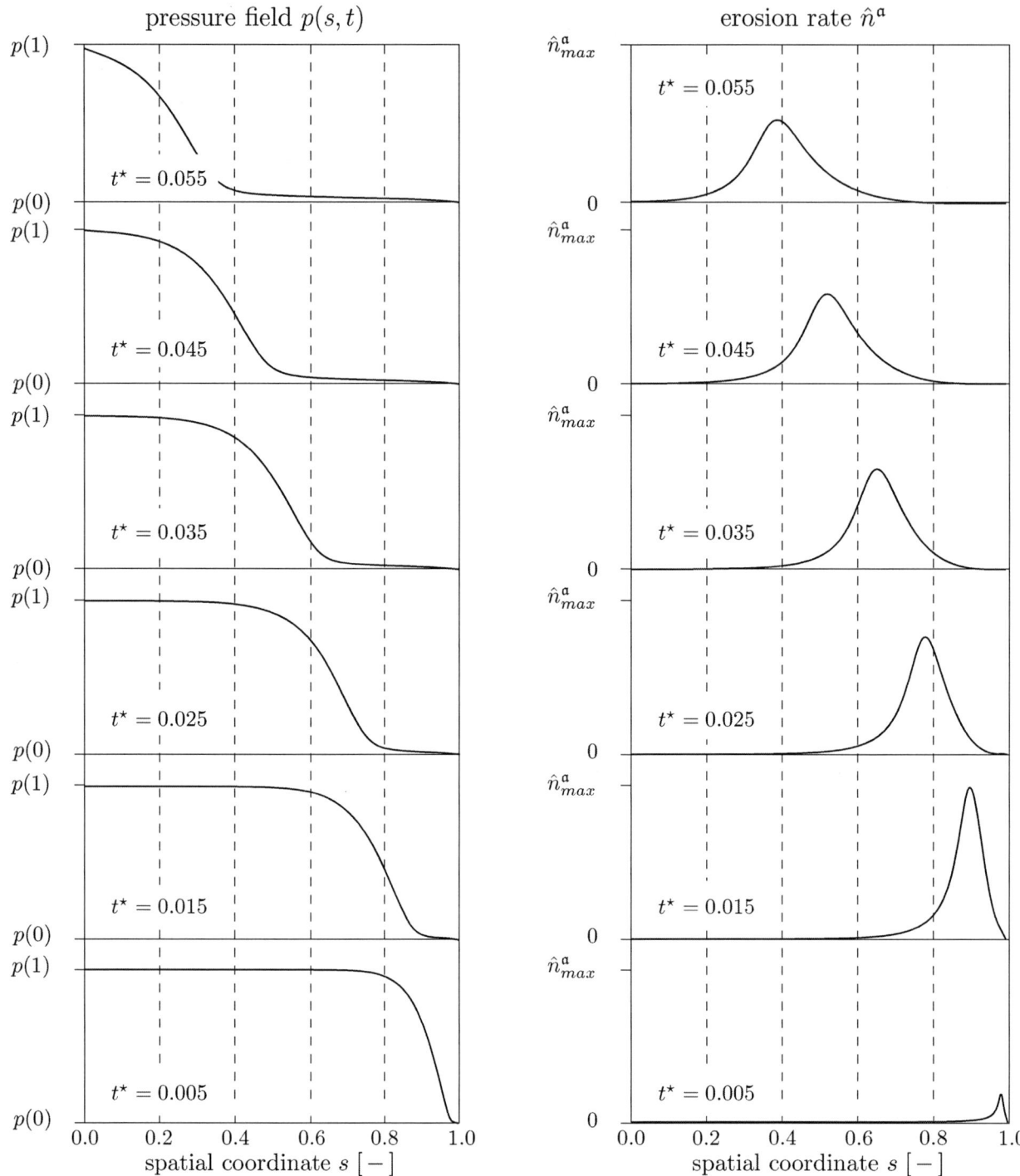

FIGURE 12. Pressure distribution $p(s,t)$ and distribution of erosion rate $\hat{n}^{\mathfrak{a}}(s,t)$.

sophisticated descriptions taking into account additional kinematical degrees of freedom or taking into account aditional coupling effects in the constitutive equations. As shown here, we have taken into account the non-linear Forchheimer equation or a modified version of the evolution equation for the volume production. In contrast to the King-Ertekin [41] and the Paterson model [31], the total problem is described within the framework of thermodynamics of closed systems. Thus, the mixture has no sinks or sources. Internal erosion, solution of fines, is treated as a mass exchange phenomenon between the solid and the abrasive phase.

In a next step, we would like to achieve that the presented modelling framework can be used to develop more realistic models e.g. to model the volumetric sand production in gas-flow weak conditions or to explain outbursts in coal on the continuum scale. Furthermore, it is obvious, that the present model can be also extended to a deformable solid skeleton. Thus, the governing constitutive equations should be questioned and eventually replaced by more sophisticated, problem-adapted constitutive assumptions. Nevertheless, before modifying and extending the present model we need to perform experiments for gas-driven erosion problems which give us realistic material parameters.

Acknowledgements: This work is dedicated to the late Professor Ioannis Vardoulakis. Without the various stimulating and motivating discussions with him about internal erosion and modelling issues, and his warm hospitality in Athens this paper would not have been possible.

REFERENCES

1. J. Ziems, *Wasserwirtschaft–Wassertechnik (WWT)* **17**, 50–55 (1967).
2. L. Wittmann, *Filtrations- und Transportphänomene in porösen Medien*, PhD-thesis, Institut für Bodenmechanik und Felsmechanik der Universität Fridericiana in Karlsruhe, 1980.
3. L. N. Reddi, *J. Infr. Systems* **3**, 78–86 (1997).
4. M. Foster, R. Fell, and M. Spannagle, *Can. Geotech. J.* **37**, 1000–1024 (2000).
5. F. Bendahmane, D. MArot, and A. Alexis, *J. Geotech. Geoenviron.* **134**, 57–67 (2008).
6. V. Chauhan, *Hydro Review Worldwide* **15**, 24–33 (2007).
7. P. Nicholson, Hurricane Katrina: Why did the levees fail, Tech. rep., Written testimony on behalf of the ASCE, before Senate Comittee on Homeland Security and Governmental Affairs, Hearing on failure of levee system surrounding New Orleans, November 2, 2005 (2005).
8. N. Sasiharan, *The failure of Teton dam - A new theory based on state based soil mechanics*, Master thesis, Washington State University, Department of Civil and Environmental Engineering, 2003.
9. I. Vardoulakis, M. Stavropoulou, and P. Papanastasiou, *Transport Porous Med.* **22**, 225–244 (1996).
10. M. B. Geilikman, and M. B. Dusseault, *Transport Porous Med.* **35**, 259–272 (1999).
11. E. Papamichos, I. Vardoulakis, J. Tronvoll, and A. Skjærstein, *Int. J. Numer. Anal. Meth. Geomech.* **25**, 789–808 (2001).
12. R. G. Wan, and J. Wang, *J. Can. Petro. Technol.* **41**, 46–52 (2002).
13. H. Steeb, S. Diebels, and I. Vardoulakis, "Sand production: A microscopically motivated multi-phase model," in *Proceedings of Powder & Grains 2005, 18th – 22nd July 2005, University of Stuttgart*, Balkema – Taylor & Francis, Leiden, The Netherlands, 2005, pp. 723–726.
14. J. Shepherd, L. K. Rixon, and L. Griffiths, *Int. J. Rock Mech. Min. Sci.* **18**, 267–283 (1981).
15. T. Xu, C. A. Tang, T. H. Yang, W. C. Zhu, and J. Liu, *Int. J. Rock Mech. Min. Sci.* **43**, 905–919 (2006).
16. I. Vardoulakis, and H. Steeb, "Sand production in gas-flown weak formations," in *Proceedings of '5th German-Greek-Polish Symposium on Advances in Mechanics', September 12-18, 2004, Bad Honnef, Germany*, 2004.
17. L. Paterson, *Int. J. Rock Mech. Min. Sci.* **23**, 327–332 (1986).
18. S. Valliappan, and S. Wohua, *Int. J. Numer. Meth. Eng.* **44**, 875–895 (1999).
19. C. A. Truesdell, Sulle basi della termomeccanica I & II, *Atti Accad Naz. Lincei. Rend. Cl. Sci. Fis. Mat. Nat.* **22**, pp. 33-38 and pp. 158-166. (1957), translated into english in: C. Truesdell (Editor), Continuum Mechanics II-IV, Gordon & Breach, New York, 1966.
20. R. M. Bowen, *Int. J. Engng. Sci.* **18**, 1129–1148 (1980).
21. R. M. Bowen, *Int. J. Engng. Sci.* **20**, 697–735 (1982).
22. W. Ehlers, "Foundations of multiphasic and porous materials," in *Porous media: Theory, experiments and numerical applications*, edited by W. Ehlers, and J. Bluhm, Springer-Verlag, Berlin, 2002, pp. 3–86.
23. K. Hutter, and K. Jöhnk, *Continuum methods and physical modeling*, Springer-Verlag, Berlin, 2004.
24. R. de Boer, *Trends in continuum mechanics of porous media*, Springer Verlag, Berlin, 2005.
25. K. Hutter, L. Laloui, and L. Vuillet, *Mech. Coh. Frict. Mat.* **4**, 295–338 (1999).
26. T. Wilhelm, and K. Wilmański, *Int. J. Multiphase Flow* **28**, 1929–1944 (2002).
27. H. Steeb, and S. Diebels, *Comp. Mat. Science* **28**, 597–607 (2003).
28. S. T. Yuster, "Theoretical considerations of multiphase flow in idealized capillary systems," in *Proceedings of the 3rd World Petroleum Congress*, 1951, vol. 2, pp. 437–445.
29. A. E. Scheidegger, *The physics of flow through porous media*, MacMillan Publishing Company, New York, 1957.
30. L. S. Bennethum, M. A. Murad, and J. H. Cushman, *Comput. Geotech.* **20**, 245–266 (1997).
31. H. P. Schlanger, and L. Paterson, "A scheme for determining gas pressure gradients in a coal seam with an advancing mine face," in *Computational Techniques and Applications: CTAC-85*, edited by J. Noye, and R. May, North-Holland, Amsterdam, 1986, pp. 775–786.

32. A. Sommerfeld, *Mechanik der Deformierbaren Medien*, vol. 2 of *Vorlesungen über Theoretische Physik*, Akademische Verlagsgesellschaft M.B.H., Leipzig, 1945.
33. H. A. Einstein, *Der Geschiebetrieb als Wahrscheinlichkeitsproblem*, Rascher & Co., Zürich, 1937.
34. R. Sakthivadivel, and S. Irmay, A review of filtration theories, Tech. Rep. HEL 15-4, University of California, Berkeley, US (1966).
35. P. Y. Julien, *Erosion and sedimentation*, Cambridge University Press, Cambridge, 1998.
36. W. Ehlers, and G. Eipper, *Transport Porous Med.* **34**, 179–191 (1999).
37. A. D. Ville, *Transport Porous Med.* **22**, 287–306 (1996).
38. B. Markert, *Porous media viscoelasticity with application to polymeric foams*, PhD-thesis, Institute of Applied Mechanics (CE), Chair II, University of Stuttgart, 2005.
39. M. A. Biot, *J. Acoust. Soc. Am.* **29**, 168–191 (1956).
40. K. Eriksson, D. Estep, P. Hansbo, and C. Johnson, *Computational differential Equations*, Cambridge University Press, Cambridge, 1996.
41. G. King, and T. Ertekin, "A comperative evaluation of vertical and horizontal drainage wells for the degasification of coal seams," in *SPE 13 091, 59th Annual Technical Conference and Exhibition of the SPE, Houston, Texas, USA, (Sept. 16–19, 2004)*, 1984.

A New Approach based on Langevin type Equation for Driven Granular Gas under Gravity

J. Wakou* and M. Isobe[†]

Miyakonojo National College of Technology, Miyakonojo-shi, Miyazaki, 885-8567
[†]*Graduate School of Engineering, Nagoya Institute of Technology, Nagoya 466-8555*

Abstract. We propose a novel approach based on a Langevin equation for fluctuating motion of the center of mass of granular media fluidized by energy injection from a bottom plate. In this framework, the analytical solution of the Langevin equation is used to derive analytic expressions for several macroscopic quantities and the power spectrum for the center of mass. In order to test our theory, we performed event-driven molecular dynamics simulations for one- and two-dimensional systems. Energy is injected from a vibrating bottom plate in the one-dimensional case and from a thermal wall at the bottom in the two-dimensional case. We found that the theoretical predictions are in good agreement with the results of those simulations under the assumption that the fluctuation-dissipation relation holds in the case of nearly elastic collisions between particles. However, as the inelasticity of the interparticle collisions increases, the power spectrum for the center of mass obtained by the simulations gradually deviates from the prediction of theoretical curve. Connection between this deviation and violation of the fluctuation-dissipation relation is discussed.

Keywords: Granular matter, Center of mass, Langevin equation, Event-driven molecular dynamics simulation
PACS: 81.05.Rm, 45.70.Qj, 47.70.Nd

INTRODUCTION

Granular materials fluidized by external driving have been widely studied in recent years in the field of nonequilibrium statistical physics. In order to fluidize granular matter, we have to supply sufficiently large energy continuously, because the kinetic energy of grains is dissipated due to inelastic collisions among grains. There are various ways to supply energy to the system: a way commonly used in experiments is use of a vibrating bottom plate. In computer simulations, a stochastic thermal wall is also often used as an ideal model of energy supply from boundaries. When the balance between energy input and dissipation is achieved, the system is in a nonequilibrium steady state.

It is known that the granular fluids under external driving show various fascinating phenomena, such as convection inside the fluids, waves and patterns that appear on the surface, size segregation like the Brazil nut phenomenon, a transition from a condensed state to a fluidized state (See Ref. [1] and references therein). It is also important to study how macroscopic quantities in the system, e.g., height of the center of mass (COM) and granular temperature, depend on the system parameters, such as the number of particles, the restitution coefficient, the amplitude and frequency of the external vibration. Experimental studies have revealed some scaling relationships for the macroscopic quantities [2, 3, 4, 5].

There have been also a lot of theoretical studies to understand these phenomena and properties. At the most microscopic scale, molecular dynamics (MD) simulations that follow the motion of all particles using computer have been commonly used [2, 3, 6, 7, 8, 9, 10, 11]. One of the most important tools to study dilute granular gas is kinetic theory, which is formulated in terms of the Boltzmann (-Enskog) equation under the assumption of "molecular-chaos". It has been used to study velocity distributions and the scaling relationships in granular fluids under external vibration [4, 12, 13, 14, 15, 16]. At more macroscopic scale, hydrodynamic equations which are derived from the Boltzmann equation have been applied to driven granular systems to study local density, flow, and temperature profiles [17, 18, 19, 20, 21].

These theoretical approaches have successfully described a variety of phenomena mentioned above, within the range of validity of each approach. It was found, however, some discrepancies in the results of these approaches, when they were applied to the scaling relationships for the macroscopic quantities. For example, as summarized in the review of previous works by McNamara and Luding [9], experimental results by Warr, Huntley, and Jacques [4], suggest a scaling law for the height of the COM $h_{c.m.}$ measured from its height at rest $h_{c.m.0}$ taking the form

$$h_{c.m.} - h_{c.m.0} \sim (A_0\omega)^\delta H^\nu, \tag{1}$$

with $\delta = 1.3 \pm 0.04$ and $v = 0.27 \pm 0.11$, where $A_0\omega$ represents the maximum velocity of the vibrating bottom, and H is the number of layers of particles at rest. Results of MD simulations of a two-dimensional system by Luding, Herrmann, and Blumen [6], however, suggest $\delta \simeq 1.5$ and $v \simeq 1$ for simulations without rotation of particles; results by Luding [7] give $\delta = 1.60 \pm 0.10$ and $v = 0.76 \pm 0.11$ for simulations including rotation of particles. On the other hand, results of kinetic theoretical approaches [4, 13, 14] give $\delta = 2$, $v = 1$. Although there are some studies that take into account an effect of the wall [9], an effect of high density [15], and an effect of nonuniformity of the hydrodynamic fields [20], this discrepancy has not yet been fully explained.

Recently we have proposed a novel theoretical approach [22, 23] which describes the system at more macroscopic level than the previous approaches mentioned above. Our theory is on the basis of a Langevin equation of motion of the COM, which is derived by focusing on the force exerted to the granular media by the bottom plate. Some macroscopic quantities can be easily obtained from the solution of this equation. In this proceedings, we will first describe how to formulate the Langevin equation for the COM, and then test the theory by comparing some important predictions derived from the Langevin equation with the result of extensive event-driven MD simulations. A detailed discussion on validity of the fluctuation-dissipation relation which connects the intensity of the random force with the friction coefficient will be one of the main points of this article.

THEORETICAL FORMULATION

In the following we explain our theoretical formulation in the case of D-dimensional system of grains on a vibrating bottom plate. The case of grains on a heat source at bottom can be discussed only by a slight modification as shown later.

We consider granular medium that consists of N inelastic particles of mass m and diameter d bouncing on a vibrating bottom plate, subjected to gravity with acceleration g. The z-direction is chosen to be opposite to the direction of gravity. The motion of particles is confined in a box of cross section S, where the quantity S is an area for $D = 3$, a length for $D = 2$, and unity for $D = 1$. The bottom plate oscillates sinusoidally with amplitude A_0 and angular frequency ω. Hence, the height of the bottom plate $z_0(t)$ at time t is

$$z_0(t) = A_0 \sin(\omega t). \qquad (2)$$

In this paper, we focus only on a fluidized state without any structure in the horizontal directions, such as convection and surface wave. Such a fluidized state can be achieved assuming sufficiently small length scales of the cross section of the box containing granular medium, so that convections and surface waves are suppressed. Moreover, we ignore any boundary effects associated with the side walls for simplicity. Indeed, in event-driven MD simulations to be described later, periodic boundary conditions in the horizontal directions have been used.

Langevin equation

We will begin by considering the equation of motion of the COM of the particles:

$$M\frac{d^2Z}{dt^2} = -Mg + F_b. \qquad (3)$$

Here, $M \equiv Nm$ is the total mass of particles and Z is the height of the COM. Ignoring the boundary effects concerning the side walls, the force acting on the COM is the sum of gravity and the force exerted by the bottom plate F_b.

There are two important timescales in the system: the oscillation period of the bottom plate $\tau(\equiv 2\pi/\omega)$ and the macroscopic relaxation time τ_{rel} to the stationary state. If τ is comparable to τ_{rel}, the energy supplied by one stroke of the bottom plate is almost dissipated during the period, and the particles will be in a condensed state. Such a condensed state is beyond the scope of the present study. In this study, we restrict ourselves to the high-frequency case $(\tau/\tau_{rel})^2 \ll 1$, in which the system is in a fluidized state.

In order to evaluate F_b, the force exerted on the particles by the bottom plate, we consider its reaction force, that is, the force exerted on the bottom plate by the particles. Here we draw an analogy between our system and a system of a Brownian particle exhibiting Brownian motion (see, for example, Ref. [24]). In our system, granular particles randomly collide with the bottom wall and exert a force on it. In Brownian motion, solvent molecules randomly collide with a Brownian particle and exert a force on it. These two forces are expected to have similar properties. The force on one side of the Brownian particle consists of a force due to the static pressure, the frictional force, and the random force. Thus, on the basis of the analogy, it is plausible to assume that F_b consists of the following four kinds of forces: average force due to the static pressure, frictional force, elastic force, and random force.

The elastic force has its origin in elastic oscillation excited by the bottom plate. An elastic oscillation mode that shows the slowest relaxation is the mode with the largest-wavelength in the system. In the stationary state, this mode plays a dominant role in deciding macroscopic property of the system. In the largest-wavelength mode, the granular fluid shows macroscopic oscillation alternating between the expansion state with the highest surface of the fluid and the contraction state with the lowest surface, which accompanies oscillation of the height of the COM with the same frequency. It must be noted that the oscillation of the bottom plate does not trigger resonances with the largest-wavelength mode and the other macroscopic modes, because the condition we assumed above, $(\tau/\tau_{rel})^2 \ll 1$, suggests that the time scale of the bottom plate oscillation is much shorter than time scales of macroscopic motions. Another important elastic oscillation is a sound wave that is directly excited by the bottom plate. Hence, we take into account two forces as the elastic force: a force due to this largest-wavelength mode and a force due to a sound wave excited by the bottom plate oscillation.

Finally, we assume the following form of $F_b(t)$:

$$F_b(t) = Mg - M\mu V_z(t) - M\Omega^2 \left(Z(t) - \overline{Z}\right) + f_s + R(t). \tag{4}$$

The first term is the average force acting on the bottom plate. The second term is the frictional force, which is assumed to be linear in the z-component of the velocity V_z of the COM, with a constant coefficient μ. The third term is the elastic force resulting from the largest-wavelength mode showing macroscopic expansion and contraction of the granular fluid. We assume that it is linear in the height of the center of mass Z measured from its long time average $\overline{Z}$. A constant coefficient Ω specifies the angular frequency of macroscopic oscillation of the granular fluid. The fourth term is the other elastic force resulting from excitation of a sound wave by the vibrating bottom plate. We will formulate this term below. The last term is the random force. Its property will also be specified later.

The period of the macroscopic slowest oscillation τ_{osc} is given by $\tau_{osc} = 2\pi/\Omega$. Similarly, the macroscopic relaxation time τ_{rel} is given by $\tau_{rel} = 1/\mu$. Since the both time scales characterize macroscopic changes that extend to the whole system, it is plausible to assume that they are on the same order as the characteristic time for a sound wave to travel along the vertical direction from the bottom to the surface of the granular gas. Let us define the thermal velocity c, by $c = \sqrt{Dk_BT/m}$, where T is the global temperature related to the mean square velocity fluctuation and k_B is the Boltzmann constant. Then, the characteristic time can be estimated as c/g because the velocity of sound is on the order of c and the height of the surface of the granular gas is, as the first order approximation, on the order of $c^2/2g$, which is the maximum height of a particle launched from the bottom with the thermal velocity c. Thus, we assume that Ω and μ are on the order of g/c:

$$\Omega = \hat{\Omega}\frac{g}{c}, \quad \mu = \hat{\mu}\frac{g}{c}, \tag{5}$$

where $\hat{\Omega}$ and $\hat{\mu}$ are numerical factors on the order of 1 that are determined curve-fitting the results of our simulation.

We estimate the elastic force $f_s(t)$ on the basis of hydrodynamic sound-wave theory [25]. In a normal fluid, sound waves propagate according to a relationship between the pressure and the velocity of the fluid. Let us denote a small change in the pressure from its equilibrium value by p', a typical velocity of the fluid particles in the wave by v, and the velocity of sound by c_s. If the condition $v \ll c_s$ is satisfied, we have a relationship $p' = \rho c_s v$ for a traveling plane wave, where ρ is the constant equilibrium density of the fluid. We assume that this relationship is also satisfied in fluidized granular media under the same condition $v \ll c_s$. The velocity of sound c_s is on the order of the thermal velocity c. The density ρ is on the order of $M/(Sc^2/2g)$, where $c^2/2g$ is the first order estimation of the surface height of the granular gas as described above, and S represents the area of the base of the system. In the vicinity of the bottom plate, v may be approximated by the velocity of the bottom plate $v_0(t) = A_0\omega\cos(\omega t)$. Since $f_s(t)$ corresponds to $p'S$ at the bottom plate, we have

$$f_s(t) = \hat{\sigma}M\frac{g}{c}A_0\omega\cos(\omega t), \tag{6}$$

where $\hat{\sigma}$ is a numerical factor on the order of 1 that is used as a curve-fit parameter when we compare our theoretical predictions with the results of simulations. The condition $v \ll c_s$ is also written as $v_0 \ll c$ in the vicinity of the bottom plate. Hence, the maximum value of $v_0(t)$, $A_0\omega$, must be much small compared to c: $A_0\omega \ll c$. Similar estimation of the pressure of sound wave has already been discussed in Ref. [8].

As property of the random force, we assume stationary Gaussian white noise:

$$\langle R(t) \rangle = 0, \quad \langle R(t)R(t') \rangle = I\delta(t-t'). \tag{7}$$

We also assume that the intensity of the random force I is determined by the fluctuation-dissipation relation, which is expected to be satisfied when the system is close to equilibrium state:

$$I = 2M\mu k_B T. \tag{8}$$

We will discuss later how the fluctuation-dissipation relation is violated in the stationary state which deviates far from equilibrium.

Substituting Eq. (4) into Eq. (3), we have the following linear Langevin equation for the COM:

$$\frac{dV_z}{dt} = -\Omega^2 (Z - \bar{Z}) - \mu V_z + \frac{f_s}{M} + \frac{R}{M}. \tag{9}$$

Analytical solution

It is straightforward to obtain the analytical solution of the Langevin equation (9) of the form

$$Z(t) - \bar{Z} = A_0 \zeta \sin(\omega t + \theta) + \int_{-\infty}^{t} G(t-t') \frac{R(t')}{M} dt' + F_{ini}(t), \tag{10}$$

where

$$\zeta = \frac{\hat{\sigma} \frac{g}{c} \omega}{\sqrt{(\Omega^2 - \omega^2)^2 + (\mu\omega)^2}}, \tag{11}$$

and

$$\tan \theta = -\frac{\omega^2 - \Omega^2}{\mu\omega} \quad \left(-\frac{\pi}{2} \leq \theta < 0\right), \tag{12}$$

respectively. The function $G(t)$ is given by

$$G(t) = \frac{e^{-\frac{\mu}{2}t}}{\omega_0} \sin(\omega_0 t), \tag{13}$$

where $\omega_0 = (\Omega^2 - (\mu/2)^2)^{1/2}$. The last term $F_{ini}(t)$ consists of those that depend on the initial conditions and vanish after a sufficient amount of time. Thus, the term is negligible when calculating long-time averages of physical quantities in the stationary state.

NUMERICAL TEST OF THE LANGEVIN APPROACH

We have performed event-driven MD simulations of one- and two-dimensional systems. The one-dimensional system consists of inelastic hard rods on a vibrating bottom plate. The two-dimensional system consists of inelastic hard disks on a thermal bottom wall. These two systems have been investigated in many previous numerical and theoretical works. In our two-dimensional simulations, we have used an efficient event-driven MD algorism developed by one of the authors [26]. Here we will demonstrate how our theory based on the Langevin equation can describe macroscopic properties of the granular fluids.

1D system on a vibrating bottom plate

We first study a one-dimensional granular fluid on a vibrating bottom plate [2, 3, 12]. We performed simulations systematically by changing the number of particles N ($N = 10, 20, 100, 1000$), the restitution coefficient r ($r = 0.80 \sim$

0.9999), and the maximum acceleration of the bottom plate $\Gamma \equiv A_0 \omega^2/g$ ($\Gamma = 10 \sim 2560$). The physical quantities are averaged over a sufficiently long period of time. When we compare our theory with simulation, we evaluate the global temperature T by $k_B T/2 = \overline{E}_K$, where $\overline{E}_K$ is the stationary value of the kinetic energy per particle defined as

$$\overline{E}_K \equiv \lim_{t_M \to \infty} \frac{1}{t_M} \int_0^{t_M} \frac{1}{N} \sum_{i=1}^{N} \frac{m}{2} v_i(t)^2 dt, \qquad (14)$$

where v_i is velocity of the i-th particle. Hence, the thermal velocity c is evaluated as $c \equiv \sqrt{k_B T/m} = \sqrt{2\overline{E}_K/m}$.

Using the solution of the linear Langevin equation, we can calculate some macroscopic quantities. Let us first consider the amplitude ζ of the oscillation of the COM. Theoretical prediction is given in Eq. (11). It must be noted here that we consider a fluidized state with the timescale $(\tau/\tau_{rel})^2 \ll 1$. This can be rewritten by substituting $\tau = 2\pi/\omega$ and $\tau_{rel} = (c/g)/\hat{\mu}$ as $\hat{\omega}^2 \gg 1$, where $\hat{\omega}$ is defined as $\hat{\omega} \equiv \omega c/g$. In this limit, we expanded ζ in Eq. (11) in terms of $\hat{\omega}^{-2}$:

$$\zeta = \frac{\hat{\sigma}}{\hat{\omega}} \left(1 + O\left(\hat{\omega}^{-2}\right)\right). \qquad (15)$$

Thus, it is predicted that the amplitude is inversely proportional to the scaled angular frequency of the vibration.

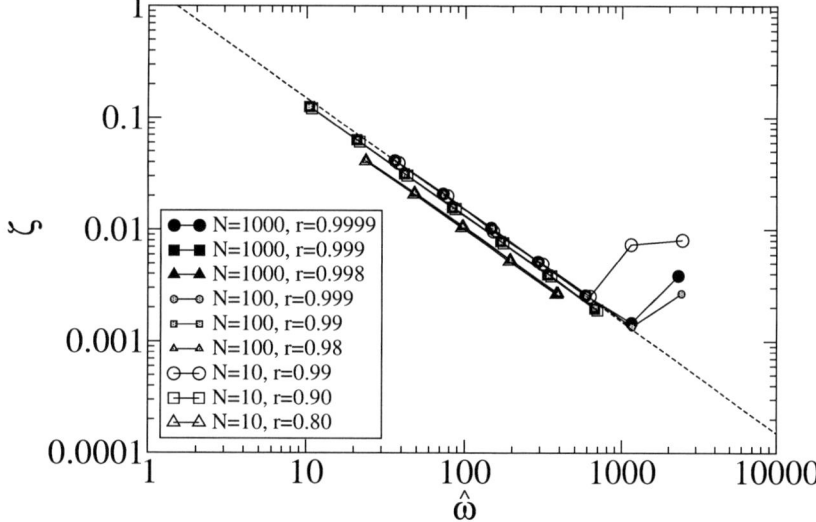

FIGURE 1. The amplitude ζ of the oscillation of the center of mass as a function of the scaled angular frequency $\hat{\omega} \equiv \omega c/g$ of the bottom plate vibration. The accelerations Γ used in the simulations are $\Gamma = 10, 20, 40, 80, 160, 320$, and 640, except for the simulations that experienced an "inelastic collapse" (an infinite number of collisions in a finite time) [27, 28]. The dashed line corresponds to the theoretical prediction $\hat{\sigma}/\hat{\omega}$ given in Eq.(15) with a curve-fit parameter $\hat{\sigma} = 1.5$.

Figure 1 gives the simulation results for ζ as a function of $\hat{\omega}$. We obtained good agreement between the theoretical prediction (15) and the simulation result when we chose the curve-fit parameter $\hat{\sigma} = 1.5$.

Secondly, we consider the power injection by the bottom plate P_b, which has been the subject of recent studies [4, 8, 13, 14, 16]. It is defined by using the force exerted by the bottom plate F_b and its velocity v_0:

$$P_b = \lim_{t_M \to \infty} \frac{1}{t_M} \int_0^{t_M} F_b(t) v_0(t) dt. \qquad (16)$$

Substituting F_b using Eq. (3) into Eq. (16), and integrating by parts twice, we have

$$P_b = \lim_{t_M \to \infty} \frac{1}{t_M} \int_0^{t_M} \left(M\frac{d^2Z}{dt^2} + Mg\right) v_0(t) dt = -M\omega^2 \lim_{t_M \to \infty} \frac{1}{t_M} \int_0^{t_M} Z(t) v_0(t) dt. \qquad (17)$$

Then, substituting Eq. (10) into Eq. (17), we obtain

$$P_b/MgA_0\omega = \frac{\hat{\sigma}}{2} \frac{A_0\omega}{c} \frac{\omega^2(\omega^2 - \Omega^2)}{(\omega^2 - \Omega^2)^2 + (\mu'\omega)^2} = \frac{\hat{\sigma}}{2} \frac{A_0\omega}{c} \left(1 + O\left(\hat{\omega}^{-2}\right)\right). \qquad (18)$$

The results obtained by neglecting terms on the order of $\hat{\omega}^{-2}$ coincide with the scaling predicted by kinetic theories [4, 13, 14]: $P_b \sim Mg(A_0\omega)^2/c$. Figure 2 shows that the scaling relationship (18) with the same curve-fit parameter as Fig.1, $\hat{\sigma} = 1.5$, agrees well with the simulations in the region where $A_0\omega/c \lesssim 1$. This result is consistent with the condition $A_0\omega/c \ll 1$ required for the formula given by Eq. (6) to be valid.

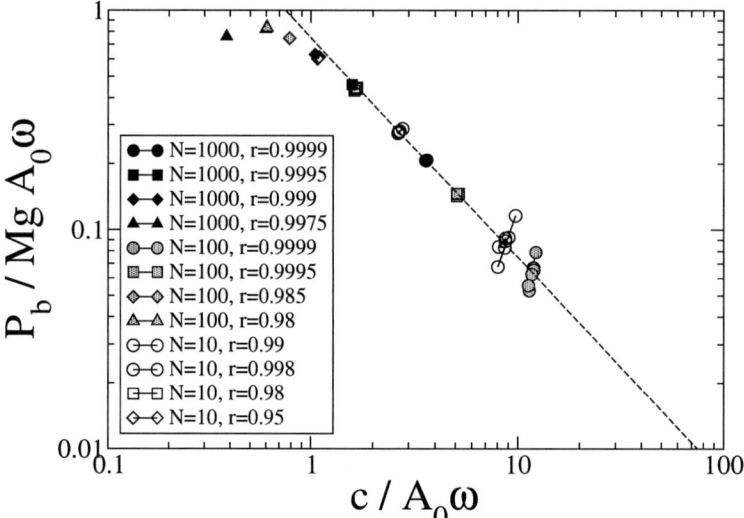

FIGURE 2. The power injection by the bottom plate P_b scaled by $MgA_0\omega$, as a function of $c/A_0\omega$. Here c is the thermal velocity evaluated as $c = \sqrt{2\overline{E}_K/m}$. The accelerations Γ used in the simulations were $\Gamma = 10, 20, 40, 80, 160, 320$, and 640, except the simulations that experienced an inelastic collapse. The dashed line corresponds to the theoretical prediction $(\hat{\sigma}/2)(A_0\omega/c)$ given by Eq. (18) with a curve-fit parameter $\hat{\sigma} = 1.5$.

Thirdly, let us consider the power spectrum I_{CM} for the height of the COM. According to the Wiener-Khinchin theorem, this can be calculated analytically from the Fourier transform of the two-time correlation function $\psi_{CM}(t)$ defined by

$$\psi_{CM}(t) = \lim_{t_M \to \infty} \frac{1}{t_M} \int_0^{t_M} \langle \delta Z(t')\delta Z(t'+t) \rangle dt', \qquad (19)$$

where $\delta Z(t) \equiv Z(t) - \overline{Z}$ and the brackets $\langle \cdots \rangle$ indicate an average over the random force $R(t)$. Substituting Eq. (10) into Eq. (19) and performing the Fourier transform, we obtain

$$I_{CM}(\hat{\omega}')/\frac{c^5}{Ng^3} = \frac{\pi}{2}N\zeta^2\left(\frac{Ag}{c^2}\right)^2 (\delta(\hat{\omega}'-\hat{\omega}) + \delta(\hat{\omega}'+\hat{\omega})) + \frac{2\hat{\mu}}{(\hat{\Omega}^2 - \hat{\omega}'^2)^2 + (\hat{\mu}\hat{\omega}')^2}, \qquad (20)$$

where $\hat{\omega}'$ is the angular frequency ω' scaled by g/c: $\hat{\omega}' = \omega'c/g$. Hence, our theory predicts that the power spectrum consists of two terms: the first term gives the delta-functional peak at the frequency of the bottom plate oscillation $\hat{\omega} \equiv \omega c/g$. The second term represents a continuous spectrum. Indeed it has already been shown in Refs. [2, 3] that the power spectrum for the motion of the COM in a fluidized state consists of a continuous spectrum and a sharp peak at the frequency of the vibration. In Fig.3, we present the scaled power spectrum for the case of $N = 100$, $r = 0.99$, and $\Gamma = 160$ obtained from our simulation. The result of simulation indeed shows the sharp peak corresponding to the first term and the continuous part corresponding to the second term. The part of continuous spectrum agrees well with our theoretical prediction when we choose the curve-fit parameters $\hat{\mu} = 2.0$ and $\hat{\Omega} = 1.5$.

The continuous part of the scaled power spectrum given by the second term of Eq. (20) predicts a universal behavior independent of the system parameters. In order to confirm this property, we first divided the logarithm of the angular frequency $\log \omega'$ into bins with a constant interval, and took an average of the power spectrum over the bins; then we scaled it as given by Eq. (20). The results of simulations with various system parameters are shown in Fig.4. The simulation data indeed collapse on a single master curve which agrees very well with the curve of the theoretical prediction.

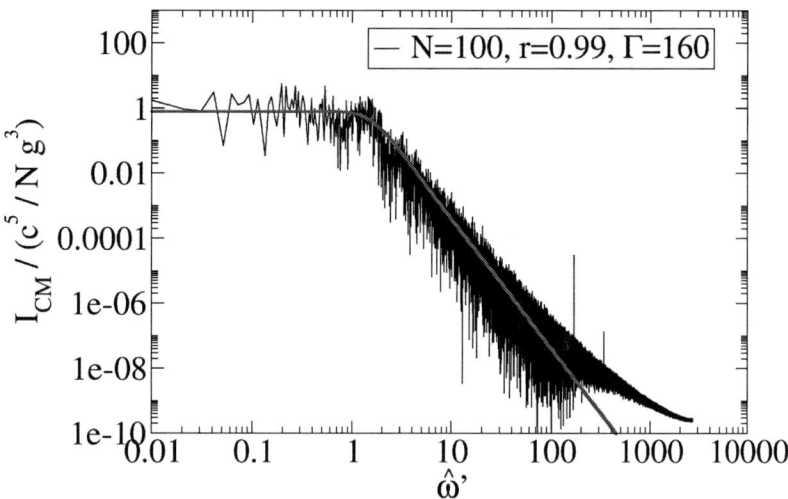

FIGURE 3. The power spectrum I_{CM} for the height of the center of mass scaled by c^5/Ng^3 obtained from simulations for $N=100$, $r=0.99$, and $\Gamma=160$. The gray solid line depicts the theoretical prediction given by the second term in Eq. (20) with curve-fit parameters $\hat{\mu}=2.0$ and $\hat{\Omega}=1.5$.

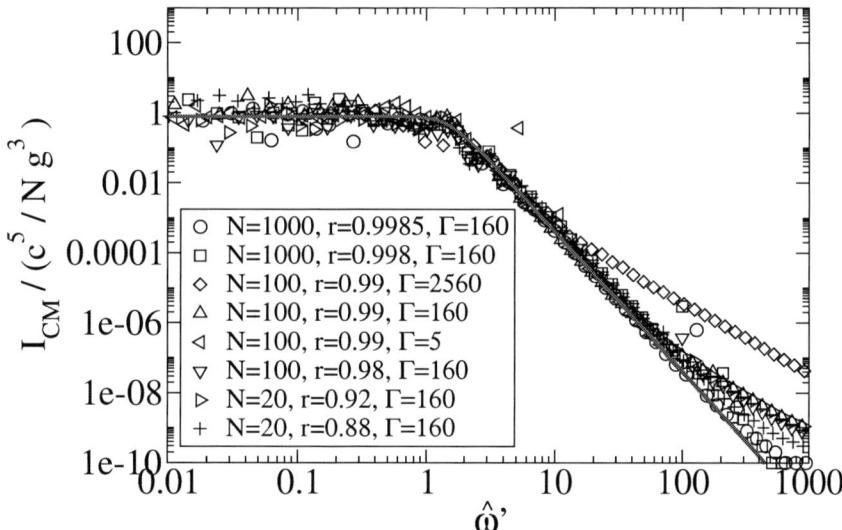

FIGURE 4. The power spectrum I_{CM} for the height of the center of mass scaled by c^5/Ng^3 for various system parameters. The gray solid line is the same as in Fig.3.

Furthermore, we study how the law of equipartition fails when the inelasticity is increased. Let us define the stationary value of the kinetic energy of the COM for the motion along the z-axis as

$$\overline{K}_{CM} \equiv \lim_{t_M \to \infty} \frac{1}{t_M} \int_0^{t_M} \left\langle \frac{1}{2} M V_z(t)^2 \right\rangle dt \qquad (21)$$

Using Eq. (10), $\overline{K}_{CM}$ can be calculated and yields

$$\frac{\overline{K}_{CM}}{\overline{E}_K} = 1 + \frac{N}{8}\left(\frac{mgA_0\hat{\sigma}}{\overline{E}_K}\right)^2. \qquad (22)$$

Figure 5 shows comparison of simulation results with the theoretical prediction (22). Here, the vertical axis shows the ratio of the kinetic energy of the COM, $\overline{K}_{CM}$, to the kinetic energy per particle $\overline{E}_K$. When the law of equipartition is satisfied, this ratio is equal to 1. The simulation data show deviation from the equipartition at $\overline{E}_K \ll 10$. This deviation can be well explained by our theory as far as it is not too large, using the same curve-fit parameter $\hat{\sigma} = 1.5$ as used in Fig.1 and Fig.2.

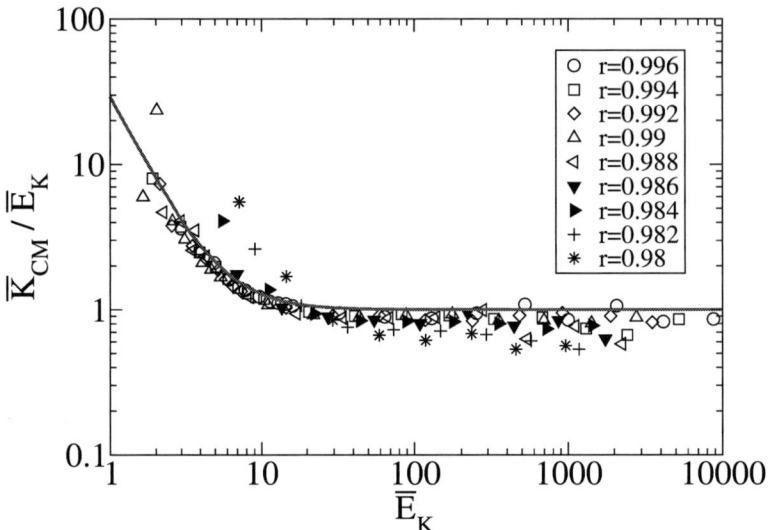

FIGURE 5. The ratio of kinetic energy of the center of mass $\overline{K}_{CM}$ and kinetic energy per particle $\overline{E}_{CM}$ for $N = 100$ and different Γ values (from 10 to 2560), except the simulations that experienced an inelastic collapse. The gray solid line gives the theoretical prediction Eq. (22) with a curve-fit parameter $\hat{\sigma} = 1.5$.

2D system on a thermal wall

We then study a two-dimensional granular fluid on a thermal bottom wall [10, 11, 21]. The system considered here consists of N inelastic hard disks with mass m and diameter d, in a two-dimensional space of a width L. We set periodic boundary condition in the horizontal direction and infinite height in the vertical direction. The stochastic thermal wall with the temperature T_0 is set at the bottom. When a disk collides with the wall, it comes off with the value of the vertical velocity component, v_n, sampled from the probability density

$$p(v_n) = \frac{mv_n}{k_B T_0} \exp\left(-\frac{mv_n^2}{2k_B T_0}\right). \tag{23}$$

In order to prevent unphysical flows in the horizontal direction which may be caused by the periodic boundary conditions, we impose that the velocity component parallel to the wall remains unchanged by the collision with the wall. The dynamics is evolved by binary inelastic collisions between disks with a constant normal restitution coefficient r. The system can be completely characterized by the number of disks N, the dimensionless width $\hat{L} \equiv L/d$, the dimensionless driving intensity $\Lambda \equiv k_B T_0/mgd$, and the restitution coefficient r [10, 11]. The value of $\hat{L}$ is chosen sufficiently small to prevent the density instability, so that the system remains homogeneous in the horizontal direction.

Our theory can be easily applied to the case of the thermal bottom wall by simply modifying $f_s(t) = 0$ in Eq. (9). The other difference from the one-dimensional system is that the thermal velocity c is defined by $c \equiv \sqrt{2k_B T/m}$. When we compare our theory with simulation, we evaluate T by $k_B T = \overline{E}_K$ and c by $c = \sqrt{2\overline{E}_K/m}$.

Then, the theoretical curve of the power spectrum I_{CM} for the height of the COM is given by the second term in Eq. (20) divided by the system dimensionality 2, which comes from the change in the relation between $k_B T$ and $\overline{E}_K$:

$$I_{CM}(\hat{\omega}')/\frac{c^5}{Ng^3} = \frac{\hat{\mu}}{(\hat{\Omega}^2 - \hat{\omega}'^2)^2 + (\hat{\mu}\hat{\omega}')^2}. \tag{24}$$

In the Figs. 6 and 7, the power spectra I_{CM} of the COM with fixed parameters $(\hat{L}, \Lambda) = (10, 10^3)$ for various (N, r) are shown. Figure 6 shows the results of the simulations in the case of nearly elastic collisions $(1 - r \ll 1)$. Here we find that the scaled power spectra concentrate into a master curve. The theoretical curve with the curve-fit numerical factors $\hat{\mu} = 2.0$ and $\hat{\Omega} = 1.5$ shows good agreement with numerical simulations. On the other hand, Fig.7 shows the result of the simulations in the large inelasticity case; it demonstrates a systematic deviation from the curve predicted by our theory. We speculate that this deviation is caused by violation of the fluctuation-dissipation relation, which will be discussed in the following sections.

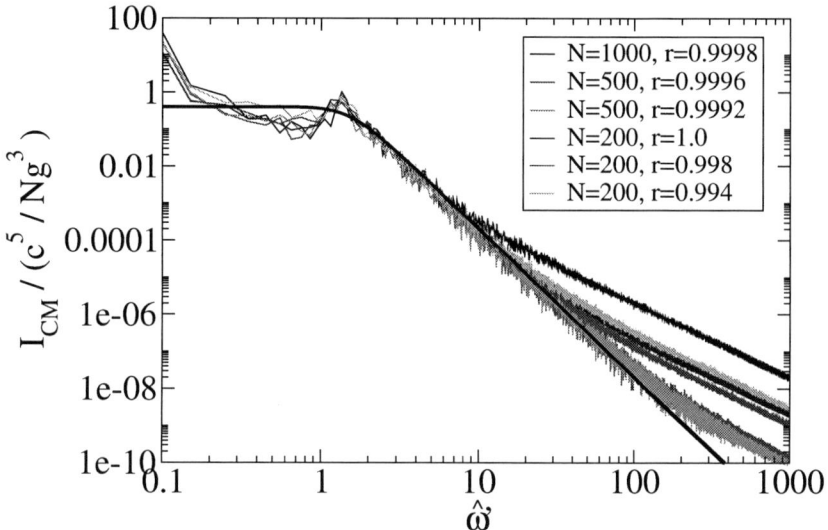

FIGURE 6. The power spectrum I_{CM} for the height of the center of mass scaled by c^5/Ng^3 obtained from simulations for various (N, r). The driving intensity is $\Lambda = 1000$. Here the restitution coefficients r used in the simulations are close to 1 (nearly elastic case). The thick solid line depicts the theoretical prediction given by Eq. (24) with curve-fit parameters $\hat{\mu} = 2.0$ and $\hat{\Omega} = 1.5$.

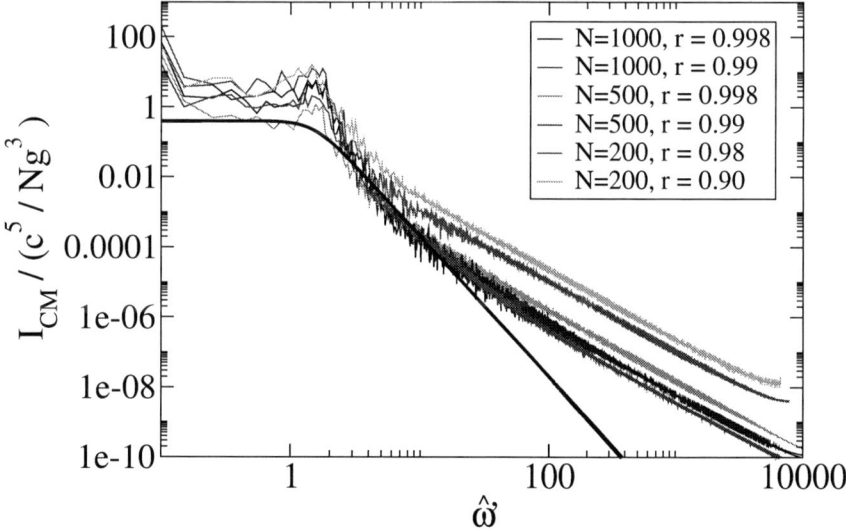

FIGURE 7. The same result as in Fig.6 but the restitution coefficients r are smaller than those used in Fig.6 (large inelasticity case). The thick solid line depicts the theoretical prediction given by Eq. (24) with curve-fit parameters $\hat{\mu} = 2.0$ and $\hat{\Omega} = 1.5$.

Failure of the law of equipartition in 2D system

We first investigate how the law of equipartition fails when inelasticity is increased. Figure 8 shows the ratio of the kinetic energy of the COM $\overline{K}_{CM}$ to $k_B T/2 (= \overline{E}_K/2)$. Since $\overline{K}_{CM}$ is defined by Eq. (21) using only the z-component of the COM velocity V_z, this ratio is to be equal to 1 when the law of equipartition is satisfied. We found a systematic deviation from the law of equipartition at low temperature; the deviation seems to be inversely proportional to T and independent of N. Our theory can not give any explanation on this behavior of the ratio of the energies.

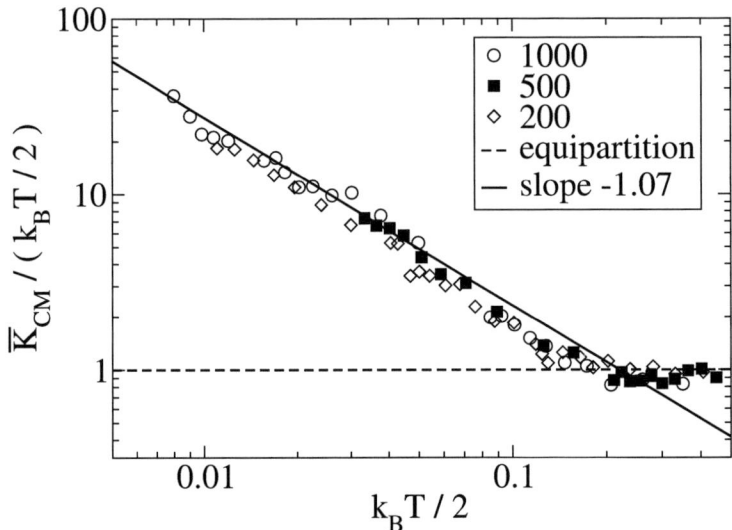

FIGURE 8. The ratio of kinetic energy of the center of mass $\overline{K}_{CM}$ and the half of kinetic energy per particle $k_B T/2$ plotted against $k_B T/2$. Here $k_B T$ is evaluated by $k_B T = \overline{E}_K$. The restitution coefficients r are from 0.964 to 0.9995 for $N = 1000$, from 0.991 to 0.9999 for $N = 500$, and from 0.90 to 0.999 for $N = 200$. The dashed line gives the law of equipartition of energy. The solid line gives a slope of -1.07.

It is known from previous studies that the system shows density inversion in a certain condition [10, 11, 20, 21]. Here we discuss on a connection between the density inversion phenomena and the failure of the energy equipartition. In the previous study based on hydrodynamic equations, Bromberg et al. [21] have shown that if a parameter λ defined by

$$\lambda \equiv \frac{\pi^{1/2}}{2} N_h \left(1 - r^2\right)^{1/2}, \qquad (25)$$

where N_h is the number of layers of disks at rest, becomes larger than the critical value $\lambda_c \simeq 1.06569...$, then density inversion develops in the system. Here we show in Fig. 9 the ratio of the energies as a function of λ. We observe indeed that the deviation from the equipartition starts at a certain value of λ near λ_c. We also confirmed that at the data points where the law of equipartition fails, the density inversion indeed develops as shown in Fig. 10.

Failure of fluctuation-dissipation relation in 2D system

The fluctuation-dissipation relation (of the second kind) (8) with respect to our Langevin equation (9) with $f_s = 0$ can be derived using the same procedure in the study of Brownian motion based on the Langevin equation [24]. To begin with, let us define the intensity of the random force I by Eq. (7). Then, we calculate $\overline{K}_{CM}$ using the solution (10) of the Langevin equation (9) with $f_s(t) = 0$, and represent it in terms of I:

$$\overline{K}_{CM} = \frac{M}{4\mu} I. \qquad (26)$$

Finally, if the equipartition of energy, that is , $\overline{K}_{CM} = k_B T/2$, is satisfied, then we obtain the fluctuation-dissipation relation as given in Eq. (8). Therefore, in the case of large inelasticity where the law of equipartition of energy fails, the fluctuation-dissipation relation also fails to be satisfied.

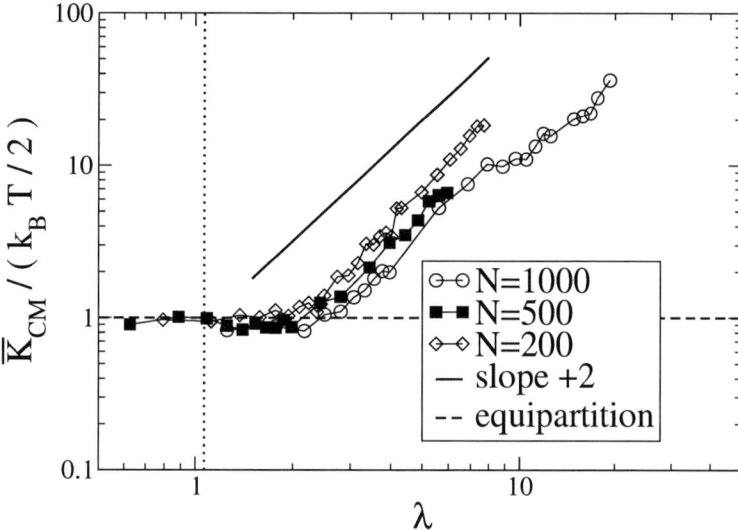

FIGURE 9. The ratio of kinetic energy of the center of mass $\overline{K}_{CM}$ and the half of kinetic energy per particle $k_B T/2$ plotted against λ defined by Eq. (25). The dashed line gives the law of equipartition of energy. The dotted line indicates λ_c, the critical value of λ for the density inversion. The solid line gives a slope of $+2$.

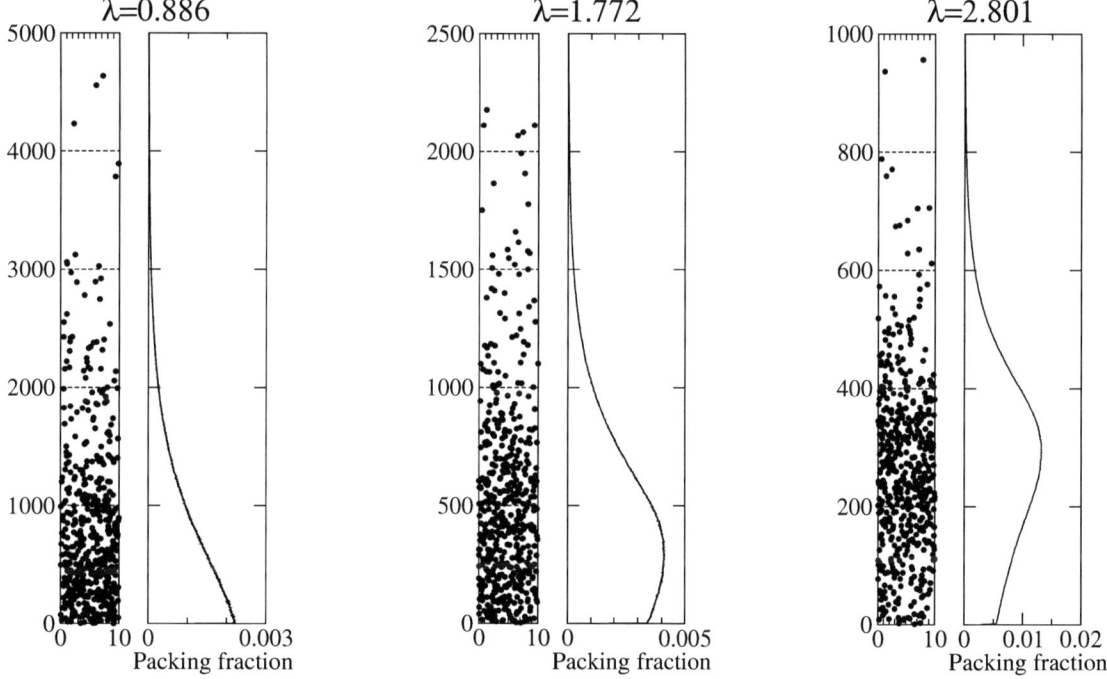

FIGURE 10. Snapshots and packing fraction profiles of simulations for $\Lambda = 1000$, $N = 500$, and different λ values.

Having observed that the fluctuation-dissipation relation is violated in the large inelasticity case, one can then go on to consider whether the linear Langevin equation itself is valid in the case of large inelasticity. In order to clarify this point, we use Eq. (26) and define I in terms of $\overline{K}_{CM}$ without use of the fluctuation-dissipation relation: $I \equiv 4\mu \overline{K}_{CM}/M$. Then, the theoretical expression of I_{CM} becomes

$$I_{CM}(\hat{\omega}')/\left[4\left(\frac{c}{g}\right)^3 \frac{\overline{K}_{CM}}{M}\right] = \frac{\hat{\mu}}{(\hat{\Omega}^2 - \hat{\omega}'^2)^2 + (\hat{\mu}\hat{\omega}')^2}. \quad (27)$$

When we compare Eq. (27) with the results of simulation, we determine $\overline{K}_{CM}$ from the simulation data. Figure 11 shows the rescaling of the power spectra successfully collapses the simulation data onto a single curve near the peak. The curve of the theoretical expression (27) with $\hat{\mu} = 0.6$ and $\hat{\Omega} = 1.5$ seems to show better fitting than the case of $\hat{\mu} = 2.0$ and $\hat{\Omega} = 1.5$. This change in the coefficient $\hat{\mu}$ in the case of large inelasticity may be attributed to the internal structure caused by the density inversion. We have not yet understand how the values of the phenomenological constants in our theory are related to the density profile of the system.

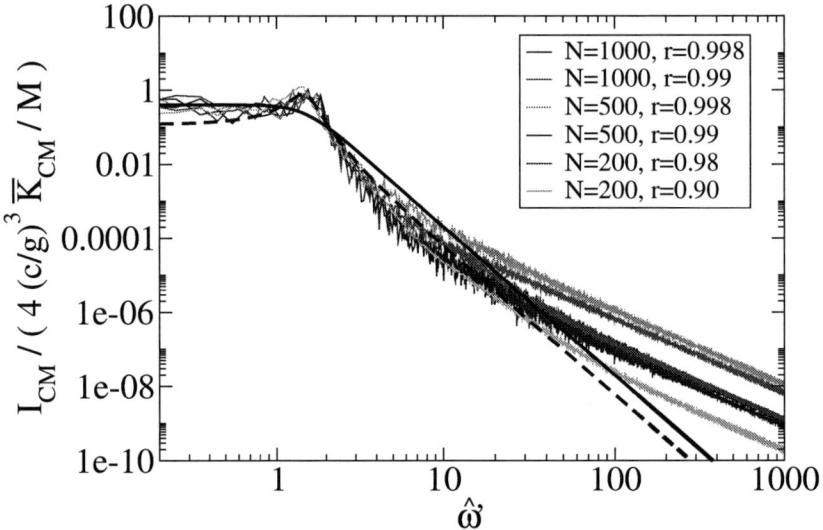

FIGURE 11. The power spectrum I_{CM} for the height of the center of mass rescaled by $4(c/g)^3 \overline{K}_{CM}/M$ for various (N, r). The parameters and lines of the simulations are the same as in Fig.7. The thick solid line gives the theoretical prediction given by Eq. (24) with curve-fit parameters $\hat{\mu} = 2.0$ and $\hat{\Omega} = 1.5$. The thick dashed line gives the theoretical prediction with $\hat{\mu} = 0.6$ and $\hat{\Omega} = 1.5$.

CONCLUDING REMARKS

A Langevin equation for the motion of the center of mass was formulated to describe macroscopic properties of the fluidized state of granular matter under gravity. The analytical expressions of some macroscopic quantities were derived and compared with the results of extensive simulations. There are three numerical factors, $\hat{\sigma}$, $\hat{\mu}$ and $\hat{\Omega}$, that can not be specified by our phenomenological consideration; they were used as curve-fit parameters when we compared our theoretical predictions with the results of simulations. In the present study, we performed event-driven molecular dynamics simulations for the following two systems: a one-dimensional system of inelastic rods on a vibrating bottom plate and a two-dimensional system of inelastic disks on a thermal bottom wall.

In the one-dimensional system, we found good agreement between our theory and simulation. The results of simulations suggest the following values of the numerical factors: $\hat{\sigma} = 1.5$, $\hat{\mu} = 2.0$ and $\hat{\Omega} = 1.5$. Deviation from the law of equipartition is observed when the kinetic energy per particle is sufficiently small. This deviation is explained well by our theory as far as it is not too large to violate the fluctuation-dissipation relation.

In the two-dimensional system, the results of simulations are in good agreement with the theoretical predictions if the collisions between disks are nearly elastic. The results of simulations suggest the following values of the numerical factors: $\hat{\mu} = 2.0$ and $\hat{\Omega} = 1.5$. As the inelasticity of the interparticle collisions increases, however, the power spectrum for the center of mass obtained by the simulations gradually deviates from the prediction of theoretical curve. We speculated that this deviation is caused by violation of the fluctuation-dissipation relation. From the systematic event-driven simulation for the wide range of parameters, we found failure of the law of equipartition of energy with respect to the kinetic energy of the center of mass, which necessarily causes violation of the fluctuation-dissipation relation. Contrary to the one-dimensional system, the deviation from the law of equipartition can not be explained within the framework of our theory. Furthermore, we obtained evidence that this failure of the law of equipartition of energy is caused by the inversion of the density profile [10, 11, 20, 21]. Finally, we showed that the theoretical predictions obtained without assuming the fluctuation-dissipation relation agree with the results of simulations by choosing the curve-fit parameters $\hat{\mu} = 0.6$, $\hat{\Omega} = 1.5$. This result suggests that the Langevin equation itself might remain valid even when the fluctuation-dissipation relation is violated, and it might be still useful to explain a universal behavior of the power spectrum in the large inelasticity case.

ACKNOWLEDGMENT

This study was supported by Grant-in-Aid for Scientific Research from the Ministry of Education, Culture, Sports, Science and Technology No. 19740236. Part of the computations for this study was performed using the facilities of the Supercomputer Center, Institute for Solid State Physics, the University of Tokyo. This study was financially supported by the Hosokawa Powder Technology Foundation.

REFERENCES

1. I. S. Aranson, and L. S. Tsimring, *Rev. Mod. Phys.*, **78**, 641–92 (2006).
2. E. Clément, S. Luding, A. Blumen, J. Rajchenbach, and J. Duran, Int. J. Mod. Phys. B **7**, 1807–27 (1993).
3. S. Luding, E. Clément, A. Blumen, J. Rajchenbach, and J. Duran, *Phys. Rev. E*, **49**, 1634–46 (1994).
4. S. Warr, J. M. Huntley, and G. T. H. Jacques, *Phys. Rev. E*, **52**, 5583–95 (1995).
5. R. D. Wildman, J. M. Huntley, and D. J. Parker, *Phys. Rev. E*, **63**, 061311-1–10 (2001).
6. S. Luding, H. J. Herrmann, and A. Blumen, *Phys. Rev. E*, **50**, 3100–08 (1994).
7. S. Luding, *Phys. Rev. E*, **52**, 4442–57 (1995).
8. S. McNamara, and J. -L. Barrat, *Phys. Rev. E*, **55**, 7767–70 (1997).
9. S. McNamara, and S. Luding, *Phys. Rev. E*, **58**, 813–22 (1998).
10. M. Isobe, and H. Nakanishi, *J. Phys. Soc. Jpn.*, **68**, 2882–85 (1999).
11. M. Isobe, *Phys. Rev. E*, **64**, 031304-1–14 (2001).
12. B. Bernu, F. Delyon, and R. Mazighi, *Phys. Rev. E*, **50** 4551–59 (1994).
13. V. Kumaran, *Phys. Rev. E*, **57**, 5660–64 (1998).
14. V. Kumaran, *J. Fluid Mech.*, **364**, 163–85 (1998).
15. P. Sunthar, and V. Kumaran, *Phys. Rev. E*, **60**, 1951–55 (1999).
16. R. Soto, *Phys. Rev. E*, **69**, 061305-1–5 (2004).
17. P. K. Haff, *J. Fluid Mech.*, **134**, 401–30 (1983).
18. J. Lee, *Physica A*, **219**, 305–26 (1995).
19. J. Lee, *Physica A*, **238**, 129–48 (1997).
20. J. J. Brey, M. J. Ruiz-Montero, and F. Moreno, *Phys. Rev. E*, **63**, 061305-1–10 (2001).
21. Y. Bromberg, E. Livne, and B. Meerson, "Development of a Density Inversion in Driven Granular Gases", in *Granular Gas Dynamics*, edited by T. Pöschel and N. Brilliantov, Springer, 2003, pp.251–66.
22. J. Wakou, A. Ochiai, and M. Isobe, *J. Phys. Soc. Jpn.*, **77**, 034402-1–5 (2008).
23. J. Wakou, and M. Isobe, "A New Approach based on Langevin type Equation for Granular Media Fluidized Vibrations", in *Powders and Grains 2009: Proceedings of the 6th International Conference on Micromechanics of Granular Media*, edited by M. Nakagawa, and S. Luding, AIP Conference Proceedings 1145, American Institute of Physics, New York, 2009, pp.717-20.
24. R. Kubo, M. Toda, and N. Hashitsume, *Statistical Physics II*, Springer, Berlin, 1985.
25. L. D. Landau and E. M. Lifshitz, *Fluid Mechanics*, Pergamon Press, New York, 1987.
26. M. Isobe, *Int. J. Mod. Phys.*, **C10**, 1281–93 (1999).
27. B. Bernu, and R. Mazighi, *J. Phys. A* **23**, 5745–54 (1990).
28. S. McNamara, and W. R. Young, *Phys. Fluids* A **4**, 496–504 (1992).

ELASTO-PLASTICITY AND MICROMECHANICS
OF RATE-INDEPENDENT FLOW

Response of a cohesionless packing to a point load

El Hadji Bouya Amar[+], D. Clamond[*], N. Fraysse[+], J. Rajchenbach[+]

[+] *Laboratoire de Physique de la Matière Condensée (CNRS-UMR 6622)*
[*] *Laboratoire Jean-Alexandre Dieudonné (CNRS UMR 6621)*
Université de Nice Sophia Antipolis, Parc Valrose, 06108 Nice Cedex 2, France.

Abstract. In order to study the mechanical behaviour of grain piles, we investigate the response of a non-cohesive two-dimensional packing of cylinders submitted to a localized force, that in the reversible regime of deformation. By means of image processing, we obtain an accurate measurements of the individual grain displacements. The measured displacement field deviates unambiguously from the predictions of linear continuum elasticity. Instead, the data reveal a localization process, as well as a partial agreement with the diffusive models of Harr (1966) or of Coppersmith et al. (1996).

Keywords: Mechanics, Granular Materials, Statics, Constitutive relation.
PACS: 45.70-n, 45.70-cc, 83.10Gr, 83.80Fg

INTRODUCTION

In civil engineering, cohesionless soils or grain piles submitted to a compressive load are usually asumed to obey linear elasticity below the Mohr-Coulomb plastic threshold [1]. Nevertheless, several important features cast doubts on the relevance of the linear elasticity to the mechanical behavior of grain packings below plastic yielding. First, if one considers a loose packing, it is obvious that a compressive load first densifies the medium (i.e. increases the number of intergranular contacts). Second, it is worth noting that the *normal* contact force between grains is essentially *nonlinear*. For two spherical grains in contact, the relation between the normal contact force F_n and the centroid displacement δ is the given by the Hertz law, which reads $F_n \propto \delta^{3/2}$ [2]. The nonlinearity of the Hertz law proceeds from the fact that the contact area between the two spheres in contact increases with the normal loading. The nonlinear character of the force-displacement relationship is featured by any point contact between two solids (for example, $F \propto \delta^2$ for a cone-plane contact). Hence it is clear that the linear elasticity cannot describe the stress-strain relation of a one-dimensional array of spheres, and its applicability to two- and three-dimensional packings remains thus highly questionable. Moreover, the *tangential* contact force between grains originates in the Coulomb dry friction, which is also intrinsically *highly* nonlinear. For a son-sliding mutual contact, the contact force between two bodies belongs to the Coulomb cone. In presence of sliding, the contact force belongs to the surface of the Coulomb cone. Hence, the set of momentum and force equilibrium equations relative to individual particles and describing the packing at rest has to be supplemented by a set of inequalities prescribing the Coulomb non-sliding criterion. Determining the ensemble of contact forces in a grain packing at rest is a difficult *nonvariational* problem, which admits in general a plurality of solutions depending upon the average coordination number [3-6]. In soil mechanics, the non-uniqueness of the solutions arising from dry friction was recognized by Rankine [7] and Darwin [8] more than one century ago. To illustrate the possibility of multivalued solutions arising from frictional contacts, we recall with all due apologies the following undergraduate-level exercise: Two identical blocks (of weight Fn), connected with a spring, rest on a horizontal frictional foundation (Fig. 1). It is clear that there is a continuous range of equilibrium solutions obeying the inequality $-\mu F_n < F_t < \mu F_n$ (where μ the friction coefficient between the blocks and the table, and F_t the force exerted by the spring). At equilibrium, the spring, whose state is macroscopically inaccessible, can be either stretched or compressed (Fig. 1a and 1b), depending upon the preparation history. Although a regularization procedure consisting of mapping the tangential friction force onto a tangential spring provides an acceptable equilibrium solution from the viewpoint of the force balance (while disregarding the relation between stress and strain), it is clear that such an *Ansatz* is an oversimplification. It reduces the initial *nonvariational* problem into a *variational* one, and the plurality of solutions into a unique solution. Since both normal and tangential contacts are strongly nonlinear, it appears that the classical framework hypothesizing a linear-elastic constitutive relation for frictional granular packings below the plastic failure deserves a critical experimental assessment.

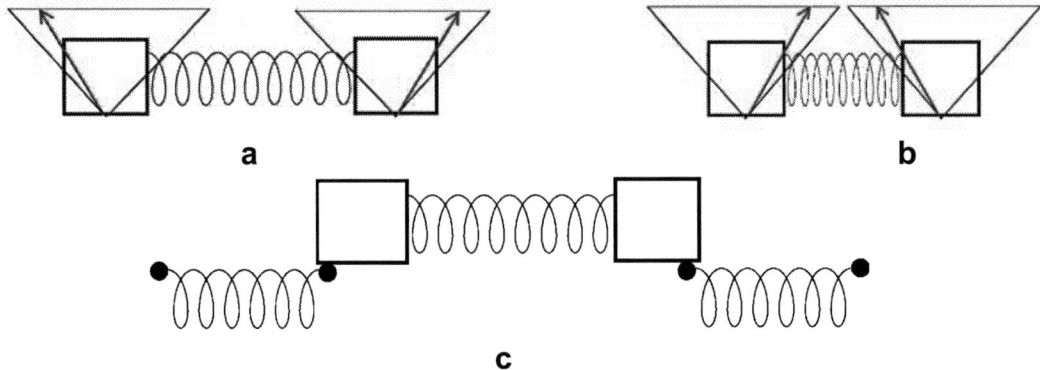

FIGURE 1 Example showing the possibility of multivalued solutions for the set of equilibrium forces in presence of dry friction. Two identical blocks, connected by a spring, rest on a horizontal frictional foundation. There is a large ensemble of equilibrium contact forces, that all satisfy the static Coulomb criterion requiring that the contact force belongs to lie within the Coulomb cone. At equilibrium, the spring can be either stretched (Fig. 1.a) or compressed (Fig. 1.b), depending upon the preparation process. On the other hand, the usual regularization procedure consists in substituting the frictional contacts by tangential springs (Fig. 1.c). In the latter case, the original nonvariational problem reduced to a variational one, which admits an unique solution for the forces. If now we consider microscopic grains in contact, note that the elastic deformations at contacts (modelled here by the spring) are inaccessible.

EXPERIMENTAL SETUP

In order to gain insight into the nature of the stress equations, we performed a point-punch test on a two-dimensional packing. The point-punch test represents the Green function for the mechanical response of the discrete medium, provided it is linear (an assumption that is also questionable [9]). The sample studied is prepared by cutting a 10-mm thick elastomeric plate into identical cylinders (8 mm diameter, 10 mm long) grains. The collection of grains is then packed into a triangular compact lattice (i.e. centered hexagonal) bounded by a rigid metal frame. The frame ensures zero-displacement boundary conditions. The Young's modulus and Poisson ration of the polyurethane elastomer are respectively $Y = 4\ 10^6$ Pa and $\nu = 0.46$. The punch consists of a steel blade 10-mm long, 3 mm-thick and with Young's modulus $Y' = 2\ 10^{11}$ Pa. The elastomeric material was chosen because it small Young's modulus allows a high relative precision in the measurement of the deformations induced by a gentle point load. Note that the 2d triangular packing can be considered as *isotropic* from the viewpoint of linear elasticity.

Previous experimental studies probing the response of a granular pile submitted to a point load led to controversial findings. Using a photoelastic visualization method and a piling constituted of square tiles, Da Silva and Rajchenbach [10] concluded that their observations cannot be interpreted with linear elastic modelling, and that they were rather consistent with a hierarchical process for stress transmission, as proposed by Harr [11] and Coppersmith et al. [12]. Indeed, in two dimensions, the elastic response of a semi-infinite medium to a point normal forcecan be described as follows [13]: if the origin (of polar coordinates r, θ) is taken as the point of application of the load P, the stress is everywhere radial, and its magnitude is given by $\sigma_r(r, \theta) = 2P\cos\theta /\pi r$. Hence the contours of constant stress magnitude are a set of circles passing through the point of application of the force. This result holds for a semi-infinite elastic medium, and the iso-stress contours are slightly modified in the presence of rigid boundary conditions, as discusssed below. By means of the same photoelastic method, but using pentagonal grains, Geng et al. [14] observed a wedge-shaped brighten domain (with a wedge angle close to 60°), which is not consistent with the circular iso-stress contours (slightly modified by the boundary conditions) expected from isotropic, linear elasticity.

In our experimental procedure, a sequence of pictures is taken while the external load is increased. Then successive pictures are processed in order to access the displacement field. For the sake of improving the accuracy, the bulk of the grains has been seeded with fine tracer particles (0.2 mm diameter) which allows for a precise tracking of local displacements. The measurement accuracy is of the order of 2 pixels, which corresponds to 0.02 cylinder diameter. The experimental load-cell is sketched in Fig. 2.

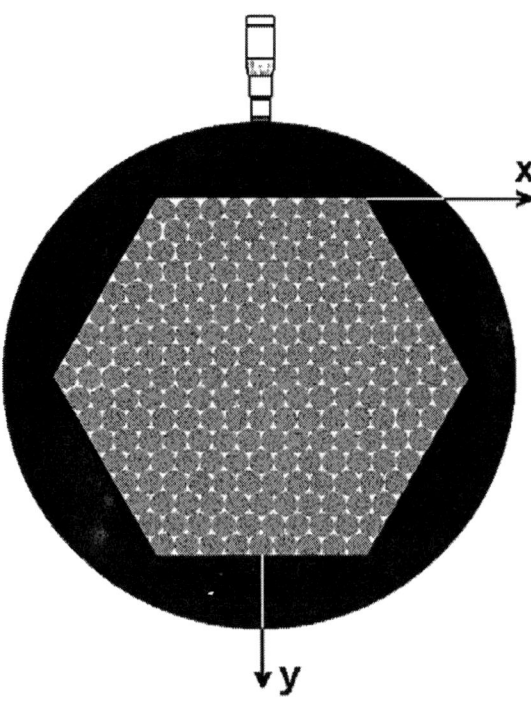

FIGURE 2. Sketch of the load-cell. The collection of cylinders is packed into a triangular (centered hexagonal) compact lattice, and confined into a rigid hexagonal frame. Cylinders are 8 mm in diameter and the ends were seeded with tracer particles. The punch is progressively moved inwards, while the displacements of the tracers are tracked.

EXPERIMENTAL RESULTS

In Fig. 3, we show the the local displacement amplitudes obtained in the discrete packing (full circles), resulting from a punch indentation of 2 mm, as a function of the position (x,y). Note that the observed fluctuations should not be attributed to limitations on the measurement accuracy, but to the inhomogeneity in the displacements on the scale of one grain: tracers located at the vicinity of a contact between two grains undergo a displacement much larger than that experienced by a tracer located near the center of the same grain. It is of interest to compare the displacement field to that obtained in the case of a continuous elastic material. With that aim, we positioned in the same hexagonal cell a continuous plate of the same thickness (10 mm) and made of the same elastomeric material, and the plate was submitted to the same punch test. The displacement obtained for the continuous plate and measured along the direction of loading, is shown in the same figure for comparison (Fig. 3, empty circles). Moreover, in order to compare the actual measured response and the theoretical elastic predictions, we show in the same figure the displacement curves obtained by solving analytically the Navier-Lamé equation (with Poisson ratio $\nu = -0.9$ and $\nu = + 0.5$) with the same boundary conditions, and in the planar strain approximation. Note that a variation in Young's modulus has no effect on displacements along the punching axis, since the present boundaries impose conditions not on stresses, but on displacements. The experimental points obtained for the continuous plate compare well with the theoretical elastic description, and thus clearly indicate an elastic response of the polymeric material.

On the other hand, the displacement field corresponding to the 2D discrete packing deviates unambiguously from the predictions of continuum elasticity. Instead, it rather follows an exponential-like decay with the distance along the direction of loading The decay along Oy suggests a localization process.

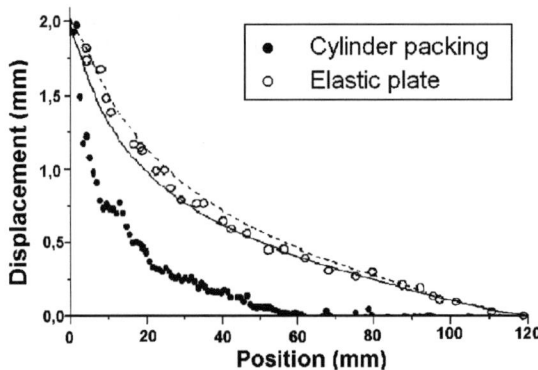

FIGURE 3. Magnitude of local displacements, measured along the punch axis Oy (● discrete piling, ○ continuous plate) and elastic predictions for various Poisson modulus (continuous line: $\nu = -0.9$, dashed line $\nu = +0.5$) in a plane strain modelling.

To sharpen our comparison of the mechanical response of a 2D granular packing, with that of a continuous elastic medium, it is interesting to examine the displacement field in locations off the axis loading. In Fig. 4, we show the map of the displacement amplitude as a function of the position *(x,y)*, coded in grey-levels. Apart from the fluctuations originating in the position of the tracers, according to their distance relative to grain centers or contacts (as described above), it is clear that the iso-displacement contours look closely like a family of ellipses.

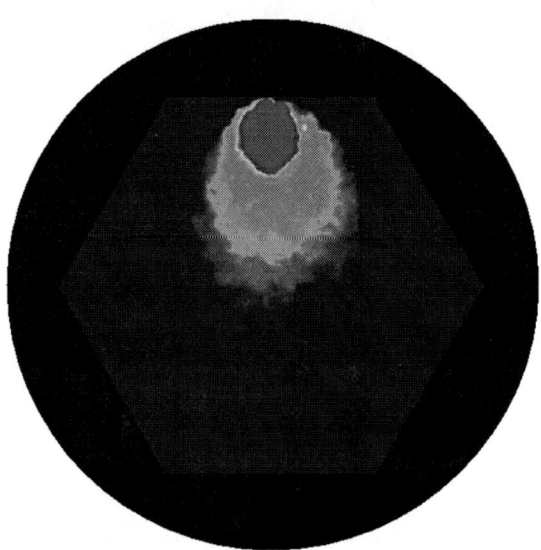

FIGURE 4. Same data as in Fig. 3. Displacement amplitude as a function of the position *(x,y)*, coded in grey-levels. The contours corresponding to displacements of equal amplitude are approximately elliptical.

From the assumption of the exponential-like decay of the displacements in the punching axis direction, and of the elliptic-like shape of the iso-displacement contours in the plane *(x,y)*, elementary geometrical considerations lead to the following form for the displacement magnitude in the discrete medium:

$$|displacement| \propto \exp(-y/a) \exp(-x^2/by) \qquad (1)$$

where the lengths *a* and *b* are here of comparable magnitude, typically 3 or 4 grain diameters. To confirm the validity of relation (1), we have plotted in Fig. 5 the displacement amplitude as a function of the transverse position *x*, for various ordinates *y*. It is clear that the set of experimental data can be reasonably fit by a family of Gaussian curves, with standard deviation varying as $\sqrt{y}$. Note that the previous data obtained by Da Silva et al. [10] can be represented by the same fitting function (1), with a (resp. b) $\cong$ 4 (resp. 0.5) grain diameters. The Gaussian widening along Ox is reminiscent of the diffusive, hierarchical models proposed previously by Harr [11] and Coppersmith et al. [12]. Nevertheless, note that the latter models predict a $1/\sqrt{y}$ decay along the direction of loading, rather than the exponential-like one found here.

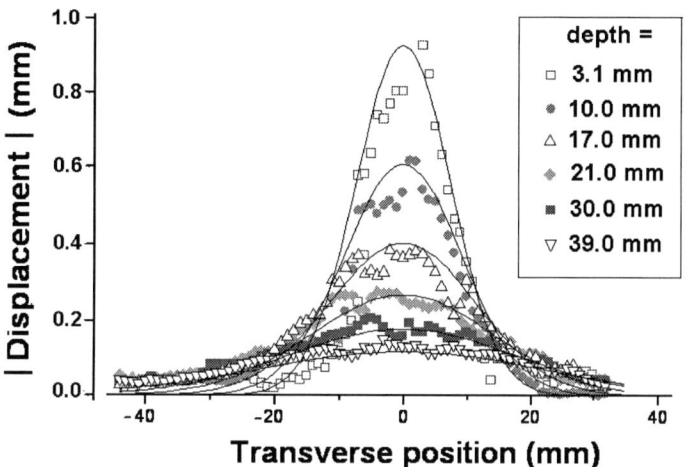

FIGURE 5. Displacement amplitude as a function of the deviation x to the axis of punching, plotted for various ordinates y ($\square$, $y=$ 3,1 mm), ($\bullet$, $y=$ 10.0 mm), (Δ, $y=$ 17.0 mm) ($\blacklozenge$, $y=$ 21.0 mm). ($\blacksquare$, $y=$ 30.0 mm). (∇, $y=$ 39.0 mm). Continuous lines correspond to Gaussian fits.

At this point, it is worth addressing the issue of sample size, i.e. the cell size compared to the grain diameter. The packing is here comprised of about 220 grains. As shown above, the localization length is typically 4 grain diameters, and the tracer displacements attain a zero value (within the experimental accuracy) at positions located far from the cell boundaries. We conclude that using larger cells (with rigid boundaries imposing a zero normal displacement) would not lead to any change both in the localization phenomenon or in the reported localization length. It is also of interest to clarify here the meaning of 'reversible deformation' in the present test. We mean that there was no measurable macroscopic irreversible deformation (or settling) remaining after the point load was removed. Note that we cannot exclude the possibility of microscopic slippage events occuring during the loading-unloading process at the level of microscopic contacts.

CONCLUSION

In summary, we have performed accurate measureements of the displacement field in a two-dimensional packing of cylindrical elastic grains, in response to a point load in the reversible regime of deformation. The coarse-grained displacement field deviates unambiguously from the predictions of standard linear elasticity. We have found that the deformation amplitude, measured along the loading axis Oy, follows closely an exponential decay as a function of the distance y to the punch, which indicates a localization process. On the other hand, the iso-displacement contours resemble a family of ellipses, and the transverse width of the strained region varies approximately as $\sqrt{y}$. We emphasize that individual grains behave elastically, and the collective response of the packing is reversible, but that the coarse-grained strain field cannot be accounted for by a linear elastic model.

REFERENCES

1. K. Terzaghi, *Theoretical Soil Mechanics,* John Wiley & Son, New York, 1943.
2. H. Hertz, *J. Reine Angew. Math.* **92**, 156 (1881).
3. J.J. Moreau, "Indeterminacy due to dry friction in multibody dynamics", in *European Congress on Computational Methods in Applied Sciences and Engineering,* edited by P. Neittaanmäki, T. Rossi, S. Korotov, E. Oñate, J. Périaux & D. Knörzer, ECCOMAS 2004, , Jyväskylä, Finland, 2004.
4. T. Unger, J. Kertész and D. Wolf, *Phys. Rev. Lett.* **94**, 178001 (2005).
5. S. Alexander, *Phys. Reports* **296**, 65-236 (1998).
6. J.J. Moreau, "Unilateral contact and dry friction in finite freedom dynamics" in *Nonsmooth Mechanics and Applications*, edited by J.J. Moreau and P.D. Panagiotopoulos , CISM Courses and Lectures 302, p. 1-82, Springer-Verlag, Wien, New-York, 1988, pp. 1-82.
7. W.J.M. Rankine, *Phil. Trans. Roy. Soc. Lond.* **147**, 9-27 (1857).
8. G. H. Darwin, *Minutes of the Proc. Inst. Civ. Eng.* **71**, 350-378 (1883).
9. J.J. Moreau, "Non linéarité de la réponse d'un granulat sec ", Conference, Carry le Rouet, 2004.
10. M. Da Silva and J. Rajchenbach, *Nature* **406**, 708-710 (2000).
11. M.E. Harr, *Foundations of Theoretical Soil Mechanics*, McGraw-Hill, New York, 1966.
12. S.N. Coppersmith, C.H. Liu, S. Majumdar, O. Narayan, T.A. Witten, *Phys. Rev. E* **53**, 4673-4885 (1996).
13. J. Boussinesq, *Application des Potentiels à l'Etude de l'Equilibre et des Mouvements des Solides Elastiques*, Gauthier-Villars, Paris, 1885.
14. J. Geng, D. Howell, E. Longhi, R.P. Behringer, G. Reydellet, L. Vanel and E. Clement, *Phys. Rev. Lett.* **87**, 035506 (2001).

kGamma distributions in granular packs

T. Aste*,**, G. W Delaney[‡] and T. Di Matteo**

*School of Physical Sciences, University of Kent, Canterbury, Kent, CT2 7NH, UK.
[†]Department of Applied Mathematics, Research School of Physical Sciences and Engineering, The Australian National University, 0200 Canberra, ACT, Australia.
**King's College Department of Mathematics, Strand London WC2R 2LS, UK.
[‡]CSIRO Mathematical and Information Sciences, Private Bag 33, Clayton South, Vic, 3168, Australia.

Abstract. It has been recently pointed out that local volume fluctuations in granular packings follow remarkably well a shifted and rescaled Gamma distribution named the *kGamma* distribution [T. Aste, T. Di Matteo, Phys. Rev. E 77 (2008) 021309]. In this paper we confirm, extend and discuss this finding by supporting it with additional experimental and simulation data.

Keywords: k-Gamma distribution; packing fraction; Voronoi Volume; local density fluctuations
PACS: 81.05.Rm,45.70.Cc,45.70.Vn,45.70.Mg

INTRODUCTION

The description and understanding of amorphous structures is very challenging because the lack of any translational symmetry makes it hard to encode the structural information into a compact form. The absence of periodicity does not exclude however repetitiveness. Indeed, in amorphous systems several local configurations, or 'motifs', are repeated often. However, one must consider that these 'repetitions' typically concern similar but non identical units with small variations between one and the other. Furthermore, the choice of the parameters that identify the local motifs is somehow arbitrary and, as a consequence, depending on the detail of the description, one can gather information about the local structure at different levels. The identification of these motifs and the study of their variations are the fundamental first step toward the understanding of the structure of granular materials and of other disordered systems such as amorphous phases. In these disordered systems, the structure is necessarily defined in statistical terms and it can be characterized by the probability of occurrence of a given motif. Statistical mechanics is the theoretical instrument to calculate such probability.

In this paper we focus on equal-sized bead packings both from experiments and simulations. For these systems we identify the local structural motifs with the Voronoi cells which are defined as the portion of space closest to a given bead center than to any other in the packing. In particular, we choose to consider the volume V of such cells as the only parameter identifying the local structural organization. The overall packing fraction ϕ of the sample is directly related to the average Voronoi volume $\langle V \rangle$ by $\phi = V_s / \langle V \rangle$ with $V_s = \pi d^3/6$ the volume of a bead with diameter d. The Voronoi cell defines a region of pertinence around each bead. The fluctuations of the Voronoi volumes are therefore a measure of the local variations in the packing fraction. Indeed, the local fraction of occupied volume is $\varphi = V_s/V$. In this paper we discuss the distribution of such volumes showing that it is very well reproduced by a kGamma distribution [1]. Remarkably, this functional form is retrieved in a wide set of very different systems from idealized hard sphere packings to glass beads in water.

EXPERIMENTS AND NUMERICAL SIMULATIONS

Acrylic beads in air

The experimental data sets of bead configurations that we analyze in this paper are from the database on disordered packings [2] which contains structural data from experimental sphere packings obtained by X-ray Computed Tomography. Specifically our study concerns 6 samples (A-F) composed of acrylic beads prepared in air within a cylindrical container with an inner diameter of 55 *mm* and filled to a height of ~ 75 *mm* [3, 4, 5]. Samples A and C contain $\sim 150,000$ beads with diameter $d = 1.00$ *mm* and polydispersity within 0.05 *mm*. Samples B, D-F contain $\sim 35,000$

beads with diameter $d = 1.59$ *mm* and polydispersity within 0.05 *mm*. The two samples at lower packing fraction (A, B) were obtained by placing a stick in the middle of the container before pouring the beads and then slowly removing the stick [6]. Sample C was prepared by gently and slowly pouring the spheres into the container. Sample D was obtained by a faster pouring. In sample E, a higher packing fraction was achieved by gently tapping the container walls. The densest sample (F) was obtained by a combined action of gentle tapping and compression from above (with the upper surface left unconfined at the end of the preparation). To reduce boundary effects, the inside of the cylinder was roughened by randomly gluing spheres to the internal surface. The packing fraction of each of the samples is estimated at: A, $\phi \sim 0.586$; B, $\phi \sim 0.596$; C, $\phi \sim 0.619$; D, $\phi \sim 0.626$; E, $\phi \sim 0.630$; and F, $\phi \sim 0.640$.

Glass beads in water

Twelve other samples (FB12-24 and FB27) containing about 150,000 glass beads with diameters 0.25 *mm* are also analysed. The packings were prepared in water by means of a fluidised bed technique [2, 7, 8] within a vertical polycarbonate tube with an inner diameter of 12.8 *mm* and a length of 230 *mm*. Packing fractions between 0.56 and 0.60 were obtained by using different flow rates with higher rates associated with lower packing fractions. After each flow pulse, the particles sediment forming a mechanically stable packing.

Identification of the grain positions by X-ray computed tomography

X-ray Computed Tomography (XCT) is used to calculate the coordinates of the bead centers. This is done by applying a convolution method to the three-dimensional XCT density map efficiently implemented by use of (parallel) Fast Fourier Transform. Furthermore a watershed method is also used to identify distinct grains. With this technique the estimation of the position of the centre of mass of each grain can be achieved with a precision better than 0.1% of the sphere diameter. This is well below the grain polydispersity that is estimated around 1 to 3 % depending on the sample.

Lubachevsky-Stillinger simulations

A set of simulated packings are produced by using a modified Lubachevsky-Stillinger (LS) algorithm [9]. The simulation is an event-driven Newtonian dynamics in which the spheres are considered perfectly elastic without any rotational degree of freedom and with no friction. The simulation is performed in a cubic box with periodic boundary conditions, without gravity. During the simulation, the radii of the spheres are gradually increased from a very loose initial state to more densely packed configurations. In these simulations the principal control parameter is the growth rate for the sphere radii. Small values of growth rates will result in crystallisation. To avoid crystallization the growth rate should be rather large, forcing the packing into "jammed" non-crystalline structures where the spheres cannot be further expanded at finite pressure [10, 11]. Simulations were performed by using the code at: http://cherrypit.princeton.edu/Packing/C++/ on $N = 10000$ spheres with initial temperature 0.1, with initial packing fraction 0.1 and with a number of event per cycle equal to 20. The spheres were expanded with different growth rates between 2e-5 to 0.5, until a maximal reduced pressure of 10^{12} was reached [12].

Discrete element method simulations

We use Discrete Element Method simulation (DEM) which integrates the Newton equation of motion with both translational and rotational degrees of freedoms for elasto-frictional spheres under gravity [13, 14, 15]. The spheres interact only when overlapping, with a normal repulsive force $F_n = k_n \xi_n^{3/2}$ where $\xi_n = d - |\vec{r}_i - \vec{r}_j|$ is the overlap between grains of diameters d with centres at $\vec{r}_i$ and $\vec{r}_j$ and k_n is the stiffness parameter ($k_n = d/2Y/(3(1-P^2))$, with Y the Young's modulus and P the Poisson ratio) [16, 17]. Tangential force under oblique loading is also considered as $F_t = -\min(|k_t \xi_n^{1/2} \xi_t|, |\mu F_n|) \cdot \text{sign}(v_t)$, with $\xi_t = \int_{t_0}^{t} v_t(t') dt'$ the displacement in the tangential direction that has taken place since the time t_0 when the two spheres first got in contact, where v_t is the relative shear velocity and μ

is the kinematic friction coefficient between the spheres and k_t the tangential stiffness parameter typically assumed $2/7k_n$ [18]. Normal visco-elastic dissipation $F_n = -\gamma_n \xi_n^{1/2} \dot{\xi}_n$ (with $\dot{\xi}_n$ the normal velocity) and a viscous friction force $F_t = -\gamma_t v_t$ [19] are also included.

Here we report data for 64 simulations prepared by pouring, into a cylinder with a rough boundary, spherical beads with diameters 3mm. The cylinder had section $\approx 22d$ and it was filled with 9614 beads to an height of $\approx 56d$ resulting in an initial packing fraction around 0.25. The beads were let sediment under gravity reaching a final mechanically stable state with packing fraction in a range between 0.55 to 0.64 depending mainly on the value of the friction coefficient (larger frictions smaller packing fractions), and also on the gravity and on the stokes constant. The time step has been set at 8.0e-6 sec, the grain mass is 0.003 kg and k_n =1.9e7, k_t =5.6e6. The simulations were performed at various k_s from 1e-3 to 1e2, various values of gravity from 1 to 10 and several friction coefficients from 1e-4 to 1e4.

EQUILIBRIUM STATISTICAL MECHANICS PREDICTION FOR THE VORONOI VOLUME DISTRIBUTION

Granular structures are disordered. This means that, differently form crystals, a unique "ideal" structure where all the grains positions are uniquely assigned does not exist. There are instead a very large number of structures that have equivalent global properties (packing fraction, mechanical properties, etc.) but differ in the way the grains are locally arranged. For these disordered packings we aim to find a relation between global functional properties and local structural properties and identify the probability of occurrence of specific local structural features for given global properties. Statistical mechanics is the theoretical framework that allows us to perform this kind of computation.

A statistical mechanics approach for granular systems was firstly proposed by Edwards in 1989 [20]. The key idea is that these non-thermal systems can be described by using a formalism very similar to the one developed for molecular gasses by substituting the constraint on the energy with a constraint on the volume occupied by the system. Although this is one of the few examples of extension of classical statistical mechanics concepts to a-thermal systems, the Edwards' approach is rather straightforward. Any reader with some familiarity with thermal physics and classical statistical mechanics will recognize that the forthcoming statistical description of granular systems is formally identical to the one for molecular gasses with 'E' substituted with 'V'. However, conceptually, the approach is not trivial because in granular systems we lack mechanisms equivalent to temperature and molecular chaos that allow thermal systems to explore homogeneously the phase space. In granular systems energy is dissipated in inelastic collision and the system soon reaches a static state with immobile grains at mechanical equilibrium. Such state can only be changed by perturbing the system, injecting energy, for instance through vibrations or fluid flow [21, 7]. For a given preparation protocol one aims to identify the probability of occurrence of some specific structural features and their related functional properties. In order to associate a probability to a given structure, one should (virtually) explore the whole set of possible structures which are attainable through a given preparation protocol and compute the frequency of occurrence of that specific structure within the ensemble of all attainable realizations. Within equilibrium statistical mechanics approach this computation is typically performed by assigning an entropy and maximizing it; finding in this way the configurations with maximum likelihood. It is beyond the propose and the possibility of this paper to fully expose the subdue issues around this kind of approach that have been debated in the literature for the last twenty years since 1989 [20, 22, 23, 24, 25, 26]. Here we are merely comparing the theoretical prediction from a statistical mechanics approach (namely Eq.4) with data from experiments and computer simulations. To have a better insight of our approach to this problem and for a formal deductive derivation of Eq.4, the interest reader can refer to [27] and references therein. Let us hereafter only briefly sketch the main ideas and the main passages of this approach.

In analogy with the Edwards' original approach here we consider the ensemble of mechanically stable configurations that can be achieved by means of a given preparation protocol resulting in a given average packing fraction over a large number of trials [20, 22, 23, 24, 25, 26]. Here we look at the statistics of the local configurations of each Voronoi cell and we maximize entropy to calculate the probability $p(V)$ for a cell with volume V in a sample with packing fraction ϕ where the average Voronoi volume is $\langle V \rangle = V_s/\phi$. A classical equilibrium statistical mechanics theory 'a la Edwards' gives:

$$p(V) = \frac{\Omega(V)e^{-\beta V}}{\sum_{V'}\Omega(V')e^{-\beta V'}} \quad , \qquad (1)$$

with β a Lagrange multiplier which is determined by the constraint on the average volume:

$$\langle V \rangle = \sum_V V p(V) \quad . \qquad (2)$$

Here the challenge is to compute $\Omega(V)$ which is the 'density of states' counting the number of microscopic configurations associated with a Voronoi volume V.

Let us note that Eq.1 is the analogous for these non-thermal systems to the Boltzmann distribution for molecular gasses where in this case the particle energy is substituted with the Voronoi cell volume.

Explicit derivation of the probability distribution from a simple hypothesis

In order to compute $\Omega(V)$ here we use a very simple hypothesis: there are k 'degrees of freedom' contributing to the volume of each Voronoi cell. The idea is that each Voronoi cell is composed of k elements each one contributing independently to the cell volume V. Each of these 'elementary cells' can have arbitrary volumes larger than v_{min} under the condition that their combination must add to a total volume V. Under this assumption $\Omega(V)$ can be computed exactly:

$$\Omega(V) = \frac{1}{\Lambda^{3k}} \int_{v_{min}}^{V} dv_1 \int_{v_{min}}^{V} dv_2 .... \int_{v_{min}}^{V} dv_k \delta(v_1 + v_2 + ... + v_k - V) = \frac{(V - kv_{min})^{k-1}}{\Lambda^{3k}(k-1)!} \quad , \tag{3}$$

with Λ a constant analogous to the Debye length. Substituting into Eq.1, and by using Eq.2 we obtain for the Lagrange multiplier $\beta = k/(\langle V \rangle - kv_{min})$ and the probability $p(V)$ takes the form:

$$p(V) = \frac{k^k}{\Gamma(k)} \frac{(V - V_{min})^{(k-1)}}{(\langle V \rangle - V_{min})^k} \exp\left(-k \frac{V - V_{min}}{\langle V \rangle - V_{min}}\right) \quad , \tag{4}$$

with $V_{min} = kv_{min}$ which is the minimum volume attainable for a Voronoi cell in the packing. For a packing of equal spheres V_{min} is exactly known being the volume of a dodecahedral cell circumscribing the sphere which is $V_{min} = 5^{(5/4)}/\sqrt{2(29+13\sqrt{5})}d^3 \simeq 0.694 d^3$ [6]. Eq.4 is a Gamma distribution in the variable $V - V_{min}$; it is characterized by a 'shape' parameter k and a 'scale' parameter $(\langle V \rangle - V_{min})/k$ [28]. We call such a function: *kGamma* distribution [1]. Interestingly, a mathematical study for the Voronoi statistics in two dimensional point processes predicts a gamma distribution for the cell area distribution [29].

The mean of the distribution $p(V)$ is $\langle V \rangle$ and its variance is

$$\sigma_v^2 = \frac{(\langle V \rangle - V_{min})^2}{k} \quad , \tag{5}$$

which implies

$$k = \frac{(\langle V \rangle - V_{min})^2}{\sigma_v^2} \quad . \tag{6}$$

This last equation gives directly the parameter k from a measure of the variance of the distribution. Therefore, there are no free fit parameters in Eq.4.

It might be of some use to compute explicitly the related distribution for the local packing fraction φ which is given by the identity $p(\varphi)d\varphi = p(V)dV$ yielding to

$$p(\varphi) = \frac{k^k}{\Gamma(k)} \frac{\varphi_{max}}{\varphi^2} \frac{(\varphi_{max}/\varphi - 1)^{(k-1)}}{(\varphi_{max}/\phi - 1)^k} \exp\left(-k \frac{\varphi_{max}/\varphi - 1}{\varphi_{max}/\phi - 1}\right) \quad , \tag{7}$$

with $\varphi_{max} = V_s/V_{min} \simeq 0.75$ the maximum attainable local packing fraction in equal sphere packings.

VORONOI VOLUME DISTRIBUTIONS FROM EXPERIMENTS AND SIMULATIONS

We have tested the validity and resilience of the kGamma behavior over a set of several hundreds numerical simulations and over 18 different experiments. The simulations consisted of both Lubachevsky-Stillinger newtonian dynamics of frictionless hard spheres and DEM simulations of elasto-frictional spheres. The experiments include dry and fluidized bead samples.

Figure 1 shows the resulting distribution of the Voronoï volumes. One can see that such distributions span a very broad interval of volumes with V between $\approx 1.3 V_s$ and $\approx 2.5 V_s$ with large differences between different samples

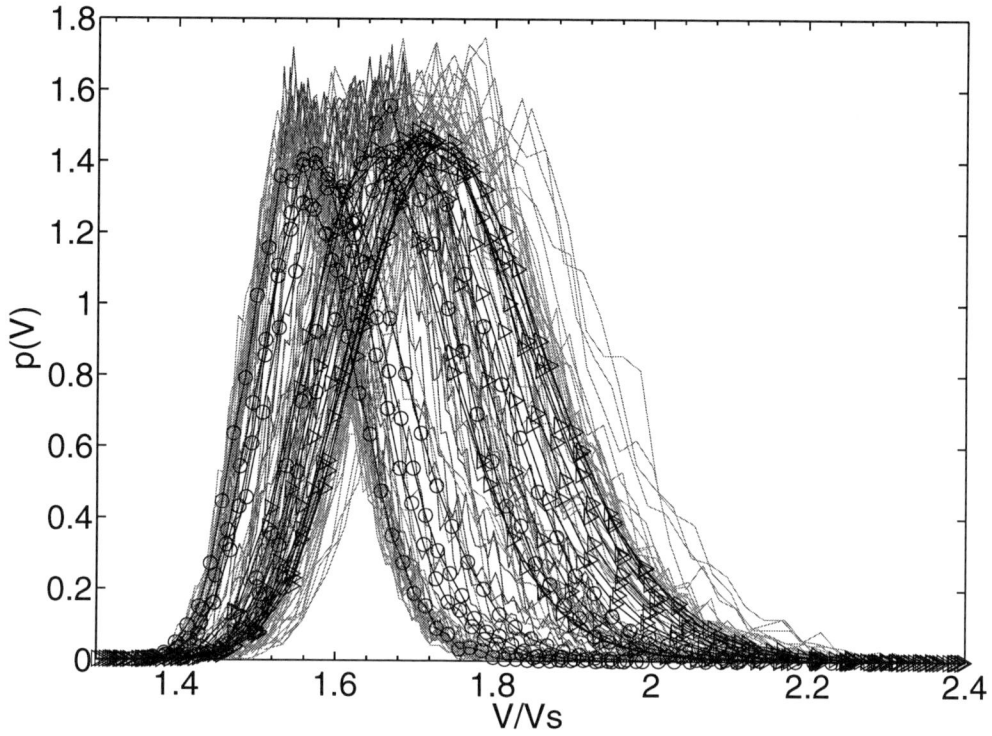

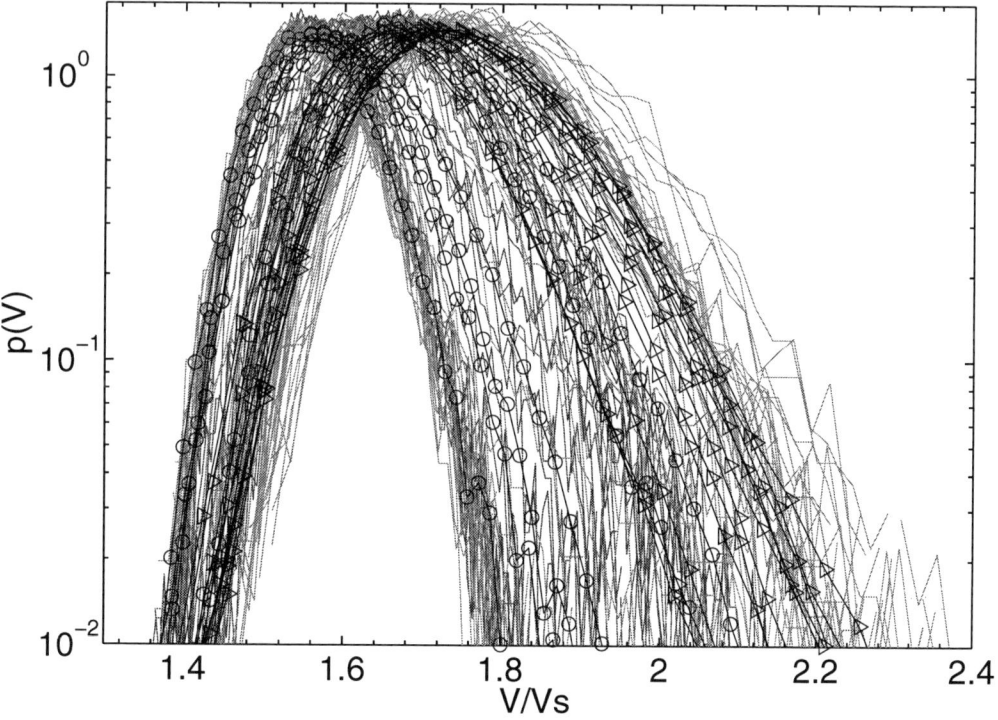

FIGURE 1. (Color online). (top) Voronoï volume distributions from all the experimental and simulation data. The ○ refer to dry acrylic beads experiments and the ▷ refer to glass beads in water. The read lines (color online) are Lubachevsky-Stillinger simulations and the green lines are DEM simulations. $V_s = \pi/6d^3$ is the volume of a spherical bead. (bottom) Same plot in semi-log Y scale.

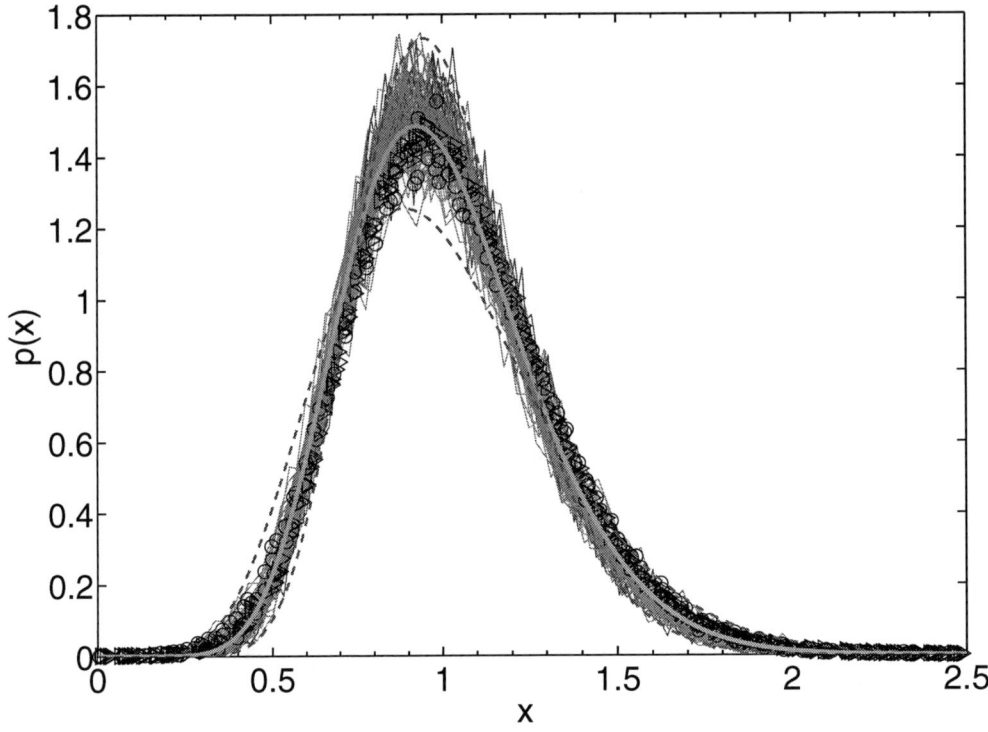

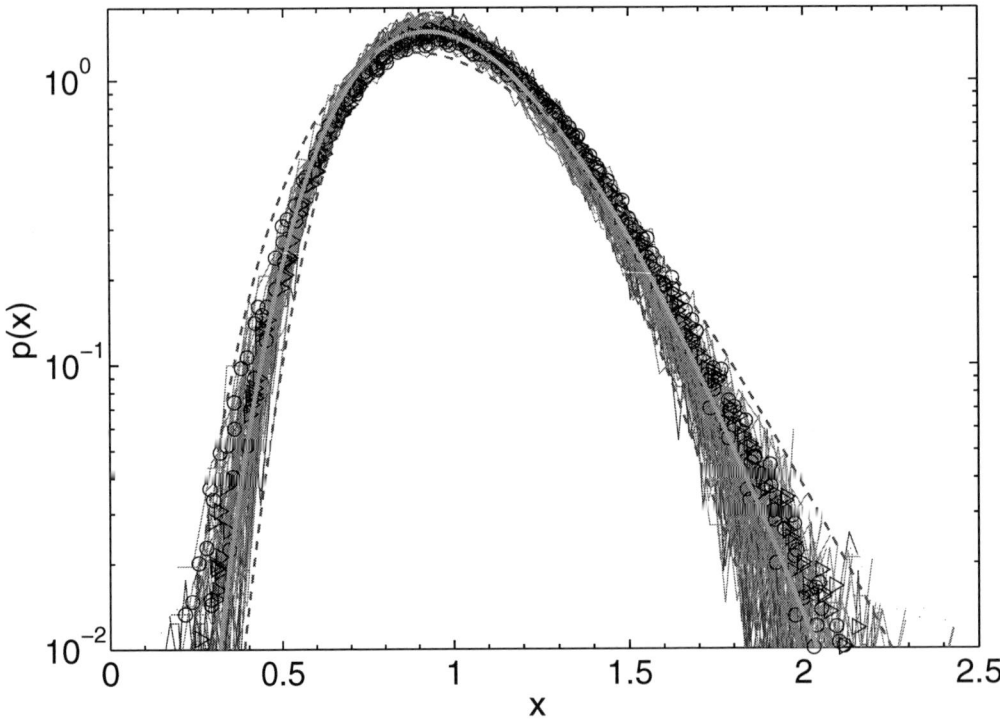

FIGURE 2. (Color online). (top) Same Voronoï volume distributions as in Fig.1 but plotted vs. $x = (V - V_{min})/(\langle V \rangle - V_{min})$. (bottom) Same plot in semi-log Y scale. The tick full line is the kGamma function $p(x) = k^k/\Gamma(k)x^{k-1}\exp(-kx)$ for $k = 13$. The two dashed lines are two kGamma functions for $k = 9$ and $k = 18$.

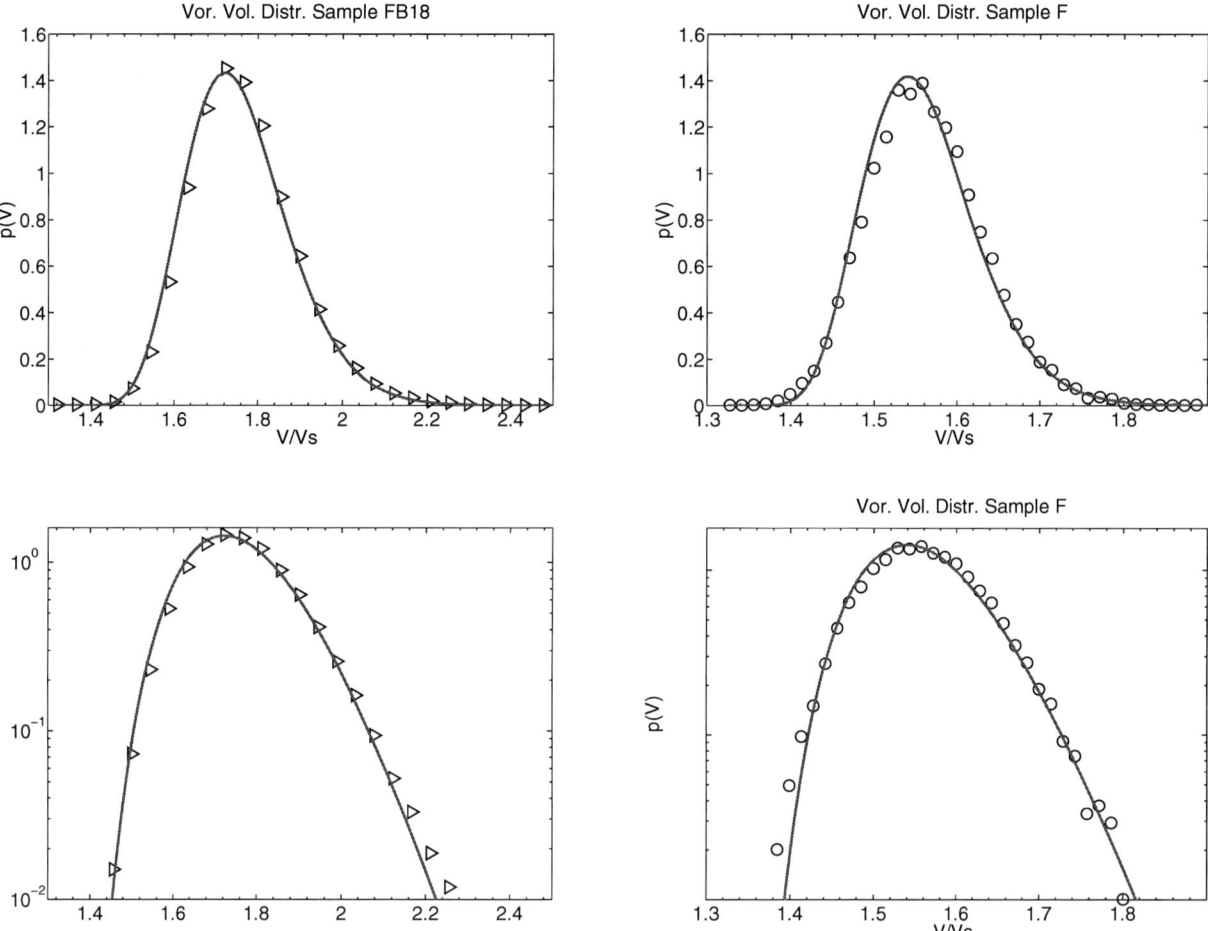

FIGURE 3. (Color online). (left) Experimental Voronoï volumes distributions for the sample with lowest packing fraction obtained by using fluidized bead technique for glass beads in water (FB18). (right) Experimental Voronoï volumes distributions for the sample with highest packing fraction obtained by pouring acrylic beads in air (F). (bottom) Same plots in semi-log Y scale. The lines are the kGamma distributions with $k = (\langle V \rangle - V_{min})^2 / \sigma_v^2$ with σ_v the measured standard deviation and $\langle V \rangle = V_s/\phi$ with ϕ the measured packing fraction. There are no adjustable parameters or fits.

shown both in the average values and in the distribution spreading. We observe that all distributions show some degree of asymmetry around the maximum with larger probabilities for large volume fluctuations.

From Eq. 4, one can see that *kGamma* distributions characterized by similar values of k must result into similar behaviors when plotted as function of $x = (V - V_{min})/(\langle V \rangle - V_{min})$. Figure 2 shows the plot of all the distributions as a function of such shifted-rescaled variable. We note that all distributions collapse into a very similar functional form which is very well described by the kGamma function $p(x) = k^k/\Gamma(k) x^{k-1} \exp(-kx)$ with k ranging in a narrow interval.

The goodness of the description of these distributions by means of the kGamma function in Eq.4 can be judged from Fig.3 which shows the agreement between the prediction from the kGamma function $p(V)$ and the measured data from experiments. Similar agreements are found across all samples from both experiments and simulations. It should be stressed that in this plot there are no adjustable parameters or fitting constants. Indeed, the only two parameters in Eq.4 are $\langle V \rangle$ and k which are uniquely determined respectively from the sample packing fraction ($\langle V \rangle = V_s/\phi$) and from the measured standard deviation of the Vorornoi volumes ($k = (\langle V \rangle - V_{min})^2/\sigma_v^2$, Eq.6).

The impressive fact of such an agreement is that these systems are very different (ideal Newtonian spheres, elasto-frictional spheres under gravity, real experimental acrylic beads in air and glass beads in water) and they are prepared in very different ways (pouring, tapping, fluid flows, molecular dynamics, shearing). The collapse of all

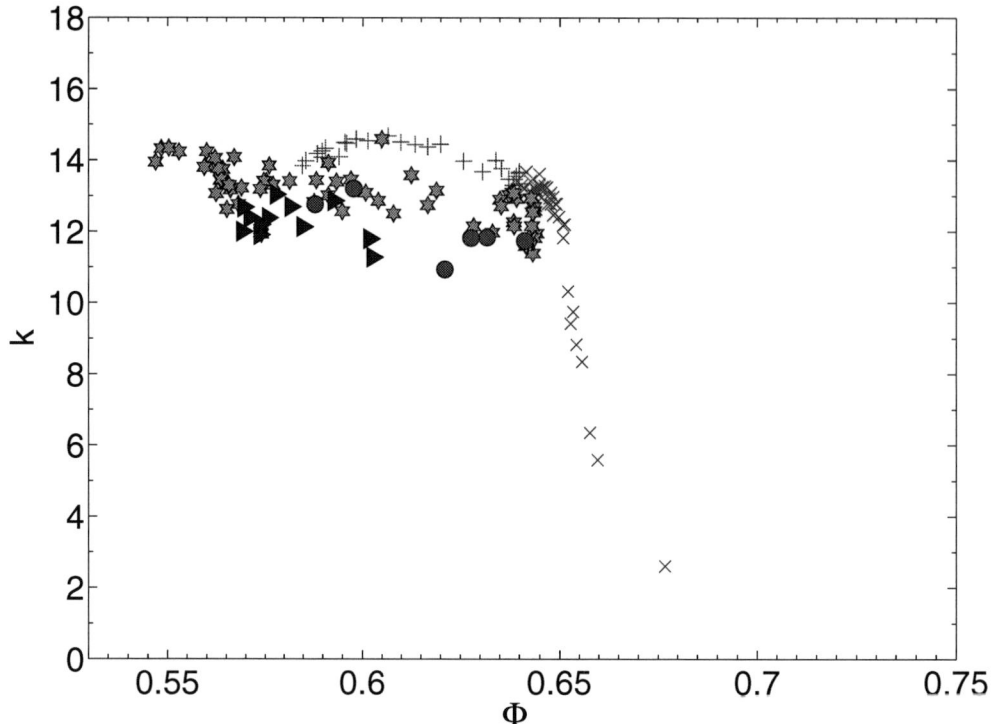

FIGURE 4. (Color online). Behavior of k calculated from $k = (\langle V \rangle - V_{min})^2/\sigma_v^2$ (Eq.6) vs. packing fraction ϕ. The $+$ refer to Lubachevsky-Stillinger simulations, the $\star$ refer to DEM simulations, the $\circ$ refer to dry acrylic beads experiments and the $\triangleright$ refer to glass beads in water. The $\times$ refer instead to Lubachevsky-Stillinger simulations above $\phi \simeq 0.64$ where a poly-crystaline phase starts to emerge.

these distributions around a unique functional form suggests that the main driving mechanism which determines these fluctuations is an exchange of volume among Voronoi cells that possesses only a small number of degrees of freedom. Accordingly with our hypothesis, such a number is given by the parameter k.

In Fig.4 we report the measured values of k as function of the sample packing fraction calculated from the relation $k = (\langle V \rangle - V_{min})^2/\sigma_v^2$ (Eq.6). We observe a rather narrow and homogeneous range of values laying between 11 and 15 for all samples. A sharp drop in k is observed above $\phi \simeq 0.64$ where in the Lubachevsky-Stillinger simulations a crystalline phase starts to nucleate and grow. Let us note that in this phase a peak at larger packing fractions appears in the volume distributions (not reported in Figs.1 and 2 to avoid confusion.).

The good predictive potential of kGamma distributions for volume fluctuations in granular assemblies can also be inferred from the simple measure of the standard deviation. Indeed, standard deviations can be easily measured and require smaller statistical sets with respect to the whole distribution. Figure 5 reports the trend of the measured standard deviation versus the packing fraction for all the samples. It is clearly evident from the figure that they all follow a common decreasing trend before the crystallization onset above $\phi \simeq 0.64$. The line in the figure is the the prediction from Eq.5: $\sigma = (V_s/\phi - V_{min})/\sqrt{k})$, for $k = 13$.

CONCLUSIONS

In this paper we have shown that kGamma distributions describe very well the observed Voronoi volume distributions in a wide variety of systems from simulated hard spheres to real experimental packings. The agreement between the predicted distribution and the measures is remarkable also considering that there are no adjustable parameters or fitting constants.

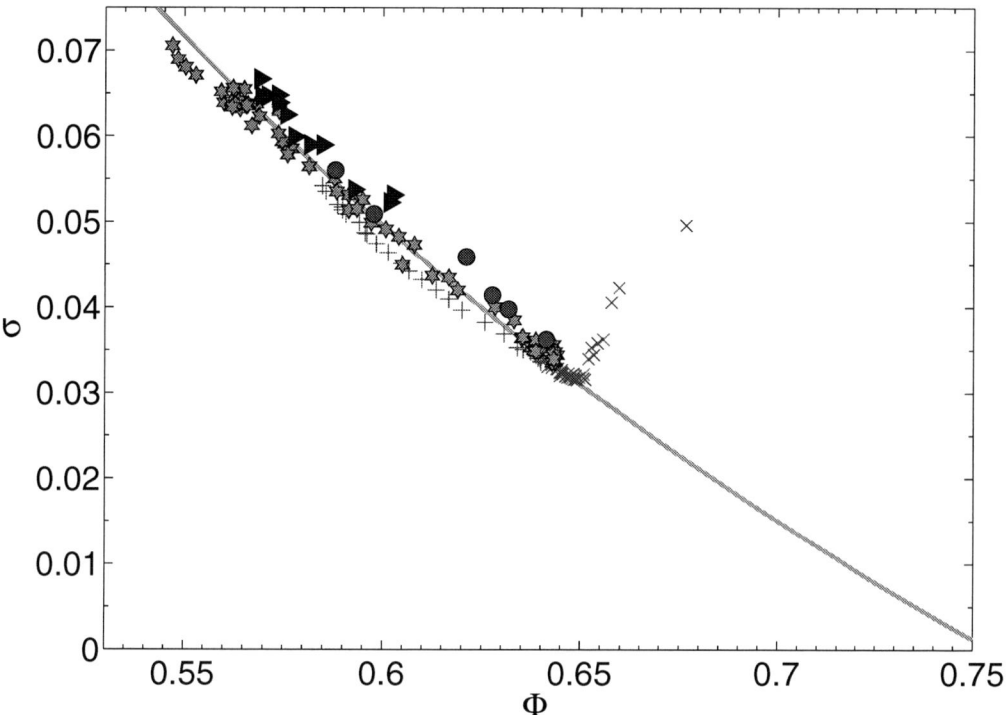

FIGURE 5. (Color online). Behavior of σ_v vs. packing fraction ϕ. The line is the prediction from Eq.5 (i.e. $\sigma = (V_s/\phi - v0)/\sqrt{k}$) for $k = 13$. The symbols are the same as in Fig.4.

We have briefly explained as a statistical mechanics approach can be used to retrieve such a distribution from a very simple hypothesis that the Voronoi cells in the systems have k degrees of freedom associated to their volumes. Despite the very good agreement with the empirical data that has been shown in this paper, this hypothesis is certainly quite strong and unrealistic. Indeed, it might be true that in average there are k degrees of freedom per Voronoi cell but it is unlikely that *each* Vorornoi cell has *exactly* k degrees of freedom. A relaxation of this hypothesis on a more realistic ground would produce different outcomes predicting for instance other kinds of Gamma distributions. However, the beauty and the strength of the present approach is that we have only one parameter that is fully determined by the measured standard deviation. In our view, the exceptional agreement of all the experimental and numerical data with the prediction from kGamma distribution does not justify, at the present, the introduction of a more complicated theoretical framework with the consequent incorporation of extra adjustable parameters.

Future studies will focus on the effect of polydispersity and shapes [30].

Acknowledgements

We thank Antonio Coniglio, Mario Nicodemi, Massimo Pica Ciamarra, Matthias Schröter and Harry Swinney for helpful discussions. Many thanks to T.J. Senden, M. Saadatfar, A. Sakellariou, A. Sheppard, A. Limaye for the help with tomographic data.

REFERENCES

1. T. Aste, and T. Di Matteo, *Phys. Rev. E* **77**, 021309 (2008).

2. See, *http://wwwrsphysse.anu.edu.au/granularmatter/* (2006-2009).
3. T. Aste, M. Saadatfar, and T. J. Senden, *Phys. Rev. E.* **71**, 061302 (2005).
4. T. Aste, M. Saadatfar, A. Sakellariou, and T. Senden, *Physica A* **339**, 16–23 (2004).
5. T. Aste, *J. Phys.: Condens. Matter* **17**, S2361–90 (2005).
6. T. Aste, and D. Weaire, *The Pursuit of Perfect Packing*, Institute of Physics, Bristol, 2000.
7. M. Schröter, D. I. Goldman, and H. L. Swinney, *Phys. Rev. E.* **71**, 30301 (R) (2005).
8. T. Aste, T. Di Matteo, M. Saadatfar, T. Senden, M. Schröter, and H. L. Swinney, *preprint, arXvi-0612063* (2006).
9. M. Skoge, A. D. F. H. Stillinger, and S. Torquato, *Phys. Rev. E.* **74**, 041127 (2006).
10. S. Torquato, T. M. Truskett, and P. G. Debenedetti, *Phys. Rev. Lett.* **84**, 2064–7 (2000).
11. M. Rintoul, and S. Torquato, *Phys. Rev. E.* **58**, 532–7 (1998).
12. A. Donev, S. Torquato, F. H. Stillinger, and R. Connelly, *Phys. Rev. E.* **71**, 011105 (2005).
13. S. Hutzler, G. Delaney, D. Weaire, and F. MacLeod, *American Journal of Physics* **72**, 1508–16 (2004),
14. G. Delaney, S. Inagaki, and T. Aste, "Fine tuning DEM simulations to perform virtual experiments with three-dimensional granular packings," in *Granular and Complex Materials*, edited by T. Aste, T. D. Matteo, and A. Trodesillas, Wold Scientific, 2007, vol. 8, pp. 169–85.
15. G. Delaney, S. Inagaki, T. D. Matteo, and T. Aste, "Virtual experiments on complex materials," in *Complex Systems II*, edited by D. Abbott, T. Aste, R. D. M. Batchelor, and T. D. M. and T. Guttmann, 2008, vol. 6802, pp. G8020-1–8.
16. L. Landau, and E. Lifshitz, *Theory of Elasticity*, Pergamon, New York, 1970.
17. H. A. Makse, N. Gland, D. L. Johnson, and L. Schwartz, *Physical Review E* **70**, 061302 (2004).
18. P. A. Cundall, and O. D. L. Strack, *Geotechnique* **29**, 47–65 (1979).
19. J. Schafer, S. Dippel, and D. Wolf, *J. Phys. I* **6**, 5 (1996).
20. S. Edwards, and R. Oakeshott, *Physica A* **157**, 1080–90 (1989).
21. H. M. Jaeger, S. R. Nagel, and R. P. Behringer, *Rev. Mod. Phys.* **68**, 1259 (1996).
22. A. Mehta, and S. F. Edwards, *Physica A* **157**, 1091–7 (1989).
23. S. Edwards, *Granular matter: an interdisciplinary approach*, Springer-Verlag, New York, 1994, pp. 121–40.
24. S. F. Edwards, and D. V. Grinev, *Phys. Rev. Lett.* **82**, 5397 (1999).
25. J. Brujić, S. F. Edwards, I. Hopkingson, and H. A. Makse, *Physica A* **327**, 201–12 (2003).
26. R. Blumenfeld, and S. F. Edwards, *Phys. Rev. Lett.* **90**, 114303 (2003).
27. T. Aste, and T. Di Matteo, *arXiv:0711.3239* (2007).
28. R. V. Hogg, and A. T. Craig, *Introduction to Mathematical Statistics*, Macmillan, New York, 1978.
29. K. Brakke, *Statistics of random plane voronoi tessellations*, unpublished (1986), http://www.susqu.edu/brakke/papers/voronoi.htm.
30. G. W. Delaney, S. Hutzler, and T. Aste, *Phys. Rev. Lett.* **101**, 120602 (2008).

Stress transmission and incipient yield flow in dense granular materials

Raphael Blumenfeld

Earth Science and Engineering, Imperial College London, London SW7 2AZ, UK
and
Cavendish Laboratory, J J Thomson Avenue, Cambridge CB3 0HE, UK

Abstract. Jammed granular matter transmits stresses non-uniformly like no conventional solid, especially when it is on the verge of failure. Jamming is caused by self-organization of granular matter under external loads, often giving rise to networks of force chains that support the loads non-uniformly. An ongoing debate in the literature concerns the correct way to model the static stress field in such media: good old elasticity theory or newcomer isostaticity theory. The two differ significantly and, in particular in 2D, isostaticity theory leads naturally to force chain solutions. More recently, it has been proposed that real granular materials are made of mixtures of regions, some behaving elastically and some isostatically. The theory to describe these systems has been named stato-elasticity.

In this paper, I first present the rationale for stato-elasticity theory. An important step towards the construction of this theory is a good understanding of stress transmission in the regions of pure isostatic states. A brief description is given of recently derived general solutions for 2D isostatic regions with nonuniform structures, which go well beyond the over-simplistic picture of force chains.

I then show how the static stress equations are related directly to incipient yield flow and derive the equations that govern yield and creep rheology of dense granular matter at the initial stages of failure. These equations are general and describe strains in granular materials of both rigid and compliant particles.

Keywords: Stress transmission, failure, yield
PACS: 45.70.-n, 83.80.Fg

INTRODUCTION

The ubiquity of granular materials in nature and their overwhelming technological significance have focused much attention on these systems. A coherent fundamental understanding of these materials in general and of their behaviours in specific regimes in particular, is yet to emerge. The lack of such understanding handicaps effective modelling for practical applications. As matters stand currently, any progress towards a first-principles modelling has significant scientific and applied benefits. Particularly problematic to model are dense granular materials (DGM). One reason that constitutive relations in the dense regime have been difficult to obtain is that, although the material may appear uniform to the eye above some scale, the internal stress distribution is not - DGM transmit stresses non-uniformly along force networks[1, 2, 3, 4]. The network re-organises as the material is loaded by external forces, with its structure continually failing and re-consolidating. The configurations of the stress networks determine, in turn, where failure occurs, as well as the dynamics as the material deforms. Thus, ultimately, stress determination and dynamics must be modelled self-consistently; we are currently still far from this goal.

Nevertheless, a significant step in this direction is a fundamental understanding of the way that granular matter transmits stresses when the structure is given. This problem has been debated much in the literature. Conventional models, based on constitutive relations involving strain or strain rate, are hard-pressed to explain the observed nonuniform stress fields. Therefore, it was suggested in the mid-nineties that these could be explained by assuming that DGM are statically determinate, or isostatic[5, 6, 7, 8]. Following this suggestion, phenomenological continuous stress-structure constitutive relations were proposed [9, 10, 11, 12, 13, 14, 15]. It was assumed that there are linear relations between the stress components, with the coefficients somehow related to the local structure. In two-dimensional systems, there is only one relation (there are three in three dimensions):

$$A\sigma_{xx} + B\sigma_{xy} + C\sigma_{yy} = 0 \,. \tag{1}$$

In the treatments that followed this suggestion, the coefficients A, B and C were taken to be constant and uniform throughout the material. It has been proposed that, if these relations lead to hyperbolic set of stress equations - in contrast to the elliptic nature of the equations of elasticity theory, for example - then the solution would propagate along force chains, thus reproducing the experimental observations in two dimensions. Since the stress coefficients in the stress-structure relations were constant, the characteristic paths, and consequently the force chains, were unavoidably straight.

This treatment remained phenomenological until the end of the twentieth century and quite controversial on many levels - a controversy that will be discussed in slightly more detail below. In 2002, this formalism received a firmer theoretical support when equation (1) was derived from first principles, relating the coefficients A, B and C directly to the local granular structure[16]. This formulation did away with the phenomenological aspect and made the two-dimensional theory parameter-free. The constitutive equation (1) was then rewritten in the form

$$Q : \sigma = 0, \qquad (2)$$

where Q is a rank two tensor that depends uniquely on the local granular structure. In a later work [17] it was shown that the geometry constrains the components of the structure tensor Q in such a way that the equations are indeed hyperbolic, as had been conjectured in previous works[8, 9].

That relation (2) can be derived from first principles is intriguing - constitutive closure relations are normally postulated. Nevertheless, the basic derivation in [16] was a not sufficient. To clinch the applicability of this relation it is necessary to be able to coarse-grain it to the macroscale. This proved to be a problem: the components of the fabric tensor Q were shown to fluctuate locally around zero mean, hindering coarse-graining by simple volume averages. However, it was observed that the fluctuations are anti-correlated between neighbouring particles, much in the same way that disordered antiferromagnetic spin systems are between neighbouring spins. This observation was then used to reformulate the problem in [18] in a way that could be coarse-grained. This resolved the problem, finally establishing equation (2) as a general closure relation for ideally isostatic granular materials of rigid particles. It should be noted that, once the fabric tensor Q is given, a unique solution exists for any arbitrary combination of boundary stress loads [19, 20].

Nevertheless, the development of isostaticity theory has been fraught with other doubts and criticism. Granular assemblies in nature are not necessarily isostatic and the usefulness of this concept to general particulate systems has been questioned[21, 22]. In the original formulation the particles were considered perfectly rigid and convex, hence making contacts at points. This led to concerns over the validity of the theory when the particles have finite compliance. This issue was examined in detail in [23], where it was shown that increasing particles compliance does not invalidate isostaticity theory altogether. Rather, the theory needs to be corrected and the corrections increase with particles compliance and contact area between particles. However, significantly, the corrections decrease rapidly with increasing system size.

To be specific, consider a granular assembly of N stiff, but not perfectly rigid, particles. The mean coordination number is such that, had the particles been perfectly rigid, the assembly would be isostatic. The analysis in [23] shows that, under a given external loading, the stress solution in the material would be the isostatic solution, but with corrections. This was shown as follows. First, the centroid point of each contact surface was identified and an alternative approximate system was considered, whose inter-granular contacts were points located at those centroid. To choose the approximate system, it was noted that the exact solution could be found by solving for an equivalent system whose contact forces acted at contact points located judiciously across the contact surfaces. The location of the equivalent contact points, however, needs to be determined by the effective surface contact force distributions, which are unknown. In the approximate system, the contact points are chosen at the centroids of the contact surfaces.

Now, the alternative approximate system is also marginally rigid and, therefore, it has an isostatic stress solution. The error between this solution and the correct one is due to the differences between the approximate and correct equivalent contact points. This error was then shown to decreases as N^α, where $\alpha = 3/4$ and $8/9$ in two and three dimensions, respectively[23]. It was therefore concluded that, on macroscopic sizes, the isostatic solutions of the approximate systems converge to the correct stress solutions of compliant-particles systems.

Another concern had been the range of applicability of isostaticity theory for materials that are not precisely at marginally rigid stress states. The problem was that the theory has been developed under the assumption that the mean coordination number is exactly right to make the material statically determinate. The contention was that the status of the theory is unclear when the mean coordination number goes above that value. This is a valid contention - it is clear that the equations can no longer hold as originally formulated. However, it turns out that the way that the solutions deteriorate, whether abruptly or gradually, is an important issue. This issue will be discussed later in the paper, following insight drawn from experimental observations [24]. That discussion will be shown to lead to two conclusions: (i) that the isostatic, or marginally rigid, state behaves as a critical point and (ii) that the stress behaviour of real materials depends on the proximity to the critical point [17, 25].

A third concern is seemingly fundamental. It had been conjectured [8, 12, 13] and then shown more rigorously [17] that (2) leads to hyperbolic stress equations, whose solutions propagate along characteristics paths (CPs) (see below). The CPs are determined by the structure, namely, the tensor Q and they define preferred directions. The question then is how can this happen in isotropic systems. Unfortunately, this question has not been discussed in great detail in the literature. The view of this author is that the structure can only be isotropic under isotropic loading. DGM respond to applied loading by local rearrangement and self-organisation. The material response to anisotropic loading cannot but be anisotropic, leading to a locally anisotropic fabric tensor. Thus, it is rare to find, either in nature or in simulations, DGM that are isotropic. However, we can consider a 'gedunken' isotropic material. To generate such a system one should start from an isotropic structure and apply to it isotropic boundary loads, e.g. purely compressive σ_{rr} along the boundary of a circular disc. Then every point along the boundary is a stress source of two CPs that propagate into the system. This gives rise to an isotropic distribution of CPs in the material and, correspondingly to an isotropic solution. Thus, isostaticity theory should not violate any common sense symmetry.

An interesting simulation to test this issue [26], claiming to the contrary, is nevertheless inconclusive. Such tests should make sure of the isotropy by: (i) loading the system isotropically (i.e. independent of direction rather than equally along the axes, which leads to isotropy in fluids but not in DGM) and (ii) that the isotropy is reflected in the statistics of the fabric tensor Q. Thus, this contention still needs to be substantiated.

It should be noted that the above argument suggests that a complete stress theory of granular materials should ultimately take into consideration the material organisation under the loading, a point that has been alluded to above.

Turning to failure and yield of DGM, it should be commented that the manner in which DGM transmit static stresses is significant to the understanding of these materials well beyond static phenomena. Not only do DGM exhibit simultaneously properties that are normally associated with solids, liquids and gases, but they also display rich behaviour that is uniquely their own and cannot be observed in other conventional phases. Thus, static stress transmission is an obvious first step to study the poorly understood (and therefore inadequately modelled) dynamics of DGM. One particular area, where modelling could benefit much from improvement in fundamental understanding, is of flow and rheology. The importance of flow of DGM to many real-life applications cannot be over-emphasized. For example, in powder metallurgy and the transport of cereals, seeds and coal, it is important that the material fail easily locally so that it can flow without blockage. Yet, in other application, such as construction, it is essential that soils be sufficiently stable so that structures do not collapse. In both these cases we need a good understanding of the basic physics in order to model: (i) the threshold to failure; (ii) the failure mechanisms and (iii) the post-failure dynamics of flow.

The main problem is that, unlike in fluid mechanics, where the continuum equations of motion are well established - the Navier-Stokes equations - there is no agreement on the continuous equations that describe rheology of DGM. A major stumbling block is that the homogenization from the particle-level dynamics to the macro-scale is sensitive both to the physics on the discrete level and to the coarse-graining procedure used to upscale to the continuum. Moreover, unlike in conventional fluids, which have no yield threshold to flow, most real DGM flow only once they have been loaded above some stresses, known generally as the 'yield surface'. Thus, the failure of DGM is affected directly by the way that the material supports static stresses. This further suggests that an essential criterion for a good model is a seamless convergence to the correct stress state when the flow stops.

In this paper, I will first review a simple experiment that leads to intriguing two conclusions: first, that isostatic states can be approached very closely, or even exactly realised, in real systems and, second, that isostatic states act as critical points in the sense that they sport a diverging lengthscale. The physical interpretation of this lengthscale and its relevance to understanding the physics of DGM will be discussed. I will then describe a recent development on modelling failure and yield of DGM. Specifically, I will present a derivation of the equations of flow as the material fails. The relation between these equations and the convergence to isostatic static stress solutions will be shown and discussed, thus establishing a seamless transition from the static to the quasi-static dynamic theories. Many of the results reported here have been obtained in collaboration with R. C. Ball and S. F. Edwards.

A TWO-DIMENSIONAL FREE-FALL PILING EXPERIMENT

Mean coordination number and marginal rigidity

In this section I describe the experiment reported in [24] and discuss an interesting interpretation of its results. The main aim of the experiment was to investigate the realizability and nature of marginally rigid states. Before getting down to the experiment, it is constructive to discuss the conditions that define such states.

It is clear that there is a minimal number of contacts per particle (called the mean coordination number), below which the system cannot be mechanically stable, except for a measure zero of metastable configurations. Marginally rigid states are mechanically stable configurations of particles that are minimally connected. In such states the intergranular forces are statically determinate, a fact that can be used to identify the condition for marginal rigidity. This condition depends on several characteristics of the particles, including whether they are frictional or frictionless, and spherical or not. To some extent, it also depends on system size, but limiting the discussion to macroscopically large assemblies, where this dependence is negligible, I will disregard it in the following.

Consider a d-dimensional assembly of N ($\gg 1$) rigid particles of sufficiently high friction coefficient, such that negligibly few particles slipped as it was constructed. Let z_g be the number of force-carrying contacts (defined here as the coordination number) of particle g. The corrections due to size effects are of order $N^{-(d-1)/d}$ in d dimensions and hence negligible. The mean coordination number per particle is

$$\bar{z} = \frac{1}{N} \sum_{g=1}^{N} z_g , \qquad (3)$$

and the total number of inter-granular contacts is therefore

$$N_{cont} = \frac{1}{2} N \bar{z} . \qquad (4)$$

At each force-carrying contacts there is an inter-granular force to be determined, giving altogether dN_{cont} unknowns. To solve for these we have at our disposal the balance conditions. For each particle there are d force, and $d(d-1)/2$ torque moment, balance equations. The latter is the number of axes of rotations in d dimensions. The total number of equations is then

$$N_{eq} = N[d + d(d-1)/2] = N \frac{d(d+1)}{2} . \qquad (5)$$

For the system to be statically determined, the number of equations should equal the number of unknowns. From expressions (4) and (5) we see that this gives a condition for the mean coordination number

$$\bar{z} = z_c^f = d+1 . \qquad (6)$$

In assemblies of frictionless particles, the direction of the force vector at each contact point is determined by the geometry (specifically, by the normal direction at the local contact), which leaves the magnitude of the force as the only unknown. It follows that the number of unknowns is $N_{cont} = \frac{1}{2} N \bar{z}$, while the number of equations is the same as in (4). Again demanding that the number of equations be equal to the number of unknowns, we find that the condition for static determinacy is that the mean coordination number is

$$\bar{z} = z_c^s = d(d+1) . \qquad (7)$$

A redundant case to be considered is when the particles are smooth and perfectly spherical. This case is rare in nature but it is used frequently for numerical simulations. In this case, the number of unknowns is again equal to the number of contacts, but the number of equation is not the same as in the previous cases. Due to the shape of the particles, when the forces are balanced on a particle, the torque balance equations for it are also automatically satisfied, since all the force direction lines pass through the particle center. This makes the torque balance equations redundant, leaving only Nd equations to solve. It follows that, for static determinacy in such systems, the mean coordination number should be

$$\bar{z} = z_c^{ss} = 2d . \qquad (8)$$

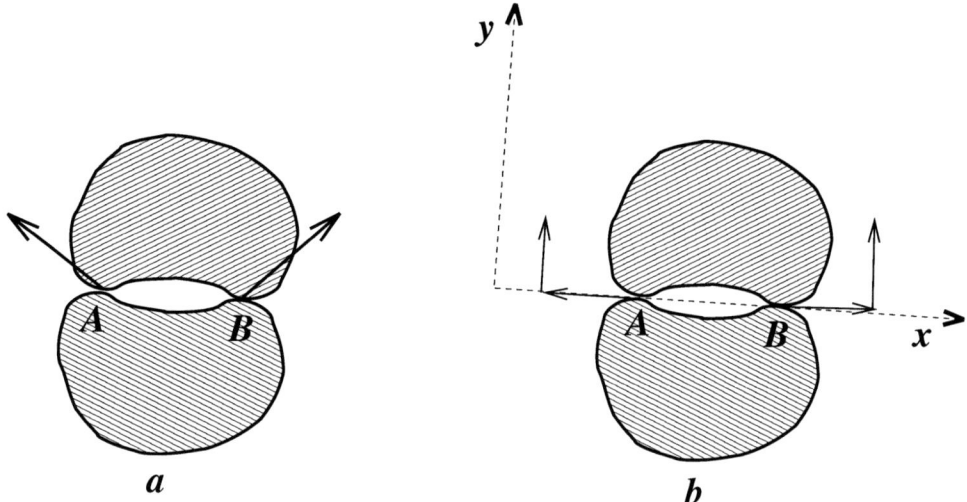

FIGURE 1. Two rigid particles in contact through two points A and B. The two contact forces in (a) can be decomposed into the four components in (b), of which only three can be resolved. Since the two x-components of the forces are equal, opposite, and act along the same line, then the stress around the two particles is unaffected by this indeterminacy. Consequently the mean coordination number is larger than for systems with only convex particles.

The above considerations have been around for a long time, in fact some of these arguments date back to Maxwell[27, 28]. Nevertheless, they do not hold for all packings. In particular, they include a couple of underlying assumptions. One is that in particles of arbitrary shape there is no redundancy of any of the torque moment equations. For example, if the contact force vectors around a particle all happen to pass through one or more of its axes of rotation then the torque around those axes ares automatically zero and there is no need to invoke the relevant couple equations. In the following I assume that such occurrences are very rare, in fact of measure zero. Another assumption is that the particles are rigid and convex and, consequently, that a particle can only make contact with any neighbour at one point. This assumption fails when particles have non-convex shapes, whereupon two particles can make contact at two points. To understand the consequence of lifting the one-contact-point assumption, consider a piling experiment, where particles with high friction coefficients 'fall' very gently, so that they come to rest to form a stationary pile without slipping and without disturbing the rest of the pile. In d dimensions, any solid, and in particular both the falling particle and the consolidated pile, has $d(d+1)/2$ degrees of freedom: motion along one of the d directions and rotation around the $d(d-1)/2$ axes of rotation. Thus, prior to the collision the particle-pile system has $d(d+1)$ degrees of freedom. Once the particle has settled, the new consolidated pile has only $d(d+1)/2$ degrees of freedom. This means that $d(d+1)/2$ degrees of freedom have been 'lost' and these provides $d(d+1)/2$ conditions. From these conditions we can solve for the forces between the recently-arrived particle and the pile. At each contact Newton's third law (action and reaction) is satisfied, which gives that there are $d+1$ unknowns per particle to solve for and, when there is only one point of contact between particles, then we could solve for the inter-granular forces if there are on average $d+1$ contacts per particle. This is an alternative derivation the above value of z_c^f.

However, when the shape of the landing particle has a non-convex part, the particle may lean stably against only one particle, in which case the above argument needs to be reconsidered. For simplicity, the situation is illustrated in two dimensions in figure 1. Particles contacting at two points transmit two inter-granular forces and hence give rise to four unknowns. This is one unknown more than the $d(d+1)/2 = 3$ conditions that can be derived from the 'loss' of the degrees of freedom. Therefore, one unknown must remain indeterminate. Fortunately, this does not affect the global stress field, as we shall see next. Let us decompose the forces along axes x and y, where x coincides with the line that joins the two contact points, as shown in figure 1. We can solve for the y-components and use Newton third law to determine that the x-component of one force is equal and opposite to the x-component of the other. The remaining one unresolved unknown is the magnitude of the x-component. The local stress on either of the two particles can be derived from the area normalized symmetric part of the force moment

$$\sigma_{ij}^g = \frac{1}{A} \sum_{g'=1}^{z_g} f_i^{g'g} \rho_j^{g'g}, \qquad (9)$$

where σ_{ij}^g is the ij-th component of the stress on particle g, g' are the touching neighbours of this particle, $f_i^{gg'}$ is the i-th component of the force that partice g' applies on g, $\rho_j^{g'g}$ is the j-th component of the position vector extending from the center of mass (say) of particle g to its contact with g', and A is the area associated with particle g. Since the unresolved x-components are equal, opposite and act along a line tangent to the contact points, then they do not contribute at all to the forces, nor to the torque moment around g. It follows that the unresolved degree of freedom does not affect the stress around this particle. This argument, which was first presented in [24], leads to two conclusions. First, the stress around two particles touching at two points may be well defined in spite of the fact that not all the inter-granular forces can be determined. Second, in counting the number of contacts we have to allow for multiple contacts. The latter means that, in a system that contains multiple inter-particle contacts, the mean coordination number for marginal rigidity is higher than the one calculated previously. To find how much higher should z_c^f be, for example, in two dimensions, we note that there is one such contact for every pair of double-contacting particles, increasing the mean coordination by $1/N$ for each one of them. It follows that the correct value for marginal rigidity in two dimensions for a system with a fraction w of double contacts out of all inter-granular contacts is

$$\bar{z} = z_c^f = 3 + w. \qquad (10)$$

This observation will be directly relevant to the experiment to be described next.

The experiment

The experiment models free fall and pile-up of particles in two dimensions (see figure 2). Model noncircular particles of approximately $1.5 \pm 0.2 \text{cm}^2$ were punched out from a cardboard sheet, using a stainless steel punch. The cardboard was fibrous, ensuring that the cut edges gave high coefficient of inter-granular friction. The friction coefficient was estimated from slipping tests to be higher than 50. The particles had a slight non-convex region. The experiment was carried out on a horizontal glass plate, with the pile building up within a U-shaped collector, whose surface was made of a similar material as the particles. The internal dimensions of the collector were $18 \times 24 \text{cm}^2$. The particles were initially placed on a thin transparent film, lying on top of the glass plate, at a notionally random distribution subject to the requirement of no contacts and that the initial spatial distribution is reasonably uniform. These particles were then conveyed towards the collector at an approximate speed of 0.5m/min. This was done either by moving the collector towards the particles (at high initial densities) or by moving the particles towards the collector via sliding the transparent plastic sheet under it (at low initial densities). Care was taken to maintain a constant advance rate and particularly to avoid relative transverse motion between the collector surface and the 'free falling' particles. The slow fall rate ensured both negligible particle deformation and minimisation of inertial effects, which could transmit vibrations through the forming pile and rearrange it subsequently to piling. Data was collected in the form of photographs of the growing pile, taken normal to the plane at regular intervals during the process, and a photograph of the 'consolidated' final pile. For each final pile we have determined its bulk and boundary particles, the contacts and the double contacts. The determination of the number of contacts was done both by eye and by taking the zero limit of the cumulative distribution of the spaces between neighbouring particles. Both methods agreed excellently. Following the discussion of the previous subsection, to determine the proximity to the marginal state, we measured coordination number as twice the total number of contacts divided by the total number of particles. This quantity, if calculating coordination from individual particles, means double counting coordinations between boundary particles and the boundary. Relatively few double contacts were observed and these were counted as two separate contacts. In the experiment the double contacts comprised 10% of the total number of inter-granular contacts. The photographs taken during each piling process were also used to identify a yield front (YF) - a 'layer' of particles that have collided with the pile, but which have not yet come to rest. The particles that belong to this layer were identified by comparing the intermediate positions of the particles to their position in the final pile. The comparison was done by superposing the photographs on top of a lightbox and determining overlaps. The force of sliding friction between the particles and the moving base played a role analogous to gravitational force on a stationary mass. Since the approach was at constant (low) velocity, the falling particles did not accelerate and the situation is indeed of free fall in two dimensions.

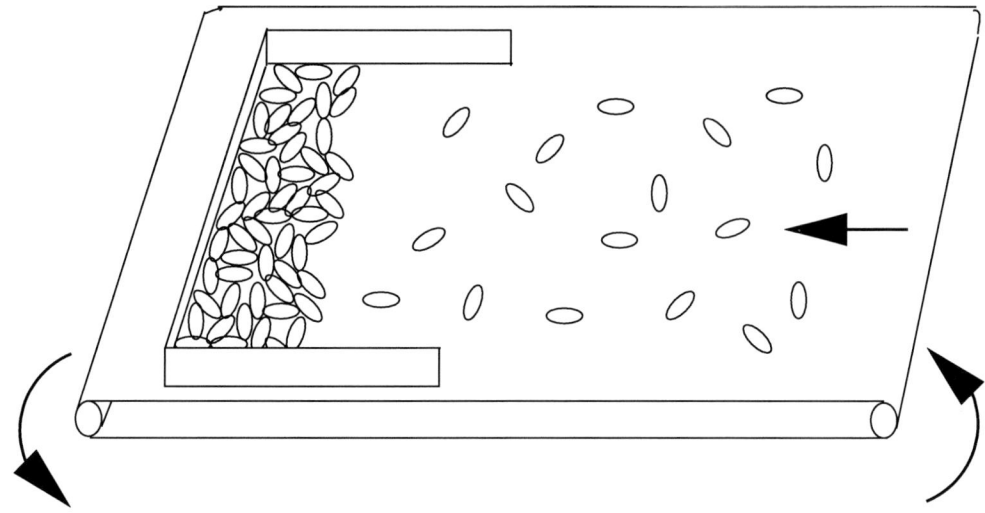

FIGURE 2. A two-dimensional free-fall piling experiment. Particles are conveyed at constant slow speed and pile up against a collector. The particle-particle and particle-collector friction coefficient is very high. The friction between particles and the underlying conveyor mimics gravity to press the pile against the collector. Photographs are taken from below to monitor the piling process.

Measurements were taken of both the density of the consolidated pile (figure 3) and its mean coordination number (figure 4) as functions of the initial density of the falling particles. It can be observed that, as the initial density increases the final pile density decreases. This is characteristic of a jamming system. As particles fall, contacts are made and broken in the YF. With very dilute initial density, particles join the pile and stabilize by making two contacts with it. As the initial density increases, falling particles interfere with one another and not all of them can make two contacts with the stationary pile. Thus, the decrease in the number of contacts, and hence in the final density, is a collective phenomenon - the evolution of the contact network is highly correlated and is governed by the cooperative stress-driven dynamics.

The experiment has a natural limit density, identified as the point where the density of the initial particles is equal to the density of the final pile (see figure 3). Clearly, there can be no experimental measurement to the right of this point since the density of the initial particles cannot exceed that of the consolidated pile that it generates. This density, ϕ_c, is found by extrapolating a fit to the the pile density measurements as a function of the initial density, and locating the intersection with the equal-densities line. It is plausible that this is the density of loose random packing, since it is the absolute limiting lowest density that one can obtain in this experiment.

One significant quantitative result is that, on extrapolating the fit of the mean coordination number to ϕ_c, $\bar{z}$ was found to approach 3.1 ± 0.1, in excellent agreement with the prediction of the marginal rigidity mean coordination number given in (10), combined with the observed value of $w = 0.1$. This leads to the conclusion that isostatic states can be approached arbitrarily closely. This observation has been supported by an independent repeat experiment, carried out with different, and more, particles[29].

Another significant observation concerns the YF. Figure 5 shows three typical experiments at three different initial densities before all the particles have come to rest. Of particular interest is the amount of particles that belong to the YF (coloured red). In the YF, particles are continuously shifting and moving relative to their neighbours en route to a stable mechanical equilibrium. Specifically, observe that the number of the particles that comprise the YF increases rapidly as the initial density increases toward the limit density ϕ_c (see figure 6). At low initial densities, the YF is about one particle thick while close to ϕ_c it is more than 50% of the pile. This suggests that, for macroscopically large packings, the number of particles participating in the YF is a finite fraction of the total number of particles. This conclusion, which needs to be checked for larger assemblies of particles, suggests that the size of the YF *diverges* as the system tends to the 'thermodynamic limit', i.e. when the total number of particles tends to infinity. These observations and conclusions have also been supported by the experiment of Sibille and Mullin[29], also shown in figure 6.

A diverging lengthscale is a fingerprint of a critical point and its occurrence in marginally rigid packings gives an important clue to the nature of the isostatic state. Specifically, diverging lengthscales are common in second order phase transitions and in critical phenomena [30]. To understand the physical phenomenon that this lengthscale

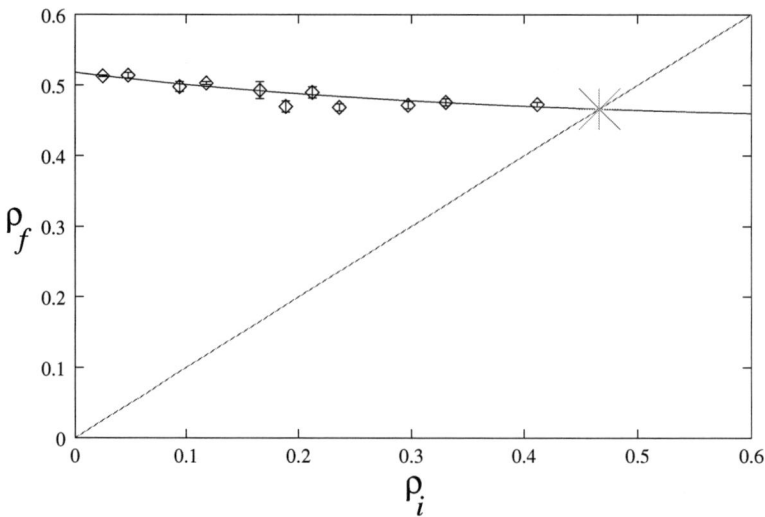

FIGURE 3. The density of the consolidated pile, as a function of the initial density of the free-falling particles. Note that the higher the initial density the lower the density of the final pile. The line fitting the result is a simple exponential and comes both as a guide to the eye and to provide an estimate for the critical density, ρ_c. The value of ρ_c is determined by the intercept of the fitting line and the equal-density line, which passes through the origin.

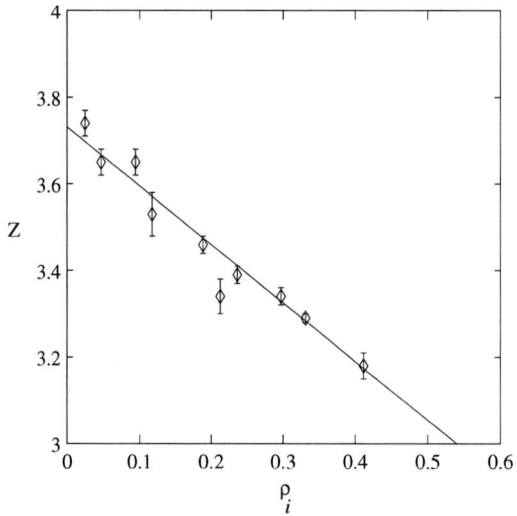

FIGURE 4. The mean coordination number of the consolidated pile, z, as a function of the initial density of the free-falling particles. the number of contacts was found both visually and by taking the zero limit of the cumulative distribution of the space probability density between neighbouring particles. The two method agreed perfectly. A straight line fit was used to extrapolate the value of the mean coordination number to ρ_c. The value of z at ρ_c was found to be 3.1, in excellent agreement with the marginally rigid value for our pile with 10% double contacts.

is associated with, recall that, at the critical point, the mean coordination number is on the verge of becoming mechanically unstable. Perturbing a particle from its position affects the particles in contact with it and they then have to shift position too. This, in turn, affect their neighbours and so on. Thus, the first interpretation is that the diverging lengthscale corresponds to the response length of the marginally rigid pile to small particle displacements. The minimal connectivity of the pile makes every contact breakage major event that typically leads to a long-range rearrangement.

Furthermore, applying a small force to a particle in a marginally rigid pile, one expects the force to be sufficient to break a contact. However, to prevent the contact from breaking we need to adjust the force on the neighbour particle. This in turn would give rise to further adjustments due to the tenuous connectivity. This suggests that the force response

FIGURE 5. The behaviour of the yield front. The particles belonging to the front (coloured red) have encountered the pile but have not yet reached the final consolidated positions. The three different pictures show measurements from three different experiments at initial densities that increase from top to bottom. At the low density the front is less than one particle deep, while at the high density it comprises more than half the pile. As the density increases the consolidation process becomes more cooperative.

length should also become comparable to the system size, i.e. diverge. Note that such a thought experiment probes nothing else but the Green function of the continuous stress theory that applies to these material.

IMPLICATIONS OF THE EXPERIMENT FOR CONTINUUM STRESS THEORY

The conclusions and interpretations of the previous section are significant since they point directly to a resolution of the long-standing debate over the correct theory for real granular materials. It is the critical nature the marginally rigid state and the divergence of the response of the system to stress perturbations, which gives us a big clue.

As mentioned, critical points are well known from the study of traditional second-order phase transitions. A system undergoing such a transition is in one phase initially and, as it approaches the critical point, regions of another phase form and develop. A well-known example is the transition in metals from a normal conducting state to a superconducting one upon reducing the temperature. The temperature where the transition occurs is the critical point. As the temperature decreases toward this point, superconducting regions form, and then grow, within the initially normal phase. At the critical temperature, the superconducting regions span the entire material.

The typical size of the superconducting clusters can be quantified by measuring the the density-density correlation function of the superconducting phase and identifying the correlation length over which this function decays. As the critical point is approached from above, the correlation length increases until it become comparable to the system size at the critical temperature. In systems that go to the thermodynamic limit, i.e. become infinitely (or macroscopically) big, the correlation length diverges. It should be emphasized that, even when the system is only close to the critical point, there are sufficiently large superconducting regions to affect experimental measurements of the conductivity.

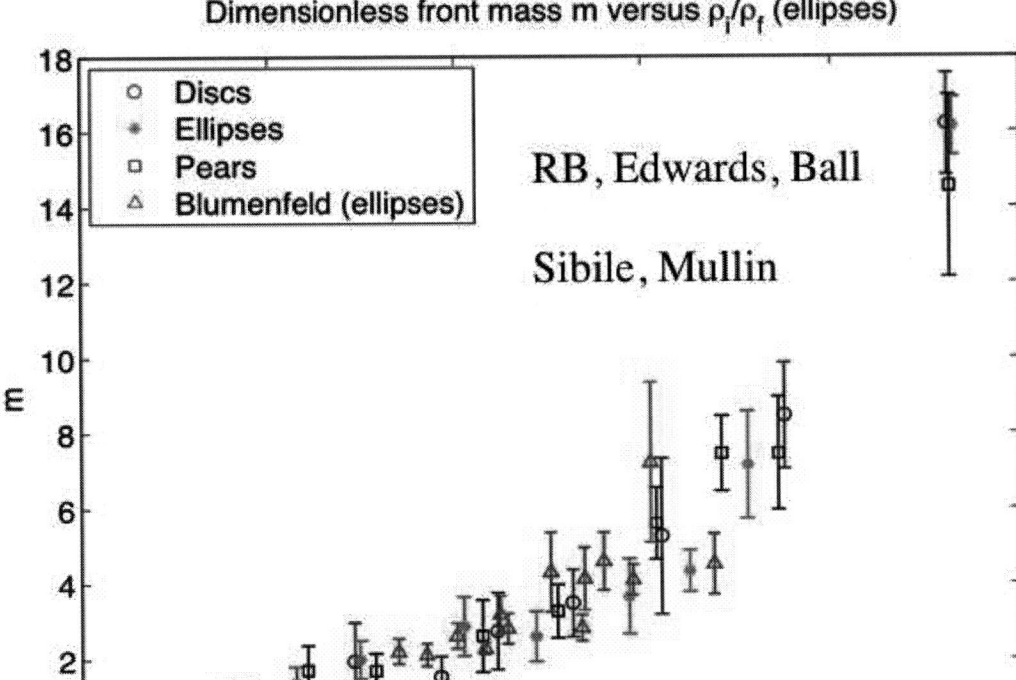

FIGURE 6. The measured front (arbitrarily normalized) mass as a function of the initial density. The plot shows a diverging trend as the ratio of the initial and final densities approach unity at the critical density. (plot courtesy of Sibille and Mullin [29])

Indeed, macroscopic measurements depend sensitively on the proximity of the material to a critical point.

This physical interpretation of the diverging lengthscale is very similar to the situation in DGM. At the critical density ρ_c, the material is minimally connected and hence fully isostatic. The proximity of the granular system to the critical point can be parameterized by $\delta z = \bar{z} - z_c$, the difference of the mean coordination number from the critical value z_c. Finite positive value of δz correspond to existence of over-connected regions in the material. A steady increase in δz from zero corresponds to growth of the over-connected regions at the expense of the isostatic medium.

Thus, we can regard real DGM as two-phase composites: part isostatic and part over-connected. The stress field in the isostatic regions is governed by the isostaticity equations, namely, the balance conditions and (2), whilst the over-connected regions follow a different closure relation.

In principle, it is straightforward to construct a stato-elasticity theory. Consider a material containing a particular distribution of isostatic and over-connected regions. To determine which phase a region belongs to and the boundaries of a region can be done by the local mean coordination number. The exact algorithm for such a determination is downstream from the main thrust of this paper. In a simple version, we can assume that elasticity theory holds in all the over-connected regions, while isostaticity theory holds only in the marginally rigid regions. Then, in each region the stress solutions are written in general form, using the different sets of equations, and the solutions in neighbouring regions are tailored to match at common boundaries. While there are practical problems in this procedure, exact solutions are currently possible for simple geometries [31]. .

This resolves the long-standing controversy in the field on the correct way to model stresses in granular materials - whether with the theory of elasticity or isostaticity. In a sense, both approaches are valid, but in different regimes. Real DGM generically contain both phases: marginally rigid regions where the hyperbolic [17] equations of isostaticity apply and over-connected regions where the elliptic equations of elasticity do. Thus, granular media should be modelled by a combined theory.

However, to implement a stato-elasticity theory, one must have a good grasp of each of the separate theories. While a good understanding of elasticity exists, that of isostaticity is still rudimentary. In particular, for a long time, isostaticity theory was associated only with force chains and long response lengths. This changed recently when a more thorough study of nonuniform isostatic media has been carried out [19, 20], showing that the idea of well isolated force chain needs to be revised when the components of the fabric tensor Q are not constant across the medium.

Force chains have been observed experimentally and they occur naturally when equations (1) and (2) are hyperbolic [8, 12, 13, 14]. That these equations are indeed hyperbolic on the smallest scale was shown from first-principles in [17]). The solutions of hyperbolic equations propagate along characteristic paths (CPs) and in the original works the forces were finite along the CPs and vanished elsewhere. Thus, the CPs were equated with the observed force chains. However, this picture is based on analyses of systems where the components of Q were uniform. The recent work of [19] and [20] showed that the situation is more complex because, when Q varies with position, the CPs of the solutions 'couple' and 'interact', with the strength of the coupling proportional to the gradients of the components of Q. The coupling led to several effects. First, and not surprisingly, the CPs meander rather than be straight. Second, the forces along the CPs decay, suggesting that force chains eventually 'evaporate'. Third, forces can 'leak' to the region between the CPs, with the leakage increasing with the magnitudes of the gradients of Q. Fourth, for locally sharp gradients of Q, the CPs can branch. This result was significant because it explained observed branching without having to invoke additional stochastic mechanism [32]. These results form a basis for better understanding of stresses in isostatic regions within the two-phase picture described above.

FAILURE AND CREEP FLOW

In this section, I focus on the physics of failure of DGM and on the initial dynamics immediately after failure. In the engineering community, the term failure is understood on the macroscopic continuum level, where the reference is to observables measured on such scales. However, failure is caused by local rearrangement of individual particles. Some such rearrangement may only lead to local deformation that do not translate to global failure and result only in irreversible behaviour, i.e. reversing the loading conditions does not reproduce the original microstructure. Some local rearrangement may spread out to span large parts of the material, in which case they are regarded as global failure. The conditions for each of these scenarios are complex and beyond the scope of this discussion. Here, I focus on the local failure of the contact network between particles and the initial motion that it results in. The main advantage of this discussion is that it allows to define an exact yield surface and criterion, as will be shown below.

Generically, a granular system fails when the external loading on it exceeds a certain threshold, commonly known as the 'yield surface'. This is a surface in the space spanned by the individual components of the stress tensor and, in natural granular systems, it normally encloses a finite region. However, one can imagine granular systems, whose yield surface degenerates to a point at the origin, namely, it would yield under the smallest external loading.

For failure to occur, regions inside the material must be at, or close to, marginally rigid states. Moreover, macroscopic failure would be preceded by an incipient marginally rigid region extending from boundary to boundary. An example of such a region is an incipient shear band. Therefore, it makes sense to focus on failures in regions that are initially in isostatic stress states. For the purpose of this discussion, I will assume that the granular system consists of convex and rigid particles of high friction coefficient to inter-granular slipping.

Generically, particle displacements occur by two mechanisms: slippage of particles in contact and rotation. In generally random granular assemblies the two mechanisms normally take place simultaneously. In the following, the displacement of a particle centroid is identified with that of a point of the continuous representative material. On the continuous level we can define a continuous displacement field, from which a strain and strain rate fields can be derived. Thus,

$$\dot{e}_{ij} \equiv \frac{1}{2}(\partial_j \dot{u}_i + \partial_i \dot{u}_j) = \dot{e}_{ij}^{rot} + \dot{e}_{ij}^{sl}, \qquad (11)$$

where u_i is the displacement in the ith direction, $\partial_i \equiv \partial/\partial x_i$ and x_i is the ith Cartesian coordinate. The quantities in this equation and in the rest of this section refer to centroids of particles, but particle indices have been omitted to avoid cluttering the notation. For analysis purposes it is useful to treat the rotational and slippage contributions independently. This is not only an exercise in idealization. It is interesting to note that there are special circumstances, where the two mechanisms are indeed separable in the sense that macroscopic strain rate can be achieved, at least momentarily, either by pure slippage or by pure particle rotations. This happens in systems where all loops have even numbers of particles around them. An example of a part of a two-dimensional such a system is shown in figure 7.

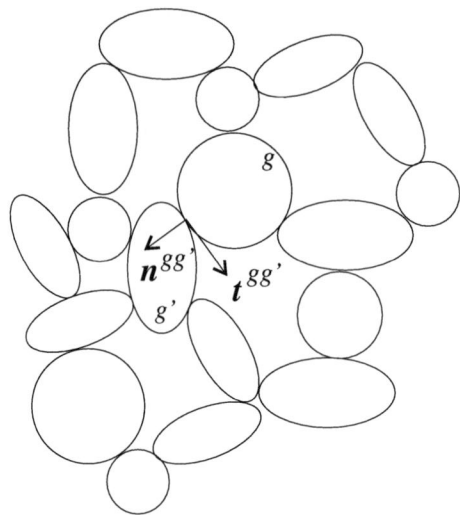

FIGURE 7. An 'unfrustrated' part of a granular assembly in two dimensions. The even number of particles around every void loop makes it possible to divide the structure into two sub-structures, black and white in the figure, such that every particle is in contact only with particles of the opposite colour. Associating spin up with one colour and down with the other, maps this exactly to an unfrustrated antiferromagnetic ground state.

In these systems, the particles can co-rotate happily, with every particle rotating oppositely to its neighbours and no slippage taking place. Such structures I term 'unfrustrated' [16], in analogy to antiferromagnetic spin systems, as explained in the caption of figure 7. It is in unfrustrated systems that the region enclosed by the yield surface shrinks to zero, for such systems can strain without dissipating energy. Exciting mechanically equilibrated unfrustrated systems by perturbative oscillatory loading, they experience what is known as 'floppy modes' [33].

Energy is dissipated when two particles in contact rotate in the same direction, whereupon they must experience mutual solid friction. The situation is analogous to two antiferromagnetically interacting spins, forced to align in parallel - hence the term frustration. As particles of unfrustrated systems rotate, they displace and contacts may be broken and new ones made. Whether the structure remains unfrustrated or not is an interesting and relevant issue that is not addressed in this presentation. To avoid this issue, I will focus on the dynamics before the connectivity changes.

A particular example of unfrustrated systems is the deformed Honeycomb structure sketched in figures 8 and 9. We can generate displacement of the entire right hand side of the structure relative to the left hand side about the shear line B in two ways. One is by making all the particles on the right hand side of B slip over those on the left hand side, as shown in figure 8. The slippage then occurs only along the shear line. Another way is to rotate all the particles on the right hand side, as shown in figure 9, and leave the particle on the left hand side stationary.

Focusing first on the slipping strain rate, let the shear line B pass through the contact between particles g and g' (figure 9) and let the tangent unit vector to their surfaces at the contact point be $\hat{t}^{gg'} = \vec{S}_B$, as in figure 7. The normal to the surfaces of the particles at the contact point is $\hat{n}^{gg'} = \varepsilon \hat{t}^{gg'}$, where ε is a $\pi/2$-rotation operator. In terms of these, the symmetric strain is proportional to

$$\gamma_{ij} = \frac{1}{2}\left(\partial_i \hat{t}_j^{gg'} + \partial_j \hat{t}_i^{gg'}\right), \tag{12}$$

where $\hat{t}_j$ is the jth component of the unit vector $\hat{t}$. It follows that the slippage strain rate at this contact is

$$\dot{e}_{ij}^{sl} = A\gamma_{ij}, \tag{13}$$

where A is the displacement rate in the tangential direction. As we shall see below, this first-principles formulation of the slippage contribution on the granular level has advantages over traditional plasticity-based models. While γ_{ij} can be determined, in principle, from the local structure, the rate A is an unknown scalar field that needs to be solved from the flow equations.

The rotation-based strain rate is interesting. Consider the local microstructure around a particle g, in contact with particles $g_1, g_2, \ldots, g_n$ (e.g. $n = 3$ in figure 10). We wish to derive the shift in the centroid of the contact points (i.e.

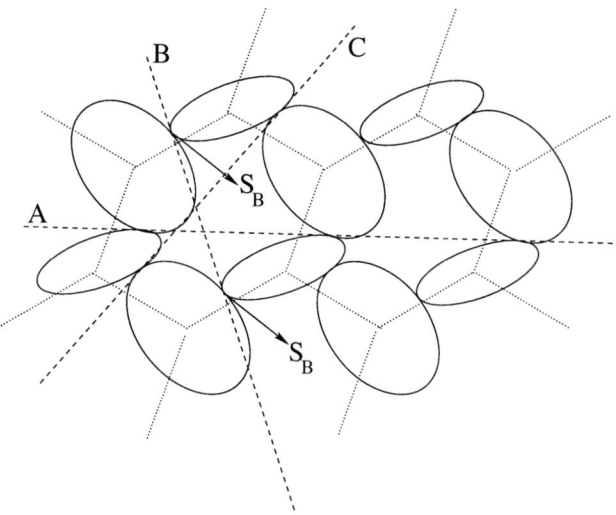

FIGURE 8. An unfrustrated deformed Honeycomb structure of particles can be sheared along three shear lines: *A*, *B*, and *C*. For example, displacing the half-system to the right of shear line *B* relative to the left hand side, can be achieved by slipping the contacts along the shear lines in the S_B direction, which is determined by the geometry at these contacts.

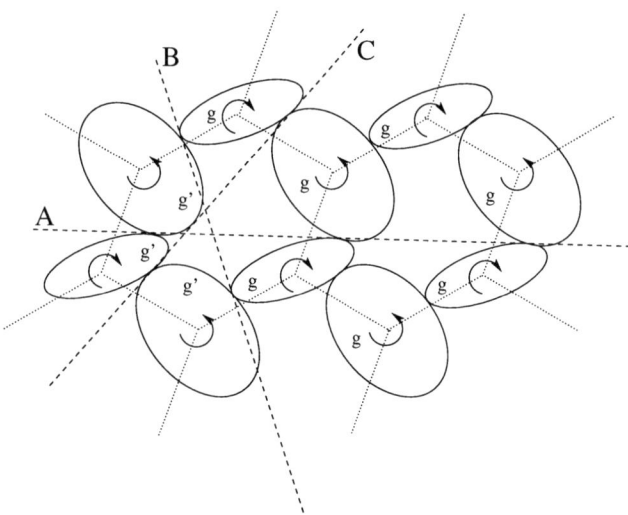

FIGURE 9. Pure rotation strain in an unfrustrated deformed Honeycomb structure of particles. The system can be strained by rotating the particles to the right of the shear line *B* (the *g* particles) by a small angle θ in the shown directions. If we keep the *g'* particles stationary, the entire right hand side of the system would displace relative to the left hand side.

the mean position vector of these points) due to an infinitesimally small rotation θ. Since for an infinitesimal rotation this is a linear-order calculation, then we can consider a small rotation around each of the contact points and superpose the contributions. This calculation leads to the result

$$e_{ij}^{rot} = Q_{ij}^T \theta . \tag{14}$$

The tensor Q^T turns out to be the transpose of the tensor Q in eq. (2) (which is symmetric anyway, $Q^T = Q$).

Thus, we can write now the strain rate for general two-dimensional systems, whether frustrated or not,

$$\dot{e}_{ij} = A\gamma_{ij} + Q_{ij}^T \omega , \tag{15}$$

where $\omega = \dot{\theta}$ is the local angular velocity field, measured through particle rotations.

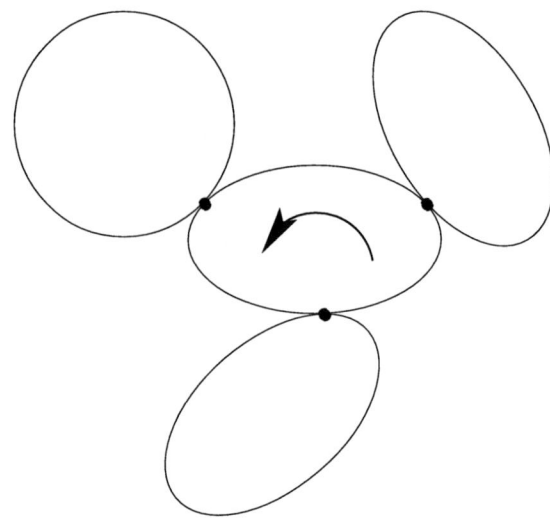

FIGURE 10. To calculate the strain that results from pure rotation, rotate the central particle by a small angle θ. To linear order this is the same as rotating the centroid of the central particle by this angle around ever contact point and superposing the results to find the displacement of that centroid. The result is that the symmetric strain is as given in equation (14).

Next, let us write explicitly the local yield criterion. Basing the condition for slippage between particles g and g' on da Vinci - Amontons - Coulomb friction threshold criterion, we have

$$|\sigma \cdot \hat{t}| > \mu \sigma \cdot \varepsilon \cdot \hat{t}, \qquad (16)$$

where we have omitted the superscripts gg' from $\hat{t}$. The macroscopic yield threshold can then be written in the exact form

$$Y(\sigma) = 1 - \prod_{gg'}\{1 - H(|\sigma \cdot \hat{t}| - \mu \sigma \cdot \varepsilon \cdot \hat{t})\} . \qquad (17)$$

When the tangential force at each and every contact is below the yield threshold $Y(\sigma) = 0$. As soon as any of these forces increases above the threshold the particular contact slips and $Y = 1$. This is then a legitimate yield function that describes failure when its value increases above zero.

The above strain rate equation is quasi-static, thus satisfying the static stress equations at all times. These are the continuous form of force balance equations

$$\mathbf{div} \cdot \sigma = F_{ext} \qquad (18)$$

and of the torque balance equations

$$\sigma = \sigma^T . \qquad (19)$$

In addition, the stress-structure relation (2) is also part of the set of equations.

It should be noted that we need no independent energy equation - the energy dissipation can be determined directly from the existing equations. It is constructive to examine the energy dissipation. To this end, multiply equation (15) from the right by the stress tensor σ and take the trace of the resulting tensor. Since the process is quasi-static, the stress tensor is static at every moment and the quantity on the left hand side, $\dot{e} : \sigma$, is exactly the energy dissipation rate. Considering the two contributions on the right hand side, we anticipate dissipation only from friction due to slippage between particles in contact; rolling of rigid particles over one another should not give rise to dissipation. Thus, we expect all the dissipation to come from the first term

$$\dot{e} : \sigma = A\gamma : \sigma . \qquad (20)$$

Note that this equation can be used to measure the scalar field A experimentally. From the left-hand-side term describing strain rate due to rotation we obtain

$$Q^T : \sigma\omega = 0 . \tag{21}$$

We now note that equation (21) should hold for any arbitrary angular velocity ω and therefore the relation should be independent of it. We then obtain that, in two dimensions $Q^T : \sigma = Q : \sigma = 0$, which, fittingly, is exactly the stress-structure relation (2) for the static stress field. Not only is this consistent with the static equations, but it also provides the bridge between the quasi-static dynamics and the static solution - it shows that, when the motion described by the above yield equation stops, the system 'lands' on an isostatic stress state - the correct state to land on because the yield surface corresponds to marginally rigid states, as discussed above.

I conclude this section by pointing out that the set of equations described above, and in particular the strain rate equation (15), can be extended to a more general description of assemblies of non-rigid particles. In such assemblies, equation (15) has to be modified with an additional term on the right hand side, describing strain rate due to the compliance of particles. This is a purely elastic term and, adding it, the equation reads

$$\dot{e}_{ij} = A\gamma_{ij} + Q^T_{ij}\omega + C_{ijkl}\sigma_{kl} , \tag{22}$$

where C_{ijkl} is a conventional compliance tensor that depends locally on the stiffness of the particles in the absence of slippage and rotation. Note that the additional term gives rise to elastic energy, in addition to the dissipation due to slippage friction.

CONCLUSIONS AND DISCUSSION

To conclude, I have reviewed in this paper recent advances on two controversial issues. One concerns the debate in the community on the correct modelling of the static stress transmission in general real dense granular materials (DGM) and the other concerns the equations that govern the failure, subsequent initiation of flow, of DGM. The recent history of isostaticity theory and its basics have been sketched briefly and, in particular, the sensitive dependence of the theory's validity on the marginally rigid nature of the material.

A key experiment has been described, which showed that: (i) marginally rigid states are realizable in DGM and (ii) that the isostatic state behaves as a critical point in the sense that a particular response length diverges in the 'thermodynamic' limit - the response to localized displacements and application of forces. The realizability of marginally rigid states gives initial support to the relevance of isostaticity theory to real systems.

The second result of the experiment has been shown to lead to a far-reaching conclusion: that real DGM should be regarded as two-phase composites - one phase made of marginally rigid regions, where stress transmission is governed by the isostaticity equations, and another phase of over-connected regions, governed by more conventional equations, possibly the elasticity equations. The diverging lengthscale at the marginally rigid state has been argued to correspond to the typical response length to stress perturbations (basically, the response length to the Green function) and, consequently, it corresponds to the typical length of force chains in two dimensions. Indeed solutions to the general isostaticity equations in two dimensions consist of *infinitely* long force chains (albeit with the caveat that force chains may eventually disperse, as shown in [19, 20]) and mentioned above.

The two-phase model of real granular materials, described here, which has been called stato-elasticity, leads to an interesting situation that has no parallel in traditional studies of composite materials. In conventional composites, e.g. of two different electrical conductors or two substances with different elastic properties, the phases satisfy the *same* field equations (e.g. Laplace's equation) but with different constitutive parameters. In the above examples, the electrical conductivity or the elastic constants would differ between the two phases. In granular materials, it is the very form of the field equations that is different. To emphasize the significance of this difference, suppose that the over-connected regions can be modelled by elasticity theory. Then the closure of the stress equations in those regions is via the compatibility conditions and a stress-strain relation, while in the isostatic regions the closure is through sterss-structure relations. This has crucial consequence - while the elasticity equations are elliptic, those of isostaticity are hyperbolic, which means that global solutions over the entire medium are difficult to obtain. This is an issue that is currently studied by this author.

Finally, I have reviewed a recent model for the failure of a DGM and the rheology that follow immediately after the failure. A set of equations has been derived for such flow. The equations are parameter-free - a significant advantage over most many models in the literature. Another advantage is that the set of equations has transparent unknown fields that need to be solved for and there are just enough equations to solve for these fields. The equations hold for DGM that are at marginal rigidity before failure and after flow has ceased and, in particular, they make sure that the system is in a correct static stress state once the motion stops. The strain rate equations, which is at the heart of the formalism, includes three mechanisms: slippage, rotation and particle compliance.

Interestingly, equation (22), with only the last two terms on the right hand side, also describes iso-auxetic materials [34]. It is expected that this equation would find more applications, another issue that this author is looking into.

REFERENCES

1. J. Smid, and J. Novosad, *Ind. Chem. Eng. Symp.* **63**, V/1–12 (1981).
2. B. Brockbank, J. M. Huntley, and R. C. Ball, *J. Phys. II (France)* **7**, 1521–1532 (1997).
3. L. Vanel, D. W. Howell, D. Clark, R. P. Behringer, and E. Clement, *Phys. Rev. E* **60**, R5040–R5043 (1999).
4. A. P. F. Atman, P. Brunet, J. Geng, G. Reydellet, P. Claudin, R. P. Behringer, and E. Clement, *Eur. Phys. J. E* **17**, 93–100 (2005).
5. S. F. Edwards, and R. B. Oakeshott, *Physica D* **38**, 88–92 (1989).
6. S. F. Edwards, and R. B. Oakeshott, *Physica A* **157**, 1080–1090 (1989).
7. A. Mehta, and S. F. Edwards, *Physica A* **157**, 1091–1100 (1989).
8. J.-P. Bouchaud, M. E. Cates, and P. Claudin, *J. Phys. I (France)* **5**, 639–656 (1995).
9. S. F. Edwards, and C. C. Mounfield, *Physica A* **226**, 1–11 (1996).
10. S. F. Edwards, and C. C. Mounfield, *Physica A* **226**, 12–24 (1996).
11. S. F. Edwards, and C. C. Mounfield, *Physica A* **226**, 25–33 (1996).
12. J. P. Wittmer, P. Claudin, and J.-P. Cates, M. E.and Bouchaud, *Nature* **382**, 336–338 (1996).
13. J. P. Wittmer, M. E. Cates, and P. Claudin, *J. Phys. I (France)* **5**, 39–80 (1995).
14. J. P. Cates, M. E.and Wittmer, J.-P. Bouchaud, and P. Claudin, *Phys. Rev. Lett.* **81**, 1841–1844 (1996).
15. S. F. Edwards, and D. V. Grinev, *Phys. Rev. Lett.* **82**, 5397–5400 (1999).
16. R. C. Ball, and R. Blumenfeld, *Phys. Rev. Lett.* **88**, 115505–115508 (2002).
17. R. Blumenfeld, *Phys. Rev. Lett.* **93**, 108301–108304 (2004).
18. R. Blumenfeld, *Physica A* **336**, 361–368 (2004).
19. M. Gerritsen, G. Kreiss, and R. Blumenfeld, *Phys. Rev. Lett.* **101**, 098001–098004 (2008).
20. M. Gerritsen, G. Kreiss, and R. Blumenfeld, *Physica A* **387**, 6263–6276 (2008).
21. S. B. Savage, "," in *Physics of Dry Granular Media*, edited by H. J. Herrmann, J. P. Hovi, and S. Luding, NATO ASI Series, Kluwer, (Amsterdam, 2002, pp. 25–95.
22. C. Goldenberg, and I. Goldhirsch, *Phys. Rev. Lett.* **89**, 084302–084305 (2002).
23. R. Blumenfeld, "Stress transmission and isostatic states of non-rigid particulate systems," in *Modeling of Soft Matter*, edited by M.-C. T. Calderer, and E. M. Terentjev, Springer-Verlag, Berlin, 2005, vol. 141 of *IMA Volume in Mathematics and its Applications*, pp. 235–246.
24. R. Blumenfeld, S. F. Edwards, and R. C. Ball, *J. Phys.: Cond. Mat.* **17**, S2481–S2487 (2005).
25. R. Blumenfeld, *New J. Phys.* **9**, 160 (2007).
26. G. Combe, and J.-N. Roux, *Phys. Rev. Lett.* **85**, 3628 (2000).
27. J. C. Maxwell, *Phil. Mag.* **27**, 250–261 (1864).
28. J. C. Maxwell, *Edinb. Roy. Soc. Trans.* **26**, 1–40 (1872).
29. L. Sibille, T. Mullin, and P. Poullain, *EuroPhys. Lett.* **86**, 44003–44008 (2009).
30. C. Domb, and M. S. Green, *Phase Transitions and Critical Phenomena*, Academic Press, London, 1972.
31. R. Blumenfeld, *in preparation* (2009).
32. J.-P. Bouchaud, P. Claudin, and M. Levine, D. E.and Otto, *Eur. Phys. J. E* **4**, 451–457 (2001).
33. M. Wyart, S. R. Nagel, and T. Witten, *Euro. Phys. Letters* **72**, 486–492 (2005).
34. R. Blumenfeld, *Molecular Simulation* **31**, 867–871 (2005).

Incremental response of a model granular material by stress probing with DEM simulations

F. Froiio [*] and J.-N. Roux [†]

[*]*Ecole Centrale de Lyon Centrale de Lyon, Laboratoire de Tribologie et Dynamique des Systèmes*
36, avenue Guy de Collongue
69134 Ecully CEDEX, FRANCE
[†] *Université Paris-Est, Laboratoire Navier*
2 Allée Kepler, Cité Descartes
77420 Champs-sur-Marne, FRANCE

Abstract. We use DEM simulations on a simple 2D model of a granular material to test for the applicability of the classical concepts of elastoplasticity (e.g., yield criterion, flow rule) to the response to stress increments of arbitrary directions. We apply stress probes in a three-dimensional stress space to various intermediate states (investigation points) along the biaxial compression path, and pay special attention to the influence of the magnitude of the increments. The elastic part of the material response is systematically identified by building the elastic stiffness matrix of well-equilibrated configurations. The influences of the contact stiffness level and of the dominant strain mechanism, contact deformation (I) or network rearrangement (II), are considered. Stress increments sharing the same principal directions as the stress state in the investigation point comply with a standard (single-mechanism) elastoplastic model with a Mohr-Coulomb criterion and well-defined flow rules and plastic moduli. Stress increments with principal axis rotation entail a response which is satisfactorily modeled by superimposing 3 plastic mechanisms, 2 of them symmetrically corresponding to shear stresses of both signs. The full dependence of strain increments on stress increments is thus parametrized with three flow rules, two of which are essentially symmetric.

Keywords: Granular materials, Discret Element Method, quasi-static deformation, incremental response, stress probing, principal stress axes rotation
PACS: 81.05.Rm, 83.80.Fg

INTRODUCTION

Elasto-plastic models are insofar the most widely spread continuum models in the literature and in the engineering practice concerned with granular materials under quasi-static loading conditions [1, 2]. These models have been insofar tested, and their parameter fitted, almost exclusively on the basis of phenomenological observation and one still counts a relatively small number of studies investigating the microscopic origin of the macroscopically observed plastic behaviour by discrete, grain-level simulations [3, 4, 5, 6].

Testing the response of representative elementary volumes (REV's) of a given material to "small" stress or strain increments, superimposed in various directions on an equilibrium state is perhaps the most appropriate procedure in order to assess the applicability of a continuum model. This procedure, known under the name of stress probing, is anyway accompanied by some remarkable practical difficulties among which the most important is that one must dispose of as many "identical" specimens as the stress increment directions to be explored (the stress probes). This is the reason why physical experiments of this kind are rare and assessing applicability of elastoplasticity for granular materials makes no exception: the only experimental work following this approach, to the author's knowledge, is the one by Royis and Doanh in 1998 [7] in which the stress probing procedure is applied on specimens issued from CD (Consolidated Drained) triaxial tests on Hostun sand. Using discrete simulations of granular material instead of physical specimens in the stress probing procedure was first proposed by Bardet [8, 9] in 1989 and offers not just an important work-around to these practical difficulties but also endows the stress probing technique with a remarkable flexibility, as we try to show in this work. A more recent study using the stress probing procedure via DEM simulations was authored by Calvetti and coworkers who focused on the elastoplastic behaviour of 3D specimens (assemblages of spheres), subjected to axisymmetric loading history (triaxial test) and stress probes with the same principal directions as the triaxial test [4, 10, 11]. Similar tests were run by Alonso-Marroquín and coworkers in 2D on polygonal particles [3, 12]. A substantial agreement can be found, among these authors, on the elastoplastic characters of the response of the tested materials, at small strains and with monotonous loading histories from virgin isotropically consolidated states. Two main features were confirmed in particular: the effectiveness, under the considered test conditions, of plastic models based on a single mechanism of plastic deformation and the "non associated" character of the flow rule.

In this work we present some preliminary results of a study which aims at assessing or clarifying other aspects of the elastoplastic response of granular materials, via a similar numerical implementation of the stress probing procedure. We use 2D specimens (assemblages of circular disks) subjected to a standard biaxial compression and then tested against stress probes in various directions of the stress space. The specificity of this work are the following: (i) the study is parametric in that we widen the range of the model parameters to access a number of significantly different classes of mechanical responses; (ii) we systematically test the dependence of the incremental response on the size of the stress increments; (iii) stress probing is performed in a three-dimensional stress space, i.e., we apply stress increments in the plane spanned by principal axes as well as stress increments inducing rotation of the latter (i.e., increments adding amounts of shear stress on principal planes). One motivation of our study is the modelling of localisation in granular media, where the applicability of such criteria as Rudnicki and Rice's may depend on *subtleties of the incremental constitutive description* [22]. As regards point (iii), in particular, let us remark that localisation appears to be crucially sensitive to the stress increments inducing rotation of principal axes, as is the case when some simple shear is superimposed on a biaxial compression [21].

In the remainder of this section we recall the basic ideas of the constitutive model to be assessed. Eventually, we describe the model material and the biaxial test procedure characterising the loading history of the specimens prior to stress probes. The next two sections present our preliminary results concerning the response to stress increments in the plane of principal stress axes and in a general three-dimensional stress space, respectively. The future steps of this study are outlined in the concluding section.

Notation

The formulas we will be needing in the following use the standard, compact notation convention of continuum mechanics. For ease of reading we will reserve boldface Greek characters to 2nd-order tensors and boldface Latin characters for Euclidean vectors. Interposed dots between two vectors or two tensors will denote the standard scalar product in the inherent linear spaces. As often is the case when dealing with frictional materials, the sign convention adopted here for the Cauchy stress tensor $\boldsymbol{\sigma}$ and for the infinitesimal strain tensor $\boldsymbol{\varepsilon}$ is such that compressive states are measured by positive values of the diagonal elements of stress and strain matrices.

Constitutive model

In most common scenarios of interest in civil engineering, granular materials are involved as large masses undergoing quasistatic deformation processes and exhibiting a variety of mechanical behaviours, from solid-like to fluid-like depending on the importance of the rearrangement of the contact network among the various microscopic ingredient of the macroscopic deformation. In this work we focus on the behaviour for small amounts of deformation from a virgin state, i.e., $\|\boldsymbol{\varepsilon}\| \sim 0.005$, where $\boldsymbol{\varepsilon}$ is the infinitesimal strain tensor. We postpone to a further publication the analyses for higher deformation levels. Despite some measurable network rearrangement can appear and even contribute substantially to the macroscopic deformation [13, 15, 16], it is commonly accepted that use of continuum models for solids is appropriate within this deformation threshold, provided the constitutive paradigm can take into account the strong irreversibilities and non-linearities that appear since the inception of deformation.

Classical plasticity models were imported in soil mechanics from metal plasticity and adapted to frictional-cohesive materials [17]. A main contribution to the understanding of frictional-cohesive materials was then given by the authors of a class of models grouped under the name of critical state soil mechanics (e.g., [18, 19]). The Cam-clay model and its ancestor the Granta-gravel model were prototypes of this family.

Only a few basic ingredients of the elastoplastic theories for granular materials need to be recalled here. *In primis* one usually mentions the hypothesis of rate-independence of the constitutive behaviour. The latter expresses the fact that the deformation process does not depend on physical time. The corresponding mathematical statement is that the strain rate is a homogeneous function of degree 1 of the stress rate, i.e.,

$$\forall \lambda > 0: \quad \dot{\boldsymbol{\varepsilon}}(\lambda \dot{\boldsymbol{\sigma}}) = \lambda \dot{\boldsymbol{\varepsilon}}(\dot{\boldsymbol{\sigma}}). \tag{1}$$

Ensuring quasi-static conditions implies the use of very low time rates, at which this hypothesis is generally satisfied. In terms of stress increments $\delta \boldsymbol{\sigma}$ and strain increments $\delta \boldsymbol{\varepsilon}$, Eq. 1 authorises to write that

$$\forall \lambda > 0: \quad \delta \boldsymbol{\varepsilon}(\lambda \, \delta \boldsymbol{\sigma}) = \lambda \, \delta \boldsymbol{\varepsilon}(\delta \boldsymbol{\sigma}), \tag{2}$$

provided increments $\delta \boldsymbol{\sigma}$ and $\delta \boldsymbol{\varepsilon}$ are small enough to be considered as infinitesimal. How small is "small" is one of the questions raised in this work and we will make use of Eq. 2 as a smallness criterion to filter out "non infinitesimal" stress and strain increments in our experimental procedure.

A second ingredient of plasticity we are interested in is the so-called *partition hypothesis*. We refer to the assumption that the strain increments $\delta \boldsymbol{\varepsilon}$ can be decomposed additively into elastic strain increments $\delta \boldsymbol{\varepsilon}^E$ and plastic strain increments $\delta \boldsymbol{\varepsilon}^P$:

$$\delta \boldsymbol{\varepsilon} = \delta \boldsymbol{\varepsilon}^E + \delta \boldsymbol{\varepsilon}^P \tag{3}$$

The former are computed according to a properly defined elastic compliance tensor $\mathbb{C}$, i.e.,

$$\delta \boldsymbol{\varepsilon}^E = \mathbb{C} \, \delta \boldsymbol{\sigma} \tag{4}$$

and relate to the amount of deformation work $\boldsymbol{\sigma} \cdot \delta \boldsymbol{\varepsilon}$ that is being reversibly stored as elastic energy. The latter relates to the amount of deformation work that is being dissipated and should fit the plastic constitutive assumption as specified here below in terms of a *yield criterion* and a *plastic flow rule*.

The yield criterion gives the recipe to compute plastic strain increments and distinguishes between "active" and "inactive" stress increments with respect to the mechanism responsible for plastic strains. With some simplification in the terminology we refer here to the yield criterion as prescription

$$\|\delta \boldsymbol{\varepsilon}^P\| = \begin{cases} \frac{1}{E^P} \delta \boldsymbol{\sigma} \cdot \boldsymbol{\xi} & \text{if } f(\boldsymbol{\sigma}) = 0 \text{ and } \delta \boldsymbol{\sigma} \cdot \boldsymbol{\xi} \geq 0 \\ 0 & \text{if } f(\boldsymbol{\sigma}) = 0 \text{ and } \delta \boldsymbol{\sigma} \cdot \boldsymbol{\xi} < 0 \\ 0 & \text{if } f(\boldsymbol{\sigma}) < 0 \end{cases} \tag{5}$$

in which the yield locus $f(\boldsymbol{\sigma}) = 0$ has outward oriented unit normal $\boldsymbol{\xi} := \frac{\partial f}{\partial \boldsymbol{\sigma}} \|\frac{\partial f}{\partial \boldsymbol{\sigma}}\|^{-1}$ and bounds the elastic domain in stress space. According to the above criterion, the only stress increments that succeed in producing plastic strain increments are those applied when the current stress state $\boldsymbol{\sigma}$ has reached the yield locus, and that point outward from the elastic domain. If these two conditions are met, the corresponding plastic strain increment will be proportional to the active part of the stress increment, i.e. the component $\delta \boldsymbol{\sigma} \cdot \boldsymbol{\xi}$, through the constant E^P called plastic modulus.

Finally, and once more loosely speaking, the plastic flow rule assigns a unique direction in stress space for all plastic strain increments, i.e., independently on the stress increment direction:

$$\forall \delta \boldsymbol{\sigma}: \quad \delta \boldsymbol{\varepsilon}^P(\delta \boldsymbol{\sigma}) = \boldsymbol{\pi}(\boldsymbol{\sigma}) \|\delta \boldsymbol{\varepsilon}^P(\delta \boldsymbol{\sigma})\| \tag{6}$$

where the tensor $\boldsymbol{\pi}$, $\|\boldsymbol{\pi}\| = 1$, is called *plastic flow direction*.

The particular yield criterion discussed in this work is of the Mohr-Coulomb type: we define function f in Eq. 5 with the expression

$$f = |\mathbf{m} \cdot \boldsymbol{\sigma} \mathbf{n}| - \mu_s \mathbf{n} \cdot \boldsymbol{\sigma} \mathbf{n}, \quad \mathbf{n} \cdot \mathbf{m} = 0 \tag{7}$$

whose terms can be described by rephrasing a few elements of Batdorf and Budiansky's plastic slip theory for polycrystalline materials [20] in the case of materials with particulate, frictional microstructure. Plastic strains are the macroscopic effect of slips along families of micro-planes inside the specimen (slip planes), characterised in Eq. 5 by an in-plane vector $\mathbf{m}$ (the slip direction) and by the unit normal $\mathbf{n}$. Plastic slip is activated when the threshold of tangential stress is reached on the inherent slip plane. At the scale of the REV this mechanism is reflected by a yield criterion $f = 0$ depending on a friction parameter μ_s. The latter is not a material constant but a parameter that evolves so to ensure $\delta f = 0$ during plastic loading (cf. Eq. 5). Implicit assumptions in Eq. 7 are that $\mathbf{n} \cdot \boldsymbol{\sigma} \mathbf{n} \geq 0$, as customary for non-cohesive granular materials, and that all activated slip planes hold nearly the same orientation.

We remark that some algebra leads to

$$\boldsymbol{\xi} = \pm \frac{1}{2}(\mathbf{m} \otimes \mathbf{n} + \mathbf{n} \otimes \mathbf{m}) - \mu_s \mathbf{n} \otimes \mathbf{n} \tag{8}$$

where the sign of the first term on the r.h.s. depends on the way the modulus operator has been resolved in Eq. 7. Using Eq. 8 with this caution we obtain, finally, a more compact form for the first equality in Eq. 7, i.e.,

$$f = \boldsymbol{\xi} \cdot \boldsymbol{\sigma}. \tag{9}$$

SPECIMEN PREPARATION

We characterise here the different types of specimens whose response under stress probes will be discussed in the following sections. The specimens are grouped into classes, based on non-dimensional control parameters, reflecting both the qualitative type of deformation response and the loading history.

Model material

The samples in use in our simulations consist of 5600 disks with diameters distributed uniformly between $0.7d$ and $1.3d$, where d is the average, representative diameter. All disks are assumed to be made of a homogeneous material, and m denotes the mass of particles of diameter d. The disks are initially arranged in rectangular cells whose wedges align along direction 1 and direction 2: the first is the "confinement" or "lateral" direction and the latter is the "axial" or "vertical" direction, in referring to the usual laboratory conditions of biaxial/triaxial testing. The cell can deform into an arbitrary parallelogram in order to accommodate the generic configuration of a two-dimensional cell undergoing homogeneous deformations at small-strains. Bi-periodicity is obtained, numerically, by an adaptation to DEM simulations of Parrinello-Rahman and Lees-Edwards techniques for molecular dynamics simulations (cf. [23, 24]). By these techniques we implement either mixed boundary conditions (for axial compression during biaxial tests, performed at constant axial strain rate and constant lateral pressure) or simple stress-rate-controlled boundary conditions (for isotropic compression during biaxial tests and for the application of stress probes). Samples are regarded in our analyses as REV's and are characterised macroscopically by the components of the stress tensor $\boldsymbol{\sigma}$ and of the infinitesimal strain tensor $\boldsymbol{\varepsilon}$. The former are computed according to the classical Love formula while the latter are retrieved, as usual, as linearised strain measures for the cell.

We use a standard linearly-elastic Coulomb-friction contact model: the normal contact force writes $F_N = K_N h_N$ where K_N is the normal contact stiffness and $h_N \geq 0$ is the (numerical) interpenetration of contacting disks; $F_T = K_T h_T$ relates the tangential contact force F_T to the relative tangential displacement h_T at contact (computed incrementally) through the tangential contact stiffness K_T; finally $|F_T|$ is bounded above by μF_N where μ is the contact friction coefficient. Here we choose $K_T = K_N$ and $\mu = 0.3$. An additional viscous force $F_N^\alpha = \alpha_N \dot{h}_N$ adds to the elastic force F_N as customary in DEM simulations, merely as a convenient means to accelerate the approach to equilibrium configurations. To this purpose we set $\alpha_N = 0.9\sqrt{2K_N m}$, where $\sqrt{2K_N m}$ is the critical value for a two-particle system with masses m interacting via a spring of stiffness K_N. Our focus being on constitutive information, gravity or other non-inertial volume actions are not considered here.

Loading history

The specimens to which stress probes are applied were first subjected to a standard procedure of strain-rate-controlled biaxial compression up to the desired stress ratio $\varsigma = Q/P$, where Q is the final value of the axial pressure σ_{22} and P is the value of confining pressure to which the lateral pressure σ_{11} is set during axial loading. Prior to axial loading the specimen were consolidated under isotropic stress conditions up to pressure P, starting from loose, randomly agitated "granular gas" configurations.

According to the loading history given above and to the previous characterisation of the model material, dimensional analysis leads to the identification of five independent dimensionless parameters that characterise separate classes of "equivalent" specimens: (i) stiffness parameter $\kappa = K_N/P$ setting the scale of contact deflections, as $h/d \propto \kappa^{-1}$; (ii) stress ratio $\varsigma = Q/P$, as an indicator of the deviatoric stress; (iii) friction coefficient μ; (iv) the damping parameter $\zeta = \alpha_N/\sqrt{2K_N m}$ and (v) inertia parameter $\gamma = \dot{\varepsilon}_{22}\sqrt{m/P}$; (vi) tangential to normal stiffness ratio K_T/K_N. We use $\zeta = 0.9$, as anticipated previously, and set $\gamma = 10^{-4}$ in order to approach quasistatic conditions with sufficient accuracy.

TABLE 1. Biaxial test families and values of variable parameters.

Biaxial tests	μ_{iso}	κ	ς (ca.)
A3		10^3	1.2, 1.4, 1.6, 1.8
A4	0	10^4	1.2, 1.4, 1.6, 1.8
A5		10^5	1.8, 1.9
B3		10^3	1.2, 1.4, 1.6, 1.8
B4	0.3	10^4	1.2, 1.4, 1.6, 1.8
B5		10^5	1.2, 1.4, 1.6, 1.8

The friction coefficient is also fixed, i.e. $\mu = 0.3$, as well as ratio $K_T/K_N = 1$, but we let κ and ς vary as detailed further on.

To further widen the spectrum of specimen classes we play with the value of contact friction coefficient μ_{iso} adopted during isotropic consolidation [15]. The only possible choice in real laboratory experiences is of course $\mu_{iso} = \mu$ but the numerical model allows us to set μ and $\mu_{iso} < \mu$ independently from each other. This possibility (supplemented with an "agitation" stage [15]) can be used as a robust procedure to obtain specimens that are "macroscopically indistinguishable", i.e. share the same solid fraction Φ, but differ markedly in terms of microstructures and deformation responses [13, 14, 16]. "Lubricated conditions" during isotropic consolidations (i.e., $\mu_{iso} = 0$) drive the material towards high values of the coordination number z; testing the material in this state gives a characteristic deformation response at small strains where the leading microscopic mechanism is the deformation at contacts (deformation response of type I). Conversely, "non-lubricated conditions" (i.e., $\mu_{iso} = \mu$), with some vibration, will result into much lower coordination numbers and lead to a deformation response dominated by a continuous network rearrangement due to microscopic instabilities (deformation response of type II). The biaxial compression tests considered in this work were all performed with friction coefficient $\mu = 0.3$, irrespective of the value μ_{iso} employed in sample preparation (i.e., during isotropic compression).

The parameters used in this work are reported in Table 1. The label A3 on the first line refers to a family of ten "equivalent" biaxial tests characterised by $\mu_{iso} = 0$ and $\kappa = 10^3$. They are equivalent in the sense that their preparation procedures differ just by the initial random velocity field at start up of the isotropic consolidation. The values of stress ratio $\varsigma = 1, 2, 1.4, 1.6, 1.8$ at the end of the same line refer to the configurations selected during axial loading, designated as specimens for the stress probing procedure. The following lines in the same table report the same information but relative to other choices of parameters κ and μ_{iso}. Fig. 1 illustrates the deformation response during axial loading for biaxial tests of families A3, A4 and A5 (type I deformation response) while Fig. 2 gives the same plots for biaxial tests of families B3, B4 and B5 (type II deformation response). A comparison between the two figures allows to visualise the macroscopic effect of the two microscopic deformation mechanisms mentioned above. The small-strain range of curves in Fig. 1 is sensitive to the stiffness parameter κ: as shown in [16], strains are actually inversely proportional to κ, for a given stress ratio ς. For the cases in Fig. 2, on the other hand, one notices that the macroscopics behaviour, already at small strains, results from microscopic instabilities and does not depend on the stiffness parameter κ.

BIAXIAL STRESS PROBES

For each test family in the Table 1, at least two of the ten equivalent biaxial tests have been considered insofar for the anlysis of the incremental response: the respective specimens were tested *via* stress probing in the above range of stress ratios. We present our results discussing case A4 ($\kappa = 10^4$, type I deformation regime, $\varsigma \simeq 1.2, 1.4, 1.6, 1.8$) claiming that the qualitative features that we observed were found repeatable in all the other cases, despite the change in control parameters. The section reports on stress probes applied in the plane of principal stress axes, which correspond in our case to a combination of increments of lateral stress σ_{11} and of axial stress σ_{22}. It will be convenient to refer to the representations in the planes σ_{11} *vs.* σ_{22} or ε_{11} *vs.* ε_{22} as to representations in the *biaxial stress plane* or in the *biaxial strain plane*, respectively.

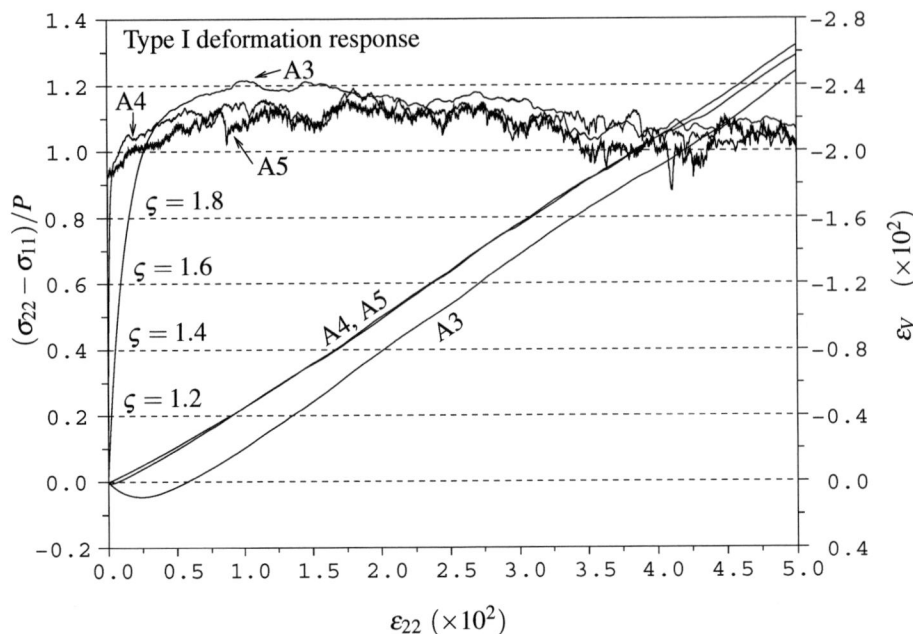

FIGURE 1. Normalised deviatoric stress *vs.* axial strain and volumetric strain *vs.* axial strain for typical biaxial tests of families A3, A4 and A5: $\mu_{iso} = 0$ (but $\mu = 0.3$) and $\kappa = 10^3$, 10^4 and 10^5, respectively.

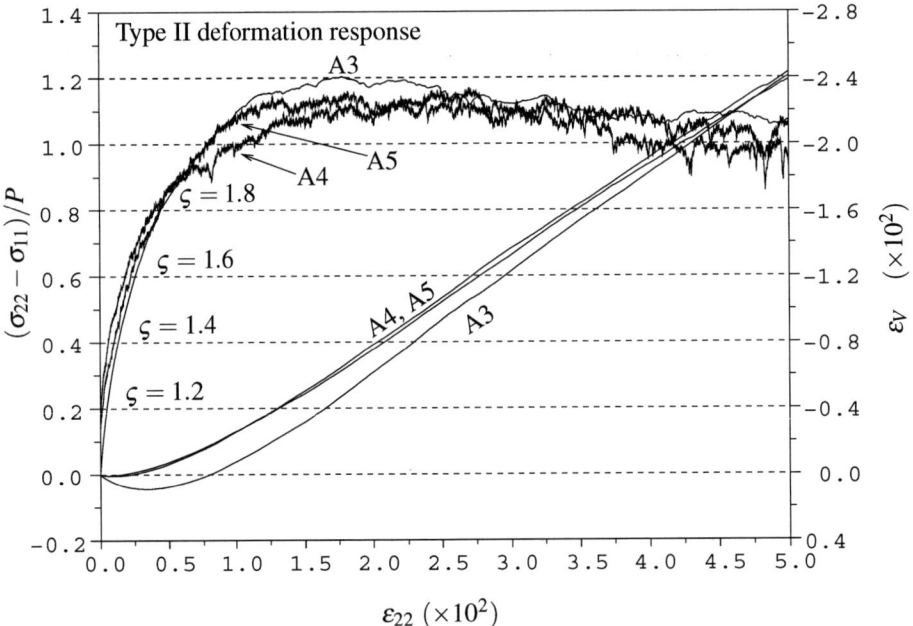

FIGURE 2. Normalised deviatoric stress *vs.* axial strain and volumetric strain *vs.* axial strain for typical biaxial tests of families B3, B4 and B5: $\mu_{iso} = \mu = 0.3$ and $\kappa = 10^3$, 10^4 and 10^5, respectively.

Incremental response

The rose of stress increments applied to the specimens is shown in Fig. 3: it consists in twelve increment levels linearly distributed from $\|\delta\boldsymbol{\sigma}\| = 2\sqrt{2}P \times 10^{-3}$ to $12 \times 2\sqrt{2}P \times 10^{-3}$ along sixteen orientations in the biaxial stress plane, labelled from 0A to 0P, with constant angular spacing $2\pi/16$. The elastic response to the increments in Fig. 3

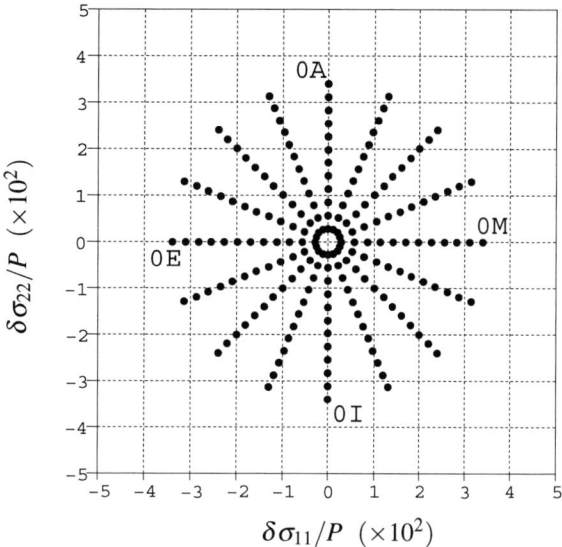

FIGURE 3. Rose of applied increments for biaxial stress probing: 16 increment directions (0A to 0P) for 12 amplitude values in the biaxial plane.

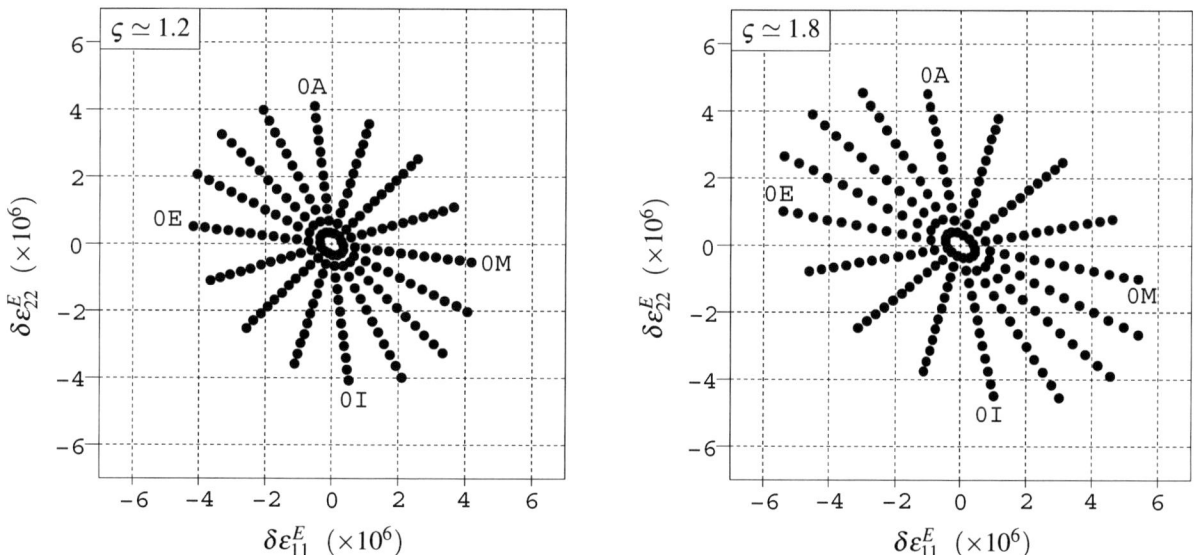

FIGURE 4. Elastic response for specimens at stress ratio $\varsigma \simeq 1.2$ (left) and $\varsigma \simeq 1.8$ (right) from a biaxial test of family A4.

is plotted in Fig. 4 for the specimen with lowest and highest values of stress ratio: the material exhibits a marked elastic anisotropy slightly evolving during the axial loading (i.e., from $\varsigma \simeq 1.2$ to 1.8). The elastic strain increments are assumed to be given by the expression in Eq. 4, where the components of the elasticity tensor $\mathbb{C}$ are computed, by assembling the contribution of the contact stiffness K_N and K_T across the contact network [14]. To test the partition hypothesis we identify plastic strain increments to the difference

$$\delta\boldsymbol{\varepsilon}^P = \delta\boldsymbol{\varepsilon} - \delta\boldsymbol{\varepsilon}^E \tag{10}$$

and check *a posteriori* whether or not this definition is effective in giving evidence of a plastic flow rule (cf. Eq 6) and yield criterion (cf. Eq 5).

Fig. 5 shows that the strain increments $\delta\boldsymbol{\varepsilon}^P$ neatly align along a direction in the biaxial strain plane, confirming the applicability of a flow rule. We measure in particular counterclockwise angles of the plastic strain increment direction

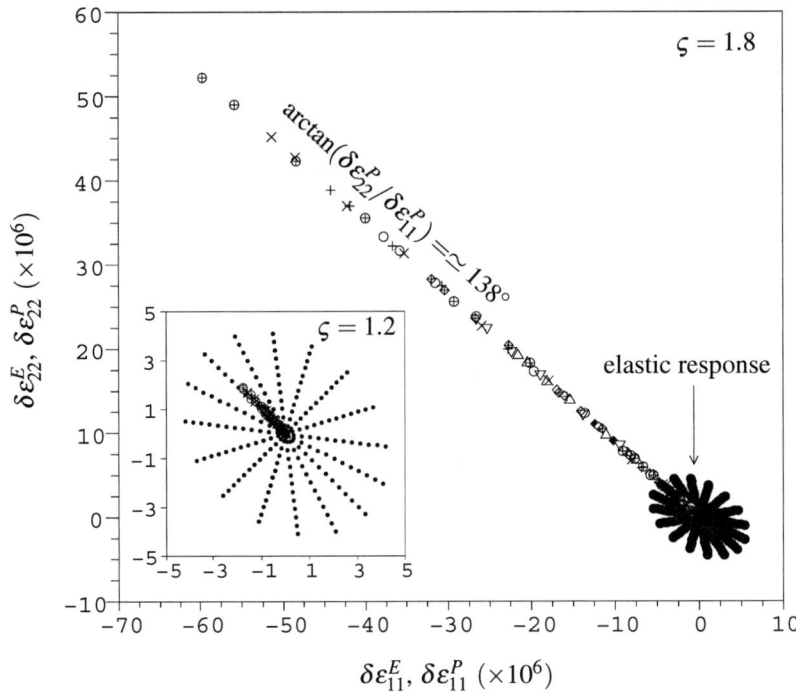

FIGURE 5. Elastic vs. plastic response for specimens at stress ratio $\varsigma = 1.2$ (inner window) and $\varsigma = 1.8$ (resp. outer) from a biaxial test of family A4.

with respect to direction ε_{11}, ranging from 132.2° to 138.3° (for $\varsigma \simeq 1.2$ and 1.8, resp.).

To investigate the existence of a yield criterion we consider the case $\varsigma \simeq 1.2$ and plot in Fig. 6 the norm $\|\delta\boldsymbol{\varepsilon}^P\|$ of the plastic strain increment against the angle of the stress increments in the biaxial stress plane. As for the plastic strain increments, stress increment angles in the biaxial plane are measured counterclockwise, with respect to direction "11". The experimental points in Fig. 6 are fitted with a truncated cosine function, expressive of the criterion in Eq 5. The corresponding phase angle (132.3° in the figure) gives the direction of the normal $\boldsymbol{\xi}$ to the supposed yield criterion in the biaxial stress space. The *load direction*, i.e., the direction associated to the current value of the stress tensor $\boldsymbol{\sigma}$ is almost orthogonal to $\boldsymbol{\xi}$, as expected (cf. Eq 9). On the other hand the plastic flow direction $\boldsymbol{\pi}$ is close but not at all coincident with the normal $\boldsymbol{\xi}$ to the yield criterion (separated of about 10°), giving evidence of the non associated character of the flow rule. These two qualitative features were systematically found in all investigated cases: for $\varsigma \simeq 1.2, 1.4, 1.6$ and 1.8, as shown it upper-right quadrant of the same figure, and over the whole range of stress increments. The angles for the direction of the normal $\boldsymbol{\xi}$ to the yield criterion and for the flow direction $\boldsymbol{\pi}$ are compared to the load direction in Table 2.

Stress increment size

An important question in the stress probing procedure concerns the appropriate size of the strain increments to apply in order to get measurements that are at the same time little affected by systematic errors and representative of the infinitesimal behaviour. To discuss this point we represent in Fig. 7 the norm $\|\delta\boldsymbol{\varepsilon}^P\|$ of the plastic strain increments versus the "active" component of the stress increment, i.e., the positive values of the scalar product $\delta\boldsymbol{\sigma} \cdot \boldsymbol{\xi}$. Fitting the yield criterion in Eq. 5 requires selecting an observation window in which the relation between the norm $\|\delta\boldsymbol{\varepsilon}^P\|$ of the plastic strain increment and the active component of the stress increment can be fitted as linear. The plot in Fig. 7 suggests that, for specimens obtained from biaxial tests of type A4, the size should not exceed the eight-level of

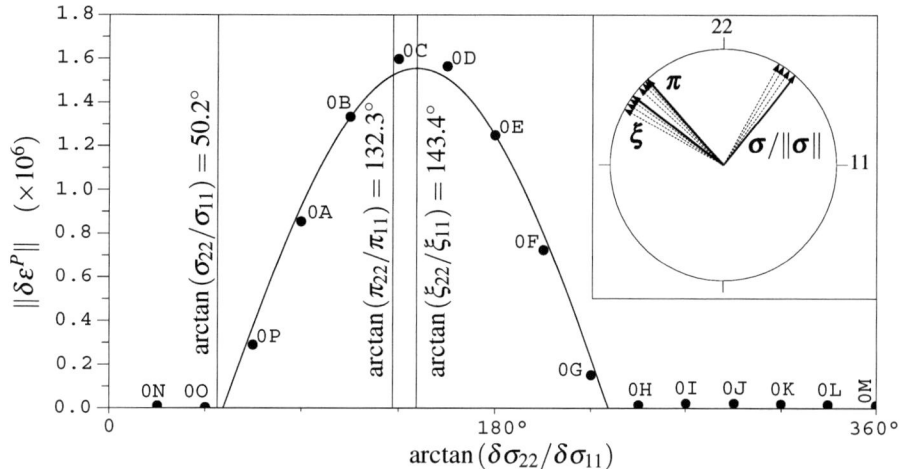

FIGURE 6. Fitting of the plastic strain increment amplitude with the truncated cosine function for specimen at stress ratio $\varsigma \simeq 1.2$ (main window) from a biaxial test of family A4. Load direction $\boldsymbol{\sigma}/\|\boldsymbol{\sigma}\|$, normal to the yield criterion $\boldsymbol{\xi}$ and plastic flow direction $\boldsymbol{\pi}$ as they evolve counterclockwise from stress ratio $\varsigma \simeq 1.2$, to 1.8 during the axial loading for the same biaxial test (upper-right window).

increment considered here (i.e., $\|\delta\boldsymbol{\sigma}\| = 8 \times 2\sqrt{2}P \times 10^{-3} \simeq 2.263P \times 10^{-2}$). According to the result analysed insofar by the authors, this range seems also to depend sensitively on the stiffness parameter κ: the higher the value of the stiffness parameter, the smaller the maximal allowed norm $\|\delta\boldsymbol{\sigma}\|/P$ of the stress increment amplitude. On the other hand it is not advisable to reduce as much as numerically possible the size of the increments, as the case $\varsigma \simeq 1.8$ in the same figure shows: evidence of a residual elastic response is given by the absence of plastic deformation increments in response to very small stress increments. This source of systematic errors can be corrected easily provided "large enough" stress increments are used. Most important this type of systematic error can be explained, mechanically, on the basis of the procedure to which the specimen were subjected prior to to stress probing. We remark in particular that, after the expected value of stress ratio ς is attained during the axial compression, the specimen is left under constant axial and lateral stresses for the time necessary to reach statical equilibrium, not to distort the incremental response during stress probing. A small parasite effect of this intermediate "creep" transition before stress probing is that part of the plastic memory, stored at contact between particles, is erased due to a slight unavoidable rearrangement of the contact network. We expect this effect to gradually fade out as slower and slower numerical tests are performed, in order to approach closer to the quasistatic limit of $\gamma \to 0$. The appreciable non-zero intercept of the (dashed) interpolation line for case $\varsigma \simeq 1.8$ in Fig. 7 can be seen as the macroscopic signature of this mechanism. In the end, an appropriate choice of the size of the increment for stress probing seems to bounded both above, by a linearity requirement, and below, due to a parasite effect of residual elastic behaviour.

ROTATION OF PRINCIPAL STRESS AXES

The same set of specimens considered in the previous section was tested under stress probes inducing rotation of principal stress axes. We discuss here a few preliminary results, and possible interpretations, that we expect to study systematically on a larger base. The stress space to which we refer is now the general stress space with coordinates σ_{11}, σ_{22} and σ_{12}, where σ_{12} is the third component of the stress tensor, dismissed insofar and corresponding to tangential stresses along the planes orthogonal to the lateral and axial direction. Analogously, the deformation response is measured in a three-dimensional strain space with coordinates ε_{11}, ε_{22} and ε_{12}.

Plastic flow direction

For this preliminary analysis, the stress increments lay in a specific plane of the stress space: the plane spanned by the direction of the normal to the yield criterion $\boldsymbol{\xi}$, detected previously, and by the direction associated to the

TABLE 2. Load direction, normal to yield criterion and flow direction for tested specimens from two biaxial tests in each family of Table 1: averaged values over the range of increment sizes.

	family A3							
	biaxial test A3-1				biaxial test A3-2			
ς	1.198	1.402	1.602	1.801	1.198	1.400	1.599	1.801
$\arctan(\sigma_{22}/\sigma_{11})$ (deg)	50.2	54.5	58.0	61.0	50.2	54.5	58.0	61.0
$\arctan(\xi_{22}/\xi_{11})$ (deg)	133.1	133.3	134.7	137.1	132.6	132.6	134.0	137.2
$\arctan(\pi_{22}/\pi_{11})$ (deg)	138.7	142.4	143.4	143.8	138.4	141.6	141.7	141.7

	family A4							
	biaxial test A4-1				biaxial test A4-2			
ς	1.228	1.404	1.605	1.803	1.228	1.403	1.604	1.802
$\arctan(\sigma_{22}/\sigma_{11})$ (deg)	50.8	54.5	58.1	61.0	50.8	54.5	58.1	61.0
$\arctan(\xi_{22}/\xi_{11})$ (deg)	138.6	143.8	148.9	151.7	146.8	147.9	149.5	150.1
$\arctan(\pi_{22}/\pi_{11})$ (deg)	132.2	133.6	135.6	138.3	135.6	136.8	138.3	141.8

	family A5							
	biaxial test A5-1				biaxial test A5-2			
ς	1.799	1.914	1.919	1.929	1.782	1.905	1.929	1.929
$\arctan(\sigma_{22}/\sigma_{11})$ (deg)	60.9	62.4	62.5	62.6	60.7	62.3	62.6	62.6
$\arctan(\xi_{22}/\xi_{11})$ (deg)	153.2	154.0	154.0	155.3	153.2	154.4	152.9	153.4
$\arctan(\pi_{22}/\pi_{11})$ (deg)	138.7	142.4	143.4	143.8	138.4	141.6	141.7	141.7

	family B3							
	biaxial test B3-1				biaxial test B3-2			
ς	1.199	1.401	1.601	1.801	1.200	1.401	1.600	1.800
$\arctan(\sigma_{22}/\sigma_{11})$ (deg)	50.2	54.5	58.0	61.0	50.2	54.5	58.0	60.9
$\arctan(\xi_{22}/\xi_{11})$ (deg)	140.8	145.4	149.4	152.8	140.7	144.9	148.9	152.8
$\arctan(\pi_{22}/\pi_{11})$ (deg)	130.3	133.9	137.9	141.1	130.8	133.9	137.5	141.9

	family B4							
	biaxial test B4-1				biaxial test B4-2			
ς	1.202	1.401	1.600	1.800	1.202	1.403	1.601	1.800
$\arctan(\sigma_{22}/\sigma_{11})$ (deg)	50.2	54.5	58.0	60.9	50.2	54.5	58.0	60.9
$\arctan(\xi_{22}/\xi_{11})$ (deg)	141.5	145.8	146.2	—	140.1	146.3	149.0	152.2
$\arctan(\pi_{22}/\pi_{11})$ (deg)	130.8	135.0	136.5	141.5	131.3	136.0	139.6	140.9

	family B5							
	biaxial test B5-1				biaxial test B5-2			
ς	1.200	1.399	1.605	1.799	1.206	1.401	1.600	1.804
$\arctan(\sigma_{22}/\sigma_{11})$ (deg)	50.2	54.4	58.1	60.9	50.3	54.5	58.0	61.0
$\arctan(\xi_{22}/\xi_{11})$ (deg)	142.1	145.1	148.9	152.5	141.2	146.4	150.0	152.0
$\arctan(\pi_{22}/\pi_{11})$ (deg)	133.4	135.2	137.4	140.7	131.9	136.1	138.9	142.2

shear stress component σ_{12}. This simplifying choice ensures anyway that we have access to the highest values of plastic deformation increments. As was the case for biaxial stress probing, the increments are applied along sixteen equally distributed directions of our stress plane, from 1A to 1P, and vary in amplitude from $\|\delta\sigma\| = 2\sqrt{2}P \times 10^{-3}$ to $\|\delta\sigma\| = 12 \times 2\sqrt{2}P \times 10^{-3}$ (see Fig 8). Points of this plane will be mapped by coordinates computed as $\sigma \cdot \xi/P$ and $\sqrt{2}\sigma_{12}$, where the factor $\sqrt{2}$ is adopted, due to the tensorial nature of $\delta\sigma$, so to visualise families of equal-norm increments as circles.

In order to discuss the validity of the partition hypothesis under rotation of principal stress axes, we refer once more to Eq. 10 as a definition for plastics strain increments. The consequence of this choice is shown in Fig. 9, for a specimen with stress ratio $\varsigma \simeq 1.8$ and loading history from a biaxial test of family A4. Elastic and plastic strain increments are plotted in coordinates of type $\sqrt{2}\delta\varepsilon \cdot \pi$ and $\delta\varepsilon_{12}$, i.e., we study exclusively the components of strain increment along a plane parallel to the plastic flow direction π, found during biaxial stress probing, and orthogonal to the biaxial strain plane. Due to π and for reasons of symmetry of the mechanical response, the plastic strain increments are expected to be confined to this plane. Fig. 10 confirms this expectation showing that the plastic response is negligible along a third plane, chosen orthogonal to the biaxial strain plane and to the one in Fig. 9.

The plot in Fig. 9 completes the one in Fig. 5 ($\varsigma \simeq 1.8$) and shows, compared to it, that stress increments inducing

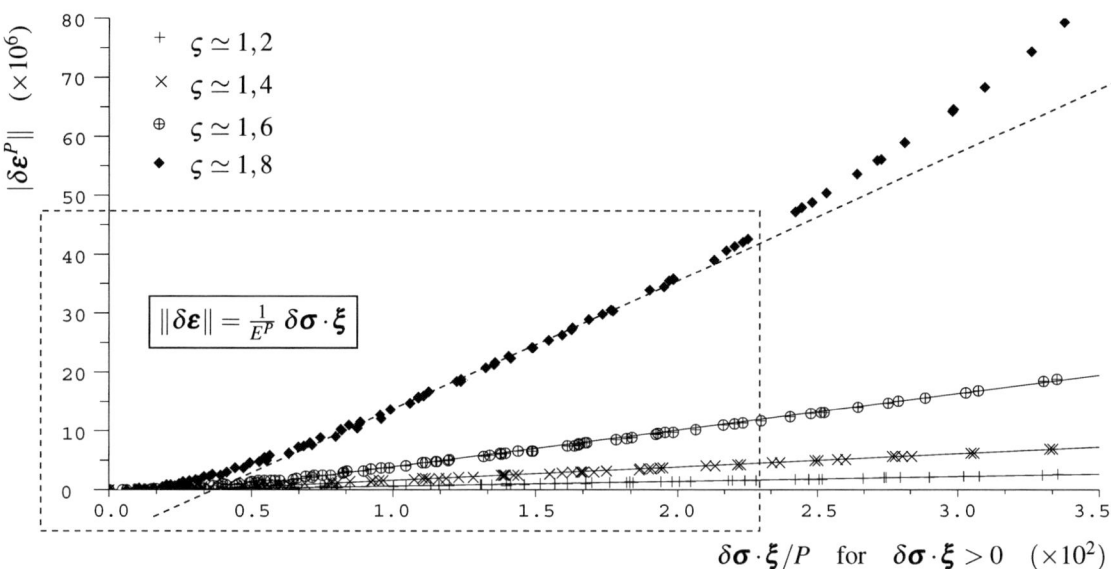

FIGURE 7. Plastic strain increment amplitude *vs.* active part of the stress increments for specimens at stress ratios $\varsigma \simeq 1.2$ to 1.8 from a biaxial test of family A4.

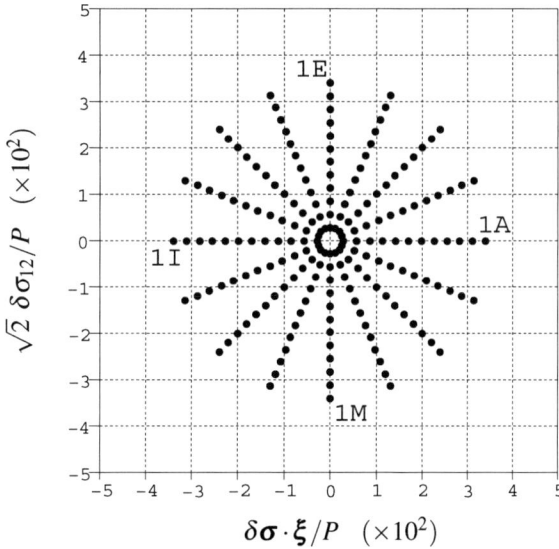

FIGURE 8. Rose of applied increments for stress probing with rotation of principal stress axes: 16 increment directions (1A to 1P) for 12 amplitude values in the plane parallel to $\boldsymbol{\xi}$ (biaxial stress probing) and orthogonal to the biaxial stress plane.

rotation of principal stress axes cause, at least, loss of uniqueness of the plastic flow direction. One notices in Fig. 9 as many plastic strain increment directions as the number of stress increment directions: dotted lines in the figure are drawn on naked-eye visible experimental point to show that plastic strain increments originated by proportional stress increments align along precise directions. The question arises whether or not this kind of incremental behaviour is representative of a non-trivial flow rule, or should be modelled in a different constitutive framework than the elastoplastic one. The issue of the flow rule cannot be treated anyway isolated from that of the yield criterion, which is considered next.

A final remark on Fig. 10 is that the envelopes of the plastic response, traced in figure for three different values of the stress increment, obey loosely the symmetry with respect to the axis corresponding to direction $\boldsymbol{\pi}$. It is questionable whether or not the appreciable deviation from symmetry would disappear for REV's of larger sizes.

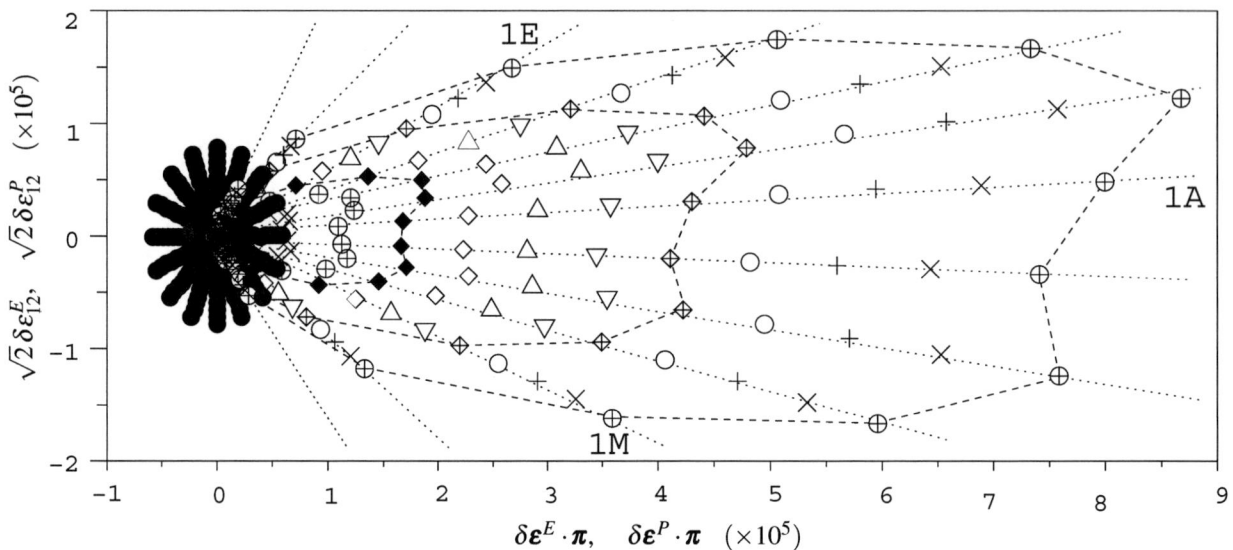

FIGURE 9. Elastic strain increments (solid circular marks) and plastic strain increments (other marks, resp.) for a specimen at stress ratio $\varsigma = 1.8$ from a biaxial test of family A4. Dotted lines show the alignment of plastic strain increments for proportional stress increments while dashed segments mark the response envelopes at increments of norm $\|\delta\boldsymbol{\sigma}\| = 4 \times 2\sqrt{2}P$, $8 \times 2\sqrt{2}P$ and $12 \times 2\sqrt{2}P$.

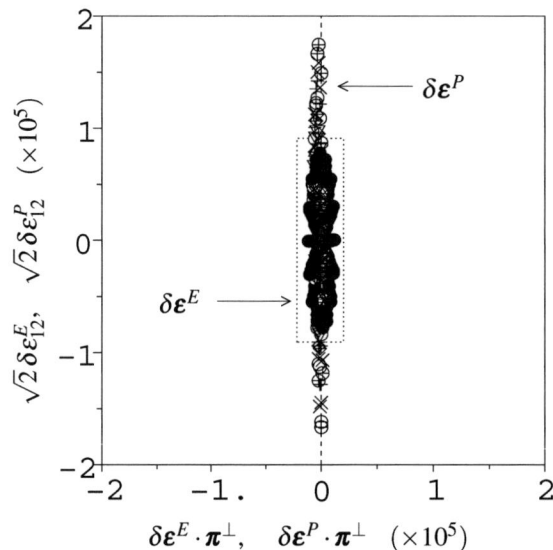

FIGURE 10. Elastic strain increments (solid circular marks) and plastic strain increments (other marks, resp.) for a specimen at stress ratio $\varsigma = 1.8$ from a biaxial test of family A4. The representation is given in a strain increment plane orthogonal to those in Fig. 9 and 5: the direction $\boldsymbol{\pi}^\perp$, $\|\boldsymbol{\pi}^\perp\| = 1$, belongs to the biaxial plane and lays orthogonal to $\boldsymbol{\pi}$ ($\arctan(\pi_{22}^\perp/\pi_{11}^\perp) = 138.3° + 90°$, cf. Table 2).

Yield criterion

The shapes of the envelopes of the elastic and plastic responses are represented in Fig. 11-a for a specimens selected at stress ratio $\varsigma \simeq 1.2$ from a biaxial test of family A5. Due to the low level of stress ratio, the plastic envelope is still bounded by the elastic one and the plastic strain increments related to stress increments of pure shear are dominant with respect to "biaxial" strain increment (parallel to $\boldsymbol{\pi}$). The open shape of the plastic strain envelope denies the existence of a uniquely defined flow rule (cf. Fig 9).

The behaviour shown in fig. 11-a can still be modelled anyway as elastoplastic, provided one drops the assumption of a unique mechanism of plastic deformation [26, 27]. We postpone a detailed exposition of our idea to a further publication and give here an example of the procedure to fit the case in the figure with a first generalisation of the classical elastoplastic framework. We consider in particular the possibility of three distinct and independent plastic mechanisms of deformation: a first mechanism detectable with biaxial stress probes and two pseudo-symmetric additional mechanisms activated by shear stress increments of positive and negative values, respectively. The partition hypothesis now writes in the form

$$\delta\boldsymbol{\varepsilon} = \delta\boldsymbol{\varepsilon}^E + \delta\boldsymbol{\varepsilon}^P_I + \delta\boldsymbol{\varepsilon}^P_{II} + \delta\boldsymbol{\varepsilon}^P_{III} \tag{11}$$

where the amplitudes of the three separate plastic increments on the r.h.s. are given by the respective yield criteria, i.e.,

$$\|\delta\boldsymbol{\varepsilon}^P_I\| = \begin{cases} \frac{1}{E^P_I}\delta\boldsymbol{\sigma}\cdot\boldsymbol{\xi}_I & \text{if} \quad f(\boldsymbol{\sigma}) = 0 \quad \text{and} \quad \delta\boldsymbol{\sigma}\cdot\boldsymbol{\xi}_I \geq 0 \\ 0 & \text{if} \quad f(\boldsymbol{\sigma}) = 0 \quad \text{and} \quad \delta\boldsymbol{\sigma}\cdot\boldsymbol{\xi}_I < 0 \\ 0 & \text{if} \quad f(\boldsymbol{\sigma}) < 0 \end{cases}, \tag{12}$$

$$\|\delta\boldsymbol{\varepsilon}^P_{II}\| = \begin{cases} \frac{1}{E^P_{II}}\delta\boldsymbol{\sigma}\cdot\boldsymbol{\xi}_{II} & \text{if} \quad f(\boldsymbol{\sigma}) = 0 \quad \text{and} \quad \delta\boldsymbol{\sigma}\cdot\boldsymbol{\xi}_{II} \geq 0 \\ 0 & \text{if} \quad f(\boldsymbol{\sigma}) = 0 \quad \text{and} \quad \delta\boldsymbol{\sigma}\cdot\boldsymbol{\xi}_{II} < 0 \\ 0 & \text{if} \quad f(\boldsymbol{\sigma}) < 0 \end{cases}, \tag{13}$$

$$\|\delta\boldsymbol{\varepsilon}^P_{III}\| = \begin{cases} \frac{1}{E^P_{III}}\delta\boldsymbol{\sigma}\cdot\boldsymbol{\xi}_{III} & \text{if} \quad f(\boldsymbol{\sigma}) = 0 \quad \text{and} \quad \delta\boldsymbol{\sigma}\cdot\boldsymbol{\xi}_{III} \geq 0 \\ 0 & \text{if} \quad f(\boldsymbol{\sigma}) = 0 \quad \text{and} \quad \delta\boldsymbol{\sigma}\cdot\boldsymbol{\xi}_{III} < 0 \\ 0 & \text{if} \quad f(\boldsymbol{\sigma}) < 0 \end{cases}, \tag{14}$$

and by the respective flow rules, grouped here below:

$$\forall \delta\boldsymbol{\sigma} : \begin{cases} \delta\boldsymbol{\varepsilon}^P_I(\delta\boldsymbol{\sigma}) = \boldsymbol{\pi}_I(\boldsymbol{\sigma})\|\delta\boldsymbol{\varepsilon}^P_I(\delta\boldsymbol{\sigma})\| \\ \delta\boldsymbol{\varepsilon}^P_{II}(\delta\boldsymbol{\sigma}) = \boldsymbol{\pi}_{III}(\boldsymbol{\sigma})\|\delta\boldsymbol{\varepsilon}^P_{II}(\delta\boldsymbol{\sigma})\| \\ \delta\boldsymbol{\varepsilon}^P_{III}(\delta\boldsymbol{\sigma}) = \boldsymbol{\pi}_{II}(\boldsymbol{\sigma})\|\delta\boldsymbol{\varepsilon}^P_{III}(\delta\boldsymbol{\sigma})\| \end{cases} \tag{15}$$

We identify in particular $\boldsymbol{\xi}_I$ to the normal to the criterion detected by the biaxial stress probing procedure, i.e. $\boldsymbol{\xi}_I \equiv \boldsymbol{\xi}$, and seek $\boldsymbol{\xi}_{II}$ and $\boldsymbol{\xi}_{III}$ in the plane of Fig. 8. Analogously we set $\boldsymbol{\pi} \equiv \boldsymbol{\pi}_I$ and pick $\boldsymbol{\pi}_{II}$ and $\boldsymbol{\pi}_{III}$ in the plane of Fig. 11-a.

As to the fitting of the flow rule, compared to the case in Fig. 6, one is now bound to use the sum of three truncated cosine functions, one for each of the three orthogonal criteria in Eqs. 12-14:

$$\delta\boldsymbol{\varepsilon} = \left(\frac{[\cos(\theta - \theta_I)]^+}{E^P_I}\boldsymbol{\pi}_I + \frac{[\cos(\theta - \theta_{II})]^+}{E^P_{II}}\boldsymbol{\pi}_{II} + \frac{[\cos(\theta - \theta_{III})]^+}{E^P_{III}}\boldsymbol{\pi}_{III}\right)\|\delta\boldsymbol{\sigma}\| \tag{16}$$

where $[\cdot]^+$ denotes the positive part of the argument function and the angles θ, θ_I, θ_{II} and θ_{III}, are measured counterclockwise in the plane of Fig. 8 starting from direction $\boldsymbol{\xi}$. Angles θ_I, θ_{II} and θ_{III}, refer to $\boldsymbol{\xi}_I$, $\boldsymbol{\xi}_{II}$ and $\boldsymbol{\xi}_{III}$, respectively (e.g. $\theta_I = 0$). Our fitting parameters are the angles θ_{II} and θ_{III}, the three plastic moduli E^P_I, E^P_{II}, E^P_{III} and the angles ω_{II} and ω_{III} referred to $\boldsymbol{\pi}_{II}$ and $\boldsymbol{\pi}_{III}$ and measured counterclockwise in the plane of Fig. 11 starting from $\boldsymbol{\pi}$ (e.g. $\omega_I = 0$). The quality of the fitting in Fig. 12 is encouraging. We remark anyway that this setting does not apply immediately to the cases with highest stress ratio (i.e., $\varsigma \simeq 1.6$ and $\varsigma \simeq 1.8$) where some degree further generality needs to be added to the model.

To conclude on the case in Fig. 11a and validate our renewed partition hypothesis, we plot in Fig. 11b the difference $\delta\boldsymbol{\varepsilon}^P - \delta\boldsymbol{\sigma}\cdot\boldsymbol{\xi}_I/E^P_I$ in order to visualise the response envelope exclusively for the plastic increments of competence of the second and third mechanisms, i.e., $\delta\boldsymbol{\varepsilon}^P_{II}$ and $\delta\boldsymbol{\varepsilon}^P_{III}$. The plastic envelop in the figure now conforms to two clearly-defined directions, i.e. the two "missing" flow directions $\boldsymbol{\pi}_{II}$ and $\boldsymbol{\pi}_{III}$.

CONCLUSION

Our concern in this work is an assessment of some features of the elastoplastic behaviour of granular materials and an evaluation on the representativity of the measurements that can be obtained from the stress probing procedure

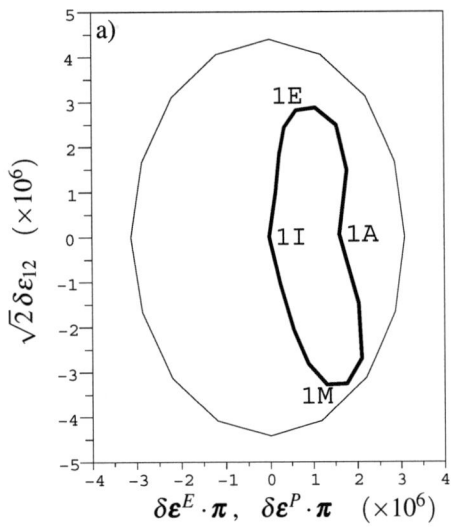
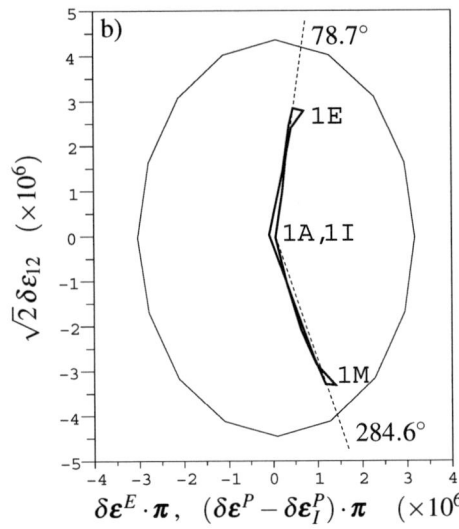

FIGURE 11. Elastic and plastic response envelopes for specimen at stress ratio $\varsigma \simeq 1.2$, from a biaxial test of family A4, under stress increments of amplitude $2 \times 2\sqrt{2}P$ in Fig. 8. Total plastic response $\delta\varepsilon^P$ (a) or plastic response for mechanisms II and III (b).

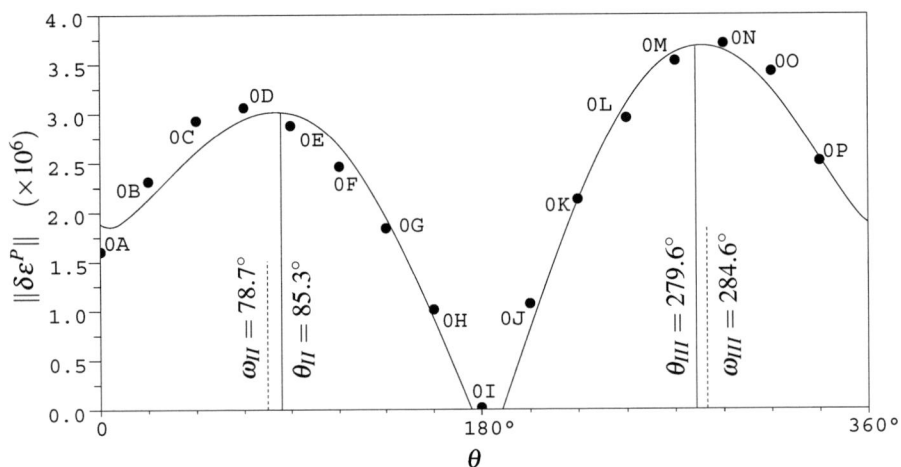

FIGURE 12. Fitting of the plastic strain increment amplitude with Eq. 16, for specimen at stress ratio $\varsigma = 1.2$, from a biaxial test of family A4, under stress increments of amplitude $2 \times 2\sqrt{2}P$ in Fig. 8.

via DEM simulations. To this extent, our study was conceived in parametrical form, and we play on the size of the stress increments, the stiffness parameter $\kappa = K_N/P$ and the stress ratio ς, within two distinct qualitative classes of deformation response (cf. Table 1 and Fig. 1-2). The results presented here were obtained from a limited number of prototype stress probing tests.

For the case of stress probes in the biaxial stress plane we observed the existence of a clear direction of accumulation for plastic strain increments, i.e. a plastic flow direction in the language of plasticity. Measurements of this quantity were robust, i.e. not affected significantly by the stress increment size. On the other hand both the normal to the yield criterion and the plastic modulus E^P were found sensitive to the increment size, especially the latter. We propose in particular that a criterion for the detection of the appropriate range of stress increments should be based on a requirement of linearity between plastic strain increments and "active" stress increments (cf. Fig. 7) with stable coefficient $1/E^P$.

The normal to the yield criterion was found systematically orthogonal, with very good approximation, to the load direction. According to the presentation in the introduction and to Eq. 9, this is the explicit signature of a yield criterion of the Mohr-Coulomb type in the sense of the pastic slip theory. All in all the response of the tested specimens to stress probes in the biaxial stress plane can be certainly ascribed to the class of elastoplastic materials with single mechanism of plastic deformation, as found in the literature [10, 11, 12]. A variable difference in angle, of the order of $10°$, was observed between the plastic flow direction and the normal to the yield criterion, confirming the non associated character of the flow rule.

As to the incremental response to stress probes with rotation of principal stress axes, i.e. with non-null tangential components, the first remark concerns the loss of a uniquely defined plastic flow direction. The stress increment were applied in a plane orthogonal to the biaxial plane and parallel to the (biaxial) normal to the yield criterion. The plastic strain increments were found exclusively in the plane orthogonal to the biaxial plane and parallel to the (biaxial) plastic flow direction. We showed that this scenario can still be modelled in the elastoplastic framework by introducing additional mechanisms of plastic deformation.

The above features were observed for the different values of parameters and classes of qualitative behaviour, but a study of their quantitative variability is part of the work to come.

ACKNOWLEDGMENTS

Laboratoire Navier is a joint research unit of Laboratoire Central des Ponts et Chaussées, Ecole Nationale des Ponts et Chaussées and Centre National de la Recherche Scientifique. PPF CEGEO is a joint programme funded by the French Ministry of Higher Education and Research.

REFERENCES

1. J. K. Mitchell, *Fundamentals of soil behavior*, Wiley, New York, 1993.
2. P. A. Vermeer, in *Physics of Dry Granular Media*, edited by H. J. Herrmann, J.-P. Hovi, and S. Luding, Balkema, Dordrecht, 1998, pp. 163–96.
3. F. Alonso-Marroquín, S. Luding, H. J. Herrmann, and I. Vardoulakis, *Phys. Rev. E* **71**, 051304 (2005).
4. C. Tamagnini, F. Calvetti, and G. Viggiani, *J. Eng. Math.* **52**, 265–91 (2005).
5. F. Darve, L. Sibille, A. Daouadji, and F. Nicot, *C. R. Mécanique* **335**, 496–515 (2007).
6. F. Radjaï, *eprint ArXiv:0801.4722v1* (2008).
7. P. Royis, and T. Doanh, *Int. Journ. Num. Anal. Methods in Geomechanics*, **22**, 34-45, (1998).
8. J. P. Bardet, and J. Proubet, in *Powders and Grains 1989*, edited by J. Biarez, and R. Gourvès, Balkema, Rotterdam, 1998, pp. 265–73.
9. J. P. Bardet, *Int. J. of Plasticity*, **10**, 879–908 (1994).
10. F. Calvetti, G. Viggiani and C. Tamagnini, *Constitutive modelling and analysis of boundary value problems in geotechnical engineering*, edited by G. Viggiani, Hevelius, Benevento, 187–216 (2003).
11. F. Calvetti, G. Viggiani and C. Tamagnini, *Rivista italiana di geotecnica*, **3**, 11–29 (2003).
12. F. Alonso-Marroquín, and H. J. Herrmann, *Phys. Rev. E* **66**, 021301 (2002).
13. J.-N. Roux, and G. Combe, *Comptes-Rendus Physique*, **3**, 131–40 (2003).
14. I. Agnolin, and J.-N. Roux, *Phys. Rev. E* **76**, 061304 (2007).
15. I. Agnolin, and J.-N. Roux, *Phys. Rev. E* **76**, 061302 (2007).
16. J.-N. Roux, and G. Combe, these proceedings.
17. W. Prager, and D. C. Drucker, *Q. Appl. Mathematics*, **2**, 157–65 (1952).
18. K. H. Roscoe, A. N. Schofield, and C. P. Wroth, *Geotechnique*, **8**, 22–53 (1958).
19. A. N. Schohield, and C. P. Wroth, *Critical State Soil Mechanics*, McGraw-Hill (1968).
20. S. B. Batdorf, and B. Budiansky, NACA TN 1871 (1949).
21. J. Desrues, and R. Chambon, *Int. J. Solid Struct.* **39**, 3757–76 (2002).
22. J. W. Rudnicki, and J. R. Rice, *J. Mech. Phys. Solids* **23**, 371–94 (1975).
23. M. Allen, and D. Tildesley, *Computer simulations of liquids*, Oxford University Press, Oxford, 1987.
24. P.-E. Peyneau, and J.-N. Roux, *Phys. Rev. E* **78**, 011307 (2008).
25. J.-N. Roux, in *Powders and Grains 2005*, edited by R. García Rojo, H. J. Herrmann, and S. McNamara, Balkema, Leiden, 2005, pp. 261–5.
26. W. T. Koiter, *Quart. Appl. Math.*, **2**, 350–4 (1953).
27. R. Baldacci, G. Ceradini, and E. Giangreco, *Plasticità*. CISIA, Milan, 1974, pp. 11–56.

On the coarse-graining of grains

I. Goldhirsch

School of Mechanical Engineering, Faculty of Engineering, Tel-Aviv University, Ramat-Aviv, Tel-Aviv 69978, Israel

Abstract. The main goal of this article is to show that there are qualitative and quantitative differences between continuum fields obtained by the coarse-graining of systems comprising discrete constituents at finite or very large ("infinite") resolution. This is demonstrated by two examples: the stress field is symmetric for infinite coarse-graining scales but can be asymmetric for finite resolution; the classical expression for the elastic energy density needs to be corrected by a term that vanishes in the limit of infinite coarse-graining scale. The article also presents a brief and somewhat biased introduction to the subject of coarse-graining (or homogenization or averaging) and a summary of part of the recent and not-so-recent work by the author and some of his collaborators. It is shown that it is possible to coarse grain discrete systems in such a way that the fields are smooth and the resolution controlled. In particular, the use of smooth coarse graining functions of pre-determined resolution ensures that the gradients of the corresponding fields are bounded by the inverse of the scale of coarse-graining. The equations of continuum mechanics are automatically satisfied in the framework of this formulation. Among the other presented results are expressions for the displacement and strain fields that are different from certain definitions in the literature. All presented results are valid for single realizations and may be used for ensemble averaging.

Keywords: coarse-graining, averaging, homogenization, granular matter, stress, strain, stress asymmetry, Cosserat theory, elasticity, heat flux
PACS: 83.80.Fg,81.05.Rm,46.65.+g,45.70.-n,46.05.+b

1. INTRODUCTION

Efforts to connect molecular scale dynamics with macroscopic continuum equations of motion (more accurately, constitutive relations) date back to classical studies by Boltzmann (for gases), Kirkwood [1], Born and Huang [2], and others [3]. Coarse graining approaches to granular matter appeared much later, perhaps starting with Weber's work [4], but are mostly based on similar principles, cf. a selection of some relatively recent studies and reviews [5, 6, 7, 8, 9, 10, 11, 12, 13, 14, 15, 16, 17, 18] and references therein; also see [19] for a review that includes historic references, and [20] concerning the coarse graining of granular gases. In some papers only spatial coarse graining was invoked, but others treated temporal coarse graining as well, see e.g., [21] and reference therein. Here we focus on spatial coarse-graining alone (which implies that instantaneous collisions are not treated here).

The classical expression for the stress tensor in terms of the constituents' degrees of freedom and their interactions, often referred to as the Born-Huang [2] or Love [22] or Voigt [23] formula, was originally derived for molecular systems and is based on the "limit" of large REV's or coarse graining scales. This limit is justified for typical molecular systems since the latter possess significant scale separation between the molecular and macroscopic scales. For instance, in gases the scales that correspond to typical gradients of the continuum fields are much larger than the corresponding mean free paths (else one deals with the Knudsen regime or shocks); this fact is among other things the basis for the notion of local equilibrium. Similar statements can be made concerning molecular solids. When the gradients (or strain) are sufficiently "small" one can employ very large REV's, i.e. much larger than the microscopic scales, e.g., the lattice constant in ordered solids, yet much smaller than the typical scales on which the physical entities of interest undergo significant changes. In the realm of nanoscale [24] or granular systems (in particular concerning the description of relatively "small" systems as in some experiments) one must not use the limit of infinite coarse graining scales. A similar statement applies when one wishes to study any system at finite (say, mesoscopic) resolution.

The formulation presented below is exact and reproduces the balance equations of continuum mechanics. The resolution of the continuum fields can be chosen at will, although certain choices are advantageous over others. Some of the results presented below are known, cf. e.g., [8, 21, 24, 25, 26, 27, 28] and references therein, and repeated here using a somewhat simpler presentation and for sake of completeness. Recent results presented below include an exact and explicit formula for the stress asymmetry, which has been a subject of debate in the literature, see [12], references therein and citations thereof, and an explicit expression for the microscopic stress field [29], which offers a possible answer to the question whether this field is unique or not [25, 26, 30, 31].

One of the goals of coarse graining is to produce macroscopic descriptions of assemblies of discrete particles. Another goal is to produce continuum equations of motion on the basis of the microscopic dynamics. There is a difference between coarse-graining as such, and the derivation of constitutive relations. One can coarse-grain systems of particles to produce fields such as mass and momentum density, as well as the balance equations of continuum mechanics, and these results would be useful for the description of the studied systems [32] but they do not comprise derivations of constitutive relations (or closures). These require additional steps which are not discussed here.

The first step in any coarse-graining procedure is to choose and define a set of continuum fields for which equations of motion are desired. The choice is usually dictated by symmetries, conservation laws, measurability and other considerations, not the least of which is the wish to obtain (at least approximately) a closed set of equations of motion ("closure"), i.e. the time derivatives of the fields should depend on the chosen set of fields alone. The fields should be defined everywhere in the space occupied by the considered system, not only e.g., at the centers of mass of the discrete particles or the interior of the particles. For instance, the strain and stress fields of a solid have to be continuous function of space, not restricted in their definitions to the positions of the centers of mass of the particles. Coarse graining involves a spatial coarse graining scale (when temporal coarse graining is invoked as well one also needs a temporal coarse graining scale, see [21]). This scale, or resolution, is important both for obtaining smooth fields and constitutive relations. As explained e.g., in [33], the density field is strongly fluctuating as a function of the coarse graining scale (or resolution) for very fine resolution (e.g., when one uses a REV whose linear dimension is smaller that the corresponding mean free path in a gas), yet its value plateaus to a space-dependent number as a function of resolution. The latter *is* what one refers to the as the macroscopic density. For very coarse resolution the density becomes space dependent again due to macroscopic variations. The same holds for practically any other continuum or macroscopic field, see e.g., [32, 34, 35] for the case of the stress field.

Coarse graining is a projection process through which information is lost. Subresolution scale information is not carried by the fields. Therefore attempts to e.g., define displacement fields that coincide with the particle displacements at their respective centers of mass clash with the notion of coarse graining both in terms of physics and underlying philosophy, even if such interpolations are mathematically feasible. Not only is information not lost this way but the fields one produces may possess large gradients (whose scale, except for the case of ordered systems under small strains, is typically an inverse particle diameter); practically all gradient expansions may fail for such fields. Such interpolation would be tantamount to requiring that the values of the momentum field of a gas at the positions of the centers of mass of the molecules equals the corresponding momenta of the molecules. As is well known the molecules of a gas (macroscopically) at rest have speeds of the order of the speed of sound in the gas, yet the macroscopic momentum and velocity fields vanish. Clearly, these issues do not arise when spatial (and often temporal) coarse graining of finite resolution is properly applied.

In statistical mechanical theories of fluids it is common [36] to define fields by first using Fourier transforms of the microscopic entities (such as momentum), then invoking a cutoff in the wave-vector space upon back-transforming. The inverse cutoff is the resolution of the coarse grained fields.

As mentioned, in early studies of elasticity (see e.g., [2]) the coarse graining scale was taken to be very large compared to the molecular scale (e.g., the lattice constant), yet still small with respect to the macroscopic scales. This is possible when the system of interest is sufficiently large compared to its microscopic scale and the gradients are "small" i.e. the properties of the system are nearly constant on the coarse graining scale of resolution. Clearly, such a choice is not possible for nanoscale systems or granular materials (in many instances) due to the lack of good scale separation. This in turn requires one to revisit the ways to coarse grain these classes of systems: due attention needs to be paid to the resolution involved. When a system is coarse grained with spatial resolution, w, no gradient involves a scale that is smaller than w, and therefore one expects to obtain smooth fields whose spatial variation is characterized by scales that exceed the resolution or coarse graining scale. When only a particle and its nearest neighbors are involved in the definition of coarse grained fields, the resolution is very fine and one may not obtain proper continuum descriptions.

2. COARSE GRAINING

Consider a finite or infinite number of particles, identified by lower case Latin indices, e.g., the mass of particle i is denoted as m_i. Vectorial and tensorial components are denoted by Greek letters. The positions of the centers of mass of the particles are denoted by $\vec{r}_i(t)$, where t denotes time, and the α's component of this vector is $r_{i\alpha}(t)$. The corresponding center of mass velocity is $\vec{v}_i(t) \equiv \dot{\vec{r}}_i(t)$. In addition, $\vec{r}_{ij} \equiv \vec{r}_i(t) - \vec{r}_j(t)$, denotes the relative positions of the centers of mass of two particles, and $\vec{v}_{ij}$, the corresponding relative center of mass velocities. The resultant force on particle i is denoted by $\vec{f}_i(t)$, and it equals $m_i\dot{\vec{v}}_i(t)$, by Newton's second law. Similarly, the angular momentum of particle i is denoted by $\vec{S}_i(t)$, and it satisfies $\dot{\vec{S}}_i(t) = \vec{M}_i(t)$ where $\vec{M}_i(t)$ is the resultant torque on particle i (this does not require the particle to be rigid). The force exerted by particle j on particle i is denoted by $\vec{f}_{ij}(t)$, and $\vec{f}_i(t) = \sum_j \vec{f}_{ij}(t)$, i.e., we assume binary and additive interactions (as mentioned, it is not very difficult to relax this assumption but we don't do it here for sake of simplicity). Bulk forces are easy to include in the formulation but they are not. By Newton's third law: $\vec{f}_{ij}(t) = -\vec{f}_{ji}(t)$; this is used below quite frequently. Nearly all factors of $\frac{1}{2}$ that appear below stem from the following trivial identity: when an expression, A, equals another expression, B, they both also equal $\frac{1}{2}(A+B)$.

A point in space, $\vec{r}$, is a point where we choose to measure a physical entity (e.g., the value of the stress tensor field) and it is not "time dependent". The fields depend on $\vec{r}$ and t, their time dependence being dictated by that of the particles' center of mass positions, $\vec{r}_i(t)$, velocities $\vec{v}_i(t)$, orientations, angular velocities, and possibly other degrees of freedom. The time dependence of the particles' degrees of freedom is often not explicitly spelled out below, nor is that of the fields, for sake of notational conciseness. The Einstein summation convention is used below only for Greek letters (i.e. vector/tensor components) but not Latin indices (that represent the particles' identities); the latter summations are explicitly presented.

First we derive the equation of continuity as a trivial example of how the formulation presented here can be used. Then we proceed to the equation of motion for the momentum density and derive an explicit formula for the stress tensor field. The latter expression is used to obtain the antisymmetric part of the stress tensor.

The most trivial and well known equation of motion for any system in which mass is conserved is the equation of continuity for the mass density, a similar statement holding for the number density (also, densities in polydisperse systems). The derivation presented below helps further set the notation and demonstrates part of the method used throughout this paper.

The microscopic mass density at a point $\vec{r}$ at time t, $\rho^{\text{mic}}(\vec{r},t)$, is defined in statistical mechanics by: $\rho^{\text{mic}}(\vec{r},t) \equiv \sum_i m_i \delta(\vec{r} - \vec{r}_i(t))$, where $\delta(\vec{r})$ is the Dirac delta function [1]. This definition complies with the basic requirement that the integral of the mass density over a volume in space equals the mass contained in this volume. It is however a singular entity. A nonsingular mass density can be defined by $\rho(\vec{r},t) = \sum_i m_i \phi(\vec{r} - \vec{r}_i(t))$, i.e., by replacing the delta-function by a (real) "coarse graining" function of space, ϕ, which possesses a pre-determined "width", w (the coarse graining scale or resolution) and which is required to be positive semi-definite and normalizable (its integral over space is unity). A simple example is $\phi(\vec{r}) \equiv \frac{1}{\Omega_d(w)} H(w - \|\vec{r}\|)$, where H represents the Heaviside function and $\Omega_d(w)$ is the volume of a sphere of radius w in d-dimensions (this would correspond to an REV of spherical shape with a uniform weight; our definition is more general since it allows for a smooth weight function and consequently fields whose derivatives are 'bounded' by the inverse of the coarse-graining scale). Often it is useful to take ϕ to be a Gaussian of width w. In the limit $w \longrightarrow 0$, the coarse-grained density field reduces to its microscopic definition. The above coarse graining procedure can also be viewed as being defined by a convolution of the microscopic density with the coarse graining function, i.e., $\rho(\vec{r},t) = \int d\vec{r}' \phi(\vec{r}-\vec{r}') \rho^{\text{mic}}(\vec{r}',t)$. In most cases known to the author the values of the fields depend mostly on w and not the precise choice of the coarse graining function as long as the latter is not chosen to be singular or highly anisotropic. Similar definitions will be used for other coarse grained fields below. Note that (chain rule): $\frac{\partial}{\partial t}\phi(\vec{r}-\vec{r}_i) = -\dot{r}_{i\beta}\frac{\partial}{\partial r_\beta}\phi(\vec{r}-\vec{r}_i) = -v_{i\beta}\frac{\partial}{\partial r_\beta}\phi(\vec{r}-\vec{r}_i)$. Also note that $\frac{\partial}{\partial r_\beta}$ can be 'moved across' all variables that describe the particles' degrees of freedom since the latter are just time dependent entities. The mass conservation (continuity) equation can be derived by taking the time derivative of the coarse grained mass density:

$$\frac{\partial \rho(\vec{r},t)}{\partial t} = \frac{\partial}{\partial t}\sum_i m_i \phi(\vec{r}-\vec{r}_i) = -\frac{\partial}{\partial r_\beta}\sum_i m_i v_{i\beta} \phi(\vec{r}-\vec{r}_i) = -\frac{\partial p_\beta(\vec{r},t)}{\partial r_\beta}, \quad (1)$$

where the coarse grained momentum density is defined by $\vec{p}(\vec{r},t) \equiv \sum_i m_i \vec{v}_i \phi(\vec{r}-\vec{r}_i)$, corresponding to the following microscopic momentum density field: $\vec{p}^{\text{mic}}(\vec{r},t) \equiv \sum_i m_i \vec{v}_i(t) \delta(\vec{r}-\vec{r}_i(t))$ (here the time dependence is explicitly presented). Note that the above result holds even for a single particle and therefore one does not need to resort to arguments involving swarms of particles or ensembles, as in some derivations in the literature. In the limit $w \to 0$ one obtains the same result for microscopic resolution. These observations hold for all fields discussed below and will not be repeated. The coarse grained velocity field is *defined* by $\vec{V}(\vec{r},t) \equiv \vec{p}(\vec{r},t)/\rho(\vec{r},t)$. Notice that the velocity field is meaningful only as a macroscopic (or mesoscopic) field, unlike the mass and momentum density fields, which

---

[1] This definition does not imply that the mass of a particle is assumed to be concentrated at its center of mass. It only represents the particle's position through its center of mass coordinates.

are coarse grained densities of well defined microscopic physical entities (from a thermodynamic point of view the velocity field is the vector chemical potential conjugate to the momentum density field). With the latter definition one now obtains the standard form of the equation of continuity: $\dot{\rho} = -\frac{\partial}{\partial r_\beta}(\rho V_\beta)$.

The momentum equation can be derived in a similar way. Taking the time derivative of the coarse grained momentum density field one obtains:

$$\frac{\partial p_\alpha(\vec{r},t)}{\partial t} = \frac{\partial}{\partial t}\sum_i m_i v_{i\alpha}\phi(\vec{r}-\vec{r}_i) = \sum_i m_i \dot{v}_{i\alpha}\phi(\vec{r}-\vec{r}_i) + \sum_i m_i v_{i\alpha}\frac{\partial \phi(\vec{r}-\vec{r}_i)}{\partial t}$$

$$= \underbrace{\sum_i m_i \dot{v}_{i\alpha}\phi(\vec{r}-\vec{r}_i)}_{A_\alpha} \underbrace{- \frac{\partial}{\partial r_\beta}\sum_i m_i v_{i\alpha} v_{i\beta}\phi(\vec{r}-\vec{r}_i)}_{B_\alpha}. \qquad (2)$$

Here the time dependence of the particles' center of mass positions and velocities is no longer spelled out. Next, each part of Eq. (2) is treated separately. The first part is: $A_\alpha \equiv \sum_i m_i \dot{v}_{i\alpha}\phi(\vec{r}-\vec{r}_i) = \sum_i f_{i\alpha}\phi(\vec{r}-\vec{r}_i) = \sum_{ij} f_{ij\alpha}\phi(\vec{r}-\vec{r}_i)$, where Newton's second law has been used and the assumption of additive binary interactions invoked. Upon exchanging the names of the dummy summation variables, i and j, one obtains: $A_\alpha = \sum_{ij} f_{ji\alpha}\phi(\vec{r}-\vec{r}_j)$. Using Newton's third law one further obtains: $A_\alpha = -\sum_{ij} f_{ij\alpha}\phi(\vec{r}-\vec{r}_j)$. Upon summing the two expressions for A_α, and dividing the sum by 2 one obtains: $A_\alpha = \frac{1}{2}\sum_{ij} f_{ij\alpha}[\phi(\vec{r}-\vec{r}_i) - \phi(\vec{r}-\vec{r}_j)]$. The following identity holds for any smooth function, ϕ:

$$\phi(\vec{r}-\vec{r}_j) - \phi(\vec{r}-\vec{r}_i) = \int_0^1 ds \frac{\partial}{\partial s}\phi(\vec{r}-\vec{r}_i+s\vec{r}_{ij}) = \int_0^1 r_{ij\beta}\, ds \frac{\partial}{\partial r_\beta}\phi(\vec{r}-\vec{r}_i+s\vec{r}_{ij}) = r_{ij\beta}\frac{\partial}{\partial r_\beta}\int_0^1 ds\,\phi(\vec{r}-\vec{r}_i+s\vec{r}_{ij}). \qquad (3)$$

Substituting this identity in the last expression for A_α one obtains: $A_\alpha = -\frac{1}{2}\frac{\partial}{\partial r_\beta}\sum_{ij} f_{ij\alpha} r_{ij\beta}\int_0^1 ds\,\phi(\vec{r}-\vec{r}_i+s\vec{r}_{ij})$. Note that Eq. (3) corresponds to a specific choice of an integration path from $\vec{r}_i$ to $\vec{r}_j$ (a straight line) and is therefore not the most general possibility for representing the difference of coarse graining functions. This issue will be further discussed below. In rewriting the expression for B_α, defined in Eq. (2), it is convenient to define the *fluctuating velocity* of particle i at time t: $\vec{v}'_i(\vec{r},t) \equiv \vec{v}_i(t) - \vec{V}(\vec{r},t)$. Note that the reference coarse grained velocity is at the "coarse graining-center", $\vec{r}$, and not at the particle's center of mass position [21]. This also corresponds to a commonly employed numerical method for defining fluctuations: one divides the system into boxes and computes the center of mass velocity of the particles in each box. The fluctuating velocity of a particle in a given box is then measured with respect to the center of mass velocity of the particles in the corresponding box and not the velocity field at the particle's center of mass position. There are other ways to define fluctuations but they do not always yield results that are consistent with the equations of continuum mechanics. Below we write $\vec{v}'_i$ for $\vec{v}'_i(\vec{r},t)$ for sake of notational brevity. The following identity follows from the definition of $\vec{v}'_i$ and that of the coarse grained velocity field:

$$\sum_i m_i \vec{v}'_i \phi(\vec{r}-\vec{r}_i) = \sum_i m_i \left[\vec{v}_i - \vec{V}\right]\phi(\vec{r}-\vec{r}_i) = \sum_i m_i \vec{v}_i \phi(\vec{r}-\vec{r}_i) - \vec{V}\sum_i m_i \phi(\vec{r}-\vec{r}_i) = \vec{p} - \rho\vec{V} = 0. \qquad (4)$$

Using this result and Eq. (2) one obtains:

$$B_\alpha = -\frac{\partial}{\partial r_\beta}\sum_i m_i v_{i\alpha} v_{i\beta}\phi(\vec{r}-\vec{r}_i) = -\frac{\partial}{\partial r_\beta}\sum_i m_i(V_\alpha + v'_{i\alpha})(V_\beta + v'_{i\beta})\phi(\vec{r}-\vec{r}_i)$$

$$= -\frac{\partial}{\partial r_\beta}\left[\sum_i m_i V_\alpha V_\beta \phi(\vec{r}-\vec{r}_i) + \sum_i m_i v'_{i\alpha} v'_{i\beta}\phi(\vec{r}-\vec{r}_i)\right] = -\frac{\partial}{\partial r_\beta}\left[\rho V_\alpha V_\beta + \sum_i m_i v'_{i\alpha} v'_{i\beta}\phi(\vec{r}-\vec{r}_i)\right]. \qquad (5)$$

Combining the above expressions for A_α and B_α one obtains the momentum conservation equation:

$$\frac{\partial p_\alpha}{\partial t} = -\frac{\partial}{\partial r_\beta}\left[\rho V_\alpha V_\beta - \sigma_{\alpha\beta}\right], \qquad (6)$$

where we *identify* the following expression for the stress tensor, σ, in terms of microscopic entities:

$$\sigma_{\alpha\beta} = -\frac{1}{2}\sum_{i,j} f_{ij\alpha} r_{ij\beta}\int_0^1 ds\,\phi(\vec{r}-\vec{r}_i+s\vec{r}_{ij}) - \sum_i m_i v'_{i\alpha} v'_{i\beta}\phi(\vec{r}-\vec{r}_i), \qquad (7)$$

The first term on the right hand side of Eq. (7) is the "contact" or "collisional" stress, while the second term is the "kinetic" or "streaming" stress, and it is negligible for quasi-static deformations.

In the field of granular matter one often replaces $\sigma_{\alpha\beta}$ by $-\sigma_{\alpha\beta}$ to render compressive stresses positive. When the coarse graining scale, w, is far larger than the typical distance between interacting particles one can neglect $s\vec{r}_{ij}$ in the integral in Eq. (7), since for most particle pairs that contribute to the sum in this equation $\|\vec{r} - \vec{r}_i\| >> \|\vec{r}_{ij}\|$. In this case: $\sigma^c_{\alpha\beta} \approx -\frac{1}{2}\sum_{i,j} f_{ij\alpha} r_{ij\beta} \phi(\vec{r} - \vec{r}_i)$, where $\vec{\sigma}^c$ stands for "the contact stress contribution". Furthermore, when $\phi(\vec{r} - \vec{r}_i)$ is chosen to equal $\frac{1}{\Omega_d(w)} H(w - \|\vec{r} - \vec{r}_i\|)$, i.e. one averages over a sphere around $\vec{r}$, the formula reduces to the well known (and frequently used) Born-Huang expression: $\sigma^c_{\alpha\beta} \approx -\frac{1}{2}\frac{1}{\Omega_d(w)}\sum_{i,j; \|(\vec{r}-\vec{r}_i)\|<w} f_{ij\alpha} r_{ij\beta}$ (the sphere can be replaced by differently shaped REV's). In other words, the standardly employed formula is only justified in the limit of large coarse graining scales and needs to be replaced by the exact expression presented above when one is concerned with mesoscopic scales. As mentioned, the original, e.g., in [2], derivation of this formula pertains to molecular systems that possess strong scale separation and therefore the use of the limit of "large" resolution compared to molecular dimensions is practically always justified. Also note that the coarse graining center in many applications of the Born-Huang formula is taken to be the center of mass of "particle i".

One may criticize [25, 26, 30, 31] the above expression for the stress field since it is not unique, in the sense that any second order tensor whose divergence vanishes can be added to this expression without changing the equation of motion. Furthermore, the choice in Eq. (3) of a straight line interpolation may seem arbitrary (other choices amount to adding a divergence free tensor to the obtained expression). However, in addition to the fact that the above choice of the integration path is simple to implement, it can be shown [31] that additional symmetry requirements (e.g., that the stress tensor is invariant to permutations of the particles' identities, symmetric for particles interacting by binary central forces, reduces to the well known formula for the equilibrium pressure and more) render this a unique choice. A stronger argument in favor of the above expression for the contact part of the stress tensor can be found in [29]. There a derivation that is directly based on the Cauchy definition of stress, rather than an identification of a term in the continuum mechanical equations of motion, is presented. The result is the same as in the above. One of the immediate consequences of this derivation [29] is that the thus derived expression provides the force transmitted through an open surface (not obvious from the term identified as stress in the standard derivation), as it should. Specifically, it is shown in [29] that the microscopic expression for the contact part of the stress tensor is given by: $\sigma^c_{\alpha\beta} = -\frac{1}{2}\sum_{ij} f_{ij\alpha} r_{\beta,ij} \int_0^1 \delta(\vec{r} - \vec{r}_i + s\vec{r}_{ij}) ds$. This comprises an exact expression for the contact stress with microscopic resolution (corresponding to $w = 0$). We shall rename the result $\sigma^{c,\text{mic}}$ to denote this fact. The coarse grained contact stress can now be obtained by convoluting the latter expression with the coarse graining function: $\int d\vec{r}' \phi(\vec{r} - \vec{r}') \sigma^{c,\text{mic}}_{\alpha\beta}(\vec{r}')$.

2.1. The antisymmetric part of the stress tensor field

The kinetic stress is manifestly symmetric and therefore it does not contribute to the antisymmetric part of the stress tensor, i.e., $\sigma_{\alpha\beta} - \sigma_{\beta\alpha} = \sigma^c_{\alpha\beta} - \sigma^c_{\beta\alpha}$, where we recall that: $\sigma^c_{\alpha\beta} = -\frac{1}{2}\sum_{i,j} f_{ij\alpha} r_{ij\beta} \int_0^1 ds\, \phi(\vec{r} - \vec{r}_i + s\vec{r}_{ij})$. It follows that:

$$\sigma_{\alpha\beta} - \sigma_{\beta\alpha} = -\frac{1}{2}\sum_{i,j} (f_{ij\alpha} r_{ij\beta} - f_{ij\beta} r_{ij\alpha}) \int_0^1 ds\, \phi(\vec{r} - \vec{r}_i + s\vec{r}_{ij}), \quad (8)$$

hence,

$$\sigma_{\alpha\beta} - \sigma_{\beta\alpha} = -\frac{1}{2}\sum_{i,j} \varepsilon_{\alpha\beta\gamma} \left[\vec{r}_{ij} \times \vec{f}_{ij}\right]_\gamma \int_0^1 ds\, \phi(\vec{r} - \vec{r}_i + s\vec{r}_{ij}), \quad (9)$$

where $\varepsilon_{\alpha\beta\gamma}$ is the antisymmetric isotropic tensor. It follows immediately that when only central forces are present and the particles are homogeneous spheres (or inhomogeneous in such a way that their centers of mass coincide with their respective geometric centers) the stress tensor is symmetric. Other possibilities are discussed below.

Denote by $\vec{M}^f_{ij}$ the torque applied on particle i by the *force* exerted by particle j on particle i. This torque can be due to frictional forces as well as normal forces e.g., when the particles are not spherical. Clearly: $\vec{M}^f_{ij} = (\vec{r}^c_{ij} - \vec{r}_i) \times \vec{f}_{ij}$, where $\vec{r}^c_{ij}$ denotes the point of contact of particles i and j; when the contact is not approximated by a point one can still define an effective point of contact but this is not further elaborated here. Also, the theory can easily be extended to the case when several points of contact between a pair of particles (e.g., for concave particles) exist. Clearly $\vec{r}^c_{ij} = \vec{r}^c_{ji}$. Similarly, the torque exerted by particle i on particle j is given by $\vec{M}^f_{ji} = (\vec{r}^c_{ij} - \vec{r}_j) \times \vec{f}_{ji} = -(\vec{r}^c_{ij} - \vec{r}_j) \times \vec{f}_{ij}$. Notice that: $\vec{r}_{ij} = -(\vec{r}^c_{ij} - \vec{r}_i) + (\vec{r}^c_{ij} - \vec{r}_j)$. It follows that: $\vec{r}_{ij} \times \vec{f}_{ij} = -\vec{M}^f_{ij} - \vec{M}^f_{ji}$. Substitution of the latter result into Eq. (9) yields:

$$\sigma_{\alpha\beta} - \sigma_{\beta\alpha} = \frac{1}{2} \sum_{i,j} \varepsilon_{\alpha\beta\gamma} \left(M^f_{ij\gamma} + M^f_{ji\gamma} \right) \int_0^1 ds\, \phi(\vec{r} - \vec{r}_i + s\vec{r}_{ij})$$

$$= \frac{1}{2} \sum_{i,j} \varepsilon_{\alpha\beta\gamma} M^f_{ij\gamma} \int_0^1 ds\, \phi(\vec{r} - \vec{r}_i + s\vec{r}_{ij}) + \frac{1}{2} \sum_{i,j} \varepsilon_{\alpha\beta\gamma} M^f_{ji\gamma} \int_0^1 ds\, \phi(\vec{r} - \vec{r}_i + s\vec{r}_{ij}). \tag{10}$$

Upon exchanging i and j in the second term in the last line of Eq. (10) and using the fact that the integral in this term is symmetric to the exchange of i and j, one obtains:

$$\sigma_{\alpha\beta} - \sigma_{\beta\alpha} = \sum_{i,j} \varepsilon_{\alpha\beta\gamma} M^f_{ij\gamma} \int_0^1 ds\, \phi(\vec{r} - \vec{r}_i + s\vec{r}_{ij}). \tag{11}$$

One can obtain further insights by rewriting this exact result as follows.

$$\sigma_{\alpha\beta} - \sigma_{\beta\alpha} = \sum_{i,j} \varepsilon_{\alpha\beta\gamma} M^f_{ij\gamma} \int_0^1 ds \left(\phi(\vec{r} - \vec{r}_i) + \int_0^s \frac{\partial}{\partial s'} \phi(\vec{r} - \vec{r}_i + s'\vec{r}_{ij}) ds' \right)$$

$$= \sum_i \varepsilon_{\alpha\beta\gamma} M^f_{i\gamma} \phi(\vec{r} - \vec{r}_i) + \varepsilon_{\alpha\beta\gamma} \frac{\partial}{\partial r_\delta} \sum_{i,j} M^f_{ij\gamma} r_{ij\delta} \int_0^1 ds \int_0^s \phi(\vec{r} - \vec{r}_i + s'\vec{r}_{ij}) ds' \tag{12}$$

where $\vec{M}^f_i \equiv \sum_j \vec{M}_{ij}$ is the resultant torque on particle i *due to forces*. A simple transformation of the above double integral yields:

$$\sigma_{\alpha\beta} - \sigma_{\beta\alpha} = \varepsilon_{\alpha\beta\gamma} \sum_i M^f_{i\gamma} \phi(\vec{r} - \vec{r}_i) + \varepsilon_{\alpha\beta\gamma} \frac{\partial}{\partial r_\delta} \sum_{i,j} M^f_{ij\gamma} r_{ij\delta} \int_0^1 ds (1-s) \phi(\vec{r} - \vec{r}_i + s\vec{r}_{ij}). \tag{13}$$

Recalling that $\dot{\vec{S}}_i = \vec{M}_i$, the resultant torque on particle i, and noting that: $\vec{M}_i = \vec{M}^f_i + \vec{M}^{rf}_i$, where $\vec{M}^{rf}_i$ denotes the contribution of rolling friction (or couples in general), one can replace $\vec{M}^f_i$ in Eq. (13) by $\dot{\vec{S}}_i - \vec{M}^{rf}_i$. The result is:

$$\sigma_{\alpha\beta} - \sigma_{\beta\alpha} = \varepsilon_{\alpha\beta\gamma} \sum_i (\dot{S}_{i\gamma} - M^{rf}_{i\gamma}) \phi(\vec{r} - \vec{r}_i) + \varepsilon_{\alpha\beta\gamma} \frac{\partial}{\partial r_\delta} \sum_{i,j} M^f_{ij\gamma} r_{ij\delta} \int_0^1 ds (1-s) \phi(\vec{r} - \vec{r}_i + \vec{r}_{ij}) \tag{14}$$

Denote by M^{rf}_{ij} the couple exerted by particle j on particle i. Clearly,

$$\sum_i M^{rf}_{i\gamma} \phi(\vec{r} - \vec{r}_i) = \sum_{ij} M^{rf}_{ij\gamma} \phi(\vec{r} - \vec{r}_i) = \sum_{ij} M^{rf}_{ji\gamma} \phi(\vec{r} - \vec{r}_j) = -\sum_{ij} M^{rf}_{ij\gamma} \phi(\vec{r} - \vec{r}_j), \tag{15}$$

where use has been made of $\vec{M}^{rf}_{ij} = -\vec{M}^{rf}_{ji}$. Therefore:

$$\sum_i M^{rf}_{i\gamma} \phi(\vec{r} - \vec{r}_i) = -\frac{1}{2} \sum_{ij} M^{rf}_{\gamma ij} (\phi(\vec{r} - \vec{r}_j) - \phi(\vec{r} - \vec{r}_i)) = -\frac{\partial}{\partial r_\delta} \frac{1}{2} \sum_{ij} M^{rf}_{ij\gamma} r_{ij\delta} \int_0^s ds\, \phi(\vec{r} - \vec{r}_i + s\vec{r}_{ij}). \tag{16}$$

Upon substituting this result in Eq. (14) one obtains:

$$\sigma_{\alpha\beta} - \sigma_{\beta\alpha} = \varepsilon_{\alpha\beta\gamma} \sum_i \dot{S}_{i\gamma} \phi(\vec{r} - \vec{r}_i) + \varepsilon_{\alpha\beta\gamma} \frac{\partial}{\partial r_\delta} \sum_{ij} \left(\frac{1}{2} M^{rf}_{ij\gamma} r_{ij\delta} \int_0^s ds\, \phi(\vec{r} - \vec{r}_i + s\vec{r}_{ij}) + M^f_{ij\gamma} r_{ij\delta} \int_0^1 ds (1-s) \phi(\vec{r} - \vec{r}_i + s\vec{r}_{ij}) \right). \tag{17}$$

Note that even in the quasistatic limit (i.e., for $\dot{\vec{S}}_i = 0$), there are contributions to the stress asymmetry that are not nominally vanishing. However these contributions take on the form of a divergence. Eq. (17) is compatible with a result by Eringen [37], but his formula is restricted to quasi-static deformations and does not display the microscale meaning of the result. Furthermore, the fact that the asymmetry disappears in the limit of large coarse-graining scales has not been known heretofore. This is shown next. One can naively estimate the various contributions to the stress asymmetry as follows (for different approaches to estimate length scales associated with Cosserat formulations, cf. e.g., [38] and references therein). The largest value of a derivative, given the resolution, w, is of the order of $\frac{1}{w}$. The typical value of the rolling friction term, up to a constant, is the product df, where f represents the norm of a typical force, and d stands for a typical particle diameter (the same holding for the force contribution to the torque). Also, the norm of $\vec{r}_{ij}$ can be represented by d. Therefore the overall contribution of this term can be roughly estimated by $\frac{1}{w\Omega_d(w)} f d^2 N_c q_1$, where $\Omega_d(w)$ is the coarse graining volume (the normalization of ϕ when it is taken to be a Heaviside function), N_c is the number of contacts in the volume and q_1 is a "reduction factor" due to cancellations stemming from possible different signs of the summands. To within an O(1) factor, $\Omega_3(w) \approx w^3$ in three dimensions. Therefore one can estimate the antisymmetric part of the stress tensor in three dimensions by $f\frac{d^2}{w^4} N_c q_1$. Compare this to a similar rough estimate of the stress: $\frac{fdN_c q_2}{w^3}$, where q_2 plays a similar role to q_1. The ratio of the former to the latter is $\frac{d}{w}$ when $q_1 \approx q_2$ (unjustified), and it is rather small for large coarse graining scales, irrespective of the neglected numerical factors. However, it can be rather significant for small coarse graining scales. Note that even in the absence of rolling friction there is a contribution to the stress asymmetry which is approximately of the same order as the above estimate. When the dynamics is not "slow" the antisymmetric part of the stress has a contribution from the time dependence of the particles' angular momenta. For frictionless spherical homogeneous particles the stress tensor is symmetric as already mentioned above. Finally, note that the above result is similar but not identical to that obtained in [12] since the authors of this paper do not use controlled resolution fields and their results are basically valid in the limit of large scale resolution; the similarity can be easily appreciated by comparing Eq. (11) with Eqs. (47-48) in ref. [12].

3. DISPLACEMENT AND STRAIN

Various definitions of the strain field using the particles' degrees of freedom have been proposed, cf. e.g., [7, 39, 40, 41]. They are usually defined on an ad-hoc basis (see e.g., the review by Bagi [7]) and restricted to quasi-static deformations. The derivation presented below is based on concepts borrowed from continuum mechanics, and not restricted to quasi-static deformations. It is worked out below for small strains only.

Recall the following elementary results from continuum mechanics. Consider a *material particle* whose position at time $t = 0$ is $\vec{R}$. It's position at time t, in the Lagrangian representation, is $\vec{r}(\vec{R},t)$. In the same representation, the displacement field is defined as: $\vec{u}^{La}(\vec{R},t) \equiv \vec{r}(\vec{R},t) - \vec{R}$. The material particle's velocity in Lagrangian coordinates is *defined* by $\vec{V}^{La}(\vec{R},t) \equiv \partial \vec{u}^{La}(\vec{R},t)/\partial t$. It therefore follows that: $\vec{u}^{La}(\vec{R},t) = \int_0^t \vec{V}^{La}(\vec{R},t')dt'$. Note that the velocity field in the Eulerian representation, $\vec{V}(\vec{r},t)$, has to satisfy: $\vec{V}(\vec{r}(\vec{R},t),t) = \vec{V}^{La}(\vec{R},t)$, i.e. the two have to equal each other, a similar statement holding for the relations between the Lagrangian and Eulerian representations of the other fields. Substituting the expression for the velocity as a ratio of momentum and mass densities, expressed in Lagrangian coordinates, yields the following expression for the displacement field:

$$\vec{u}^{La}(\vec{R},t) \equiv \int_0^t \frac{\sum_i m_i \vec{v}_i(t')\phi[\vec{r}(\vec{R},t') - \vec{r}_i(t')]}{\sum_j m_j \phi[\vec{r}(\vec{R},t') - \vec{r}_j(t')]} dt'. \qquad (18)$$

Let $\vec{u}_i(t) \equiv \vec{r}_i(t) - \vec{r}_i(0)$ denote the displacement of particle i in time t. Clearly: $\dot{\vec{u}}_i(t) = \vec{v}_i(t)$. Therefore one can perform integration by parts in Eq. (18) to obtain:

$$u_\alpha^{La}(\vec{R},t) = u_\alpha^{La,\text{lin}}(\vec{R},t) - \sum_i \int_0^t m_i \bar{u}_{i\alpha}(t') \frac{\partial}{\partial t'} \frac{\phi[\vec{r}(\vec{R},t') - \vec{r}_i(t')]}{\sum_j m_j \phi[\vec{r}(\vec{R},t') - \vec{r}_j(t')]} dt', \qquad (19)$$

where:

$$u_\alpha^{La,\text{lin}}(\vec{R},t) \equiv \frac{\sum_i m_i u_{i\alpha}(t)\phi[\vec{r}(\vec{R},t) - \vec{r}_i(t)]}{\sum_j m_j \phi[\vec{r}(\vec{R},t) - \vec{r}_j(t)]}$$

Simple manipulations of the second term on the right hand side of Eq. (19) give:

$$u_\alpha^{La}(\vec{R},t) = u_\alpha^{La,\text{lin}}(\vec{R},t) + \int_0^t \frac{1}{\rho} \frac{\partial}{\partial r_\beta} \sum_i m_i v'_{i\beta}(\vec{R},t') u'_{i\alpha}(\vec{R},t')\phi[\vec{r}(\vec{R},t) - \vec{r}_i(t')]dt', \qquad (20)$$

where ρ denotes $\rho^{La}(\vec{R},t)$, $\vec{u}'_i(\vec{R},t) \equiv \vec{u}_i(t) - \vec{u}^{La,\text{lin}}(\vec{R},t)$ defines the fluctuating displacement of particle i, and $\frac{\partial}{\partial r_\beta} \equiv \frac{\partial R_\gamma}{\partial r_\beta}\frac{\partial}{\partial R_\gamma}$. It is easy to see that the second term on the right hand side of Eq. (20) is quadratic in the strain. Therefore to linear order in the strain, the displacement field equals $\vec{u}^{La,\text{lin}}$. Upon reverting to the Eulerian representation one obtains that the displacement field, to linear order in the strain, is given by:

$$\vec{u}^{\text{lin}}(\vec{r},t) \equiv \frac{\sum_i m_i \vec{u}_i(t) \phi[\vec{r}-\vec{r}_i(t)]}{\sum_j m_j \phi[\vec{r}-\vec{r}_j(t)]} = \frac{\sum_i m_i \vec{u}_i(t) \phi[\vec{r}-\vec{r}_i(t)]}{\rho(\vec{r},t)}. \tag{21}$$

Using the notation, $\phi_i \equiv \phi(\vec{r}-\vec{r}_i(t))$ and $\phi_{i\alpha} \equiv \frac{\partial \phi_i}{\partial r_\alpha}$, one obtains from Eq. (21):

$$\frac{\partial u_\alpha^{\text{lin}}(\vec{r},t)}{\partial r_\beta} = \frac{\sum_i m_i u_{i\alpha}\phi_{i\beta}\sum_j m_j \phi_j - \sum_i m_i u_{i\alpha}\phi_i \sum_j m_j \phi_{j\beta}}{\rho^2}$$

Exchanging the dummy indices i and j in the second sum in the numerator of the last equation and rearranging terms one obtains: $\frac{\partial u_\alpha^{\text{lin}}(\vec{r},t)}{\partial r_\beta} = \frac{\sum_{ij} m_i m_j u_{ij\alpha} \phi_{i\beta} \phi_j}{\rho^2}$, where $u_{ij\alpha} \equiv u_{i\alpha} - u_{j\alpha}$. It follows that the linear strain tensor is given by:

$$\varepsilon_{\alpha\beta}^{\text{lin}} = \frac{1}{2}\left(\frac{\partial u_\alpha^{\text{lin}}}{\partial r_\beta} + \frac{\partial u_\beta^{\text{lin}}}{\partial r_\alpha}\right) = \frac{\sum_{ij} m_i m_j \phi_j (u_{ij\alpha}\phi_{i\beta} + u_{ij\beta}\phi_{i\alpha})}{2\rho^2}. \tag{22}$$

Note that no numerical derivatives are necessary in the implementation of these expressions because the spatial derivatives act only on the coarse graining function, which can be chosen e.g., to be a Gaussian.

3.1. Elastic energy density and work density

It is shown in refs. [27, 42] that the elastic energy density of a system of frictionless particles, whose interactions are represented by forces linearized around a stress-free reference state is given by:

$$e^{\text{el}}(\vec{r},t) = \frac{1}{2}\sigma_{\alpha\beta}^{\text{lin}}\varepsilon_{\alpha\beta}^{\text{lin}} - \frac{1}{8}\frac{\partial}{\partial r_\beta}\sum_{ij} f_{ij\alpha}\left(u'_{i\alpha} + u'_{j\alpha}\right) r_{ij\beta}^0 \int_0^1 ds\, \phi[\vec{r}-\vec{r}_i^0 + s\vec{r}_{ij}^0], \tag{23}$$

where the superscript 'lin' denotes that the fields were calculated to linear order in the strain, a zero superscript refers to the reference state, and $\vec{u}'_i(\vec{r},t) \equiv \vec{u}_i - \vec{u}^{\text{lin}}(\vec{r},t)$ is the fluctuating displacement of particle i, defined in a similar way to the fluctuating velocity defined above. This result was shown in [27] to follow from relatively straightforward algebraic manipulations of the classical expression for the elastic energy, $\frac{1}{2}\sigma:\varepsilon$. Another derivation of the same result, which is of more general nature, was presented in ref. [42]: there, the continuum mechanical energy balance equation was integrated in time for a "virtual quasi-static deformation", revealing that the above correction to the classical expression for the elastic energy density stems from an integration of the divergence of the heat flux term that appears in the energy equation (please note that even in athermal systems - the (granular) heat flux is the part of the energy flux that is not resolved by the macroscopic field, i.e. it is neither macroscopic work nor convective transport of energy). One can directly see the meaning of the correction from Eq. (23): it is the work associated with the fluctuating displacements. The latter are not accounted for in classical elasticity. Another extension of this result is presented below.

Consider the differential work density associated with an incremental strain, $\delta\varepsilon$ (the corresponding displacement field being $\delta\vec{u}^{\text{lin}}$) in a quasi-static deformation: $\sigma:\delta\varepsilon$. Also specialize to the case of symmetric stress. It follows that: $\sigma_{\alpha\beta}\delta\varepsilon_{\alpha\beta} = \sigma_{\alpha\beta}\frac{\partial \delta u_\alpha^{\text{lin}}}{\partial r_\beta} = \frac{\partial}{\partial r_\beta}\left(\sigma_{\alpha\beta}\delta u_\alpha^{\text{lin}}\right)$, where use has been made by the static equilibrium relation $\frac{\partial \sigma_{\alpha\beta}}{\partial r_\beta} = 0$. Substituting the above expression for the stress field one obtains: $\sigma_{\alpha\beta}\delta\varepsilon_{\alpha\beta} = -\frac{1}{2}\frac{\partial}{\partial r_\beta}\sum_{ij}\delta u_\alpha^{\text{lin}} f_{ij\alpha} r_{ij\beta}^0 \int_0^1 ds\, \phi[\vec{r}-\vec{r}_i^0 + s\vec{r}_{ij}^0]$, to lowest nonvanishing order in the strain. Using the above definition of a fluctuating displacement, it follows that $\vec{u}^{\text{lin}}(\vec{r},t) = \vec{u}_i(t) - \vec{u}'_i(\vec{r},t)$. Hence:

$$\sigma_{\alpha\beta}\delta\varepsilon_{\alpha\beta} = -\frac{1}{2}\frac{\partial}{\partial r_\beta}\sum_{ij}\left[\delta u_{i\alpha} - \delta u'_{i\alpha}\right] f_{ij\alpha} r_{ij\beta}^0 \int_0^1 ds\, \phi[\vec{r}-\vec{r}_i^0 + s\vec{r}_{ij}^0]$$

$$= -\frac{1}{2}\sum_{ij}\delta u_{i\alpha}f_{ij\alpha}r^0_{ij\beta}\frac{\partial}{\partial r_\beta}\int_0^1 ds\,\phi[\vec{r}-\vec{r}^0_i+s\vec{r}^0_{ij}] + \frac{1}{2}\frac{\partial}{\partial r_\beta}\sum_{ij}\delta u'_{i\alpha}f_{ij\alpha}r^0_{ij\beta}\int_0^1 ds\,\phi[\vec{r}-\vec{r}^0_i+s\vec{r}^0_{ij}]$$

$$= \frac{1}{2}\sum_{ij}\delta u_{i\alpha}f_{ij\alpha}\left(\phi\left[\vec{r}-\vec{r}^0_i\right]-\phi\left[\vec{r}-\vec{r}^0_j\right]\right) + \frac{1}{2}\frac{\partial}{\partial r_\beta}\sum_{ij}\delta u'_{i\alpha}f_{ij\alpha}r^0_{ij\beta}\int_0^1 ds\,\phi[\vec{r}-\vec{r}^0_i+s\vec{r}^0_{ij}]$$

In equilibrium $\sum_j f_{ij\alpha} = 0$, hence $\sum_j f_{ij\alpha}\phi\left(\vec{r}-\vec{r}_i(t)\right) = 0$ and it can be replaced by $-\sum_j f_{ij\alpha}\phi\left(\vec{r}-\vec{r}_i(t)\right)$. Therefore

$$\sigma_{\alpha\beta}\delta\varepsilon_{\alpha\beta} = -\frac{1}{2}\sum_{ij}\delta u_{i\alpha}f_{ij\alpha}\left(\phi\left[\vec{r}-\vec{r}^0_i\right]+\phi\left[\vec{r}-\vec{r}^0_j\right]\right) + \frac{1}{2}\frac{\partial}{\partial r_\beta}\sum_{ij}\delta u'_{i\alpha}f_{ij\alpha}r^0_{ij\beta}\int_0^1 ds\,\phi[\vec{r}-\vec{r}^0_i+s\vec{r}^0_{ij}]$$

$$= -\frac{1}{2}\sum_{ij}\delta u_{ij\alpha}f_{ij\alpha}\phi\left[\vec{r}-\vec{r}_i(t)\right] + \frac{1}{4}\frac{\partial}{\partial r_\beta}\sum_{ij}(\delta u'_{i\alpha}+\delta u'_{j\alpha})f_{ij\alpha}r^0_{ij\beta}\int_0^1 ds\,\phi[\vec{r}-\vec{r}^0_i+s\vec{r}^0_{ij}],$$

where use has been made of the symmetry of the integral in the above expression to the exchange of i and j. This is the differential version of the result presented above; it does not assume a specific interaction (or form of $\vec{f}_{ij}$) and is therefore quite general. An extension to non-symmetric stress tensors will be considered elsewhere. As mentioned, the new extra term represents the work associated with the fluctuating displacements, $\vec{u}'_i$, which is not accounted for in classical elasticity. As the correction term is a divergence of a flux, its average over a "large" volume becomes negligible with respect to the average of the classical expression for the energy density. One could therefore say the average of the classical expression over a sufficiently large volume provides the correct energy density. A simpler and more straightforward interpretation is obtained by noting that the correction term is of the order $\frac{d}{w}$ compared to the classical expression for the elastic energy density (same re incremental work density) and it becomes relatively negligible in the limit of large coarse graining scales. It should be emphasized that both contributions to the elastic energy density are of the same (second) order in the strain, i.e. the additional term is not a "higher order" correction but rather a correction that appears when the deformation is not affine (i.e., when $\vec{u}'_i \neq 0$) and the coarse graining scale, w, is not 'sufficiently' large with respect to the typical particle diameter, d.

4. CONCLUDING REMARKS

In the realm of granular solids there is still an active debate concerning the "correct" definitions of the pertinent fields. The formulation presented above summarizes the approach of the author and his collaborators. It has been shown that when coarse graining is done at finite resolution one obtains results that may differ from those obtained for infinite REV's. With these methods one can actually derive elasticity [27] for disordered systems, and obtain corrections to classical elasticity (which vanish for large coarse-graining scales). The result concerning the stress asymmetry should be reflected in the corresponding constitutive relations, whether derived or measured.

An interesting question is why does the above expression for the strain field depend on the masses of the particles even when applied to quasi-static deformations. Following reference [27], see also Eq. (22), any expression for the linear (in strain) displacement field that is a weighted average of the particle displacements (not necessarily by the particles' masses) gives rise to a strain field that depends on the particle displacements only through their relative displacements. These are shown in [27] to linearly depend on the strain field for "sufficiently large" coarse-graining scales. This is shown in [27] to lead to a linear elastic constitutive relation. However, the mass weighted average is also compatible with dynamic elasticity (where the second derivative of the displacement field is the first derivative of the velocity field defined as in the above) and therefore this choice is valid for both statics and dynamics. Admittedly, this problem requires further study. Also note that recent results based on the analysis of an experiment [32] indicate that the correlation length of the displacement fluctuations is rather short, of the order of a few particle diameters. This means that in practice the "sufficiently large" coarse graining scale mentioned above may be of mesoscopic dimensions.

It is important to reiterate the issue of plateaus. Experimental results concerning purely sheared two dimensional slabs [32], as well as numerical findings [24, 34, 35] show that even when the gradients of various fields are relatively large one can observe plateaus e.g., in the stress as a function of resolution, or w. This allows one to define the continuum fields in an 'objective' way.

ACKNOWLEDGMENTS

Support from the US-Israel Binational Science Foundation, grant no. 2004391 and the Israel Science Foundation, grant no. 412/08 are gratefully acknowledged. The author is indebted to Jens Boberski, Chay Goldenberg, Hans Herrmann, Niels Kruyt, Stefan Luding, Thorsten Pöschel, (the late) Ioannis Vardoulakis and Dietrich Wolf for helpful discussions.

REFERENCES

1. J. H. Irving and J. G. Kirkwood, J. Chem. Phys. **18**, 817 (1950).
2. M. Born and K. Huang, "Dynamical theory of crystal lattices" (Oxford, 1954, reissue 2000).
3. cf. e.g., L. E. Reichl, "A modern course in statistical physics" (Wiley, 1998) and numerous references therein.
4. J. Weber, Bull. de Liais, Ponts et Chausees, **20**, 1 (1966).
5. B. Cambou, P. Dubujet, F. Emeriault and F. Sidoroff, Eur. J. Mech. A – Solids, **14**, 255 (1995).
6. C. S. Chang and J. Gao, Acta Mech. **115**, 213 (1996).
7. K. Bagi, Mech. Mat. **22** (3), 165 (1996).
8. M. Babic, Int. J. Eng. Sci., **35**, 523 (1997).
9. J. D. Goddard, in Physics of Dry Granular Media, H. J. Herrmann, J.-P. Hovi and S. Luding, Eds., p. 1(Kluwer, 1998).
10. S. Nemat-Nasser, J. Mech. Phys. Solids, **48**, 1541 (2000).
11. M. Lätzel, S. Luding and H. J. Herrmann, in P. A. Vermeer, S. Diebels, W. Ehlers, H. J. Herrmann, S. Luding and E. Ramm, Eds., "Continuous and discontinuous models of cohesive-frictional materials", Lecture notes in physics 568, p. 215 (Springer, 2001).
12. J. P. Bardet and I. Vardoulakis, Int. J. Solids Struct. **38**, 353 (2001).
13. H. P. Zhu and A. B. Yu, Phys. Rev. **E 66**, 021302 (2002).
14. R. C. Ball and R. Blumenfeld, Phys. Rev. Lett. **88**, 115505 (2002).
15. W. Ehlers, E. Ramm. S. Diebels and G. A. D'Addetta, Int. J. Solids and Struct. **40**, 6681 (2003).
16. N. P. Kruyt and L. Rothenburg, Mechanics of Materials **36** (22), 1157 (2004).
17. F. Froiio, G. Tomassetti and I. Vardoulakis, Int. J. Sol. Struct. **43**, 7684–7720 (2006).
18. J. D. Goddard, in "Mathematical models of granular matter", G. Capriz, P. Giovine and P. M. Mariano, Eds., Lecture Notes in Applied Mathematics, Vol. 1937, Ch. 1, p. 1 (Springer, 2008).
19. J. D. Goddard, in D. DeKee and P. N. Kaloni, Eds., "Recent Developments in Structured Continua", Pitman Research Notes in Mathematics, No. 143, p. 179 (Longman/J. Wiley, New-York, 1986).
20. I. Goldhirsch, Annual Reviews of Fluid Mechanics, **35**, 267 (2003).
21. B. J. Glasser and I. Goldhirsch, Phys. Fluids **13**, 407 (2001).
22. A. E. H. Love, "Treatise of mathematical theory of elasticity" (Cambridge UP, 1927).
23. W. Voigt, "Theoretische Studien über die Elasticitätverhhältnisse der Krystalle", Abhindt. Ges. Wiss. Göttingen, **34** (1887).
24. C. Goldenberg and I. Goldhirsch, in "Handbook of Theoretical and Computational Nanotechnology", M. Rieth and W. Schommers, Eds., p. 330 (American Scientific, 2006).
25. A. I. Murdoch, J. Elasticity **71** (1–3), 105 (2003).
26. A. I. Murdoch, J. Elasticity **88** (2), 113 (2007).
27. I. Goldhirsch and C. Goldenberg, Eur. Phys. J. **E 9** (3), 245 (2002).
28. C. Goldenberg and I. Goldhirsch, in "The Physics of Granular Media", D. E. Wolf and H. Hinrichsen, Eds., p. 3 (Wiley, 2004).
29. I. Goldhirsch, Gran. Matt., in press (2010).
30. P. Schofield and J. R. Henderson, Prof. R. Soc. Lond. Ser. A **379** (1776) 231 (1982).
31. E. Wajnryb, A. R. Altenberger and J. S. Dahler, J. Chem. Phys. **103**, 9782 (1995).
32. J. Zhang, R. P. Behringer and I. Goldhirsch, Prog. Theor. Phys. Suppl., in press (2010).
33. G. K. Batchelor, "An introduction to fluid dynamics" (Cambridge UP, 1967).
34. C. Goldenberg, A. P. F. Atman, P. Claudin, G. Combe and I. Goldhirsch, Phys. Rev. Lett. **96** (16), 168001 (2006).
35. C. Goldenberg and I. Goldhirsch, Phys. Rev. **E 77** (4), 041303 (2008).
36. J. Machta and I. Oppenheim, Physica **A 112**, 361 (1982).
37. e.g., A. C. Eringen, "Theory of micropolar elasticity", in H. Leibowitz, Ed., Fracture: An advanced treatise, Vol. I, Mathematical fundamentals", p. 621 (Academic, 1968).
38. H. Arslan and S. Sture, Compu. Mat. Sci. **42**, 535 (2008).
39. N. P. Kruyt and L. Rothenburg, Mech. Mat., **36**, 1157 (2004).
40. F. Nicot and F. Darve, Gran. Matt., **8**, 221(2006), and refs. therein.
41. P. Y. Hicher and C. S. Chang, Int. J. Solids and Struct., **42**, 4258 (2005), and refs. therein.
42. D. Serero, C. Goldenberg, S. H. Noskowicz and I. Goldhirsch, Powder. Tech. **182**, 257 (2008).

Macroscopic stress from dynamic, rotating granular media

S. Luding

Multi Scale Mechanics, TS, CTW, UTwente,
POBox 217, 7500 AE Enschede, Netherlands;
e-mail: `s.luding@utwente.nl`

Abstract. In this brief study, the stress is averaged over a two-dimensional rigid particle (disk) that is in contact with other objects via localized (point) contacts. In contrast to previous studies, the rotation dynamics of the particles is also taken into account here The stress contains four terms, (i) the static stress due to forces, (ii) the dynamic stress due to translational velocity fluctuations, (iii) the dynamic stress due to rotational velocity fluctuations, and (iv) the stress due to changes of angular velocities due to torques.

Keywords: dynamic stress, rotations, averaging, fluctuations, non-equilibrium
PACS: 45.70, 47.50+d, 51.10.+y

INTRODUCTION

One of todays great challenges in material science and physics is the macroscopic description of the material behavior of granular media which are inhomogeneous, nonlinear, and disordered on a "microscopic" scale [1, 2, 3, 4]. This is due to the intrinsically inhomogeneous stress distribution in granular assemblies and the corresponding stress-networks involving large fluctuations of contact forces. Furthermore, the material is nonlinear due to the nature of the contact forces: If a contact opens, the interaction force vanishes or, with other words, no tensile but only compressive forces are possible in the absence of cohesion. This nonlinearity allows for a reorganization of the contact-network due to deformation.

While there is quite broad consensus on how stress should be averaged [5, 6, 7, 8, 9, 10, 11, 12, 13, 14, 15, 16, 17, 18, 19, 20], the averaging of strain is still a subject of discussion, see Ref. [21] and references therein. In many traditional studies the stress is averaged under the assumption of static equilibrium [8, 9, 22, 11, 13, 23]. However, in this study we allow for a dynamic stress, where the particles are not in equilibrium and thus moving and rotating. The stress is first averaged over single particles, for which several simplifications can be applied. Finally, the resulting stress tensor is interpreted term by term.

STRESS AVERAGING

In a first step, the usual volume average of stress is replaced by a sum over particle averaged internal stresses

$$\bar{\bar{\sigma}} = \frac{1}{V} \int_V \underline{\underline{\sigma}} dV' = \frac{1}{V} \sum_p V^p \bar{\underline{\underline{\sigma}}}^p := \frac{1}{V} \sum_p \int_{V_p} \underline{\underline{\sigma}} dV', \qquad (1)$$

where p denotes the particle with volume V^p and the sum runs over all particles within the volume V. [1] The integral on the right hand side is carried out only over a single particle (where the stress in the free volume between particles is assumed to vanish and thus neglected). This operation will allow us to deal with particle averaged stresses $\underline{\underline{\sigma}}^p$ instead of volume averages. For the sake of simplicity we skip the stress-superscript p in the following.

---

[1] Note that the weight factor V^p in the sum cancels the prefactor $1/V^p$ in the implicit definition of the particle stress, where the particle stress is averaged over points strictly inside the particle only. This is matter of choice and V^p could be replaced by V_c^p at all instances, which would define the particle stress in its Voronoi cell, with $\sum_p V_c^p = V$. We do not apply this latter definition – which would not change the averaged stress anyway – since the volumes V_c^p are a-priori unknown and in general different per particle in a disordered packing.
Assume a single particle p, with $V^p = \pi a^2$, and constant stress σ_0 in its inside. When it occupies, e.g., a square cell with volume $V = 4a^2$, the averaged internal particle stress is $\sigma^p = \sigma_0$, while the average stress over the volume is $\bar{\sigma} = \nu \sigma_0$, with volume fraction $\nu = V^p/V$

In order to proceed, we first introduce the identity for the transposed [2] stress

$$\underline{\underline{\sigma}}^T = (\text{grad}\,\vec{x}) \cdot \underline{\underline{\sigma}}^T = \text{div}(\vec{x} \otimes \underline{\underline{\sigma}}) - \vec{x} \otimes \text{div}\,\underline{\underline{\sigma}}\,, \tag{2}$$

where grad and div are the usual vector operator derivatives with respect to $\vec{x}$, grad$\vec{x}$ is the unit tensor, and $\otimes$ is the dyadic product of vectos or tensors. Combinig Eqs. (1) and (2) yields

$$\underline{\underline{\bar{\sigma}}}^T = (\underline{\underline{\bar{\sigma}}}^p)^T = \frac{1}{V^p} \int_{V^p} \underline{\underline{\sigma}}^T dV' = \frac{1}{V^p} \int_{V^p} [\text{div}(\vec{x} \otimes \underline{\underline{\sigma}}) - \vec{x} \otimes \text{div}\,\underline{\underline{\sigma}}] dV'\,, \tag{3}$$

The law of momentum balance in Eulerian reference frame, for the volume occupied by particle p at time t, reads

$$\rho \ddot{\vec{x}} + \rho \vec{v} \cdot \text{grad}\,\vec{v} = \text{div}\,\underline{\underline{\sigma}} + \rho \vec{b}\,, \tag{4}$$

where the double-dots are partial derivatives with respect to time and $\vec{b}$ is an external acceleration, e.g. gravity. Inserting Eq. (4) in Eq. (3) (and restricting to one particle) leads to

$$\underline{\underline{\bar{\sigma}}}^T = \frac{1}{V^p}\left[\underbrace{\int_{\partial V^p} \vec{x} \otimes \underline{\underline{\sigma}} \cdot d\vec{A}}_{\underline{\underline{\bar{\sigma}}}_s^T} - \underbrace{\int_{V^p} \vec{x} \otimes \rho(\ddot{\vec{x}} - \vec{b}) dV'}_{-\underline{\underline{\bar{\sigma}}}_v^T} - \underbrace{\int_{V^p} \vec{x} \otimes \rho(\vec{v} \cdot \text{grad}\,\vec{v}) dV'}_{-\underline{\underline{\bar{\sigma}}}_d^T}\right], \tag{5}$$

where the first term comes from Gauss' theorem that allows to translate a volume integral into a surface integral. The term that involves the constant acceleration is merged into the second term. The three stress contributions will be addressed seperately below.

The surface integral of Eq. (5)

Using the Cauchy theorem $\vec{t} = \underline{\underline{\sigma}} \cdot \vec{n}$ and the definition $d\vec{A} = \vec{n}dA$, with the normal $\vec{n}$ to the boundary ∂V^p of particle p, allows to transform the first part of Eq. (5) into a sum [23]:

$$\underline{\underline{\bar{\sigma}}}_s^T = \frac{1}{V^p} \int_{\partial V^p} (\vec{x} \otimes \vec{t})\,dA = \frac{1}{V^p} \sum_c \vec{x}^c \otimes \vec{f}^c\,, \tag{6}$$

after replacing the surface stresses active at the contacts by the corresponding forces $\vec{f}^c$. Note that here very small contact areas are assumed, so that the integral can be transformed into the sum in Eq. (6), which represents the transition from the Eulerian (continuum) to the Lagrangian (discrete) picture, respectively.

Introducing the branch vector $\vec{l}^{pc}$ by the vector addition $\vec{x}^c = \vec{x}^p + \vec{l}^{pc}$, as shown in Fig. 1, leads to

$$\underline{\underline{\bar{\sigma}}}_s^T = \frac{1}{V^p}\left[\vec{x}^p \otimes \sum_c \vec{f}^c + \sum_c \vec{l}^{pc} \otimes \vec{f}^c\right]. \tag{7}$$

With Newtons law for the motion of particle p with mass m,

$$m\ddot{\vec{x}}^p = \sum_c \vec{f}^c + m\vec{b}\,, \tag{8}$$

one finally gets

$$\underline{\underline{\bar{\sigma}}}_s^T = \frac{1}{V^p}\left[m\vec{x}^p \otimes (\ddot{\vec{x}}^p - \vec{b}) + \sum_c \vec{l}^{pc} \otimes \vec{f}^c\right] \tag{9}$$

for the first integral in Eq. (5).

---

[2] The transposed stress is introduced on the left hand side of Eq. (2) just to keep the T-symbol out of the integrals for convenience. The stress in Eq. (1) is then just the transposed of the transposed stress. For symmetric stresses, the T could be dropped completely, however, our result is non-symmetric in general, dynamic situations.

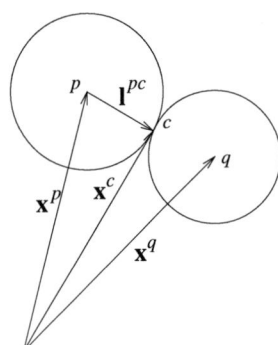

FIGURE 1. Schematic plot of two particles p and q with their common contact c.

The volume integral of Eq. (5)

The volume integral

$$\underline{\underline{\bar{\sigma}}}_v^T = -\frac{1}{V^p}\int_{V^p}\left(\vec{x}\otimes\rho\ddot{\vec{x}} - \vec{x}\otimes\rho\vec{b}\right)dV' \qquad (10)$$

contains those terms acting on all material points of particle p. Therefore, one has to introduce a vector $\vec{l}$ which points from the center of mass of the particle to the material points inside so that $\vec{x} = \vec{x}^p + \vec{l}$, see Fig. 2.

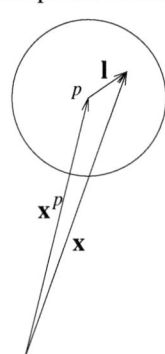

FIGURE 2. Schematic plot of a particle p and its material point $\vec{x}$.

This leads to

$$\underline{\underline{\bar{\sigma}}}_v^T = -\frac{1}{V^p}\int_{V^p}(\vec{x}^p + \vec{l})\otimes\rho\left(\ddot{\vec{x}}^p + \ddot{\vec{l}} - \vec{b}\right)dV' , \qquad (11)$$

where the vectors $\vec{x}^p$ and $\vec{b}$ are constant, so that one can draw them out of the integral. The integral $\int_{V^p}\rho dV'$ is the mass m of the particle as implied in the following. In separate terms the stress reads

$$\underline{\underline{\bar{\sigma}}}_v^T = -\quad (1/V^p)[\quad +m\vec{x}^p\otimes\ddot{\vec{x}}^p \qquad (12)$$

$$+\vec{x}^p\otimes\int_{V^p}\rho\ddot{\vec{l}}dV' \qquad (13)$$

$$-m\vec{x}^p\otimes\vec{b} \qquad (14)$$

$$+\left(\int_{V^p}\rho\vec{l}dV'\right)\otimes\ddot{\vec{x}}^p \qquad (15)$$

$$+\int_{V^p}\rho\vec{l}\otimes\ddot{\vec{l}}dV' \qquad (16)$$

$$-\left(\int_{V^p}\rho\vec{l}dV'\right)\otimes\vec{b}\ ] \qquad (17)$$

The fourth term, Eq. (15), and the sixth term, Eq. (17), vanish due to the fact that $\int_{V^p} \rho \vec{l} dV'$ is the definition of the center of mass and $\vec{l}$ is defined relative to the center of mass. For the rotational motion of a rigid body with angular velocity ω around its center of mass one has

$$\begin{aligned} \dot{\vec{l}} &= \vec{\omega} \times \vec{l}, \text{ and} \\ \ddot{\vec{l}} &= \dot{\vec{\omega}} \times \vec{l} + \vec{\omega} \times \dot{\vec{l}} \\ &= \dot{\vec{\omega}} \times \vec{l} + \vec{\omega} \times (\vec{\omega} \times \vec{l}) \end{aligned} \qquad (18)$$

so that also the second term, Eq. (13), equals zero because both $\vec{\omega}$ and $\dot{\vec{\omega}}$ are constant over the rigid particle and thus can be drawn out of the integral. Finally, using $\vec{\omega} \times (\vec{\omega} \times \vec{l}) = \vec{\omega}(\vec{\omega} \cdot \vec{l}) - \vec{l}(\omega^2) = -\vec{l}(\omega^2)$, since $\vec{\omega}$ and $\vec{l}$ are perpendicular in 2D disks rotating around their axis of rotational symmetry, one gets

$$-\frac{1}{V^p} \left[m\vec{x}^p \otimes (\ddot{\vec{x}}^p - \vec{b}) + \int_{V^p} \rho \vec{l} \otimes (\dot{\vec{\omega}} \times \vec{l} - \omega^2 \vec{l}) dV' \right] . \qquad (19)$$

Using the identity $\vec{l} \otimes (\dot{\vec{\omega}} \times \vec{l}) = -(\vec{l} \otimes \vec{l}) \times \dot{\vec{\omega}}$ and drawing the constants out of the integrals, yields

$$\underline{\underline{\bar{\sigma}}}_v^T = -\frac{1}{V^p} \left[m\vec{x}^p \otimes (\ddot{\vec{x}}^p - \vec{b}) + \underline{\underline{\theta}} \times \dot{\vec{\omega}} + \omega^2 \underline{\underline{\theta}} \right] , \qquad (20)$$

after introducing the symmetric tensor $\underline{\underline{\theta}} := -\int_{V^p} \rho \vec{l} \otimes \vec{l}$.

The dynamic stress in Eq. (5)

The integral

$$\underline{\underline{\bar{\sigma}}}_d^T = -\frac{1}{V^p} \int_{V^p} (\vec{x} \otimes \rho \vec{v} \cdot \operatorname{grad} \vec{v}) dV' \qquad (21)$$

can be simplified by transforming the components of the term

$$-x_\alpha v_\gamma v_{\beta,\gamma} = -(x_\alpha v_\gamma v_\beta)_{,\gamma} + x_{\alpha,\gamma} v_\gamma v_\beta + x_\alpha v_{\gamma,\gamma} v_\beta , \qquad (22)$$

where the $,\gamma$ replaces the gradient. The last term on the r.h.s. vanishes due to the incompressibility of the particles $v_{\gamma,\gamma} = 0$. The first term leads to a surface integral using the Cauchy theorem, however, it vanishes since $\vec{n} \cdot \dot{\vec{l}} = 0$, and the surface integral over terms $\vec{v}^p \cdot \vec{n}$ also vanishes due to the symmetric particle shape. The second integral survives and, after replacing $\operatorname{grad} \vec{x}$ by the unit tensor, has to be treated in a way similar to the volume integral in the previous subsection. Therefore, we replace the vector $\vec{v}$ by $\dot{\vec{x}} = \dot{\vec{x}}^p + \dot{\vec{l}}$, so that

$$\underline{\underline{\bar{\sigma}}}_d^T = \frac{1}{V^p} \int_{V^p} \left(\rho (\dot{\vec{x}}^p + \dot{\vec{l}}) \otimes (\dot{\vec{x}}^p + \dot{\vec{l}}) \right) dV' . \qquad (23)$$

Since the mixed terms contain $\dot{\vec{x}}^p \otimes \dot{\vec{l}}$ and thus vanish due to the definition of $\vec{l}$. The dyadic velocity tensor $\rho \dot{\vec{x}}^p \otimes \dot{\vec{x}}^p$ can be easily integrated so that the remaining integral contains

$$\dot{\vec{l}} \otimes \dot{\vec{l}} = (\vec{\omega} \times \vec{l}) \otimes (\vec{\omega} \times \vec{l}) = \omega^2 (l^2 \underline{\underline{I}} - \vec{l} \otimes \vec{l}) . \qquad (24)$$

The integral over the first term leads to $ma^2\omega^2 \underline{\underline{I}}/2$, with the particle radius a, and the second term is again $\underline{\underline{\theta}}$, so that the dynamic stress is

$$\underline{\underline{\bar{\sigma}}}_d^T = \rho \left[\vec{v}^p \otimes \vec{v}^p + \frac{1}{2} a^2 \omega^2 \underline{\underline{I}} \right] + \frac{1}{V^p} \omega^2 \underline{\underline{\theta}} . \qquad (25)$$

The combined particle stress

Inserting Eqs. (9), (20), and (25) in Eq. (5), finally leads to

$$(\bar{\bar{\sigma}}^p)^T = \bar{\bar{\sigma}}^T = \frac{1}{V^p}\left[\sum_c \vec{l}^{pc} \otimes \vec{f}^c + m\vec{v}^p \otimes \vec{v}^p - \underline{\underline{\theta}} \times \dot{\vec{\omega}} + \underline{\underline{\theta}}_0 \omega^2\right], \tag{26}$$

with the moment of inertia tensor for rotation around the disk center $\underline{\underline{\theta}}_0 = (m/2)a^2 \underline{\underline{I}}$.

The many-particle macroscopic stress

Inserting the internal averaged particle stress from Eq. (26) into Eq. (1) finally leads to the macroscopic averaged many particle stress:

$$\bar{\bar{\sigma}} = \frac{1}{V}\sum_p V^p \bar{\bar{\sigma}}^p = \frac{1}{V}\sum_{c\in V} \vec{f}^c \otimes \vec{l}^{pc} + \frac{1}{V}\sum_p \left[m\vec{u}^p \otimes \vec{u}^p - \dot{\vec{\omega}}^p \times \underline{\underline{\theta}} + \underline{\underline{\theta}}_0 (\omega^p)^2\right], \tag{27}$$

where $\vec{v}^p = \vec{v}_{cm} + \vec{u}^p$ in Eq. (26) contained still both the average center of mass (cm) velocity and the fluctuation velocities u. The constant, moving reference frame "stress" $\bar{\bar{\sigma}}_{cm} = \frac{M}{V}\vec{v}_{cm} \otimes \vec{v}_{cm}$ is divergence-free in Eq. (4) and therefore can be disregarded in order to gain objectivity and frame independence of the stress.

DISCUSSION

The first stress term in Eq. (26) is the well-known, static contribution to the stress tensor and the second term is the dynamic (kinetic) contribution due to the fluctuating particle motion with respect to the Eulerian reference frame ($\rightarrow$ for details see the kinetic theory of gases), i.e. a kinetic energy density. The third, asymmetric term is related to the change of angular velocity and, thus, couples the non-equilibrium translational degrees of freedom to the non-equilibrium rotational motion. Finally, the last term is the rotational energy density, paralleling the dynamic, translational contribution to the stress tensor. Note that our formulation is not identical with the improved stress formulation as recently proposed by Goldhirsch [20], which in principle allows for non-symmetric stress also in the case of static equilibrium.

Comparison of the different analytical stress predictions among each other, with particle simulation results and with experiments is the next step to be done in order to gain improved insight about the effects of dynamics on the stress in granular media.

OUTLOOK

After having performed the above calculations with various assumptions (rigid disk in a two dimensional system) one might wish to compute the stress tensor for the more general case of three-dimensional, possibly non-spherical, objects with internal degrees of freedom like vibrational modes.

While the generalization to 3D spheres sees straightforward, the consequences of a non-sperical geometry and some non-rigidity might complicate the integrals too much to allow a comparatively straight approach.

ACKNOWLEDGMENTS

Helpful discussions with M. Lätzel, N. P. Kruyt, S. Diebels, W. Volk, W. Ehlers, and H. J. Herrmann are gratefully acknowledged. This work was made possible by the financial support of the Deutsche Forschungsgemeinschaft (DFG) and the Stichting voor Fundamenteel Onderzoek der Materie (FOM), financially supported by the Nederlandse Organisatie voor Wetenschappelijk Onderzoek (NWO).

REFERENCES

1. H. J. Herrmann, J.-P. Hovi, and S. Luding, editors, *Physics of dry granular media - NATO ASI Series E 350*, Kluwer Academic Publishers, Dordrecht, 1998.
2. P. A. Vermeer, S. Diebels, W. Ehlers, H. J. Herrmann, S. Luding, and E. Ramm, editors, *Continuous and Discontinuous Modelling of Cohesive Frictional Materials*, Springer, Berlin, 2001, lecture Notes in Physics 568.
3. Y. Kishino, editor, *Powders & Grains 2001*, Balkema, Rotterdam, 2001.
4. M. Nakagawa, and S. Luding, editors, *Powders and Grains 2009*, American Institute of Physics (AIP), Golden, Colorado, USA, 2009, confernce Proceedings #1145, ISBN 978-0-7354-0682-7.
5. A. Drescher, and G. de Josselin de Jong, *J. Mech. Phys. Solids* **20**, 337–351 (1972).
6. L. Rothenburg, and A. P. S. Selvadurai, "A micromechanical definition of the Cauchy stress tensor for particulate media," in *Mechanics of Structured Media*, edited by A. P. S. Selvadurai, Elsevier, Amsterdam, 1981, pp. 469–486.
7. S. B. Savage, and D. J. Jeffrey, *J. Fluid. Mech.* **110**, 255 (1981).
8. P. A. Cundall, A. Drescher, and O. D. L. Strack, "Numerical experiments on granular assemblies; Measurements and observations," in *IUTAM Conference on Deformation and Failure of Granular Materials*, Delft, 1982, pp. 355–370.
9. J. D. Goddard, "Microstructural Origins of Continuum Stress Fields - a Brief History and Some unresolved Issues," in *Recent Developments in Structered Continua. Pitman Research Notes in Mathematics No. 143*, edited by D. DeKee, and P. N. Kaloni, Longman, J. Wiley, New York, 1986, p. 179.
10. R. J. Bathurst, and L. Rothenburg, *J. Appl. Mech.* **55**, 17–23 (1988).
11. N. P. Kruyt, and L. Rothenburg, *ASME Journal of Applied Mechanics* **118**, 706–711 (1996).
12. J. J. Moreau, "Numerical Investigation of Shear Zones in Granular Materials," in *Friction, Arching and Contact Dynamics*, edited by D. E. Wolf, and P. Grassberger, World Scientific, Singapore, 1997.
13. C.-L. Liao, T.-P. Chang, D.-H. Young, and C. S. Chang, *Int. J. Solids & Structures* **34**, 4087–4100 (1997).
14. I. Goldhirsch, and C. Goldenberg, *Eur. Phys. J. E* **9**, 245–251 (2002).
15. C. Goldenberg, and I. Goldhirsch, *Phys. Rev. Lett.* **89**, 084302 (2002).
16. C. Goldenberg, and I. Goldhirsch, *Granular Matter* **6**, 87–96 (2004).
17. L. Staron, F. Radjai, and J.-P. Vilotte, *Eur. Phys. J. E* **18**, 311–320 (2005).
18. C. Goldenberg, and I. Goldhirsch, *Nature* **435**, 188–191 (2005).
19. C. Goldenberg, A. P. F. Atman, P. Claudin, G. Combe, and I. Goldhirsch, Scale separation in granular packings: stress plateaus and fluctuations (2005), submitted to Physical Review Letters; e-print: cond-mat/0511610.
20. I. Goldhirsch, *Granular Matter* **12** (2010), submitted.
21. O. Duran, N. P. Kruyt, and S. Luding, *Int. J. of Solids and Structures* **47**, 251–260 (2010).
22. C. S. Chang, "Micromechanical modelling of constitutive relations for granular media," in *Micromechanics of granular media*, Elsevier, Amsterdam, 1988.
23. M. Lätzel, S. Luding, and H. J. Herrmann, *Granular Matter* **2**, 123–135 (2000), e-print cond-mat/0003180.

Simple Interaction Model for Partially Wet Granular Materials

Namiko Mitarai and Hiizu Nakanishi[†]

Niels Bohr Institute, Blegdamsvej 17, DK-2100, Copenhagen, Denmark.
[†]*Department of Physics, Kyushu University 33, 812-8581, Fukuoka, Japan.*

Abstract. It is well known that just adding some liquid to dry granular materials changes their behaviors very much. Liquid forms bridges between grains, and the bridge induces cohesion between grains due to the surface tension. When the liquid content is small, the liquid forms a bridge at each contact point (the pendular state), which induces two-body cohesive force. As the liquid content increases, some liquid bridges merge, and more than two grains interact through a single liquid cluster (the funicular state). We propose a simple phenomenological model for wet granular media to take into account many particle interaction through a liquid cluster in the funicular state as well as two-body cohesive force by a liquid bridge in the pendular state. In our model, the cohesive force acts among the grains connected by a liquid-gas interface. As the liquid content is increased, the number of grains that interact through the liquid increase, but the liquid-gas interface may decrease when liquid clusters are formed. Due to this competition, the shear stress shows a maximum as a function of the liquid-content.

Keywords: granular flow, cohesion, wet granular materials, capillary force, surface tension
PACS: 45.70-n, 82.70-y, 83.80.Fg

INTRODUCTION

When the granular material is partially wet, the surface tension results in the cohesive force among grains by forming liquid bridges and/or liquid clusters connecting grains. This cohesion dramatically changes the behavior of the granular materials. For example, we can make a tunnel going through a wet sand pile, but it is impossible to do with a sand pile with dry non-cohesive grains[1].

We also know that the mechanical property of the partially wet granular materials depends on the liquid content [1, 2, 3, 4]. The partially wet granular systems are often categorized in the following five regimes depending on the liquid content (Fig.1):

(a) The dry state: No liquid in the system. Cohesive interactions between grains are negligible[1].

(b) The pendular state: Small amount of liquid in the system. The liquid bridges are formed at the particle-particle contact point, and the cohesive force acts between a pair of particles connected by a liquid bridge.

(c) The funicular state: More liquid in the system. Some liquid bridges merge to form liquid clusters, which connects multiple particles. The cohesive force acts through liquid bridges and liquid clusters.

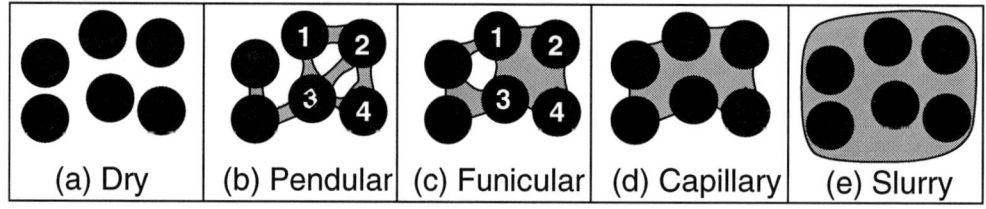

FIGURE 1. Schematic description of wet granular materials with various liquid content. (a) the dry state, (b) the pendular state, (c) the funicular state, (d) the capillary state, and (e) the slurry state.

---

[1] Here we consider the case where the electrostatic effects are negligible

(d) The capillary state: The pores are almost saturated by liquid, but still the pressure in the liquid-phase is lower than that of the air-phase. Thus the particles at the liquid-air interface feel the cohesive force to keep the particles together.

(e) The slurry state: The pores are completely saturated by liquid, and no air in the system. The liquid pressure is the same as the air pressure, and there is no cohesive force between grains.

The states with the cohesive effects, i.e., the pendular, the funicular, and the capillary states, are also often referred to as the "unsaturated" states, to be distinguished from the wet but non-cohesive slurry state, which is sometimes called as the "saturated" state [5].

The research on unsaturated granular materials has been done intensively in engineering fields, from civil engineering to pharmaceuticals and food processing, because of its important and broad applications [3, 5]. The subject is also attracting more and more research interests in physics [1, 4, 6] One of the interesting points to understand would be how the system behavior changes as a function of the liquid content. It is clear that the macroscopic cohesive effect increases with the liquid content at the very small liquid content regime, and decreases with the liquid content at the large, almost saturated liquid content regime, but what happens in between? There are several experiments that suggests existence of a maximum or a few maxima in the cohesive effects, but the behavior has not been fully understood yet and still under active research [7, 8, 9, 10, 11, 12].

The most well-understood state in the unsaturated regimes is the pendular state. The force exerted by a liquid bridge between a pair of particles has been extensively studied [13], and the collective behaviors are studied from the static case where cohesion is the primary effect [14, 15, 16, 17, 18, 19, 20, 21, 22, 23, 24] to the dynamic case where the viscous effect also becomes significant [25, 26, 27]. Recently some experiments start to clarify the behavior also with liquid content beyond the pendular state under well-controlled conditions. It has been shown in several experiments that, upon increasing the liquid content, there is a maximum in the material yield stress at rather low liquid content, in the pendular state or the funicular state [7, 9, 8, 10]. Especially, Møller et al. [9] showed a clear increase of the modulus with the liquid content in the pendular state and a drop in the funicular state by the measurement of the shear modulus.

To understand the fundamental properties of partially wet granular materials, not only more well-controlled experiments but also numerical simulations that reproduce some properties should be useful. The significance of the numerical simulation has been shown in the research on dry granular systems, where the development of the molecular dynamics or discrete element methods using simplified interactions between grains contribute a lot to understand the fundamental physics of the system [28, 29, 30]. It is easy to imagine that numerical simulation of plausible interaction model of partially wet granular materials would also be a great help to understand the system. One way to do this is to simulate the solid granular particle moving in the space filled with two kinds of fluids (e.g., the water and the air) that obeys the full hydrodynamic equations, which may not impossible for small system (cf. [31]) but computationally very heavy. Attempts to model the particle-particle interactions of partially wet granular materials in a simple way have also been made. In most of the case, one considers the pendular state, where the main additional interactions is the two-body attractive force [1, 3, 6, 32, 33, 23, 24, 34, 35]. There are, however, not many interaction models which takes into account the many-particle interaction via a liquid cluster. One example is a model proposed by Grof et al. [11, 12], which assumes that the granular particle motion is rather slow, and the fluid motion is simulated by minimizing the liquid-gas interface for the given grain configuration. However, it is still quite a heavy interaction model to simulate large systems.

We recently proposed a simple phenomenological model that takes into account the many-body interaction through the liquid clusters in the funicular state [36]. In our model, the cohesive force acts among the grains connected by the liquid-gas interface. As the liquid content increases, the number of grains that interact through the liquid increases, but the liquid-gas interface may decrease when the liquid clusters are formed. In this paper we briefly summarize the behavior of this model, where the competition results in a maximum shear stress in the liquid-content dependence.

MODEL

Our model is based on the soft sphere model of dry grains with elastic repulsive force and dissipation (Discrete Element Method, DEM [28]). To consider the partially wet case, we add a cohesive interaction which corresponds to the liquid bridge and a rule to define the liquid cluster. In this section, we first introduce a model for the pendular state with a history-dependent two-body cohesion, as proposed by Schultz et al. [32]. We then extend the model to the funicular state by introducing a rule to define a "liquid cluster" based on the interaction network.

Model P for the pendular state with two-body interaction

For simplicity, we consider a frictionless two-dimensional system, where grains are modeled by polydisperse disks. Let us consider the interaction between particles i and j of diameters σ_i and σ_j and masses m_i and m_j, at positions $\boldsymbol{r}_i$ and $\boldsymbol{r}_j$ with velocities $\boldsymbol{v}_i$ and $\boldsymbol{v}_j$, respectively. There is no interaction between them before they collide. Once they are in contact, i.e., $|\boldsymbol{r}_{ij}| = |\boldsymbol{r}_i - \boldsymbol{r}_j|$ becomes less than $(\sigma_i + \sigma_j)/2$, the force on the particle i by j is given as a function of $\Delta \equiv (\sigma_i + \sigma_j)/2 - |\boldsymbol{r}_{ij}|$ by

$$\boldsymbol{F}_{ij} = f(\Delta)\boldsymbol{n}_{ij} - \eta(\Delta)\boldsymbol{v}_{ij} \qquad (1)$$

with the normal unit vector $\boldsymbol{n}_{ij} = \boldsymbol{r}_{ij}/|\boldsymbol{r}_{ij}|$ and the relative velocity $\boldsymbol{v}_{ij} = \boldsymbol{v}_i - \boldsymbol{v}_j$, as long as $|\boldsymbol{r}_{ij}| < \alpha(\sigma_i + \sigma_j)/2$; The parameter $\alpha (\geq 1)$ determines the distance between grains at which the liquid bridge breaks, and once $|\boldsymbol{r}_{ij}|$ exceeds $\alpha(\sigma_i + \sigma_j)/2$, $\boldsymbol{F}_{ij} = 0$ and the pair i, j do not interact until they are in contact again.

The functions $f(\Delta)$ and $\eta(\Delta)$ in the interaction of Eq.(1) are given by

$$f(\Delta) = \max(k\Delta, -F_0) \qquad (2)$$

and

$$\eta(\Delta) = \begin{cases} \eta_{\text{grain}} & \text{for } \Delta > 0, \\ \eta_{\text{liquid}} & \text{for } \Delta < 0. \end{cases} \qquad (3)$$

Note that Δ characterizes the overlap (for $\Delta > 0$) or the spacing (for $\Delta < 0$) between the grains. The parameter k characterizes the elastic constant, $\eta(\Delta)$ is the damping parameter, and a positive constant F_0 characterizes the cohesion due to a liquid bridge. The damping parameter $\eta(\Delta)$ in eq.(3) depends on Δ because the dissipation when grains are in contact (characterized by η_{grain}) and the dissipation via liquid viscosity (characterized by η_{liquid}) has different physical origin and their values can be quite different. We call this model as Model P.

Notice that the model with $\alpha = 1$ corresponds to dry granular materials with neither cohesive force nor hysteresis, and α is greater than unity for wet cohesive grains. When $\alpha > 1$, there is a hysteresis in the interaction [32], and this hysteresis causes dissipation in addition to the viscous force. It should also be noted that there is no friction in the present model for simplification, which may have strong effect in the behavior of the cohesive grains [37, 34, 38, 39]

Model F for the funicular state with many-body interaction

When the amount of liquid is increased enough for the system to be in the funicular state, some pores are filled by liquid, and more than two grains are connected by a single liquid-filled region. This is schematically shown in Fig. 1(c), where four particles (with numbers 1 to 4) are connected by a liquid cluster. In this case, there is no liquid-air interface that connects particles 2 and 3 in contrast to the pendular state (Fig. 1(b)). There is no direct cohesive force between the particles 2 and 3, though the cohesive force between the particles connected by interfaces in Fig. 1(b) (between 1 and 2, 2 and 4, 4 and 3, and 1 and 3) still tend to hold all the particles together.

We model such a multi-particle effect by making a bond inactive when the bond connects the two particles between which a liquid cluster exists. The presence of the liquid cluster between two grains is determined by the simple criterion: The grains i and j are connected by a liquid cluster without liquid-gas interface if they are connected by a liquid bridge and the number of grains $N_{i,j}$ that are connected to both i and j by liquid bridges is larger or equal to a certain threshold value n_{th}. The liquid bridge is the same as defined in the previous subsection, and we take $n_{th} = 2$ for the 2-dimensional system (see Fig.1(b)). If two grains are connected by a big cluster without liquid-gas interface, the liquid bridge between them is inactive and no cohesive force acts through it.

Namely, the force on grain i by j is given not by Eq. (1) but by

$$\boldsymbol{F}_{ji} = f(\Delta)[\Theta(f(\Delta)) + (1 - \Theta(f(\Delta)))\Theta(n_{th} - N_{i,j})]\boldsymbol{n}_{ij} - \eta(\Delta)\boldsymbol{v}_{ij}, \qquad (4)$$

where $f(\Delta)$ is given by Eq. (2) and $\Theta(x)$ is the Heaviside step function defined as $\Theta(x) = 1$ for $x > 0$ and $\Theta(x) = 0$ for $x \leq 0$. We call this model for the funicular state as Model F here. Note that it is straightforward to extend this model to a three-dimensional system with choosing $n_{th} = 3$.

In this model (Model F), the particle-particle interactions between grains i and j can be classified into the following categories: (i) the repulsive interaction with $N_{i,j} < n_{th}$ and $\Delta > 0$, (ii) the cohesive interaction with $N_{i,j} < n_{th}$ and

$\Delta < 0$, (iii) the repulsive interaction with $N_{i,j} \geq n_{th}$ and $\Delta > 0$ (overlapping particles in the liquid cluster), and (iv) the non-cohesive viscous interaction with $N_{i,j} \geq n_{th}$ and $\Delta < 0$ (particles connected by the liquid cluster without interface between them).

The key parameter in the models is the parameter α to determine the interaction range. In Model P, increasing α just increases the number of cohesive bonds. In Model F, on the other hand, not only the number of interacting particles increases, but also the fraction of bonds that are surrounded by other bonds ($N_{i,j} \geq n_{th}$) increases; such bonds exert no cohesion. Thus, the larger α corresponds to the case with more liquid content. We examine how the competition between these effects of increasing α in Model F affects the response of the wet granular assembly.

SIMULATION RESULTS

Simulation setup

Here we focus on the behavior under steady shear [36], which is easier setup to obtain the reproducible results than the static simulations that are sensitive to the initial packing procedure [38].

The grains are modeled by polydisperse disks whose diameter are uniformly distributed between 0.8σ to σ, and all the grains have the same mass m. The system is periodic in the x direction. In the y direction, there are two parallel rough walls of mass M, each of which consists of N_w particles with diameter σ glued without spacing; the total length of the wall $L = \sigma N_w$).

The shear is applied by the moving top (bottom) wall in the x direction at a velocity U ($-U$). A constant pressure P_w is applied to the walls in the y direction, and the y coordinate of the top (bottom) wall Y_t (Y_b) obeys the equation of motion $M\ddot{Y}_t = F_{w,t}(t) - P_w L$ ($M\ddot{Y}_b = F_{w,b}(t) + P_w L$), where $F_{w,t}(t)$ ($F_{w,b}(t)$) is the y-component of the force exerted from particles on the upper (bottom) wall at time t.

For Model F, we also need to decide whether there are liquid bridges between grains in the wall; That affects the formation of the liquid cluster near the wall. We initially assign liquid bridges between neighboring pair of grains in the wall with the probability $1/2$, and the bridges among wall grains do not disappear or newly created during the simulation.

The following data are collected in the steady states, where both the average kinetic energy of grains and the the y-coordinates of the walls become almost constant.

The parameters are given in the dimensionless form with the unit length σ, unit mass m, and the unit time scale $\sqrt{100m\sigma^2/F_0}$ in the following. In this unit, we choose the elastic constant k to be 5×10^4, the viscous coefficient due to grain contact η_{grain} to be 10 (which gives the restitution coefficient e_n for dry grains ($\alpha=1$) around 0.9), and the viscous coefficient due to liquid bridge η_{liquid} to be 1. We simulate in the following the shear rate $\dot{\gamma} \sim 1$ or less in this unit, where the viscous effect much smaller than the cohesive force. We set the wall mass M to be $M = L$.

The system size are mostly $L = 40$ and the number of particles in the system $N = 1600$ unless otherwise noted. Results with various values of wall pressure P_w has been reported elsewhere [36], thus here we focus on $P_w = 100$ the shear rate $\dot{\gamma}$ dependence, which is adjusted by changing the wall velocity U. The liquid-content dependence is studied by changing the interaction range α. The second-order Adams-Bashforth method and the trapezoidal rule are used to integrate the equations for the velocity and position, respectively, with a time step for integration $dt = 10^{-4}$.

Results

We found that the model under the given boundary condition sometimes solidify and stick to one of the walls, if the interaction range α is too big and/or the system size L is too large [36]. This effect is smaller in Model F than Model P, because in Model F, the bond surrounded by others are non-cohesive, hence easier to rearrange the particle positions. In the following, we present the data with $L = 20$ for model P where the uniform shear flow has been approximately obtained; we confirmed that the values of shear stress for the case with $L = 40$ and $L = 20$ matches as long as the uniform shear flow is obtained.

It should be noted that, even in the case where the uniform shear flow is obtained on average, the system experiences rather large fluctuations because of the finite system size. Figure 2(a) shows the x-component of the instantaneous force on the wall per unit length, of which time average is the shear stress S.

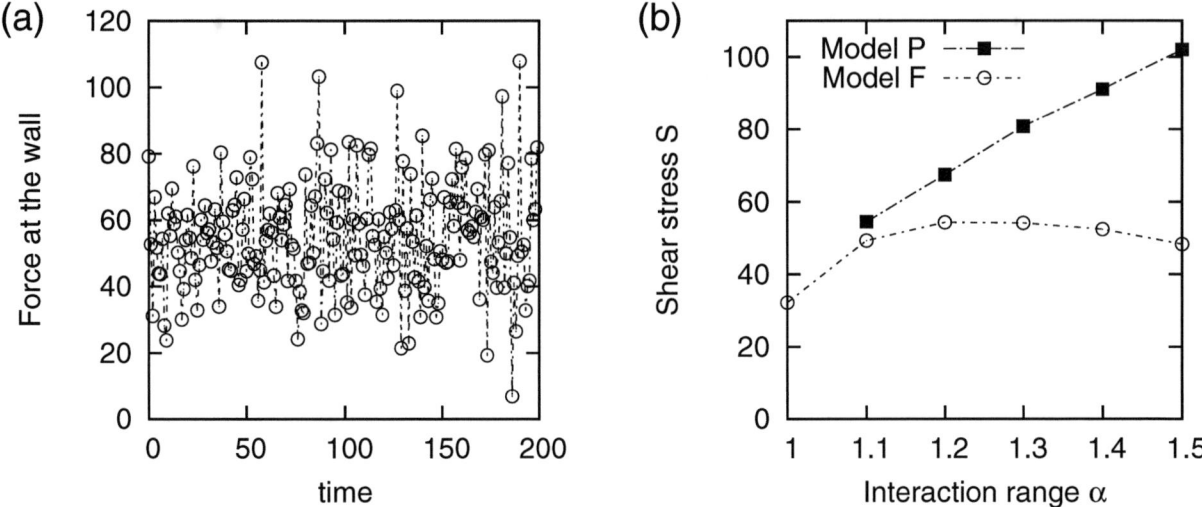

FIGURE 2. (a) The time series of the *x*-component of the instantaneous force at the wall per wall length in the steady state with $\alpha = 1.3$, $P_w = 100$, $\dot{\gamma}$ 1.0, and $L = 40$ for Model F. One can see that the fluctuation is large even in the steady state. The shear stress is calculated from the time average of the force. (b) The interaction range α dependence of the shear stress. Comparison of Model F (open circles) and Model P (solid squares), under the pressure $P_w = 100$ and the shear rate $\dot{\gamma}$ 1.0 [36].

Figure 2(b) shows the value of the shear stress S as a function of the interaction range α with $P_w = 100$ and $\dot{\gamma}$ 1 [36]. Here, the results for Model F is with the system size $L = 40$, while for Model P the system size is $L = 20$, in order to realize the uniform shear flow. We can clearly see that, in the case of Model P, S increases linearly with α, while S has a maximum around α 1.2 in Model F. The lower value of S in Model F is understood by the "liquid cluster" effect, namely, the surrounded bond becomes non-cohesive.

The maximum in the shear stress S as a function of α can be seen in Model F with other sets of parameters. In Fig. 3, the results for model F under $P_w = 100$ with two values of shear rate, $\dot{\gamma}$ 1.0 and $\dot{\gamma}$ 0.1, are shown. The shear stress is smaller for smaller shear rate but maxima can be seen in both cases, though the location of the maximum seems to be shifted towards smaller value of α for $\dot{\gamma}$ 0.1. It has also been confirmed that the maxima can be seen with various values of P_w with $\dot{\gamma}$ 1 [36].

The value of the packing fraction ϕ at the middle of the system is shown in Fig. 3(b). The smaller shear rate shows larger packing fraction, which can be understood as a consequence of the less fluctuation in the smaller shear rate. We see that the number of interacting particles per particle N_{int} is larger for smaller $\dot{\gamma}$ (Fig. 3(c)), which is possibly due to the higher packing fraction. At the same time, in both $\dot{\gamma}$, we see N_{int} increase with α.

N_{int} includes all kinds of interactions, namely, the repulsive interactions with $\Delta > 0$, the cohesive interactions with $\Delta < 0$ and $N_{i,j} < n_{th}$, and the non-cohesive viscous interactions with $\Delta < 0$ and $N_{i,j} \geq n_{th}$ coming from the "liquid cluster" effect. Out of these, the number of the cohesive interactions per particle N_c is shown in Fig. 3(d) as a function of α, which should show the direct effect of the competition between the increasing interaction and increasing non-cohesive viscous interactions with increasing α. We see clear maxima in N_c in both $\dot{\gamma}$, though the location of maxima are different from those of the shear stress S (Fig. 3(a)). The maxima in N_c, however, has tendency to be shifted towards smaller α for smaller shear rate $\dot{\gamma}$, which is consistent with the behavior of the shear stress. Similar behaviors have again been seen with various values of P_w [36].

SUMMARY AND DISCUSSION

In this paper, we introduced a simple model for the funicular state where the liquid clusters can result in many-body interactions. We have shown that, upon increasing the interaction range α, there exist a maximum in the number of particles cohesive interaction and the shear stress in the steady shear flow [36]. These maxima should be due to the competition between the number of interacting particles and the number of particles connected by liquid cluster.

In order to compare this model with experiments, we need to simulate the 3-dimensional case, and also find out

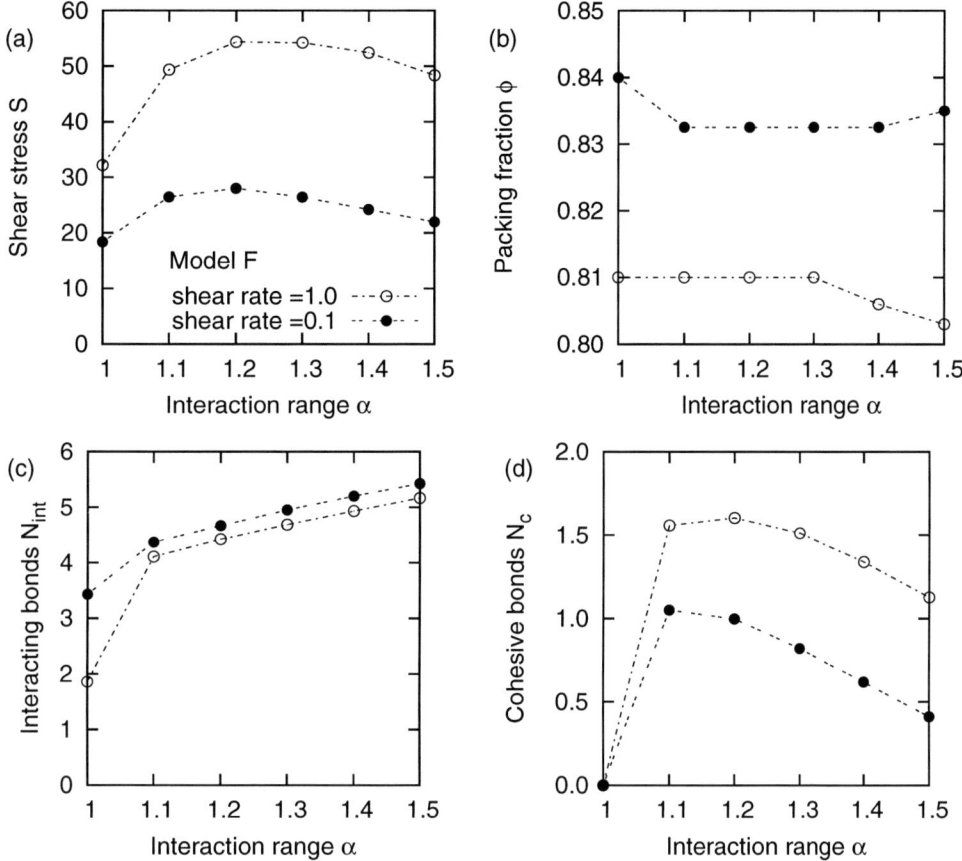

FIGURE 3. Comparison of the shear rate $\dot{\gamma}$ 1.0 (open circles) and the shear rate $\dot{\gamma}$ 0.1 (solid circles) for Model F under the pressure $P_w = 100$. (a) The shear stress S, (b) the packing fraction ϕ, (c) the number of bonds per particle N_{int}, and (d) the number of cohesive interactions per particle N_c.

the relationship between the interaction range α and the liquid volume V_l, since V_l is the actual control parameter in experiments. The simplest estimation would be $V_l \propto [\sigma(\alpha\ 1)]^d$, where d is the spatial dimension, because α characterizes the distance between interacting particles. It is our future project to see the correspondence in three-dimension.

In addition, one should note that the large fluctuation in the stress exists for the systems of the size studied here. This large fluctuation should be due to the existence of large liquid clusters, whose distributions we have studied elsewhere [36], and found that the system-wide cluster are being formed. These facts suggests that a careful analysis on the system size effect and the boundary effect is necessary for the quantitative comparison of the simulation results with those of experiments.

In addition, there are many effects that have not been taken into account in this model for simplification. One of them is the lubrication effect due to viscosity, which would be significant in high-shear rates [40]. The friction is another important effect, which should especially be relevant for small liquid content case (α 1) [34, 37, 38, 39]. Another point is the conservation of liquid, which will affect the configuration of the interaction network. The model proposed by Richefeu et al. [23, 24] is one of the possible approach to include liquid conservation effectively, where a liquid volume dependent interaction has been adopted and the liquid is redistributed to each liquid bridge according to a simple rule along the time evolution. It would be interesting to see how the behavior changes if we add the liquid cluster criterion to this.

ACKNOWLEDGMENTS

This work is partially supported by the grant-in-aid (21540418) by Japan Society for the Promotion of Science (JSPS).

REFERENCES

1. N. Mitarai, and F. Nori, *Adv. Phys.* **55**, 1–45 (2006).
2. D. Newitt, and J. Conway-Jones, *Trans. I. Chem. Eng.* **36**, 422–441 (1958).
3. S. Iveson, J. Litster, K. Hapgood, and B. J. Ennis, *Powder Technol.* **117**, 3–39 (2001).
4. A. Kudrolli, *Nature materials* **7**, 174–175 (2008).
5. N. Lu, and W. Likos, *Unsaturated Soil Mechanics*, John Wiley & Sons, Hoboken, 2004.
6. S. Herminghaus, *Advances in Physics* **54**, 221–261 (2005).
7. Z. Fournier, D. Geromichalos, S. Heminghaus, M. Kohonen, F. Mugele, M. Scheel, M.Schulz, B. Schulz, C. Schier, R. Seemann, and A. Skudelny, *J. Phys.: Condens. Matter* **17**, S477–S502 (2005).
8. N. Lu, B. Wu, and C. Tan, *J. Geotech. Geoenv. Eng.* **133**, 144 (2007).
9. P. C. F. Møller, and D. Bonn, *Europhys. Lett.* **80**, 38002 (2007).
10. M. Scheel, R. Seemann, M. Brinkmann, M. D. Michiel, A. Sheppard, B. Breidenbach, and S. Herminghaus, *Nature Mat.* **7**, 189 (2008).
11. A. Grof, C. Lawrence, and F. Stepanek, *J. Colloid and Interface Sci.* **319**, 182 (2008).
12. A. Grof, C. Lawrence, and F. Stepanek, *Gran. Matt.* **10**, 93 (2008).
13. C. Willett, M. Adams, S. Johnson, and J. Seville, *Langmuir* **16**, 9396–9405 (2000).
14. L. Bocquet, E. Charlaix, S. Ciliberto, and J. Crassous, *Nature* **396**, 735–737 (1998).
15. D. Hornbaker, R. Albert, I. Albert, A.-L. Barabási, and P. Schiffer, *Nature* **387**, 765 (1997).
16. R. Albert, I. Albert, D. Hornbaker, P. Schiffer, and A.-L. Barabási, *Phys. Rev. E* **56**, R6271–R6274 (1997).
17. N. Fraysse, H. Thomé, and L. Petit, *Euro. Phys. J. B* **11**, 615–619 (1999).
18. N. Olivi-Tran, N. Fraysse, P. Girard, M. Ramonda, and D. Chatain, *Eur. Phys. J. B* **25**, 217–222 (2002).
19. T. Halsey, and A. Levine, *Phys. Rev. Lett.* **80**, 3141–3144 (1998).
20. S. Nase, W. Vargas, A. Abatan, and J. McCarthy, *Powder Technol.* **116**, 214–223 (2001).
21. F. Restagno, C. Ursini, H. Gayvallet, and E. Charlaix, *Phys. Rev. E* **66**, 021304 (2002).
22. S. Nowak, A. Samadani, and A. Kudrolli, *Nature Physics* **1**, 50 (2005).
23. V. Richefeu, M. E. Youssoufi, and F. Radjai, *Phys. Rev. E* **73**, 051304 (2006).
24. V. Richefeu, F. Radjai, and M. E. Youssoufi, *Eur. Phys. J. E* **21**, 359 (2006).
25. P. Tegzes, T. Vicsek, and P. Schiffer, *Phys. Rev. Lett.* **89**, 094301 (2002).
26. P. Tegzes, T. Vicsek, and P. Schiffer, *Phys. Rev. E* **67**, 051303 (2003).
27. Q. Xu, A. V. Orpe, and A. Kudrolli, *Phys. Rev. E* **76** (2007).
28. P. A. Cundall, and O. D. L. Strack, *Geotechnique* **29**, 47 (1979).
29. H. Jaeger, S. Nagel, and R. Behringer, *Rev. Mod. Phys.* **68**, 1259–1273 (1996).
30. J. Duran, *Sands, Powders, and Grains: An Introduction to the Physics of Granular Materials*, Springer, New York, 1997.
31. M. Cates, R. Adhikari, and K. Stratford, *J. Phys.: Condens. Matter* **17**, S2771–S2778 (2005).
32. M. Schulz, B. Schulz, and S. Herminghaus, *Phys. Rev. E* **67**, 052301 (2003).
33. R. Brewster, G. Grest, J. Landry, and A. Levine, *Phys. Rev. E* **72**, 061301 (2005).
34. P. G. Rognon, J.-N. Roux, D. Wolf, M. Naaïm, and F. Chevoir, *Europhys. Lett.* **74**, 644 (2006).
35. P. G. Rognon, J.-N. Roux, M. Naaïm, and F. Chevoir, *J. Fluid Mech* **596**, 21–47 (2008).
36. N. Mitarai, and H. Nakanishi, *arXiv:0908.1477v1, to appear in Eurphys. Lett.* (2009).
37. G. Bartels, T. Unger, D. Kadau, D. Wolf, and J. Kertész, *Gran. Matt.* **7**, 139–143 (2005).
38. F. A. Gilabert, J.-N. Roux, and A. Castellanos, *Phys. Rev. E* **75**, 011303 (2007).
39. F. A. Gilabert, J.-N. Roux, and A. Castellanos, *Phys. Rev. E* **78**, 031305 (2008).
40. P. G. Rognon, I. Einav, and C. Gay, "Thixotropy in immersed granular materials," in *these proceedings*, 2009.

Shear induced diffusion in dense granular flows

Ashish V. Orpe*,†, Chris H. Rycroft**,‡ and Arshad A. Kudrolli†

*Chemical Engineering Division, National Chemical Laboratory, Pune 411008, India
†Department of Physics, Clark University, Worcester, MA 01610 USA
**Department of Mathematics, Lawrence Berkeley Laboratory, Berkeley, CA 94720, USA
‡Department of Mathematics, University of California, Berkeley, CA 94720, USA

Abstract. The dynamics of dense granular flows subjected to gravity induced shear are investigated experimentally using the refractive index matching technique. The system consists of grains flowing inside a bin with a rectangular cross-section and sheared by a rough boundary on one side and smooth boundaries on the remaining sides. The particles flow within a viscous interstitial liquid having the same refractive index as particles and are imaged in the bulk using laser fluorescence. The particle positions are identified very accurately and tracked over long durations to obtain the mean and fluctuating properties. The shear is observed to be non-linear and localized in a region of 3 to 4 particles near the boundary. The boundary imposes a packing order, and the grains are observed to flow in layers, parallel to the shearing boundary, which get progressively more disordered with distance from the walls. We have also carried out soft particle simulations in a equivalent system incorporating the Cundall-Strack contact model between the particles and ignoring the hydrodynamic effects of the interstitial liquid to understand the effect of particle friction coefficient, elasticity, contact model and polydispersity on the mean and fluctuating flow properties. We find the mean velocity and the number density of the particles as a function of flow cross-section and the particle fluctuation properties observed in the experiments and the simulations to in very good agreement after appropriate scaling.

Keywords: granular, diffusion, index matching, laser fluorescence, LAMMPS, DEM
PACS: 45.70.Mg,07.05.Tp

INTRODUCTION

Sheared granular systems are prevalent in various natural instances and several industrial operations. Many of these applications have densely packed grains exhibiting curious features like jamming, shear banding, solid-liquid coexisting behavior etc. In spite of the prevalence of dense granular systems, a hydrodynamic theory and a length scale other than the particle diameter has not been consistently identified leading to a large amount of research in the past two decades and a variety of theoretical advances [1, 2, 3, 4, 5, 6]. The key to successful hydrodynamic modeling is a clearer understanding of the fluctuating behavior of grains which can be described accurately by quantifying the particle diffusion behavior. Indeed, particle diffusion for dense granular flows has been measured previously, in strictly two-dimensional systems [7] and quasi-two-dimensional systems [8, 9, 10]. While these studies have provided significant insights, it is not clear how the walls affect the particle diffusion and so there is a need to measure the diffusion behavior of flowing grains in the bulk. The need for this is recognized within the research community and work has started on this front [11, 12], though there is still a lot to achieve in this area. For instance, a detailed knowledge similar to that established for non-Brownian suspensions [13, 14] does not exists for dense granular flows.

Several innovative experimental techniques have been devised over the past two decades to study the dynamics of granular systems in the bulk like MRI [15, 16, 17, 18], PEPT [19, 20, 21], X-ray tomography [22, 23, 24, 25]. However, one technique that has received significant attention in the recent years is the refractive index matching technique [11, 12, 26, 27, 28, 29, 30, 31, 32]. The technique has been used previously to study flow details of non-Brownian density matched suspensions [14, 33] and allows for direct tracking of the particles in three dimensions and is very suitable to measure microscopic quantities like particle diffusion, local structure, velocity auto-correlations etc. The technique, described in detail later, involves particles flowing through a liquid having the same refractive index and are visualized by a laser sheet fluorescing the liquid and the particles appearing as dark shadows on a bright background which can be tracked using standard available routines. However, one biggest question mark over this technique has been the utility of the results for the dry systems as the technique involves a viscous refractive index matched liquid. Further, despite its significant usage the effects of interstitial liquid on the dynamics of particle flow have not been studied systematically, neither have been the results from this technique validated thoroughly with simulations or theory predictions.

In addition to several innovative experimental techniques, computational models have been an invaluable tool in understanding three dimensional granular flow, because they can provide complete information about a particle system without the possible difficulties in experiments. One such approach incorporates soft particles which are able to overlap and interact via contact force model. The simulations are then carried out using fixed-time step integration via discrete element method (DEM) [34]. Here we are referring to the simulations which employ a modified version of a contact model originally employed by Cundall and Strack [35] for simulation of cohensionless particles. The model is simulated by the Large-scale Atomic/Molecular Massively parallel simulator (LAMMPS) developed at Sandia National Laboratories. This simulation code is well-developed and has been used by many research groups for various granular problems like avalanching flow [36, 37, 38], static granular packings [39, 40, 41], granular drainage [42, 43, 44], granular shear and its relation to constitutive modeling [45, 46, 47, 48], and granular segregation [49]. The simulation model consists of several free parameters which determine the relative strength of the different forces and many of the studies listed above use the same values. These parameters have been established for relatively smaller systems and so it is not very clear if these parameters remain valid for large system sizes. Further, to enable the simulations to be computationally feasible for a problem of practical size, some approximations are made in the contact models which has led to skepticism about the capability of these contact models to capture the underlying physics of granular flows.

In this paper we attempt to address the issues outlined above through simultaneous (index-matching) experiments and (DEM) simulations. This work has been published earlier in its entirety [50] and over here we summarize the key results from that work. Many details are omitted to restrict the length of the paper and at appropriate locations the more interested reader is referred to the original published article. The paper proceeds as follows: We first describe the experimental and simulation methodology. Then we discuss the mean flow results and their dependence on several particle properties and flow features. A more detailed discussion of the particle diffusion behavior is presented next followed by conclusions.

EXPERIMENTS

The schematic of the experimental shear cell made out of optically smooth transparent glass plates is shown in Fig. 1. Glass beads of diameter $d = 1.0$ mm flow from the top chamber into the bottom chamber through a narrow slit. The shear in the flow relative to other surfaces is generated by gluing a layer of glass beads to one of the sides in the top chamber while keeping the remaining sides optically smooth. The interstitial space between the particles is filled with a liquid of viscosity $v = 2.2 \times 10^{-2}$ kg m^{-1} s, density $\rho = 1.0 \times 10^3$ kg m^{-3} and has the same refractive index as the particles. Two side chambers, completely filled with liquid, are provided along the entire height of the system to allow the liquid to redistribute (shown by blue arrows) when the particles drain from the orifice. The liquid motion in and out of the side chambers takes place through a wired mesh fixed between the chambers which is small enough to prevent the migration of particles. Such arrangement of side chambers ensures a steady flow by preventing liquid forcing upwards through the orifice when the particles drain down. A dye added to the liquid is illuminated by a light sheet of thickness less than $0.1d$ and imaged from an orthogonal direction using a 512×480 pixel resolution CCD camera, where 20 pixels corresponds to one d. The particles in an image appear dark against a bright background with a flat intensity profile across each particle. We then make use of convolution procedure [29] to convert each image into a 2D map consisting of bright sharp peaks of intensity corresponding to particle centers which are then obtained using a centroid algorithm [51]. This procedure yields particle position in every image to within a twentieth of a particle diameter. Because of small variations of refractive index within the glass beads and defects, the accuracy with which we can determine the position of the particle diminishes with optical length within the index matched sample. Therefore, we restrict our data acquisition to a window which is within $30d$ from a side wall. A sequence of images is recorded in the region of interest at a frame rate of 60 Hz and particle trajectories are obtained by comparing the particle centers in consecutive images. The particle trajectory data is further analyzed to obtain the mean and fluctuating properties of the flow. The images of the particle flow in the $x-y$ plane were obtained by using the laser orientation shown in Fig. 1a. The images in the $x-z$ plane were obtained by positioning the laser sheet orthogonal to the $x-y$ plane.

The mean flow velocity measured in experiments as shown in the next section, has a maximum of approximately $0.6d$ s^{-1} [12]. For these values of velocities, the Reynolds number is about 10^{-2} and the ratio of viscous drag on the particle to the gravitational force is estimated to be less that 10^{-2}. For the amount of shear observed, the maximum strain rate is about 0.25s^{-1} which is comparable in magnitude to that in experiments with a plate dragged on a granular bed [30], where the drag friction experienced by the plate was measured to be unchanged when the index matching

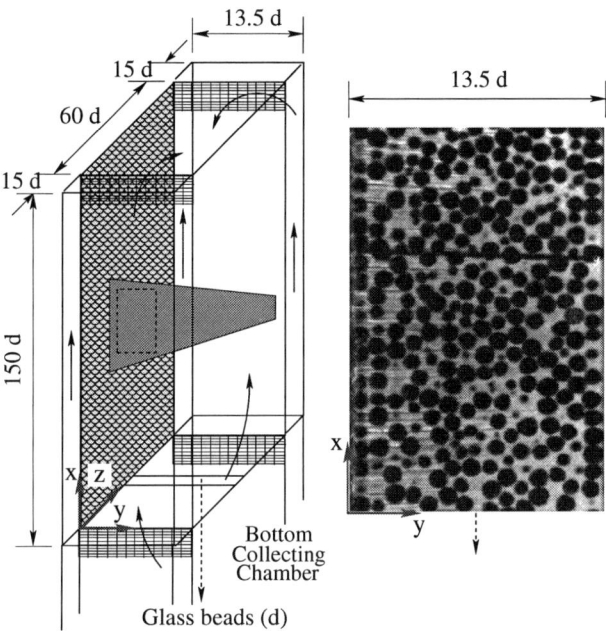

FIGURE 1. (a) Schematic diagram of the shear cell. Blue colored arrows show the liquid distribution in and out of the side chambers. (b) A typical image of particles observed in a thin slice in the shear plane obtained with a fluorescent particle index matching technique. The co-ordinate system used is also shown.

liquid was used. Based on these observations we anticipate that the interstitial liquid does not affect the motion of the particles and so neglect its presence in the simulation model.

SIMULATION MODEL

In the simulations, we define lengths in terms of a particle diameter d, and we define a natural mass unit m. Particles have unit density, and thus the mass of a particle is $m_p = 4\pi(0.5)^3/3 = 0.524m$. The results from simulations are expressed in terms of a natural time unit $\tau = \sqrt{d/g}$. For any two particles separated by $\mathbf{r}$, and which are in compression such that $\delta = d - |\mathbf{r}| > 0$, where δ is the overlap distance, they experience a force $\mathbf{F} = \mathbf{F}_n + \mathbf{F}_t$ where the normal and tangential components are given by

$$\mathbf{F}_n = f(\delta/d)\left(k_n \delta \mathbf{n} - \frac{\gamma_n \mathbf{v}_n}{2}\right) \qquad (1)$$

$$\mathbf{F}_t = f(\delta/d)\left(-k_t \Delta \mathbf{s}_t - \frac{\gamma_t \mathbf{v}_t}{2}\right) \qquad (2)$$

Here, $\mathbf{n} = \mathbf{r}/|\mathbf{r}|$. $\mathbf{v}_n$ and $\mathbf{v}_t$ are the normal and tangential components of the relative surface velocity, and $k_{n,t}$ and $\gamma_{n,t}$ are the elastic and viscous constants respectively of the viscoelastic force model. We primarily consider the force model for Hookean contacts wherein $f(\zeta) = 1$. The normal damping term is set to $\gamma_n = 50\sqrt{(g/d)}$ and the tangential term is set to zero. $\Delta \mathbf{s}_t$ is the elastic tangential displacement between spheres, obtained by integrating tangential relative velocities during elastic deformation for the lifetime of the contact. If $|\mathbf{F}_t| > \mu|\mathbf{F}_n|$, so that a local Coulomb yield criterion is exceeded, then $\mathbf{F}_t$ is rescaled so that it has magnitude $\mu|\mathbf{F}_n|$ and $\Delta \mathbf{s}_t$ is modified so that equation 2 is upheld. The reader is referred to our earlier work [50] and Silbert [36] for details of the above parameters and their underlying assumptions. These parameters have been established for small system sizes (typically 24,000 particles) and we test their validity for a much larger system size (147,000 particles) used in our simulations. The biggest assumption of the model is the choice of the normal elastic constant, which is set to $k_n = 2 \times 10^5 mg/d$. As noted in the original calibration and in subsequent studies [36, 41, 43], this is significantly lower than what would be realistic for typical hard materials such as glass, where $k_n = O(10^{10}mg/d)$ would be more reasonable. However, such a constant would be

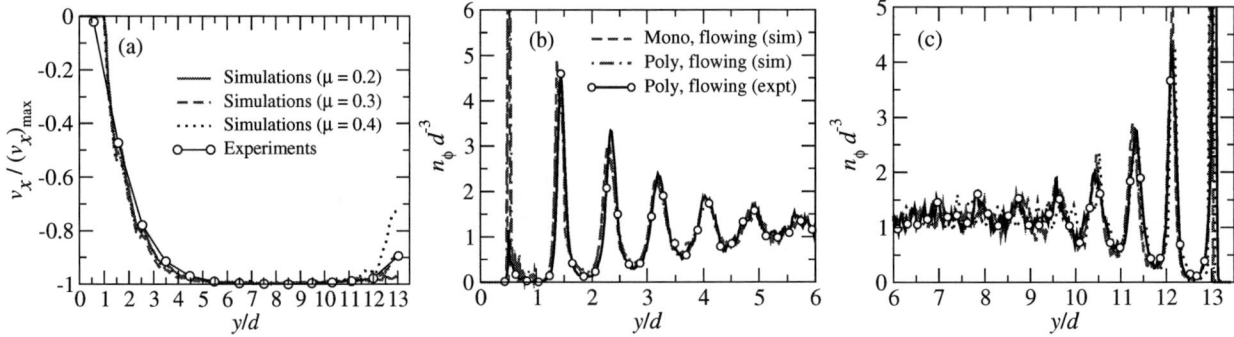

FIGURE 2. (a) Vertical component of downward velocity v_x as a function of y co-ordinate. (b) & (c) Number density (number of particles per unit volume) variation as a function of y co-ordinate. Simulation results in (b) and (c) are obtained for a friction coefficient $\mu = 0.3$. Open circles represent experimental results while lines represent simulation results. Number density for simulations is shown for flowing and static particles and for mono- and poly-dispersed particles.

prohibitively computationally expensive, since the time step required must have the form $\delta t \propto k_n^{-1/2}$. Both Silbert [36] and Landry [41] discuss that the chosen value of k_n is a reasonable compromise, which is small enough to feasibly simulate, but large enough to avoid the system exhibiting excessive elastic effects. Here, we make use of $\delta t = 10^{-4}\tau$. Further details of the simulation regarding the creation of initial packings, creation of the rough wall as in experiments and the details of the drainage process can be found here [50].

RESULTS AND DISCUSSION

We first discuss the mean flow behavior in sheared region while making a systematic comparison between experiments and simulations. Following that we present a more detailed discussion about the particle diffusion behavior as exhibited in the sheared region. Discussion about some of the less important results have been omitted due to constraints on the length of the manuscript and the readers are referred to the published work [50] for more details.

All the data in experiments and simulations have been obtained in a test region ($0d \leq y \leq 13.5d$, $20d \leq z \leq 40d$, $80d \leq x \leq 100d$) which is high above the orifice to avoid any exit effects and also away from the side walls to avoid any wall effects. The experimental results are averaged over ten different runs with the standard deviation over these runs being less than the symbol size shown in Fig. 2a. The simulation results are obtained over a time window ($300\tau \leq t \leq 900\tau$) in which the flow is steady. In both, experiments and simulations, plotted results represent time windows which avoid free surface effects as the particles move downwards through the test region.

Structure and Mean flow

The mean flow as seen from Fig. 2a consists of a shear layer localized to within $4-5$ particle diameters from the rough wall, while the remaining region moves as a plug with a slight slip near the smooth wall. The experimental and simulation profiles match very closely over most of the flow region after normalization with the maximum velocity. The simulation results show a very weak dependence on friction coefficient which is in agreement with previous results carried out for much wider ranges of μ [46]. The differences exhibited by simulation profiles near the smooth wall suggests potentially subtle effects in particle dynamics, arising out of particle sliding or rolling against the smooth wall and was found very difficult to match with the experiments. However, the profiles match very closely in the shearing region which is of primary interest over here and so in subsequent discussions we concentrate on $\mu = 0.3$, as using $\mu = 0.2$ and 0.4 had negligible effect on the presented results.

The flow near the walls, occurs in layers moving past one another. The particle number density plotted in Fig. 2b,c show the formation and positions of the layers. The peaks in profile correspond to the layers of particles. Particle size polydispersity does not seem to affect the density profile as seen from the simulation results which are in very close agreement with the experimental profiles obtained for polydisperse particle size distribution. This behavior is a bit surprising as it can be expected that the peaks would have been smeared due to presence of polydisperse

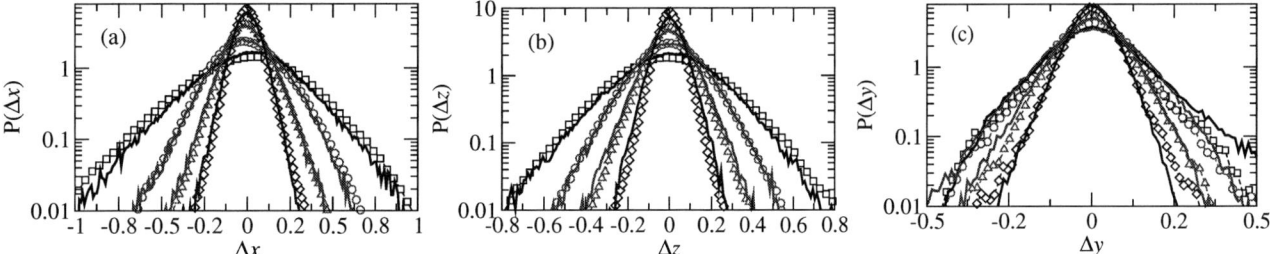

FIGURE 3. Distribution of components of particle displacements in the three directions. The displacements are calculated using a time interval corresponding to a mean distance drop of $1d$. Symbols represent experimental data while lines of same color represent simulation results. Layers 1 to 4 are shown with black squares, red circles, green triangles and blue diamonds respectively.

particle size distribution. Similar lack of dependence on the particle size polydispersity was also observed for vertical velocity profiles and many other measurements (not shown). To probe the structure further, we also computed the two dimensional order parameter $q_6 = <\exp(i6\theta_p)>$ from the experimental data, where $<.>$ is the average over all the nearest neighbor bonds and θ_p is the angle between the particle bonds. For disordered packing, $q_6 = 0$. Evaluating q_6 for the experimental data in the $x-z$ plane, we find that q_6 decreases from about 0.075 near the side walls to less than 0.03, a few grain diameters away from the side wall, indicating disordered packing. q_6 in the $x-y$ plane is observed to be ~ 0.1 but still significantly smaller than $q_6 = 1$ which is typically observed for a hexagonal packing.

The temporal variations in the mean velocity components were examined next. In the experiments, the velocity components in all the three directions obtained using lowest possible time interval showed a lot of fluctuations which appear to be consistent with the random noise. Indeed, averaging over longer time intervals resulted in a near constant velocity. The results showed no evidence of any kind of oscillations or correlation between layers for any time intervals. The simulation results, however, showed significant unsteady oscillations only in v_x which were found to be correlated between the layers. The velocity magnitude varied over a large range of timescale τ and so any suitable time averaging was not able to eliminate these oscillations. A change in the integration time step by a factor of 2 was not able to decrease the magnitude of these oscillations. The absence of these oscillations in experiments suggests that they are most likely attributed to the approximations made in the simulation model, such as the value of the normal spring constant being smaller than the realistic one. To examine this, simulations were carried out for an order of magnitude higher spring constant ($k_n = 2 \times 10^6 mg/s$). This resulted in the waves which were much smaller and the variations in the velocity occurred over much faster time scale τ so that through suitable time averaging the variations could be smoothened out significantly. Complete elimination of these waves would, however, need a much higher spring constant which is not possible with the current numerical capabilities. A more detailed discussion of this oscillatory behavior in simulations can be found here [50]. All the particle diffusion results presented in the next section have been obtained by appropriate subtraction of these temporal and spatial variations in the mean velocity.

Particle diffusion

In this section, we examine particle rearrangements and diffusion in the experiments and simulations. While in previous section, we found that the overall flow characteristics obtained from simulations and experiments are in good agreement, over here we would like to explore whether this agreement persists for microscopic behavior. The simulation results shown in subsequent paragraphs were obtained for monodisperse particles with friction coefficient (μ) of 0.3 and spring constant (k_n) of $2 \times 10^5 mg/d$ unless stated otherwise.

The particle diffusion in a slow, dense granular flow arises out of particle rearrangements caused due to outside forcing, in this instance the gravity induced shear. Experimental measurements of the diffusion behavior were carried out in a rectangular silo by Choi et al. [10], but with measurements near the wall. Their results pointed to a rate-independence in granular mixing, allowing time to be scaled out, and mean squared displacements to be expressed in terms of total deformation only. Quasi-2D measurements in a rectangular silo by Natarajan et al. [8] showed that the self-diffusion coefficients in the flow direction were an order of magnitude higher than those in the direction transverse to the flow. Utter et al. [7] measured self-diffusion in dense, sheared granular flows in a two dimensional Couette assembly. They showed that the measured diffusivities are proportional to the local shear rate with diffusivities along the direction of mean flow approximately twice as large as those in perpendicular direction. In three dimensions

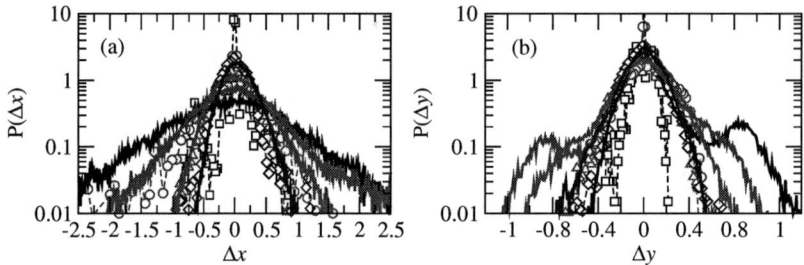

FIGURE 4. Distribution of displacements in x and y direction calculated for a time interval corresponding to a mean distance drop of $8d$. Symbols represent experimental data while lines of same color represent simulation results. Layers 1 to 4 are shown with black squares, red circles, green triangles and blue diamonds respectively.

Campbell [52] has considered particle diffusion in a cell with Lees-Edward boundary conditions, although for a smaller system and much rapid flow. Very recently, Orpe *et al.* [12] measured diffusion behavior in three dimensions in gravity induced shear and showed that the velocity auto-correlations in dense sheared granular flows decay faster than that for elastic fluids at equilibrium.

We first compute the probability distribution functions (PDFs) of particle displacements in three co-ordinate directions which arise out of particle fluctuations. These particle displacements were obtained by suitable subtraction of the mean component of the flow. Details about accurate accounting for the spatial and temporal variations in the mean velocity while computing particle fluctuations can be found elsewhere [50]. Consider first the measurements in the $x-z$ plane made for each flowing layer of particles. Within each layer, the displacements are obtained from successive positions of a particle over a time interval corresponding to a mean distance drop of $1d$, where d is the particle diameter. The distributions of these displacements obtained from all the tracked positions of all the particles is shown in Fig. 3a and b for first four layers from the rough shearing surface. A very good agreement is seen between the experiments and simulations. The fluctuations increase for the layers closer to the wall as expected. For a normal diffusive process, the curves would be expected to appear quadratic on a semi-log plot, but here we see a slower decay and larger tails corresponding to ballistic motion on a sub-particle length scale. Also visible in Fig. 3a is a small skewness with more large steps in the negative x direction than in the positive x direction. This anisotropy is caused by gravity: when a gap opens up in the particle packing, it is possible that particles will fall downward to fill it resulting in a large downward displacement even after the mean velocity is subtracted.

To obtain particle displacements in the y direction we make use of measurements in the $x-y$ plane. In the experimental measurements, trajectories of the tracked particles are limited to within a layer $1d$ wide due to limitations in tracking the particles traversing across the layers. Thus, any large displacements encountered by a particle while moving between the layer are discounted. No such limitation exists in simulation and particles moving across the layers in y direction are tracked very accurately. Fig. 3c shows the comparison of the y displacements PDFx in the four layers between experiments and simulations obtained for mean distance drop of $1d$. A very good agreement is obtained. However, significant differences are observed for the distributions of displacements obtained over a time interval corresponding to a mean distance drop of of $8d$ as shown in Fig. 4a,b. The PDFs in x direction for experiments are much noisier compared to those in simulations. This is due to insufficient data for averaging caused due to the inability in tracking particles over longer times which can move in and out of the viewing plane and also between the layers. In Fig. 4b, while the central peaks for the PDFs in y direction are similar in size, additional peaks can be seen in the simulation data corresponding to the particles that have moved between the layers over much longer times. These displacements of particles over very long times is not accessible through experimental measurements. Such complicated behavior makes it hard to assign a meaningful diffusion constant in y direction. On the length scales that can be observed, the PDFs do not appear to tend anything resembling a Gaussian, and particle motion is a combination of stochastic behavior and layer confinement.

Fig. 5 shows a logarithmic plot of the mean squared displacements as a function of mean distance dropped for three coordinate directions and four layers. The x and z plots are based on measurements in the $x-z$ plane and we see quite good agreement in all the four layers. As noted by Choi *et al.* [10], a transition from superdiffusive behavior characterized by slopes greater than one, to a normal diffusion with slopes closer to one can be seen at a mean distance drop of approximately $1d$. A slight disagreement is seen in the z direction, particularly for layers 1 and 2 for very large displacements (*ie.* more than $4d$), wherein we see the experimental mean squared displacements to be slightly larger than in simulations. We believe that this bias may be introduced perhaps due to loss of particles which have moved

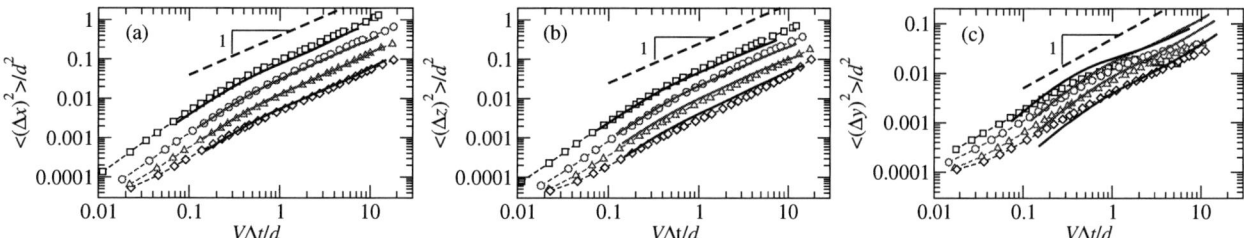

FIGURE 5. Mean squared displacements for the three directions plotted as a function of mean distance drop. Symbols represent experimental data while lines of same color represent simulation results. Layers 1 to 4 are shown with black squares, red circles, green triangles and blue diamonds respectively. Here V represents the mean velocity within each layer.

between the layers over long displacements since the tracking was done only within one layer. The effect is not seen for layers 3 and 4 wherein the particle hopping between the layers is almost negligible. These results probably state the lack of reliability in experiments to determine fluctuations over very long time scales and displacements, thus stressing the importance of simulation data over these time scales. The y plot is based on the measurements in the $x-y$ plane. We see good agreement for small distances but the curves begin to diverge for larger distances, due to the simulation methods counting those particles moving between the layers. The plot for layer 1 exhibits fundamentally different behavior since particles can only jump to layers in one direction as opposed to both. The curves do not exhibit slopes close to one, confirming that a diffusion constant cannot be meaningfully defined.

The diffusion constants were obtained only for x and z direction from the mean squared displacements shown in Fig. 5a,b. The shear rates were obtained numerically from the mean velocity profile shown in Fig. 2a. The diffusion constants are observed to vary roughly linearly with the shear rate, though the relationship cannot be said to be conclusive owing to only four data points for the four shear rates corresponding to the four layers. The diffusion constant in x direction is found to be greater than that in the z direction. Such anisotropy has been anticipated in sheared athermal suspensions [13] and also has been observed in two-dimensional granular experiments [7, 8]. The anisotropy is observed in both experimental and simulation data which points to the importance of particle geometry and local packing in determining local rearrangement and diffusion of particles rather than the details of interactions between particles; experiments were performed in presence of interstitial liquid and simulations in the absence of it. Some differences in the fluctuations of particles in experiments and simulations is observed which seem to arise owing to a systematic loss of particles in experiments due to limitation in tracking particles over long times rather than due to physical differences.

All the above results for particle diffusion from simulations were re-calculated for higher spring constant ($k_n = 2 \times 10^6 mg/d$) to determine the effect of wave-like behavior in the mean velocity mentioned in the previous section. However, we did not find appreciable differences between the results as long as the temporal variations in the mean velocity arising out of wave-like behavior were accounted accurately. The simulation data for particle diffusion obtained from both values of the spring constant agreed very well with the experiments.

Another quantity which describes the diffusion behavior of particles is the velocity autocorrelation function for any time and is based upon finding collection of velocity pairs separated by time t and then computing the product-moment correlation coefficient. We computed the three components of the velocity autocorrelation function from experiments and simulations. The methodology to subtract the spatial and temporal mean drift was the same as used while computing PDFs of displacements. The velocities for computing correlation function were obtained for a time interval corresponding to a mean distance drop of $0.01d$.

The auto correlations for the simulations were first obtained using the spring constant of $k_n = 2 \times 10^5 mg/d$. The correlations exhibited chaotic oscillations at large times which appears unphysical since after undergoing several collisions over long times, the velocities should be uncorrelated. The reason for this behavior is possibly due to the velocity waves mentioned in the earlier section. The correlations were then computed using a higher spring constant $k_n = 2 \times 10^6 mg/d$. The results showed significant improvement. Though some noise was still visible, the correlations do decay to zero at long times. Such sensitive dependence on spring constant was not observed for other diffusion measurements mentioned above. The reason for this may be because autocorrelations involve calculation of velocities from the difference of two positions which is more sensitive to velocity waves than diffusion measurements. No chaotic oscillations were observed in the correlation function obtained experimentally. We obtained very good agreement between experiments and simulation data (not shown here) particularly for the first and the second layer where the velocity waves had lesser effect. The initial decay of correlations was found to be closer to an exponential decay observed by Orpe *et al.* [12] as opposed to a long time tail that has been observed for the unsheared granular systems [31]. Further details about the exact form of decay of correlation was not possible based on the current levels of accuracy in the experimental and simulation technique.

SUMMARY

To summarize, we have shown very good quantitative agreement between the index-matched experiments and DEM simulations employing the Cundall-Strack granular contact model with typical values of the contact parameters used in several granular flow problems by different research groups. The close agreement between various macroscopic and microscopic flow features provides validation that both techniques can be used reliably to study granular flows, albeit both having their limitations: interstitial fluid effects at high flow rates in experiments and approximations in the simulation contact model. The observed successful matching can be partially attributed to the fact that the contact models employed in discrete-element simulation reproduce the physics in the index-matched flow experiments very closely. At the same time the results indicate that many key features of slow, dense granular flows exhibit some degree of universality across various situations. The results show that velocity profiles are similar for friction coefficients varying from 0.2 to 0.4. A small amount of polydispersity appears to have minimal effect on the velocity profiles and packing structure.

Despite the success, the results show potential areas of concern. In experiments, the measurements are made within a thin laser sheet and so a complete three dimensional map is not available which prevents us from measuring several properties. Further, very long time dynamics are not possible to measure due to lack of information of particles which move in and out of the laser sheet. In simulations, the biggest concerning factor was the presence of velocity waves. Our results indicate that increasing the normal contact stiffness by a factor of ten results in the waves which are smaller in magnitude and the mean velocity fluctuations which occur at much higher frequencies. While this did not result in complete elimination of the waves, it allowed for significant smoothening of the velocity variations through suitable time-averaging. The waves appear to have no analog in experiments. At the same time, we note that our study does not provide evidence to show that these waves are completely unphysical in all situations like granular material made up of softer composition (e.g. acrylic glass) wherein such waves can be observed and can be an interesting direction for further study.

ACKNOWLEDGMENTS

This work was supported by the Director, Office of Science, Computational and Technology Research, U.S. Department of Energy under Contract Nos. DE-AC02-05CH11231 and the National Science Foundation under grants DMS-0410110 and DMS-070590, Clark University under Grant No. DMR-0605664 and National Chemical Laboratory under Institute Start-Up Grant No. MLP013626. We are also grateful to the Scientific Cluster Support (SCS) program at the Lawrence Berkeley Laboratory. Ashish V. Orpe expresses his gratitude to the IUTAM-ISIMM symposium organizers for the invitation to give a lecture and also for the financial support provided for the travel.

REFERENCES

1. J. T. Jenkins, and S. B. Savage, *J. Fluid Mech.* **130**, 187–202 (1983).
2. H. M. Jaeger, and S. R. Nagel, *Science* **255**, 1523–1531 (1992).
3. H. M. Jaeger, S. R. Nagel, and R. P. Behringer, *Rev. Mod. Phys.* **68**, 1259–1273 (1996).
4. T. Halsey, and A. Mehta, editors, *Challenges in Granular Physics*, World Scientific, 2002.
5. P. G. de Gennes, *Rev. Mod. Phys.* **71**, S374–S382 (1999).
6. I. S. Aranson, and L. S. Tsimring, *Rev. Mod. Phys.* **78**, 641–692 (2006).
7. B. Utter, and R. P. Behringer, *Phys. Rev. E* **69**, 031308 (2004).
8. V. V. R. Natarajan, M. L. Hunt, and E. D. Taylor, *J. Fluid Mech.* **304**, 1 (1995).
9. K. M. Hill, G. Gioia, and V. V. Tota, *Phys. Rev. Lett.* **91**, 064302 (2003).
10. J. Choi, A. Kudrolli, R. R. Rosales, and M. Z. Bazant, *Phys. Rev. Lett.* **92**, 174301 (2004).
11. O. Pouliquen, M. Belzons, and M. Nicolas, *Phys. Rev. Lett.* **91**, 014301 (2003).
12. A. Orpe, V. Kumaran, A. Reddy, and A. Kudrolli, *Europhys. Lett.* **84**, 64003 (2008).

13. D. R. Foss, and J. F. Brady, *J. Fluid Mech.* **401**, 243–274 (2006).
14. V. Breedveld, D. van den Ende, M. Bosscher, R. J. J. Jongschaap, and J. Mellema, *J. Chem. Phys.* **116**, 10529 (2002).
15. M. Nakagawa, S. A. Altobelli, A. Caprihan, E. Fukushima, and E. K. Jeong, *Exp. Fluids* **16**, 54 (1993).
16. K. Yamane, M. Nakagawa, S. A. Altobelli, T. Tanaka, and Y. Tsuji, *Phys. Fluids* **10**, 1419 (1998).
17. L. Sanfratello, A. Caprihan, and E. Fukushima, *Gran. Mat.* **9**, 1 (2007).
18. K. Sakaie, D. Fenistein, T. J. Caroll, M. van Hecke, and P. Umbanhowar, *Europhys. Lett.* **84**, 38001 (2008).
19. D. J. Parker, C. J. Broadbent, P. Fowles, M. R. Hawkesworth, and P. A. McNeil, *Nucl. Instrum. Meth.* **236**, 592 (1993).
20. Y. L. Ding, J. P. K. Seville, R. Forster, and D. J. Parker, *Chem. Eng. Sci.* **56**, 1769 (2001).
21. A. Ingram, J. P. K. Seville, D. J. Parker, X. Fan, and R. G. Forster, *Powder Technol.* **158**, 76 (2005).
22. A. Sakellariou, T. J. Sawkins, T. J. Senden, and A. Limaye, *Physica A* **339**, 152 (2004).
23. D. M. Mueth, G. F. Debregeas, G. S. Karczmar, P. J. Eng, S. R. Nagel, and H. M. Jaeger, *Nature* **406**, 385 (2000).
24. T. Aste, M. Saadatfar, A. Sakellariou, and T. J. Senden, *Physica A* **339**, 16 (2004).
25. T. Aste, *Phys. Rev. Lett.* **96**, 018002 (2006).
26. N. Jain, D. V. Khakhar, R. M. Lueptow, and J. M. Ottino, *Phys. Rev. Lett.* **86**, 3771 (2001).
27. J.-C. Tsai, G. A. Voth, and J. P. Gollub, *Phys. Rev. Lett.* **91**, 064301 (2003).
28. S. J. Fiedor, and J. M. Ottino, *Phys. Rev. Lett.* **91**, 244301 (2003).
29. J.-C. Tsai, and J. P. Gollub, *Phys. Rev. E* **70**, 031303 (2004).
30. S. Siavoshi, A. V. Orpe, and A. Kudrolli, *Phys. Rev. E* **73**, 010301(R) (2006).
31. A. Orpe, and A. Kudrolli, *Phys. Rev. Lett.* **98**, 238001 (2007).
32. S. Slotterback, M. Toiya, L. Goff, J. F. Douglas, and W. Losert, *Phys. Rev. Lett.* **101**, 258001 (2008).
33. C. J. Koh, P. Hookham, and L. G. Leal, *J. Fluid Mech.* **266**, 1 (1994).
34. P. A. Cundall, "A computer model for simulating progressive large scale movements in blocky rock systems," in *Proc. Symp. Int. Soc. for Rock Mechanics, Nancy, France*, 1971, pp. 129–136.
35. P. A. Cundall, and O. D. L. Strack, *Geotechnique* **29**, 47 (1979).
36. L. E. Silbert, D. Ertaş, G. S. Grest, T. C. Halsey, D. Levine, and S. J. Plimpton, *Phys. Rev. E* **64**, 051302 (2001).
37. L. E. Silbert, J. W. Landry, and G. S. Grest, *Phys. Fluids* **15**, 1 (2003).
38. R. C. Brewster, J. W. Landry, G. S. Grest, and A. J. Levine, *Phys. Rev. E* **72**, 061301 (2005).
39. L. E. Silbert, D. Ertaş, G. S. Grest, T. C. Halsey, and D. Levine, *Phys. Rev. E* **65**, 031304 (2002).
40. L. E. Silbert, G. S. Grest, and J. W. Landry, *Phys. Rev. E* **66**, 061303 (2002).
41. J. W. Landry, G. S. Grest, L. E. Silbert, and S. J. Plimpton, *Phys. Rev. E* **67**, 041303 (2003).
42. H. P. Zhu, and A. B. Yu, *J. Phys. D* **37**, 1497 (2004).
43. C. H. Rycroft, M. Z. Bazant, G. S. Grest, and J. W. Landry, *Phys. Rev. E* **73**, 051306 (2006).
44. C. H. Rycroft, G. S. Grest, J. W. Landry, and M. Z. Bazant, *Phys. Rev. E* **74**, 021306 (2006).
45. X. Cheng, J. B. Lechman, A. Fernandez-Barbero, G. S. Grest, H. M. Jaeger, G. S. Karczmar, M. E. Mobius, and S. R. Nagel, *Phys. Rev. Lett.* **96**, 038001 (2006).
46. K. Kamrin, C. H. Rycroft, and M. Z. Bazant, *Modelling Simul. Mater. Sci. Eng.* **15**, S449–S464 (2007).
47. M. Depken, J. B. Lechman, M. van Hecke, W. van Saarloos, and G. S. Grest, *Europhys. Lett.* **78**, 58001 (2007).
48. C. H. Rycroft, K. Kamrin, and M. Z. Bazant (2009), (accepted in J. Mech. Phys. Solids).
49. J. Sun, F. Battaglia, and S. Subramaniam, *Phys. Rev. E* **74**, 061307 (2006).
50. C. H. Rycroft, A. V. Orpe, and A. A. Kudrolli, *Phys. Rev. E* **80**, 031305 (2009).
51. J. C. Crocker, and D. G. Grier, *Journal of Colloid and Interface Science* **179**, 298–310 (1996).
52. C. S. Campbell, *J. Fluid Mech.* **348**, 85 (1997).

Equilibrium of granular clusters: influence of boundary curvature and contact properties

Rossella Pignatelli

Department of Structural Engineering, Politecnico di Milano
Piazza Leonardo da Vinci, 32, I-20133 Milan, Italy
`pignatelli@stru.polimi.it`

Abstract. The total stiffness matrix of a heap of packed grains is determined through the analysis of grain-to-grain contact in incremental deformation regime. The obtained stiffness takes into account possible frictional sliding, relative rotation, contact deformability and curvature of grain boundaries. The spectral analysis of the stiffness matrix permits to evaluate the stability of the heap: examples on two specific grain clusters are presented.

Keywords: Grain boundary curvature, tangential stiffness, granular matter
PACS: 45.70.-n

INTRODUCTION

The equilibrium of dry and dense heaps of grains with arbitrary shape and dimension is analysed here. A discrete approach is followed. The aim is to determine a global stiffness matrix of the heap taking into account the curvature of the grain boundary at the contact points and the type of contact. Transmission of torques between pairs of granules in contact is considered. The determination of a global stiffness matrix for a granular heap permits stability analyses performed by varying parameters involved in that matrix. It furnishes information for establishing continuum models in which constitutive relations are given through homogenization of discrete schemes.

The analysis developed here is of incremental type. Perturbation of the external loads is prescribed in such a way that they induce just infinitesimal displacements. Inertial effects are neglected. Both translational and rotational degrees of freedom of the granules are involved. Infinitesimal deformation of the grains at the contact points is included in the analyses. For each contact only the resultant of the local stress distribution is considered and is applied at a single point called the *contact point*: it is identified by the virtual intersection of the original surfaces of grains before the contact. Hertz's law is used and the model by Mindlin and Deresiewicz [1] characterizes the contact. Frictional sliding and cohesion are accounted for. Tangential stiffness of granular heaps is determined in [2] with the method used here. However, in contrast with the model proposed in [2], friction, cohesion, rotational stiffness and rotational friction are involved in the present analysis.

Numerical analyses of eigenvalues and eigenvectors of the stiffness matrix of special grain clusters are developed. They permit to evaluate the influence on the total heap stiffness of type of contact, curvature of grain boundary, grain characteristic dimension, contact forces already present in the actual state, rolling, sliding and contact deformation. Analyses of the influence of particle shape and force distribution on the stability of grain clusters have been developed in other works (see e.g. [3, 4], where the simulations are based on discrete element method (DEM)) from other points of view. Various are, in fact, the paths that can be followed in describing the mechanical behavior of granular matter (see e.g. [5, 6]).

The general treatment is specified in two simple cases. They permit to clarify further the physical meaning of the steps followed in developing the proposal presented here. So, although they are simple, in this sense they are not simplistic.

The results collected here can be useful in principle for homogenization purposes (the related procedures are a non trivial task above all when referred to granular matter - this aspect has been discussed variously in literature, see e.g. [7, 8, 9, 10, 11, 12, 13]).

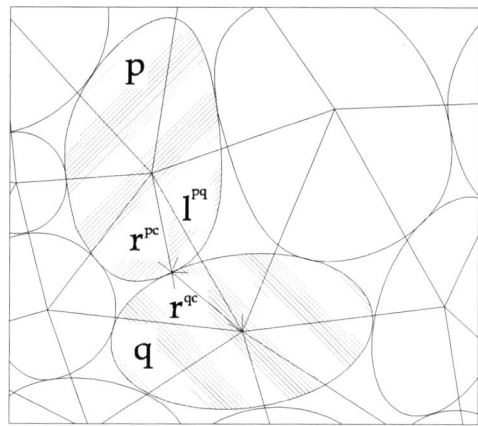

FIGURE 1. The grains p and q; the contact vectors $\mathbf{r}^{pc}$ and $\mathbf{r}^{qc}$; the branch vector $\mathbf{l}^{pq}$.

GEOMETRY AND KINEMATICS

A rather detailed representation of granular assemblies can be obtained by using graphs with nodes at the grain barycentres and edges which represent nearest-neighbors and define contacts (see [14, 15]), as shown in Figure 1. Consider a generic couple of grains p and q in contact at point $\mathbf{x}^c$. For the sake of simplicity, hereafter the contact point $\mathbf{x}^c$ will be often indicated only by c. $\mathbf{x}^p$ and $\mathbf{x}^q$ denote the grain barycentres, the *contact vector* $\mathbf{r}^{pc}$ connects the centroid of the grain p and the point c, the so-called *branch vector* $\mathbf{l}^{pq}$ links the centroids of neighboring grains p and q [15]:

$$\mathbf{r}^{pc} = \mathbf{x}^p - \mathbf{x}^c, \; \mathbf{l}^{pq} = \mathbf{x}^p - \mathbf{x}^q = \mathbf{r}^{pc} - \mathbf{r}^{qc} = -\mathbf{l}^{qp}. \qquad (1)$$

When the contact is not punctual, the barycentre of the contact area can be taken as representative point. Both translational and rotational degrees of freedom are associated with each grain barycentre, the latter ones describing the global rotation of the grain. A related infinitesimal 'extended' displacement vector $d\mathbf{u}^p$, pertaining to the barycentre of the generic grain p, can be then defined in terms of the infinitesimal displacements $d\mathbf{w}^p$ of the barycentre itself and the infinitesimal rigid-body rotation $d\varphi_i^p$ of the grain itself. The infinitesimal 'extended' displacement vectors of all grains can be then collected in the *global infinitesimal extended displacement vector* $d\mathbf{u}$ of the entire assembly. The relevant expressions, in components, are given by

$$[d\mathbf{u}^p] := \begin{bmatrix} dw_1^p \\ dw_2^p \\ dw_3^p \\ d\varphi_1^p \\ d\varphi_2^p \\ d\varphi_3^p \end{bmatrix}, \; [d\mathbf{u}] := \begin{bmatrix} d\mathbf{u}^1 \\ d\mathbf{u}^2 \\ \vdots \\ d\mathbf{u}^N \end{bmatrix}. \qquad (2)$$

Infinitesimal deformation of the grains at the contact regions can be included in the present analysis in an approximate way. Without essential loss of generality, by assuming that the contact area between generic pairs of granules is small, it is possible to approximate locally the shape of the contact region *before* the contact by a spherical surface. *After* the contact, by considering the contact area as planar, its barycentre c can be always taken as representative of the contact itself. The point c, denoted as *contact point*, is determined by the virtual intersection of the original boundaries of the two grains before the deformation (see Figure 2 for the 2D case, where the overlap is accentuated and c is the middle point of the segment connecting the two surface points where the normals to the boundary are aligned along the same direction)[2]. A vector $\mathbf{n}^c$ represents the normal to the contact region at c. For grains p and q, it is assumed that the orientation of $\mathbf{n}^c$ is outward the boundary of the grain p when $p < q$ in the indexing chosen for the whole assembly. The choice of $\mathbf{n}^c$ implies the assumption that the contact surface is sufficiently smooth at least in a neighborhood of c. At each contact point a natural local frame is defined by the normal $\mathbf{n}^c$ and the two vectors $\mathbf{t}^c$ and $\mathbf{z}^c$, of unit length, belonging to the plane orthogonal to $\mathbf{n}^c$ at c. If a global frame (x_1, x_2, x_3) is selected, transformation matrices $\mathbf{T}^{pc}$ link the local frame at the contact point c of the grain p with the global one. The (symmetric) curvature tensor at c of the

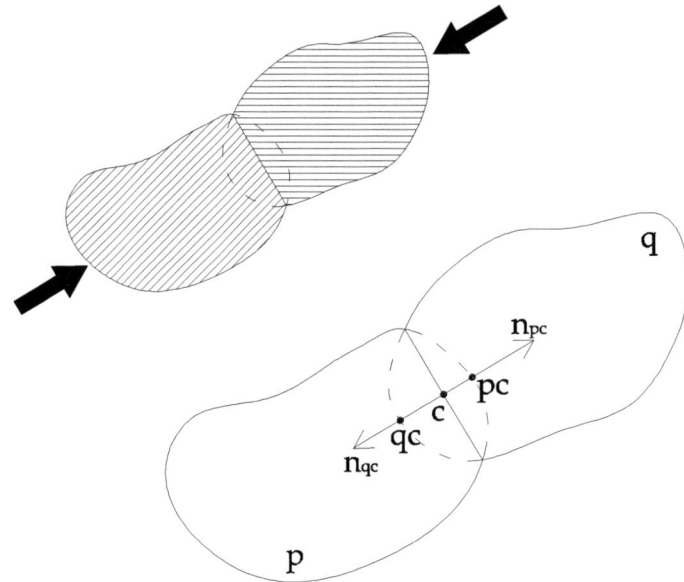

FIGURE 2. Definition of the contact point amid grains

boundary of the grain p is indicated by $\hat{\mathbf{C}}^{pc}$ and is expressed in the local frame (n^c, t^c, z^c), as suggested in [2]. In the global frame it becomes

$$\mathbf{C}^{pc} = \mathbf{T}^{pc} \hat{\mathbf{C}}^{pc} \mathbf{T}^{pcT}. \tag{3}$$

KINEMATICAL EFFECTS DUE TO PARTICLE ROLLING AND CONTACT DEFORMATION

Contact deformation

The contact deformation is described here in an approximate way. Essentially, one imagines a fictitious (non-physical) rigid-body motion of the grain, allowing grains overlap that defines the contact region through the imaginary intersection. Figure 2 pictures what has been just described: the (non-physical) virtual interpenetration of the granules is described by the dashed lines, the solid line in the middle defines the physical contact region with barycentre c, the point considered as representative of the entire region, as pointed out previously. The point c can be considered as belonging to the grain p and the grain q. As depicted in Figure 2 c itself can be considered as the dual image of two points belonging to the boundary of the grains p and q when the grains were in contact only at a point. These two points at the boundaries are indicated by pc and qc respectively. The infinitesimal displacement $d\hat{\mathbf{v}}^{c,def}$ of the point c can be written as the difference between the infinitesimal displacements of the points pc and qc, denoted respectively as $d\mathbf{u}^{pc}$ and $d\mathbf{u}^{qc}$. In the local frame of the contact point it can be expressed by writing

$$d\hat{\mathbf{v}}^{c,def} := (\hat{\mathbf{T}}^{qcT} d\mathbf{u}^{qc} - \hat{\mathbf{T}}^{pcT} d\mathbf{u}^{pc}), \tag{4}$$

where $\hat{\mathbf{T}}^{pcT}$ and $\hat{\mathbf{T}}^{qcT}$ are *transformation matrices* between local and global frames, written in a way that allows the representation of the rotations. The displacement vectors $d\mathbf{u}^{pc}$ and $d\mathbf{u}^{qc}$ of the contact points pc and qc can be expressed in terms of the displacements of the grain barycentres $d\mathbf{u}^p$ and $d\mathbf{u}^q$, then the infinitesimal displacement

$d\hat{\mathbf{v}}^{c,def}$ assumes the form

$$d\hat{\mathbf{v}}^{c,def} := (\hat{\mathbf{T}}^{qcT}\hat{\mathbf{B}}^{qcT}d\mathbf{u}^q - \hat{\mathbf{T}}^{pcT}\hat{\mathbf{B}}^{pcT}d\mathbf{u}^p) = \left(\hat{\mathbf{T}}^{qcT}\hat{\mathbf{B}}^{qcT}\begin{bmatrix}dw_1^q\\dw_2^q\\dw_3^q\\d\varphi_n^q\\d\varphi_t^q\\d\varphi_w^q\end{bmatrix} - \hat{\mathbf{T}}^{pcT}\hat{\mathbf{B}}^{pcT}\begin{bmatrix}dw_1^p\\dw_2^p\\dw_3^p\\d\varphi_n^p\\d\varphi_t^p\\d\varphi_w^p\end{bmatrix}\right), \tag{5}$$

where the *transition matrices* $\hat{\mathbf{B}}^{pc}$ and $\hat{\mathbf{B}}^{qc}$ contain the components of the contact vectors $\mathbf{r}^{pc}$ and $\mathbf{r}^{qc}$:

$$\hat{\mathbf{B}}^{pc} = \begin{bmatrix} 1 & 0 & 0 & 0 & 0 & 0 \\ 0 & 1 & 0 & 0 & 0 & 0 \\ 0 & 0 & 1 & 0 & 0 & 0 \\ 0 & -r_3^{pc} & +r_2^{pc} & 1 & 0 & 0 \\ +r_3^{pc} & 0 & -r_1^{pc} & 0 & 1 & 0 \\ -r_2^{pc} & +r_1^{pc} & 0 & 0 & 0 & 1 \end{bmatrix} ; \hat{\mathbf{B}}^{qc} = \begin{bmatrix} 1 & 0 & 0 & 0 & 0 & 0 \\ 0 & 1 & 0 & 0 & 0 & 0 \\ 0 & 0 & 1 & 0 & 0 & 0 \\ 0 & -r_3^{qc} & +r_2^{qc} & 1 & 0 & 0 \\ +r_3^{qc} & 0 & -r_1^{qc} & 0 & 1 & 0 \\ -r_2^{qc} & +r_1^{qc} & 0 & 0 & 0 & 1 \end{bmatrix}. \tag{6}$$

The expression in (5) - which can be also taken as a definition - is similar to that proposed in [2], but here also the infinitesimal rotations of the points *pc* and *qc* at the contact are accounted for.

Particle rolling

Experiments on model assemblies of oval disks show that *particle rolling* is a dominant characteristic of the deformation of granular assemblies [16] and it admits various definitions in literature (see [17, 18] for an overview). Here interparticle rolling is defined as that part of the common translation of the contact material points *pc* and *qc* (Figure 2) that originates from the particle rotation. This form of rolling requires the knowledge of the local surface curvatures of the two particles at the contact, as proposed in [19]. A first effect of the interparticle rolling is represented by the increments of the contact vectors, denoted by $d\mathbf{r}^{pc}$ and $d\mathbf{r}^{qc}$ respectively, obtainable by considering the shifts of *pc* and *qc*, namely $d\mathbf{s}^{pc}$ and $d\mathbf{s}^{qc}$ along the surface and taking into account the results in [20]. A second effect is the infinitesimal rotation of the contact point around the barycentre of the grain, denoted as $d\psi^{pc}$, considered as a combination of a twist component about the normal axis and a component that is the rotation about an axis in the plane orthogonal to the normal at the contact point. The increments of the contact vectors and the infinitesimal rotation of the contact point can be written in terms of the infinitesimal grain displacements, by following [2]:

$$d\mathbf{r}^{pc} = \mathbf{A}^{pc}d\mathbf{u}^p + \mathbf{G}^{pc}d\mathbf{u}^q, \quad d\mathbf{r}^{qc} = \mathbf{H}^{pc}d\mathbf{u}^p + \mathbf{L}^{pc}d\mathbf{u}^q, \tag{7}$$

$$d\psi^{pc} = d\psi^{qc} = \mathbf{R}^{pc}d\mathbf{u}^p + \mathbf{V}^{pc}d\mathbf{u}^q, \tag{8}$$

where

$$\mathbf{A}^{pc} = \frac{1}{2}\left[\mathbf{B}^{pcT} - \check{\mathbf{I}} - (\mathbf{C}^p + \mathbf{C}^q)^{-1}\mathbf{N}^{pc} + \frac{1}{2}(\mathbf{C}^p + \mathbf{C}^q)^{-1}\mathbf{E}^{pc}(\mathbf{C}^p - \mathbf{C}^q)\mathbf{B}^{pcT}\right], \tag{9}$$

$$\mathbf{G}^{pc} = \frac{1}{2}\left[\mathbf{B}^{qcT} + (\mathbf{C}^p + \mathbf{C}^q)^{-1}\mathbf{N}^{pc} - \frac{1}{2}(\mathbf{C}^p + \mathbf{C}^q)^{-1}\mathbf{E}^{pc}(\mathbf{C}^p - \mathbf{C}^q)\mathbf{B}^{qcT}\right], \tag{10}$$

$$\mathbf{H}^{pc} = \frac{1}{2}\left[\mathbf{B}^{pcT} + (\mathbf{C}^p + \mathbf{C}^q)^{-1}\mathbf{N}^{pc} + \frac{1}{2}(\mathbf{C}^p + \mathbf{C}^q)^{-1}\mathbf{E}^{pc}(\mathbf{C}^p - \mathbf{C}^q)\mathbf{B}^{pcT}\right], \tag{11}$$

$$\mathbf{L}^{pc} = \frac{1}{2}\left[\mathbf{B}^{qcT} - \check{\mathbf{I}} - (\mathbf{C}^p + \mathbf{C}^q)^{-1}\mathbf{N}^{pc} - \frac{1}{2}(\mathbf{C}^p + \mathbf{C}^q)^{-1}[\mathbf{E}^{pc}(\mathbf{C}^p - \mathbf{C}^q)\mathbf{B}^{qcT}\right], \tag{12}$$

$$\mathbf{R}^{pc} = \left\{\mathbf{M}^{pc} - \mathbf{en}^{pc}[\mathbf{C}^q(\mathbf{C}^p + \mathbf{C}^q)^{-1}\mathbf{N}^{pc} + \mathbf{W}^{pq}\mathbf{E}^{pc}\mathbf{B}^{pcT}]\right\}, \tag{13}$$

$$\mathbf{V}^{pc} = \left\{\mathbf{M}^{pc} - \mathbf{en}^{pc}[\mathbf{C}^p(\mathbf{C}^p + \mathbf{C}^q)^{-1}\mathbf{N}^{pc} - \mathbf{W}^{pq}\mathbf{E}^{pc}\mathbf{B}^{qcT}]\right\}. \tag{14}$$

In the previous relations, e is Ricci's permutation symbol, $\check{\mathbf{I}}$ a 3×6 matrix whose left 3×3 block is the *unit matrix* and its right 3×3 block is zero, $\mathbf{C}^p$ and $\mathbf{C}^q$ the *curvature tensors* introduced in (3), $\mathbf{B}^{pcT}$ and $\mathbf{B}^{qcT}$ the *transition matrices* defined in [2] and corresponding to the first three columns of the matrices introduced in (6), $\mathbf{N}^{pc}$, $\mathbf{E}^{pc}$, $\mathbf{W}^{pq}$ and $\mathbf{M}^{pc}$ are defined respectively by

$$\mathbf{N}^{pc} := \begin{bmatrix} 0 & 0 & 0 & 0 & +n_3^{pc} & -n_2^{pc} \\ 0 & 0 & 0 & -n_3^{pc} & 0 & +n_1^{pc} \\ 0 & 0 & 0 & +n_2^{pc} & -n_1^{pc} & 0 \end{bmatrix}, \quad \mathbf{E}^{pc} := \mathbf{I} - \mathbf{n}^{pc} \otimes \mathbf{n}^{pc}, \tag{15}$$

$$\mathbf{W}^{pq} := -\frac{1}{2}\mathbf{C}^p(\mathbf{C}^p + \mathbf{C}^q)^{-1}\mathbf{C}^q + \mathbf{C}^q(\mathbf{C}^p + \mathbf{C}^q)^{-1}\mathbf{C}^p, \tag{16}$$

$$\mathbf{M}^{pc} := \frac{1}{2} \begin{bmatrix} 0 & 0 & 0 & n_1^c n_1^c & n_1^c n_2^c & n_1^c n_3^c \\ 0 & 0 & 0 & n_2^c n_1^c & n_2^c n_2^c & n_2^c n_3^c \\ 0 & 0 & 0 & n_3^c n_1^c & n_3^c n_2^c & n_3^c n_3^c \end{bmatrix}. \tag{17}$$

LOADS AND CONTACT FORCES

In the actual state any part of the grain heap is assumed to be in equilibrium. The external forces and moments acting on the generic grain p, expressed in the global frame, can be gathered in a global load vector denoted by $\mathbf{q}^p$. Then all the $\mathbf{q}^p$ vectors can be collected into the global load vector of the assembly $\mathbf{q}$. The contact forces, acting on the contact point c between grains p and q (with index $p < q$), are expressed in the local coordinate system (n^c, t^c, z^c) of the contact region and are denoted by $\mathbf{S}^c$. In writing the balance equations of the grains, it is useful to reduce the contact forces $\mathbf{S}^c$ to the barycentre of each grain. For this reason the *reduced force vectors*, defined by

$$\mathbf{f}_{red}^{pc} := \mathbf{B}^{pc}\mathbf{f}^{pc} = \mathbf{B}^{pc}\mathbf{T}^{pc}\mathbf{S}^c, \mathbf{f}_{red}^{qc} := \mathbf{B}^{qc}\mathbf{f}^{qc} = -\mathbf{B}^{qc}\mathbf{T}^{qc}\mathbf{S}^c, \tag{18}$$

are introduced. In (18) $\mathbf{B}^{pc}$ and $\mathbf{B}^{qc}$ are the 6×3 *transition matrices* corresponding to the first three columns of the matrices defined in (6), $\mathbf{T}^{pc}$ and $\mathbf{T}^{qc}$ the 3×3 *transformation matrices* between the local frame at the contact point c and the global one - they contain direction cosines.

THE CONTACT MODEL

The contact model is a simplified description of the relationship between the increments of the contact forces and the contact deformation. In the standard contact model contacts transmit only normal and transversal forces. The normal stiffness k_N^c is commonly considered Hertzian and the transversal stiffness k_T^c is described by the simplified Mindlin and Deresiewicz's relation [1] that considers a sort of Poisson effect without frictional sliding. Here the interparticle contact points are assumed to transmit also bending and twisting moments and the contact law appearing in common models (see, e.g., [2]) can be modified in the following way

$$d\hat{\mathbf{S}}^c = \hat{\mathbf{k}}^c d\hat{\mathbf{v}}^{c,def}, \tag{19}$$

where $d\hat{\mathbf{S}}^c$ is the contact action vector in the local frame and $d\hat{\mathbf{v}}^{c,def}$ the contact deformation defined in (5) in terms of interparticle rotations. The *contact stiffness matrix* $\hat{\mathbf{k}}^c$ can be written as

$$\hat{\mathbf{k}}^c = \begin{bmatrix} k_N^c & 0 & 0 & 0 & 0 & 0 \\ 0 & k_T^c & 0 & 0 & 0 & 0 \\ 0 & 0 & k_T^c & 0 & 0 & 0 \\ 0 & 0 & 0 & k_\varphi^c & 0 & 0 \\ 0 & 0 & 0 & 0 & k_\varphi^c & 0 \\ 0 & 0 & 0 & 0 & 0 & k_\varphi^c \end{bmatrix}, \tag{20}$$

where k_φ^c is the rotational stiffness term [21]. Also frictional sliding is included. The yield condition is described by the Coulomb law with addition of cohesion:

$$Q := |\mathbf{S}^c - (\mathbf{n}^c \cdot \mathbf{S}^c)\mathbf{n}^c| + \mu \mathbf{S}^c \cdot \mathbf{n}^c + c = 0, \tag{21}$$

where μ is the friction coefficient, $\mathbf{n}^c$ the unit normal vector and c the cohesion amid grains, if the granular assembly admits cohesion [22]. This yield condition depends upon the current contact force $\mathbf{S}^c$, which is known a priori [23]. A deep treatment of interparticle friction and its relation with macroscopic friction can be found in [24].

THE STIFFNESS MATRIX OF GRANULAR ASSEMBLIES

Commonly, the total stiffness matrix of the assembly is compiled starting from the infinitesimal equilibrium equation of a single grain p, expressed by the six scalar equations

$$d\mathbf{q}^p + \sum_{(c)} d\mathbf{f}_{red}^{pc} = \mathbf{0}, \qquad (22)$$

where $d\mathbf{q}^p$ is the increment of the external load vector and $d\mathbf{f}_{red}^{pc}$ the increment of the reduced contact force vector. c runs along all intergranular contacts. The increment of the reduced contact force vector $d\mathbf{f}_{red}^{pc}$ is affected by translation, rotation and deformation of the grain at contact. The re-location of the contact point due to the relative rotation and translation of grains may change the direction of the contact force causing a variation in the force components. The deformation induces a variation in the contact forces, depending on the stiffness of the contact, as described in (20). By following [2], the total increment of the reduced contact force $d\hat{\mathbf{f}}_{red}^{pc}$ in (22) can be written in terms of infinitesimal displacements of grains:

$$d\hat{\mathbf{f}}_{red}^{pc} = ((\mathbf{D}^{pc,trans} + \mathbf{D}^{pc,rot}) - \hat{\mathbf{k}}^{pc})d\mathbf{u}^p + ((\mathbf{D}^{qc,trans} + \mathbf{D}^{qc,rot}) - \hat{\mathbf{k}}^{qc})d\mathbf{u}^q, \qquad (23)$$

where the matrices involved take into account grain boundary curvature and contact forces already present in the actual state, namely

$$\mathbf{D}^{pc,trans} \;:\; = \begin{bmatrix} \mathbf{0} \\ \mathbf{eF}^{pc} \end{bmatrix} \left\{ \frac{1}{2}\mathbf{B}^{pcT} - \check{\mathbf{I}} + (\mathbf{C}^p + \mathbf{C}^q)^{-1}\mathbf{N}^{pc} + \right.$$
$$\left. + \frac{1}{2}(\mathbf{C}^p + \mathbf{C}^q)^{-1}(\mathbf{C}^p - \mathbf{C}^q)\mathbf{E}^{pc}\mathbf{B}^{pcT} \right\}, \qquad (24)$$

$$\mathbf{D}^{qc,trans} \;:\; = \begin{bmatrix} \mathbf{0} \\ \mathbf{eF}^{pc} \end{bmatrix} \left\{ \frac{1}{2}\mathbf{B}^{qcT} - (\mathbf{C}^p + \mathbf{C}^q)^{-1}\mathbf{N}^{pc} + \right.$$
$$\left. - \frac{1}{2}(\mathbf{C}^p + \mathbf{C}^q)^{-1}(\mathbf{C}^p - \mathbf{C}^q)\mathbf{E}^{pc}\mathbf{B}^{qcT} \right\}, \qquad (25)$$

$$\mathbf{D}^{pc,rot} \;:\; = \mathbf{B}^{pc}(\mathbf{ef}^{pc})\left\{ \mathbf{M}^{pc} - \mathbf{en}^{pc}[\mathbf{C}^q(\mathbf{C}^p + \mathbf{C}^q)^{-1}\mathbf{N}^{pc} + \mathbf{W}^{pq}\mathbf{E}^{pc}\mathbf{B}^{pcT}] \right\}, \qquad (26)$$

$$\mathbf{D}^{qc,rot} \;:\; = \mathbf{B}^{pc}(\mathbf{ef}^{pc})\left\{ \mathbf{M}^{pc} - \mathbf{en}^{pc}[\mathbf{C}^p(\mathbf{C}^p + \mathbf{C}^q)^{-1}\mathbf{N}^{pc} - \mathbf{W}^{pq}\mathbf{E}^{pc}\mathbf{B}^{qcT}] \right\}. \qquad (27)$$

In previous equations, e is Ricci's permutation index, $\check{\mathbf{I}}$ a 3×6 matrix whose left 3×3 block is the *unit matrix* and its right 3×3 block is zero, $\mathbf{C}^p$ and $\mathbf{C}^q$ the *curvature tensors* in (3), $\mathbf{N}^{pc}$, $\mathbf{E}^{pc}$, $\mathbf{W}^{pq}$ and $\mathbf{M}^{pc}$ have been defined in (15)-(17), $\mathbf{B}^{pc}$ and $\mathbf{B}^{qc}$ the 6×3 *transition matrices* corresponding to the first three columns of the matrices defined in (6). Since the effect of contact deformation is evaluated starting from the contact model introduced above, which is valid for contacts transmitting couples, the coefficient matrices $\hat{\mathbf{k}}^{pc}$ and $\hat{\mathbf{k}}^{qp}$ in (23) can be written in terms of the contact stiffness $\hat{\mathbf{k}}^c$ defined in (20)), namely

$$\hat{\mathbf{k}}^{pc} = \hat{\mathbf{B}}^{pc}\hat{\mathbf{T}}^{pc}\hat{\mathbf{k}}^c\hat{\mathbf{T}}^{pcT}\hat{\mathbf{B}}^{pcT}, \quad \hat{\mathbf{k}}^{qc} = -\hat{\mathbf{B}}^{qc}\hat{\mathbf{T}}^{qc}\hat{\mathbf{k}}^c\hat{\mathbf{T}}^{qcT}\hat{\mathbf{B}}^{qcT}. \qquad (28)$$

In (28), $\hat{\mathbf{B}}^{pc}$ and $\hat{\mathbf{B}}^{qc}$ are *transition matrices* defined in (6), $\hat{\mathbf{T}}^{pcT}$ and $\hat{\mathbf{T}}^{qcT}$ *transformation matrices* between local and global frames, written in a way that allows the representation of the rotations. The balance equations of a single grain p in (22) can be written in terms of infinitesimal grain displacements. The collection of the balance equations of all the m grains in the heap is then written as

$$d\mathbf{q} - \mathbf{K}^{TOT}d\mathbf{u} = \mathbf{0}. \qquad (29)$$

The total stiffness matrix $\mathbf{K}^{TOT}$ contains information on the geometrical features of the system, the contact model and the actual grain-to-grain and grain-to-supporting wall contact forces in the actual state. Its pq-block, K^{pq}, can be

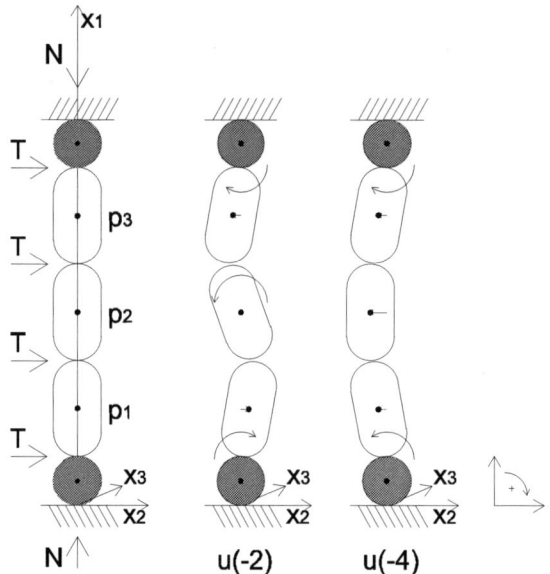

FIGURE 3. Column made of three grains with spherical curvature at the contact and their motions described by the eigenvectors u^{-2} and u^{-4}.

determined as follows: if the two neighboring grains p and q are not in contact, then the block K^{pq} is zero. Take $p \neq q$. Assume that grains p and q are in contact at c. It is then possible to write

$$\mathbf{K}^{pq} = -\mathbf{B}^{pc}\mathbf{T}^{pc}\mathbf{k}^{pc}\mathbf{T}^{pcT}\mathbf{B}^{pcT} - \mathbf{D}^{pc,trans} - \mathbf{D}^{pc,rot}, \qquad (30)$$

if $p > q$ and

$$\mathbf{K}^{qp} = -\mathbf{B}^{qc}\mathbf{T}^{qc}\mathbf{k}^{qc}\mathbf{T}^{qcT}\mathbf{B}^{qcT} - \mathbf{D}^{qc,trans} - \mathbf{D}^{qc,rot}, \qquad (31)$$

if $p < q$. In (30) and (31), $\mathbf{B}^{pc}$ and $\mathbf{B}^{qc}$ are 6×3 *transition matrices* corresponding to the first three columns of the matrices defined in (6), $\mathbf{T}^{pc}$ and $\mathbf{T}^{qc}$ 3×3 *transformation matrices* between the local frame at the contact point c and the global one, containing the direction cosines, $\mathbf{k}^{pc}$ and $\mathbf{k}^{qc}$ 3×3 *stiffness matrices* containing only normal and transversal stiffness terms. When $p = q$, the block $\mathbf{K}^{pp}$ is given by the sum, over all the contact points c between a grain p and its neighbours, of the terms in (30) (the contribution of each contact is taken into account).

The analysis of the spectral properties of the global stiffness matrix in (29) permits to analyse the stability of the equilibrium state of the whole assembly. In fact, the positive definiteness of $\mathbf{K}^{TOT}$ assures equilibrium. If $\mathbf{K}^{TOT}$ has zero or negative eigenvalues, the corresponding eigenvectors have to be analysed; if they are *consistent* with the directions of their solution vectors $d\mathbf{u}$, then the equilibrium state is not stable, while if they are *not consistent*, then the presence of zero or negative eigenvalues in the total stiffness matrix is not sufficient to declare the instability of the system. The word *consistent* means that the product $\mathbf{K}^i d\mathbf{u}$ involves 'motions' $d\mathbf{u}$ that lie within the domain of the branch $\mathbf{K}^i$ [23].

COLUMN AND CELL OF GRAINS

The example proposed by K. Bagi in [2] of a chain of grains is here extended to the case of contacts transmitting also bending and twisting moments. Also frictional sliding is taken into account. Grains made of granite, with Young modulus $E = 4160 kN/cm^2$ and Poisson ratio $v = 0.055$, are considered. The column is composed by grains with translational and rotational degrees of freedom, subjected to axial and transversal forces. The three *free* grains are limited by two *fixed* grains (Figure 3). This simple example helps to evaluate the physical meaning of the numerical analysis of the total stiffness matrix. At first the column of grains is subject only to a normal force equal to $1kN$. Grains have a contact radius equal to $2cm$ and spherical curvature. The degrees of freedom are the translations and rotations of grains $p1$, $p2$ and $p3$ in Figure 3. The resultant total stiffness matrix has four negative eigenvalues (the rest

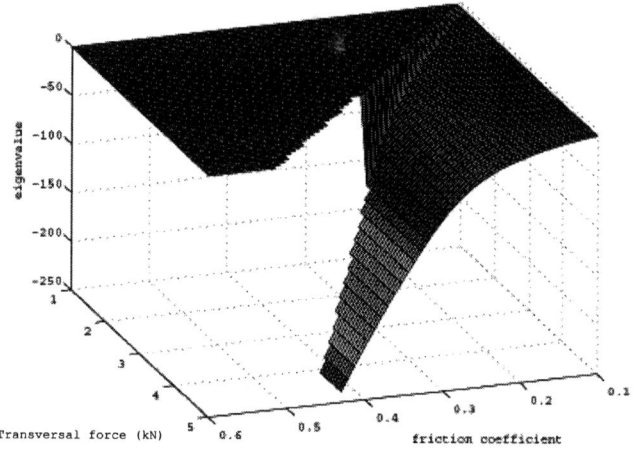

FIGURE 4. The effect of the friction coefficient and of the transversal force on the largest eigenvalue of the total stiffness matrix.

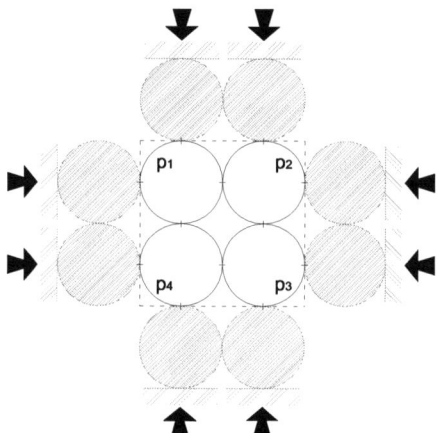

FIGURE 5. A cell of four spherical grains surrounded by fixed grains exerting a uniform compression on it.

of the eigenvalues are positive). The four corresponding eigenvectors are determined and two of them, labelled here as $\mathbf{u}^{-2}$ and $\mathbf{u}^{-4}$, are *consistent*. The *consistent* eigenvectors show that the system can buckle according to the motions in Figure 3.

Figure 4 illustrates the influence on the largest eigenvalue of the total stiffness matrix of the transversal force and the friction coefficient, for a fixed normal force equal to $10kN$. The plateau on the left (where the eigenvalue is zero) shows that the increment of the friction coefficient results in a larger domain of stability. In order to highlight the effect of interparticle friction the contacts are assumed to have no rotational stiffness in this case (while the transversal stiffness is taken into account).

The second example of a cell containing four identical spherical grains lying on the horizontal plane (x,y) and surrounded by fixed grains that exert a uniform compression on it is analysed (Figure 5). The influence of contact friction, rotational stiffness, dimension and shape of the grains is also investigated in this case. Figure 6 shows the role of curvature and the destabilizing effect of the compression force on the largest eigenvalue of the stiffness matrix, since when the grains are pressed together, they can bulk in the orthogonal direction z. The stabilizing effect of rotational contact stiffness and destabilizing role of dimension are shown in Figure 7.

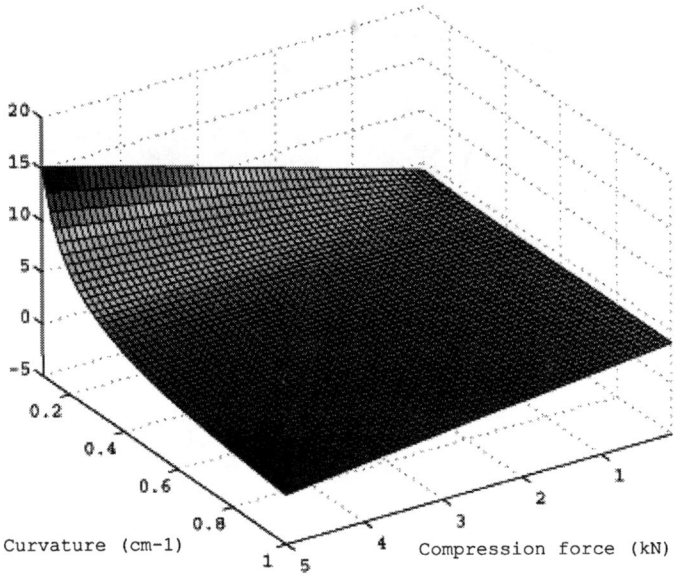

FIGURE 6. Effect of grains surface curvature and compression force on the largest eigenvalue of the stiffness matrix of the cell.

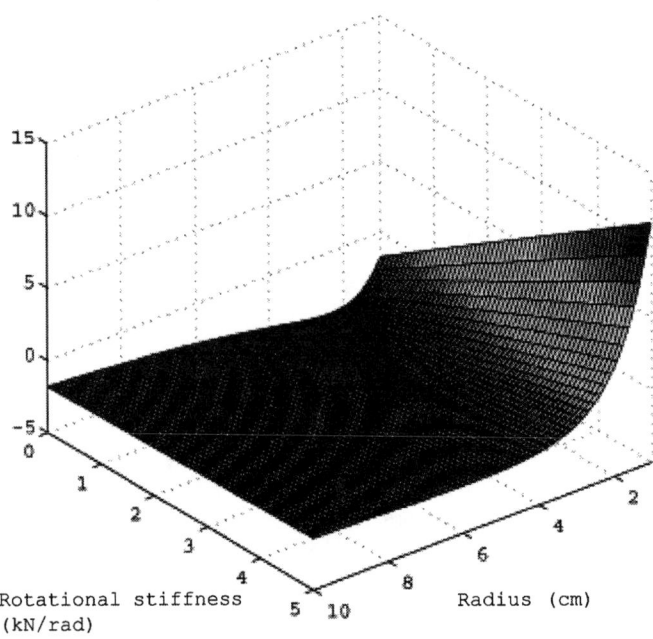

FIGURE 7. Effect of rotational stiffness and dimension of grains on the largest eigenvalue of the stiffness matrix of the cell.

DISCUSSION

The equilibrium of granular clusters has been investigated by means of the analysis of the eigenvalues and eigenvectors of the total stiffness matrix, following a method widely diffused in structural analysis. The results previously proposed in [2] are extended here by taking into account friction, cohesion and rotational contact stiffness. Two examples are shown. Although they are simple, they allow a clear view of the essential features of the analyses developed here. Buckling phenomena can be identified through the determination of the eigenvectors corresponding to negative eigenvalues.

The results obtained here can be also used for homogenization procedures involving continuum models. Relating work is in progress.

REFERENCES

1. I. Agnolin, and J. N. Roux, *Phys. Rev. E* **76**, 061304 (2007).
2. K. Bagi, *Granular Matter* **9**, 109–34 (2007).
3. S. J. Antony, R. O. Momoh, and M. R. Kuhn, *Computational Materials Science* **29**, 494–98 (2004).
4. N. P. Kruyt, and S. J. Antony, *Physical Review E* **75**, 051308 (2007).
5. J. D. Goddard, *Int. J. Solids Structures* **41**, 5851–61 (2004).
6. J. D. Goddard, and A. K. Didwania, *Mechanics of Materials* **41**, 637–51 (2009).
7. B. Cambou, P. Dubujet, F. Emeriault, and F. Sidoroff, *European J. Mech. - A Solids* **14**, 255–76 (1995).
8. K. C. Chen, J. Y. Lan, and Y. C. Tai, *Int. J. Solids Structures* **46**, 3882–93 (2009).
9. P. Dubujet, and F. Dedecker, *Granular Matter* **1**, 129–36 (2004).
10. W. Elhers, E. Ramm, S. Diebels, and G. A. D'Addetta, *Int. J. Solids Structures* **40**, 6681–6702 (2003).
11. H. M. Jaeger, S. R. Nagel, and R. P. Behringer, *Rev. Mod. Phys.* **68**, 12591273 (2001).
12. E. Kuhl, G. A. D'Addetta, H. J. Herrmann, and E. Ramm, *Granular Matter* **2**, 113–21 (2000).
13. F. Nicota, and F. Darve, *Mechanics of Materials* **37**, 980–1006 (2005).
14. M. Satake, "Graph-theoretical approach to the mechanics of granular materials," in *Continuum Models of Discrete Systems*, edited by A. J. M. Spencer, Taylor and Francis, 1987, pp. 163–73.
15. J. D. Goddard, "From granular matter to generalized continuum," in *Mathematical Models of Granular Matter*, edited by G. Capriz, P. Giovine, and P. M. Mariano, Springer, 2008, vol. 1937 of *Lect. Notes on Appl. Math.*, chap. 1, pp. 1–20.
16. K. Iwashita, T. Kakiuchi, and M. Oda, *Powders & Grains* **97**, 207–10 (1997).
17. M. R. Kuhn, and K. Bagi, "Particle rolling and its effects in granular materials," in *Proc. of QuaDPM'03 Workshop*, Budapest, Hungary, 2003, pp. 151–58.
18. M. R. Kuhn, and B. K., *ASME J. Appl. Mech.* **71**, 493–501 (2004).
19. K. Iwashita, and M. Oda, *ASCE J. Eng. Mech.* **124**, 285–92 (1998).
20. I. S. Sokolnikoff, *Tensor Analysis: Theory and Applications to Geometry and mechanics of Continua*, Wiley, New York, 1964.
21. A. Tordesillas, J. Peters, and M. Muthuswamy, *ANZIAM Journal* **46**, 260–75 (2005).
22. R. M. Neddermann, *Statics and kinematics of granular materials*, Cambridge University Press, 1992.
23. M. R. Kuhn, and C. S. Chang, *Int. J. Solids Structures* **43**, 6026–51 (2006).
24. N. P. Kruyt, and L. Rothenburg, *J. Stat. Mech.* p. P07021 (2006).

Force transmission in cohesive granular media

Farhang Radjai*,†, Vincent Topin*,†, Vincent Richefeu**, Charles Voivret‡,
Jean-Yves Delenne*,†, Emilien Azéma* and Said El Youssoufi*

*Laboratoire de Mécanique et Génie Civil, UMR 5508 CNRS-Université Montpellier 2, case 048, Pl. E. Bataillon, F-34095 Montpellier cedex 5, France
†Laboratoire de Micromécanique et Intégrité des Structures, IRSN DPAM-CNRS, case 048, Pl. E. Bataillon, F-34095 Montpellier cedex 5, France
**Laboratoire Sols Solides Structures Risques, Université Joseph Fourier, Domaine Universitaire, B.P. 53, F-38041 Grenoble, Cedex 9, France.
‡Laboratoire Surface du Verre et Interfaces, UMR 125 CNRS/Saint-Gobain, 39 Quai Lucien Lefranc, F-93303 Aubervilliers Cedex, France

Abstract.
We use numerical simulations to investigate force and stress transmission in cohesive granular media covering a wide class of materials encountered in nature and industrial processing. The cohesion results either from capillary bridges between particles or from the presence of a solid binding matrix filling fully or partially the interstitial space. The liquid bonding is treated by implementing a capillary force law within a debonding distance between particles and simulated by the discrete element method. The solid binding matrix is treated by means of the Lattice Element Method (LEM) based on a lattice-type discretization of the particles and matrix. Our data indicate that the exponential fall-off of strong compressive forces is a generic feature of both cohesive and noncohesive granular media both for liquid and solid bonding. The tensile forces exhibit a similar decreasing exponential distribution, suggesting that this form basically reflects granular disorder. This is consistent with the finding that not only the contact forces but also the stress components in the bulk of the particles and matrix, accessible from LEM simulations in the case of solid bonding, show an exponential fall-off. We also find that the distribution of weak compressive forces is sensitive to packing anisotropy, particle shape and particle size distribution. In the case of wet packings, we analyze the self-equilibrated forces induced by liquid bonds and show that the positive and negative particle pressures form a bi-percolating structure.

Keywords: granular media, force chain, granular disorder, cohesion, discrete element method, lattice element method, capillary bond, binding matrix
PACS: 45.70.-n, 81.05.Rm, 61.43.Hv

INTRODUCTION

A considerable amount of experimental and numerical work has been devoted to force transmission in model granular media such as glass bead packs [1, 2, 3, 4, 5, 6]. The force transmission in granular materials is essential for microscopic modeling of constitutive behavior and for many industrial processes that involve a better understanding of the static or dynamic forces experienced by the particles. The force distributions are found to be broad and highly heterogeneous. This heterogeneity is often described in terms of *force chains* and linked with the concept of *jamming*.

The issue that we would like to address in this paper, is to which extent the well-known features of force distributions in noncohesive granular media apply to cohesive granular media. The latter covers a wide class of materials encountered in nature and industry. Well-known examples are sedimentary rocks, wet soils, and fine and sintered powders. In contrast to noncohesive granular media, all these materials are endowed with *cohesion* resulting either from direct surface forces between particles or from the presence of a binding phase filling fully or partially the interstitial space. The effect of surface forces or a binder is to freeze or restrict the relative degrees of freedom (separation, sliding, rolling) between particles up to a threshold. Hence, depending on the boundary conditions, tensile forces can develop in cohesive granular media and their distributions are dictated by the conditions of force balance and granular disorder as in the case of compressive forces. Obviously, the distribution of tensile forces is of particular relevance to the stress intensity factor which controls the initiation and propagation of cracks.

In this paper, we investigate force and stress distributions in granular media involving either liquid bridges or a solid binding matrix between particles. The presence of liquid bridges will be treated by implementing a capillary force law within a debonding distance between particles. The simulations are performed by means of the Discrete Element

Method (DEM) using Molecular Dynamics (MD) and Contact Dynamics approaches. For solid binding, we adopt a broad framework allowing for the numerical treatment of a binding matrix with variable volume fraction. The effect of small amounts of the matrix localized at the contact zones between particles can be assimilated to that of a surface force. The force transmission in this limit is correlated with the packing structure. The other limit of high matrix volume fractions corresponds to a *cemented* granular material in which the particles are fully or partially embedded in the binding matrix. The force transmission is thus mediated both by the particles and matrix and governed by the details of the composition (phase volume fractions) and the material properties of each phase (relative stiffness, particle-matrix interface adherence).

The treatment of the matrix, as a continuous phase, requires a numerical method capable of resolving the matrix. We use the Lattice Element Method (LEM) which is found to be numerically efficient. It is based on a lattice-type discretization of all phases including the particles, matrix and their interface [7, 8, 9]. The elastic deformations of the particles are taken into account not only at their contacts with other particles or with the matrix, as in the DEM, but also in their bulk. The matrix can be introduced with the desired volume at the contact zones between the particles and in the pores with its elastic properties and adhesion with the particles. An advantage of the LEM is to give us access to stresses in the bulk of the particles and binding matrix. Hence, the forces at the contact zones can be estimated by coarse-graining from the stresses and compared to the DEM predictions for the same granular configuration.

In the following, we first focus on some important features of force transmission in noncohesive granular media. We consider both 2D and 3D granular samples and the effect of particle shape and size distribution. One section is devoted to granular media with solid bridging. The LEM is briefly introduced together with numerical procedures for sample preparation. Our main numerical results will be presented by considering the force distributions in 2D packings simulated alternatively by LEM and DEM in the limit of low matrix volume fractions, the stresses in a 3D packing and the effect of matrix volume fraction, particle stiffness and particle volume fraction on stress distributions. In another section we consider liquid bridging. We first introduce the capillary force law implemented in MD simulations. Then, we analyze the force distributions with and without a confining pressure. We also consider the tensile and compressive stresses supported by the particles. We conclude the paper with a summary of the most salient features of force transmission in cohesive granular media.

FORCE DISTRIBUTIONS IN NONCOHESIVE GRANULAR MEDIA

We study in this section the normal force distributions from numerical simulations by CD and MD methods in 2D and 3D. We consider the effect of packing anisotropy, particle shape and particle size distribution (PSD). Some of these features will be revisited in the next sections in the presence of liquid or solid binding between particles.

Background

Granular disorder and steric exclusions lead to an unexpectedly inhomogeneous distribution of contact forces under quasistatic loading [1, 10, 3, 11, 5, 12, 13, 14, 15, 6]. These force inhomogeneities in granular assemblies were first observed by means of photoelastic experiments [16, 17]. The carbon paper technique was used later to record the force prints at the boundaries of a granular packing [3]. It was found that the forces have a nearly decreasing exponential distribution. Numerical simulations by the contact dynamics (CD) method provided detailed evidence for force chains, the organization of the force network into strong and weak networks, and the exponential distribution of strong forces [18, 4]. Moreover, the force probability density functions (PDF's) from simulations showed that the weak forces (below the average force) in a sheared granular system have a nearly uniform or decreasing power law shape in agreement with refined carbon paper experiments [10, 5].

Further experiments and numerical simulations have shown that the exponential falloff of strong forces is a robust feature of force distribution in granular media both in two and three dimensions. In contrast, the weak forces are sensitive to the details of the preparation method or the internal state of the packing [19, 20, 15, 6]. A remarkable aspect of weak forces is the fact that their number does not vanish as the force falls to zero [18, 21]. Several theoretical models have been proposed allowing to relate the exponential distribution of forces to granular disorder combined with the condition of force balance for each particle [1, 22]. Recently, the force PDF's were derived for an isotropic system of frictionless particles in two dimensions from a statistical approach assuming a first shell approximation (one particle with its contact neighbors) [21].

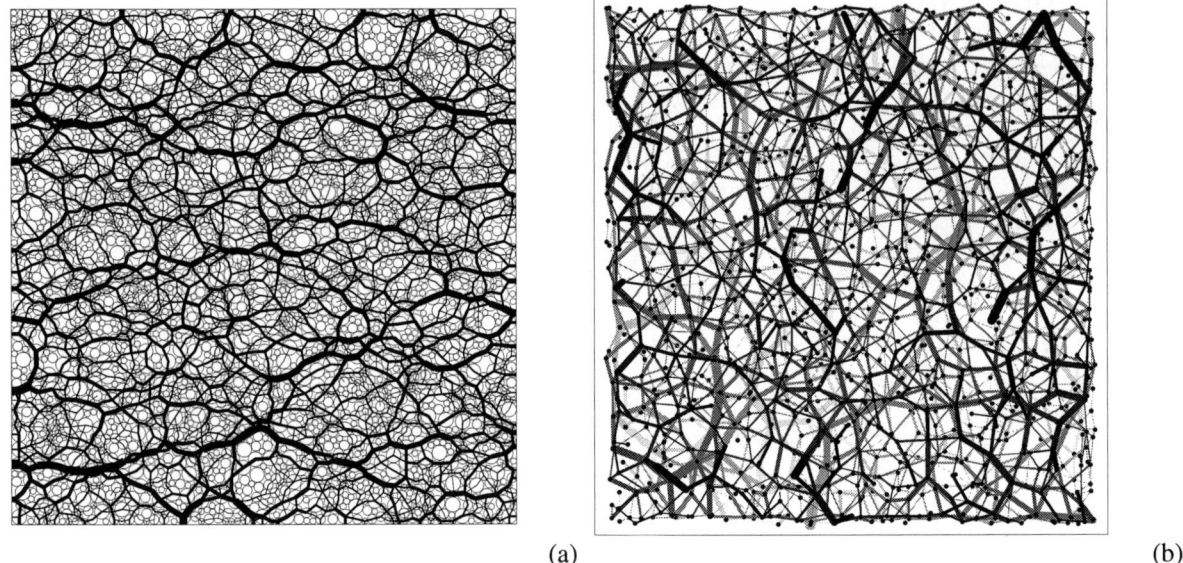

(a) (b)

FIGURE 1. The force network in a 2D packing of disks (a) and in a thin layer cut inside a 3D packing of spherical particles (b). The line thickness is proportional to the normal force. The gray level in the 3D system represents the field depth.

Figure 1 displays a 2D packing simulated by the CD method. The normal forces are encoded as the thickness of branch vectors (joining particle centers). In the same figure, the force network in a thin layer for a 3D packing of spherical particles subjected to axial compression is shown. Strong force chains are easily distinguished in both cases. The strongest chains have a linear aspect and they are mostly parallel to the axis of compression (vertical).

Discrete Element Method

The Discrete Element Method (DEM) has been extensively used since the pioneering work of Cundall for the simulation of granular materials [23]. In this method, the equations of motion are integrated for all particles by taking into account their contact interactions. In its original version, commonly used also today, the particles are treated as rigid elements but the interactions are modeled by means of visco-elastic force laws expressed in terms of the relative displacements between particles as in classical Molecular Dynamics (MD) simulations. In these MD-type approaches, the simulation of mutual exclusions between particles requires a stiff repulsive potential and thus high time resolution. In the same way, the Coulomb law for dry friction needs to be regularized such that the friction force can be expressed as a (mono-valued) function of relative tangential displacement.

The Contact Dynamics (CD) method, introduced later, provides an alternative approach based on *nonsmooth* formulation of mutual exclusion and dry friction between particles [24, 25, 26]. In this method, the equations of motion are expressed as differential inclusions and the accelerations are replaced by velocity jumps. At each time step, all kinematic constraints implied by enduring contacts and possible rolling of particles over one another are simultaneously taken into account in order to determine all velocities and contact forces. In the generic CD algorithm, an iterative process is used to solve this problem. It consists of solving a single contact problem with all other contact forces kept constant, and iteratively updating the forces until a given convergence criterion is fulfilled. Due to the implicit time integration scheme inherent in the CD method, the solution is unconditionally stable. The particle positions are updated from the calculated particle velocities before a new detection of the contacts between particles is performed.

Schematically, the MD method is based on a description of particle interactions in terms of *force laws*, i.e. bijective force-displacement relations, whereas the CD method is based on a formulation of kinematic constraints in terms of *contact laws*. Independently of particle deformability, the impenetrability of the particles and the Coulomb friction at the contact zones can be formulated in the form of contact laws expressing the contact actions as set-valued functions of particle positions. The uniqueness of the solution is not guaranteed by CD approach for perfectly rigid particles in absolute terms. However, by initializing each step of calculation with the forces calculated in the preceding step, the set of admissible solutions shrinks to a small variability basically of the same order of magnitude as the numerical resolution. In the MD method this 'force history' is by definition encoded in the particle positions.

Since the CD method handles the kinematic constraints without resorting to force laws, the particles are often treated as perfectly rigid although finite stiffness can be introduced in the same framework. This is the case of the CD simulations carried out for the analysis of force distributions in this paper. Hence, the only material parameter of the simulated static packings by the CD method is the coefficient of friction μ between the particles. On the other hand, the MD-generated packings are characterized by normal and tangential stiffnesses k_n and k_t as well as the coefficient of friction μ. The mean deformation of the particles is given by the ratio p/k_n of the average stress p to k_n.

Normal force distributions

Different numerical packings were prepared by isotropic compaction and then deformed under either slow triaxial loading in 3D or in simple shear in 2D. The particle inertia are negligibly small compared to the static confining pressure so that the packings can be considered in a *quasi-static state*. As we shall see below, the general shape of force distributions is robust with respect to the details of preparation or the microstructure. But the distribution parameters do depend on the preparation. In all examples considered below the packings are sheared until a steady or *critical* state, in the sense of soil mechanics, is reached. In this state, the shear deformation is isochoric on the average, and the memory of the preparation process is erased as a result of shearing so that the microstructure is a function only of the material parameters. The force distributions will be analyzed either in the initial isotropic state prepared by isotropic compaction with zero coefficient of friction or in the critical state.

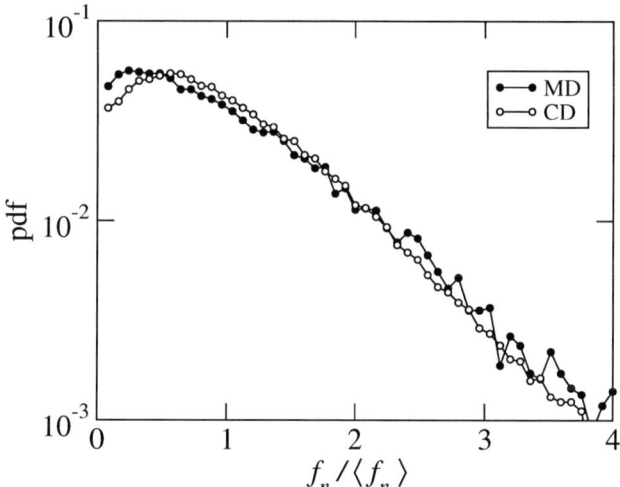

FIGURE 2. Probability density functions of normal forces in two isotropic samples of spherical particles simulated by MD and CD methods.

Figure 2 shows the PDF's of normal forces for two isotropic samples of spherical particles simulated by MD (8000 particles) and CD (20000 particles) methods. The PSD is not the same in the samples but they represent rather weakly polydisperse distributions with a ratio of 2 between the largest and smallest particle diameters. The coefficient of friction is $\mu = 0.4$ between particles and 0 with the walls. The forces have been normalized by the average force in each sample. Although the two samples are not exactly identical, the two PDF's have the same shape characterized by an exponential falloff for large forces, a small peak for a force slightly below the average force and a finite value at zero force. The position of the peak is not the same in the two distributions but the exponents of the exponential falloff are the same within statistical precision of the data:

$$P(f_n) \propto e^{-\beta f_n/\langle f_n \rangle}, \tag{1}$$

with $\beta \simeq 1.4$. This similarity between the two distributions indicates that the statics of a granular system is statistically robust with respect to the numerical approach and, in particular, the small elastic deformation at contact points in MD simulations has negligible effect on the force inhomogeneity. In other words, the physics of a static granular packing

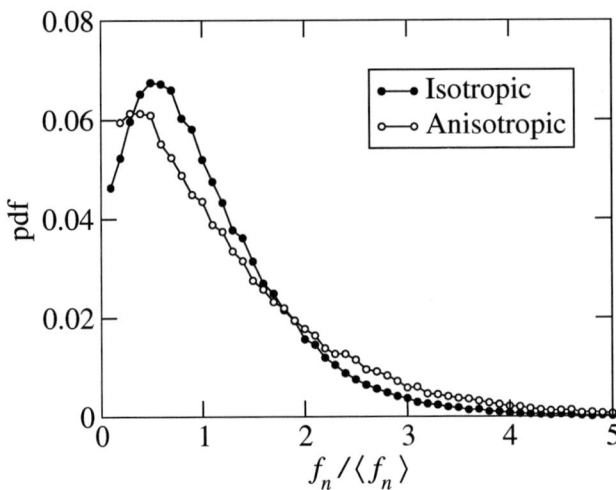

FIGURE 3. Probability density functions of normal forces in a sample of spherical particles after isotropic compaction (isotropic state) and following triaxial compression (anisotropic state).

can be approximated by considering undeformable particles as in the CD method as far as the the ratio p/k_n of the confining pressure p to the normal stiffness k_n of the particles is small (here $\simeq 10^3$).

The observed shape of force PDF's is unique in two respects: (1) the exponential part reflects the presence of very strong forces in the system often appearing in a correlated manner in the form of force chains; (2) the nonvanishing class of weak forces, with a fraction of more than 60% of contact forces below the average force, means that the stability of force chains is ensured by a large number of vanishingly small forces [4, 19]. This is a signature of the *arching effect*. Hence, the average force is a physically poor representative of the broad spectrum of forces in a granular system.

Figure 3 displays the normal force PDF's in CD simulations for the same system of spherical particles both at the isotropic state and at the critical state where the fabric and force chains are anisotropic. The effect of anisotropy is to reinforce the force inhomogeneity by increasing the relative density of weak forces [20, 27, 28]. The exponent

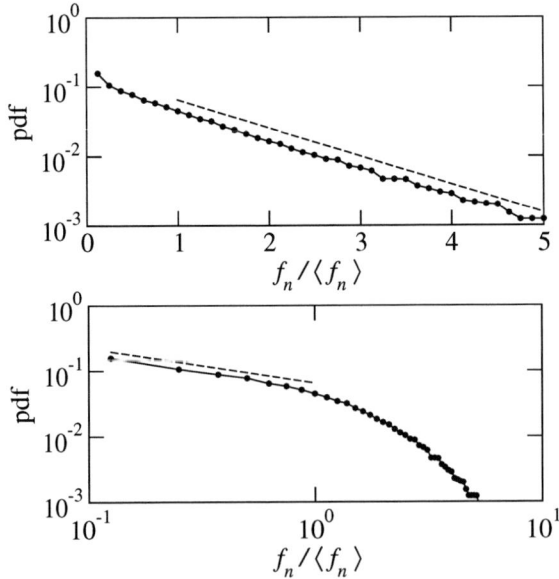

FIGURE 4. Probability density function of normal forces in an isotropic sample of irregular polyhedral particles on log-linear and log-log scales.

β remains nearly unchanged whereas the small peak near the average force disappears and the distribution of weak forces tends to become nearly uniform.

Effect of particle shape

The force distributions are sensitive to particle shapes. Fig. 4 shows the distribution of normal forces in an isotropic 3D sample of 20000 irregular polyhedral particles with $\mu = 0.5$ simulated by the CD method. We again observe the exponential tail of strong forces together with a decreasing power law distribution for weak forces.

The angular particle shape increases considerably the number of very weak forces by enhancing the arching effect. The latter is also reflected in the value of the exponent β reduced to 0.97 compared to 1.4 for spherical particles. In other words, the force chains are stronger but less in number. A detailed analysis of force and fabric anisotropies in this packing reveals the special role of face-to-face contacts in enhancing force anisotropy and thus the overall shear strength as compared to packings of spherical particles [29]. Similar trends are observed in packings of polygonal particles (in 2D simulations) [30].

Effect of particle size distribution

Figure 5 shows the normal force PDF's for increasingly broader particle size span s in a 2D sheared packing of 10000 circular particles simulated by the CD method [31]. The size span is defined by $s = (d_{max} - d_{min})/(d_{max} + d_{min})$ where d_{min} and d_{max} are the smallest and largest diameters, respectively. A monodisperse distribution corresponds to $s = 0$ and the limit $s \simeq 1$ corresponds to an infinitely polydisperse system [32]. The PSD is uniform by particle volume fractions.

The PDF becomes broader with increasing s. The weak forces have a clear power law behavior with increasing exponent α whereas the strong forces fall off exponentially with a decreasing exponent β. The power-law behavior of strong forces can be attributed to a "cascade" mechanism from the largest particles "capturing" strongest force chains down to smaller forces carried by smaller particles [31]. A map of normal forces in a highly polydisperse packing ($s = 0.96$) is shown in Fig. 6. A large number of rattlers, i.e. particles not engaged in the force network, can be observed. Although these particles represent a small volume fraction of the sample, their absence from the force-bearing network contributes to force inhomogeneity.

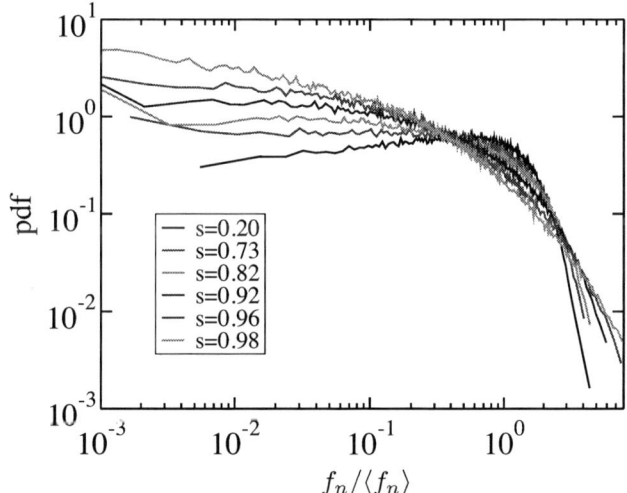

FIGURE 5. Probability density functions of normal forces for increasing span s of particle diameters.

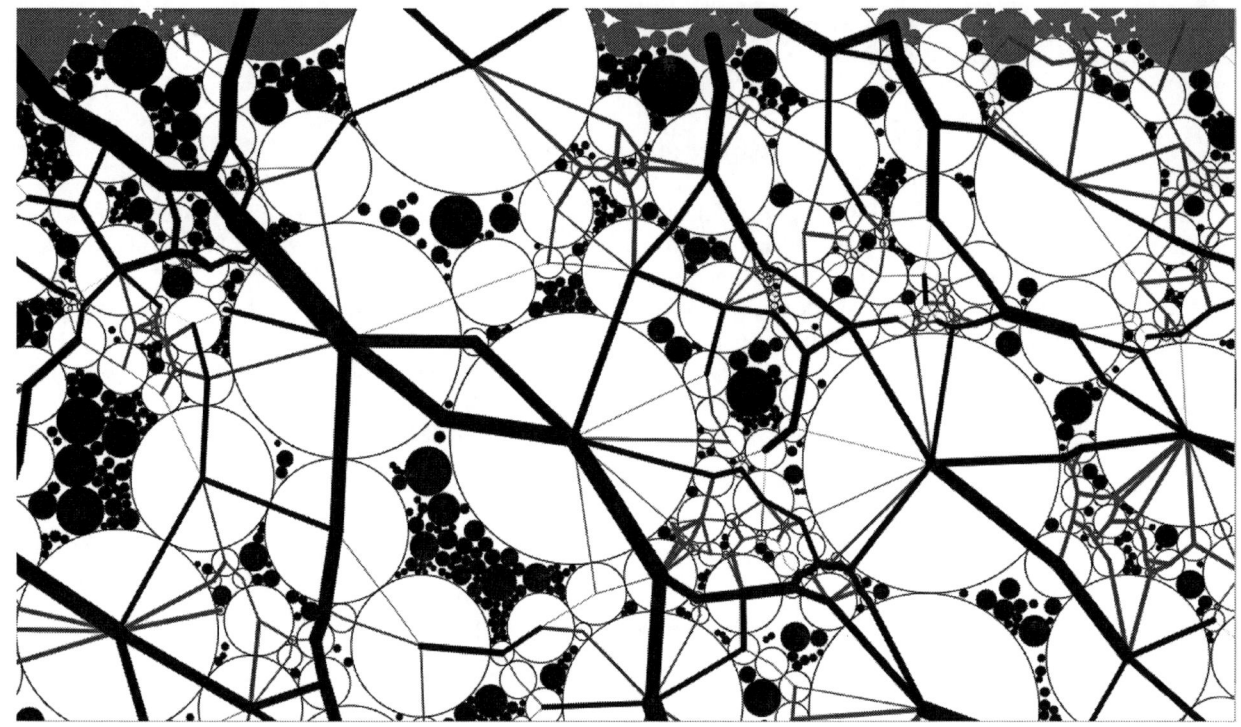

FIGURE 6. A map of normal forces in a highly polydisperse system with a uniform size distribution by particle volume fractions. The black particles are "rattlers" excluded from the force-bearing network.

A generic functional form

The above examples point to a generic PDF of normal forces in a granular packing that can be approximated by the following form [19]:

$$P(f_n) = \begin{cases} A \left(\frac{f_n}{\langle f_n \rangle}\right)^{-\alpha} & f_n/\langle f_n \rangle < 1 \\ A\, e^{\beta(1-f_n/\langle f_n \rangle)} & f_n/\langle f_n \rangle > 1 \end{cases} \quad (2)$$

where A is the normalization factor given by

$$\frac{1}{A} = \frac{1}{1-\alpha} + \frac{1}{\beta} \quad (3)$$

Considering the mean force $\langle f_n \rangle$ as the point of cross-over between the two parts of the distribution, we get the following relation between the exponents:

$$\beta^2 = (1-\alpha)(2-\alpha) \quad (4)$$

Note that the nearly uniform distribution of static forces in the case of sheared circular particles is recovered by setting $\alpha = 0$ in equation (2). Then, from equation [4] we get $\beta = \sqrt{2} \simeq 1.4$ which is the value found for the distribution of forces in sheared packings of weakly polydisperse spheres. For this system, the following fitting form was also proposed by Mueth et al. [5]:

$$P(f) = a\,(1 - b e^{-f^2}) e^{-\beta f} \quad (5)$$

where $f = f_n/\langle f_n \rangle$. As argued by Mueth et al., the above function for the range of weak forces provides a fit essentially indistinguishable from a power law $f_n^{-\alpha}$ as long as α is positive and close to zero [5].

SOLID BONDING

In this section, we consider cemented granular media in which the local cohesion is a consequence of the presence of a binding phase between the particles.

Numerical method and sample preparation

The LEM is based on a discretization of the phases on a regular or irregular lattice. Hence, the space is represented by a grid of points (nodes) interconnected by one-dimensional elements (bonds). Each bond can transfer normal force, shear force and bending moment up to a threshold in force or energy, representing the cohesion of the phase or its interface with another phase. Each phase (particle, matrix) and its boundaries are materialized by the bonds sharing the same properties. The samples are deformed by imposing displacements or forces to the nodes belonging to the contour. The total elastic energy of the system is a convex function of node displacements and thus finding the unique equilibrium configuration of the nodes amounts to a minimization problem. Performing this minimization for stepwise loading corresponds to subjecting the system to a quasistatic deformation process. The details of this method can be found in Ref. [8].

The samples are constructed either by geometric methods or by isotropic compaction of disk-like particles by DEM simulations by setting the friction coefficient between the particles to zero in order to get a dense packing. The samples are then discretized on a lattice. The matrix is introduced in the form of bridges of variable thickness, depending on the overall matrix volume fraction and the particle sizes, between neighboring particles throughout the system; see Fig. 7. As the matrix volume fraction is increased, the thickness of the bridges increases and eventually they merge to fill the interstitial space.

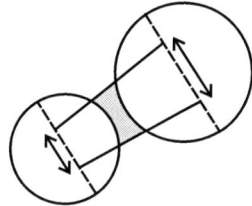

FIGURE 7. Numerical model of cementing bridge between particles. The width is increased for all pairs in a sample until the required matrix volume fraction is reached.

The elastic properties of each phase are controlled by the linear elastic properties of the bonds. The main elastic parameters that will be considered here are the Hooke constants k^p and k^m of the bonds belonging to the particles and matrix, respectively. The initial state is the reference (unstressed) configuration. When the sample is loaded, bond forces develop inside the sample. A stress tensor σ^a can be attributed to each node a of the lattice network: $\sigma_{ij}^a = \frac{1}{V^a} \sum_b r_i^{ab} f_j^{ab}$ where the summation runs over all neighboring nodes j, r_i^{ab} is the i component of the vector joining the node a to the midpoint of the bond ab and f_j^{ab} is the j component of the bond force [33, 8].

The resolution of the stresses depends on the particle size compared to the lattice element lengths. The discretization should be sufficiently fine for the particle contours to be correctly represented. The macroscopic elastic moduli might crucially depend on the discretization as more generally in porous materials. In practice, however, the resolution is set as a result of compromise between the necessary number of particles for statistical representativity and total number of nodes accessible to computer simulation. In the simulations reported in this paper, we generally favored high resolution both in 2D and 3D simulations such that the results for stress transmission reliably reflect the configuration of the particle phase.

In the following, we mainly consider node stresses in rectangular and cubic samples subjected to vertical loading with free lateral boundaries. At low matrix volume fractions, for comparison with DEM we will also evaluate the contact forces between particles from bond forces. During loading, the bond forces increase with the applied vertical stress at the boundary. Hence, the mean bond force increases linearly with the external load whereas the bond force PDF's and stresses do not evolve as long as no bond breaks. We focus here only on force distributions in the undamaged samples, i.e. in the purely elastic domain. The damage and fracture properties have been extensively studied elsewhere [8].

Sub-particle stresses and contact forces

In order to obtain fine statistics of node stresses and contact forces between particles, we simulated a large sample of about 5000 particles with a particle volume fraction of $\rho^p \simeq 0.8$. This corresponds to a packing with a dense contact network of coordination number $z = 4$. The particle diameter d varies between d_{min} and $d_{max} = 3d_{min}$ with a uniform

FIGURE 8. Vertical stress field σ_{yy} represented in color level in a cemented packing. The solid bridges and voids are in white and gray, respectively.

distribution by volume fractions ($P(d) \propto d^{-2}$). We would like to compare the contact forces in this system, simulated by the LEM, with those in a similar system simulated by the DEM. This can be done only in the limit of a small matrix volume fraction where the matrix is found in the form of small solid bridges between the particles such that its effect can be represented by a cohesion law. We used a matrix volume fraction of $\rho^m \simeq 0.01$. The DEM code is based on the standard molecular dynamics method with cohesive bonding between the particles. The sample is subjected to vertical compression.

Figure 8 shows the vertical stress field σ_{yy}. The node stresses are represented by proportional color levels over the elementary hexagonal cells centered on each node. We observe chains of highly stressed particles and higher concentration at the contact zones between the particles. In order to compare the LEM simulated packing with DEM simulations of the same packing, for which only contact forces are accessible, we compute the contact forces $\vec{f}$ by summing up the bond forces $\vec{f}^{ab}$ for all bonds ab crossing the contact plane S: $\vec{f} = \sum_{ab \in S} \vec{f}^{ab}$.

Figure 9 shows the map of normal forces between particles for the LEM and DEM packings. We observe very similar force chains despite the fact that radically different methods were used to simulate them. The Pearson product-moment correlation coefficient between the two force networks is $r = 0.92$, which indicates high similarity. The PDF's of normal and tangential forces from LEM and DEM simulations are shown in Fig. 10. We observe that the two PDF's coincide over nearly the whole range of forces. The distribution of normal forces involves an exponential fall-off in the ranges of strong compressive and for the whole range of tensile forces. The exponent in the range of tensile forces is larger than that for the compressive forces. Remark that the largest tensile forces are far below the breaking threshold. The distribution is uniform in the range of weak compressive and forces as also observed in most simulations of sheared packings composed of circular weakly polydisperse particles (see section). This excellent agreement between the force PDF's with $\beta \simeq 1.35$ may be considered as a validation of DEM results for the force networks in the sense that the contact forces in LEM simulations are calculated from a finer scale [2, 5, 6].

Having access to the node stresses, it is interesting to evaluate their PDF's in order to see whether they carry a signature of the composition. One example of the PDF of vertical stresses σ_{yy} is displayed in Fig. 11(a) for a packing under vertical compression. Since the sample is under axial compression, only 4% of vertical stresses are tensile and are thus not shown in Fig. 11(a). Interestingly, the strong stresses fall off exponentially as contact forces (see Fig. 10), $P_\sigma(\sigma_{yy}) \propto e^{-\beta \sigma_{yy}/\langle \sigma_{yy} \rangle}$ with $\beta \simeq 0.95$, and they mostly concentrate at the contact zones. The weak stresses have a nonzero PDF, much the same as weak contact forces, reflecting the arching effect. Since the contact force distributions reflect the granular disorder, i.e. the structure of the network of contiguous particles, the observed

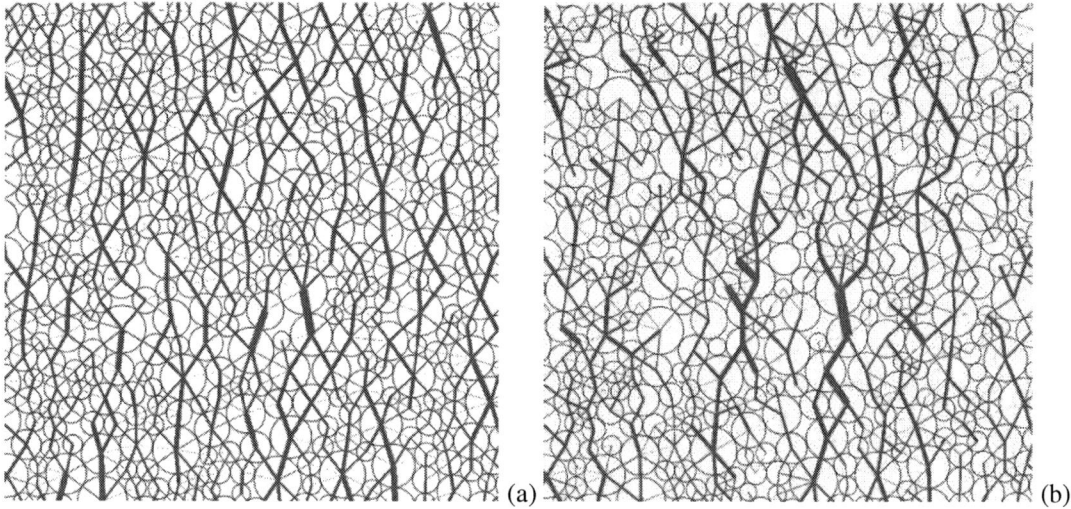

FIGURE 9. A map of normal forces in a portion of a sample under vertical compression simulated by DEM (a) and LEM (b). Line thickness is proportional to the normal force. Very weak and tangential forces are not shown.

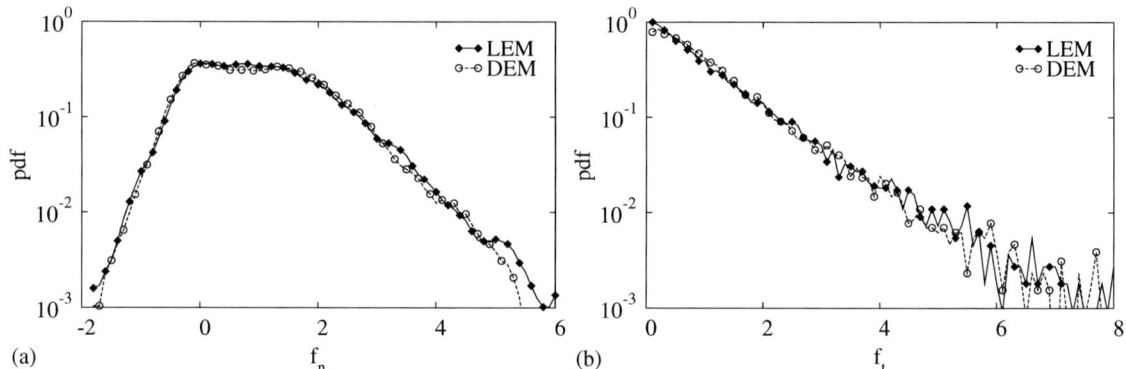

FIGURE 10. Probability density function of normal forces (a) and tangential forces (b) in a sample axially compressed by LEM and DEM simulations. The forces are normalized by the mean normal force.

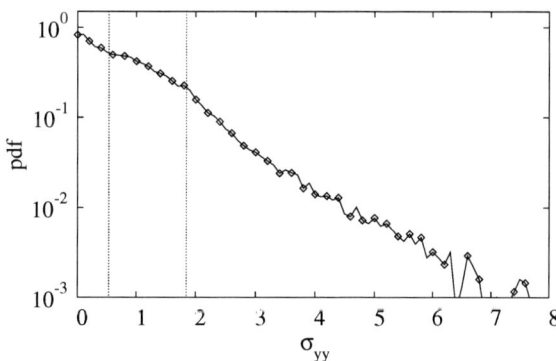

FIGURE 11. Probability density function of vertical stresses normalized by the average stress in compression.

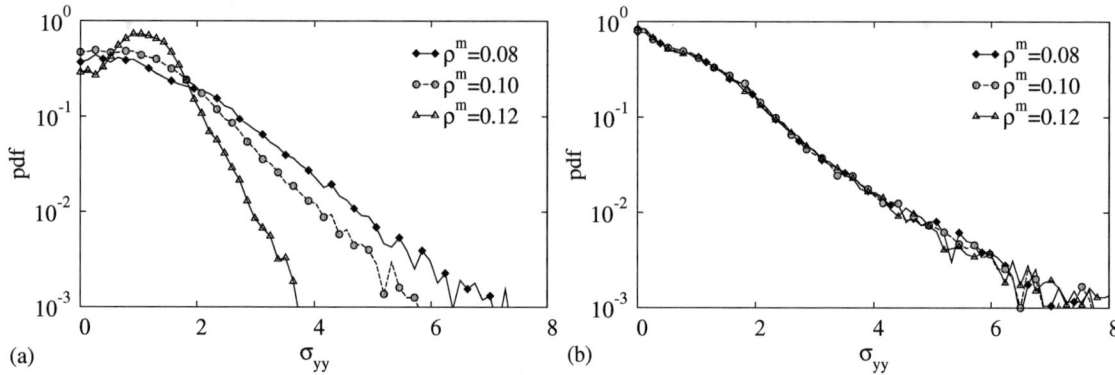

FIGURE 12. Probability density functions of normalized vertical stresses for three values of the matrix volume fraction (a) in tension and (b) in compression.

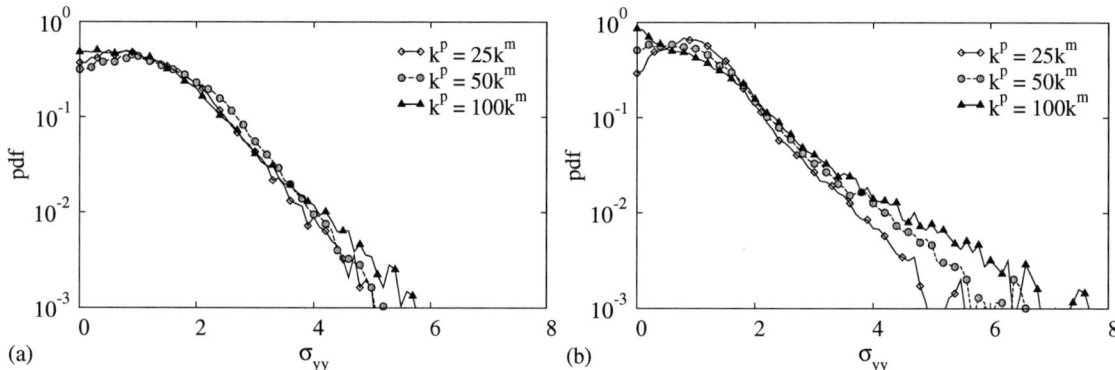

FIGURE 13. Probability density functions of normalized vertical stresses for three values of the relative stiffness k^p/k^m (a) in tension and (b) in compression.

similarity between the distributions of stresses and forces means that the sub-particle stresses are strongly affected by the granular disorder.

Effect of matrix volume fraction

It is expected that at higher matrix contents the stress is more homogeneously redistributed inside the packing due to load transfer between the particles and the matrix. Fig. 12 shows P_σ for three values of ρ^m in tension and compression for $k^p = 100k^m$. Interestingly, the exponential tail persists both in tension and in compression, but for equal matrix volume fractions, the PDF of strong stresses is broader in compression than in tension. In other words, the stress redistribution is more homogeneous in tension than in compression.

It is also interesting to observe that the stress PDF is not affected by the matrix volume fraction in compression but it is increasingly broader in tension for decreasing matrix content so that the stresses are more and more concentrated in the bridges between the particles. In tension, the exponent β varies from 1.10 to 2.55 as ρ^m varies from 0.08 to 0.12 whereas in compression we have $\beta \simeq 0.95$ for all ρ^m. As ρ^m increases, the gaussian peaked on the mean stress, corresponding mainly to the stresses in the bulk of the particles, becomes more and more pronounced.

Particle/matrix stiffness ratio

We now consider the influence of the particle/matrix stiffness ratio k^p/k^m on stress distribution. Fig. 13 displays the vertical stress PDF's for three values of k^p/k^m in tension and compression for $\rho^m = 0.10$. It is remarkable that

FIGURE 14. (Color online) Representation of a cemented granular sample composed of particles (in red), interfaces (in green) and matrix (in blue) discretized on a 3D irregular lattice.

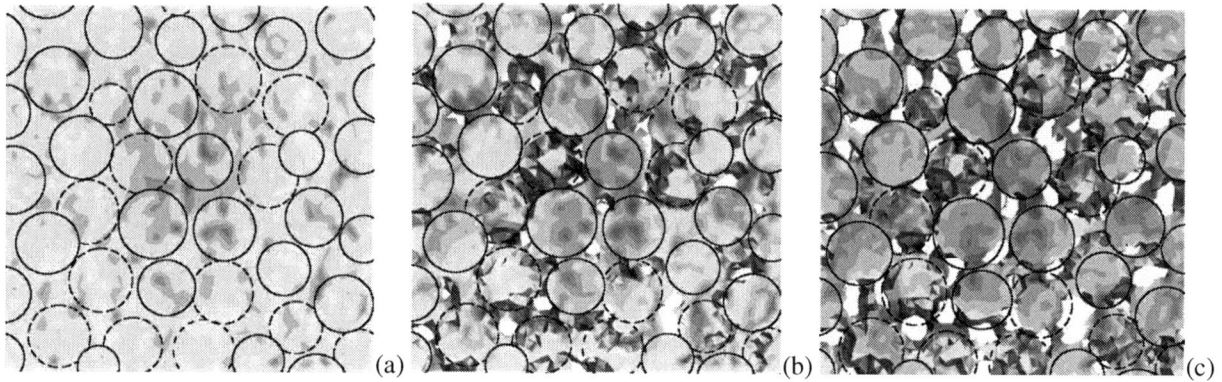

FIGURE 15. (Color on line) Vertical stresses field σ_{yy} in the 3D packing on a cut plane in color level for (a) $\rho^m = 0.37$, (b) $\rho^m = 0.23$, (c) $\rho^m = 0.10$.

in tension the particle stiffness has little influence on the pdf whereas in compression the pdf becomes increasingly broader for an increasing particle stiffness. The respective effects of particle stiffness and matrix volume fraction can be understood by remarking that, due to the presence of a granular backbone, the stress chains are essentially guided by the cementing matrix in tension and by the particle phase in compression. Therefore, the stress transmission is not affected by the matrix volume fraction in compression and only slightly influenced by particle stiffness in tension.

Effect of composition in 3D

We briefly extend here our studies to 3D cemented granular solids. We generated a dense packing of 300 particles discretized over an irregular 3D lattice containing about 500 000 elements. The particle diameters d vary between d_{min} and $d_{max} = 2d_{min}$ with a uniform distribution by volume fractions. The particle volume fraction is $\rho^p \simeq 0.63$. As in our 2D LEM simulations, the matrix is distributed uniformly in the form of bridges of varying thickness and section between neighboring particles. The filling fraction depends on the cross section of the bridges. This protocol allows us to vary the matrix volume fraction continuously from 0 to 0.37. The sample is displayed in Fig. 14.

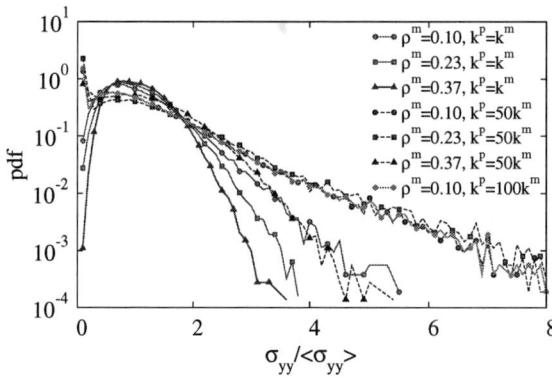

FIGURE 16. Probability density functions of normalized vertical stresses for different values of stiffness ratio k^p/k^m and values of the matrix volume fraction ρ^m in compression.

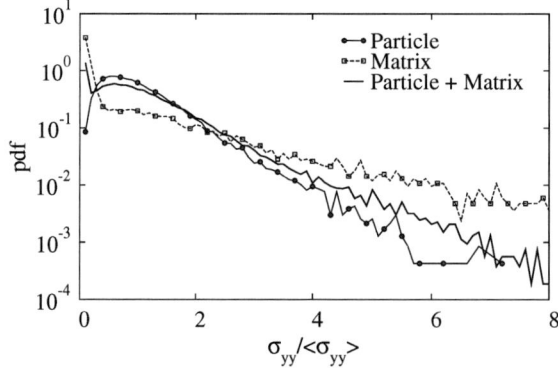

FIGURE 17. Probability density functions of normalized vertical stresses for $\rho^m = 0.1$ and $k^p/k^m = 50$ in the particle and matrix phases in comparison with that in the whole sample.

Fig. 15 displays a map of vertical stresses on a cut plane for three values of ρ^m. We observe that the stresses are more and more localized in the matrix bridges as the matrix volume fraction is reduced. Figure 16 shows the vertical stress pdf's for three values of k^p/k^m and three values of ρ^m under vertical compression with free lateral boundaries. Two limits can be distinguished: (1) The *homogeneous limit* characterized by $\rho^m = 0.37$ and $k^p = k^m$, corresponding to a homogeneous material with no void and no particle (absence of elastic contrast between particles and matrix); (2) The *granular limit* characterized by large k^p and weak amount of matrix (here $\rho^m = 0.1$) basically distributed in the form of small solid bonds between particles. The latter corresponds to a granular material with stiff particles as generally assumed in DEM simulations. We see that, as expected, the stress distribution in the homogeneous limit is the less broad one with a nearly gaussian shape. The stress variability in this system reflects the metric disorder of the underlying lattice. The distribution for $\rho^m = 0.1$ and $k^p/k^m = 100$ corresponds to the granular limit.

The strong stresses have a decreasing exponential distribution as in 2D packings in the granular limit with as exponent increasing with matrix volume fraction. A secondary peak is observed in the range of very weak stresses in all cases where the particles are stiffer than the matrix. This peak reflects the weak stresses inside the matrix bridges, as suggested by Fig. 17 where the distributions are separately plotted for the stresses in the matrix and inside the particles in the case $\rho^m = 0.1$ and $k^p/k^m = 100$. We see that the particles involve no stress peak. This peaks is thus a consequence of the low stiffness of the binding phase.

The distribution in the granular limit is practically the broadest one, and hence all distributions for all parameters lie between those for the granular and homogeneous limits. For $\rho^m = 0.23$ and $k^p/k^m = 1$ we have a porous material with no mechanical contrast between the matrix and particles. For $\rho^m = 0.37$ and $k^p/k^m = 50$ we have a granular phase embedded in a matrix with no voids. In both these cases, the stress distribution is broader than that in the homogeneous limit although the physical origins of this enhanced inhomogeneity are different. We remark that, for $\rho^m = 0.1$, increasing k^p/k^m from 50 to 100 has little influence on the stress distribution. In the same way, for

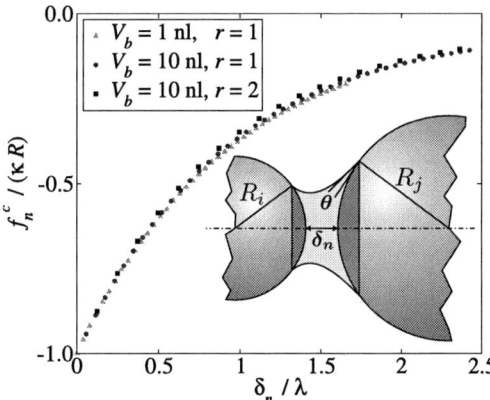

FIGURE 18. Scaled plot of the capillary force as a function of the gap between two particles for different values of the local liquid volume V_b and size ratio r according to the model proposed in this paper. Inset: Geometry of a capillary bridge.

$k^p/k^m = 50$, increasing ρ^m from 0.1 to 0.23 has practically no impact on the distribution.

LIQUID BONDING

In this section, we investigate force transmission in wet granular media composed of rigid particles interconnected by capillary bridges. The action of the capillary bridges is modeled by a capillary force law and implemented in a 3D MD code.

Numerical method

For the simulations of wet granular materials, we used the MD method with spherical particles and a capillary force law. The total normal force f_n at each contact is the sum of a repulsive force f_n^r and an attractive capillary force f_n^c. The latter is a function of the liquid bond parameters, namely the gap δ_n, the liquid bond volume V_b, the liquid surface tension γ_s, and the particle-liquid-gas contact angle θ; see inset in Fig. 18. The capillary force can be calculated by integrating the Laplace-Young equation [34, 35, 36]. However, for efficient MD simulations, we need an explicit expression of f_n^c as a function of the liquid bond parameters.

We used an analytical form for the capillary force which is well fitted by the data from direct integration of the Laplace-Young equation both for polydisperse particles [37]. At leading order, the capillary force f_0 at contact, i.e. for $\delta_n \leq 0$, is

$$f_0 = -\kappa R, \tag{6}$$

where R is a length depending on the particle radii R_i and R_j and κ is given by [38, 39, 40]

$$\kappa = 2\pi \gamma_s \cos \theta. \tag{7}$$

A negative value of δ_n corresponds to an overlap between the particles. The assumption is that the overlap is small compared to the particle diameters. The data obtained from direct integration of the Laplace-Young equation show that the geometric mean $R = \sqrt{R_i R_j}$ is more suited than the harmonic mean $2R_i R_j/(R_i + R_j)$ proposed by Derjaguin for polydisperse particles in the limit of small gaps (see below) [41]. We also note that f_0 in Eq. (6) is independent of the bond liquid volume V_b.

The adhesion force f_0 at contact is the highest level of the capillary force. The latter declines as the gap δ_n increases. The capillary bridge is stable as long as $\delta_n < \delta_n^{max}$, where δ_n^{max} is the debonding distance given by [42]

$$\delta_n^{max} = \left(1 + \frac{\theta}{2}\right) V_b^{1/3}. \tag{8}$$

Between these two limits, the capillary force falls off exponentially with δ_n:

$$f_n^c = f_0 e^{-\delta_n/\lambda}, \tag{9}$$

where λ is a length scale which should be a function of V_b and the particle radii. The asymmetry due to unequal particle sizes is taken into account through a function of the ratio between particle radii. We set

$$r = \max(R_i/R_j; R_j/R_i). \tag{10}$$

Dimensionally, a plausible expression of λ is

$$\lambda = c\, h(r) \left(\frac{V_b}{R'}\right)^{1/2}, \tag{11}$$

where c is a constant and h is a function only of r. When introduced in Equations (11) and (9), this form yields a nice fit for the capillary force obtained from direct integration of the Laplace-Young equation by setting $R' = 2R_i R_j/(R_i + R_j)$, $h(r) = r^{-1/2}$ and $c \simeq 0.9$.

Figure 18 shows the plots of Eq. 9 for three different values of the liquid volume V_b and size ratio r together with the corresponding data from direct integration. The forces are normalized by κR and the lengths by λ. The data collapse on the same cruve, indicating again that the force κR and the expression of λ in Eq. (11) characterize correctly the behavior of the capillary bridge.

Finally, the capillary force can be expressed in the following form:

$$f_n^c = \begin{cases} -\kappa R & \text{for } \delta_n < 0 \\ -\kappa R\, e^{-\delta_n/\lambda} & \text{for } 0 \leq \delta_n \leq \delta_n^{max} \\ 0 & \text{for } \delta_n > \delta_n^{max} \end{cases}, \tag{12}$$

with

$$\lambda = \frac{c}{\sqrt{2}} \left\{ \frac{1/R_i + 1/R_j}{\max(R_i/R_j; R_j/R_i)} V_b \right\}^{\frac{1}{2}}. \tag{13}$$

In the simulations, the total liquid volume is distributed among all eligible particle pairs (the pairs with a gap below the debonding distance, including the contact points) in proportion to the reduced diameter of each pair. We also assume that the particles are perfectly wettable, i.e. $\theta = 0$. The choice of the liquid volume has no influence on the value of the largest capillary force in the pendular state [43]. For our simulations, we chose a gravimetric water content of 0.007 so that the material is in the pendular state. The coefficient of friction is $\mu = 0.4$ for all simulations.

Distributions of bond forces

We consider force PDF's in a wet packing of 8000 spheres simulated by the MD method for $p_m = 0$ Pa and $p_m = 100$ Pa. The confined sample was obtained by isotropic compaction of a wet packing initially prepared with $p_m = 0$. The packing was then allowed to relax to equilibrium under the action of the applied pressure. This level of confinement is high compared to the reference pressure $p_0 = f_0/\langle d \rangle$ ($p_m/p_0 \simeq 0.5$), yet not too high to mask fully the manifestations of capillary cohesion.

Figure 19 shows the force networks in a narrow slice nearly three particle diameters thick in both samples. The tensile and compressive forces are represented by segments of different colors joining particle centers. As in dry granular media, we observe a highly inhomogeneous distribution both for tensile and compressive forces. The effect of external compressive pressure is to reduce the fraction of tensile bonds. In the unconfined packing, the bond coordination number z (average number of liquid bonds per particle) is $\simeq 6.1$ including nearly 2.97 compressive bonds and 3.13 tensile bonds. As we shall see below, these wet samples involve also a large number of weak forces ($f_n \simeq 0$) corresponding to the contacts where capillary attraction is balanced by elastic repulsion, i.e. $k_n \delta_n + f_0 \simeq 0$.

Figure 20 displays the PDF of normal forces in tensile (negative) and compressive (positive) ranges in the unconfined packing ($p_m = 0$ Pa). We observe two nearly symmetrical parts decaying exponentially from the center:

$$P(f_n) \propto e^{-\alpha_w |f_n|/f_0}, \tag{14}$$

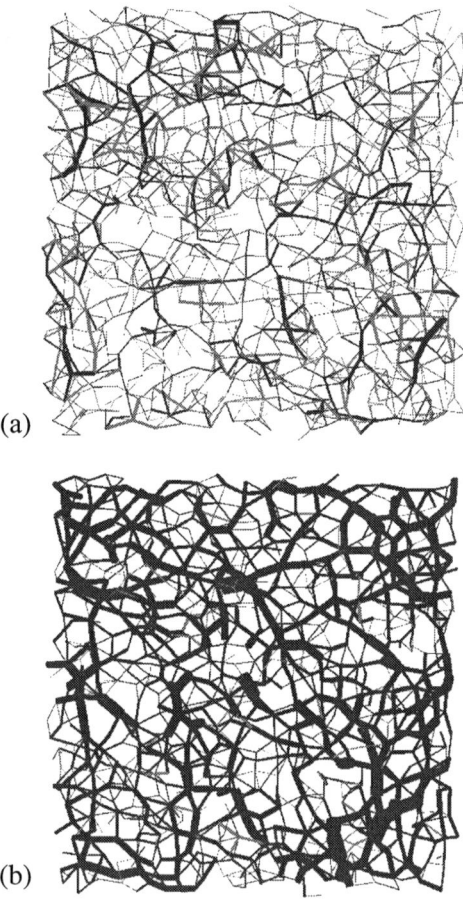

FIGURE 19. Maps of tensile (green) and compressive (red) forces in a thin layer in samples S_6 ($p_m = 0$ Pa) (a) and S_7 ($p_m = 100$ Pa) (b). Line thickness is proportional to the magnitude of the force.

with $\alpha_w \simeq 4$ for both negative and positive forces, and $f_0 = \kappa R_{max}$, where R_{max} is the largest particle radius. In contrast to dry granular media, where the distribution deviates from a purely exponential behavior for weak forces (section), here the exponential behavior extends to the center of the distribution. The tensile range is cut off at $f_n = -f_0$ corresponding to the largest capillary force. Although the confining stress is zero, positive forces as large as $2f_0$ can be found in the system. We also observe in Fig. 20 a distinct peak centered on $f_n = 0$ which is the average force for zero confining pressure. The presence of this peak, resulting from the balance between capillary attraction and elastic repulsion, suggests that a large number of weak forces play a special role with respect to the statics and stability of wet granular materials.

Figure 21 shows the PDF of normal forces in the confined packing. The symmetry of the distribution around $f_n = 0$ is now broken compared to the unconfined case in Fig. 20. The distribution is roughly exponential for both tensile and compressive forces but the exponents are different as in the case of solid cohesion at low matrix volume fraction (section). In the same figure, the PDF of normal forces in a sample without capillary cohesion is shown. We see that the exponent for compressive forces is nearly the same as in the dry packing. Another feature of force distribution observed in Fig. 21 is the presence of a distinct peak centered on zero force which was observed also for the case of unconfined packing in Fig. 20. Hence, this peak reflects a feature of force transmission in wet granular materials that will be analyzed below.

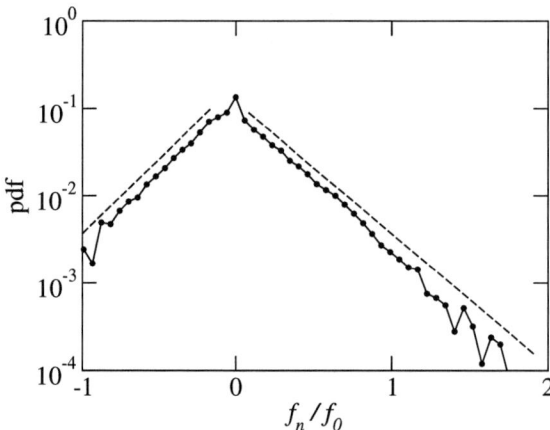

FIGURE 20. Probability density function of normal forces normalized by the largest capillary force f_0 at zero confining pressure.

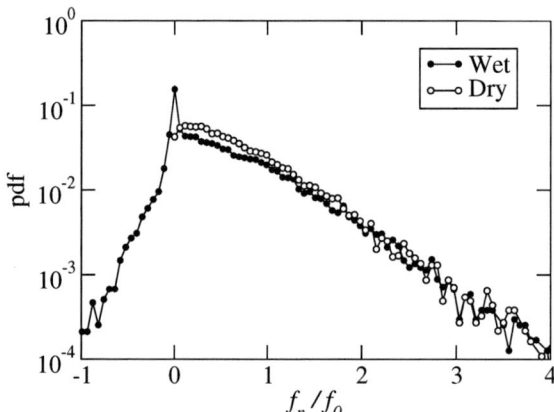

FIGURE 21. Probability density functions of normal forces normalized by the largest capillary force f_0 in the wet and dry confined packings.

Particle pressures

In an unconfined assembly of dry rigid particles, no self-stresses occur and the forces vanish at all contacts. However, the presence of liquid bonds in a wet granular material induces tensile and compressive forces whilst the average force is zero. In other words, the grains keep together to form a self-sustained structure in the absence of confining stresses. In general, various loading histories such as consolidation or differential particle swelling can induce self-stresses in a cohesive packing [44]. In our system, the self-stresses appear during relaxation. This is obviously a consequence of the tensile action of capillary bonds bridging the gaps between neighboring particles within the debonding distance.

For a local description of self-stresses we need to characterize the stress transmission at the particle scale as the smallest scale at which the force balance condition is defined for rigid particles. Although the stress tensor is by definition a macroscopic quantity, it can be shown that an equivalent particle stress σ_i can be defined for each particle i of a granular packing in static equilibrium [33, 45, 46]:

$$(\sigma_i)_{ab} = \frac{1}{V_i} \sum_{j \neq i} f_a^{ij} r_b^{ij}, \tag{15}$$

where r_{ij} is the position of the contact-point of the force f_{ij} of particle j on particle i, and a and b design the Cartesian components. V_i is the free volume of particle i, the sum of the particle volume and a fraction of the pore space:

$$V_i = \frac{\pi d_i^3}{6\nu}, \tag{16}$$

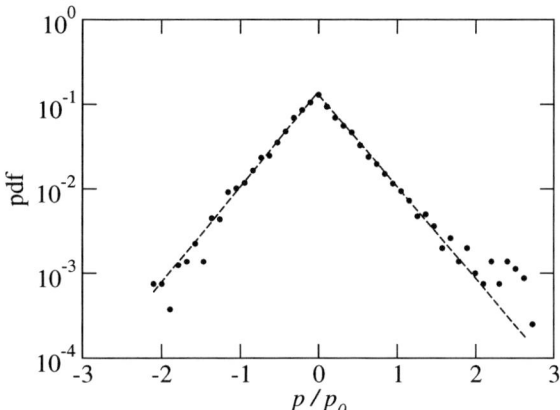

FIGURE 22. Probability density function of particle pressures normalized by reference pressure p_0 (see text) in the unconfined wet packing.

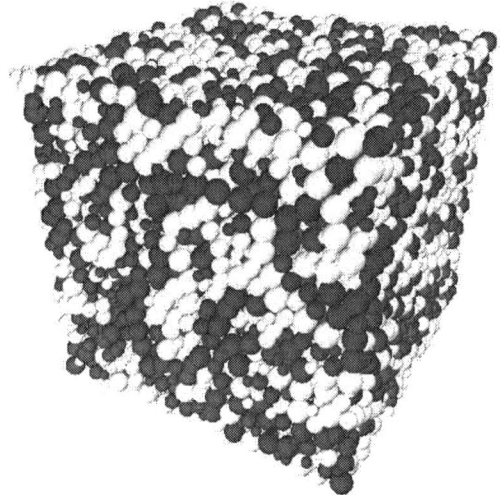

FIGURE 23. The unconfined wet packing with negative (white) and positive (black) particle pressures.

where d_i is the particle diameter, and ν is the solid fraction of the packing. The sum of particle stresses σ_i weighted by the corresponding relative free volumes V_i/V tends to the Cauchy stress tensor as the number of particles in a control volume V increases.

From particle stresses we get particle pressures:

$$p_i = \frac{1}{3} \sum_{\alpha=1}^{3} (\sigma_i)_{\alpha\alpha}. \qquad (17)$$

Each particle can take on positive or negative pressures according to the forces exerted by neighboring particles. The PDF of particle pressures is displayed in Fig. 22 for the unconfined sample. The pressures have been normalized by a reference pressure $p_0 = f_0/\langle d \rangle^2$. The distribution is symmetric around and peaked on zero pressure, and each part is well fit by an exponential form. This symmetry in the structure of self-stresses must be contrasted with the asymmetric distribution of forces (Fig. 20) due to the cutoff on tensile forces. Obviously, the exponential shape of particle pressure distributions reflects statistically that of bond forces. This distribution extends to the center $p_i = 0$.

Zero particle pressure corresponds to a state where a particle is balanced under the combined action of tensile and compressive forces. Such particle states are not marginal here and they reflect a particular stress transmission in a wet packing. The positive and negative particle pressures form separate phases as observed in Fig. 23 where positive and negative pressures are represented in black and white, respectively. Each phase percolates throughout the system. The morphology of each phase is approximately filamentary with variable thickness and a large interface between them. A detailed analysis of this structure shows that the particles at the interface between the two phases have a weak pressure and the largest negative or positive pressures are located at the heart of each phase [46].

CONCLUSION

In this paper, the distributions of contact forces and stresses were investigated in cohesive and noncohesive granular media by means of different numerical methods. The exponential fall-off of the number of strong forces and stresses is a robust feature of the distributions in packings of different particle shapes and size distributions with both liquid and solid bonding. In contrast, the force probability density in the range of weak forces and stresses was found to depend on system parameters, taking different shapes from a peaked distribution to a decreasing power law distribution. For wet granular media with a homogeneous distribution of liquid bonds, we showed the nontrivial organization of particle pressures in two separate percolating phases of tensile and compressive particle pressures with an interphase at zero pressure.

For the simulation of solid bonding, we used the lattice element method which provides a suitable framework for the investigation of stress fields in complex granular solids involving a solid matrix sticking to the particles. By coarse-graining the sub-particle stresses, we arrived at the same contact force distributions as in DEM simulations and experiments. Our data are consistent with the fact that the decreasing exponential distribution of strong forces is a signature of *granular disorder*, i.e. the disorder induced by a contiguous network of stiff particles. This signature disappears in the homogeneous limit where there is no stiffness contrast between the particle and matrix phases and the porosity vanishes or when the particles are interposed everywhere by the binding matrix. Our 3D simulations evidence the two limits of homogeneous and granular distributions. For different values of the matrix volume fraction and particle/matrix stiffness ratio, the distributions vary between these two limits.

REFERENCES

1. C. Liu, and H. M. Jaeger, *Phys. Rev. Lett.* **74** (1995).
2. F. Radjai, L. Brendel, and S. Roux, *Phys. Rev. E* **54**, 861 (1996).
3. H. Jaeger, and S. Nagel, *Reviews of Modern Physics* **68**, 1259–1273 (1996).
4. F. Radjai, D. E. Wolf, M. Jean, and J. Moreau, *Phys. Rev. Letter* **80**, 61–64 (1998).
5. D. M. Mueth, H. M. Jaeger, and S. R. Nagel, *Phys. Rev. E* **57**, 3164 (1998).
6. T. S. Majmudar, and R. P. Behringer, *Nature* **435**, 1079–1082 (2005).
7. E. Schlangen, and E. J. Garboczi, *Engineering Fracture Mechanics* **57**, 319–332 (1997).
8. V. Topin, J.-Y. Delenne, F. Radjai, L. Brendel, and F. Mabille, *The European Physical Journal E* **23**, 413–429 (2007).
9. V. Topin, F. Radjai, J.-Y. Delenne, A. Sadoudi, and F. Mabille, *Journal of Cereal Science* **47**, 347–356 (2008).
10. F. Radjai, M. Jean, J.-J. Moreau, and S. Roux, *Phys. Rev. Lett.* **77**, 274– (1996).
11. H. J. Herrmann, and S. Luding, *Continuum Mechanics and Thermodynamics* **10**, 189–231 (1998).
12. G. Lovol, K. Maloy, and E. Flekkoy, *Phys. Rev. E* **60**, 5872–5878 (1999).
13. S. G. Bardenhagen, J. U. Brackbill, and D. Sulsky, *Phys. Rev. E* **62**, 3882–3890 (2000).
14. S. Roux, and F. Radjai, "Statistical approach to the mechanical behavior of granular media," in *Mechanics for a New Millennium*, edited by H. Aref, and J. Philips, Kluwer Acad. Pub., Netherlands, 2001, pp. 181–196.
15. L. E. Silbert, D. Erta, G. S. Grest, T. C. Halsey, and D. Levine, *Phys. Rev. E* **65**, 051307 (2002).
16. P. Dantu, *Ann. Ponts Chauss.* **IV**, 193–202 (1967).
17. A. Drescher, and G. de Josselin de Jong, *J. Mech. Phys. Solids* **20**, 337–351 (1972).
18. F. Radjai and S. Roux, *Phys. Rev. E* **51**, 6177 6187 (1995).
19. F. Radjai, S. Roux, and J. J. Moreau, *Chaos* **9**, 544–550 (1999).
20. S. J. Antony, *Phys Rev E* **63**, 011302 (2001).
21. P. T. Metzger, *Phys. Rev. E* **69**, 053301; discussion 053302 (2004).
22. S. N. Coppersmith, C. Liu, S. Majumdar, O. Narayan, and T. A. Witten, *Phys. Rev. E* **53**, 4673–4685 (1996).
23. P. A. Cundall, and O. D. L. Strack, *Géotechnique* **29**, 47–65 (1979).
24. J. Moreau, *European Journal of Mechanics A/Solids* **supp.**, 93–114 (1994).
25. M. Jean, "The Non Smooth Contact Dynamics method," in *Special issue on modeling contact and friction*, 1999.
26. F. Radjai, and V. Richefeu, *Mechanics of Materials* **41**, 715–728 (2009), ISSN 0167-6636.
27. F. Radjai, I. Preechawuttipong, and R. Peyroux, "Cohesive granular texture," in *Continuous and discontinuous modelling of cohesive frictional materials*, edited by P. Vermeer, S. Diebels, W. Ehlers, H. Herrmann, S. Luding, and E. Ramm, Springer Verlag, Berlin, 2001, pp. 148–159.

28. R. C. Youngquist, P. T. Metzger, and K. N. Kilts, *SIAM J. App. Math.* **65**, 1855 (2005).
29. E. Azéma, G. Saussine, and F. Radjai, *Mechanics of Materials* **41**, 729-741 (2009).
30. E. Azéma, F. Radjai, R. Peyroux, and G. Saussine, *Phys. Rev. E* **76**, 011301 (2007).
31. C. Voivret, F. Radjai, J.-Y. Delenne, and M. S. E. Youssoufi, *Phys. Rev. Lett.* **102**, 178001 (2009).
32. C. Voivret, F. Radjai, J.-Y. Delenne, and M. S. E. Youssoufi, *Phys. Rev. E* **76**, 021301 (2007).
33. J. J. Moreau, "Numerical Investigation of Shear Zones in Granular Materials," in *Friction, Arching, Contact Dynamics*, edited by D. E. Wolf, and P. Grassberger, World Scientific, Singapore, pp. 233–247 (1997).
34. G. Lian, C. Thornton, and M. Adams, *Journal of Colloid and Interface Science* **161**, 138–147 (1993).
35. T. Mikami, H. Kamiya, and M. Horio, *Chemical Engineering Science* **53**, 1927–1940 (1998).
36. F. Soulié, M. S. E. Youssoufi, F. Cherblanc, and C. Saix, *Eur. Phys. J. E* **21**, 349–357 (2006).
37. V. Richefeu, F. Radjai, and M. S. E. Youssoufi, *Eur. Phys. J. E* **21**, 359–369 (2007).
38. C. Willett, M. Adans, S. Johnson, and J. Seville, *Langmuir* **16**, 9396–9405 (2000).
39. L. Bocquet, E. Charlaix, and F. Restagno, *Comptes Rendus Physique* **3**, 207–215 (2002).
40. M. Kohonen, D. Geromichalos, M. Scheel, C. Schier, and S. Herminghaus, *Physica A* **339**, 7–15 (2004).
41. J. N. Israelachvili, *Intermolecular and surface forces*, Academic Press, Cambridge University, 1993.
42. G. Lian, C. Thornton, and M. J. Adams, *Chemical Engineering Science* **53**, 3381–3391 (1998).
43. V. Richefeu, M. S. d El Youssoufi, and F. Radjai, *Phys. Rev. E* **73**, 051304 (2006).
44. M. S. El Youssoufi, J.-Y. Delenne, and F. Radjaï, *Phys. Rev. E* **71**, 051307 (2005).
45. L. Staron, and F. Radjai, *Phys Rev E* **72**, 041308 (2005).
46. V. Richefeu, M. S. El Youssoufi, R. Peyroux, and F. Radjai, *International Journal for Numerical and Analytical Methods in Geomechanics* **32**, 1365–1383 (2008).

How granular materials deform in quasistatic conditions

J.-N. Roux [*] and G. Combe [†]

[*] *Université Paris-Est, Laboratoire Navier, 2 Allée Kepler, Cité Descartes, 77420 Champs-sur-Marne, France*
[†] *Laboratoire 3SR, Université Joseph Fourier, 38041 Grenoble, France*

Abstract. Based on numerical simulations of quasistatic deformation of model granular materials, two rheological regimes are distinguished, according to whether macroscopic strains merely reflect microscopic material strains within the grains in their contact regions (type I strains), or result from instabilities and contact network rearrangements at the microscopic level (type II strains). We discuss the occurrence of regimes I and II in simulations of model materials made of disks (2D) or spheres (3D). The transition from regime I to regime II in monotonic tests such as triaxial compression is different from both the elastic limit and from the yield threshold. The distinction between both types of response is shown to be crucial for the sensitivity to contact-level mechanics, the relevant variables and scales to be considered in micromechanical approaches, the energy balance and the possible occurrence of macroscopic instabilities

Keywords: granular materials, elastoplasticity, stress-strain behavior
PACS: 81.05.Rm, 83.80.Fg

INTRODUCTION

The quasistatic limit, the rigid limit and the macroscopic limit

Although they are modeled, at the macroscopic level, with constitutive laws in which physical time and inertia play no part [1, 2], granular materials are most often investigated at the microscopic level by "discrete element" numerical methods (DEM) in which the motion of the solid bodies is determined through integration of dynamical equations involving masses and accelerations. Fully quasistatic approaches, in which the system evolution in configuration space, as some loading parameter is varied, is regarded as a continuous set of mechanical equilibrium states, are quite rare in the numerical literature [3, 4, 5]. It is regarded as a natural starting point, on the other hand, to perform suitable averages of the mechanical response of the elements of a contact network to derive the macroscopic material response [6]. Whether and in which cases it is possible to dispense with dynamical ingredients of the model at the granular level and how the quasistatic limit is approached are fundamental issues that still need clarification.

Another set of open questions are related to the role of particle deformability. Most DEM studies include contact elasticity in the numerical model. Experimentally, elastic behavior is routinely measured in quasistatic tests [7] and sound propagation. Yet, most often, contact deflections are quite negligible in comparison with grain diameters. In the "contact dynamics" method [8, 9], which is used to simulate quasistatic granular rheology [10, 11, 12], grains are modeled as rigid, undeformable solid bodies. The influence of contact deformability on the macroscopic behavior, the existence of a well-defined rigid limit are thus other basic issues calling for further investigations.

Small granular samples, as the ones used in DEM studies, often exhibit quite noisy mechanical properties. The approach to a macroscopic behavior expressed with smooth stress-strain curve might seem problematic, especially in the presence of rearrangement events, associated with instabilities at the microscopic level [13, 14].

The origins of strain

The present communication shows how one may shed light on the interplay between the quasistatic, rigid and macroscopic limits on distinguishing two different rheological regimes and delineating their conditions of occurrence, in simple model materials. Macroscopic strain in solidlike granular materials has two obvious physical origins: first, grains deform near their contacts, where stresses concentrate (so that one models the grain interaction with a point force); then, grain packs rearrange as contact networks break, and then repair in a different stable configuration. We refer here respectively to the two different kinds of strains as type I and II. The present paper, based on numerical simulations of simple materials, identifies the regimes, denoted as I and II accordingly, within which one mechanism or the other dominates, and discusses the consequences on the quasistatic rheology of granular materials.

NUMERICAL MODEL MATERIALS AND SIMULATION PROCEDURES

Two sets of numerical simulation results are exploited below. Two-dimensional (2D) assemblies of polydisperse disks, as in Refs. [15, 3, 16], are subjected to fully stress-controlled biaxial tests, for which a quasistatic computation method [15, 3, 17] is exploited, in addition to standard DEM simulations. The behavior of three-dimensional (3D) packs of monosized spherical particles, as in Refs. [18, 19, 20] is studied in simulated triaxial compression tests, with special attention to strains in the quasistatic limit. Part of the results are presented in the references (mostly in some conference proceedings) cited just above, pending the publication of a more comprehensive study.

Two-dimensional material and stress-controlled tests

2D systems are simulated in order to investigate basic rheophysical mechanisms with good accuracy, in the simplest conceivable, yet representative, model material. Samples made of polydisperse disks in 2D, with a uniform diameter distribution between $0.5a$ and a, are first assembled on isotropically compressing frictionless particles, thus producing dense packs with solid fraction $\Phi = 0.8434 \pm 3 \times 10^{-4}$ and coordination number z close to 4 in the large system limit. Those values are extrapolated from data averaged on sets of samples with $N = 1024, 3025$ and 4900 disks. The samples are enclosed in a rectangular cell framed by solid walls, 2 of which are mobile orthogonally to their direction, which enables us to carry out biaxial compression tests (Fig. 1). Finite system effects on Φ and z are mainly due to the surrounding walls and can be eliminated (they are proportional to perimeter to area ratio).

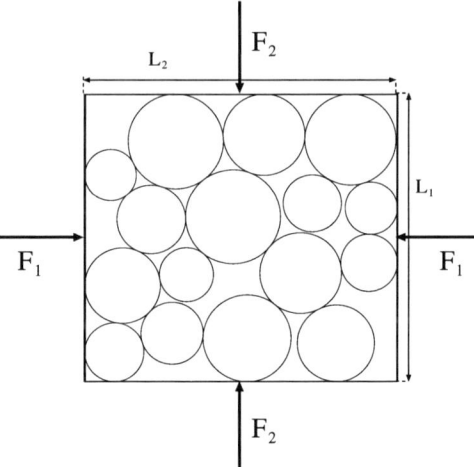

FIGURE 1. Schematic representation of the biaxial tests simulated on 2D disk samples. $\sigma_2 = F_2/L_2$ is kept constant, equal to the initial isotropic pressure P, while $\sigma_1 = F_1/L_1$ is stepwise increased.

Stress-increment controlled DEM simulations

Once prepared in mechanical equilibrium under an isotropic pressure P, disk samples, in which contacts are now regarded as frictional, with friction coefficient $\mu = 0.25$, are subjected to biaxial tests as sketched in Fig. 1. Strains $\varepsilon_1 = -\Delta L_2/L_2$, $\varepsilon_2 = -\Delta L_1/L_1$, "volumetric" strain $\varepsilon_v = \varepsilon_1 + \varepsilon_2$ are measured in equilibrium configurations, while the stress deviator $q = \sigma_2 - \sigma_1$ is the control parameter. We use soil mechanics conventions, for which compressive stresses and shrinking strains are positive. q is stepwise increased by small intervals $\Delta q = 10^{-3}P$. The contact model is the standard (Cundall-Strack [21]) one with normal (K_N) and tangential (K_T) stiffness constants such that $K_N = 2K_T = 10^5 P$. A normal viscous force is also introduced in the contact, in order to reach equilibrium configurations faster. After each deviator step, one waits for the next equilibrium configuration, in which forces and moments are balanced with good accuracy (with a tolerance below $10^{-5} Pa$ for forces on grains, below $10^{-5} PL$ for forces on walls). We refer to this procedure as *stress-increment controlled* (SIC) DEM.

The stricly quasistatic approach: SEM calculations

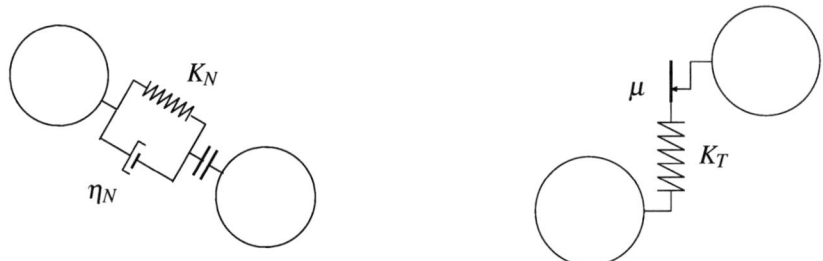

FIGURE 2. Normal (left) and tangential (right) contact behavior in 2D disk samples, as schematized with rheological elements: springs with stiffness constants K_N, K_T, dashpot with damping constant η_N, plastic slider with threshold related to normal force by coefficient μ.

The *static elastoplastic method* (hereafter referred to as SEM), amounts to dealing with the initial sample configuration as a network of springs and plastic sliders corresponding to contact behavior, as in Fig. 2 – with the dashpots ignored, as they play no role in statics. The evolution of the system under varying load is determined as a continuous trajectory in configuration space, each point of which is an equilibrium state. It has been implemented in [15, 3], and a similar approach was used in [4]. The algorithm will not be described here, as it is presented in [17]. It relies on resolution of linear system of equations, with the form of the matrix (the elastoplastic stiffness matrix) depending on contact status (nonsliding, sliding, open). The bases of the approach are also discussed in [5].

SEM calculations are possible as long as only type I strains are obtained, and the results reported here [15] correspond to the deviator interval $0 \leq q \leq q_1$ in biaxial compressions from the chosen initial state (in which all tangential forces are equal to zero), in which a type I response is obtained.

Triaxial compression of 3D bead assemblies

Triaxial compression tests of assemblies of $N = 4000$ single-sized spherical beads of diameter a are simulated by DEM, with the more standard procedure in which the *axial strain rate* $\dot{\varepsilon}_a$ is kept constant (hereafter *strain-rate controlled* or SRC DEM). The deviator stress, q, is measured, as a function of axial strain $\varepsilon_a = \varepsilon_1$, as $q = \sigma_1 - \sigma_3$, where σ_1 is the major ("axial") principal stress conjugate to ε_a, while the other two (lateral) principal stresses $\sigma_2 = \sigma_3$ are kept equal to the initial isotropic pressure P. To allow for comparisons with laboratory experiments, the beads are attributed the elastic properties of glass (Young modulus $E = 70$ GPa, Poisson ratio $\nu = 0.3$) and friction coefficient $\mu = 0.3$. The contact law is a somewhat simplified version of the Hertz-Mindlin ones [22], as in Ref. [19], which might be consulted for more details. It leads to favorable comparisons of elastic moduli [20] obtained in simulations and measured in experiments on glass beads. Preparation of cuboidal samples with periodic boundaries in all three directions (and thus statistically homogeneous and devoid of wall effects) under prescribed pressures in the range 10 kPa $\leq P \leq$ 1 MPa is detailed in Refs. [19, 23]. It is shown [19] that one may obtain, depending on the assembling procedure, for densities close to the random close packing limit $\Phi \simeq 0.64$ under low pressure, coordination numbers ranging from $z \simeq 4$ (or $z^* \simeq 4.5$ if the "rattlers", i.e., grains carrying no force, are excluded from the count) to $z = 6$ (more exactly $z^* = 6$, with 1 or 2% of rattlers) in the limit of $P \to 0$. As in [19], the low-coordination systems ($z^* \simeq 4.5$) are referred to as "C samples" in the sequel, while those with $z^* \simeq 6$ are called "A samples". We can thus assess the influence of initial coordination number on the small strain (pre-peak) behavior of a dense material.

Dimensionless control parameters

The contact law and the simulated mechanical test lead to the definition of useful dimensionless numbers. The *inertia parameter* $I = \dot{\varepsilon}_a \sqrt{m/aP}$ (in 3D) or $I = \dot{\varepsilon}_a \sqrt{m/P}$ (in 2D) characterizes the importance of inertial effects in strain-rate controlled tests under pressure P (m is the grain mass). The parameter I has been used repeatedly to describe the state of granular materials in steady flow, both in experiments [24] and in simulations [25, 26, 27], or the departure from equilibrium in a slow compression [23]. It also plays a central role in the recent formulation of a successful constitutive law for dense granular flows [28].

The importance of contact deflections, relative to grain diameter a, is expressed by the stiffness number, κ, which is defined as $\kappa = K_N/P$ in 2D models with linear contact elasticity, and as $\kappa = \left(\dfrac{E}{(1-\nu^2)P}\right)^{2/3}$ with 3D beads and Hertzian contacts. In both cases, typical contact deflections h satisfy $h/a \propto \kappa^{-1}$ [19].

The three limits mentioned in the introduction can be defined as $I \to 0$ (quasistatic limit), $\kappa \to \infty$ (rigid limit), $N \to \infty$ (macroscopic limit).

SIMULATION RESULTS

Biaxial tests in 2D

Type I response interval, quasistatic approach

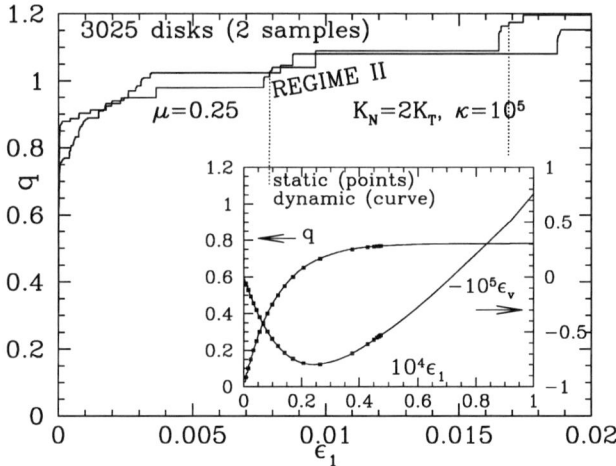

FIGURE 3. q (normalized by P) versus ε_a in SIC tests on 2 samples of 3025 disks, showing a very stiff increase (confused with vertical axis), and then a staircase regime. Inset: detail of very small strains, with comparison of SIC DEM and SEM calculations.

In Fig. 3, $q(\varepsilon_a)$ curves as obtained by SIC DEM are shown for two samples of 3025 disks. The curves first exhibit a very sharp increase of deviator q, which, as revealed once the strain scale is blown up by a factor of 10^4 in the insert, is in fact an interval of type I response: direct SEM calculation are possible, and coincide with DEM results. The smoothness of the stress variations versus strain in that range is characteristic of a continuous trajectory of equilibrium states in configuration space. Beyond the transition to type II strain regimes, a staircase-shaped deviator curve (Fig. 3) is observed, exhibiting intervals of stability (nearly vertical parts of curve in Fig. 3), separated by rearrangement events (horizontal parts of curve in Fig. 3) in which the system gains kinetic energy before a new stable contact network is formed. We could check that the SEM procedure is able to reproduce the stability intervals obtained with SIC DEM. On reversing the load (stepwise decreasing q), a considerably larger deviator range is accessible to SEM calculations, and thus in regime I, as illustrated by the two (quasi-vertical) dotted lines on the main plot of Fig. 3.

TABLE 1. Average and standard deviation of q_1 as obtained over 26 samples with N=1024, 10 samples with N=3025 and 6 samples with N=4900.

SEM	N=1024	N=3025	N=4900
$\langle q_1 \rangle$	0.750 ± 0.050	0.774 ± 0.033	0.786 ± 0.024

Role of contact stiffness

As the system, in regime I, is equivalent to a network of springs and plastic sliders (Fig. 2), type I strains are all inversely proportional to stiffness level κ, provided the compression that decreases κ does not significantly affect the sample geometry. The curves pertaining to different κ values coincide if expressed with stress ratios and variables $\kappa \varepsilon$, as shown in Fig 4.

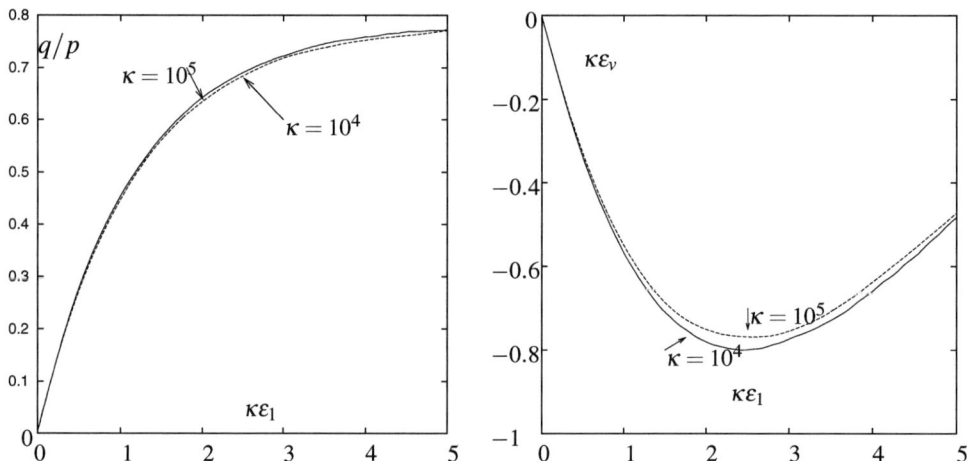

FIGURE 4. Stress ratio q/p (left) and rescaled volumetric strain $\kappa \varepsilon_v$ (right) vs. rescaled axial strain $\kappa \varepsilon_1$ for $\kappa = 10^5$ and $\kappa = 10^4$.

Approach to the macroscopic limit

The staircase-shaped loading curves in regime II should approach in the large sample limit a smooth stress-strain curve, as observed in very slow laboratory tests. To check for the approach of such a macroscopic limit, the average $\langle q(\varepsilon_1) \rangle$ and the standard deviation $\sigma(q)(\varepsilon_1)$ are computed as functions of axial strain for sets of samples of three different sizes, and the region of the $\varepsilon_1 - q$ plane corresponding to $\langle q \rangle (\varepsilon_1) - \sigma(q) \leq q \leq \langle q \rangle (\varepsilon_1) + \sigma(q)$ is shaded on Fig. 5, darker zones corresponding to larger N. Fluctuations about the average curve decrease as the system size increases, and the insert shows that the standard deviation, as averaged over the interval $0 \leq \varepsilon_1 \leq 0.02$, regresses as $N^{-1/2}$. Thus staircase curves get smoothed in the large system limit, which implies that the "stairs" become increasingly small and numerous: as N increases rearrangement events (microscopic instabilities) become smaller and smaller, but more and more frequent. A similar regression is observed for the volumetric strain curve.

Unlike the small type I response intervals observed within regime II, the stability range $q \leq q_1$ of the initial, isotropic structure does not dwindle as the system size increases. As shown in Table 1, the initial regime I deviator interval even increases a little, approaching a finite limit as $N \to \infty$.

Our implementation of SEM involves no creation of new contacts (although this could be taken into account in a more refined version). This approximation becomes exact in the rigid limit of $\kappa \to \infty$, because a finite strain increment is necessary for additional contacts to close, while type I strains scale as κ^{-1}. The near coincidence of SEM and DEM approaches, the latter involving contact creations, shows that new contacts are indeed negligible for $\kappa = 10^5$. q_1 thus represents the maximum deviator stress supported by the initial contact network, beyond which [5], due to contact sliding and opening, an instability or a "floppy mode" appears. The hallmark of such instabilities is the negativity of

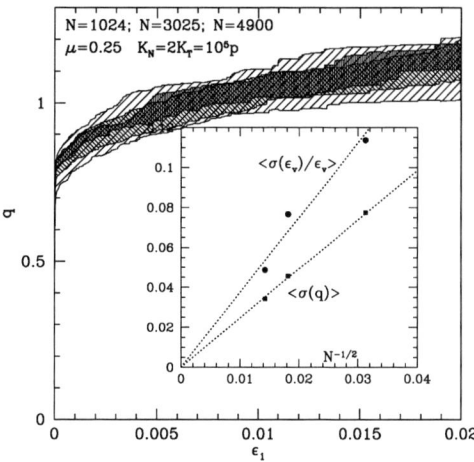

FIGURE 5. Main plot: sample to sample average of q versus ε_1. Shaded regions extend to one standard deviation about the average, with, in this order, darker and darker shades of gray for N=1024, 3025, 4900. Insert: regression of fluctuations, proportional to $N^{-1/2}$. Average standard deviations $\sigma(q)$ and $\sigma(\varepsilon_v)$ over interval $0 \leq \varepsilon_1 \leq 0.02$.

the second-order work [5, 17], *viz.*

$$\Delta^2 W(\Delta \mathbf{U}) = \Delta \mathbf{U} \cdot \underline{\underline{\mathbf{K}}} \cdot \Delta \mathbf{U}$$

for some direction of displacement increment vector $\Delta \mathbf{U}$. Vector $\Delta \mathbf{U}$ comprises all increments of grain displacements and rotations, and $\underline{\underline{\mathbf{K}}}$ is the stiffness matrix, which depends, *via* the status of contacts, on the direction of $\Delta \mathbf{U}$. $\Delta^2 W(\Delta \mathbf{U}) < 0$ implies that the increment of contact forces resulting from a small perturbation $\Delta \mathbf{U}$ will accelerate the resulting motion, whence a spontaneous increase of kinetic energy.

Transition stress q_1 and the "critical yield analysis" approach

One may wonder whether q_1 marks the upper bound q_u of the deviator interval for which contact forces balancing the external load (i. e., statically admissible) and satisfying Coulomb's inequality (i. e., plastically admissible) can be found in the network. This is the "critical yield analysis" approach to failure in structural mechanics. It is known that q_1 and q_u would coincide if the sliding in contacts where friction is fully mobilized implied dilatancy, with an angle equal to the friction angle (the discrete analog of an "associated" flow rule). Fig. 6 shows that q_1 is well below q_u. With a dilatant sliding rule, the material response in biaxial compression would be stiffer, and initially (rather paradoxically) more contractant, and the deviator would reach 1.3P (instead of about 0.8P) before failure of the initial contact network.

Evolution of microscopic state variables

As mentioned above, contact creation is negligible in regime I and the fabric evolution is essentially due to contacts opening, mostly in the direction of extension (direction 2). As the initial coordination number is maximal, because of the absence of friction in the assembling process, very few contacts are gained in the direction of compression. In the initial state, all contacts only bear normal force components. Friction mobilization is gradual, but the proportion of sliding contacts, as shown in Fig. 7 steadily increases from zero in regime I, and reaches an apparent plateau in regime II. This means that the interval of elastic response is, strictly speaking, reduced to naught, even though the stress-strain curve can *approximately* be described as elastic in a very small range. The appearance of sliding contacts can effectively reduce the degee of static indeterminacy in the system. If the status is assumed to be fixed for all contacts, the Coulomb condition, satisfied as an equality, reduces the number of independent contact components from $2dN_c$ (in 2D) to $2N_c - \chi_s$, with χ_s the number of sliding contacts among a total of N_c. It has been speculated [29, 30] that failing

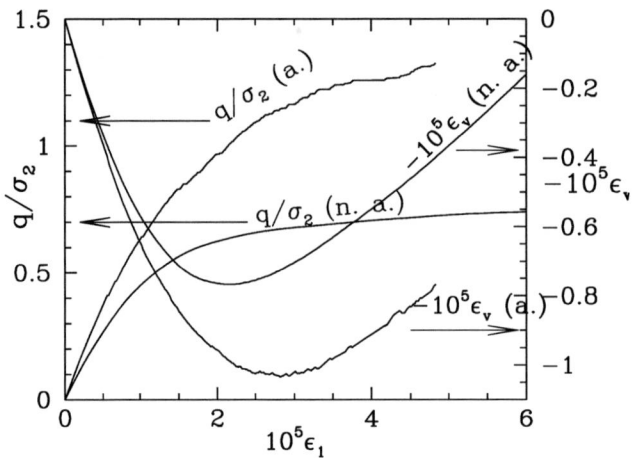

FIGURE 6. Comparison of SEM calculation with the normal (parallel, curves marked "n. a." for non associated) and the "associated" (dilatant, curves marked "a.") sliding rule in contacts in sample with 1024 disks.

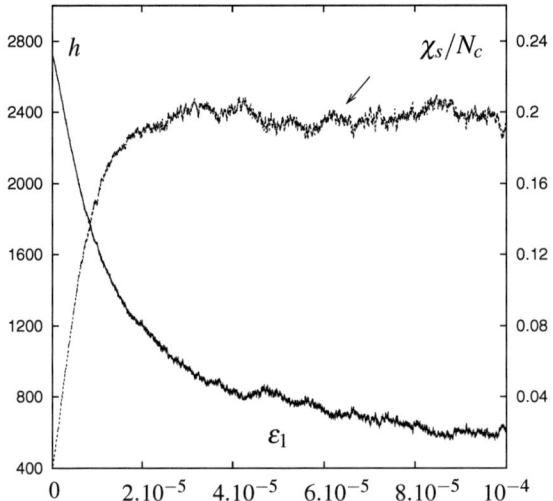

FIGURE 7. In a sample with N=3025, degree of force indeterminacy h (solid line) and proportion of sliding intergranular contacts χ_s/N_c (dotted line), versus axial strain ε_1.

contact networks (regime II) should correspond to vanishing force indeterminacy. The data of Fig. 7 provide evidence against such a prediction, as h stabilizes to about 600, a moderate (10% of the total number of degrees of freedom in a sample of 3025 disks), yet finite value.

3D triaxial tests

The simulations reported here compare dense states A (high coordination number) and C (low coordination number). State A is similar to the dense disk sample studied in the previous section, as both were initially assembled with frictionless grains. (A samples, once packed under low pressure, are nevertheless compressed to the desired confining pressure with the value $\mu = 0.3$ of the friction coefficient used in the triaxial tests [19]). Pressure values correspond to glass beads, and vary between 10 kPa ($\kappa = 39000$) and 1 MPa ($\kappa = 1800$). We first check for the approach of the quasistatic and the macroscopic limit in 3D, strain-rate controlled DEM simulations, then discuss the influence of coordination number, and regimes I and II, in the light of the previous 2D study.

Reproducibility, quasistatic limit

As the system size increases, sample to sample fluctuations should regress, as checked in 2D (Fig. 5). Our 3D results are based on 5 samples of 4000 beads of each type, and Fig. 8 checks for stress-strain curve reproducibility in both A and C cases, for small axial strains. Thanks to the fully periodic boundary conditions [19], the macroscopic mechanical behavior is quite well defined with $N = 4000$. The approach to the quasistatic limit, in SRC tests can be assessed on checking for the innocuousness of the dynamical parameters, i.e., inertial number I, and reduced damping parameter ζ. ζ is defined as the ratio of the viscous damping constant in a contact to its critical level, given the instantaneous value of the stiffness constant. We found it convenient to use a constant value of ζ in our simulations, as in [19]. Fig. 8 also shows that provided inertial number I, characterizing dynamical effects, is small enough, both I and ζ become irrelevant. Fig. 8 shows that the quasistatic limit is correctly approached for $I \leq 10^{-3}$, quite a satisfactory result, given

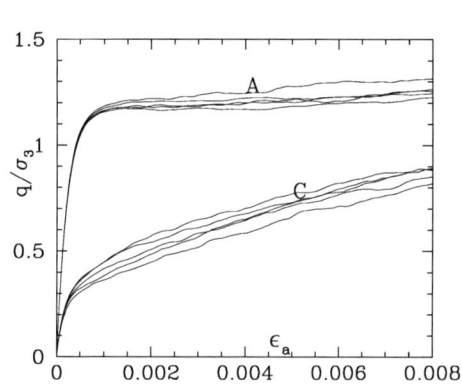
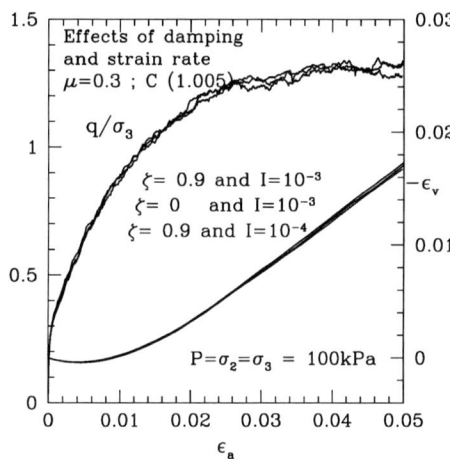

FIGURE 8. Left: small strain part of $q(\varepsilon_a)$ curves for 5 different samples of each type, A (top curves) and C (bottom ones) with $N = 4000$ beads. Right: $q(\varepsilon_a)$ and $\varepsilon_v(\varepsilon_a)$ curves in one type C sample for the different values of ζ and I indicated.

that usual laboratory tests with $\dot{\varepsilon}_a \sim 10^{-5}\,\mathrm{s}^{-1}$ correspond to $I \leq 10^{-8}$.

Influence of initial coordination number

Fig. 9 compares the behavior of initial states A and C, in triaxial compression with $P = 100$ kPa ($\kappa \simeq 6000$). Although, conforming to the traditional view that the peak deviator stress is determined by the initial sample density,

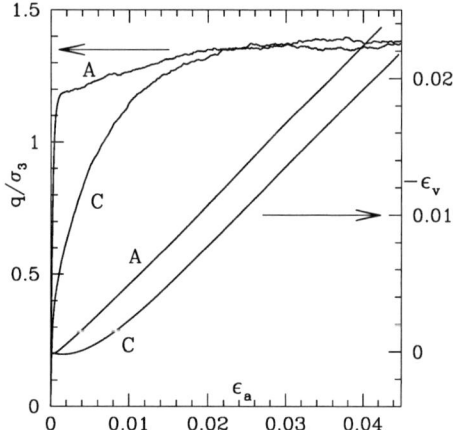

FIGURE 9. $q(\varepsilon_a)$ (left scale) and $\varepsilon_v(\varepsilon_a)$ (right scale) curves for A and C states under $P = 100$ kPa. Averages over 5 samples of 4000 spherical grains.

maximum q values are very nearly identical in systems A and C, the mobilization of internal friction is much more gradual for C. For A, the initial rise of deviator q for small axial strain is quite steep, and the volumetric strain variation becomes dilatant almost immediately, for $\varepsilon_a \sim 10^{-3}$. In [20] it was shown that measurements of elastic moduli provide information on coordination numbers. It is thus conceivable to infer the rate of deviator increase as a function of axial strain from very small strain ($\sim 10^{-5}$ or below [7, 20]) elasticity. Most experimental curves obtained on sands, which do not exhibit q maxima or dilatancy before $\varepsilon_a \sim 0.01$, are closer to C ones. However, some measurements on glass bead samples [31] do show fast rises of q at small strains, somewhat intermediate between numerical results of types A and C.

Influence of contact stiffness

The small strain (say $\varepsilon_a \leq 5.10^{-4}$) interval for A samples, with its fast q increase, is in regime I, as one might expect from 2D results on disks. This is readily checked on changing the confining pressure. Fig. 10 shows the curves for triaxial compressions at different P values (separated by a factor $\sqrt{10}$) from 10 kPa to 1 MPa, with a rescaling of the strains by the stiffness parameter κ, in one A sample. Their coincidence for $q/P \leq 1$ evidences a wide deviator range in regime I. For larger strains, curves separate on this scale, and tend to collapse together if

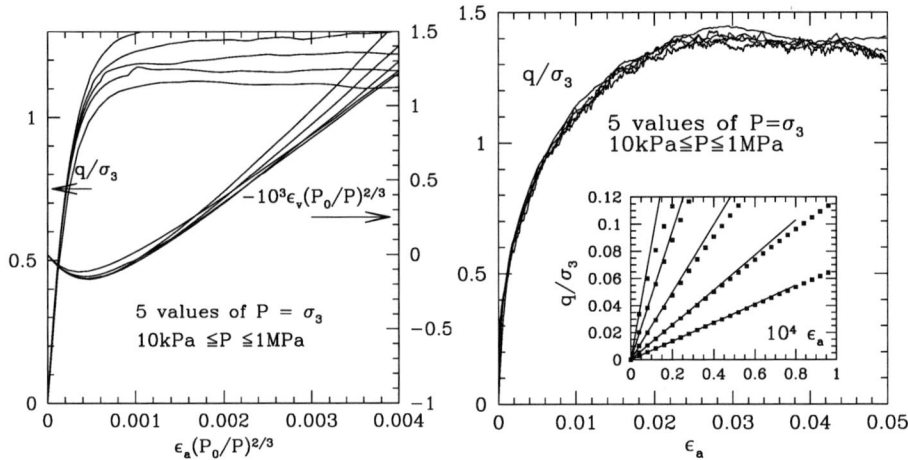

FIGURE 10. Left: $q(\varepsilon_a)/P$ and $\varepsilon_v(\varepsilon_a)$ curves for one A sample and different P values. Strains on scale $(P/P_0)^{2/3} \propto \kappa^{-1}$, $P_0 = 100$ kPa. Right: $q(\varepsilon_a)/P$ for the same P values in one C sample. Inset: detail with blown-up ε scale, straight lines corresponding to Young moduli in isotropic state.

q/P, ε_v are simply plotted versus ε_a. The strain dependence on stress ratio is independent from contact stiffness. This different sensitivity to pressure is characteristic of regime II. Fig 10 also shows that it applies to C samples almost throughout the investigated range, down to small deviators (a behavior closer to usual experimental results than type A configurations). At the origin (close to the initial isotropic state, see inset on fig. 10, right plot), the tangent to the curve is given by the elastic (Young) modulus of the granular material, E_m, and therefore q/P scales with κ, but curves quickly depart from this behaviour (around $q = 0.2P$). The approximately elastic range [20] is quite small, as observed in experiments [7, 32, 33].

Calculations with a fixed contact list

Within regime I, the mechanical properties of the material can be successfully predicted on studying the response of one given set of contacts. Those might slide or open, but the very few new contacts that are created can be neglected. To check this in simulations, one may restrict at each time step the search for interacting grains to the list of initially contacting pairs. Fig. 11 compares such a procedure to the complete calculation. The curve marked "NCC" for *no contact creation* is indistinguishable from the other one for $q \geq 0.8$. We thus check that, in regime I, the macroscopic behavior is essentially determined by the response of a fixed contact network.

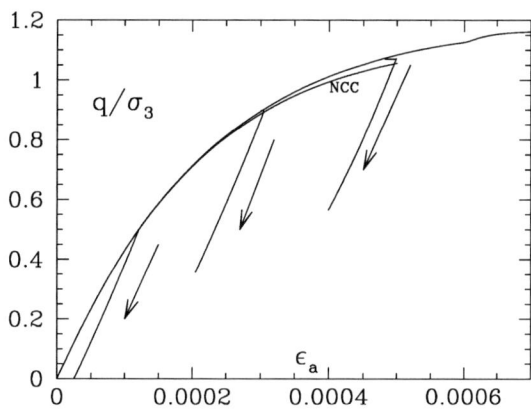

FIGURE 11. Very small strain part of $q(\varepsilon_a)$ curve in one A sample, showing beginning of unloading curves (arrows). Curve marked NCC was obtained on calculating the evolution of the same sample without any contact creation.

Type I strains and elastic reponse

Fig. 11 also shows that the small strain response of A samples, within regime I, close to the initial state, is already irreversible: type I strains are not elastic. An approximately elastic behavior is only observed for very small strains, as depicted in the inset of Fig. 10 (right part). In this small interval near the initial equilibrium configuration, the stress-strain curve is close to its initial tangent, defined by the elastic modulus. Moduli [20] can be calculated from the stiffness matrix of contact networks. One may also check that the unloading curves shown on Fig. 11 (and the ones of Fig. 3 in 2D as well) comprise a small, approximately elastic part, with the relevant elastic modulus (the Young modulus for a triaxial test at constant lateral stress) defining the initial slope. At the microscopic level, a small elastic response is retrieved upon reversing the loading direction because contacts stop sliding. The elastic range is strictly included in the larger range of type I behavior.

Fluctuations and length scale

Finally, let us note that regimes I and II also differ by the importance of sample to sample fluctuations: curves in Fig. 8 (left plot) pertaining to the different samples of type A or C are confused as long as $q \leq 1.1P$ (case A) or $q \leq 0.3P$ (case C), which roughly corresponds to the transition from regime I to regime II. Larger fluctuations imply that the characteristic length scale associated with the displacement field (correlation length) is larger in regime II. Whether and in what sense rearrangements triggered by instabilities in regime II, in a material close to the rigid limit (large κ), can be regarded as local events is still an open issue.

CONCLUSION

Numerical studies thus reveal that the two regimes, in which the origins of strain differ, exhibit contrasting properties. Although the reported studies in 2D and 3D differ in many respects (linear versus Hertzian contacts, wall versus periodic boundaries, SIC versus SRC DEM), the same phenomena were observed in both cases. Regime I corresponds to the stability range of a given contact structure. It is larger in highly coordinated systems. It is observed in the beginning of monotonic loading tests, in which the deviator stress increases from an initial isotropic configuration, and also after changes in the direction of load increments (hence a loss in friction mobilization). Strains, for a given stress level, are then inversely proportional to contact stiffnesses. The deviator range in regime I, $q \leq q_1$, in usual monotonic tests, is stricly larger than the small elastic range, but strictly smaller than the maximum deviator. It does not vanish in the limit of large systems, unlike in the singular case of rigid, frictionless particle assemblies [13, 34]. Regime I is limited by the occurrence of elastoplastic instabilities in the contact network and does not coincide with the prediction of the critical yield approach. In regime I, the work of the externally applied load is constantly balanced by the one of contact forces, so that the kinetic energy approaches zero in the limit of slow loading rates. A remarkable consequence is that the instability condition based on the negativity of the macroscopic second-order work [35] is never fulfilled, as macroscopic and microscopic works coincide, and the latter is positive. In regime II, network rearrangements are triggered by instabilities and some bursts of kinetic energy are observed [14]. Larger fluctuations witness longer-ranged correlations in the displacements. The microscopic origin of macroscopic strains, which are independent on contact elasticity for usual stiffness levels κ, lies in the geometry of grain packings.

On attempting to predict a macroscopic mechanical response from packing geometry and contact laws, the information about which kind of strain should dominate is crucial.

A promising perspective is the study of correlated motions associated with rearrangement events.

REFERENCES

1. G. Gudehus, F. Darve, and I. Vardoulakis, editors, *Constitutive Relations for Soils*, Balkema, Rotterdam, 1984.
2. D. M. Wood, *Soil Behaviour and Critical State Soil Mechanics*, Cambridge University Press, 1990.
3. J.-N. Roux, and G. Combe, *Comptes Rendus Physique* **3**, 131–40 (2002).
4. K. Kaneko, K. Terada, T. Kyoya, and Y. Kishino, *Int. J. Solid Struct.* **40**, 4043–69 (2003).
5. S. McNamara, and H. J. Herrmann, *Phys. Rev. E* **74**, 061303 (2006).
6. L. La Ragione, V. Magnanimo, and J. T. Jenkins, "Constant Pressure Axisymmetric Compression of an Aggregate of Identical Elastic, Frictional Spheres," in [36], pp. 1100–3.
7. S. Shibuya, F. Tatsuoka, S. Teachavorasinskun, X.-J. Kong, F. Abe, Y.-S. Kim, and C.-S. Park, *Soils Found.* **32**, 26–46 (1992).
8. M. Jean, *Computational Methods in Applied Mechanics and Engineering* **177**, 235–257 (1999).
9. F. Radjaï, and V. Richefeu, *Mechanics of Materials* **41**, 715–28 (2009).
10. F. Radjaï, and S. Roux, "Contact dynamics study of 2D granular media : critical states and relevant internal variables," in *The Physics of Granular Media*, edited by H. Hinrichsen, and D. E. Wolf, Wiley-VCH, Berlin, 2004, pp. 165–87.
11. F. Radjaï, H. Troadec, and S. Roux, "Basic features of granular plasticity," in *Granular Materials: Fundamentals and Applications*, edited by S. J. Antony, W. Hoyle, and Y. Ding, Royal Society of Chemistry, Cambridge, 2004, pp. 157–83.
12. F. Radjaï, *ArXiv e-prints* (2008), 0801.4722.
13. G. Combe, and J.-N. Roux, *Phys. Rev. Lett.* **85**, 3628–31 (2000).
14. L. Staron, J.-P. Vilotte, and F. Radjaï, *Phys. Rev. Lett.* **89**, 204302 (2002).
15. G. Combe, *Mécanique des matériaux granulaires et origines microscopiques de la déformation*, Presses du Laboratoire Central des Ponts et Chaussées, Paris, 2002.
16. G. Combe, and J.-N. Roux, "Discrete numerical simuations, quasistatic deformation and the origins of strain in granular materials," in *Deformation characteristics of geomaterials*, edited by H. di Benedetto, T. Doanh, H. Geoffroy, and C. Sauzéat, Swets and Zeitlinger, Lisse, 2003, pp. 1071–8.
17. J.-N. Roux, and G. Combe, "Méthodes quasi-statiques," in *Modélisation numérique discrète des matériaux granulaires*, edited by F. Radjaï, and F. Dubois, Hermes, 2009.
18. J.-N. Roux, "The nature of quasistatic deformation in granular materials," in *Powders and Grains 2005*, edited by R. García Rojo, H. J. Herrmann, and S. McNamara, Balkema, Leiden, 2005, pp. 261–5.
19. I. Agnolin, and J.-N. Roux, *Phys. Rev. E* **76**, 061302 (2007).
20. I. Agnolin, and J.-N. Roux, *Phys. Rev. E* **76**, 061304 (2007).
21. P. A. Cundall, and O. D. L. Strack, *Géotechnique* **29**, 47–65 (1979).
22. K. L. Johnson, *Contact Mechanics*, Cambridge University Press, 1985.
23. I. Agnolin, and J.-N. Roux, *Phys. Rev. E* **76**, 061303 (2007).
24. GDR MiDi, *European Physical Journal E* **14**, 341–65 (2004).
25. F. da Cruz, S. Emam, M. Prochnow, J.-N. Roux, and F. Chevoir, *Phys. Rev. E* **72**, 021309 (2005).
26. T. Hatano, *Phys. Rev. E* **75**, 060301(R) (2007).
27. P.-E. Peyneau, and J.-N. Roux, *Phys. Rev. E* **78**, 011307 (2008).
28. P. Jop, Y. Forterre, and O. Pouliquen, *Nature* **441**, 727–30 (2006).
29. J. Lanier, and M. Jean, *Powder Technology* **109**, 206–21 (2000).
30. N. P. Kruyt, and L. Rothenburg, "Plasticity of Granular Materials: a structural mechanics view," in [36], pp. 1073–9.
31. S. Emam, J. Canou, A. Corfdir, J.-C. Dupla, and J.-N. Roux, "Élaboration et comportement mécanique de matériaux granulaires solides modèles : expériences et simulations numériques," in *Rhélogie des pâtes et des matériaux granulaires*, edited by B. Cazacliu, and J.-N. Roux, Presses du Laboratoire Central des Ponts et Chaussées, Paris, 2006, vol. SI12 of *Etudes et Recherches des Laboratoires des Ponts et Chaussées*, pp. 105–45.
32. H. di Benedetto, H. Geoffroy, C. Sauzéat, and B. Cazacliu, "Sand behaviour in very small to medium strain domains," in *Pre-failure Deformation Characteristics of Geomaterials*, edited by M. Jamiolkowski, R. Lancellotta, and D. Lo Presti, Balkema, Rotterdam, 1999, pp. 89–96.
33. R. Kuwano, and R. J. Jardine, *Géotechnique* **52**, 727–49 (2002).
34. P.-E. Peyneau, and J.-N. Roux, *Phys. Rev. E* **78**, 041307 (2008).
35. F. Prunier, F. Laoufa, and F. Darve, *European Journal of Environmental and Civil Engineering* **13**, 135–47 (2009).
36. M. Nakagawa, and S. Luding, editors, *Powders and Grains '09*, vol. 1145 of *AIP Conference Proceedings*, 2009.

Theory of random packings

Chaoming Song, Ping Wang, Hernán A. Makse

Levich Institute and Physics Department, City College of New York, New York, NY 10031, US

Abstract. We review a recently proposed theory of random packings. We describe the volume fluctuations in jammed matter through a volume function, amenable to analytical and numerical calculations. We combine an extended statistical mechanics approach 'a la Edwards' (where the role traditionally played by the energy and temperature in thermal systems is substituted by the volume and compactivity) with a constraint on mechanical stability imposed by the isostatic condition. We show how such approaches can bring results that can be compared to experiments and allow for an exploitation of the statistical mechanics framework. The key result is the use of a relation between the local Voronoi volume of the constituent grains and the number of neighbors in contact that permits a simple combination of the two approaches to develop a theory of random packings. We predict the density of random loose packing (RLP) and random close packing (RCP) in close agreement with experiments and develop a phase diagram of jammed matter that provides a unifying view of the disordered hard sphere packing problem and further shedding light on a diverse spectrum of data, including the RLP state. Theoretical results are well reproduced by numerical simulations that confirm the essential role played by friction in determining both the RLP and RCP limits. Finally we present an extended discussion on the existence of geometrical and mechanical coordination numbers and how to measure both quantities in experiments and computer simulations.

Keywords: granular matter, random close packing, statistical mechanics
PACS: 81.05.Rm, 83.80.Fg

I. STATISTICAL MECHANICS OF JAMMED MATTER

Conventional Statistical Mechanics uses the ergodic hypothesis to derive the microcanonical and canonical ensembles, based on the quantities conserved, typically the energy E [1]. Thus the entropy in the microcanonical ensemble is $S(E) = k_B \log \int \delta(E - \mathcal{H}(p,q)) dp dq$, where $\mathcal{H}(p,q)$ is the Hamiltonian. This becomes the canonical ensemble with $\exp[-\mathcal{H}(\partial S/\partial E)]$. Experiments [2, 3, 4, 5] indicate that systematically shaken granular materials show reversible behavior, and the analogue of the conserved quantity is the volume V, thus the micro-canonical ensemble or V-ensemble is [6, 7, 8, 9, 10, 11, 12]:

$$\Omega(V) = \exp[S(V)/\lambda] = \int \delta(V - \mathcal{W}(\vec{r}_i)) \, \Theta_{\text{jam}}(\vec{r}_i) \, d\vec{r}_i, \qquad (1)$$

where $\Theta_{\text{jam}}(\vec{r}_i)$ is a function that defines the jammed configuration. As a minimum requirement the jamming function $\Theta_{\text{jam}}(\vec{r}_i)$ should ensure touching grains, and obedience to Newton's force laws. $\vec{r}_i$ denotes the particle positions in the system and $\mathcal{W}(\vec{r}_i)$ is the volume function defining the volume associated with each grain (see below). This gives a canonical ensemble of $\exp[-\mathcal{W}(\partial S/\partial V)]$. Just as $\partial E/\partial S = T$ is the temperature in equilibrium system, the temperature-like variable in granular systems is the compactivity $X = \partial V/\partial S$. In Eq. (1), λ is the analogue of the Boltzmann constant.

Thermodynamic analogies may illuminate methods for attempting to solve certain problems, but inevitably fail at some point in their application. The mode of this failure is an interesting phenomenon, illustrated by the compaction experiments of the groups of Chicago, Texas, Paris and Schlumberger [2, 3, 4, 5]. They have shown that reversible states exist along a branch of compaction curve where statistical mechanics is more likely to work. Conversely, experiments also showed a branch of irreversibility where the statistical framework is not expected to work. Poorly consolidated formations, such as a sandpile, are irreversible and a new "out-of-equilibrium" theory is required to describe them. Below we focus on a theoretical description of the reversible branch of the compaction curve focusing on a theory of the random close packed state.

The canonical partition function in the V-ensemble is the starting point of the statistical analysis of jamming:

$$\mathcal{Z}(X) = \int g(\mathcal{W}) e^{-\mathcal{W}/X} \Theta_{\text{jam}} d\mathcal{W}, \qquad (2)$$

where $g(\mathcal{W})$ is the density of states for a given volume $\mathcal{W}$.

From Eq. (2) we identify three minimal steps in developing analytical solutions which are discussed in the next sections. Section II discusses the need for a volume function in terms of the contact network. Section III discusses the need for a proper definition of jammed state that allows one to define Θ_{jam}. Section IV discusses the density of states. Finally in Section V we explain the geometrical and mechanical coordination numbers and how to measure them in Section V.A, and we conclude in Section VI.

II. VOLUME FUNCTION

While it is always possible to quantify the total volume of the system, it is unclear how to treat the volume fluctuations at the grain level. The first step to study the V-ensemble is to find the volume $\mathscr{W}(\vec{r}_i)$ associated to each particle $\vec{r}_i$ that successfully tiles the system. This is analogous to the additive property of energy in equilibrium statistical mechanics.

Initial attempts included a model volume function under mean-field approximation [6], the work of Ball and Blumenfeld [13] and simpler versions in terms of the first coordination shell [14]. These definitions are problematic since some are not additive, others present problems in polydisperse systems or are proportional to coordination contrary to expectation. In Ref. [15, 16] we have found an analytical form of the volume function in three-dimensions and demonstrated that it is the Voronoi volume of a particle i:

$$\mathscr{W}_i^{\text{vor}} = \frac{1}{3} \oint \left(\min_{\hat{s} \cdot \hat{r}_{ij} > 0} \left(\frac{r_{ij}}{2\hat{s} \cdot \hat{r}_{ij}} \right) \right)^3 ds, \tag{3}$$

where $\vec{r}_{ij}$ is the vector from the position of particle i to that of particle j, the integration is done over all the directions $\hat{s}$ forming an angle θ_{ij} with $\vec{r}_{ij}$ as in Fig. 1a, and R is the radius of the grain. R will be set to unity for simplicity. While this formula may seem complicated, it has a simple interpretation depicted in Fig. 1a.

The Voronoi construction is additive and successfully tiles the total volume. Prior to this result, there was no analytical formula to calculate the Voronoi volume in terms of the contact network r_{ij}. A further simplification arises when we consider isotropic systems. Then the volume function reduces to the orientation volume, without the average over $\hat{s}$. We define the reduced free orientational volume function as

$$w^s \equiv \frac{\mathscr{W}_i^s - V_g}{V_g}, \tag{4}$$

with $\mathscr{W}_i^s \equiv V_g \left(\frac{1}{2R} \min_{\hat{s} \cdot \hat{r}_{ij} > 0} \frac{r_{ij}}{\hat{s} \cdot \hat{r}_{ij}} \right)^3$, see Fig. 1a ($V_g$ is the particle volume). This equation allows theoretical analysis in the V-ensemble since it reduces the complicated definition (3) to a more amenable "one-dimensional" volume which can be treated analytically.

The next step is to develop a theory of volume fluctuations to coarse grain $\mathscr{W}_i^s$ over a mesoscopic length scale. We call this the quasi-particle approximation. It could be considered as well as a mean-field approximation, although mean field is supposed to be exact in infinite dimensions. The approximations used in the present theory are supposed to get better as the dimension increases, but we cannot claim that the theory is exact in large dimensions. Thus, we prefer to call our approximation "quasi-particle" in the spirit of Landau and the quasiparticles as "coordinons".

The coarsening reduces the degrees of freedom to one variable, the coordination number of each grain, and defines an average volume function which is more amenable to statistical calculations than Eq. (3) as shown in [16]:

$$w(z) = \langle w^s \rangle_i = \frac{2\sqrt{3}}{z}, \tag{5}$$

valid for monodisperse hard spheres where z is the geometrical coordination number. For now on we assume $V_g = 1$ for simplicity. The available volume per grain is inversely proportional with the coordination number, in agreement with the X-ray tomography experiments (see Fig. 6 in [17] where the volume fraction is $\phi^{-1} = w + 1$).

III. DEFINITION OF JAMMING VIA Θ_{jam}: ISOSTATIC ENSEMBLE

The definition of the constraint function Θ_{jam} is intimately related to the proper definition of a jammed state, with a minimum requirement of mechanical equilibrium. In an attempt to define the jammed states in a rigorous mathematical

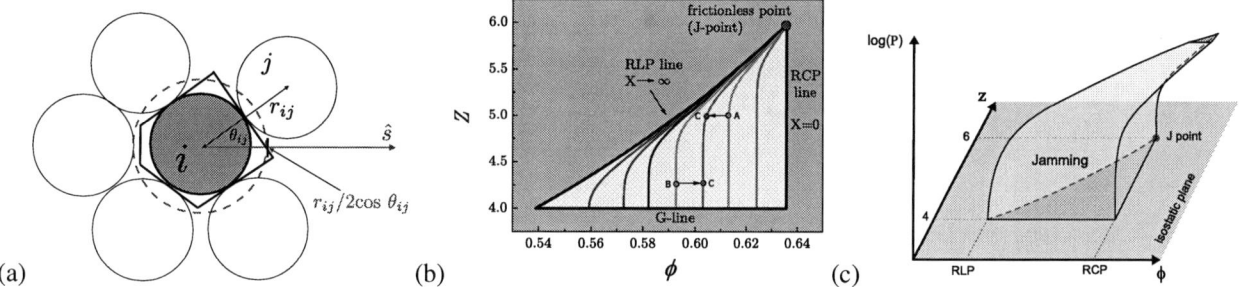

FIGURE 1. (a) Schematics of the Voronoi volume and the orientational volume associated with particle i. The boundary of the Voronoi cell (shown in two-dimensions for simplicity) corresponds to the irregular pentagon in black which defines $\mathscr{W}_i^{\text{vor}}$. The limit of the Voronoi cell of particle i in the direction $\hat{s}$ is the minimum of $r_{ij}/2\cos\theta_{ij}$ over all the particles in the packing, as indicated. This defines the orientational volume $\mathscr{W}_i^s$ which is the volume of the sphere of radius $r_{ij}/2\cos\theta_{ij}$ defined by the dash red circle in the figure. (b) Phase diagram of jamming in the hard sphere plane under the isostatic assumption. All the disordered packings fall within the yellow triangle demarcated by the RCP and RLP lines and the G-line. The isocompactivity lines are in color. (c) Generalization of the phase diagram to the space (Z, ϕ, p).

way, Torquato and coworkers have proposed three categories of jamming [18]: locally jammed, collectively jammed and strictly jammed based on geometrical constraints. Unfortunately this definition cannot be easily extended to frictional systems since it is based on geometry and does not include the contact forces. Other approaches based on minima of the potential energy landscape also fail since such a potential does not exist for frictional grains due to their path-dependency.

Then we define an alternative approach to characterize jamming for the general case of frictional granular matter. In [16] we propose the isostatic condition [19, 20] as a possible formulation of jamming. The isostatic condition implies a mechanical coordination number to be $Z = 2d$ for frictionless spherical particles and $Z = d+1$ for infinitely rough particles (with interparticle friction coefficient $\mu \to \infty$). Numerical simulations [21, 22], experiments [23] and theoretical work [20] suggest that at the jamming transition the system becomes exactly isostatic. However, no rigorous proof of this statement exist. It should be noted that $Z = 6$ is a necessary but not sufficient condition for isostaticity [20]. A kinematic condition has to be satisfied as well, which refers to the location of the center of the particles that are determined only by the length of the vectors that join the center of contacting particles. In other words the equations of equilibrium needs to be independent of each other. While this condition has been proved to exist in frictionless packings [20], the problem of frictional packings, even with infinite friction, remains open. Thus, to be precise, the only point where we can claim isostaticity is the frictionless point. In any case, we consider that we are able to extend the isostatic condition to infinite frictional packings as well at $Z = 4$. Interpolating between the two limits, there exist packings of finite μ; the coordination number smoothly varies between $Z(\mu = 0) = 6$ and $Z(\mu \to \infty) \to 4$ [24, 16].

We notice that while the relationship between friction and Z may not be unique, the theory is only based on Z. Thus, given the mechanical coordination number we predict the state of the packing with the compactivity. Additionally, we conjecture that μ determines $Z(\mu)$ if we follow certain protocols as discussed in [16]. These protocols imply compression of a packing from an unjammed state to jamming by following one single continuous path. That is, the path dependent shear forces are not reset by a sudden change in the preparation path. It should be stated that there are other protocols, like shear cycling, that start with a given Z and can produce packings that continuously compactify until RCP (and even beyond). This is done by effectively changing the path followed by the shear forces at every cycle. Thus, for shear cycling protocols, a unique relation may not be expected between Z and μ. However, the theoretical results as derived below are still valid in shear cycling experiment since they concern the relation between Z and volume. In fact, a shear experiment may be the easiest way to obtain the packings at the RLP line as explained below.

Assuming that a system of hard spheres is isostatic at the jamming transition, Eq. (2) can be written in terms of z and the mean-field Eq. (5) can be used in the single-particle (or more precisely quasi-particle) partition function:

$$\mathscr{Z}_{\text{iso}}(X) = \int_Z^6 e^{-w(z)/X} g(z) dz. \tag{6}$$

This ensemble is referred to as the *Isostatic-ensemble*. Note that the upper limit of integration is $z = 6$ [16]. This implies that only disordered packings are included. The solution of such a partition function for monodisperse hard spheres has been done in [16] revealing the phase diagram depicted in Fig. 1b.

This phase diagram predicts a series of important results, such as the value of RCP at $X = 0$,

$$\phi_{\text{RCP}} = 6/(6 + 2\sqrt{3}), \tag{7}$$

and the lowest density of the RLP at $X = \infty$,

$$\phi_{\text{RLP}} = 4/(4 + 2\sqrt{3}), \tag{8}$$

in close agreement with experiments. The diagram restricts the possible packings to the yellow triangle in Fig. 1b, ranging from frictionless systems with $Z = 6$, to infinitely rough grains in the $Z = 4$ granular line or G-line.

IV. DENSITY OF STATES

A difficult problem is the determination of the density of states $g(\mathscr{W})$ in Eq. (2). For the simplest case of the Iso-ensemble from Eq (6), the density of states reduces to

$$g(z) = (h_z)^z, \tag{9}$$

where h_z is a small microscopic constant arising due to the discrete volume space of configurations [16]. The situation is analogous to the discreteness of the configuration space imposed by the Heisenberg uncertainty principle in quantum mechanics. The formula is analogous to the factor h^{-d} for the density of states in equilibrium statistical mechanics. While the degrees of freedom $\{p_i, q_i\}$ are continuous, the uncertainty principle imposes the discreteness $(\Delta p, \Delta q)$ in the configurational space given by $\Delta p \Delta q \sim h$. This consideration allows for the approximate solution explained in the above section and depicted in the phase diagram of Fig. 1b.

V. GEOMETRICAL VERSUS MECHANICAL COORDINATION NUMBER

It is important to note that the derivation of the volume function in Section II implies nothing about the value of the contact forces; the volume function represents the contribution arising purely from the geometry of the packing. Thus, the coordination number z appearing in Eq. (5) is the *geometrical* coordination number related to volume, which is different from the *mechanical* coordination number Z that counts the number of contacts per particle with non-zero force related to the isostatic condition and force network.

Having acknowledged a difference between the geometrical coordination number z in Eq. (5) and the mechanical coordination number Z which counts only the contacts with non-zero forces, below we discuss the bounds of z and how to measure it.

Since some geometrical contacts may carry no force, then we have:

$$Z \leq z. \tag{10}$$

To show this, imagine a packing of infinitely rough ($\mu \to \infty$) spheres with volume fraction close to 0.64. There must be $z = 6$ nearest neighbors around each particle on the average. However, the mechanical balance law requires only $Z = 4$ contacts per particle on average, implying that 2 contacts have zero force and do not contribute to the contact force network.

Such a situation is possible as shown in Fig. 2: starting with the contact network of an isostatic packing of frictionless spheres having $Z = 6$ and all contacts carrying forces (then $z = 6$ also as shown in Fig. 2a), we simply allow the existence of tangential forces between the particles and switch the friction coefficient to infinity. Subsequently, we solve the force and torque balance equations again for this modified packing of infinitely rough spheres but same geometrical network, as shown in Fig. 2b [Notice that the shear force is composed of an elastic Mindlin component plus the Coulomb condition determined by μ. Thus when $\mu \to \infty$, the elastic Mindlin component still remains].

The resulting packing is mechanically stable and is obtained by setting to zero the forces of two contacts per ball, on average, to satisfy the new force and torque balance condition for the additional tangential force at the contact. Such a solution is guaranteed to exist due to the isostatic condition: at $Z = 4$ the number of equations equals the number of force variables. Despite mechanical equilibrium, giving $Z = 4$, there are still $z = 6$ geometrical contacts contributing to the volume function.

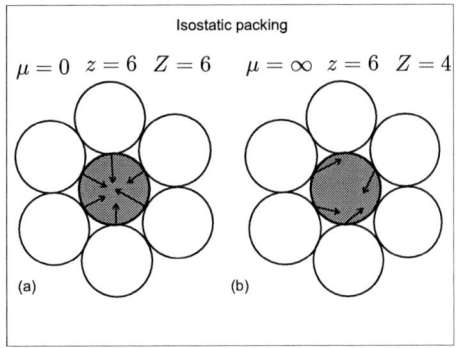

FIGURE 2. (a) Consider a frictionless packing at the isostatic limit with $z = 6$. In this case the isostatic condition implies also $Z = 6$ mechanical forces from the surrounding particles. (b) If we now switch on the tangential forces using the same packing as in (a) by setting $\mu \to \infty$, the particle requires only $Z = 4$ contacts to be rigid. Such a solution is guaranteed by the isostatic condition for $\mu \to \infty$. Thus, the particle still have $z = 6$ geometrical neighbors but only $Z = 4$ mechanical ones.

Therefore, we identify two types of coordination number: the geometrical coordination number, z, contributing to the volume function and the mechanical coordination number, Z, measuring the contacts that carry forces only. This distinction is crucial to understand the sum over the states and the bounds in the partition function.

We have established a lower bound of the geometrical coordination in Eq. (10). The upper bound arises from considering the constraints in the positions of the rigid hard spheres. For hard spheres, the Nd positions of the particles are constrained by the $Nz/2$ geometrical constraints, $|\vec{r}_{ij}| = 2R$, of rigidity. Here d is the dimension. Thus, the number of contacts satisfies $Nz/2 \leq Nd$, and z is bounded by:

$$z \leq 2d. \tag{11}$$

Notice that this upper bound applies to the geometrical coordination, z and not to the mechanical one, Z, and it is valid for any system irrespective of the friction coefficient, from $\mu = 0 \to \infty$.

In conclusion, the mechanical coordination number, Z, ranges from 4 to 6 as a function of μ, and provides a lower bound to the geometrical coordination number, while the upper bound is $2d$. A granular system is specified by the interparticle friction which determines the average mechanical coordination at which the system is equilibrated, $Z(\mu)$. The possible microstates in the ensemble available for this system follow a Boltzmann distribution Eq. (2) for states satisfying the following bounds:

$$Z(\mu) \leq z \leq 2d = 6. \tag{12}$$

V.A. How to measure the geometrical coordination number

Measuring the geometrical coordination number can be a tricky task, in principle. At the onset, it is the coordination number of a single quasiparticle. What we measure in a real packing (numerically or experimentally generated) is an ensemble average of many quasiparticle according to the partition function Eq. (6). Thus, rigorously speaking, it is not possible to isolate a quasi-particle and measure its properties in a real packing. Beyond this caustic and somehow pessimistic remark, yet rigorous, below we offer light at the end of the tunnel by using the theoretical predictions to define an approximative, yet accurate, way to measure the geometrical coordination.

Figure 3 summarizes the predictions of the theory regarding the behaviour of z and Z for the packings in the phase diagram. First, we can think that the quasiparticle behaviour is revealed when the system has infinite compactivity and behaves like a non-interacting "gas" of quasiparticles. This is the behaviour found along the RLP line in the phase diagram of Fig. 1b. Indeed, along this line, the packings have the highest entropy [25] and therefore they are the most likely to be found numerically and experimentally. Indeed, this is what is found in our numerical simulations [16, 25]. Along this line we expect that the mechanical and geometrical coordinations are the same, $z \approx Z$. This result comes about since at infinite compactivity we are exploring the states with the highest volume or lower volume fraction. In the partition function Eq. (6), these are states with $z \approx Z$. It should be noticed that the density of states plays a role in this argument. We are assuming that the density decays very rapidly owning to a small constant $h_z \to 0$ in Eq.

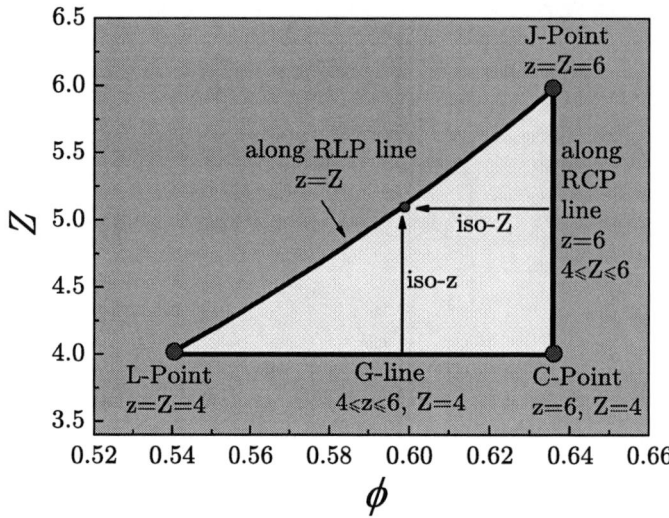

FIGURE 3. Summary of the theoretical findings regarding the range of z and Z along the different iso-z, iso-Z, and iso-X lines and the J, C, and L-points in the phase diagram.

(9). This is of course a very reasonable assumption, since h_z is related to the discretization of the volume space of configurations, and it is like a Planck constant of granular matter. However, we may even relax this consideration and allow this constant to be of the order 1. The above argument is still correct, but the only difference is that the RLP limit is not at infinite compactivity but in the limit of $X \to 0^-$. That is, the highest volume corresponds to negative temperature states, which are even hotter than $X \to +\infty$. An extended explanation of this point is given in [25] and its follow up paper in cond-mat.

The second interesting aspect of the theory is the prediction that along the RCP line in Fig. 3, the geometrical coordination number is constant and equal to the maximum coordination, which produces the minimum Voronoi volume: $z = 6$. At the same time, the mechanical coordination number varies from 4 to 6 as we reduce friction to zero. This result is explained with the analysis offered in Fig. 2 and explained above. The conclusion is that the packings along the RCP line are geometrically the same (with $z = 6$ all of them) but they differ in the value of the forces between the particles as Z varies from 6 to 4. This prediction can be also applied to other packings with lower ϕ. For a given volume fraction, the packings along the iso-z vertical line in Fig. 3 should have approximately the same z but different Z from 4 to the maximum Z given by the RLP line.

Using this theoretical result, it is easy then to define the geometrical coordination number for real packings and propose a clear way to measure it. The idea is to inflate the particles infinitesimally by a Δr value and measure the contacting particles. By setting $\Delta r = 0$, we clearly measure the mechanical coordination. By considering an infinitesimally small Δr we should measure the geometrical one. The question is to know what value of Δr to use. Here is where the theory comes handy. We know that along the RCP line the packings are the same geometrically. So, whatever the definition of coordination number we use, it should satisfy that after a given Δr, we should find the same packing structure for the packings along the RCP line and below the given Δr, the structure should change reflecting the different values of Z for different packings in the RCP line.

Using these considerations, we identify the geometrical coordination as follows. Two particles that may not be in contact (giving rise to a zero force) may be close enough to be considered as contributing to the geometrical coordination. Indeed, it is known that the radial distribution function $g(r)$ has a singularity, $g(r) \sim (r - 0.5)^{-0.5}$ [24], implying that there are many particles almost touching. We introduce a modified radial distribution function (RDF) $g_z(r)$ in order to approximately identify z and Z from real packings:

$$g_z(r) = \frac{1}{N}\frac{R^2}{r^2}\sum_i^N\sum_{j\neq i}^N \Theta\left(\frac{r_{ij}}{r-R}-1\right)\Theta\left(\frac{r+R}{r_{ij}}-1\right), \quad r > R \tag{13}$$

where R is the radius of particle, N is the number of particles, r_{ij} is the distance of two particle's centers, $r_{ij} = |\vec{r}_i - \vec{r}_j|$, and Θ is the Heaviside step function. The RDF describes the average value of the number of grains in contact with a virtual particle which has been inflated up to a radius $r \geq R$, and the factor of R^2/r^2 is the ratio of a real sphere's area

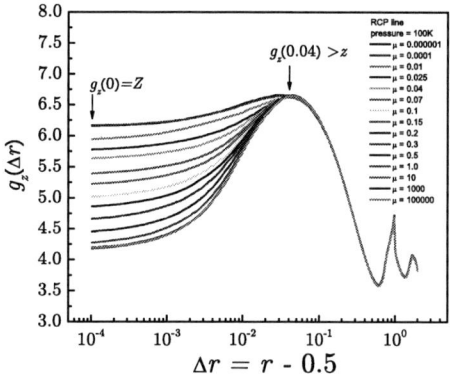

FIGURE 4. $g_z(\Delta r)$ of packings with various friction coefficient μ along RCP line. We set $2R = 1$.

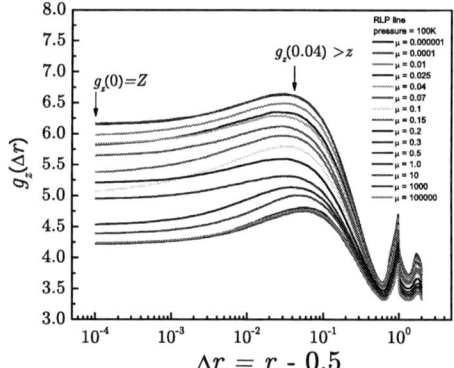

FIGURE 5. $g_z(\Delta r)$ of packings with various friction coefficient μ along RLP line.

and the virtual one's. Without the normalization factor R^2/r^2, Eq. (13) is the same definition of coordination as used by Torquato and Zamponi in their analysis of infinite pressure jammed hard sphere glasses [26, 27]. This factor is not crucial for our analysis (as the main constraint is that when $r = R$ we should get Z) as explained below, but we argue that it is useful since we need to proper normalize by the fact that we are inflating the balls. Please notice that Eq. (29) in Supplementary Information Section of [16] contains a typo. The correct definition is Eq. (13).

$g_z(r)$ measures the number of balls with their volume intersecting the surface of a sphere of radius r measured from the center of a given ball. When $r = R$ in (13) we obtain the mechanical coordination number while the geometrical one is obtained for a small value $\Delta r = \frac{r-R}{2R} \neq 0$ for which we distinctly find a signature from computer simulations, unambiguously defining it at $\Delta r = 0.04$ for the system size used by following the packings along the RCP line.

Figures 4 and 5 plot the $g_z(\Delta r)$ of packings with various friction coefficient μ along the RCP and RLP lines respectively. Following the definition of Eq. (13), g_z, with $\Delta r = 0$, should be directly equal to the mechanical coordination number, Z, and should range from 4 to 6 along both RCP and RLP (if $h_z \ll 1$) lines which is confirmed by our numerical simulations in Figs. 4 and 5, respectively.

More importantly, as shown in the figures, we find that $g_z(\Delta r)$ along the RCP line is exactly the same for all the packings when $\Delta r > 0.04$ as shown in Fig. 4. For $\Delta r < 0.04$, Fig. 4 shows that the packings have different mechanical coordination numbers from 4 to 6. This is exactly what the theory predicted. All these packings have actually the same geometrical structure evidenced when $\Delta r > 0.04$ but with different mechanical coordinations which appears only in a difference observed for $\Delta r < 0.04$. Based on this analysis, we then define the geometrical coordination number as the one appearing at $\Delta r = 0.04$. We identify the geometrical coordination number as $z = g_z(0.04)$ under the accuracy of the simulations and for this particular system size $N = 10,000$ (we notice though that this value may depend on system size).

It is important to note that in terms of the radial distribution function, $g(r)$, nothing really happens at $\Delta r = 0.04$: that is, there is no peak in $g(r)$ at 0.04 and the second peak after the first coordination shell appears for larger Δr in $g(r)$.

On the other hand, we clearly see a peak in Fig. 4 at this value. We point out that the peak at 0.04 is the byproduct of the normalization factor R^2/r^2 in Eq. (13). This factor is suggested since we need to renormalize by the area of the virtual sphere. However, other factors, for instance R^3/r^3 in Eq. (13) would produce a peak in Fig. 4 located in another position. The important fact is not the location of the peak, but the fact that above 0.04 all the functions in Fig. 4 coincide. This is the basis of the definition of 0.04 as the location to define the geometrical coordination number. At this position, there is no peak in the $g(r)$. Indeed, the second peak in $g(r)$ beyond the first coordination shell appears much further around $\Delta r \approx 1$ and are indeed also identified by $g_z(\Delta r)$ as can be seen in Fig. 4 and Fig. 5 as well. Therefore the peak associated with the geometrical coordination number is not revealed from the structure in $g(r)$. It has a more subtle meaning as explained above.

It is also important to note that in experiments there is always an uncertainty in measuring the position of the particles. According to our analysis an small uncertainty of a few percent will render the mechanical coordination into the geometrical one. Thus experiments will be very difficult to differentiate between z and Z. A possible solution to this problem is to use complementary fluorescent techniques [23, 28, 29] to obtained a signal when the particles are carrying a force and not to rely on geometrical reconstruction of index-matched images or X-ray tomography. Another route would be to obtain approximate coordinates from experiments and then use them as input into a Molecular Dynamics simulations to obtain the exact force balance for each particle. This last approach may provide the final way to accurately measure the coordinations of the particles with accuracy. In our website http://www.jamlab.org we offer the computer codes to perform MD simulations with Hertz-Mindlin forces as well the code to calculate the entropy of the packings.

The theoretical analysis is also confirmed in the RLP packings. Along the RLP line, Fig. 5, we find that the geometrical coordination number as extracted from $g_z(0.04)$ is very close to the mechanical one. Since the RLP line is at $X \to \infty$ and $h_z \ll 1$, all the states along RLP have $z \approx Z$ as we move along the line varying the friction coefficient. Thus, the numerical results confirm the theory.

In conclusion, a prescription to measure the geometrical coordination is the following: First we identify a theoretical way to define it. For instance, here we use the theoretical prediction that all the packings along RCP have the same z. With the proper definition of coordination number for an inflated particle, Eq. (13), we calculate the coordination as a function of Δr. This identifies $\Delta r = 0.04$ as the position to obtain the geometrical z. We then explore any packing (not only at RCP line) and apply Eq. (13) at $\Delta r = 0.04$ and obtain z. Figure 3 summarizes the theoretical predictions of values of z and Z for all the packings in the phase diagram. If the reader still has doubts about the difference between z and Z we offer a final more vivid way to understand it in terms of the famous kissing number conjecture from Newton and Gregory due to a remark of A. Coniglio (private communication): a mechanical contact is like a French kiss while a geometrical contact is any other inconsequential kiss.

VI. CONCLUSIONS

In conclusion, using Edwards statistical mechanics we have elucidated some aspects of RLP and RCP in the disordered spherical packing problem. The phase diagram introduced here serves as a beginning to understand how random packings fill space in three dimensions. The comparative advantage of the present approach over extensive work done in the past, is in the classification of all packings through X, Z and ϕ in the theoretical phase diagram from where these studies could be systematically performed. This classification guides the search for indications of jamming from a systematic point of view, through the exploration of all jammed states from $\mu = 0$ to $\mu \to \infty$. Our results not only apply to packings at the jamming transition in the limit of hard spheres, but may also be extended to a general phase diagram as sketched in Fig. 1c to include states with finite nonzero pressure. Such states are described by an angoricity in addition to the compactivity as developed here. Extensions to other dimensions, polydisperse systems and other shapes of particles like ellipsoids and sphero-cylinders are being worked out with the goal of developing a unifying thermodynamic view of the physics of packings.

ACKNOWLEDGMENTS

This work is supported by the National Science Foundation, CMMT Division and the Department of Energy, Office of Basic Energy Sciences, Geosciences Division.

REFERENCES

1. L. D. Landau and E. M. Lifshitz, *Statistical Physics* (Pergamon, NY, 1970).
2. E. R. Nowak, J. B. Knight, M. L. Povinelli, H. M. Jaeger and S. R. Nagel, *Powder Technol.* **94**, 79 (1997).
3. P. Philippe, and D. Bideau, *Europhys. Lett.* **60**, 677 (2002).
4. J. Brujić, P. Wang, C. Song, D. L. Johnson, O. Sindt, and H. A. Makse, *Phys. Rev. Lett.* **95**, 128001 (2005).
5. M. Schröter, D. I. Goldman, and H. L. Swinney, *Phys. Rev. E* **71**, 030301(R) (2005).
6. S. F. Edwards, in *Granular matter: an interdisciplinary approach* (ed A. Mehta) 121-140 (Springer-Verlag, New York, 1994).
7. R. Blumenfeld and S. F. Edwards, *Phys. Rev. Lett.* **90**, 114303-1 (2003)
8. F. Lechenault, F. da Cruz, O. Dauchot, and E. Bertin, *J. Stat. Mech.*, P07009 (2006).
9. M. P. Ciamarra, A. Coniglio, and M. Nicodemi, *Phys. Rev. Lett.* **97**, 158001 (2006).
10. H. A. Makse and J. Kurchan, *Nature* **415**, 614 (2002).
11. P. Wang, C. Song, and H. A. Makse, *Nature Physics* **2**, 526-531 (2006).
12. C. Song, P. Wang, and H. A. Makse, *Proc. Nat. Acad. Sci.* **102**, 2299-2304 (2005).
13. R. C. Ball and R. Blumenfeld, *Phys. Rev. Lett.* **88**, 115505-1 (2002).
14. H. A. Makse, J. Brujić, and S. F. Edwards, in *The Physics of Granular Media*, edited by H. Hinrichsen and D. E. Wolf (Wiley-VCH, 2004).
15. C. Song, P. Wang, and H. A. Makse, http://arxiv.org/abs/0808.2186
16. P. Wang, C. Song, and H. A. Makse, *Nature* **453**, 629-632 (2008).
17. T. Aste, M. Saadatfar, and T. J. Senden, *J. Stat. Mech.*, P07010 (2006).
18. S. Torquato and F. H. Stillinger, *J. Phys. Chem B* **105**, 11849 (2001).
19. S. Alexander, *Phys. Rep.* **296**, 65 (1998).
20. C. F. Moukarzel, *Phys.Rev. Lett* **81**, 1634 (1998); *ibid*, *Gran. Matter* **3**, 41 (2001).
21. H. A. Makse, D. L. Johnson, and L. M. Schwartz, *Phys. Rev. Lett.* **84**, 4160 (2000).
22. C. S. O'Hern, S. A. Langer, A. J. Liu, and S. R. Nagel, *Phys Rev. Lett.* **88**, 075507 (2002).
23. J. Brujić, C. Song, P. Wang, C. Briscoe, G. Marty, and H. A. Makse, *Phys. Rev. Lett.* **98**, 248001 (2007).
24. L. E. Silbert, D. Ertas, G. S. Grest, T. C. Halsey, and D. Levine, *Phys. Rev. E* **65**, 031304 (2002).
25. C. Briscoe, C. Song, P. Wang, and H. A. Makse. *Phys. Rev. Lett.* **101**, 188001 (2008).
26. M. Skoge, A. Donev, F. H. Stillinger, and S. Torquato, *Phys. Rev. E* **74**, 041127 (2006).
27. G. Parisi, and F. Zamponi, to be published in *Rev. Mod. Phys.* Arxiv preprint at www.arxiv.org/abs/0802.2180 (2008).
28. J. Brujić, S. F. Edwards, D. V. Grinev, I. Hopkinson, D. Brujić, and H. A. Makse, *Faraday Discuss.*, **123**, 207 (2003).
29. J. Brujić, S. F. Edwards, I. Hopkinson, H. A. Makse, *Physica* A **327**, 201 (2003).

A plasticity model with microstructure evolution for quasi-static granular flows

Jin Sun and Sankaran Sundaresan

Department of Chemical Engineering, Princeton University, Princeton, New Jersey 08544 USA

Abstract. We develop a plasticity model to predict complex rheological behaviors of quasi-static granular flows. The stress is decomposed to a pressure term and a deviatoric term with a macroscopic friction coefficient. The closures to the pressure and the friction coefficient are linked to the microstructure evolution, which is modeled by the coordination number and fabric evolution equations. The material constants in the model are functions of particle-level properties and are calibrated using the data from simulations of steady and unsteady simple shear using the discrete element method (DEM).

Keywords: Dense granular flows, constitutive modeling, discrete element method
PACS: 83.10.Gr, 83.10.Rs, 83.50.Ax, 45.70.-n, 46.35+z, 62.20.fq

INTRODUCTION

Continuum models play an important role in scientific understanding and design of processes handling dense granular materials, e.g., hopper discharge or dense-phase pneumatic conveying [1]. However, developing an accurate constitutive model for dense flows remains an open problem. Dense flow characteristics change as the deformation rate varies. We focus our attention to the quasi-static regime, where granular materials flow so slowly that their stresses approach constant values and thus exhibit the characteristic of rate-independence.

In the quasi-static regime, particles make enduring contacts with each other and often deform irreversibly under external loads. Due to these physical features, rate-independent plasticity models for metals have been adapted for granular materials. In plasticity theory, concepts of a yield function, a plastic potential and a flow rule are used to construct constitutive models [2, 3]. For example, a Mohr-Coulomb yield criterion has been used to derive a rigid plasticity model for flow down a rough inclined plane [4]. A von Mises-type yield function was used in derivation of another rigid-perfectly plastic model for analysis of instability of hopper flows [5]. These model were applicable only to incompressible flows. Critical state soil mechanics models were formulated to predict volume fraction changes associated with loading histories [6]. The models, including both rigid plastic and elastic-plastic models, employ a volume fraction dependent elliptical yield function and the associative flow rule. They have been shown to be very useful to illuminate soil behaviors. However, they do have their limitations, mainly that they are not applicable to cyclic loading conditions and that they do not adequately model stress-induced anisotropy.

Experiments have shown that volume fraction is insufficient in describing granular flow behaviors and structural quantities such as coordination number and fabric tensor have been measured [7, 8] using photoelastic particles. These experiments of dense assemblies deformed under biaxial loads and in simple shear, have shown that particle contacts distributed non-uniformly and a fabric tensor can be defined to characterize the anisotropy. Analogous microstructure anisotropy and rearrangement have also been speculated by Gadala-Maria and Acrivos [9] in their experiments of unsteady shear of non-colloidal suspensions. Stress measurements in these experiments have all shown significant transitions when loading conditions changed even though they did not depend on the magnitudes of deformation rates.

To account for the anisotropy and these complex behaviors, description of microstructure has been introduced to the constitutive model. For example, an elastic-plastic model has been proposed for soil by incorporating a structural tensor in a von Mises-type yield function [10]. Fabric tensor has been employed for an elastic-plastic model based on the double-sliding mechanism [11, 12]. While these types of models had a certain degree of success in predicting complex rheological behaviors, they also substantially increased the model complexity. The more serious deficiencies are that their microstructure evolution equations are based on no [10] or very little [11, 12] micromechanical data and that their material constants are not directly connected to particle-level properties, such as friction coefficients. Recent theoretical and computational results have indicated the importance of particle friction coefficients on granular jamming transitions [13]. These shortcomings limit the applicability and accuracy of the models.

In this paper, we present the development of a simple plasticity model based on the micromechanical data obtained from extensive simulations using the discrete element method (DEM). We attempt to use the simplest form to capture the most of complex rheological behaviors of interest, which emphasizes applications to quasi-static shear flows. The very small elastic strain [14] will be neglected in the model due to the focus on continuous flows instated of statics. We will show the macroscopic rheological responses and the microstructure characteristics and evolutions under both steady and unsteady shear. The correspondence between the macroscopic responses and the microstructure variations will be demonstrated. Based on these physical insights, We will then present how the constitutive model can be closed by relations involving the microstructural internal variables. The evolution equations will also be developed based on the DEM data to form a complete constitutive model. As particle-level properties, such as particle elasticity and friction coefficient, are known in the DEM simulations, we correlate the material constants in the model to these properties. The correctness of the model will be demonstrated. We note that similar modeling effort have been reported recently for dense suspension flows [15, 16].

After we describe the DEM method and simulation setups in the next section, we will present simulation results of the rheological responses and microstructural evolutions during unsteady shear. In the next section, we will construct our model within proper mathematical constraints using the physical insights gained from the DEM simulations. We will summarize the attributes of our model in the conclusion section.

As for notation, we employ the regular Roman font for scalars, the Roman boldface for vectors and the San Serif font for the second order tensors.

COMPUTATIONAL METHODS AND SIMULATION DETAILS

In this paper, we present the results of DEM simulations carried out using the large-scale atomic/molecular massive parallel simulator (LAMMPS) developed at Sandia National Laboratories [17]. In the simulations, particles interact only at contact. For two particles $\{i, j\}$ with radii $\{a_i, a_j\}$ separated by r_{ij}, they experience a force, $\mathbf{F}_{ij} = \mathbf{F}_{n_{ij}} + \mathbf{F}_{t_{ij}}$, when $\delta_{ij} = d - r_{ij} > 0$, where $d = a_i + a_j$, $\mathbf{n}_{ij}$ is the contact normal unit vector pointing from the center of particle j to that of particle i, and $\mathbf{t}_{ij}$ is the tangential unit vector. The force is calculated using a spring-dashpot model, which has been well tested and used in many other studies [18, 19, 20]. The normal and tangential components of the interact force acting on particle i are

$$\mathbf{F}_{n_{ij}} = f(\delta_{ij}/d)(k_n \delta_{ij} \mathbf{n}_{ij} - \gamma_n m_{\text{eff}} \mathbf{v}_{n_{ij}}), \tag{1}$$

$$\mathbf{F}_{t_{ij}} = f(\delta_{ij}/d)(-k_t \mathbf{u}_{t_{ij}} - \gamma_t m_{\text{eff}} \mathbf{v}_{t_{ij}}), \tag{2}$$

where $k_{n,t}$ and $\gamma_{n,t}$ are the spring stiffness and viscous damping constants, respectively, and $m_{\text{eff}} = m_i m_j/(m_i + m_j)$ is the effective mass of spheres with masses m_i and m_j. The corresponding contact force on particle j is simply given by Newton's third law, i.e., $\mathbf{F}_{ji} = -\mathbf{F}_{ij}$. The function $f(\delta_{ij}/d) = 1$ is for the Hookean (linear spring-dashpot) model, and $f(\delta_{ij}/d) = \sqrt{\delta_{ij}/d}$ is for Hertzian contacts with viscoelastic damping between spheres. In the simulations using the linear spring-dashpot model, we set $k_t = 2/7 k_n$, $\gamma_t = 0$, and γ_n is chosen to satisfy the normal restitution coefficient $e = \exp\left(-\frac{\gamma_n \pi}{\sqrt{4k_n/m_{\text{eff}} - \gamma_n^2}}\right) = 0.7$. In the simulations using the Hertzian model, the restitution coefficient depends on the initial approaching velocity. For the results presented here, only Hookean model was employed.

The value of the spring stiffness constant is chosen to be large enough to minimize particle interpenetration, yet not so large as to require an unreasonably small simulation time step. The tangential force at each contact is computed by keeping track of the elastic shear displacement, $\mathbf{u}_{t_{ij}}$, throughout the lifetime of a contact. The rate of change of the elastic shear displacement is set to zero at the initiation of a contact and given by

$$\frac{d\mathbf{u}_{t_{ij}}}{dt} = \mathbf{v}_{t_{ij}} - \frac{(\mathbf{u}_{t_{ij}} \cdot \mathbf{v}_{ij})\mathbf{r}_{ij}}{r_{ij}^2}. \tag{3}$$

The last term in Eq. 3 arises from the rigid body rotation around the contact point and ensures that $\mathbf{u}_{t_{ij}}$ always lies in the local tangent plane of contact. As the shear displacement increases, the tangential force reaches the limit imposed by a static yield criterion, characterized by a local particle friction coefficient. The tangential force is limited by truncating the magnitude of $\mathbf{u}_{t_{ij}}$ to satisfy the Coulomb criterion $|\mathbf{F}_{t_{ij}}| < \mu |\mathbf{F}_{n_{ij}}|$.

In order to maintain homogeneous deformations over large strain scales, we performed the simulations in three-dimensional (3D) periodic domains without gravity. Simple shear flow was induced via the Lees-Edwards boundary conditions [21], with the stream velocity in the x-direction and the velocity gradient in the z-direction. The shearing motion induced by this boundary driven algorithm takes time to occur so that the flow would not be homogeneous immediately after a shear rate change, which renders the algorithm not suitable to study time dependent flows. This disadvantage can be greatly alleviated through the use of the SLLOD algorithm [22]. The SLLOD algorithm implies that a change in shear rate is not achieved by simply moving the boundaries of the system faster or slower, but by applying a force to the entire system. Thus the SLLOD algorithm was applied to all the simulations presented in this paper. Homogeneous stress and strain can be extracted from this type of flows, which facilitates the constitutive modeling. The macroscopic stress is calculated by

$$\sigma = \frac{1}{V} \sum_i \left[\sum_{j \neq i} \frac{1}{2} \mathbf{r}_{ij} \mathbf{F}_{ij} + m_i (\mathbf{v}'_i)(\mathbf{v}'_i) \right], \qquad (4)$$

where V is the total volume of the simulation domain and $\mathbf{v}'_i$ is the fluctuating velocity of a particle relative to its mean streaming velocity in the shear flow.

Steady state is reached after an assembly is sheared over a large strain (of order unity at least) and the stress and microstructural quantities cease to evolve. The macroscopic strain rate tensor at the steady state then follows

$$D = \frac{1}{2} \dot{\gamma} \begin{pmatrix} 0 & 1 & 0 \\ 1 & 0 & 0 \\ 0 & 0 & 0 \end{pmatrix}, \qquad (5)$$

where the $\dot{\gamma}$ is the flow velocity gradient and also referred to as the shear rate. The type of unsteady flows studied here is the shear reversal flow, where the flow direction is reversed after the steady state is reached and maintained for a certain strain. In this case, the $\dot{\gamma}$ changes its sign after reversal. The shear rate is maintained low in the sense that the inertia number [23] (or Weissenberg number [24]), $I = \frac{\dot{\gamma} d}{\sqrt{p/\rho}}$, is smaller than 10^{-3}. So that the quasi-static response is guaranteed according to previous experimental and simulation data from our and other researchers' work [23, 25, 26].

Rheological responses of the assembly may be probed by examining either the stress responses under strain constraints or the strain responses under stress constraints. Simulations were performed in these two ways using both constant volume and constant confining pressure conditions. Under constant volume condition, the volumetric strain is kept zero so that dilation of the assembly is avoided and the strain rate tensor will be exact as in Eq. 5. Under constant confining pressure condition, the difference between normal stress components is taken into account and the volumetric strain is non-zero during transient states. The caveat of constant pressure condition is that this type of control is difficult to achieve in physical experiments. However, the common constant-loading type of shearing experiments produce essentially the same rheological responses as the constant pressure simulations performed in this paper. The rheological behaviors of these shear flows serve as the physical basis to the constitutive modeling. The data from both steady and unsteady shear simulations are used to calibrate the coefficients in the model.

UNSTEADY SHEAR RHEOLOGICAL BEHAVIORS

In the following, simulations of an assembly of 2000 mono-disperse spheres under different flow conditions will be presented and analyzed. The dependence on system sizes, i.e., particle numbers, has been tested using assemblies with particle number varying from 1000 to 10000 particles. It has been found that the variations of variables decrease with increasing number of particles and have little difference after 2000 particles. In all cases, particle friction coefficient μ equals 0.5 and the inertial number $I \approx 0.0003$ at steady state. The Hookean contact model is used by default in these simulations unless otherwise indicated. All the results are ensemble averages of 10-20 realizations with different initial configurations.

In the case presented in Fig. 1, an assembly with particle volume fraction ϕ of 0.6 was sheared at $\dot{\gamma}_0$ to reach a steady state under the constant volume condition. The data were plotted against accumulated shear strain $\dot{\gamma}_0 t$ and started with the steady state. The shearing was stopped at $\dot{\gamma}_0 t = 1$, i.e., the shear strain equals one. The pressure scaled by particle diameter and stiffness, and the shear stress to pressure ratio are shown in Figs. 1(a) and (b) respectively. They clearly exhibited the rate-independent response as the stress level was retaining during the no-shear period when $1 < \dot{\gamma}_0 t < 2$ with little change from the steady state. The rate-independent characteristic has also been verified by collapsing the stress data from simulations with a variety of shear rates. Those results add no significant information and are not shown in the figures. At $\dot{\gamma}_0 t = 2$, the flow direction was reversed and the shear rate became $\dot{\gamma} = -\dot{\gamma}_0$. The pressure and stress ratio had significant transitions spanning a strain larger than unity after the reversal. The pressure dropped to a

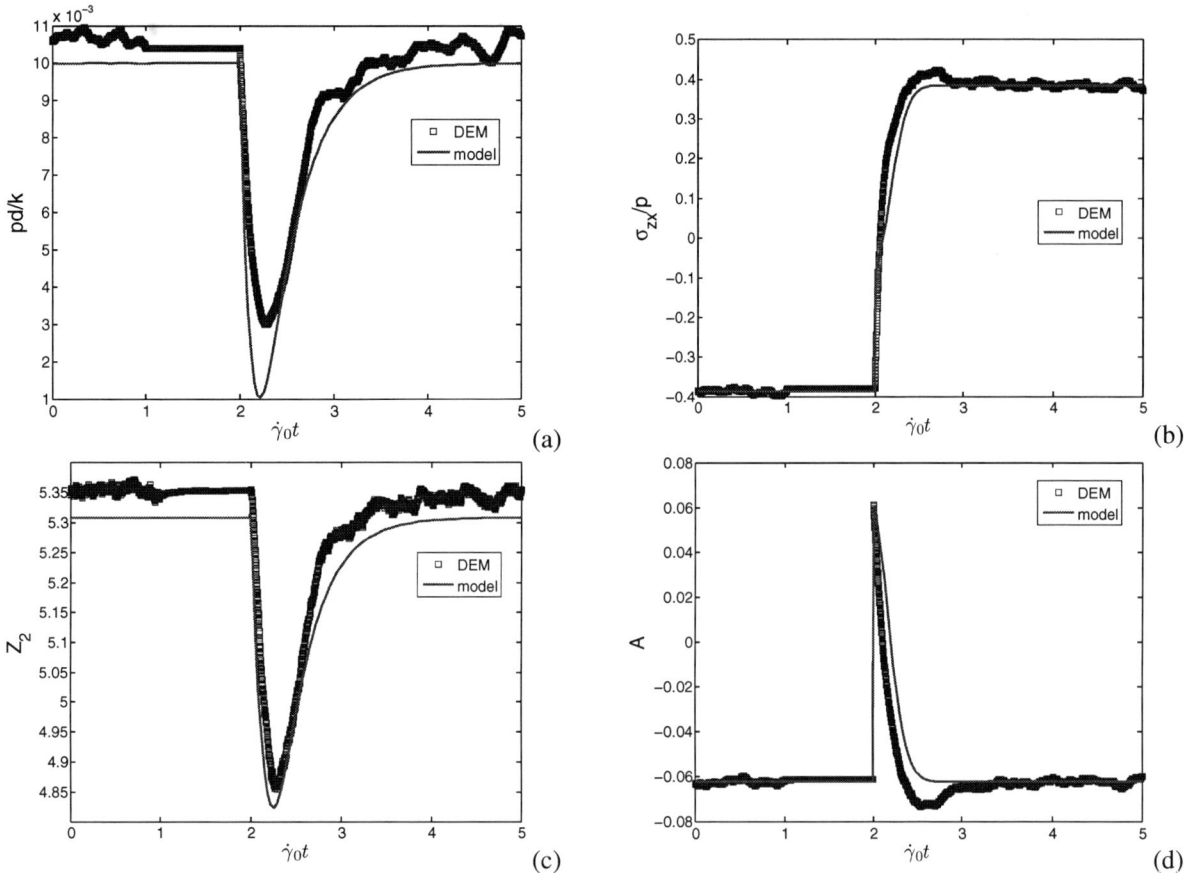

FIGURE 1. Evolutions of (a) pressure, (b) stress ratio, (c) coordination number and (d) anisotropy for an assembly subjected to unsteady shear under constant volume condition with $\phi = 0.60$. Blue square symbols denote the data from DEM simulations and the red solid curves are the constitutive model results.

lower value and slowly returned to the steady state. As the volumetric strain was kept at zero, this behavior rules out the volume fraction change as a necessary cause to the pressure transition. The steady state magnitude of the stress ratio is about 0.4, which is close to the measured quasi-static value in a shear cell experiment using polystyrene beads [4]. Upon shear reversal, the stress ratio slowly evolved to the new steady state value, which was same in magnitude as the initial state, but had an opposite sign. Similar stress transitions after shear reversal has also been observed in experiments with dense granular materials [27] and with dense suspensions [9].

For the case where the assembly was sheared under constant pressure condition, we observe that the volume fraction and stress ratio (see Figs. 2(a) and (b) respectively) also experienced transitions after shear reversal at $\dot{\gamma}_0 t = 1$. The volume fraction increased at first (the assembly compacted) and then decreased (the assembly dilated) after a shear strain of about 0.5 back to the steady state. The compaction behavior was also reported in the Couette shear cell experiment after shear reversal [27]. The volume fraction evolution under constant pressure is another evidence that the pressure is not a unique function of volume function. The stress ratio evolution had a similar trend as that in the constant volume case. At this point, we conclude that the macroscopic rheological responses have a significant transition with a strain scale of order unity after shear reversal, under both constant volume and constant pressure constraints. We note that a constitutive model without history effect would predict an instantaneous recovery to the steady state and not be able to capture this gradual transition.

To further investigate the mechanism of this transition due to shear reversal, we study the microstructure evolution during this unsteady shear. We quantify the microstructure using two variables, the average coordination number Z and the fabric tensor A. The average coordination number is defined as the mean contacts per particle in the contact network, $Z = 2N_c/N$, where N_c is the total number of contacts and N is the total number of particles in the contact network [28]. It characterizes the connectivity of a granular assembly. When the coordination number is equal to a

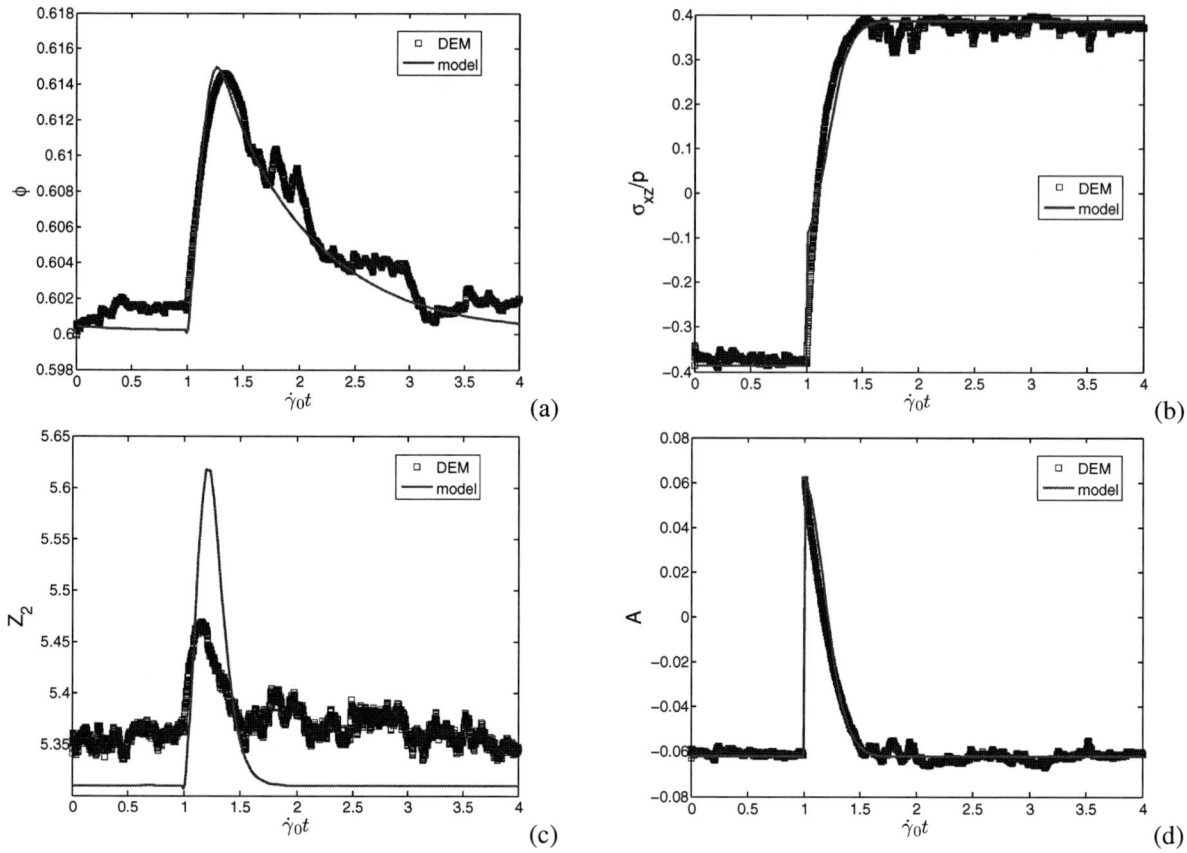

FIGURE 2. Evolutions of (a) pressure, (b) stress ratio, (c) coordination number and (d) anisotropy for an assembly subjected to unsteady shear under constant pressure condition. Blue square symbols denote the data from DEM simulations and the red solid curves are the constitutive model results.

critical value Z_c, the granular assembly is at an isostatic state, where the number of degrees of freedom is matched by number of constraints between particles. Z_c varies from 4 to 6 as particle friction coefficient changes from infinite to zero in three dimensions (3D) [13]. Z_c is also identified with the jamming point, above which granular flows can transit to the quasi-static regime [29, 13]. The value of Z_c will be determined for the granular systems as shown later in the paper. In the computation of Z from our DEM particle data, we neglect any particle with one or zero contact, i.e., do not count the floaters or rattlers. The rationale is that only particles with two and more contacts participate in the contact network. The fact is that the floaters and rattlers make up only a very small portion of the total particles in our simulations. When presenting our data, we use the symbol Z_2 to distinguish from those including floaters and rattlers; in the text, we drop the subscript 2 with the understanding that we only use Z_2 to calibrate our model constants.

Fabric tensor is used to characterize the anisotropy of microstructure. It can be understood as a statistical moment with respect to the probability distribution function of an orientational vector **n**. With different choices of the vector **n** and weighting factors, there have been various definitions [7, 30, 31, 32]. In this paper we identify **n** as the unit contact normal vector pointing from center to center of two particles in contact. We define the fabric tensor as

$$\mathsf{A} = \frac{1}{N_c} \sum_{\alpha=1}^{N_c} \mathbf{n}^\alpha \mathbf{n}^\alpha - \frac{1}{3}\mathsf{I}, \tag{6}$$

where I is the unitary tensor. By this definition, A is a symmetric traceless second rank tensor. The eigenvectors of the fabric tensor give the principal directions of the mean contact orientations. The eigenvalues, in turn, provide a measure of the extent of contact orientations along the principal directions. We may use the difference between the largest (major) and smallest (minor) eigenvalues or the second invariant as a measure of the anisotropy intensity. That being said, the structural anisotropy can easily be related to the shear (xz) component of the fabric tensor for simple shear flows. Our DEM data shows that the numerical values of the diagonal components of A are close to zero and so are the A_{xy} and A_{yz} off-diagonal components. Only the A_{xz} component is non-trivial. Approximately, the difference between the major and minor eigenvalues is $2A_{xz}$ and the corresponding eigenvectors lie in the x-z plan at an angle of $\pm 45°$ to the x-axis. Thus we study the magnitude of A_{xz} for the intensity of the anisotropy and the sign of A_{xz} for the orientation of the principle direction in simple shear flows.

The evolutions of Z and A_{xz} are plotted in Figs. 1(c) and (d) for the constant-volume case. It can be immediately recognized that the Z and A_{xz} have similar evolution patterns as the pressure and stress ratio, respectively. The Z values held up during the shear-stop phase, decreased significantly after the shear reversal and returned to the steady state value after the same strain scale as the pressure did. The transition of the coordination number after shear reversal indicates that the contacts were broken at first and then built up. The A_{xz} variation shows that the microstructure lost its anisotropy and then regained slowly the anisotropy with an opposite principle direction over a comparable strain scale as the shear stress ratio did. The evolutions of the two variables together revealed the microstructural rearrangement after shear reversal, which involves both connectivity and anisotropy changes and has a large strain scale of order unity. During the constant-pressure shear, large scale microstructural rearrangement was also indicated by the evolutions of Z and A_{xz} in Figs. 2(c) and (d). The Z increased after the shear reversal, whose magnitude is non-negligible, albeit small. Since the pressure was maintained constant for the entire shearing process, this increase suggests that pressure may be influence by the anisotropy in addition to the coordination number. The anisotropy effect on pressure will be further specified in the next section. The evolution of A_{xz} (see Fig. 2(d)) again mirrors that of the stress ratio.

To summarize our findings from the unsteady shear simulations, we have demonstrated the rate-independent pressure and stress ratio have a significant transition after shear reversal. This transition is well correlated to the microstructure rearrangement caused by the reversal, which suggests that the microstructural variables, Z and A can be used as internal variables for constitutive models.

PLASTICITY MODEL FORMULATION AND CALIBRATION

The micromechanical analyses of the DEM results in section show the microstructural rearrangement during unsteady flows. This dissipative change of structure calls for a plasticity model to describe the quasi-static rheological behaviors. A constitutive framework for purely dissipative materials [33, 34] is compatible to the physical instances revealed by our simulations and thus applicable to the current modeling work. For such material, the local Cauchy stress $\sigma(t)$ is given by

$$\sigma(t) = \eta : \mathsf{D}(t) \quad (\text{i.e., } \sigma_{ij} = \eta_{ijkl} D_{kl}(t)), \tag{7}$$

where η is a positive-definite viscosity depending generally on $\mathsf{D}(t)$ and on local deformation history. For plastic material response at the quasi-static regime, $\mathsf{D} \to 0$, we have

$$\eta = \mathcal{O}(|\mathsf{D}|^{-1}), \text{ for } |\mathsf{D}| \to 0, \tag{8}$$

where $|\mathsf{D}| = \sqrt{\frac{1}{2}\mathsf{D}^\mathsf{T} : \mathsf{D}}$ denotes the modulus of the strain rate tensor. The stress can be re-written as

$$\sigma(t) = \mu_\mathrm{p} : \hat{\mathsf{D}}(t), \tag{9}$$

where $\mu_\mathrm{p} = \lim_{|\mathsf{D}| \to 0}(|\mathsf{D}|\eta)$ is a plastic modulus and $\hat{\mathsf{D}} = \mathsf{D}/|\mathsf{D}|$ denotes a versor of a second rank tensor. With this general framework, the rate-independent plastic response of dense granular materials can be described.

The fabric evolutions were shown in section to be able to effectively represent the deformation history and clearly established the fabric tensor as an internal variable. The Eq. 9 can be simplified by assuming μ_p is given uniquely as an isotropic function of the fabric tensor A as derived in [35]. The stress can then be expressed as an isotropic function of A and $\hat{\mathsf{D}}$, given in a general expression obeying the representation theorem [43]

$$\begin{aligned}\sigma = &\psi_0 \mathsf{I} + \psi_1 \hat{\mathsf{D}} + \psi_2 \mathsf{A} + \psi_3 \hat{\mathsf{D}}^2 + \psi_4 \mathsf{A}^2 + \psi_5(\mathsf{A}\hat{\mathsf{D}} + \hat{\mathsf{D}}\mathsf{A}) \\ &+ \psi_6(\mathsf{A}^2\hat{\mathsf{D}} + \hat{\mathsf{D}}\mathsf{A}^2) + \psi_7(\mathsf{A}\hat{\mathsf{D}}^2 + \hat{\mathsf{D}}^2\mathsf{A}) + \psi_8(\mathsf{A}^2\hat{\mathsf{D}}^2 + \hat{\mathsf{D}}^2\mathsf{A}^2),\end{aligned} \tag{10}$$

where the ψ coefficients are scalar polynomials of joint invariants of A and $\hat{\mathsf{D}}$ and I is the unit tensor. It should be noted that this equation is in the same form as the Hand anisotropic fluid model [36].

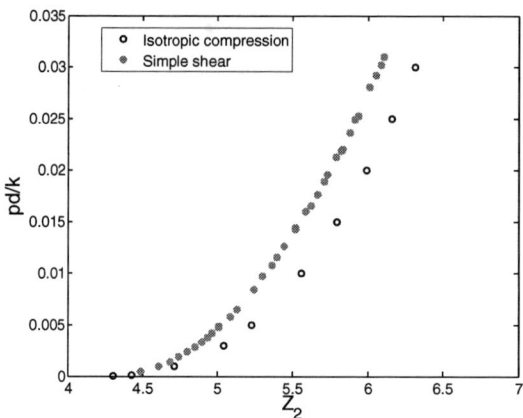

FIGURE 3. Pressure variations against coordination number from simple shear and isotropic compression simulations. The inter-particle friction coefficient is 0.5.

There are numerous combinations that may be used for the coefficients in this general expression, which renders it very complicated to determine and calibrate them. We attempt to determine a simple stress expression within this framework that is able to describe the essential physical features of the steady and unsteady quasi-static shear flows as probed in the last section. For this purpose, we retain only the constant and linear terms in $\hat{D}$, i.e., the ψ_0 and ψ_1 terms in equation 10. We identify the ψ_0 with the confining pressure p and make the ψ_1 dependent on p too. Our simplified stress equation reads

$$\sigma = p\mathsf{I} - p\eta \frac{\mathsf{S}}{|\mathsf{D}|}, \qquad (11)$$

where S is the deviatoric strain rate tensor and η is a scalar macroscopic friction coefficient. This stress equation is consistent with the granular plasticity proposed by Goddard [34], whose friction tensor is reduced to the current scalar friction coefficient. It also bears a similar form as previous rigid-plastic models [4, 5] and a recent visco-plastic model [37] but has an important distinction that η has history dependence built in using the fabric tensor A. This attribute will become clear as we detail the equations for pressure and the friction coefficient η in the following section.

The pressure and η relations

Granular pressure is conventionally modeled as a function of particle volume fraction. The elastic contribution can be understood from DEM simulation to have a power law relation [38], i.e., $p \sim (\phi - \phi_c)^\alpha$, where ϕ_c is a critical volume fraction of transition to jamming or quasi-static state. The pressure in plastic flow at critical state is usually proposed according to the critical state soil mechanics [6] as a logarithmic relation with the volume fraction: $\frac{1}{\phi} \sim \log p$. Similar pressure-volume fraction relations have been widely used in plastic flows [39, 40, 41, 42]. However, both the elastic pressure and the critical state pressure relations neglect the microstructural rearrangements brought by the disturbance in flow conditions. As shown in Figs. 1 and 2, the microstructural rearrangements as shown by the coordination number and anisotropy evolutions after shear reversal decoupled the apparent correspondence between the pressure and the volume fraction variations. To take account of the pressure variation due to this microstructural rearrangement, we model the pressure as a function of Z and A consistent with our stress equation (see Eq. 10). We model the anisotropy effect on pressure by decomposing the pressure into one part at the isotropic state and the other due to the anisotropy

$$pd/k = a_1(Z - Z_c)^{a_2} + a_3(\mathsf{A}:\mathsf{A})(Z - Z_c)^{a_4}, \qquad (12)$$

where Z_c is the critical coordination number for the transition to quasi-static regime and the a's are material constants. The pressure data at the isotropic state was generated from stress-controlled isotropic compression of the assembly. The pressure is plotted as a function of Z in Fig. 3 for both isotropic compression and steady state simple shear. We are able to calibrate the first term on the right hand side in Eq. 12 using the isotropic compression data and the second term using the pressure and A : A difference from the simple shear data.

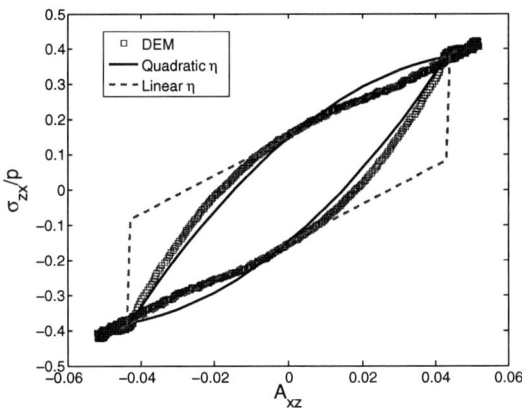

FIGURE 4. Stress ratio variations against the shear component of fabric tensor for unsteady simple shear. The assembly is subjected to unsteady shear under constant volume condition with $\phi = 0.60$ and $\mu = 0.5$.

The shear stress to pressure ratio has been shown in Figs. 1 and 2 to have a similar evolution pattern as the shear component of the fabric tensor A_{xz}. In Fig. 4, we further look at the functional dependence of the stress ratio. The stress ratios during a shear reversal under constant volume condition (the same data as in Fig. 1(b)) were plotted against A_{xz} in Fig. 4. The data for unsteady shear under constant pressure would collapse essentially into the same curve and is thus not plotted for clarity. The two branches in Fig. 4 show that the stress ratio evolves from one steady state to the other and back. Figure 4 clearly reveals a functional dependence of the stress ratio on A_{xz}. We therefore model η using the following expression

$$\eta = b_1 + b_2 \frac{\mathbf{A}:\mathbf{S}}{|\mathbf{D}|}, \tag{13}$$

where the b's are are material constants. The stress ratios calculated from Eq. 13 are plotted in the dashed line and compared with the DEM data in Fig. 4. This equation yields only linear approximation of the DEM results but captures most of the stress ratio variation except immediately after the stress ratio starts to change. A quadratic relation,

$$\eta = b_1 + b_2 \frac{\mathbf{A}:\mathbf{S}}{|\mathbf{D}|} + b_3 \frac{(\mathbf{A}:\mathbf{S})^2}{|\mathbf{D}|^2}, \tag{14}$$

shown in the solid lines in Fig. 4 better approximates the stress ratio variation. However, the strain, during which the differences resulted from Eqs. 13 and 14 are most significant, is very small. We therefore employ Eq. 13 for the constitutive model in this paper.

Microstructure evolution equations

To complete the constitutive model, appropriate evolution equations for the microstructural variables are required. The stress can then be evolved through the microstructural evolutions during unsteady flows. Based on our DEM data of the microstructure evolutions, we model the rate of change of both A and Z as functions of A and D. The reason for this choice will become evident when a comparison is made with the DEM data.

In order to be frame indifferent, the time derivative of a tensor must take account of the frame rotation [43]. We use the Jaumann derivative here, i.e., $\mathring{\mathbf{A}} = \dot{\mathbf{A}} + \mathbf{A} \cdot \mathbf{W} - \mathbf{W} \cdot \mathbf{A}$, where $\mathbf{W}$ is the spin tensor and $\dot{\mathbf{A}}$ denotes the total time derivative. The most general expression for $\mathring{\mathbf{A}}$ has been shown according to the representation theorem to be [36]

$$\mathring{\mathbf{A}} = \alpha_0 \mathbf{I} + \alpha_1 \mathbf{D} + \alpha_2 \mathbf{A} + \alpha_3 \mathbf{D}^2 + \alpha_4 \mathbf{A}^2 + \alpha_5 (\mathbf{AD} + \mathbf{DA}) + \\ \alpha_6 (\mathbf{A}^2 \mathbf{D} + \mathbf{DA}^2) + \alpha_7 (\mathbf{AD}^2 + \mathbf{D}^2 \mathbf{A}) + \alpha_8 (\mathbf{A}^2 \mathbf{D}^2 + \mathbf{D}^2 \mathbf{A}^2), \tag{15}$$

where the α coefficients are scalar polynomials of joint invariants of A and D. We again simply this expression according to the physical behaviors observed in our DEM simulations shown in Figs. 1 and 2. The expression obtained

reads
$$\mathring{\mathsf{A}} = c_1\mathsf{S} + c_2|\mathsf{D}|\mathsf{A} + c_3(\mathsf{A}:\mathsf{D})\mathsf{A}, \tag{16}$$

which only retains the α_1 and α_2 terms from Eq. 15 with linearity in D. We performed linear stability analysis of this equation and found this form to be able to return to steady state correctly. We solved Eq. 16 for the homogeneous simple shear and compared to the DEM results in Figs. 1(d) and 2(d). It can be seen that Eq. 16 is able to capture the steady state and the evolution of A_{xz} to a good degree.

The evolution equation for the coordination number Z follows the same general form as Eq. 15. Since Z is a scalar, the equation will reduce to a scalar form. In fact, Z can be treated as the trace of a generalized structure tensor with deviatoric part as A. The evolution equation after simplification takes the following form

$$\dot{Z} = d_1(\mathsf{A}:\mathsf{D} + d_2|\mathsf{D}|) + d_3|\mathsf{D}|(f(\phi) - Z) + d_4\mathrm{tr}(D), \tag{17}$$

where d_1-d_4 are material constants and tr(D) is the trace of D. The function $f(\phi)$ dictates how the coordination number varies with volume fraction during the steady shear. The equation for $f(\phi)$, established by fitting the Z variation against ϕ at steady state, is

$$f(\phi) = Z_c + \beta_1(\phi - \phi_c)^{\beta_2}. \tag{18}$$

The β coefficients are material constants calibrated using the DEM data. The tr(D) term is used to take account of the effects of compaction and dilation on increasing and decreasing the coordination number. Therefore the Reynolds dilatancy constraint is implicitly incorporated in Eq. 17, which can be re-written in an explicit form proposed by Goddard and Didwania [44]

$$\mathrm{tr}(D) = \alpha|\mathsf{D}|, \tag{19}$$

where the coefficient of dilatancy $\alpha = \frac{1}{d_4}\left[d_1(\frac{\mathsf{A}:\mathsf{S}}{|\mathsf{D}|} + d_2) + d_3(f(\phi) - Z) - \frac{\dot{Z}}{|\mathsf{D}|}\right]$, is dynamically changing according to the evolutions of Z and A. The computational results from Eq. 17 were compared to the DEM data in Figs. 1(c) and 2(c). For both conditions, the evolution strains scales after shear reversal were correctly produced. There were inaccuracies in the steady state and minimum and maximum values calculations, which do not affect the pressure calculation significantly as will be shown next.

Equations 11, 12, 13, 16 and 17 compose the complete constitutive model, which takes a simple form of plasticity model for stress equation but incorporates the microstructure evolution. We now study if the macroscopic quantities can be correctly described by this model. With the Z and components of A solved from Eqs. 12 and 16, we feed these quantities to Eqs. 12 and 13 to calculate the pressure and stress ratio, respectively for shear flows under the constant volume condition. The results are shown in Figs. 1(a) and (b) to compare with the DEM data. The rate-independence characteristic of the model is obvious as the calculated pressure and stress ratio did not change when the shearing motion was stopped at $\dot{\gamma}t = 1$. The large-strain evolutions of the pressure and stress ratio after the shear reversal at $\dot{\gamma}t = 2$ were correctly captured. There were small quantitative differences between the model and the DEM results. For the shear under constant pressure condition, we adjusted Z to match the target pressure value, which was constant and not shown in the figure. The modeled evolution of the volume fraction and stress ratio, as shown in Figs. 2(a) and (b), matched the DEM data well. It can be concluded from these comparisons that the constitutive model is self-consistent and able to correctly describe the rheological behaviors of the steady and unsteady quasi-static shear flows.

CONCLUSIONS

We have developed a plasticity model that is able to predict complex rheological behaviors of quasi-static granular flows. The stress equation simply contains a pressure term and a deviatoric term with a macroscopic friction coefficient. The closures to the pressure and the friction coefficient are linked to the microstructure evolution, which is modeled by the coordination number and fabric evolution equations. The material constants in the model are functions of particle-level properties and can be calibrated using the shear flow data from the DEM simulations.

We note that the current model is intended to capture the most salient rheological features of the quasi-static granular flows using the simplest model. Therefore, some rheological features are neglected in favor of the model simplicity. For example, the normal stress differences, which are observed in our DEM simulations, are not captured in this model. However, the model can be extended to include such subtle features by using additional terms in Eq. 10. The model can be refined in a similar way when applied to triaxial deformations. These types of extension work will be interesting to pursue in the future.

ACKNOWLEDGMENTS

The authors are grateful to professors Joe Goddard, Prabhu Nott and Stefan Lunding for their stimulating discussions on continuum modeling of dense granular materials. We also thank Sebastian Chialvo for providing some DEM data. This work is supported by a DOE-UCR grant DE-FG26-07NT43070.

REFERENCES

1. S. Sundaresan, *Powder Technology* **115**, 2–7 (2001).
2. R. Hill, *The Mathematical Theory of plasticity*, Oxford University Press, Oxford, 1950.
3. R. M. Nedderman, *Statics and Kinematics of Granular Materials*, Cambridge University Press, 1992.
4. S. B. Savage, *Mechanics of granular materials: new models and constitutive relations*, Elsevier Ltd, 1983, chap. Granular flow down rough inclines-review and extension.
5. D. G. Schaeffer, *Journal of Differential Equations* **66**, 19–50 (1987).
6. A. Schofield, and P. Wroth, *Critical State Soil Mechanics*, McGraw-Hill Inc., 1968.
7. M. Oda, J. Konishi, and S. Nemat-Nasser, *Géotechnique* (1980).
8. G. Subhash, S. Nemat-Nasser, M. M. Mehrabadi, and H. M. Shodj, *Mechanics of Materials* **11**, 87–106 (1991).
9. F. Gadala-Maria, and A. Acrivos, *Journal of Rheology* **24**, 799–814 (1980).
10. J. H. Prevost, *Soil Dynamics and Earthquake Engineering* **4**, 9–17 (1985).
11. S. Nemat-Nasser, *Journal of the Mechanics and Physics of Solids* **48**, 1541–1563 (2000).
12. S. Nemat-Nasser, and J. Zhang, *International Journal of Plasticity* **18**, 531–547 (2002).
13. C. Song, P. Wang, and H. A. Makse, *Nature* **453**, 629–632 (2008).
14. R. Kuwano, and R. J. Jardine, *Géotechnique* **52**, 727–749 (2002).
15. J. J. Stickel, R. J. Phillips, and R. L. Powell, *Journal of Rheology* **50**, 379–413 (2006).
16. J. J. Stickel, R. J. Phillips, and R. L. Powell, *Journal of Rheology* **51**, 1271–1302 (2007).
17. S. Plimpton, *Journal of Computational Physics* **117**, 1–19 (1995/3/1).
18. L. E. Silbert, D. Ertas, G. S. Grest, T. C. Halsey, D. Levine, and S. J. Plimpton, *Physical Review E* **64**, 051302 (2001).
19. L. E. Silbert, G. S. Grest, R. Brewster, and A. J. Levine, *Physical Review Letters* **99**, 068002–4 (2007).
20. C. H. Rycroft, K. Kamrin, and M. Z. Bazant, *Journal of the Mechanics and Physics of Solids* **57**, 828–839 (2009/5).
21. A. W. Lees, and S. F. Edwards, *Journal of Physics C: Solid State Physics* **5**, 1921–1928 (1972).
22. D. J. Evans, and G. P. Morriss, *Statistical Mechanics of Nonequilibrium Liquids*, Academic Press, San Diego, 1990.
23. G. D. R. MiDi, *European Physical Journal E – Soft Matter* **14**, 341–305 (2004).
24. J. Goddard, *Acta Mechanica* **63**, 3–13 (1986).
25. F. da Cruz, S. Emam, M. Prochnow, J.-N. Roux, and F. Chevoir, *Physical Review E (Statistical, Nonlinear, and Soft Matter Physics)* **72**, 021309–17 (2005).
26. L. Aarons, and S. Sundaresan, *Powder Technology* **169**, 10–21 (2006).
27. M. Toiya, J. Stambaugh, and W. Losert, *Physical Review Letters* **93**, 088001–4 (2004).
28. H. P. Zhang, and H. A. Makse, *Physical Review E (Statistical, Nonlinear, and Soft Matter Physics)* **72**, 011301–12 (2005).
29. C. S. O'Hern, L. E. Silbert, A. J. Liu, and S. R. Nagel, *Physical Review E* **68** (2003).
30. R. J. Bathurst, and L. Rothenburg, *Mechanics of Materials* **9**, 65–80 (1990).
31. F. Radjai, D. E. Wolf, M. Jean, and J.-J. Moreau, *Physical Review Letters* **80** (1998).
32. J. D. Goddard, *Physics of dry granular media*, Kluwer, 1998.
33. J. D. Goddard, *Journal of Non-Newtonian Fluid Mechanics* **14**, 141–160 (1984).
34. J. D. Goddard, *Journal of Fluid Mechanics* **568**, 1–17 (2006).
35. S. C. Cowin, *Mechanics of Materials* **4**, 137–147 (1985).
36. G. L. Hand, *Journal of Fluid Mechanics* **13**, 33–46 (1962).
37. P. Jop, Y. Forterre, and O. Pouliquen, *Nature* **441**, 727–730 (2006).
38. H. A. Makse, D. L. Johnson, and L. M. Schwartz, *Phys. Rev. Lett.* **84**, 4160–4163 (2000).
39. S. C. Cowin, *Journal of Applied Mechanics* **44**, 409–412 (1977).
40. N. P. Kruyt, *Journal of Applied Mechanics* **57**, 1032–1035 (1990).
41. S. B. Savage, *Journal of Fluid Mechanics* **377**, 1–26 (1998).
42. A. Srivastava, and S. Sundaresan, *Powder Technology* **129**, 72–85 (2003).
43. C. Truesdell, and W. Noll, *The Non-Lindear Field Theories of Mechanics*, Springer, 2004, 3rd edn.
44. J. D. Goddard, and A. K. Didwania, *The Quarterly Journal of Mechanics & Applied mathematics* **51**, 15–41 (1998).
45. T. L. Youd, *Highway Research Record* **374**, 1–6 (1971).
46. D. Wood, and M. Budhu, *Soilds under Cyclic and Transient Loading*, Balkema, Rotterdam, The Netherlands, 1980, chap. The behaviour of Leighton Buzzard sand in cyclic simple shear tests, p. 9.
47. O. Reynolds, *Philosophical Magazine* **20**, 469–481 (1885).
48. N. Okada, and N.-N. S., *Géotechnique* **44**, 1–19 (1994).

Modelling Limit States Within the Framework of Hypoplasticity

Erich Bauer

Institute of Applied Mechanics, Graz University of Technology, 8010 Graz, Austria

Abstract. The focus of this paper is on modelling limit stress states or so-called critical states in which a cohesionless granular body can be deformed continuously at a constant stress and a constant volume. The constitutive equations are based on the framework of hypoplasticity, which is an alternative concept to elasto-plasticity. The requirements for modelling limit stress states are studied for both a standard and a micro-polar hypoplastic model. As in a micro-polar description the stress tensor is usually non-symmetric, the question arises as to whether the result for the limit stress ratio based on the symmetric Cauchy stress tensor in the standard continuum is the same as the one obtained from the micro-polar continuum description. To this end the limit stress state is analysed under plane shearing of a lateral infinite granular strip located between parallel rough platens under a constant normal stress. It is shown that the stress ratio in the limit state obtained from the micro-polar hypoplastic model can only be the same as for the standard hypoplastic continuum model under certain conditions on the centre line of the shear band.

Keywords: Granular materials, critical states, limit stress states, micro-polar continuum, hypoplasticity
PACS: 81.05.Rm;46.35.+z

INTRODUCTION

Limit stress states or so-called critical stress states of cohesionless granular materials are defined for a simultaneous vanishing of the stress rate and the volume strain rate. Starting from an arbitrary stress state granular materials can reach critical stress states for instance by monotonic shearing. For the same orientation of the stress deviator in the space of principal stresses, the stress ratio in critical stress states is independent of the mean stress and the initial stress ratio, i.e. the memory of the material of the initial stress ratio is swept out. Such states are well defined for granular materials and therefore play an important role for the constitutive modelling and the calibration. An example is the so-called critical friction angle defined on the critical stress ratio and a fundamental material parameter in soil mechanics.

In the present paper critical stress states are modelled using a hypoplastic continuum description. Originally, the concept of hypoplasticity was developed based on a local or so-called standard continuum, cf. Kolymbas [1], where the constitutive equation for the evolution of the stress tensor is described by an inherently non-linear tensor valued function depending in the simplest case on the current stress and the rate of deformation. In particular, the evolution equation is represented by the sum of a tensor function which is linear in the rate of deformation and by a tensor function which is non-linear in the rate of deformation. Thus an intrinsic inelastic material behaviour is modelled without a decomposition of the deformation into elastic and plastic parts. The flow rule and the stress limit condition are not described by separate functions as they are included in the evolution equation for the state quantities. In this respect the concept of hypoplasticity differs fundamentally from the concept of elasto-plasticity. In the last two decades various hypoplastic material models have been proposed. More sophisticated ones include additional state variables like pressure dependent limit void ratios, cf. [2, 3, 4]. Creep and stress relaxation of weathered grains can be taken into account with an evolution equation for the solid hardness depending on the state of weathering, cf. [5]. With additional structure tensors initial anisotropy, e.g. [6, 7], cohesion, cf. [8], and a so-called inter-granular strain, cf. [9], can be modelled. Recently, hypoplastic models have been extended also to clay soil, e.g. [10], to viscous material properties, e.g. [11], and to a micro-polar description, e.g. [12, 13, 14]. Extended hypoplastic concepts are also proposed for instance by Goddard [15] and an embedding of hypoplasticity into the framework of thermodynamics was shown by Schneider and Hutter [16].

Constitutive models based on a standard continuum fail to describe micro-polar properties of granular materials which are experimentally evident when shear deformation takes place. In a ring shear apparatus, for instance, the displacement field across the height of a granular layer under shearing is non-linear and combined with particle rotation, e.g. [17]. Under large monotonic shearing the deformation localizes within a zone of a thickness of a few grain diameters, e.g. [18, 19, 20, 21]. In order to model such properties, a micro-polar continuum description can be used, which allows relating the characteristic length to the mean grain size in a physically natural manner, e.g. [22, 13, 23]. Analytical solutions are rare and can only be derived for certain boundary value problems using simplified versions of hypoplastic models as shown for a micro-polar hypoplastic model by Bauer and Huang [24], Huang [25] and Bauer [26]. These papers give the analytical solution for the initial response of shearing of an infinite granular layer. The present paper also presents an analytical solution for the stresses and couple stresses in the limit state. In contrast to the micro-polar hypoplastic versions by Huang et al. [14], a simplified micro-polar hypoplastic model is employed for the present study, where the influence of the rate of deformation and the rate of curvature is decoupled. In particular, the evolution equation for the non-symmetric stress tensor and the couple stress tensor is described by tensor-valued functions which are non-linear and homogeneous of first order in the rate of deformation and the rate of curvature, respectively.

The focus of the the present paper is on modelling limit states for both a standard and a micro-polar hypoplastic continuum model. As in a micro-polar description the stress tensor is usually non-symmetric, the question arises as to whether the critical stress ratio defined for steady flow states in the standard continuum description is the same as the one obtained in the micro-polar hypoplastic continuum description. It is shown that this is only the case under certain conditions on the centre line of the localized zone.

The paper is organized as follows: First the concept of hypoplasticity is briefly outlined based on the standard continuum. The requirements for modelling limit stress states, which are characterized by a simultaneous vanishing of the stress rate and the volume strain rate, are given and analysed for a particular hypoplastic equation. Then an extension to a micro-polar hypoplastic continuum is presented. For the particular case of plane shearing of an infinite granular layer under a constant pressure perpendicular to the direction of shearing analytical solutions for the initial response and for the limit state are presented. Finally the solutions for the limit stress state obtained from the micro-polar model and the standard model are compared and discussed.

Throughout the paper, bold lower case, bold upper case and calligraphic letters denote vectors, tensors of second order and of fourth order, respectively. In particular, the identity tensor of second order is denoted by $\mathbf{I}$ and the identity tensor of fourth order is denoted by $\mathscr{I}$. Indices on vector and tensor components refer to an orthonormal Cartesian basis $\mathbf{e}_i$ ($i = 1, 2, 3$). Operations and symbols are defined as: $\mathscr{I} = \delta_{ik}\delta_{jl}\mathbf{e}_i \otimes \mathbf{e}_j \otimes \mathbf{e}_k \otimes \mathbf{e}_l$, $\mathbf{I} = \delta_{ij}\mathbf{e}_i \otimes \mathbf{e}_j$, $\mathbf{A} \otimes \mathbf{B} = A_{ij}B_{kl}\mathbf{e}_i \otimes \mathbf{e}_j \otimes \mathbf{e}_k \otimes \mathbf{e}_l$, $\mathbf{AB} = A_{ik}B_{kj}\mathbf{e}_i \otimes \mathbf{e}_j$, $\mathscr{A} : \mathbf{B} = A_{ijkl}B_{kl}\mathbf{e}_i \otimes \mathbf{e}_j$, $\mathbf{A} : \mathbf{B} = A_{ij}B_{ij}$, $\mathrm{tr}\mathbf{A} = \mathbf{I} : \mathbf{A} = A_{ii}$ and $A_{ij}^T = A_{ji}$. Herein δ_{ik} denotes the Kronecker delta and the summation convention over repeated indices is employed. A superimposed dot indicates the material time derivative, and the symbol $\|.\|$ the Euclidian norm of a tensor, i.e. $\|\mathbf{A}\| = \sqrt{A_{ij}A_{ij}}$. Compressive stress and strain and their rates are negative as in the sign convention of continuum mechanics.

THE CONCEPT OF HYPOPLASTICITY FOR A NON-POLAR CONTINUUM

Within the framework of a standard hypoplastic continuum, e.g. [1, 27], the objective stress rate tensor $\mathring{\mathbf{T}}$ is described by an tensor-valued function $\mathbf{H}$ depending in the simplest case on the current Cauchy stress tensor $\mathbf{T}$ and the symmetric part of the velocity gradient or so-called strain rate tensor $\mathbf{D}$, i.e.

$$\mathring{\mathbf{T}} = \mathbf{H}(\mathbf{T}, \mathbf{D}) . \tag{1}$$

For the objective stress rate the time derivative given by Green-Naghdi [28] is adopted, i.e.

$$\mathring{\mathbf{T}} = \dot{\mathbf{T}} - \mathbf{\Omega}\mathbf{T} + \mathbf{T}\mathbf{\Omega}. \tag{2}$$

Herein $\dot{\mathbf{T}}$ denotes the material time derivative of the Cauchy stress tensor $\mathbf{T}$, and the angular velocity tensor $\mathbf{\Omega}$ is related to the rotation tensor $\mathbf{R}$ and to the rate of the rotation tensor, $\dot{\mathbf{R}}$, as $\mathbf{\Omega} = \dot{\mathbf{R}}\mathbf{R}^T$. The Green-Naghdi stress rate is chosen because, in contrast to the Jaumann stress rate, it can appropriate describe steady stress states under shearing, cf. [29]. In particular, the Jaumann stress rate can show stress oscillation, while in the Green-Naghdi stress rate the term $\mathbf{T}\mathbf{\Omega} - \mathbf{\Omega}\mathbf{T}$ vanishes for large shearing so that in the limit state the objective stress rate is equal to the material time derivative of the Cauchy stress tensor, i.e. $\mathring{\mathbf{T}} = \dot{\mathbf{T}}$.

To specify the tensor function $\mathbf{H}$ in equation (1) several requirements must be fulfilled which are based on continuum mechanics and on general mechanical properties of granular materials detected in experiments. For instance frame indifference requires that **H** is an isotropic tensor-valued function of its arguments, cf. [30, 31]. This requirement is satisfied for instance for terms taken from the general representation theorem for a tensor function of two symmetric tensors, cf. [32]. A further benefit of this representation theorem is that tensors of higher order than two are expressed by lower order tensors. Thus function **H** can be written in the most general case as

$$\mathbf{H}(\mathbf{T},\mathbf{D}) = \alpha_1 \mathbf{I} + \alpha_2 \mathbf{T} + \alpha_3 \mathbf{D} + \alpha_4 \mathbf{T}^2 + \alpha_5 \mathbf{D}^2 + \alpha_6 (\mathbf{TD}+\mathbf{DT}) + \\ + \alpha_7 (\mathbf{TD}^2 + \mathbf{D}^2\mathbf{T}) + \alpha_8 (\mathbf{T}^2\mathbf{D} + \mathbf{DT}^2) + \alpha_9 (\mathbf{T}^2\mathbf{D}^2 + \mathbf{D}^2\mathbf{T}^2) , \tag{3}$$

where **I** denotes the identity tensor and the scalar quantities α_i ($i = 1,\ldots,9$) are functions of the invariants and joint invariants of **T** and **D**. In order to further specify the constitutive relation (1) it is assumed that the material behaviour to be described is rate independent. This requirement is fulfilled if function **H** is positively homogeneous of the first order in **D**, i.e.

$$\mathbf{H}(\lambda\mathbf{D}) = \lambda\,\mathbf{H}(\mathbf{D}) \quad \text{for any scalar}\ \lambda > 0. \tag{4}$$

To describe inelastic behaviour function **H** must be non-linear in **D** so that

$$\mathbf{H}(\mathbf{D}) \neq -\mathbf{H}(-\mathbf{D}). \tag{5}$$

The requirements of rate independent behaviour and inelastic behaviour are fulfilled by a decomposition of the isotropic tensor valued function $\mathbf{H}(\mathbf{T},\mathbf{D})$ in equation (1) into the sum of the following two parts, cf. [1, 27]:

$$\mathbf{H}(\mathbf{T},\mathbf{D}) = \mathscr{A}(\mathbf{T}) : \mathbf{D} + \mathbf{B}(\mathbf{T})\,||\mathbf{D}||. \tag{6}$$

Herein the function $\mathscr{A}(\mathbf{T}) : \mathbf{D}$ is linear in **D** and the function $\mathbf{B}(\mathbf{T})\,||\mathbf{D}||$ is non-linear in **D**. The constitutive equation (6) is homogeneous of the first order in **D** and therefore it describes a rate independent material behaviour. It is easy to prove that the constitutive equation (6) is generally inherently non-linear, e.g. for two particular strain rates $\mathbf{D}_a$ and $\mathbf{D}_b$ with the same Euclidean norm, i.e. $||\mathbf{D}_a|| = ||\mathbf{D}_b||$, but opposite signs, i.e. $\mathbf{D}_b = -\mathbf{D}_a$, the corresponding stress rates are:

$$\mathbf{H}(\mathbf{T},\mathbf{D}_a) \neq -\mathbf{H}(\mathbf{T},\mathbf{D}_b) \quad \text{or equivalently} \quad \mathbf{H}(\mathbf{T},\mathbf{D}_a) \neq -\mathbf{H}(\mathbf{T},-\mathbf{D}_a).$$

It can be noted that for $\mathbf{T} \neq \mathbf{0}$ and $\mathbf{D} \neq \mathbf{0}$ the linear part and the non-linear part of the hypoplastic constitutive equation (6) do not vanish so that an intrinsic inelastic material behaviour is described without a decomposition of the deformation into elastic and plastic parts. In other words, a splitting of the strain rate into elastic and plastic parts in the sense of the concept of elasto-plasticity cannot be carried out for equation (6) as outlined in the appendix. In some extended hypoplastic models, however, relation (6) is modified in such a way that under certain condition the response of the constitutive equation becomes hyperelastic or hypoelastic. Such models basically deviate from the standard concept of hypoplasticity outlined in equation (6) and they are not considered in the present paper.

One of the fundamental properties of cohesionless granular materials is reflected by the experimental finding, e.g. Goldscheider [33], that for drained deformation with fixed components of the strain rate, i.e. for a constant **D**, the corresponding stress path asymptotically lead to an almost linear path in which the orientation of the stress path is independent of the initial stress state. In other words, under deformation with fixed components of the strain rate the memory of the material of the initial stress state declines with continuous deformation, cf. [34, 35]. In order to model this experimental finding the tensor function **H** must be homogeneous in the stress. In particular the tensorial function $\mathscr{A}(\mathbf{T})$ of rank four and the tensorial function $\mathbf{B}(\mathbf{T})$ of rank two of the constitutive equation (6) must be homogeneous of the same order in **T**. Then the terms of equation (6) can, for instance, be factorized using the dimensionless tensor $\hat{\mathbf{T}} = \mathbf{T}/\mathrm{tr}\mathbf{T}$ and equation (6) can be represented as

$$\mathbf{H}(\mathbf{T},\mathbf{D}) = (\mathrm{tr}\mathbf{T})^m\,\bar{f}_c\left[\mathscr{L}(\hat{\mathbf{T}}) : \mathbf{D} + \mathbf{N}(\hat{\mathbf{T}})\,||\mathbf{D}||\right]. \tag{7}$$

Herein m denotes the order of homogeneity in **T** and $\bar{f}_c$ is a constitutive constant with the dimension of $[stress]^{1-m}$. Equations (6) and (7) represent the basic concept of hypoplasticity. Based on equation (7) various specific representations of the functions $\mathscr{L}$ and **N** have been proposed in the literature. A comprehensive historical review can be found for instance in Wu and Kolymbas [27] as well as in Bauer and Herle [36].

Constitutive equations of the type of equation (7) are apt to describe limit stress states. In particular for certain pairs of $\hat{\mathbf{T}}$ and **D**, where $\hat{\mathbf{D}}$ is kept constant, the stress rate vanishes, i.e. $\mathring{\mathbf{T}} = \mathbf{0}$, so that

$$\mathscr{L}(\hat{\mathbf{T}}) : \mathbf{D} + \mathbf{N}(\hat{\mathbf{T}})\,||\mathbf{D}|| = \mathbf{0}. \tag{8}$$

From equation (8) a kind of a flow rule can be derived, i.e.

$$\hat{\mathbf{D}} = \frac{\mathbf{D}}{||\mathbf{D}||} = -\mathscr{L}^{-1}(\hat{\mathbf{T}}) : \mathbf{N}(\hat{\mathbf{T}}), \qquad (9)$$

where $\hat{\mathbf{D}}$ is a non-vanishing dimensionless tensor. Inserting equation (9) into the identity $\hat{\mathbf{D}} : \hat{\mathbf{D}} = 1$ leads to a scalar equation depending only on the quantity $\hat{\mathbf{T}}$, i.e.

$$\left[\mathscr{L}^{-1}(\hat{\mathbf{T}}) : \mathbf{N}(\hat{\mathbf{T}}) \right] : \left[\mathscr{L}^{-1}(\hat{\mathbf{T}}) : \mathbf{N}(\hat{\mathbf{T}}) \right] - 1 = 0 \,. \qquad (10)$$

The set of all stresses which fulfil this condition can be represented by a surface in the stress space which is called the limit stress surface, cf. [37, 38]. While in the concept of elasto-plasticity the stress limit condition is prescribed by a separate equation in hypoplasticity the flow rule and the limit stress state are embedded in the constitutive equation. It is worth noting that equation (8) first only fulfils the requirement for a vanishing stress rate, but it does not a priori restrict the set of possible strain rates $\mathbf{D}$ to critical states. In other words, limit stress states are generally included in hypoplastic constitutive equations of the form of equation (7) and are therefore independent of the specific representation of the tensor functions $\mathscr{L}(\hat{\mathbf{T}})$ and $\mathbf{N}(\hat{\mathbf{T}})$ but a vanishing stress rate need not be simultaneously related to a vanishing volume strain rate. An example are peak stress states in which the stress rate vanishes, but usually not the volume strain rate. In order to model critical states the constitutive equation must simultaneously satisfy for certain pairs of $\hat{\mathbf{T}}$ and $\mathbf{D}$ the conditions

$$\mathring{\mathbf{T}} = \mathbf{0} \quad \text{and} \quad \text{tr}\,\mathbf{D} = 0. \qquad (11)$$

In addition to relation (8) the following condition for a vanishing volume strain rate must therefore be satisfied as well:

$$\text{tr}\hat{\mathbf{D}} = \text{tr}\left[-\mathscr{L}^{-1}(\hat{\mathbf{T}}) : \mathbf{N}(\hat{\mathbf{T}}) \right] = 0. \qquad (12)$$

It can be noted that the total strain rate in (9) is equal to the plastic strain rate, i.e. $\mathbf{D} = \mathbf{D}^p$ and $\hat{\mathbf{D}} = \hat{\mathbf{D}}^p$, only for states where (11) is fulfilled. For modelling critical states requirements (10) and (12) are necessary conditions for the representation of the tensorial functions $\mathscr{L}(\hat{\mathbf{T}})$ and $\mathbf{N}(\hat{\mathbf{T}})$, which are fulfilled for the following specific relations proposed by Bauer [39]:

$$\mathscr{L}(\hat{\mathbf{T}}) = \hat{a}^2 \mathscr{I} + \hat{\mathbf{T}} \otimes \hat{\mathbf{T}} \qquad (13)$$

and

$$\mathbf{N}(\hat{\mathbf{T}}) = \hat{a}\, (\hat{\mathbf{T}} + \hat{\mathbf{T}}^*) \,. \qquad (14)$$

Relations (13) and (14) are also apt to model essential properties of granular materials which are not only related to critical states as outlined in more detail by Bauer [4] and Gudehus [3]. Herein the normalized tensor $\hat{\mathbf{T}}^*$ is the deviatoric part of the normalized tensor $\hat{\mathbf{T}}$, i.e. $\hat{\mathbf{T}}^* = \hat{\mathbf{T}} - \mathbf{I}/3$. For the specific representation of $\mathscr{L}(\hat{\mathbf{T}})$ in equation (13) and $\mathbf{N}(\hat{\mathbf{T}})$ in equation (14) the homogeneity of the function $\mathbf{H}$ in (7) is $m = 1$. Thus the hypoplastic constitutive equation can be represented as:

$$\mathring{\mathbf{T}} = \bar{f}_c\, \text{tr}\mathbf{T} \left[\hat{a}^2 \mathbf{D} + \text{tr}(\hat{\mathbf{T}}\mathbf{D})\hat{\mathbf{T}} + \hat{a}(\hat{\mathbf{T}} + \hat{\mathbf{T}}^*)||\mathbf{D}|| \right] . \qquad (15)$$

In the present version $\bar{f}_c$ is a dimensionless constant which can be calibrated based on an isotropic compression test. Factor $\hat{a}$ in equation (15) is related to critical stress states which can be easily shown by applying the definition for critical states to the hypoplastic constitutive equation. In particular, for a vanishing stress rate, i.e. $\mathring{\mathbf{T}} = \mathbf{0}$, the constitutive equation reduces to

$$\hat{a}^2 \mathbf{D} + \hat{\mathbf{T}}\,\text{tr}(\hat{\mathbf{T}}\mathbf{D}) + \hat{a}(\hat{\mathbf{T}} + \hat{\mathbf{T}}^*)||\mathbf{D}|| = \mathbf{0}. \qquad (16)$$

It follows from equation (16) that factor $\bar{f}_c$ does not influence states with a vanishing stress rate. Dividing relation (16) by $||\mathbf{D}||$ and using the dimensionless tensor $\hat{\mathbf{D}} = \mathbf{D}/||\mathbf{D}||$ leads to

$$\hat{\mathbf{D}} = -\frac{1}{\hat{a}^2}\hat{\mathbf{T}}\,\text{tr}(\hat{\mathbf{T}}\hat{\mathbf{D}}) - \frac{1}{\hat{a}}[\hat{\mathbf{T}} + \hat{\mathbf{T}}^*]. \qquad (17)$$

For critical states the volume strain rate must also vanish, i.e.

$$\mathrm{tr}\mathbf{D} = \mathrm{tr}\hat{\mathbf{D}} = -\frac{1}{\hat{a}^2}\mathrm{tr}\hat{\mathbf{T}}\mathrm{tr}(\hat{\mathbf{T}}\hat{\mathbf{D}}) - \frac{1}{\hat{a}}[\mathrm{tr}\hat{\mathbf{T}} + \mathrm{tr}\hat{\mathbf{T}}^*] = 0,$$

so that with respect to $\mathrm{tr}\hat{\mathbf{T}} = \mathrm{tr}(\hat{\mathbf{T}}/\mathrm{tr}\hat{\mathbf{T}}) = 1$ and $\mathrm{tr}\hat{\mathbf{T}}^* = 0$ the equation reduces to

$$\mathrm{tr}(\hat{\mathbf{T}}\hat{\mathbf{D}}) + \hat{a} = 0.$$

Substituting this relation into equation (17) yields

$$\hat{\mathbf{D}} = -\frac{1}{\hat{a}}\hat{\mathbf{T}}^*. \tag{18}$$

Since for the deviatoric part of the stress tensor $\mathrm{tr}\mathbf{T}^* = \mathrm{tr}\hat{\mathbf{T}}^* = 0$, it can easily be shown that relation (18) also fulfils the requirement (11) for a vanishing volume strain rate, i.e.

$$\mathrm{tr}\hat{\mathbf{D}} = -\frac{1}{\hat{a}}\mathrm{tr}\hat{\mathbf{T}}^* = 0.$$

Taking into account that $\hat{\mathbf{D}}^* = \mathbf{D}/\|\mathbf{D}\| - (1/3)\mathbf{1}\,\mathrm{tr}(\hat{\mathbf{D}})/\|\mathbf{D}\|$ and $\mathrm{tr}(\hat{\mathbf{D}}) = 0$, one obtains

$$\hat{\mathbf{D}}^* = \hat{\mathbf{D}},$$

so that for plastic flow equation (18) can be represented as

$$\hat{\mathbf{D}}^* = -\frac{1}{\hat{a}}\hat{\mathbf{T}}^*. \tag{19}$$

Relation (19) reflects coaxiality between $\hat{\mathbf{D}}^*$ and $\hat{\mathbf{T}}^*$ for critical states. Inserting (19) into the identity $\|\hat{\mathbf{D}}^*\|^2 = \hat{\mathbf{D}}^* : \hat{\mathbf{D}}^* = 1$ leads to an equation for the critical stress surface [4], i.e.

$$\hat{a} = \|\hat{\mathbf{T}}^*\|. \tag{20}$$

With respect to $\hat{\mathbf{T}}^* = \mathbf{T}/\mathrm{tr}(\mathbf{T}) - (1/3)\mathbf{1}$ it is obvious that the critical stress surface is a cone with its apex at the origin of the principal stress component space (Figure 1.a). Herein the dimensionless factor $\hat{a}$ can be interpreted as the radius of the intersection of the critical stress surface with the π-plane (Figure 1.b). In particular $\hat{a}$ determines the shape of the critical stress surface and can be formulated in terms of the invariants of the stress tensor. It was shown by Bauer [29] that factor $\hat{a}$ can be adapted to various stress limit conditions without loss of the representation of the proposed tensor function (15). For the sake of simplicity $\hat{a}$ is assumed to be constant in the present paper, which means that the trace of the limit surface in the π-plane is a circle.

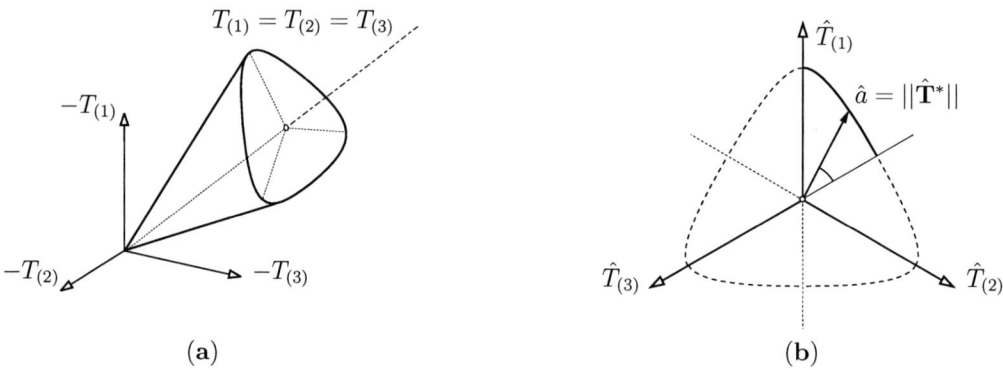

FIGURE 1. Critical stress limit states: (a) in the space of principal stresses; (b) representation of $\hat{a}$ in the π-plane.

As the constitutive equation (15) only takes into account the two constants $\bar{f}_c$ and $\hat{a}$ and the current stress as the only history dependent state variable, the prediction of general stress-strain paths is limited. For a refined modelling of the mechanical properties of granular materials the tensor functions $\mathscr{L}$ and $\mathbf{N}$ in (13) and (14) may also depend on additional scalar- and tensor-valued state variables. For instance, the influence of density on the incremental stiffness can be taken into account by the current void ratio as an additional state variable, i.e. functions $\mathscr{L}$ and $\mathbf{N}$ can be scaled with a pressure dependent stiffness factor and density factor [2, 40, 3, 4]. In order to also take into account particle rotation and couple stresses the non-polar constitutive relation was extended to a micro-polar continuum, e.g. [13, 14], which is discussed in more detail in the next section.

MICRO-POLAR EXTENSION OF THE HYPOPLASTIC MODEL

In a micro-polar continuum, e.g. [41], independent degrees of freedom ω_i^c ($i = 1,2,3$) for the micro-rotations are introduced besides the displacements u_i in the standard continuum. With respect to the rate of the macro-spin vector $\dot{\boldsymbol{\omega}}$, the rate of the micro-spin vector $\dot{\boldsymbol{\omega}}^c$ and the permutation tensor $\boldsymbol{\varepsilon}$, the macro-spin tensor $\mathbf{W}$, the micro-spin tensor $\mathbf{W}^c$, the rate of deformation $\mathbf{D}^c$ and the rate of curvature $\mathbf{K}$ are defined as:

$$\mathbf{W} = -\boldsymbol{\varepsilon}\,\dot{\boldsymbol{\omega}}, \quad \mathbf{W}^c = -\boldsymbol{\varepsilon}\,\dot{\boldsymbol{\omega}}^c, \quad \mathbf{D}^c = \mathbf{L} - \mathbf{W}^c, \quad \mathbf{K} = \operatorname{grad}\dot{\boldsymbol{\omega}}^c. \tag{21}$$

The velocity gradient $\mathbf{L} = \operatorname{grad}\dot{\mathbf{u}}$ in equation (21) is related to the macro-motion and can be decomposed into the symmetric part $\mathbf{D} = (\mathbf{L} + \mathbf{L}^T)/2$ and the skew-symmetric part $\mathbf{W} = (\mathbf{L} - \mathbf{L}^T)/2$, i.e. the definition of tensor $\mathbf{D}$ and tensor $\mathbf{W}$ is the same as in the standard continuum. Hence definition (21) for the rate of deformation can alternatively be written as

$$\mathbf{D}^c = \mathbf{D} + \mathbf{W} - \mathbf{W}^c. \tag{22}$$

Representation (22) indicates that in the case where the macro-spin is equal to the micro-spin, i.e. $\dot{\boldsymbol{\omega}} = \dot{\boldsymbol{\omega}}^c$, the rate of deformation $\mathbf{D}^c$ reduces to the tensor $\mathbf{D}$ of the standard non-polar continuum. The kinematic quantities $\mathbf{D}^c$ and $\mathbf{K}$ are associated with the stress tensor $\mathbf{T}$ and the couple stress tensor $\mathbf{M}$ defined for the current configuration. For quasi-static processes the local equilibrium equations read:

$$\operatorname{div}\mathbf{T} + \rho\,\tilde{\mathbf{b}} = \mathbf{0}, \tag{23}$$

$$\operatorname{div}\mathbf{M} - \boldsymbol{\varepsilon} : \mathbf{T} + \rho\,\tilde{\mathbf{m}} = \mathbf{0}. \tag{24}$$

Herein ρ denotes the bulk density of the material, $\tilde{\mathbf{b}}$ and $\tilde{\mathbf{m}}$ represent the body force and body couple, respectively. Equation (24) indicates that the stress tensor in a micro-polar continuum is usually non-symmetric with the exception of states with $\operatorname{div}\mathbf{M} = \mathbf{0}$ and $\tilde{\mathbf{m}} = \mathbf{0}$. With respect to the time derivative given by Green and Naghdi [28] the objective measures for the stress rate and couple stress rate read:

$$\overset{\circ}{\mathbf{T}} = \dot{\mathbf{T}} - \boldsymbol{\Omega}\mathbf{T} + \mathbf{T}\boldsymbol{\Omega}, \tag{25}$$

$$\overset{\circ}{\mathbf{M}} = \dot{\mathbf{M}} - \boldsymbol{\Omega}\mathbf{M} + \mathbf{M}\boldsymbol{\Omega}. \tag{26}$$

For the stress tensor and couple stress tensor the following evolution equations are considered:

$$\overset{\circ}{\mathbf{T}} = \bar{f}_c \operatorname{tr}\mathbf{T} \left[\hat{a}^2 \mathbf{D}^c + (\hat{\mathbf{T}} : \mathbf{D}^c)\hat{\mathbf{T}} + \hat{a}(\hat{\mathbf{T}} + \hat{\mathbf{T}}^*)\|\mathbf{D}^c\|\right], \tag{27}$$

$$\overset{\circ}{\mathbf{M}} = d_{50}\bar{f}_c \operatorname{tr}\mathbf{T} \left[a_m^2 \bar{\mathbf{K}} + (\hat{\mathbf{M}} : \bar{\mathbf{K}})\hat{\mathbf{M}} + 2a_m \hat{\mathbf{M}}\|\bar{\mathbf{K}}\|\right]. \tag{28}$$

Herein $\hat{\mathbf{T}} = \mathbf{T}/\operatorname{tr}\mathbf{T}$, $\hat{\mathbf{T}}^* = \hat{\mathbf{T}} - \mathbf{I}/3$, $\hat{\mathbf{M}} = \mathbf{M}/(d_{50}\operatorname{tr}\mathbf{T})$ and $\bar{\mathbf{K}} = d_{50}\mathbf{K}$ are normalized quantities. The parameter d_{50} denotes the mean grain diameter which enters the constitutive equations as the internal length. Factors $\hat{a}$ and a_m are related to critical states which will be investigated in more detail in the following.

For the present micro-polar model critical states are defined for a simultaneous vanishing of the stress rate the volume strain rate and the couple stress rate, i.e.

$$\overset{\circ}{\mathbf{T}} = \mathbf{0} \quad ; \quad \operatorname{tr}\mathbf{D}^c = \operatorname{tr}\mathbf{D} = 0 \quad \text{and} \quad \overset{\circ}{\mathbf{M}} = \mathbf{0}, \tag{29}$$

for certain monotonic deformations with $\|\mathbf{D}^c\| \neq 0$ and $\|\mathbf{K}\| \neq 0$. The stress limit condition and couple stress limit condition can be derived in a similar way as shown for the standard hypoplastic model in the foregoing section. As $\overset{\circ}{\mathbf{T}} = \mathbf{0}$ also $\operatorname{tr}\overset{\circ}{\mathbf{T}} = 0$, the condition for the vanishing stress rate is equivalent to $\operatorname{tr}\overset{\circ}{\mathbf{T}} = 0$ and $\overset{\circ}{\mathbf{T}}^* = \overset{\circ}{\mathbf{T}} - (\operatorname{tr}\overset{\circ}{\mathbf{T}})\mathbf{I}/3 = \mathbf{0}$. In particular equation (27) yields for $\operatorname{tr}\mathbf{D}^c = 0$ and $\operatorname{tr}\overset{\circ}{\mathbf{T}} = 0$:

$$\hat{\mathbf{T}} : \mathbf{D}^c = -\hat{a}\|\mathbf{D}^c\|, \tag{30}$$

With respect to $\hat{\mathbf{D}}^c = \mathbf{D}^c/\|\mathbf{D}^c\|$ and $\overset{\circ}{\mathbf{T}}^* = \mathbf{0}$, the following relation can be derived:

$$\hat{a}^2 \hat{\mathbf{D}}^c + \left(\hat{\mathbf{T}} : \hat{\mathbf{D}}^c\right)\hat{\mathbf{T}}^* + 2\hat{a}\hat{\mathbf{T}}^* = \mathbf{0}, \tag{31}$$

Substituting (30) into (31) the flow rule for the stress state is obtained, i.e.

$$\hat{\mathbf{D}}^c = -\frac{1}{\hat{a}}\hat{\mathbf{T}}^*. \tag{32}$$

Substituting (32) into $\|\hat{\mathbf{D}}^c\| = \sqrt{\hat{\mathbf{D}}^c : \hat{\mathbf{D}}^c} = 1$ one obtains:

$$\hat{a} = \|\hat{\mathbf{T}}^*\|. \tag{33}$$

For a vanishing couple stress rate, i.e. $\overset{\circ}{\mathbf{M}} = \mathbf{0}$, equation (28) reduces to:

$$a_m^2 \bar{\mathbf{K}} + (\hat{\mathbf{M}} : \bar{\mathbf{K}})\hat{\mathbf{M}} + 2a_m \hat{\mathbf{M}} \|\bar{\mathbf{K}}\| = \mathbf{0}. \tag{34}$$

Double contraction of equation (34) with $\bar{\mathbf{K}}$ leads to a quadratic equation for $(\hat{\mathbf{M}} : \bar{\mathbf{K}})$, i.e.

$$(\hat{\mathbf{M}} : \bar{\mathbf{K}})^2 + 2a_m \|\bar{\mathbf{K}}\| (\hat{\mathbf{M}} : \bar{\mathbf{K}}) + a_m^2 \bar{\mathbf{K}} : \bar{\mathbf{K}} = 0, \tag{35}$$

with the unique solution

$$\hat{\mathbf{M}} : \bar{\mathbf{K}} = -a_m \sqrt{\bar{\mathbf{K}} : \bar{\mathbf{K}}}. \tag{36}$$

Substituting (36) into (34) and with $= \sqrt{\bar{\mathbf{K}} : \bar{\mathbf{K}}} = \|\bar{\mathbf{K}}\|$ the flow rule for the couple stress state is obtained, i.e.

$$\hat{\bar{\mathbf{K}}} = \frac{\bar{\mathbf{K}}}{\|\bar{\mathbf{K}}\|} = -\frac{1}{a_m}\hat{\mathbf{M}}. \tag{37}$$

Substituting (37) into $\|\hat{\bar{\mathbf{K}}}\| = \sqrt{\hat{\bar{\mathbf{K}}} : \hat{\bar{\mathbf{K}}}} = 1$ one obtains

$$a_m = \|\hat{\mathbf{M}}\|. \tag{38}$$

The analysis of critical states shows that $\hat{a}$ is related to the normalized limit stress deviator and a_m is related to the limit couple stress. Factors $\hat{a}$ and a_m reflect the intergranular slide resistance and rotation resistance of particles, respectively. In contrast to more sophisticated micro-polar hypoplastic models, cf. [12, 24, 13, 23, 14], some simplification are assumed in the present constitutive relations (27) and (28). In particular, the stress rate is homogeneous of first order in the rate of deformation only and the couple stress rate is homogeneous of first order in the rate of curvature only, so that in critical states factors $\hat{a}$ and a_m are not coupled. Furthermore $\hat{a}$ and a_m are assumed to be constant in the present paper. A comparison of equation (33) with equation (20) shows that in contrast to the standard continuum model factor $\hat{a}$ is related to a non-symmetric tensor in the micro-polar model. In this context the question arises as to whether factor $\hat{a}$ can be calibrated for both continuum descriptions in a unified manner. In order to investigate this question the behaviour of a granular strip under monotonic shearing is considered in the following section.

SHEARING WITH DILATANCY

If micro-polar continua are considered, the simple shear test is no longer an element test, which means that numerical results obtained from finite element calculations are influenced by the size of the specimen. Consequently the shearing of an infinite granular layer is considered instead, where the results obtained are independent of the co-ordinate in the direction of shearing, cf. [23]. In particular the monotonic shearing of an infinite granular layer located between parallel rough platens is investigated. With respect to a constant normal stress prescribed on the top platen the shearing of granular materials can lead to a decrease or increase of the volume, which is usually termed dilatancy. For an infinite layer the field quantities are independent of the co-ordinate in the direction of shearing, i.e. $\partial(.)/\partial x_1 = 0$ with respect to the co-ordinate system in Figure (2). Furthermore for the assumption of plane strain conditions the field quantities are also independent of the co-ordinate x_3. Then the macro-motion for plane shearing with dilatancy can be described by

$$x_1 = X_1 + f_1(X_2,t), \quad x_2 = X_2 + f_2(X_2,t) \quad \text{and} \quad x_3 = X_3, \tag{39}$$

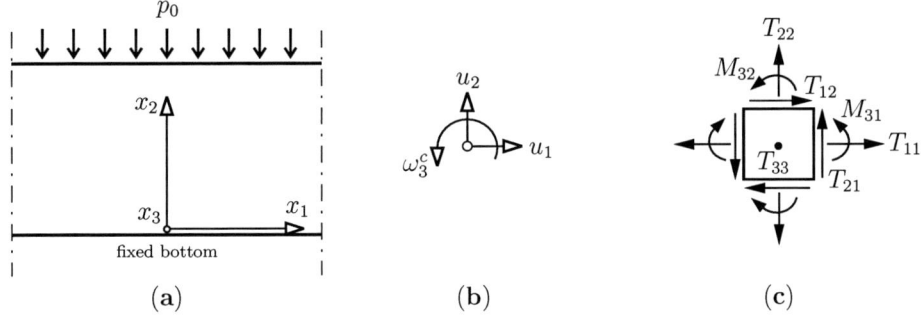

FIGURE 2. Modelling of plane shearing under constant vertical pressure $p_0 = -T_{22}$: (a) section of the infinite granular layer between parallel plates with rough surfaces, (b) relevant kinematics of plane shearing with dilatancy and degrees of freedom u_1, u_2 and ω_3^c, (c) stress components T_{11}, T_{22}, T_{33}, T_{12}, T_{21} and couple stress components M_{31} and M_{32}.

where the spatial co-ordinates x_i ($i = 1,2,3$) represent the current position of a particle and X_i ($i = 1,2,3$) are the corresponding co-ordinates in the initial configuration. $f_i(X_2,t)$ ($i = 1,2$) are time dependent functions with respect to $f_i(X_2, t = 0) = 0$ in the initial configuration at time $t = 0$. The only possible micro-motion is the micro-rotation ω_3^c, which is orientated perpendicular to the (x_1,x_2)–plane as shown in Figure (2). The displacements $u_i = x_i - X_i$ read

$$u_1 = f_1(X_2,t) , \quad u_2 = f_2(X_2,t) \quad \text{and} \quad u_3 = 0. \tag{40}$$

Since u_2 depends on X_2, the kinematics of shearing with dilatancy is taken into account. With respect to

$$g_i(X_2,t) = \frac{\partial f_i(X_2,t)}{\partial X_2} \quad \text{and} \quad \dot{g}_i(X_2,t) = \frac{d g_i(X_2,t)}{dt}, \quad (i=1,2), \tag{41}$$

for any field variable ϕ defined in the infinite layer, the following relations are valid:

$$\frac{\partial \phi}{\partial X_1} = \frac{\partial \phi}{\partial x_1} = 0, \quad \frac{\partial \phi}{\partial X_2} = (1+g_2)\frac{\partial \phi}{\partial x_2}, \quad \frac{\partial \phi}{\partial X_3} = \frac{\partial \phi}{\partial x_3} = 0. \tag{42}$$

Therefore the non-vanishing components of the velocity gradient and macro-spin read:

$$L_{12} = \frac{\partial \dot{u}_1}{\partial x_2} = \frac{\dot{g}_1}{1+g_2}, \quad L_{22} = \frac{\partial \dot{u}_2}{\partial x_2} = \frac{\dot{g}_2}{1+g_2}, \quad W_{12} = \frac{1}{2}\frac{\partial \dot{u}_1}{\partial x_2} = -\dot{\omega}_3. \tag{43}$$

It follows from L_{22} in (43) that $\dot{g}_2/(1+g_2)$ is equal to the volume strain rate and therefore a measure of the dilatancy behaviour of the granular layer. As body forces, body couples and inertia forces are neglected, the equilibrium equations (23) and (24) reduce to

$$\frac{\partial T_{ij}}{\partial x_j} = 0, \tag{44}$$

and

$$\frac{\partial M_{ij}}{\partial x_j} - \varepsilon_{ikl} T_{kl} = 0. \tag{45}$$

With respect to the relations in (42) the time derivatives of equations (44) and (45) reads

$$\frac{\partial \dot{T}_{ij}}{\partial x_j} - \frac{\partial T_{ij}}{\partial x_m}\frac{\partial \dot{u}_m}{\partial x_j} = 0, \tag{46}$$

and

$$\frac{\partial \dot{M}_{ij}}{\partial x_j} - \frac{\partial M_{ij}}{\partial x_m}\frac{\partial \dot{u}_m}{\partial x_j} - \varepsilon_{ikl}\dot{T}_{kl} = 0. \tag{47}$$

For the infinite shear layer equation(46) and equation(47) yield:

$$\frac{\partial \dot{T}_{12}}{\partial x_2} = 0, \quad \frac{\partial \dot{T}_{22}}{\partial x_2} = 0, \tag{48}$$

$$\frac{\partial \dot{M}_{32}}{\partial x_2} - \frac{\dot{g}_2}{1+g_2}(T_{12} - T_{21}) - (\dot{T}_{12} - \dot{T}_{21}) = 0. \tag{49}$$

The material time derivatives $\dot{T}_{ij}$ and $\dot{M}_{ij}$ in (48) and (49) are related to the constitutive equations using the corresponding objective derivatives defined in (25) and (26). With respect to the non-zero stresses and couple stresses shown for plane shearing in Figure (2) the components of the objective stress rate and couple stress rate read:

$$\begin{aligned}
\overset{\circ}{T}_{11} &= \bar{f}_c \operatorname{tr}\mathbf{T}\left[\psi_1 \hat{T}_{11} + \hat{a}(2\hat{T}_{11} - 1/3)\sqrt{\psi_2}\right], \\
\overset{\circ}{T}_{22} &= \bar{f}_c \operatorname{tr}\mathbf{T}\left[\hat{a}^2 D^c_{22} + \psi_1 \hat{T}_{22} + \hat{a}(2\hat{T}_{22} - 1/3)\sqrt{\psi_2}\right], \\
\overset{\circ}{T}_{33} &= \bar{f}_c \operatorname{tr}\mathbf{T}\left[\psi_1 \hat{T}_{33} + \hat{a}(2\hat{T}_{33} - 1/3)\sqrt{\psi_2}\right], \\
\overset{\circ}{T}_{12} &= \bar{f}_c \operatorname{tr}\mathbf{T}\left[\hat{a}^2 D^c_{12} + \left(\psi_1 + 2\hat{a}\sqrt{\psi_2}\right)\hat{T}_{12}\right], \\
\overset{\circ}{T}_{21} &= \bar{f}_c \operatorname{tr}\mathbf{T}\left[\hat{a}^2 D^c_{21} + \left(\psi_1 + 2\hat{a}\sqrt{\psi_2}\right)\hat{T}_{21}\right], \\
\overset{\circ}{M}_{31} &= d_{50} \bar{f}_c \operatorname{tr}\mathbf{T}\left[\left(\hat{M}_{32}\bar{K}_{32} + 2a_m\sqrt{\bar{K}^2_{32}}\right)\hat{M}_{31}\right], \\
\overset{\circ}{M}_{32} &= d_{50} \bar{f}_c \operatorname{tr}\mathbf{T}\left[a_m^2 \bar{K}_{32} + \left(\hat{M}_{32}\bar{K}_{32} + 2a_m\sqrt{\bar{K}^2_{32}}\right)\hat{M}_{32}\right],
\end{aligned} \tag{50}$$

with the abbreviations:

$$\psi_1 = \hat{T}_{12}D^c_{12} + \hat{T}_{21}D^c_{21} + \hat{T}_{22}D^c_{22}, \quad \psi_2 = D^{c\,2}_{12} + D^{c\,2}_{21} + D^{c\,2}_{22},$$

and the kinematic quantities:

$$D^c_{12} = \dot{\omega}^c_3 + \frac{\dot{g}_1}{1+g_2}, \quad D^c_{21} = -\dot{\omega}^c_3, \quad D^c_{22} = \frac{\dot{g}_2}{1+g_2}, \quad \bar{K}_{32} = d_{50}\frac{\partial \dot{\omega}^c_3}{\partial x_2}.$$

In the following it assumed that at the top of the layer a vertical pressure is applied and kept constant, i.e. $T_{22} = -p_0 = \text{const}$. Shearing is initiated by a prescribed horizontal displacement of the top layer with the constant velocity $\dot{u}_{1T}$. The bottom is fixed. While the symmetry condition for the lateral infinite shear layer has been already considered above, the boundary conditions at the bottom and top of the layer are still to be specified. Herein the micropolar boundary conditions allow the modelling of the influence of the rotational resistance of particles in contact with the bounding structure, e.g. [23]. For the present investigation very rough bottom and top surfaces of the bounding structure are assumed, so that neither sliding nor rotation may occur, i.e. the layer of grains in contact with the rough bounding structure become part of the boundary. Thus the boundary conditions at the bottom and at the top of the layer read:

$$\begin{aligned}
x_2 = 0 &: \quad \dot{u}_1 = 0, \quad \dot{u}_2 = 0, \quad \dot{\omega}^c_3 = 0 & (51) \\
x_2 = h &: \quad \dot{u}_1 = \dot{u}_{1T}, \quad \dot{T}_{22} = 0, \quad \dot{\omega}^c_3 = 0. & (52)
\end{aligned}$$

Initial response

In this section an analytical solution to the initial response of the shear layer is derived, in which for the initial state an isotropic stress state is considered, i.e.

$$T_{ij} = -p_0\, \delta_{ij}\ , \quad M_{ij} = 0, \tag{53}$$

where p_0 denotes the isotropic initial pressure. It follows from the relations in (53) that for the initial state the shear stresses and couple stresses are assumed to be zero, thus the material time derivatives of the stresses and the couple stresses coincide with the objective rates at the beginning of shearing. With respect to $g_1 = g_2 = 0$, $f_c = -\bar{f}_c\, p_0$ and the initial state assumed in (53), the relations for the non-zero rate of the stress and couple stress components by (50) reduce to:

$$\dot{T}_{11} = \mathring{T}_{11} = f_c \left[\frac{\dot{g}_2}{9} + \frac{\hat{a}}{3} \sqrt{\dot{g}_2^2 + (\dot{\omega}_3^c + \dot{g}_1)^2 + \dot{\omega}_3^{c\,2}} \right], \tag{54}$$

$$\dot{T}_{22} = \mathring{T}_{22} = f_c \left[\hat{a}^2\, \dot{g}_2 + \frac{\dot{g}_2}{9} + \frac{\hat{a}}{3} \sqrt{\dot{g}_2^2 + (\dot{\omega}_3^c + \dot{g}_1)^2 + \dot{\omega}_3^{c\,2}} \right], \tag{55}$$

$$\dot{T}_{33} = \mathring{T}_{33} = \dot{T}_{11}, \tag{56}$$

$$\dot{T}_{12} = \mathring{T}_{12} = f_c\, \hat{a}^2\, (\dot{\omega}_3^c + \dot{g}_1), \tag{57}$$

$$\dot{T}_{21} = \mathring{T}_{21} = -f_c\, \hat{a}^2\, \dot{\omega}_3^c, \tag{58}$$

$$\dot{M}_{32} = \mathring{M}_{32} = f_c\, a_m^2\, d_{50}^2\, K_{32}. \tag{59}$$

Herein the quantities:

$$\dot{g}_1 = \frac{d\dot{u}_1}{dx_2} = -2\,\dot{\omega}_3, \quad \dot{g}_2 = \frac{d\dot{u}_2}{dx_2}, \quad \dot{\omega}_3^c \quad \text{and} \quad K_{32} = \frac{d\dot{\omega}_3^c}{dx_2} \tag{60}$$

have to be determined for the given boundary value problem. With respect to the requirement in (48) the quantity $\dot{T}_{12}$ in (57) is constant across the height of the layer, i.e.

$$\dot{T}_{12} = f_c\, \hat{a}^2\, (\dot{\omega}_3^c + \dot{g}_1) = \chi, \tag{61}$$

where χ denotes a constant. By substituting the relations (57, 58, 59) into (49) one obtains

$$\frac{\partial}{\partial x_2}\left(f_c\, a_m^2\, d_{50}^2\, \frac{\partial \dot{\omega}_3^c}{\partial x_2} \right) - f_c\, \hat{a}^2\, (2\,\dot{\omega}_3^c + \dot{g}_1) = 0. \tag{62}$$

For an initially homogeneous layer f_c, $\hat{a}$ and d_{50} are independent of the co-ordinates $x_i\,(i=1,2,3)$ so that (62) and (61) lead to the following differential equation for $\dot{\omega}_3^c$:

$$\eta^2\, \frac{d^2 \dot{\omega}_3^c}{dx_2^2} - \dot{\omega}_3^c = C_1, \tag{63}$$

with the abbreviations $\eta = (a_m/\hat{a})\, d_{50}$ and $C_1 = \chi/(f_c\, \hat{a}^2) = \dot{\omega}_3^c + \dot{g}_1$.

The differential equation (63) has the general solution

$$\dot{\omega}_3^c = C_2\, \cosh(x_2/\eta) + C_3\, \sinh(x_2/\eta) - C_1, \tag{64}$$

where C_2 and C_3 are integration constants. With the solution for $\dot{\omega}_3^c$ all other state quantities can be derived as outlined in detail by Bauer [26]. With respect to the given boundary conditions the following results for the initial response can be obtained:

$$K_{32} = \frac{d\dot{\omega}_3^c}{dx_2} = \frac{1}{\eta}\left[C_2 \sinh(x_2/\eta) + C_3 \cosh(x_2/\eta)\right], \tag{65}$$

$$\dot{\omega}_3 = \frac{1}{2}[\dot{\omega}_3^c - C_1] = \frac{1}{2}[C_2 \cosh(x_2/\eta) + C_3 \sinh(x_2/\eta)] - C_1, \tag{66}$$

$$\dot{g}_1 = -2\dot{\omega}_3 = 2C_1 - [C_2 \cosh(x_2/\eta) + C_3 \sinh(x_2/\eta)], \tag{67}$$

$$\dot{u}_1 = 2C_1 x_2 - C_2 \eta \sinh(x_2/\eta) - C_3 \eta \cosh(x_2/\eta) + C_4, \tag{68}$$

$$\dot{g}_2 = -\sqrt{\frac{C_1^2 + [C_2 \cosh(x_2/\eta) + C_3 \sinh(x_2/\eta) - C_1]^2}{9\hat{a}^2 + 1/(9\hat{a}^2) + 1}}, \tag{69}$$

$$\dot{u}_2 = -\int \sqrt{\frac{C_1^2 + [C_2 \cosh(x_2/\eta) + C_3 \sinh(x_2/\eta) - C_1]^2}{9\hat{a}^2 + 1/(9\hat{a}^2) + 1}}\, dx_2 + C_5, \tag{70}$$

$$\frac{\dot{T}_{11}}{f_c \hat{a}^2} = -\frac{C_1^2 + [C_2 \cosh(x_2/\eta) + C_3 \sinh(x_2/\eta) - C_1]^2}{81\hat{a}^4 + 9\hat{a}^2 + 1} + $$
$$+ \left[\frac{C_1^2 + [C_2 \cosh(x_2/\eta) + C_3 \sinh(x_2/\eta) - C_1]^2}{(81\hat{a}^4 + 9\hat{a}^2 + 1)/(9\hat{a}^2 + 1/(9\hat{a}^2) + 2)}\right]^{\frac{1}{2}}, \tag{71}$$

$$\dot{T}_{22} = 0, \tag{72}$$

$$\frac{\dot{T}_{33}}{f_c \hat{a}^2} = \frac{\dot{T}_{11}}{f_c \hat{a}^2}, \tag{73}$$

$$\frac{\dot{T}_{12}}{f_c \hat{a}^2} = C_1, \tag{74}$$

$$\frac{\dot{T}_{21}}{f_c \hat{a}^2} = -C_2 \cosh(x_2/\eta) - C_3 \sinh(x_2/\eta) + C_1, \tag{75}$$

$$\frac{\dot{M}_{32}}{f_c \hat{a} a_m d_{50}} = C_2 \sinh(x_2/\eta) + C_3 \cosh(x_2/\eta), \tag{76}$$

$$\tag{77}$$

with the constants C_i ($i = 1..5$):

$$C_1 = \frac{\dot{u}_{1T}}{2[h - \eta \tanh(h/(2\eta))]}, \tag{78}$$

$$C_2 = C_1, \tag{79}$$

$$C_3 = -\frac{\dot{u}_{1T} \tanh(h/(2\eta))}{2[h - \eta \tanh(h/(2\eta))]}, \tag{80}$$

$$C_4 = \eta C_3, \tag{81}$$

$$C_5 = \left[\int \dot{g}_2 \, dx_2\right]_{x_2=0}. \tag{82}$$

In Figure (3) the initial response is shown for a shear layer with a height of $h = 1$ [cm], a prescribed horizontal shear velocity of $\dot{u}_{1T} = 1$ [cm/s], and the constitutive constants $d_{50} = 0.5$ [mm], $\hat{a} = 0.33$ and $a_m = 1$. It is obvious that even for the case of an initially isotropic stress state and zero couple stresses the initial rate of the state quantities

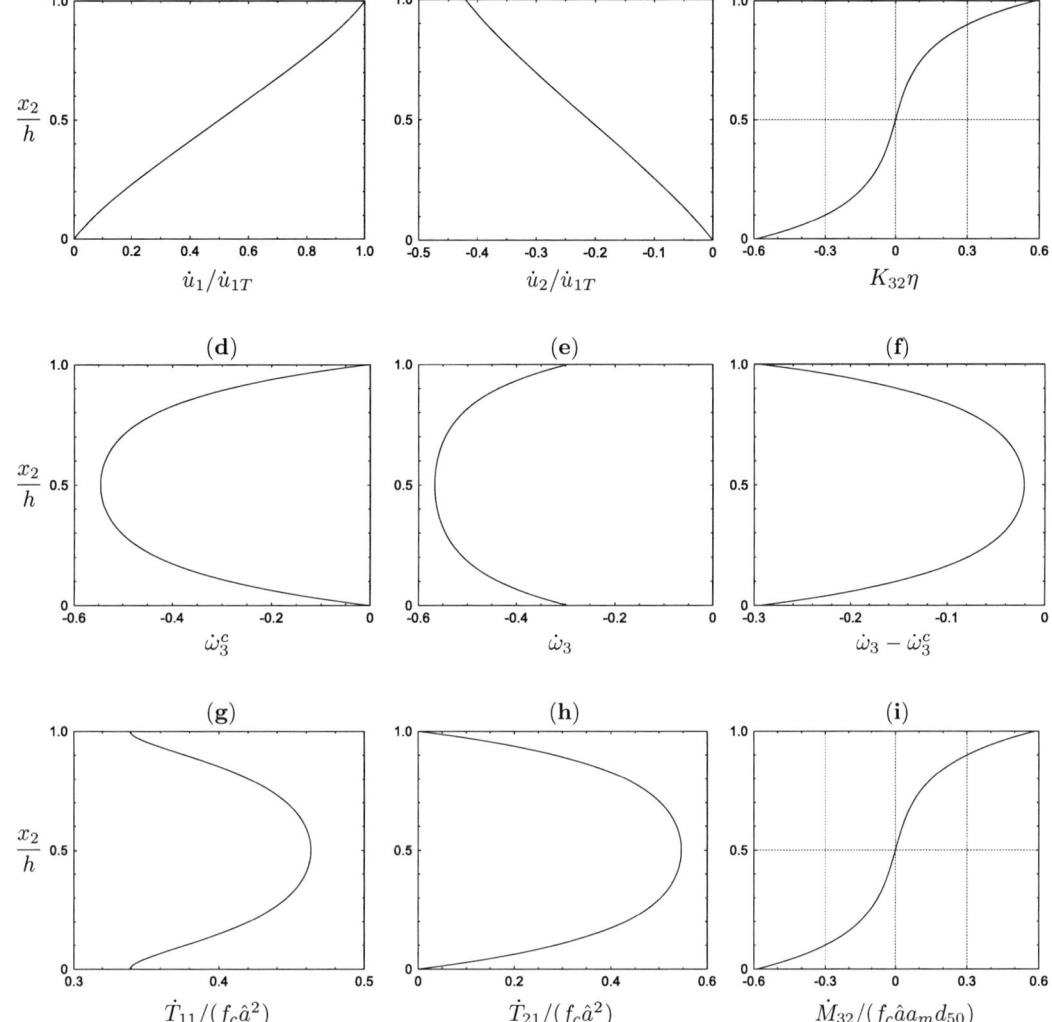

FIGURE 3. Distribution of the initial response across the height of a sheared layer [26]: (a) horizontal velocity $\dot{u}_1$, (b) vertical velocity $\dot{u}_2$, (c) micro-rotation rate $\dot{\omega}_3^c$, (d) macro-rotation rate $\dot{\omega}_3$, (e) difference of macro- and micro-rotation rates $\dot{\omega}_3 - \dot{\omega}_3^c$, and (f) curvature rate K_{32} (a) shear stress rate $\dot{T}_{12}$, (b) shear stress rate $\dot{T}_{21}$, (c) normal stress rate $\dot{T}_{11} = \dot{T}_{33}$, and (d) couple stress rate $\dot{M}_{32}$.

under shearing is different from the standard non-polar continuum. With the exception of $\dot{T}_{22}$ and $\dot{T}_{12}$ the rates of the state quantities are non-linear functions of the co-ordinate x_2 as a result of the micro-polar quantities contained in the present micro-polar hypoplastic model.

Figures (3.a) and (3.b) show that the velocities $\dot{u}_1(x_2)$ and $\dot{u}_2(x_2)$ are non-linearly distributed across the height of the shear layer. Thus, in a micro-polar continuum the deformation is inhomogeneous from the beginning of shearing. The gradient of the micro-rotation is termed rate of curvature, K_{32}, and shown in Figure (3.c). The distribution of the rate of the micro-rotation (Figure 3.d) and the rate of the macro-rotation (Figure 3.e) are also non-linear and they are different from each other (Figure 3.f). With respect to the rule of signs in Figure (2) a negative micro-rotation rate means clockwise rotation for a movement of the top surface to the right. The extreme values of $\dot{\omega}_3^c$ and of $\dot{\omega}_3$ occur on the centre line of the layer. From Figure (3.c) and Figure (3.i) it can be concluded that at the beginning of shearing the influence of micro-polar properties is more pronounced close to the bottom and top boundaries of the layer. The shear stress rate $\dot{T}_{12}$ is constant as necessary for equilibrium and it is different from the shear stress rate $\dot{T}_{21}$ (Figure 3.h). Therefore the stress tensor also becomes non-symmetric in the case of an initially isotropic stress state. The distribution of the normal stress rate $\dot{T}_{11}$ is non-linear (Figure 3.g) and it is equal to the normal stress rate $\dot{T}_{33}$

but different to $\dot{T}_{22}$. The normalized quantity of the couple stress rate $\dot{M}_{32}$ (Figure 3.i) coincides with the normalized curvature rate $\dot{K}_{32}$ shown in Figure (3.c), which is due to relation (59). Although the couple stress rate is zero on the centre line of the layer, i.e. $\dot{M}_{32}(x_2 = h/2) = 0$, the shear stress rate $\dot{T}_{12}(x_2 = h/2)$ is different to the shear stress rate $\dot{T}_{21}(x_2 = h/2)$. The results indicate that micro-polar effects appear across the entire height of the shear layer and a localization of the deformation is not manifest at the beginning of shearing. Numerical simulations of large shearing show that shear localization takes place with advanced shearing in the area of the maximum magnitude of the rate of the micro-rotations, cf. [25, 23]. Thus the extreme value of the rate of micro-rotations obtained for the initial response (Figure 3.d) is an indicator of where shear strain localization will develop under continuous shearing. Independent of the amount of shearing the rate of curvature and the rate of the couple stresses remain zero on the centre line of the shear layer because of the symmetry.

Stationary stress and couple stress states

According to hypoplasticity, stationary stress states and couple stress states or so-called critical states are reached asymptotically for monotonic shearing in the finial state. While at the beginning of shearing the granular material can show dilatancy the volume change becomes smaller for larger shear deformation and vanishes in the steady state. Thus, at least in the asymptotic state the deformation must be isochoric to fulfil the requirement for critical states, i.e. for isochoric deformation $\text{tr}\,\mathbf{D}^c = D^c_{22} = 0$, so that $\dot{g}_2 = 0$. With respect to $\mathring{\mathbf{T}} = \mathbf{0}$, $\text{tr}\,\mathbf{D}^c = 0$ and $\mathring{\mathbf{M}} = \mathbf{0}$ the constitutive equations (50) reduces to:

$$\mathring{T}_{11} = 0 \quad : \quad \bar{\psi}_1 \hat{T}_{11} + \hat{a}(2\hat{T}_{11} - 1/3)\sqrt{\bar{\psi}_2} = 0, \tag{83}$$

$$\mathring{T}_{22} = 0 \quad : \quad \bar{\psi}_1 \hat{T}_{22} + \hat{a}(2\hat{T}_{22} - 1/3)\sqrt{\bar{\psi}_2} = 0, \tag{84}$$

$$\mathring{T}_{33} = 0 \quad : \quad \bar{\psi}_1 \hat{T}_{33} + \hat{a}(2\hat{T}_{33} - 1/3)\sqrt{\bar{\psi}_2} = 0, \tag{85}$$

$$\mathring{T}_{12} = 0 \quad : \quad \hat{a}^2 D^c_{12} + \left(\bar{\psi}_1 + 2\hat{a}\sqrt{\bar{\psi}_2}\right) \hat{T}_{12} = 0, \tag{86}$$

$$\mathring{T}_{21} = 0 \quad : \quad \hat{a}^2 D^c_{21} + \left(\bar{\psi}_1 + 2\hat{a}\sqrt{\bar{\psi}_2}\right) \hat{T}_{21} = 0, \tag{87}$$

$$\mathring{M}_{31} = 0 \quad : \quad \left(\hat{M}_{32}\bar{K}_{32} + 2a_m\sqrt{\bar{K}^2_{32}}\right) \hat{M}_{31} = 0, \tag{88}$$

$$\mathring{M}_{32} = 0 \quad : \quad a^2_m \bar{K}_{32} + \left(\hat{M}_{32}\bar{K}_{32} + 2a_m\sqrt{\bar{K}^2_{32}}\right) \hat{M}_{32} = 0, \tag{89}$$

with the abbreviations:

$$\bar{\psi}_1 = \hat{T}_{12} D^c_{12} + \hat{T}_{21} D^c_{21}, \quad \bar{\psi}_2 = D^{c\,2}_{12} + D^{c\,2}_{21}.$$

The solutions to this equation system read:

$$T_{11} = T_{22} = T_{33}, \tag{90}$$

$$T_{12} = -3\hat{a}\frac{D^c_{12}}{\sqrt{D^{c\,2}_{12} + D^{c\,2}_{21}}} T_{22}, \tag{91}$$

$$T_{21} = -3\hat{a}\frac{D^c_{21}}{\sqrt{D^{c\,2}_{12} + D^{c\,2}_{21}}} T_{22}, \tag{92}$$

$$M_{31} = 0, \tag{93}$$

$$M_{32} = \pm a_m. \tag{94}$$

These solutions are based on the assumption that $K_{32} \neq 0$. With respect to the fact that the rate of curvature changes the sign within the shear zone, i.e. the value of $K_{32} = 0$ on the centre line of the shear layer, the analytical solutions for the limit stress and limit couple stress are not unique and they can be different from the ones given above. In particular on the centre line of the shear layer the rate of curvature, K_{32}, and the couple stress M_{32} remain zero from the beginning of shearing, which is also in accordance with the results obtained from numerical simulations, cf. [23, 14]. It must be noted that the numerical investigation of the evolution of the state quantities using the finite element method is limited to a moderate distortion of the elements so that the asymptotic state cannot be reached in the simulation. Thus numerical results obtained from finite element calculations for medium large shearing can only show a certain tendency towards the limit state. The solutions (91) and (92) for the non-symmetric shear stresses depend on the shear strain rates, which are usually distributed in a non-linear way across the shear layer. Numerical simulations indicate however that on the centre line of the layer the shear stresses become symmetric. In particular with the assumption that on the centre line of the shear layer $D_{12}^c = D_{21}^c$ one obtains:

$$T_{12} = T_{21} = -\frac{3}{\sqrt{2}} \hat{a} T_{22}, \tag{95}$$

which is the same result as obtained for the standard continuum description with the hypoplastic constitutive equation (15), cf. [29]. Equation (45) indicates that as a consequence of the symmetry of the shear stresses in (95) the gradient of the couple stress M_{32} becomes zero, which does not contradict with solution (94). Thus parameter $\hat{a}$ can directly be related to the normalized shear stress, $(-T_{12}/T_{22})$, and therefore also to the so-called critical friction angle φ_c, i.e.

$$\hat{a} = -\frac{\sqrt{2}}{3} \frac{T_{12}}{T_{22}} = \frac{\sqrt{2}}{3} \tan \varphi_c. \tag{96}$$

CONCLUSION

In this paper the modelling of limit stress states in hypoplasticity is demonstrated for both a standard and a micro-polar continuum description. In hypoplasticity the evolution equations for the stress tensor and the couple stress tensor are based on isotropic tensor-valued functions, which are incrementally non-linear. Thus an intrinsic inelastic material behaviour is modelled without a decomposition of the deformation into elastic and plastic parts. Limit stress states or so-called critical states are embedded in the evolution equations for the case of a simultaneous vanishing of the stress rate, the couple stress rate and the volume strain rate. Analytical relations for the limit states and the flow rules can be predicted by the model. In limit states the stress limit condition is related to the normalized stress deviator, which is a non-symmetric tensor in the case of a micro-polar continuum. For plane shearing between parallel plates with rough surfaces it is shown that on the centre line of the shear zone the rate of curvature and the couple stress is zero and that under certain conditions the shear stresses become symmetric. In this case the limit stress ratio is the same as the one obtained for the standard hypoplastic model and the micro-polar hypoplastic model.

APPENDIX

In the concept of elasto-plasticity it is usually postulated that the total strain rate $\mathbf{D}$ can be decomposed into the elastic part $\mathbf{D}^e$ and the plastic part $\mathbf{D}^p$ as

$$\mathbf{D} = \mathbf{D}^e + \mathbf{D}^p. \tag{97}$$

The objective stress rate $\mathring{\mathbf{T}}$ is related to the elastic part of the strain rate according to

$$\mathring{\mathbf{T}} = \mathscr{C} : \mathbf{D}^e, \tag{98}$$

where the fourth-order tensor $\mathscr{C}$ may also depend on the current stress state. With respect to (97) rewriting equation (98) yields

$$\mathscr{C}^{-1} : \mathring{\mathbf{T}} = \mathbf{D} - \mathbf{D}^p. \tag{99}$$

In hypoplasticity the constitutive equation

$$\mathring{\mathbf{T}} = \mathscr{A} : \mathbf{D} + \mathbf{B} \|\mathbf{D}\| \tag{100}$$

can be rewritten as

$$\mathscr{A}^{-1} : \mathring{\mathbf{T}} = \mathbf{D} + \mathscr{A}^{-1} : \mathbf{B} \|\mathbf{D}\|, \tag{101}$$

where $\mathscr{A}$ and **B** are non-vanishing tensor functions of the current stress state only. With the tensor

$$\mathbf{D}^L = -\mathscr{A}^{-1} : \mathbf{B} ||\mathbf{D}|| \tag{102}$$

equation (101) can be represented as

$$\mathscr{A}^{-1} : \mathring{\mathbf{T}} = \mathbf{D} - \mathbf{D}^L . \tag{103}$$

Equation (103) obtained from the hypoplastic constitutive equation has a similar structure as equation (99) obtained from the concept of elasto-plasticity provided $\mathbf{D}^L$ in (103) can be interpreted as the plastic strain rate according to (97), i.e. if $\mathbf{D}^L = \mathbf{D}^p$. However, $\mathbf{D}^L$ in (102) depends on the Euclidean norm of the total strain rate **D**. Thus $\mathbf{D}^L$ is independent of the sign of **D** and in general it includes both reversible and irreversible parts of the strain rate. The relation $\mathbf{D}^L = \mathbf{D}^p$ only holds in critical states. Although some similarities between elasto-plasticity and hypoplasticity could be shown by Wu and Niemunis [38] the splitting of the total strain rate into elastic and plastic parts as postulated in (97) cannot be applied to constitutive equations of type (100).

REFERENCES

1. D. Kolymbas, *Arch. Appl. Mech.* **61**, 143–51 (1991).
2. W. Wu, and E. Bauer, "A hypoplastic model for barotropy and pyknotropy of granular soils", in *Proc. of the Int. Workshop on Modern Approaches to Plasticity*, edited by D. Kolymbas, Elsevier press 1993, pp. 225-45 (1992).
3. G. Gudehus, *Soils Found. (J.Jp. Geo. Soc.)* **36**, 1-12 (1996).
4. E. Bauer, *Soils Found.(J.Jp. Geo. Soc.)* **36**, 13-26 (1996).
5. E. Bauer, *Acta Geotechnica* **4**, 261-72 (2009).
6. W. Wu, *Int. J. Num. Anal. Meth.Geomech.* **22**, 921-40 (1998).
7. E. Bauer, W. Huang, and W. Wu, *Int J. Solids Struct.* **41**, 5903-19 (2004).
8. E. Bauer, and W. Wu, *Powder Technology* **85**, 1-9 (1995).
9. A. Niemunis, and I. Herle, *Mech. Cohes.-Frict. Mater.* **2**, 279-299 (1997).
10. D. Masin, and I. Herle, *Acta Geotechnica* **2**, 261-68 (2007).
11. A. Niemunis, C.E. Grandas-Tavera, and L.F. Prada-Sarmiento, *Acta Geotechnica* **4**, 293-314 (2009).
12. J. Tejchman, and E. Bauer, *Comput. Geotech.* **19**, 221-44 (1996).
13. G. Gudehus, and K. Nübel, *Géotechnique* **54**, 187-201 (2004).
14. W. Huang, K. Nübel, and E. Bauer, *Mech. Mat.* **34**, 563-76 (2002).
15. J.D. Goddard, *Granular Mat.*, accepted and to appear (2009).
16. L. Schneider, and K. Hutter, *Solid-Fluid Mixtures of Frictional Materials in Geophysical and Geotechnical Context*, Springer Press, (2009).
17. V.K. Garga, and J.A. Infante Sedano, *Geotechnical Testing Journal* **25**, 414-21 (2002).
18. K. H. Roscoe, *Géotechnique* **20**, 129-70 (1970).
19. H.B. Mühlhaus, and I. Vardoulakis, "The thickness of shear bands in granular materials", *Geotechnique* **37**, 271-283 (1987).
20. M. Oda, "Micro-fabric and couple stress in shear bands of granular materials", in *Powders and Grains* 3, edited by C. Thornton, 1993, pp. 161-167.
21. J. Desrues, R. Chambon, M. Mokni, and F. Mazerolle, *Géotechnique* **46**, 529-46 (1996).
22. G. Gudehus, "Shear localization in simple grain skeleton with polar effect", in *Proc. of the 4th Int. Workshop on Localization and Bifurcation Theory for Soils and Rocks*, edited by T. Adachi, F. Oka, and A. Yashima, Balkema press 1998, pp. 3-10.
23. W. Huang, and E. Bauer, *Int. J. for Numer. Anal. Meth. Geomech.* **27**, 325-52 (2003).
24. E. Bauer, and W. Huang, "Evolution of polar quantities in a granular Cosserat material under shearing," in *Proc. 5th Int. Workshop on bifurcation and Localization Theory in Geomechanics*, edited by H.-B. Mühlhaus, A.V. Dyskin, and E. Pasternak, Balkema press 2001, pp. 227-38.
25. W. Huang, *Hypoplastic modelling of shear localisation in granular materials*, Ph.D. thesis, Graz University of Technology, Austria (2000).
26. E. Bauer, *J. Engin. Mathematics* **52**, 35-51 (2005).
27. W. Wu, and D. Kolymbas, "Hypoplasticity then and now", in *Constitutive Modelling of Granular Materials*, edited by D. Kolymbas, Springer press, 2000, pp. 57-105.
28. A.E. Green, and P.M. Naghdi, *Arch. Rat. Mech. Anal.* **18**, 251-81 (1965).
29. E. Bauer, *Mech. Cohes.-Frict. Mater.* **5**, 125-48 (2000).
30. R. Rivlin, and J. Ericksen, *J. Rat. Mech. Anal.* **4**, 323-425 (1955).
31. C. Truesdell, and W. Noll, "The Non-Linear Field Theories of Mechanics", *Encyclopedia of Physics* **III c**, Springer press (1965).
32. C.C. Wang, *Rat. Mech. Anal.* **36**, 166-223 (1970).
33. M. Goldscheider, *Mech. Res. Comm.* **3**, 463-68 (1976).
34. G. Gudehus, M. Goldscheider, and H. Winter, "Mechanical Properties of Sand and Clay and Numerical Integration Methods: Some Sources of Errors and Bounds of Accuracy", in *Finite Elements of Geomechanics*, edited by G. Gudehus, John Wiley, New York, 1977, pp. 121-50.

35. G. Gudehus, "Attractors,percolation thresholds and phase limits of granular soils", in *Powders and Grains*, edited by R.P. Behringer and J.T. Jenkins, Balkema press, 1997, pp. 169-83.
36. E. Bauer, and I. Herle, "Stationary states in hypoplasticity", in *Constitutive Modelling of Granular Materials*, edited by D. Kolymbas, Springer press, 2000, 167-92.
37. W. Wu, and D. Kolymbas, *Mech. Mater.* **9**, 245-53 (1990).
38. W. Wu, and A. Niemunis, *Mech. Cohes.-Frict. Mater.* **1**, 145-63 (1996).
39. E. Bauer, "Constitutive Modelling of Critical States in Hypoplasticity", in *Proceedings of the Fifth International Symposium onNumerical Models in Geomechanics, Davos, Switzerland*, Balkema, 15-20 (1995).
40. W. Wu, E. Bauer, and D. Kolymbas, *Mech. Mat.* **23**, 45-69 (1996).
41. A.C. Eringen, "Polar and Nonlocal Field Theories", *Continuum Physics*, **IV**, Academic Press, New York, San Francisco, London (1976).

Homogenisation of Discrete Media towards Micropolar Continua: A Computational Approach

Wolfgang Ehlers

Institute of Applied Mechanics (CE)
Chair of Continuum Mechanics
University of Stuttgart
70550 Stuttgart, Germany

Abstract. The present article aims at linking particle dynamics to continuum mechanics by use of homogenisation techniques. By use of homogenisation strategies, it can be shown that particle dynamics corresponds to the micropolar Cosserat continuum rather than to the standard Cauchy one.

Keywords: particle dynamics, Cosserat effects, stresses and couple stresses, homogenisation techniques
PACS: 81.05.Rm, 83.80.Fg

INTRODUCTION

The mechanical behaviour of natural systems like soil as well as general granular aggregates can be observed on different scales. While the macroscopic, continuum-mechanical view on natural systems with elastic, elasto-plastic or elasto-viscoplastic material properties is generally very convenient for the description of large-scale problems, there are some features like localisation, fracturing and granular flow that cannot be described by simple continuum models. These effects stemming from the microstructure of the material can be integrated in the overall description on the basis of extended continuum models like those given by micropolar theories, as for example, the Cosserat theory [1][1]. On the other hand, a purely microscopic view on soil or granular matter considering these materials as an ensemble of rigid particles interacting through a set of contact laws leads to huge numerical costs and is, therefore, not generally desirable. The present contribution exhibits the description of particle ensembles by means of Newton's equations of momentum and angular momentum in combination with contact laws and neighbouring lists. As a result, one obtains a tool for the computation of both dynamical problems like the outflow of a hopper and quasi-static problems like the biaxial test of soil mechanics including the development of shear bands. Proceeding from volume averaging techniques, one does not only succeed in transferring contact forces towards stresses but also in transforming contact moments towards couple stresses. These results clearly demonstrate that a macroscopic description of granular material should include microscopic information through the consideration of micropolar media. The numerical examples address these effects by exhibiting the computation of dynamical and quasi-static problems including the description of shear bands.

THE PARTICLE MODEL

Based on Newton's equations and convenient neighbouring lists, a set of rigid spherical particles can be described and numerically treated on the basis of the following equations:

---

[1] The reader, who is interested in a closer view on micropolar continua, is referred to the early work by Günther [2] and Schaefer [3] and to the later work by the author [4]. Extended work on microcontinua, in general, has been provided by Eringen, cf. e. g. [5]. Micropolar theories, in general, split into approaches including constrained and unconstrained microrotations. The approaches by Günther and Schaefer proceed from unconstrained rotations as well as the present article.

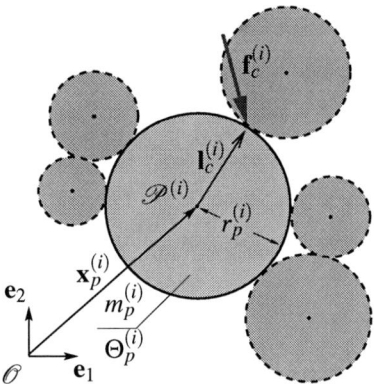

FIGURE 1. Particle $\mathscr{P}^{(i)}$ in contact.

- balance of linear momentum:

$$m_p^{(i)} \ddot{\mathbf{x}}_p^{(i)} = \sum_{c=1}^{N} \mathbf{f}_c^{(i)} + m_p^{(i)} \mathbf{g}, \tag{1}$$

- balance of angular momentum:

$$(\boldsymbol{\Theta}_p^{(i)} \boldsymbol{\omega}_p^{(i)})^\cdot = \sum_{c=1}^{N} \mathbf{l}_c^{(i)} \times \mathbf{f}_c^{(i)}. \tag{2}$$

Therein, $m_p^{(i)}$ and $\boldsymbol{\Theta}_p^{(i)}$ are the mass and the tensor of inertia of a particle $\mathscr{P}^{(i)}$, $\ddot{\mathbf{x}}_p^{(i)}$ is the acceleration of its particle centre and $\boldsymbol{\omega}_p^{(i)}$ its rotational velocity vector, while $(\cdot)^\cdot$ characterises the material time derivative. Furthermore, $\mathbf{g}$ is the gravitational force per unit mass, and $\mathbf{f}_c^{(i)}$ is an external contact force acting at one of the N contact points of the particle surface with the corresponding branch vector $\mathbf{l}_c^{(i)}$, cf. Figure 1.

Splitting the contact forces $\mathbf{f}_c^{(i)}$ into normal and tangential parts, $\mathbf{f}_{cn}^{(i)}$ and $\mathbf{f}_{ct}^{(i)}$, yields

$$\begin{aligned}
\mathbf{f}_c^{(i)} &= \mathbf{f}_{cn}^{(i)} + \mathbf{f}_{ct}^{(i)} \quad \text{with} \\
\mathbf{f}_{cn}^{(i)} &= [E_n^{(ij)} \Delta^{(ij)} + D_n^{(ij)} \dot{\Delta}^{(ij)}] \mathbf{n}^{(ij)}, \\
\mathbf{f}_{ct}^{(i)} &= E_t^{(ij)} [\bar{\mathbf{u}}_t^{(ij)} - \bar{\mathbf{u}}_{tp}^{(ij)}] + D_t^{(ij)} [\dot{\bar{\mathbf{u}}}_t^{(ij)} - \dot{\bar{\mathbf{u}}}_{tp}^{(ij)}].
\end{aligned} \tag{3}$$

Here, it has been assumed that the normal contact force follows a viscoelastic contact law, while the tangential one behaves viscoelastic-plastically. With respect to the viscoelastic contacts, $E_n^{(ij)}$ and $D_n^{(ij)}$ as well as $E_t^{(ij)}$ and $D_t^{(ij)}$ are the normal and tangential elastic and viscous moduli. Furthermore,

$$\Delta^{(ij)} = \langle \|\mathbf{x}_p^{(i)} - \mathbf{x}_p^{(j)}\| - r_p^{(i)} - r_p^{(j)} \rangle \tag{4}$$

is the so-called overlap, and $\mathbf{n}^{(ij)}$ is the outward-oriented unit surface normal directed from particle $\mathscr{P}^{(i)}$ towards $\mathscr{P}^{(j)}$. Regarding the relative motion between the particles in contact, $\dot{\bar{\mathbf{u}}}_t^{(ij)}$ is the relative tangential velocity at the contact point, and $\dot{\bar{\mathbf{u}}}_{tp}^{(ij)}$ is its frictional part. Finally, $\bar{\mathbf{u}}_t^{(ij)}$ defines the relative tangential displacement between $\mathscr{P}^{(i)}$ and $\mathscr{P}^{(j)}$ obtained from integration over the contact time, while $\bar{\mathbf{u}}_{tp}^{(ij)}$ is obtained as its frictional part. To explain these terms in detail, consider the equations of rigid body motion of particles $\mathscr{P}^{(i)}$, where

$$\mathbf{v}_c^{(i)} = \dot{\mathbf{x}}_p^{(i)} + \boldsymbol{\omega}_p^{(i)} \times \mathbf{l}_c^{(i)} \tag{5}$$

defines the velocity at a contact point of the particle's boundary and

$$v_{ct}^{(i)} = \mathbf{t}^{(ij)} \cdot \mathbf{v}_c^{(i)} \tag{6}$$

its tangential projection with $\mathbf{t}^{(ij)}$ as the tangential unit vector pointing in the direction of $\mathbf{f}_{ct}^{(i)}$. Following this yields

$$\dot{\mathbf{u}}_t^{(ij)} = \mathbf{t}^{(ij)} \left(v_{ct}^{(i)} - v_{ct}^{(j)} \right) \quad \text{and} \quad \bar{\mathbf{u}}_t^{(ij)}(t) = \mathbf{t}^{(ij)}(t) \int_{\tau_c}^{t} [v_{ct}^{(i)}(\tau) - v_{ct}^{(j)}(\tau)] \, d\tau, \qquad (7)$$

where the time span between τ_c and t defines the contact time. Since $\mathbf{t}^{(ij)}$ is varying during the contact time, $\bar{\mathbf{u}}_t^{(ij)}$, as part of the constitutive setting (3), is not obtained as the time integral of $\dot{\mathbf{u}}_t^{(ij)}$.

Splitting $\bar{\mathbf{u}}_t^{(ij)}$ in viscoelastic and plastic parts, $\bar{\mathbf{u}}_t^{(ij)} - \bar{\mathbf{u}}_{tp}^{(ij)}$ and $\bar{\mathbf{u}}_{tp}^{(ij)}$, requires the following analogs of plasticity formulae, where $\mu^{(ij)}$ is the friction coefficient [6]:

- yield condition: $\qquad F := \sqrt{\mathbf{f}_{ct}^{(i)} \cdot \mathbf{f}_{ct}^{(i)}} - \mu^{(ij)} \|\mathbf{f}_{cn}^{(i)}\| \leq 0,$

- plastic flow rule: $\qquad \dot{\mathbf{u}}_{tp}^{(ij)} = \Lambda \mathbf{t}^{(ij)} \longrightarrow \bar{\mathbf{u}}_{tp}^{(ij)} = \mathbf{t}^{(ij)}(t) \int_{\tau_c}^{t} \mathbf{t}^{(ij)}(\tau) \cdot \dot{\mathbf{u}}_{tp}^{(ij)}(\tau) \, d\tau,$ $\qquad (8)$

- Kuhn-Tucker conditions: $\quad F \leq 0, \quad \Lambda \geq 0, \quad \Lambda F = 0.$

Once the above set of equations is given, the numerical scheme for the computation of initial-boundary-value problems within the Discrete Element Method (DEM) proceeds from convenient neighbouring lists and the Verlet algorithm obtained from (1) and (2) for an explicit time integration, viz.:

- translational and rotational velocities at $t + \Delta t$:

$$\ddot{\mathbf{x}}_p^{(i)}(t+\Delta t) = \frac{\mathbf{f}^{(i)}(t)}{m_p^{(i)}} \Delta t + \dot{\mathbf{x}}_p^{(i)}(t), \qquad \text{where} \quad \mathbf{f}^{(i)}(t) = \sum_{c=1}^{N} \mathbf{f}_c^{(i)}(t) + m_p^{(i)} \mathbf{g},$$
$$\dot{\boldsymbol{\varphi}}_p^{(i)}(t+\Delta t) = (\boldsymbol{\Theta}_p^{(i)})^{-1} \bar{\mathbf{m}}^{(i)}(t) \Delta t + \dot{\boldsymbol{\varphi}}_p^{(i)}(t), \quad \text{where} \quad \bar{\mathbf{m}}^{(i)}(t) = \sum_{c=1}^{N} \mathbf{l}_c^{(i)} \times \mathbf{f}_c^{(i)}(t). \qquad (9)$$

- translational and rotational displacements at $t + \Delta t$:

$$\mathbf{x}_p^{(i)}(t+\Delta t) = \frac{\mathbf{f}^{(i)}(t)}{m_p^{(i)}} \Delta t^2 + 2 \mathbf{x}_p^{(i)}(t) - \mathbf{x}_p^{(i)}(t - \Delta t),$$
$$\boldsymbol{\varphi}_p^{(i)}(t+\Delta t) = (\boldsymbol{\Theta}_p^{(i)})^{-1} \bar{\mathbf{m}}^{(i)}(t) \Delta t^2 + 2 \boldsymbol{\varphi}_p^{(i)}(t) - \boldsymbol{\varphi}_p^{(i)}(t - \Delta t). \qquad (10)$$

To exhibit the possibilities of the above scheme, Figure 2 displays the numerical computation of an outflow of spherical particles out of a hopper.

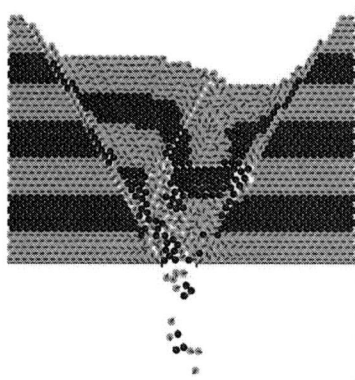

FIGURE 2. Granular flow – outflow of a hopper.

In this particular computation, use was made of an initial setting of mono-disperse particles in a triangular lattice. As a result, the particle ensemble forms shear bands under 60° while particles are falling away. It is furthermore seen from the red pointers with a horizontal initial position that the particles rotate at the shear band and while falling. As a result of this finding, it is expected that a homogenisation over a convenient representative elementary volume (REV) will rather exhibit a Cosserat continuum than a Cauchy one, where the latter is also known after Boltzmann[2]. It is furthermore expected that the rotational degrees of freedom will not be activated in the overall domain but in the shear band zone.

HOMOGENISATION TECHNIQUE

Based on the Hashin's MMM principle [7], any homogenisation process proceeds from the assumption that the REV (volume $\mathscr{R}$, surface $\partial \mathscr{R}$) as the homogenisation volume is small compared to the overall domain. As a result, volume integrals are negligible in comparison to surface integrals:

$$\int_{\mathscr{R}} (\cdot) \, \mathrm{d}v \ll \int_{\partial \mathscr{R}} (\cdot) \, \mathrm{d}a. \tag{11}$$

Corresponding to (1) and (2), the respective continuum-mechanical balances read [4]:
- balance of linear momentum:

$$\int_{\mathscr{V}} \rho \, \ddot{\mathbf{x}} \, \mathrm{d}v = \int_{\partial \mathscr{V}} \mathbf{t} \, \mathrm{d}a + \int_{\mathscr{V}} \rho \, \mathbf{g} \, \mathrm{d}v, \tag{12}$$

- balance of angular momentum (Cauchy continuum):

$$\int_{\mathscr{V}} \mathbf{x} \times \rho \, \ddot{\mathbf{x}} \, \mathrm{d}v = \int_{\partial \mathscr{V}} \mathbf{x} \times \mathbf{t} \, \mathrm{d}a + \int_{\mathscr{V}} \mathbf{x} \times \rho \, \mathbf{g} \, \mathrm{d}v, \tag{13}$$

- balance of angular momentum (Cosserat continuum):

$$\int_{\mathscr{V}} [\mathbf{x} \times \rho \, \ddot{\mathbf{x}} + \rho (\bar{\mathbf{\Theta}} \bar{\boldsymbol{\omega}})^{\bullet}] \, \mathrm{d}v = \int_{\partial \mathscr{V}} [\mathbf{x} \times \mathbf{t} + \bar{\mathbf{m}}] \, \mathrm{d}a + \int_{\mathscr{V}} [\mathbf{x} \times \rho \, \mathbf{g} + \rho \bar{\mathbf{c}}] \, \mathrm{d}v. \tag{14}$$

In (12)-(14), $\mathscr{V}$ and $\partial \mathscr{V}$ are the volume and the surface of the considered continuum, ρ is the mass density, $\mathbf{t}$ is the surface traction, $\bar{\mathbf{\Theta}}$ is the mass-specific tensor of microinertia, $\bar{\boldsymbol{\omega}}$ is the local rotational velocity, $\bar{\mathbf{m}}$ is the Cosserat surface couple, and $\bar{\mathbf{c}}$ is the volume couple.

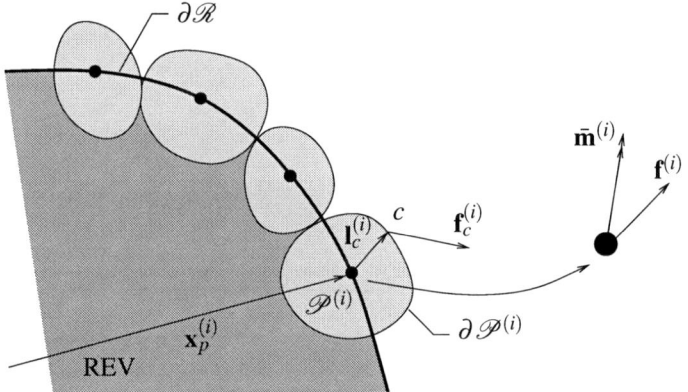

FIGURE 3. REV bounded by the curve connecting the boundary particle centres.

Combining particle mechanics and continuum mechanics on the basis of an REV, cf. Figure 3, the MMM principle (11) yields the following statement with respect to the momentum balances (1) and (12):

---

[2] In the continuum-mechanical literature, the standard continuum with symmetric Cauchy stresses is historically either called after Cauchy or after Boltzmann. However, the following statement can be found in the "Classical Field Theories" by Truesdell and Toupin, §205, footnote 4 after stating Cauchy's second law of motion (Eq. 205.11) yielding the symmetry of the Cauchy stress: "The German literature persists in attributing credit to Boltzmann here."

$$\int_{\partial \mathscr{R}} \mathbf{t}\, da = \sum_{i=1}^{B} \mathbf{f}^{(i)} = \mathbf{0}, \quad \text{where} \quad \mathbf{f}^{(i)} = \sum_{c=1}^{N_0} \mathbf{f}_c^{(i)} \tag{15}$$

is the resultant contact force acting at the centre of $\mathscr{P}^{(i)}$. Furthermore, B is the number of boundary particles, while N_0 is the number of external contacts at $\mathscr{P}^{(i)}$. Once the contact force resultant is placed at the particle centre, it is necessary to introduce the additional moment $\bar{\mathbf{m}}^{(i)}$ summing up the offsets of the individual contact forces, viz.:

$$\bar{\mathbf{m}}^{(i)} = \sum_{c=1}^{N_0} \mathbf{l}_c^{(i)} \times \mathbf{f}_c^{(i)}. \tag{16}$$

Comparing the balance equations of angular momentum (2), (13) and (14), it is obvious in analogy to the procedure to obtain (15) that one concludes to

$$\int_{\partial \mathscr{R}} [\mathbf{x} \times \mathbf{t} + \bar{\mathbf{m}}]\, da = \sum_{i=1}^{B} [\mathbf{x}_p^{(i)} \times \mathbf{f}^{(i)} + \bar{\mathbf{m}}^{(i)}] = \mathbf{0}. \tag{17}$$

Thus, an ensemble of particles obviously always behaves like a Cosserat continuum. On the other hand, shrinking the particle radii $r_p^{(i)}$ towards zero decreases the Cosserat effect until $\bar{\mathbf{m}}$ and $\bar{\mathbf{m}}^{(i)}$ are vanishing as a result of vanishing branch vectors $\mathbf{l}_c^{(i)}$. This incorporates the standard continuum approach (Cauchy continuum) into a micropolar theory (Cosserat continuum). The question whether or not Cosserat effects become apparent in a numerical setting depends on two conditions: (1) the granular microstructure must exhibit non-vanishing $r_p^{(i)}$ such that couple stresses are possible; (2) the problem under consideration must exhibit tangential contact forces which are usually activated by local rotations as they occur, for example, during shear banding of granular material. As a result, one observes couple stresses.

Proceeding from standard arguments of continuum mechanics together with Cauchy's theorem yields the Cauchy stress tensor $\mathbf{T}$ and the corresponding tensorial stress moment $\mathbf{M}$ as

$$\mathbf{t} = \mathbf{T}\mathbf{n} \quad \text{and} \quad \mathbf{x} \times \mathbf{t} = (\mathbf{x} \times \mathbf{T})\mathbf{n} =: \mathbf{M}\mathbf{n} =: \mathbf{m}. \tag{18}$$

In addition, the couple stress tensor $\bar{\mathbf{M}}$ reads

$$\bar{\mathbf{m}} = \bar{\mathbf{M}}\mathbf{n}. \tag{19}$$

In case that a continuous REV is considered, the homogenisation of stresses, stress moments and couple stresses yields the average

$$\langle \mathbf{A} \rangle = \frac{1}{V_R} \int_{\mathscr{R}} \mathbf{A}\, dv, \tag{20}$$

where $\mathbf{A}$ is the substitute for $\mathbf{T}$, $\mathbf{M}$ and $\bar{\mathbf{M}}$. It is furthermore concluded from [8] that the following identities hold:

$$\begin{aligned}\mathbf{A}^T &= \mathbf{I}\mathbf{A}^T = (\operatorname{grad}\mathbf{x})\mathbf{A}^T, \\ \operatorname{div}(\mathbf{x} \otimes \mathbf{A}) &= (\operatorname{grad}\mathbf{x})\mathbf{A}^T + \mathbf{x} \otimes \operatorname{div}\mathbf{A}.\end{aligned} \tag{21}$$

Thus, it is easily seen that

$$\begin{aligned}\mathbf{A}^T &= \operatorname{div}(\mathbf{x} \otimes \mathbf{A}) - \mathbf{x} \otimes \operatorname{div}\mathbf{A}, \\ \langle \mathbf{A} \rangle^T &= \frac{1}{V_R} \int_{\mathscr{R}} [\operatorname{div}(\mathbf{x} \otimes \mathbf{A}) - \mathbf{x} \otimes \operatorname{div}\mathbf{A}]\, dv,\end{aligned} \tag{22}$$

where $\langle \mathbf{A} \rangle^T = \langle \mathbf{A}^T \rangle$ has been used. With the aid of the Gaussian integral theorem and Hashin's MMM principle, $(22)_2$ finally results in

$$\langle \mathbf{A} \rangle^T = \frac{1}{V_R} \int_{\partial \mathscr{R}} (\mathbf{x} \otimes \mathbf{A})\mathbf{n}\, da. \tag{23}$$

From (18), (19) and (23), the volume averages of **T**, **M** and **M̄** read:

$$\langle \mathbf{T} \rangle = \frac{1}{V_R} \int_{\partial \mathscr{R}} \mathbf{t} \otimes \mathbf{x}\, da,$$
$$\langle \mathbf{M} \rangle = \frac{1}{V_R} \int_{\partial \mathscr{R}} \mathbf{m} \otimes \mathbf{x}\, da, \quad (24)$$
$$\langle \mathbf{\bar{M}} \rangle = \frac{1}{V_R} \int_{\partial \mathscr{R}} \mathbf{\bar{m}} \otimes \mathbf{x}\, da.$$

The final step in the present homogenisation procedure stems from the comparison of (24) representing the homogenisation over continuous REV and the respective terms governing particle ensembles. Thus, $(24)_1$ and (15) combine to the homogenised Cauchy stress

$$\langle \mathbf{T} \rangle = \frac{1}{V_R} \sum_{i=1}^{B} \mathbf{f}^{(i)} \otimes \mathbf{x}_p^{(i)} \quad (25)$$

obtained from contact forces at the boundary particles of the REV and the location vectors of the corresponding particle centres. Analogously, $(24)_2$, (17) and (18) combine to

$$\langle \mathbf{M} \rangle = \frac{1}{V_R} \sum_{i=1}^{B} \mathbf{m}^{(i)} \otimes \mathbf{x}_p^{(i)} = \frac{1}{V_R} \sum_{i=1}^{B} (\mathbf{x}_p^{(i)} \times \mathbf{f}^{(i)}) \otimes \mathbf{x}_p^{(i)} = \frac{1}{V_R} \sum_{i=1}^{B} \mathbf{x}_p^{(i)} \times (\mathbf{f}^{(i)} \otimes \mathbf{x}_p^{(i)}), \quad (26)$$

while the homogenised couple stress yields

$$\langle \mathbf{\bar{M}} \rangle = \frac{1}{V_R} \sum_{i=1}^{B} \mathbf{\bar{m}}^{(i)} \otimes \mathbf{x}_p^{(i)}, \quad (27)$$

where $(24)_3$ together with (16) and (19) has been used.

From (25)-(27), it is seen that the REV of a particle ensemble basically behaves like a Cosserat continuum. This includes $\langle \mathbf{\bar{M}} \rangle \neq \mathbf{0}$ and, as a result, $\langle \mathbf{T} \rangle \neq \langle \mathbf{T} \rangle^T$ [4, 8, 9].

Numerical computations carried out on the basis of the DEM will clearly demonstrate this effect after homogenisation.

NUMERICAL EXAMPLE

The following example concerns a biaxial test carried out on a dry Hostun sand specimen at the Laboratoire 3S at Grenoble. In the experiment, a rectangular sand specimen is pressed between top and bottom loading plates, while the sides are stabilised by a lateral hydraulic pressure. During the experiment, the specimen is covered by a latex membrane such that the lateral load can be applied by a fluid pressure.

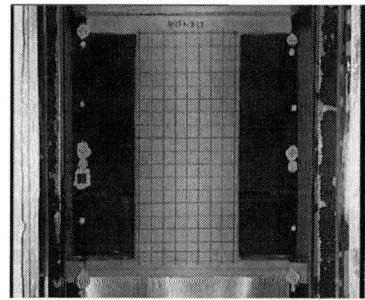 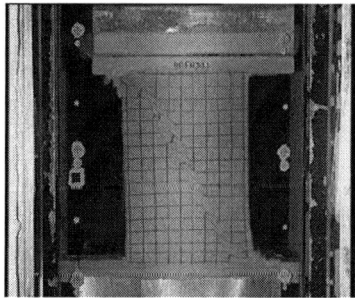

FIGURE 4. Biaxial test – initial configuration (left) and final configuration (right) exhibiting a shear band after loading.

By use of the DEM, the present example is computed with the equations described above, where, in addition, the latex membrane is modelled by elastic springs between the centres of the boundary particles. The present computation is based on an initial configuration of mono-disperse spherical particles in their densest packing. After a certain amount of loading, a shear band occurs and the right upper triangle starts to slide over the left lower one, thereby forming the typical "noses" at both ends of the shear band. It can furthermore be observed that the shear band width is growing during the sliding process until it reaches its final value which, in the present example, reaches approximately five particle diameters.

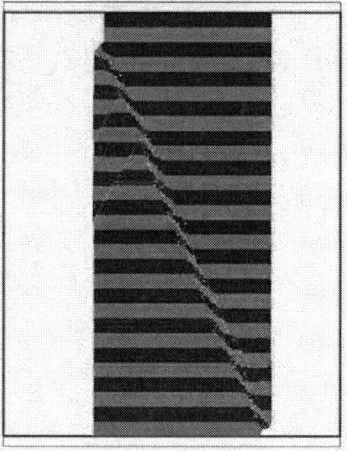

FIGURE 5. Biaxial test – DEM computation exhibiting a definite primary shear band and two further secondary ones.

Proceeding from the homogenisation technique presented in (25) and (27), the present computation makes use of a particle-centre-based procedure such that each particle can be understood as the centre of an REV, cf. [8, 9]. Following this yields the results exhibited in Figure 6. To obtain these results, small REV of only a few particles around the centre particles have been used. Larger REV would include, also across the shear band domain, areas without micropolar rotations and, therefore, would not be able to precisely predict the shear band activity. This is also seen from Figure 6, where the couple stresses only occur in the shear zone, while the remainder of the specimen does not show any Cosserat effect. Once $\langle \bar{\mathbf{M}} \rangle \neq \mathbf{0}$, the stresses $\langle \mathbf{T} \rangle$ are no longer symmetric. This means that the skew-symmetric part $skw \langle \mathbf{T} \rangle$ is non-zero in the same domain as $\langle \bar{\mathbf{M}} \rangle$. Obviously, this domain is the shear zone, where the particles exhibit distinct rotations.

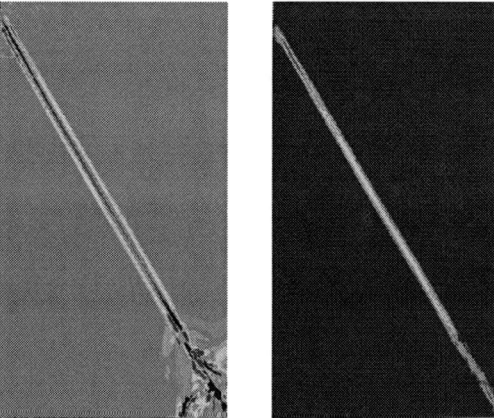

FIGURE 6. Homogenised skew-symmetric stresses $skw \langle \mathbf{T} \rangle$ (left) and homogenised couple stresses $\langle \bar{\mathbf{M}} \rangle$ (right) displayed at the reference geometry.

A continuum-mechanical investigation of both Cauchy and Cosserat continua and various numerical applications of non-polar and micropolar materials can furthermore be found in [4]. The examples presented there show the same result, namely, that micropolar degrees of freedom and with them couple stresses and skew-symmetric stress parts only occur in zones with distinct micro-information (micro-rotation). Such zones obviously include shear bands as they have been presented here.

Finally, it should be noted once again that a micropolar continuum always includes the possibility of exhibiting micropolar rotations and, as a result, the possibility of exhibiting couple stresses. However, micro-rotations are not always active, since the appearance of micro-rotations depends on the considered initial-boundary-value problem.

CONCLUSION

In the present article, it has been shown that the discrete element method is capable to describe particle-dynamical problems such as the outflow of a hopper as well as geomechanical problems such as the quasi-static biaxial test. Local information of stress and couple stress tensors can be obtained by use of homogenisation techniques combined with a particle-centre-based strategy. The homogenised values of stresses and couple stresses directly correspond to respective results obtained by use of the micropolar Cosserat continuum approach. Concerning the determination of material parameters of micropolar elasto-plastic material, the interested reader is finally referred to [10].

REFERENCES

1. E. Cosserat and F. Cosserat, *Théorie des corps déformables*. A. Hermann et fils, Paris 1909. (Theory of Deformable Bodies, NASA TT F-11 561, 1968).
2. W. Günther, *Abhandlungen der Braunschweigischen Wissenschaftlichen Gesellschaft* **10** (1958), 195–213.
3. H. Schaefer, *Zeitschrift für Angewandte Mathematik und Mechanik (ZAMM)* **47** (1967), 485–498.
4. W. Ehlers, "Foundations of Multiphasic and Porous Materials," in *Porous Media: Theory, Experiments and Numerical Applications*, edited by W. Ehlers and J. Bluhm, Springer-Verlag, Berlin, 2002, pp. 3–86.
5. A. C. Eringen, *Microcontinuum Field Theories I: Foundation and Solids*, Springer-Verlag, New York 1999.
6. B. Scholz, *Application of a Micropolar Model to the Localization Phenomena in Granular Materials*. Dissertation thesis, Report No. II-15 of the Institute of Applied Mechanics (CE), University of Stuttgart 2007.
7. Z. Hashin, *ASME Journal of Applied Mechanics* **50**, 481–505 (1983).
8. W. Ehlers, E. Ramm, S. Diebels, and G. A. D'Addetta, *International Journal of Solids and Structures* **40**, 6681–6702 (2003).
9. G. A. D'Addetta, E. Ramm, S. Diebels, and W. Ehlers, *Engineering Computations* **21**, 360–383 (2004).
10. W. Ehlers and B. Scholz, *Archive of Applied Mechanics*, **77**, 911–931 (2007).

Remarks on Constitutive Laws for Dry Granular Materials

Pasquale Giovine

Dipartimento di Meccanica e Materiali, Facoltà di Ingegneria, Università Mediterranea
Via Graziella n.1, Località Feo di Vito, I-89122, Reggio Calabria, Italy
email: giovine@unirc.it

Abstract. In this paper we resume some remarks upon the model for dry granular materials formulated in [19] which generalized the distributed theory of Goodman & Cowin [24] by allowing the possibility of rotations of grains. Firstly, we reconsider the balance equations for micromomentum and energy by bearing the bulk and rotational effects of granules in mind; secondly, we derive the constitutive relations and define the dissipation function; thirdly, we specify an appropriate equilibrium state for dry granular materials in order to obtain additional informations on constitutive fields.

Keywords: Continua with microstructure, granular materials, constitutive equations, dissipation.
PACS: 81.05.Rm, 47.57.Gc

INTRODUCTION

Granular materials partake of the properties of solids and, under different circumstances, of some properties of liquids, if not gases, so the search of an overall model for them could be quite ambitious.

The case of slow, or quasi-static, motion was studied in [19] where dry granular materials were considered as a generalization of continua with constrained affine microstructure (see [5]). This concept was useful in some context in offering an interpretation of constitutive prescriptions, involving the displacement gradients of higher order than the first, which allows one to circumvent certain apparent inconsistencies with the second law of thermodynamics; on the other hand, the refinement of the Cauchy theory was also necessary to characterize the more complex structure of the medium, even if some problem of physical concreteness [26] or of mathematical complexity could arise [12].

The extended model was suggested by the dynamical approach in [20], where a Hamiltonian variational principle of local type in the conservative case was stated and a particular expression for the density of kinetic energy was obtained; it was composed of three terms: the classical one, one modelling the relative motion between grains and one pertaining to dilatation of compressible grains themselves.

It is interesting to note that the theory of Goodman & Cowin [24], that describes continua with 'small' spherical grains which only may contract and expand homogeneously, is recovered by that theory if the microdeformation is constrained to be spherical (see also [9]).

Moreover, it is observed in [22] an advantage of the model [20] with respect to usual purely propagative ones for the study of seismic waves spreading through a sediment filled basin; in fact, if one considers rigid grains, he obtains the reproduction of a nonlinear effect experimentally observed for real seismic waves: site amplification decreases as the amplitude of the incident wave increases.

The additional kinematic parameters introduced in [19] take into account not only the dilatational motions of the individual (compressible) grains and of the grains relative to one another, but also the rotation of granules; hence a new equation for the balance of rotational micromomentum is presented in the model, while the Cauchy stress tensor is not necessarily symmetric, in general, and its spherical part includes a term which accounts for the dilatancy of Reynolds [34].

In this paper we reformulate the balance equations for the micromomentum and those for granular and internal energy in order to put in evidence peculiar microstructural effects. Moreover, we assume that the entropy flux is not equal to the heat flux divided by the temperature, as suggested in [32] and [24]. After we propose the appropriate constitutive relations, along with the thermodynamic restrictions and the invariance principles, in order to obtain the constitutive equations for a non-conducting dry granular material with rotating grains in presence of dissipation. At the end we define an equilibrium process to evaluate equilibrium values of the stress and of internal bulk and spin micro–forces.

MODELING AND DEVELOPMENT OF THE THEORY

The equations of motions for a dry cohesionless granular material proposed in [19] were obtained by using the following assumptions:

i) the material element of our body $\mathscr{B}$ is a sort of quasi–particle which consists of a grain and its immediate neighbours (see [4, 2]) and that is called a 'chunk' of material; the chunks are thought of as envelopes which fill the body without voids between them [7];

ii) the quasi-static motion of the body admits only dilatation, or contraction, of the individual (compressible) grain and of the grains relative to one another as well as the rotation of the granule itself (see, also, [1, 11]); therefore the continuum model can depict the granular material as a medium with constrained affine microstructure [17, 10, 5];

iii) The total kinetic energy per unit mass κ of the medium is then the sum of four components ($\kappa = \kappa_t + \kappa_f + \kappa_d + \kappa_r$):

a) the usual translational kinetic energy κ_t $\left(:= \frac{1}{2}\mathbf{v}^2\right)$ related to the velocity $\mathbf{v}$ of the centre of mass $\mathbf{x}$ of the material element;

b) the 'fluctuation' kinetic energy κ_f associated with the dilatancy of the chunk, that is the homogeneous expansions of it represented by the motion of individual grains relative to the centre of mass and expressed in terms of the rate of change of the chunk mass density ρ by $\frac{1}{2}\gamma(\rho)\dot{\rho}^2$, with $\gamma(\rho)$ a constitutive coefficient (see [23] for the rigid case); the dot denotes the material time derivative, i.e., $(\dot{\cdot}) := \frac{\partial(\cdot)}{\partial\tau} + \mathbf{v}\cdot\text{grad}(\cdot)$, with τ the time. We observe that the chunk mass density ρ is equal to ν times the proper mass density ρ_m: $\rho = \rho_m\nu$, where $\nu \in (0,1)$ is the volume fraction of the grains;

c) the dilatational kinetic energy κ_d related to local expansions (or contractions) of the compressible inclusions in the chunk and written in terms of the rate of change of the proper mass density of the grains ρ_m by $\frac{1}{2}\alpha(\rho_m)\dot{\rho}_m^2$, with $\alpha(\rho_m)$ another constitutive coefficient;

d) the rotational kinetic energy κ_r $\left(:= \frac{1}{2}\mu^2(\rho_m)\dot{\mathbf{R}}\cdot\dot{\mathbf{R}}\right)$ due to grain rotations, with $\mathbf{R}$ a proper orthogonal tensor and $\mu(\rho_m)$ a third constitutive coefficient.

The expressions of the constitutive functions $\gamma(\rho)$, $\alpha(\rho_m)$ and $\mu(\rho_m)$ can be very general because they depend on geometric configurations of the grains and of the chunks and on the admissible micro-motions (see §2 of [2]). In §3 and Appendix of [19] simple microstructural motions and peculiar geometrical shapes for chunks and granules were assumed in order to explicitely compute them.

In particular, one can imagine the chunk consisting, in a mental magnification, of a spherical grain and its immediate spherical neighbours, and the envelope of the chunk as a spherical surface of variable radius ς containing all these spherical compressible inclusions of variable radius φ, with interstices filled with a fluid of negligible mass; moreover, the chunks and the compressible grains are supposed to expand and/or contract homogeneously with independent, more or less violent, purely radial pulsations, while rotations of single grains are not necessarily related to macro-rotations of the own chunk.

Therefore, if the envelopes and the grains have the radius ς_* and φ_*, respectively, in a reference placement $\mathscr{B}_*$ of the material, one can calculate the following expressions for functions γ, α and μ:

$$\gamma(\rho) = \gamma_* \rho^{-\frac{8}{3}}, \quad \alpha(\rho_m) = \alpha_* \rho_m^{-\frac{8}{3}}, \quad \mu^2(\rho_m) = 3\alpha_* \rho_m^{-\frac{2}{3}} \left(= 3\alpha(\rho_m)\rho_m^2\right), \tag{1}$$

with

$$\gamma_* = \frac{16}{351}\rho_*^{\frac{2}{3}}\varsigma_*^2, \quad \alpha_* = \frac{1}{15}\rho_{m*}^{\frac{2}{3}}\varphi_*^2. \tag{2}$$

These postulates were used in [19] to add evidence that a dry granular medium with rotating grains can be viewed as a peculiar continuum with a spherical microstructure, i.e., for which the microstructural tensor field $\mathbf{G}$ of Lin$^+$, describing the changes in the affine structure, is conformal (see [9]):

$$\mathbf{G}(\mathbf{x},\tau) = \beta(\mathbf{x},\tau)\mathbf{R}(\mathbf{x},\tau), \tag{3}$$

with $\beta(\mathbf{x},\tau) = \left(\frac{\rho_{m*}}{\rho_m}\right)^{\frac{1}{3}}$ and $\mathbf{R}(\mathbf{x},\tau) \in \text{Orth}^+$. Lin$^+$ and Orth$^+$ are the collection of second–order tensors with positive determinant and proper orthogonal, respectively.

Moreover, the following set of balance equations, that govern the flow of the dry granular medium which is a non-conductor of heat, was obtained:

$$\dot{\rho} + \rho \operatorname{tr} \mathbf{L} = 0, \tag{4}$$

$$\rho \dot{\mathbf{v}} = \rho \mathbf{f} + \operatorname{div} \mathbf{T}, \tag{5}$$

$$3\rho\rho_m \left[\alpha(\rho_m)\left(\ddot{\rho}_m + \rho_m \tilde{\mathbf{Y}} \cdot \tilde{\mathbf{Y}}\right) + \frac{1}{2}\alpha'(\rho_m)\dot{\rho}_m^2 \right] = \beta \mathbf{R} \cdot (\mathbf{Z} - \operatorname{div}\Sigma - \rho \mathbf{B}), \tag{6}$$

$$3\rho \operatorname{skw}\left[\rho_m^2 \alpha(\rho_m)\tilde{\mathbf{Y}}\right]^{\cdot} = \operatorname{skw}\left[\beta \mathbf{R}(\operatorname{div}\Sigma - \mathbf{Z} + \rho \mathbf{B})^T\right], \tag{7}$$

$$\rho \dot{\varepsilon} = \left(\dot{\beta}\beta^{-1}\right)[\mathbf{Z} \cdot (\beta \mathbf{R}) + \Sigma \cdot \operatorname{grad}(\beta \mathbf{R})] + (\mathbf{Y} - \tilde{\mathbf{Y}}) \cdot \operatorname{skw}\left[\beta \mathbf{Z} \mathbf{R}^T + \Sigma \odot \operatorname{grad}(\beta \mathbf{R})\right] +$$
$$+ \left(\beta \Sigma^T \mathbf{R}^T\right) \cdot \operatorname{grad}\left(\dot{\beta}\beta^{-1}\right) + \left(\beta \mathbf{R} \oslash {}^t\Sigma\right) \cdot \operatorname{grad}\tilde{\mathbf{Y}} + \mathbf{D} \cdot \operatorname{sym} \mathbf{T} + \rho \lambda, \tag{8}$$

Equation (4) is the conservation law of mass with $\mathbf{L}$ the usual velocity gradient: $\mathbf{L} := \operatorname{grad} \mathbf{v}$, while its symmetric and skew parts are the stretching tensor $\mathbf{D}\,(:=\operatorname{sym}\mathbf{L})$ and the spin tensor $\mathbf{Y}\,(:=-\operatorname{skw}\mathbf{L})$, respectively. Equation (5) is the standard law of Cauchy's balance, where $\mathbf{f}$ is the vector density per unit mass of external body actions and $\mathbf{T}$ the Cauchy stress tensor. Equations (6) and (7) are the balances of microstructural interactions due to the compressibility and the rotations of granules, respectively: $\tilde{\mathbf{Y}} = -\dot{\mathbf{R}}\mathbf{R}^T$ is the micro-spin tensor; $\rho \mathbf{B}$ and $-\mathbf{Z}$ are the resultant tensor densities per unit volume of external body interactions on the microstructure and internal self–force, respectively, while Σ is the third–order microstress tensor: they account for interactive forces between the gross and fine structures or between chunks as well as internal dissipative contributions due to the stir of the chunks' surface or grain–boundary collisions or more exchange of granules through the chunk boundary itself. Moreover, Σ, in general, is not necessarily related to a sort of boundary microtractions, unless it is possible to define a physically significant connection on the manifold of values of the microstructure by which the gradient on it may be evaluated in covariant manner, rather it expresses weakly nonlocal effects which cannot be separated, in the physical meaning, by the microactions $\mathbf{Z}$ and hence only the sum $(\operatorname{div}\Sigma - \mathbf{Z})$ has sense (see [6]). Equation (8) is the reaction free balance of total energy in which ε is the specific internal energy per unit mass and λ the scalar rate of heat generation per unit mass due to irradiation.

The tensor product $\odot$ between third–order tensors is so defined: $(\operatorname{grad}\mathbf{G} \odot \Sigma)_{ij} := \mathbf{G}_{ih,k}\Sigma_{jhk}$; the minor left transposition (of exponent t) on the third order tensor Σ has the following meaning: $(({}^t\Sigma \mathbf{a})\mathbf{b})\mathbf{c} = ((\Sigma \mathbf{a})\mathbf{c})\mathbf{b}$, for each triple of vectors $\mathbf{a}$, $\mathbf{b}$ and $\mathbf{c}$; the tensor product $\oslash$ between tensors of the second and the third order is so defined: $(\mathbf{A} \oslash \Sigma)_{ijl} := \mathbf{A}_{ih}\Sigma_{hjl}$.

At the end, we wish to observe that the expression for the Cauchy stress tensor $\mathbf{T}$ obtained in [19], by considering the balance of moment of momentum and assumption iii b), is the sum of three terms:

$$\mathbf{T} = \operatorname{sym} \mathbf{T}_c + \operatorname{skw}\left[\beta \mathbf{R} \mathbf{Z}^T + \operatorname{grad}(\beta \mathbf{R}) \odot \Sigma\right] + \mathbf{M}; \tag{9}$$

the first symmetric one is the object of a constitutive prescription; the second one is clearly not symmetric due to the presence of the microstructure (see [5]); the third contribution comes from an influx of linear momentum described by a tensor of inertia flux $\mathbf{M}$ which is the Lagrangian derivative, times $\rho \mathbf{I}$, of the fluctuation kinetic energy κ_f and measures the agitation within a chunk of material, that is the rate of dilatancy of Reynolds [34]:

$$\mathbf{M} := -\rho^2 \left[\gamma(\rho)\ddot{\rho} + \frac{1}{2}\gamma'(\rho)\dot{\rho}^2\right] \mathbf{I} \tag{10}$$

(see the review paper [28] where many granular theories which split the stress tensor are examined, as in particular, e.g., the spherical part of the Reynolds stress type tensor of the turbulence theory in their equation (3.14), or, also, the collisional–translational contribution to the total stress tensor in equation (2.6) of [29]).

Remark: A relation of evolution for the granular temperature of the body, the so called granular heat transfer equation (4.6) of [7] or balance of pseudo–thermal energy (2.7) of [29], can be easily recovered as a direct consequence of our equations for micromomentum balance. If we define as granular temperature ϑ the extra kinetic energy (multiplied by $\frac{2}{3}$) due to chunk dilatancy and grain agitation (which is the trace of the Reynolds tensor which measures the momentum flux in fluid dynamics), that is:

$$\vartheta := \frac{2}{3}(\kappa_f + \kappa_s) = \frac{1}{3}\gamma(\rho)\dot{\rho}^2 + \frac{1}{3}\alpha(\rho_m)\dot{\rho}_m^2 + \rho_m^2 \alpha(\rho_m)\tilde{\mathbf{Y}} \cdot \tilde{\mathbf{Y}}, \tag{11}$$

where relation (1)$_3$ was used.

After, differentiating equation (11) with respect to the time and using equations (6) and (7), we obtain the following balance of granular energy:

$$\frac{3}{2}\rho\dot{\vartheta} = \text{div}\,\mathbf{u} + \mathbf{M}\cdot\mathbf{L} + \rho\mathbf{B}\cdot(\beta\mathbf{R})^{\cdot} + \iota, \tag{12}$$

where $\mathbf{u} := \left[\Sigma^T(\beta\mathbf{R}^T)^{\cdot}\right]$ is the granular heat flux vector, an interstitial work flux of mechanical nature, in excess of the usual flux due to surface tractions, owing to collisions between grains and the chunk boundary or exchange of granules through the chunk boundary itself (see [16], [15] and [21]); $(\mathbf{M}\cdot\mathbf{L})$ is the rate of working of the inertia component due to dilatancy of the Cauchy stress tensor; $[\rho\mathbf{B}\cdot(\beta\mathbf{R})^{\cdot}]$ is the 'stir' due to external actions, a granular heat source; $\iota := -[\mathbf{Z}\cdot(\beta\mathbf{R})^{\cdot} + \Sigma\cdot\text{grad}\,(\beta\mathbf{R})^{\cdot}]$ is the local rate of dissipation due to the inelastic nature of collisions between particles (see, also, equation (2.4) of [29]).

MICROSTRUCTURAL EFFECTS

The micromomentum balances (6) and (7) are not immediately recognized to have the classical expression of laws of balance, but, modulo some innocuous changes in notations, we can give them a more identifiable aspect. Therefore, it is convenient to introduce the external and internal bulk forces related to grain compressibility ϕ and δ, respectively, along with the stirring microstress vector $\mathbf{h}$ as follows:

$$\phi := -\frac{1}{3}\beta\mathbf{R}\cdot\mathbf{B}, \quad \delta := -\frac{1}{3}[\mathbf{Z}\cdot(\beta\mathbf{R}) + \Sigma\cdot\text{grad}\,(\beta\mathbf{R})], \quad \mathbf{h} = -\frac{1}{3}\beta\Sigma^T\mathbf{R}^T. \tag{13}$$

Moreover, we define all the axial vectors (multiplied by a minus sign) corresponding to skew tensors of equation (7), that is the inertia term $\mathbf{w}$, the contributions to the spin due to actions external to the body $\mathbf{b}$, such as assigned couples on a internal grain, those due to internal dissipative interactions $\mathbf{z}$, like frictional grain rotations, and the spinning hyperstress tensor $\mathbf{S}$:

$$\mathbf{w} := \frac{1}{2}\mathcal{E}\tilde{\mathbf{Y}}, \quad \mathbf{b} := \frac{1}{6}\mathcal{E}\left(\beta\mathbf{R}\mathbf{B}^T\right), \quad \mathbf{z} := \frac{1}{6}\mathcal{E}\left[\beta\mathbf{R}\mathbf{Z}^T + \text{grad}\,(\beta\mathbf{R})\odot\Sigma\right], \quad \mathbf{S} := \frac{1}{6}\mathcal{E}\left(\beta\mathbf{R}\oslash{}^t\Sigma\right), \tag{14}$$

where $\mathcal{E}$ is the Ricci's permutation tensor. We remark that the following relations apply:

$$\tilde{\mathbf{Y}} = \mathcal{E}\mathbf{w}, \quad \text{skw}\,(\beta\mathbf{R}\mathbf{B}^T) = 3\mathcal{E}\mathbf{b}, \quad \text{skw}\left[\beta\mathbf{R}(\text{div}\,\Sigma - \mathbf{Z})^T\right] = 3\mathcal{E}(\text{div}\,\mathbf{S} - \mathbf{z}). \tag{15}$$

As a consequence the balances of dilatational and rotational micromomentum (6)-(7), respectively, assume now a more classical form, namely

$$\rho\rho_m\left[\alpha(\rho_m)(\ddot{\rho}_m + 2\rho_m\mathbf{w}\cdot\mathbf{w}) + \frac{1}{2}\alpha'(\rho_m)\dot{\rho}_m^2\right] = \text{div}\,\mathbf{h} - \delta + \rho\,\phi, \tag{16}$$

$$\rho\left[\rho_m^2\alpha(\rho_m)\mathbf{w}\right]^{\cdot} = \text{div}\,\mathbf{S} - \mathbf{z} + \rho\mathbf{b}, \tag{17}$$

while, for the definition (14)$_3$, the Cauchy stress tensor (9) reduces to the following simple expression:

$$\mathbf{T} = \text{sym}\,\mathbf{T}_c + 3\mathcal{E}\mathbf{z} + \mathbf{M}. \tag{18}$$

In particular, we are able to assert that our theory for dry granular material with rotating granules is a generalization of the continuum theory of Goodman and Cowin [24] as well as of the Cosserat brothers' medium [14].

In fact, if the rotation $\mathbf{R}$ reduces to the identity tensor $\mathbf{I}$, then $\mathbf{w}$ and $\mathbf{z}$ are null and, with the appropriate identifications and when the restrictive hypothesis of constant equilibrated inertia is abandoned (see comments to equation (2.13) of [33]), equation (16) reduces to the balance of equilibrated force (4.10) of [24]; nevertheless, a clear difference remains in the stress tensor $\mathbf{T} = \text{sym}\,\mathbf{T}_c + \mathbf{M}$, that is now symmetric as in [24], but here, the presence of the tensor of inertia flux $\mathbf{M}$, related to the effects of dilatancy of Reynolds, appears explicitly.

If instead ρ_m is constant, as it is the case for rigid granules, like pebbles, the vectorial equation for the microstructure (17) represents the pure evolution equation for the free micro–rotations, as that present in the Cosserat theories (see, e.g., [25] or [5]), with a generally nonsymmetric stress tensor $\mathbf{T}$.

Now we wish to put in evidence that the hyperstress tensor $\mathbf{S}$ and the microstress vector $\mathbf{h}$ defined in $(14)_4$ and $(13)_3$, respectively, are peculiar examples of the twisting and the stirring hyperstress tensors defined in [8], but we are not in agreement with those authors about the possibility to assign prescribed boundary conditions to both of them, the twister ($\mathbf{Sm}$) and the stirrer ($\mathbf{h}\cdot\mathbf{m}$), with $\mathbf{m}$ the exterior unit normal to the boundary surface.

In fact, as we observed in comments to micromomentum balances (6) and (7), they are of different physical nature: while for the twister the boundary distribution of the external couples could be assigned in analogy to the microrigid Cosserat continua [14], for the stirrer, it appears difficult instead to imagine a direct way to act on the proper grain compressibility through the boundary itself; rather, only the sum $(\operatorname{div}\mathbf{h}-\delta)$ makes sense, as it has the right properties of covariance and can express weakly non–local effects (see [3], page 21 of [30] and [26]). This is a consequence of a close observation, presented in [6], about the manifold of values of microstructural parameters with or without a physically significant connection and the ensuing possible presence or absence, respectively, of related boundary conditions. In conclusion, when the weakly non–local effects are not considered in the model, the microstress $\mathbf{h}$ could be neglected, as usual in many granular theories.

Finally, we must also adapt the expression of the balance of energy (8) to the present circumstances by introducing $(13)_{2,3}$ and $(14)_{1,3,4}$ in it to obtain

$$\rho\dot{\varepsilon} = \operatorname{sym}\mathbf{T}\cdot\mathbf{D} - 3\left(\mathscr{E}\mathbf{z}\right)\cdot\mathbf{Y} - 3\delta\left(\dot{\beta}\beta^{-1}\right) - 3\mathbf{h}\cdot\operatorname{grad}\left(\dot{\beta}\beta^{-1}\right) + 6\mathbf{z}\cdot\mathbf{w} + 6\mathbf{S}\cdot\operatorname{grad}\mathbf{w} + \rho\lambda. \tag{19}$$

Remark: In the same context, the mechanical interstitial work flux $\mathbf{u}$, introduced in the balance of granular energy (12) at the end of previous section, is now written as

$$\mathbf{u} = 6\mathbf{S}^T\mathbf{w} - 3\dot{\beta}\beta^{-1}\mathbf{h}, \tag{20}$$

where contributions related to fluctuations and rotations of granules appear explicitely.

DISSIPATION IN DRY GRANULAR MATERIALS

In presence of dissipation, we accept here the modified imbalance of entropy as suggested by Müller [32] and used by Goodman and Cowin [24] to investigate the constitutive postulates for a granular material. They assume that the entropy flux is not equal to the heat flux divided by the temperature, hence the second law for a granular material, which is a non-conductor of heat, applies in the following form

$$\rho\theta\dot{\eta} + \theta\operatorname{div}\mathbf{k} \geq \rho\lambda, \tag{21}$$

wherein η is the density of entropy per unit mass, θ the absolute positive temperature and $\mathbf{k}$ the entropy flux.

Now we can follow the approach of Coleman and Noll [13], appropriately adapted to this more general context, to study the thermodynamic consequences of the imbalance (21), which must be identically satisfied during every admissible thermodynamic process. We observe that, for the presence of the entropy flux $\mathbf{k}$ in (21), we can regard all body forces, here represented by $\rho\mathbf{f}$, $\rho\phi$ and $\rho\mathbf{b}$, and the energy supply $\rho\lambda$ to be external fields that can be assigned arbitrarily and which guarantee the balance laws of macro- and micro-momentum and of energy to be identically satisfied. Therefore, in exploiting the entropy inequality as an identity, only the balance of mass must be taken care of (see, also, §3.3 a) of [27] and §26 of [5]).

Introducing, as usual, the Helmholtz free energy per unit mass $\psi := \varepsilon - \theta\eta$ and eliminating $\rho\lambda$ with the use of the energy balance (19), we readily obtain the reduced version of the Clausius-Duhem inequality (21), that is,

$$\rho\left(\dot{\psi} + \dot{\theta}\eta\right) \leq \operatorname{sym}\mathbf{T}\cdot\mathbf{D} - 3\left(\mathscr{E}\mathbf{z}\right)\cdot\mathbf{Y} - 3\delta\left(\dot{\beta}\beta^{-1}\right) - 3\mathbf{h}\cdot\operatorname{grad}\left(\dot{\beta}\beta^{-1}\right) + 6\mathbf{z}\cdot\mathbf{w} + 6\mathbf{S}\cdot\operatorname{grad}\mathbf{w} + \theta\operatorname{div}\mathbf{k}. \tag{22}$$

It is well known that the flow behavior of granular materials is essentially fluid-like, except that it exhibits dependence upon the distribution of the volume fraction (or of the chunk mass density) and the temperature in the reference configuration (see, *e.g.*, [24]). Therefore, in the present work, we postulate that the response of a dry granular material with rotating grains, in presence of dissipation, depends on the set $\mathscr{U} := \{\mathscr{S}, \omega := \dot{\beta}\beta^{-1}, \mathbf{w}, \mathbf{D}, \mathbf{Y}\}$ of constitutive variables, where $\mathscr{S}$ is the following array of variables which describe the elastic state of the material:

$$\mathscr{S} \equiv \{\rho_*, \rho, \mathbf{n} := \operatorname{grad}\rho, \mathbf{N} := \operatorname{grad}(\operatorname{grad}\rho), \beta, \mathbf{R}, \theta_*, \theta\}. \tag{23}$$

We observe that the symmetric tensor $\mathbf{N}$ is inserted among the variables for consistency with the results obtained in [20] from a Hamiltonian variational principle in the conservative case and in the absence of rotations: it is the likely appropriate second geometric measure of local structure, namely, a sort of rough measurements of anisotropy of grains and chunk distributions (see, also, [31]).

By invoking the principles of equipresence and material objectivity, which imply that each dependent constitutive field is given by a smooth isotropic function of the set $\mathscr{U}$, that is

$$\{\psi, \eta, \operatorname{sym}\mathbf{T}_c, \mathbf{z}, \delta, \mathbf{h}, \mathbf{S}, \mathbf{k}\} = \{\hat{\psi}, \hat{\eta}, \hat{\mathbf{T}}, \hat{\mathbf{z}}, \hat{\delta}, \hat{\mathbf{h}}, \hat{\mathbf{S}}, \hat{\mathbf{k}}\}(\mathscr{U}), \tag{24}$$

let us check the compatibility of these prescriptions with the reduced Clausius–Duhem inequality (22). In order to perform the computation of the material time derivative of the free energy ψ, we need the following identities:

$$\left(\overline{\operatorname{grad}\rho =}\right)\dot{\mathbf{n}} = \operatorname{grad}(\dot{\rho}) - \mathbf{L}^T\mathbf{n} = -(\operatorname{tr}\mathbf{D})\mathbf{n} - \rho\operatorname{grad}(\operatorname{tr}\mathbf{D}) - \mathbf{L}^T\mathbf{n},$$
$$\hat{\psi}_{\mathbf{R}} \cdot \dot{\mathbf{R}} = -\hat{\psi}_{\mathbf{R}}\mathbf{R}^T \cdot \tilde{\mathbf{Y}} = \left[-\mathscr{E}\left(\hat{\psi}_{\mathbf{R}}\mathbf{R}^T\right)\right] \cdot \mathbf{w}, \tag{25}$$

where we used the conservation of mass (4) in $(25)_1$, the definition of micro-spin $\tilde{\mathbf{Y}}$ and relation $(15)_1$ in $(25)_2$; moreover, the subscript in $(25)_2$ denote partial differentiation with respect to the shown field, e.g., $\hat{\psi}_{\mathbf{R}} := \frac{\partial \hat{\psi}}{\partial \mathbf{R}}$.

Therefore, by using in (22) the functional dependence $(24)_{1,8}$ of the free energy ψ and the entropy flux $\mathbf{k}$ and the differentiation chain rule, along with relations (25) and $(14)_4$, we obtain the following inequality:

$$\{\hat{\mathbf{T}} + \rho\left[(\rho\hat{\psi}_\rho + \hat{\psi}_{\mathbf{n}} \cdot \mathbf{n})\mathbf{I} + \operatorname{sym}(\mathbf{n} \otimes \hat{\psi}_{\mathbf{n}})\right] + \mathbf{M}\} \cdot \mathbf{D} - [3\mathscr{E}\hat{\mathbf{z}} + \rho\operatorname{skw}(\mathbf{n} \otimes \hat{\psi}_{\mathbf{n}})] \cdot \mathbf{Y} - \rho(\hat{\eta} + \hat{\psi}_\theta)\dot{\theta} -$$
$$- \left(3\hat{\delta} + \rho\beta\hat{\psi}_\beta\right)\omega + \left[6\hat{\mathbf{z}} + \rho\mathscr{E}\left(\hat{\psi}_{\mathbf{R}}\mathbf{R}^T\right)\right] \cdot \mathbf{w} - (3\hat{\mathbf{h}} - \theta\hat{\mathbf{k}}_\omega) \cdot \operatorname{grad}\omega + (6\hat{\mathbf{S}} + \theta\operatorname{skw}\hat{\mathbf{k}}_{\mathbf{w}}) \cdot \operatorname{skw}(\operatorname{grad}\mathbf{w}) + \tag{26}$$
$$+ [\rho^2(\mathbf{I} \otimes \hat{\psi}_{\mathbf{n}}) + \theta\,^t(\hat{\mathbf{k}}_{\mathbf{D}}^T)] \cdot \operatorname{grad}\mathbf{D} - \rho\left(\hat{\psi}_{\mathbf{N}} \cdot \dot{\mathbf{N}} + \hat{\psi}_\omega\dot{\omega} + \hat{\psi}_{\mathbf{w}} \cdot \dot{\mathbf{w}} + \hat{\psi}_{\mathbf{D}} \cdot \dot{\mathbf{D}} + \hat{\psi}_{\mathbf{Y}} \cdot \dot{\mathbf{Y}}\right) +$$
$$+ \theta\left[(\mathbf{F}^{-1}\hat{\mathbf{k}}_{\rho_*}) \cdot \operatorname{Grad}\rho_* + \hat{\mathbf{k}}_\rho \cdot \mathbf{n} + \hat{\mathbf{k}}_{\mathbf{n}} \cdot \mathbf{N} + \hat{\mathbf{k}}_{\mathbf{N}} \cdot \operatorname{grad}\mathbf{N} + \hat{\mathbf{k}}_\beta \cdot \operatorname{grad}\beta + \,^t(\hat{\mathbf{k}}_{\mathbf{R}}^T) \cdot \operatorname{grad}\mathbf{R} +$$
$$+ (\mathbf{F}^{-1}\hat{\mathbf{k}}_{\theta_*}) \cdot \operatorname{Grad}\theta_* + \hat{\mathbf{k}}_\theta \cdot \operatorname{grad}\theta + \theta\operatorname{sym}\hat{\mathbf{k}}_{\mathbf{w}} \cdot \operatorname{sym}(\operatorname{grad}\mathbf{w}) + \,^t(\hat{\mathbf{k}}_{\mathbf{Y}}^T) \cdot \operatorname{grad}\mathbf{Y}] \geq 0,$$

where $\mathbf{F}$ is the deformation gradient and Grad denotes differentiation in the reference configuration. It must be identically satisfied in every admissible thermodynamic process for any dry granular materials with rotating grains.

In relation (26) we can specify the values of scalars $\dot{\theta}$, $\dot{\omega}$, of vectors $\operatorname{Grad}\rho_*$, $\operatorname{grad}\beta$, $\operatorname{Grad}\theta_*$, $\operatorname{grad}\theta$, $\operatorname{grad}\omega$, $\dot{\mathbf{w}}$, of tensors $\operatorname{grad}\mathbf{w}$, $\dot{\mathbf{N}}$, $\dot{\mathbf{D}}$, $\dot{\mathbf{Y}}$ and of third-order tensors $\operatorname{grad}\mathbf{D}$, $\operatorname{grad}\mathbf{N}$, $\operatorname{grad}\mathbf{R}$ and $\operatorname{grad}\mathbf{Y}$, independently of the set of constitutive variables $\mathscr{U}$ and arbitrarily; hence (26) is linear in those quantities and its fulfillment is assured only if respective coefficients vanish. It follows that

$$\psi = \hat{\psi}(\rho_*, \rho, \mathbf{n}, \beta, \mathbf{R}, \theta_*, \theta), \quad \mathbf{k} = \hat{\mathbf{k}}(\rho, \mathbf{n}, \mathbf{N}, \omega, \mathbf{w}, \mathbf{D}), \tag{27}$$

$$\eta = -\hat{\psi}_\theta, \quad \mathbf{h} = \tfrac{1}{3}\theta\,\hat{\mathbf{k}}_\omega, \quad \mathbf{S} = -\tfrac{1}{6}\theta\operatorname{skw}\hat{\mathbf{k}}_{\mathbf{w}}, \quad \theta\,\hat{\mathbf{k}}_{\mathbf{D}} = -\rho^2\hat{\psi}_{\mathbf{n}} \otimes \mathbf{I}, \quad \hat{\mathbf{k}}_{\mathbf{w}} \cdot \mathbf{G}, \quad \hat{\mathbf{k}}_{\mathbf{N}} \cdot \Gamma = 0, \tag{28}$$

for every symmetric tensor $\mathbf{G}$ and completely symmetric third order tensor Γ. Therefore, the constitutive equations for η, $\mathbf{h}$ and $\mathbf{S}$ are determined as soon as those for ψ and $\mathbf{k}$ are known; besides (and as a consequence of (27)), η is also independent of $\mathbf{N}$, ω, $\mathbf{w}$, $\dot{\rho}$, $\mathbf{D}$ and $\mathbf{Y}$, while $\mathbf{h}$ and $\mathbf{S}$ are independent of ρ_*, β, $\mathbf{R}$, θ_*, θ, $\dot{\rho}$ and $\mathbf{Y}$. The classical Gibbs relation $(28)_1$ is a familiar result in thermostatics, but we wish to note that here it is valid for a dry granular material in non-equilibrium.

Thus the reduced entropy inequality (26) is now

$$\{\hat{\mathbf{T}} + \rho\left[(\rho\hat{\psi}_\rho + \hat{\psi}_{\mathbf{n}} \cdot \mathbf{n})\mathbf{I} + \operatorname{sym}(\mathbf{n} \otimes \hat{\psi}_{\mathbf{n}})\right] + \mathbf{M}\} \cdot \mathbf{D} - \left(3\hat{\delta} + \rho\beta\hat{\psi}_\beta\right)\omega +$$
$$+ \left[6\hat{\mathbf{z}} + \rho\mathscr{E}\left(\hat{\psi}_{\mathbf{R}}\mathbf{R}^T\right)\right] \cdot \left(\mathbf{w} - \tfrac{1}{2}\mathscr{E}\mathbf{Y}\right) + \theta\left(\hat{\mathbf{k}}_\rho \cdot \mathbf{n} + \hat{\mathbf{k}}_{\mathbf{n}} \cdot \mathbf{N}\right) \geq 0, \tag{29}$$

where the condition of frame–indifference for $\hat{\psi}$ with respect to vector and tensor arguments was used, that is

$$\operatorname{skw}\left(\mathbf{n} \otimes \hat{\psi}_{\mathbf{n}} + \mathbf{R}\,\hat{\psi}_{\mathbf{R}}^T\right) = \mathbf{0}. \tag{30}$$

We investigate constitutive relations $(28)_{4-6}$ by observing that they are similar, with the appropriate identifications, to those proposed by Dunn and Serrin [16] for materials of Korteweg type and in [18] for granular materials with inelastic grains. By imposing the objectivity of **k** under a frame change, from Appendices in [16] it follows that **k** is linear in **D** and is the sum of two distinct parts:

$$\mathbf{k} = \mathbf{c}(\rho, \mathbf{n}, \mathbf{N}, \omega, \mathbf{w}) - \rho^2 (\mathrm{tr}\,\mathbf{D})\,\mathbf{a}(\rho, \mathbf{n}), \tag{31}$$

where $\mathbf{c} = (\sigma_1 \mathbf{n} + \sigma_2 \mathbf{N}\mathbf{n}) \times \mathbf{w} + \sigma_3 \mathbf{n} \times (\mathbf{N}\mathbf{n})$, with σ_i, $i = 1, 2, 3$, scalar-valued functions of ρ, ω and $\mathbf{n} \cdot \mathbf{n}$; the constitutive vector $\mathbf{a}$ $(:= \theta^{-1} \psi_\mathbf{n})$ depends only on ρ and $\mathbf{n}$ in order to assure the compatibility between relations $(28)_4$ and (27). We see at once that, if the material possesses a center of simmetry for **c**, *i.e.*, if

$$\mathbf{c}(\rho, -\mathbf{n}, \mathbf{N}, \omega, -\mathbf{w}) = -\mathbf{c}(\rho, \mathbf{n}, \mathbf{N}, \omega, \mathbf{w}), \tag{32}$$

then all coefficients σ_i must vanish, and so the entropy flux **k** reduces to its second part alone

$$\mathbf{k} = -\rho^2 (\mathrm{tr}\,\mathbf{D})\,\mathbf{a}(\rho, \mathbf{n}). \tag{33}$$

In this case the microstress vector **h** and the hyperstress tensor **S** vanish, for $(28)_{2,3}$, and, by using (33) in (29), there remains the following residual entropy inequality which defines the dissipation density function $\mathscr{D}$ of the thermo-kinetic process

$$\mathscr{D} := \left\{ \hat{\mathbf{T}} + \rho \left[\rho \hat{\psi}_\rho - (\hat{\psi}_\mathbf{n} + \rho \hat{\psi}_{\rho\mathbf{n}}) \cdot \mathbf{n} - \rho \hat{\psi}_{\mathbf{nn}} \cdot \mathbf{N} \right] \mathbf{I} + \rho \mathrm{sym}\,(\mathbf{n} \otimes \hat{\psi}_\mathbf{n}) + \mathbf{M} \right\} \cdot \mathbf{D} - \tag{34}$$
$$- \left(3\hat{\delta} + \rho\beta \hat{\psi}_\beta \right) \omega + \left[6\hat{\mathbf{z}} + \rho \mathscr{E} \left(\hat{\psi}_\mathbf{R} \mathbf{R}^T \right) \right] \cdot \left(\mathbf{w} - \tfrac{1}{2} \mathscr{E} \mathbf{Y} \right) \geq 0,$$

where all terms are indifferent to changes in observer, as requested (see, also, §5 of [19]).

When the material does not possess a center of simmetry for **c**, the analysis is more subtle and it will presented in a forthcoming paper.

EQUILIBRIUM

We can obtain additional informations on constitutive relations for a non-conducting dry granular material from the residual entropy inequality (34) by assuming that processes do not deviate far from thermodynamic equilibrium; we define such an *equilibrium* state for our medium as one in which the independent dissipative variables $\Xi := \{\mathbf{D}, \omega, \mathbf{w} - \tfrac{1}{2}\mathscr{E}\mathbf{Y}\}$ all vanish. Hence the dissipation $\mathscr{D}$ has a minimum with respect to Ξ in equilibrium; it follows then that

$$\left.\frac{\partial \mathscr{D}}{\partial \Xi}\right|_{\Xi=0} = 0 \quad \text{and} \quad \left.\frac{\partial^2 \mathscr{D}}{\partial \Xi^2}\right|_{\Xi=0} \cdot (\Xi \otimes \Xi) \geq 0, \quad \forall\, \Xi \neq \mathbf{0}. \tag{35}$$

Condition $(35)_1$ yields the following equilibrium values of partial stress tensor $(\mathrm{sym}\,\mathbf{T}_c)$ and of internal bulk and spin microforces δ and **z**, respectively:

$$\mathrm{sym}\,\mathbf{T}_e := \mathrm{sym}\,\mathbf{T}|_{\Xi=0} = \rho \left[(\hat{\psi}_\mathbf{n} + \rho \hat{\psi}_{\rho\mathbf{n}}) \cdot \mathbf{n} + \rho \hat{\psi}_{\mathbf{nn}} \cdot \mathbf{N} - \rho \hat{\psi}_\rho \right] \mathbf{I} - \rho \mathrm{sym}\,(\mathbf{n} \otimes \hat{\psi}_\mathbf{n}), \tag{36}$$

$$\delta_e := \delta|_{\Xi=0} = -\tfrac{1}{3}\rho\beta \hat{\psi}_\beta, \quad \mathbf{z}_e := \mathbf{z}|_{\Xi=0} = -\tfrac{1}{6}\rho \mathscr{E} \left(\hat{\psi}_\mathbf{R} \mathbf{R}^T \right). \tag{37}$$

Thus the equilibrium values $\mathrm{sym}\,\mathbf{T}_e$, δ_e and $\mathbf{z}_e$ are specified when the Helmholtz free energy ψ is given as a function of $\mathscr{V} := (\rho_*, \rho, \mathbf{n}, \beta, \mathbf{R}, \theta_*, \theta)$, like the entropy density η and the entropy flux **k**.

By inserting relations (36) and $(37)_2$ and using (30) in the expression for the complete Cauchy stress tensor (18), we obtain its value at equilibrium:

$$\mathbf{T}_e := \mathbf{T}|_{\Xi=0} = \rho \left[(\hat{\psi}_\mathbf{n} + \rho \hat{\psi}_{\rho\mathbf{n}}) \cdot \mathbf{n} + \rho \hat{\psi}_{\mathbf{nn}} \cdot \mathbf{N} - \rho \hat{\psi}_\rho \right] \mathbf{I} - \rho \mathbf{n} \otimes \hat{\psi}_\mathbf{n} \tag{38}$$

and demonstrate the property of granular materials to support shear stresses at equilibrium due to the presence of the stress tensors of Ericksen type $(-\rho \mathbf{n} \otimes \hat{\psi}_\mathbf{n})$, either in the absence of the microrotations of the granules or when the grains are rigid; moreover, the pressure $\pi := \rho^2 \hat{\psi}_\rho$ associated with the chunk mass density ρ appears similar to the usual thermodynamic pressure in a compressible fluid, except that, now, it depends on all variables $\mathscr{V}$ just listed.

The other condition at equilibrium $(35)_2$ furnishes the following inequality:

$$6\mathbf{z}^e_\mathbf{y} \cdot (\mathbf{y} \otimes \mathbf{y}) - 3\delta^e_\omega \omega^2 + (6\mathbf{z}^e_\omega - 3\delta^e_\mathbf{y}) \cdot \omega \mathbf{y} + \mathbf{T}^e_\mathbf{D} \cdot (\mathbf{D} \otimes \mathbf{D}) + \left[\mathbf{T}^e_\mathbf{y} + 6\left(\mathbf{z}^e_\mathbf{D}\right)^T \right] \cdot (\mathbf{D} \otimes \mathbf{y}) + (\mathbf{T}^e_\omega - 3\delta^e_\mathbf{D}) \cdot \omega \mathbf{D} \geq 0. \tag{39}$$

It must be satisfied for each scalar ω, vector $\mathbf{y}$ $\left(:= \mathbf{w} - \frac{1}{2}\mathscr{E}\mathbf{Y}\right)$ and tensor $\mathbf{D}$ (the exponent e denotes values of functions in equilibrium). The quadratic form (39) in ω, $\mathbf{y}$ and $\mathbf{D}$ is satisfied if and only if

$$\delta_\omega^e \leq 0, \quad -8\delta_\omega^e \mathbf{z}_\mathbf{y}^e \cdot (\mathbf{y} \otimes \mathbf{y}) \geq \left[(3\mathbf{z}_\omega^e - \delta_\mathbf{y}^e) \cdot \mathbf{y}\right]^2 \tag{40}$$

plus a third very elaborate inequality, hard to analyze in any generality.

These three limitations assure that

$$\mathbf{z}_\mathbf{y}^e \cdot (\mathbf{y} \otimes \mathbf{y}) \geq 0 \quad \text{and} \quad \mathbf{T}_\mathbf{D}^e \cdot (\mathbf{D} \otimes \mathbf{D}) \geq 0, \tag{41}$$

for all vectors $\mathbf{y}$ and tensors $\mathbf{D}$. Condition $(41)_2$ is simply the familiar assertions that, at equilibrium, the viscosity tensor $\mathbf{T}_\mathbf{D}^e$ is positive semi–definite; it can depend on the list $\mathscr{V}$.

For definition $(13)_2$ and $(14)_2$, the positiveness of the swelly scalar $(-\delta_\omega)$ at equilibrium in $(40)_1$, as well as the fact that the twisty tensor $\mathbf{z}_\mathbf{y}^e$ is positive semi–definite for $(41)_1$, guarantees that terms related to internal bulk and spin microforces in the dissipation $\mathscr{D}$ are purely dissipative.

CONCLUSIONS

The continuum theory for non-conducting dry granular materials with rotating grains presented in this study has the potential for application in the area of soil mechanics: it seems to be suitable to analyze the dynamics and statics of sand layers. We demonstrated that the fluctuation kinetic energy associated with the dilatancy of the chunks gives rise to the generation of a tensor of inertia flux in the Cauchy stress, while the presence of the bulk and rotational micromotions allows the property of granular materials to support shear stresses at equilibrium due to the presence of a term of Ericksen type in the stress tensor, which is now not necessarily symmetric, in general. We put in evidence microstructural effects in the expression of micromomentum and energy balances in order to recognize peculiar contributes of micromotions. We obtained from constitutive relations that the entropy flux depends linearly on the trace of the velocity gradient and that equilibrium values of Cauchy tensor and of internal microactions are determined as soon as that for the free energy is known.

ACKNOWLEDGMENTS

I wish to thank Prof. K. Hutter for some helpful suggestions. The support of the Italian "Gruppo Nazionale di Fisica Matematica" of the "Istituto Nazionale di Alta Matematica" and of the Department of Mechanics and Materials, Faculty of Engineering of the "Mediterranean" University of Reggio Calabria, Italy, is gratefully acknowledged.

REFERENCES

1. G. Ahmadi, A Generalized Continuum Theory for Granular Materials, *Int. J. Non–Linear Mech.* **17**, 21–33 (1982).
2. A. Bedford, and D.S. Drumheller, On Volume Fraction Theories for Discretized Materials, *Acta Mechanica* **48**, 173–184 (1983).
3. Bridgman, P.W.: Reflections on Thermodynamics. Proc. Am. Acad. Arts Sci., **82**, 301–309 (1953)
4. Y.A. Buyevich, and I.N. Shchelchkova, Flow of Dense Suspensions, *Prog. Aerospace Sci.* **18**, 121–150 (1978).
5. G. Capriz, *Continua with Microstructure*. Springer Tracts in Natural Philosophy. Springer-Verlag, Berlin Heidelberg New York, **35** 1989.
6. G. Capriz, and P. Giovine, Remedy to Omissions in a Tract on Continua with Microstructure, *Atti XIII Congresso AIMETA '97*, Siena, Meccanica Generale, **I**, 1–6 (1997).
7. G. Capriz, and G., Mullenger, Extended Continuum Mechanics for the Study of Granular Flows, *Rend. Acc. Lincei, Matematica* **6**, 275–284 (1995).
8. G. Capriz, and G., Mullenger, Dynamics of Granular Fluids, *Rend. Sem. Mat. Univ. Padova* **111**, 247–264 (2004).
9. G. Capriz, and P. Podio–Guidugli, Materials with Spherical Structure, *Arch. Rational Mech. Anal.* **75**, 269–279 (1981).
10. G. Capriz, P. Podio–Guidugli, and W. Williams: Balance Equations for Materials with Affine Structure, *Meccanica* **17**, 80–84 (1982).
11. K.C. Chen, J.Y. Lan, and Y.C. Tai, Description of Local Dilatancy and Local Rotation of Granular Assemblies by Microstretch Modeling, *Int. J. Sol.Structures* **46**, 3882–3893 (2009).

12. M. Cieszko, Extended Description of Pore-Space Structure and Fluid Flow in Porous Materials. Application of Minkowski Space, *Proceed. of the XIV Int. Symp. Trends Appl. Math. Mech., STAMM'04*, Shaker Verlag, 93–102 (2004).
13. B.D. Coleman, and W. Noll, The Thermodynamics of Elastic Materials with Heat Conduction and Viscosity, *Arch. Rational Mech. Anal.* **13**, 167–178 (1963).
14. E. and F. Cosserat, *Théorie des Corps Déformables*, Hermann, 1909.
15. F. Dell'Isola, and P. Seppecher, Edge Contact Forces and Quasi–Balanced Power, *Meccanica* **32**, 33–52 (1997).
16. J.E. Dunn, and J. Serrin, On the Thermomechanics of Interstitial Working, *Arch. Rat. Mech. Anal.* **88**, 95–133 (1985).
17. A.C. Eringen, "Mechanics of Micromorphic Continua", in *Proc. IUTAM Symposium on Mechanics of Generalized Continua Freudenstadt and Stuttgart (1967)*, edited by E. Kröner, Springer, Berlin Heidelberg New York, 1968, pp.18–35.
18. P. Giovine, Nonclassical Thermomechanics of Granular Materials, *Mathematical Physics, Analysis and Geometry* **2**, 179–196 (1999).
19. P. Giovine, "An Extended Continuum Theory for Granular Media", in *Mathematical Models of Granular Matter*, edited by G. Capriz, P. Giovine, and P. Mariano, Series: Lecture Notes in Mathematics **1937**, Springer Verlag, Berlin Heidelberg, 2008, pp.167–192.
20. P. Giovine, and F. Oliveri, Dynamics and Wave Propagation in Dilatant Granular Materials, *Meccanica* **30**, 341–357 (1995).
21. P. Giovine, and M.P. Speciale, "On Interstitial Working in Granular Continuous Media", in *Proceed. 10th Int. Conf. on Waves and Stability in Continuous Media (WASCOM'99), Vulcano (Me)*, edited by V. Ciancio, A. Donato, F. Oliveri, and S. Rionero, World Scientific, Singapore, 2001, pp.196–208.
22. C. Godano, and F. Oliveri, Nonlinear seismic waves: a model for site effects. *Int. J. Non-linear Mech.* **34**, 457–468 (1999).
23. J.D. Goddard, and A.K. Didwania, Computations of Dilatancy and Yield Surfaces for Assemblies of Rigid Frictional Spheres, *Q. J. Mechanics Appl. Math.* **51**, 15–43 (1998).
24. M.A. Goodman, and S.C. Cowin, A Continuum Theory for Granular Materials, *Arch. Rational Mech. An.* **44**, 249–266 (1972).
25. G. Grioli, Sui Continui di Cosserat con Rotazioni Libere, *Rend. Semin. Mat. Univ. Padova* **47**, 299–312 (1972).
26. G. Grioli, Microstructures as a Refinement of Cauchy Theory. Problems of Physical Concreteness, *Continuum Mech. Thermodyn.* **15**, 441–450 (2003).
27. K. Hutter, The Foundations of Thermodynamics, Its Basic Postulates and Implications. A Review of Modern Thermodynamics, *Acta Mechanica* **27**, 1–54 (1977).
28. K. Hutter, and K.R. Rajagopal, On Flows of Granular Materials, *Continuum Mech. Thermodyn.* **6**, 81–139 (1994).
29. P.C. Johnson, and R. Jackson, Frictional–Collisional Constitutive Relations for Granular Materials, with Applications to Plane Shearing, *J. Fluid Mech.* **176**, 67–93 (1987).
30. I.A. Kunin, *Elastic Media with Microstructure II*, Springer Series in Solid–State Sciences **44**, Springer–Verlag, Berlin, 1983.
31. N. Mitarai, H. Hayakawa, and H. Nakanishi, Collisional Granular Flow as a Micropolar Fluid, *Phys. Rev. Lett.* **88**, 174301 (2002).
32. I. Müller, On the entropy inequality, *Arch. Rational Mech. Anal.* **26**, 118–141 (1967).
33. J.W. Nunziato, and S.C. Cowin, A Nonlinear Theory of Elastic Materials with Voids, *Arch. Rational Mech. Anal.* **72**, 175–201 (1979).
34. O. Reynolds, On the Dilatancy of Media Composed of Rigid Particles in Contact, *Phil. Magazine* **20**, 469–481 (1885).

Granular hypoplasticity with Cosserat effects

J. D. Goddard

Department of Mechanical and Aerospace Engineering
University of California, San Diego,
9500 Gilman Drive
La Jolla, CA 92093-0411 USA

Abstract. This paper provides an extension to Cosserat mechanics of a recently proposed version of hypoplasticity [1], and this extension is achieved economically by means of a novel complex-variable formulation of Cosserat theory.

The present work represents a compact synthesis and theoretical framework for both non-polar and polar hypoplasticity, and it encompasses various special cases considered in the literature, as discussed in the recent monograph by Tejchman [2].

The current approach offers a perspective on granular dilatancy, elastoplastic yield, and energy dissipation which differs from the standard hypoplasticity and which serves to establish a connection to classical incremental plasticity. In contrast to the classical theory, the present approach, based entirely on the concept of pseudo-linear forms, admits but does require elastoplastic potentials to describe plastic flow. When such potentials are assumed, it is shown that they can be related to the plastic moduli of the present formulation.

It is also shown that hypoplasticity allows for a distinction between *active* and *passive* internal variables, with the latter serving to define *parameters*. Finally, the known forms for linear isotropic Cosserat elasticity are employed to represent isotropic hypoplasticity, and the resulting formulae appear to encompass several empiricisms found in the literature.

Keywords: micropolar, Cosserat continuum, parametric hypoplasticity, stiff elastoplasticity, couple stress, dilatancy
PACS: 81.05.Rm, 83.80.Fg

INTRODUCTION

Under the rubric of *parametric hypoplasticity*, the author has proposed in a recent work [1], hereinafter denoted by "Ref. I", a generalized version of conventional hypoplasticity [3] for assemblies of non-cohesive, nearly rigid particles. Starting from the notion of strictly dissipative materials and proceeding through incremental elastoplasticity, it provides a set of rate-independent ODEs, representing a minimal mathematical framework for the continuum modeling of the elastoplasticity of granular media, including:

1. elastoplastic yield with small-strain elastic response,
2. pressure-sensitive dilatant plasticity,
3. history-dependence, including the development of anisotropy based on evolutionary granular "texture" or "fabric", and
4. non-negative dissipation, based on stiff elastoplasticity.

Following a brief summary of the previous work, we consider the incorporation of Cosserat effects. As mentioned in the Abstract, many special cases of the Cosserat model are to be found in the literature [4, 5, 6, 7, 8, 9, 10, 11]. Much of the history and many theoretical concepts are covered in the recent monograph of Tejchman [2], which is focused on numerical simulations of soil mechanics based on standard hypoplastic modeling. As evident from this monograph, the exhaustive studies of Bauer and co-workers, e.g. [12], provide a compelling mathematical approach to the largely empirical "critical-state" soil mechanics.

As is also evident from the previous works, the Cosserat model provides one of the simplest of the numerous multi-polar continuum models, which we recall serve to regularize simple-continuum models, e.g. in granular shear bands, by the introduction of additional kinematic gradients involving material length scales.

The present article aims to provide a systematic and concise treatment of the underlying theory. Although space does not allow for a comprehensive discussion of the various constitutive models presented in the above-cited literature, the present treatment may suggest a more systematic approach to such modeling.

The style of the exposition reflects the opinion, inspired by several treatises [13, 14, 3], that a compact formulation in direct tensor notation serves to elucidate otherwise complex expressions couched in tensor components, notwithstanding the computational merits of the latter. To those who share this view, the present work may serve as a useful overview of the literature.

Nonpolar hypoplasticity

It is worthwhile summarizing briefly the approach to hypoplasticity that is presented in more detail in Ref. I. With a view to thermodynamically admissibility, and in contrast to the standard expositions [15, 3], we proceed from the special case of a strictly dissipative material, in which the local Cauchy stress $\mathbf{T}(t)$ is given, *modulo* work-free reactions to any internal constraints, by the linear form in the local deformation rate $\mathbf{D}(t)$:

$$\mathbf{T}(t) = \boldsymbol{\eta} : \mathbf{D}(t), \text{ i.e. } T_{ij}(t) = \eta_{ijkl} D^{kl}(t), \tag{1}$$

where $\boldsymbol{\eta}$ is a positive-definite *viscosity* depending on $\mathbf{D}(t)$ or $\mathbf{T}(t)$ and local deformation history. With the modification described below, (1) describes rigid plasticity, and, within the framework of parametric hypoplasticity, the history dependence is described by a set $\mathcal{A}$ of internal variables which satisfy rate-independent[1] ODEs.

We generally adopt the notation of Ref. I, with vectors and second-rank tensors denoted, respectively, by bold lowercase Roman and uppercase Roman or Greek symbols, and with fourth-rank tensors, regarded as linear operators on second-rank tensors, denoted by bold lowercase Greek, $\boldsymbol{\alpha}, \boldsymbol{\beta}, \ldots$. As an exception, lowercase Greek is used for vectors representing Cosserat rotations, and Blackboard Bold is employed for complex variables, with real and imaginary parts denoted, respectively, by $\mathfrak{R}, \mathfrak{I}$, and overbars denoting complex conjugates.

The respective idemfactors are denoted by $\mathbf{I}$, with components δ_{ij}, and by $\boldsymbol{\delta}$, with components $\delta_{ijkl} = \delta_{ik}\delta_{jl}$, and we employ standard tensor notation and summation convention for components on arbitrary curvilinear coordinates, with colons denoting linear operations on tensors represented by ordered pair-wise contraction on the trailing indices of prefactors with leading indices of postfactors.

Primes denote deviators:

$$\mathbf{A}' := \mathbf{A} - \frac{1}{3}\mathbf{I}(\mathbf{I} : \mathbf{A}), \ldots \; \boldsymbol{\alpha}' := \boldsymbol{\alpha} - \frac{1}{3}\mathbf{I}(\mathbf{I} : \boldsymbol{\alpha}), \ldots \tag{2}$$

and superposed carat denotes the *versor* (or director) of real second rank tensors[2]:

$$\hat{\mathbf{A}} \stackrel{\text{def}}{=} \frac{\mathbf{A}}{|\mathbf{A}|}, \text{ where } |\mathbf{A}| \stackrel{\text{def}}{=} (\mathbf{A} : \mathbf{A}^{\mathrm{T}})^{1/2} \equiv A_{ij}A^{ij}, \tag{3}$$

while superscripts T and * denotes respective transpose or tensorial duals, with $A_{ij}^{\mathrm{T}} \stackrel{\text{def}}{=} \overline{A}_{ji}$ and $\alpha_{ijkl}^{*} \stackrel{\text{def}}{=} \overline{\alpha}_{klij}$, etc.

We denote various 4th-rank tensor moduli by the symbol $\boldsymbol{\mu}$ and the corresponding compliances by $\boldsymbol{\kappa} = \boldsymbol{\mu}^{-1}$, such that, whenever these represent invertible linear transformations on the space of second-rank tensors,

$$\boldsymbol{\kappa} : \boldsymbol{\mu} = \boldsymbol{\delta}, \; (\text{i.e. } \kappa_{ijkl}\mu^{kl}_{..mn} = \delta_{ijmn}), \tag{4}$$

The symbols $(\dot{\,}) \stackrel{\text{def}}{=} d(\,)/dt$ denote material time-derivatives defined by (18) below.

The relation (1) provides a theoretical framework for the rheology of idealized rigid-particle suspensions in viscous fluids [1], and a model of rate-independent rigid plasticity is obtained on replacing $\boldsymbol{\eta}$ by $\boldsymbol{\mu}_{\mathrm{P}}/|\mathbf{D}|$, where $\boldsymbol{\mu}_{\mathrm{P}} = \boldsymbol{\mu}_{\mathrm{P}}(\mathcal{A}, \hat{\mathbf{D}})$ defines a positive-definite plastic modulus. A dissipative *flow rule* and *yield condition* are then given, respectively, by

$$\hat{\mathbf{D}} = \boldsymbol{\kappa}_{\mathrm{P}} : \mathbf{T}, \text{ and } Y = \|\mathbf{T}\|^2_{\boldsymbol{\zeta}} \stackrel{\text{def}}{=} \mathbf{T} : \boldsymbol{\zeta} : \mathbf{T} = 1, \text{ where } \boldsymbol{\zeta} = \boldsymbol{\kappa}_{\mathrm{P}}^{*} : \boldsymbol{\kappa}_{\mathrm{P}}, \tag{5}$$

with $Y < 1$ defining rigid states. In the standard granular model, $\boldsymbol{\kappa}_{\mathrm{P}} = p\boldsymbol{\kappa}_{\mathrm{C}}$, where $p = -\mathrm{tr}(\mathbf{T})/3 > 0$, and $\boldsymbol{\kappa}_{\mathrm{C}}$ is a non-dimensional Coulomb compliance.

By the extension of an analysis relating elastic moduli and elastic potentials in Ref. I, Appendix A establishes a relation between plastic moduli and plastic potentials, on the presumption that the latter exist.

---

[1] As an historical note, it is recalled that rate-independence reflects the absence of a characteristic material time scale, as pointed out in an article on particulate mechanics[16] that cites the work of Pipkin and Rivlin [17] but overlooks a germane work by Gudehus [18] on granular materials.
[2] to be distinguished from the use of carats in certain works [8] to denote stress normalized by isotropic pressure

An accounting for elastic effects, based on the elastic-plastic decomposition $\mathbf{D} = \mathbf{D}_E + \mathbf{D}_P$, where $\mathbf{D}_P$ satisfies (5), and deformation rate $\mathbf{D}_E$ is given in terms of stress rate by an incremental elastic modulus $\boldsymbol{\mu}_E$, leads to the hypoplastic form [1]

$$\overset{\circ}{\mathbf{T}} = \boldsymbol{\mu}_H : \mathbf{D} - |\mathbf{D}|\mathbf{N}, \quad \text{with} \quad \boldsymbol{\mu}_H \equiv \boldsymbol{\mu}_E, \quad \text{and} \quad \mathbf{N} = \beta \boldsymbol{\mu}_E : \boldsymbol{\kappa}_P : \mathbf{T}, \tag{6}$$

where $\overset{\circ}{\mathbf{T}}$ represents an objective (e.g. Jaumann) rate, and β an "inelastic-clock" function (denoted by ϑ in Ref. I) that vanishes inside the yield surface, e.g., in terms of a particular Heaviside function:

$$\beta = H(Y-1), \quad \text{with} \quad H(0) \overset{\text{def}}{=} H(0+) = 1 \tag{7}$$

The conventional hypoplastic model [15, 3, 2] takes $\beta \equiv 1$, with no distinction between plastic and total deformation rates, and assumes $\boldsymbol{\mu}_H$ and $\mathbf{N}$ to be functions of $\mathbf{T}(t)$, without symmetry or positivity restrictions on $\boldsymbol{\mu}_H$. While the model does not identify a yield surface *per se*, the "stationary state" where stress rate vanishes represents the dissipative yield surface identified in the present work, subject to the interpretation of $\boldsymbol{\mu}_H$ and $\mathbf{N}$ in (6). Furthermore, and contrary to occasional remarks in the literature, hypoplasticity *does* allow for elastic response, provided that term $\mathbf{N}$ become negligible for small $|\mathbf{T}|$, which is suggested immediately by the form of (6).

Without pursuing the details [1] here, we note that the interpretation of dilatancy as an internal constraint requires that the expression for $\mathbf{N}$ in (6) be replaced by

$$\mathbf{N} = \frac{\beta}{\sqrt{1 + \alpha^2/3}} \boldsymbol{\mu}_E : (\boldsymbol{\kappa}'_P : \mathbf{T} + \tfrac{1}{3}\alpha\mathbf{I}), \tag{8}$$

where $\boldsymbol{\kappa}'_P$ is a deviatoric plastic compliance and α is a coefficient of dilatancy [1]. Furthermore, in the standard granular model, one has $\boldsymbol{\kappa}'_P = \boldsymbol{\kappa}'_C/p$, where p is confining pressure and $\boldsymbol{\kappa}'_C$ a nondimensional Coulomb compliance that is independent of p.

As a final observation, we note that (6) is a special case of the rate-independent *pseudo-linear*[3] form:

$$\overset{\circ}{\mathbf{T}} = \boldsymbol{\pi}(\mathbf{T}, \hat{\mathbf{D}}) : \mathbf{D}, \quad \text{with} \quad \boldsymbol{\pi} \overset{\text{def}}{=} \boldsymbol{\mu}_H - \mathbf{N}_H \otimes \hat{\mathbf{D}}, \quad \text{i.e.} \quad \pi_{ijkl} = \mu_{ijkl} - N_{ij}\hat{D}_{kl}, \tag{9}$$

which obviously reduces elastoplasticity to a set of rate-independent ODEs with $\mathbf{D}$ as control variable.

Unfortunately, there is no guarantee of thermodynamic admissibility of the general form (9), which, for example, may fail to exhibit non-negative work on cycles. In contrast, the interpretation (6) in terms of a physically-motivated incremental elastoplasticity leads to a weak form of Il'yushin's postulate discussed in Ref. I[4], and this provides yet another motivation for the use of evolutionary internal variables.

Parametric Hypoplasticity

This more general version of hypoplasticity, anticipated in early works on the subject [19, 15] has been computationally implemented in numerous works by Tejchman, Bauer and co-workers [20, 7, 12, 8, 21, 22] for polar as well as non-polar media. In the non-polar form, these models involve internal variables whose history dependence is determined by ODEs having the pseudo-linear form:

$$\overset{\circ}{\mathcal{T}} = \mathcal{P}(\mathcal{T}, \hat{\mathbf{D}})[\mathbf{D}], \quad \text{with} \quad \mathcal{T} = \{\mathbf{T}, \mathcal{A}\}, \tag{10}$$

where $\mathcal{P}[\mathbf{D}]$ is linear in $\mathbf{D}$, and $\mathcal{A}$ denotes the set of internal variables. While the unqualified term "hypoplasticity" appears to enjoy general usage in the current literature, the adjective *parametric* is employed here to emphasize the special nature of parameters regarded as internal variables. As discussed below, the term *parameter* refers to any internal variable whose power (work-rate) can be derived from the kinematics represented by $\mathbf{D}$. This is made more precise below, after consideration of some typical examples of parameters.

---

[3] As $\mathbf{Y} = \boldsymbol{\pi}(\hat{\mathbf{X}}) : \mathbf{X} = \boldsymbol{\Pi}(\hat{\mathbf{X}})|\mathbf{X}|$, i.e. $Y_{ij} = \pi_{ij}{}^{kl}X_{kl} = \Pi_{ij}(\hat{\mathbf{X}})|\mathbf{X}|$, reflects the degree-one homogeneity of *cones*, other terminology might be suggested.

[4] which also overlooks the older work of Gudehus [18], who considers the relevance of Il'yushin's postulate to granular plasticity.

In the simplest example [12], $\mathcal{A}$ is given solely by the void ratio e, determined through plastic volume change by:

$$\overset{\circ}{e} \equiv \dot{e} = (1+e)\text{tr}(\mathbf{D}_\text{P}) = \frac{(1+e)\alpha\beta}{\sqrt{1+\alpha^2/3}}|\mathbf{D}| \tag{11}$$

Then, with $\boldsymbol{\mu}_\text{E}, \boldsymbol{\kappa}'_\text{P}, \alpha$ in (6) depending on $\mathcal{X}$, (10) is represented by (6), (7) and (11). We recall that the standard hypoplastic model [12] without dilatancy constraint assumes $\beta = 1, \alpha = \sqrt{3/2}$.

A discussion of various limiting forms of this model is given in Ref. I, including "stiff" elastoplasticity, where

$$\epsilon = \|\boldsymbol{\kappa}_\text{E}\|/\|\boldsymbol{\kappa}_\text{P}\| = p\|\boldsymbol{\kappa}_\text{E}\|/\|\boldsymbol{\kappa}_\text{C}\| \ll 1, \text{ with } \alpha = O(\epsilon^{-1}) \tag{12}$$

In this limit, (6) reduces to a stiff ODE, with $\mathbf{D} \approx \mathbf{D}_\text{P}$, except on small strain scales $O(\epsilon)$ near points of elastic loading or unloading. As pointed out in Ref. I, the relation (12) is particularly apt for stiff geomaterials such as sand.

Another candidate for scalar parameter is the accumulated plastic shear strain γ_P common to numerous plasticity theories, which satisfies the further ODE:

$$\dot{\gamma}_\text{P} \overset{\text{def}}{=} |\mathbf{D}'_\text{P}| = \frac{\beta}{\sqrt{1+\alpha^2/3}}|\mathbf{D}| \tag{13}$$

Since α and β may generally depend on both e and γ_P, both ODEs (11) and (13) are needed.

For isotropic materials, the moduli and compliances can be represented as isotropic polynomials in $\mathbf{T}$ or $\widehat{\mathbf{D}'}$, with scalar coefficients depending on their joint isotropic scalar invariants and on a set of scalar parameters such as e and γ_P cited above. However, in order to describe strain-induced anisotropy in a material which is isotropic in some virgin state, one needs additional tensorial variables [18], such as a 2nd-rank *fabric*[5] tensor $\mathbf{A}$. In this case, a general form of (10) is given by the higher-order ODE for parametric evolution [16, 1] obtained by taking $\mathcal{A} = \{\mathbf{A}_1, \mathbf{A}_2, \ldots, \mathbf{A}_n\}$, $\mathbf{A}_1 \equiv \mathbf{A}$, with

$$\overset{\circ}{\mathbf{A}}_k = \beta|\mathbf{D}|\mathbf{A}_{k+1}, \text{ for } k = 1, \ldots, n-1,$$
$$\overset{\circ}{\mathbf{A}}_n = \boldsymbol{\pi}(\mathcal{X}, \hat{\mathbf{D}}) : \mathbf{D}, \quad \mathcal{X} = \{\mathbf{T}, \gamma_\text{P}, e, \mathcal{A}\}, \tag{14}$$

together with (11) and (13), where $\boldsymbol{\pi}$ represents a pseudo-linear form like that in (9), and all coefficients $\alpha, \beta, \ldots$ depend on $\mathcal{X}$, with $\boldsymbol{\pi}$ given generally as a tensor polynomial in $\mathbf{T}, \hat{\mathbf{D}}, \mathbf{A}$. The associated anisotropy is to be distinguished from inherent or "initial" anisotropy [22] associated with certain virgin states of the material.

Recent work on concentrated fluid-particle suspensions discussed in Ref. I suggests that a low-order model, with $n \leq 2$ and with a low-order tensor polynomial in $\mathbf{A}$, may suffice to describe induced anisotropy in simple shear. It remains to be seen whether a similar simplification applies to granular media subject to more complex deformation histories, but recent numerical simulations [25] are encouraging.

Passive vs. active variables

Underlying the continuum-mechanical *principle of virtual work*, internal variables represent configurational degree of freedom endowed with specific mechanical power and work-conjugate stress, the latter representing the dual space to the configurational tangent space. However, within the framework of parametric hypoplasticity, parameters represent a special class of internal variables whose kinematics are given as pseudo-linear forms in $\mathbf{D}$, implying that their stress and stress power are not independent of the usual mechanical quantities.

For example in the case of void ratio e, the power is given by $p\dot{e}/(1+e)$, where $p = -\text{tr}(\mathbf{T})/3$, whereas for accumulated plastic shear γ_P defined by (13)[6], it is easy to show that the power is given by $\tau_\text{P}\dot{\gamma}_\text{P}$, with $\tau_\text{P} = \mathbf{T}:\boldsymbol{\kappa}_\text{P}:\mathbf{T}$.

In the case of fabric, with conjugate stress $\mathbf{S}_\text{A}$,

$$\overset{\circ}{\mathbf{A}} = \boldsymbol{\pi}:\mathbf{D} \Rightarrow \mathbf{S}_\text{A}:\overset{\circ}{\mathbf{A}} = \mathbf{T}_\text{A}:\mathbf{D}, \text{ where } \mathbf{T}_\text{A} = \mathbf{S}_\text{A}:\boldsymbol{\pi}(\mathbf{T}, \mathcal{A}, \hat{\mathbf{D}}) \tag{15}$$

---

[5] "Texture" is often employed in the metal-plasticity literature to denote a similar concept, a term dating at least as far back as the early work of Hill [23]. That work does not provide a complete description of strain-induced evolution of anisotropy, and a cursory survey of subsequent literature suggests that, with notable exceptions [24], attempts are rarely made to relate micro-level to continuum-level descriptions of texture evolution.

[6] Conversely, given any norm $\tau = |\mathbf{T}|$, one can define $\dot{\gamma}_\text{P} = \hat{\mathbf{D}}:\boldsymbol{\mu}_\text{P}:\hat{\mathbf{D}}/\tau$, which appears equivalent to Hill's [26] duality based on plastic potentials.

In this sense, the nominal hypoplastic parameters may be treated as *passive* internal variables, enslaved to **D**. On the other hand, the Cosserat rotations considered next do not qualify as parameters, since they involve independent kinematics, with conjugate stresses and work effects that are not derivable from $\{\mathbf{T}, \mathcal{A}, \hat{\mathbf{D}}\}$.

COSSERAT EFFECTS

The following is a highly condensed restatement of Cosserat mechanics, which is treated at much greater length in many previous works [27, 13, 28, 29, 30, 4, 5]. From the conventional continuum-mechanical viewpoint, a Cosserat continuum[7] is defined by a differentiable map assigning spatial position $\mathbf{x}(\mathbf{x}°, t)$ and microstructural rotation $\mathbf{P}(\mathbf{x}°, t)$ to each material particle, with $\mathbf{x} = \mathbf{x}°$, $\mathbf{P} = \mathbf{I}$ in a given reference configuration, where $\mathbf{P} \in \mathbf{SO}(3)$ denotes a real, proper orthogonal tensor (cf. [34]).

Kinematics - rotation and spin

Following Mindlin [28, 29], in a slight departure from several previous works [30, 8], we employ a rotation relative to the material rather than to a fixed spatial frame. In either case, we can express the kinematics concisely in terms of the map $\mathbb{R}^3 \to \mathbb{R}^3 \times \mathbf{SO}(3)$ given by

$$\mathbf{x}° \to \{\mathbf{x}, \boldsymbol{\theta}\}, \text{ where } \boldsymbol{\theta} = -\tfrac{1}{2}\boldsymbol{\epsilon}:\boldsymbol{\Theta}, \text{ and } \boldsymbol{\Theta} = -\boldsymbol{\epsilon}\cdot\boldsymbol{\theta},$$
$$\text{i.e. } \theta_i = -\tfrac{1}{2}\epsilon_{ijk}\Theta^{jk}, \text{ and } \Theta_{ij} = -\epsilon_{ijk}\theta^k, \tag{16}$$

and the *Cayley-Gibbs-Rodrigues* relation [13, 35, 36, 37, 38, 39]

$$\mathbf{P} = \exp\boldsymbol{\Theta} = \mathbf{I} + \left(\frac{\sin\vartheta}{\vartheta}\right)\boldsymbol{\Theta} + \left(\frac{1-\cos\vartheta}{\vartheta^2}\right)\boldsymbol{\Theta}^2, \quad \vartheta = \{-\mathrm{tr}(\boldsymbol{\Theta}^2)/2\}^{1/2} \tag{17}$$

Here $\hat{\boldsymbol{\theta}} = \boldsymbol{\theta}/|\boldsymbol{\theta}|$ represents the axis of rotation and $\vartheta = |\boldsymbol{\theta}| = (\theta^i \theta_i)^{1/2}$ the angle of rotation about the axis.

If $\mathbf{P}$ is taken as primary variable, the skew-symmetric tensor $\boldsymbol{\Theta} = \log\mathbf{P} \in \mathfrak{so}(3)$ (*Lie algebra*) represents an inverse of the map $\mathfrak{so}(3) \to \mathbf{SO}(3)$ (*Lie group*), and it can be defined uniquely and computed by various methods [40]. Alternatively, and more conveniently, we may regard $\boldsymbol{\theta}$ as the primary variable, with (16) defining the associated map or *Cosserat placement* $\mathbb{R}^3 \to \mathbb{R}^6$.

To connect the vector of the *logarithmic spin* $\boldsymbol{\Omega} = d\boldsymbol{\Theta}/dt$ to that of the *instantaneous spin* $\boldsymbol{\Upsilon} = (d\mathbf{P}/dt)\mathbf{P}^\mathrm{T}$, where

$$\frac{d}{dt} \stackrel{\text{def}}{=} \left(\frac{\partial}{\partial t}\right)_{\mathbf{x}°}, \tag{18}$$

we recall the rather remarkable result of Kafadar and Eringen [30, Eqs.(2)-(9)], which can be expressed as:

$$\boldsymbol{\nu} \stackrel{\text{def}}{=} -\tfrac{1}{2}\boldsymbol{\epsilon}:\boldsymbol{\Upsilon} = \boldsymbol{\Gamma}\boldsymbol{\omega}, \text{ where } \boldsymbol{\omega} = -\tfrac{1}{2}\boldsymbol{\epsilon}:\boldsymbol{\Omega},$$
and
$$\boldsymbol{\Gamma} = \mathbf{I} + \left(\frac{1-\cos\vartheta}{\vartheta^2}\right)\boldsymbol{\Theta} + \left(\frac{\vartheta - \sin\vartheta}{\vartheta^3}\right)\boldsymbol{\Theta}^2, \text{ with } \boldsymbol{\Gamma}^{-1} = \mathbf{I} - \tfrac{1}{2}\boldsymbol{\Theta} + \frac{1}{\vartheta^2}\left(1 - \frac{\vartheta}{2}\cot\frac{\vartheta}{2}\right)\boldsymbol{\Theta}^2 \tag{19}$$

Either definition of spin is acceptable, and this relation makes it easy to connect their work-conjugate stresses[8].

It should be noted that (19) has also been derived by Shuster [42, Eqs. (276)-279][9], and that Iserle et al. [37, Eqs. B.10-B.11] present it as the differential of the Lie-group exponential, cf. [36, Eqs. 17-19]. It can also be obtained from the representation of rotations by quaternions [43].

---

[7] More general continua [31] are obtained by assigning additional kinematic variables to material points, the simplest example being Eringen's *microstretch* continuum [30] involving dilatation as well as rotation. Although sometimes proposed as a model of granular dilatancy [32, 33], the physical significance is not clear. In the typical granular system, the dilatation of individual grains is small compared to that of void space, which is already represented by a simple continuum with variable density or void ratio e.

[8] Owing no doubt to the group character of rotations, this relation is much simpler than that connecting the conjugate stress for logarithmic (*Hencky*) strain to Cauchy stress, as illustrated by the work Xiao et al. [41].

[9] pointed out to me by Professor Reuven Segev

Complex-variable representation

Starting from (16), one can describe Cosserat kinematics by means of a *complex placement*

$$\varkappa(\mathbf{x}^\circ) = \mathbf{x} + \imath\boldsymbol{\theta}, \tag{20}$$

leading to complex displacement and velocity field defined, respectively, by

$$\mathbb{u} = \mathbf{u} + \imath\boldsymbol{\theta}, \text{ with } \mathbf{u} = \mathbf{x} - \mathbf{x}^\circ \text{ and } \mathbb{v} = \frac{d\mathbb{u}}{dt} = \mathbf{v} + \imath\boldsymbol{\omega}, \text{ where } \mathbf{v} = \frac{d\mathbf{u}}{dt}, \text{ and } \boldsymbol{\omega} = \frac{d\boldsymbol{\theta}}{dt}, \tag{21}$$

More physically relevant quantities, particularly for small-strain theories, would be obtained by replacing $\boldsymbol{\theta}$ with $\xi\boldsymbol{\theta}$, where ξ is a material length scale, e.g. a particle diameter. However, in an abstract mathematical sense, the form (20) is equally valid, with the proviso that physically meaningful norms must involve a length scale, e.g.

$$|\mathbb{u}| = \sqrt{|\mathbf{u}|^2 + \xi^2|\boldsymbol{\theta}|^2} \tag{22}$$

which of course differs from the usual complex-variable modulus.

As a generalization of the standard deformation gradient $\mathbf{F}$ [14], we have:

$$\mathbb{F} = \partial\varkappa/\partial\mathbf{x}^\circ = \mathbf{F} + \imath\mathbf{G}, \text{ where } \mathbf{F} = \partial\mathbf{x}/\partial\mathbf{x}^\circ, \mathbf{G} = \partial\boldsymbol{\theta}/\partial\mathbf{x}^\circ \tag{23}$$

with corresponding rate[10]

$$\frac{d\mathbb{F}}{dt} = \mathbb{L}\mathbb{F}, \text{ with } \mathbb{L} = (\boldsymbol{\nabla}\mathbb{v})^T = (\mathbf{L} + \imath\mathbf{K}), \mathbf{L} = (\boldsymbol{\nabla}\mathbf{v})^T, \mathbf{K} = (\boldsymbol{\nabla}\boldsymbol{\omega})^T, \tag{24}$$

where the deformation gradient and Cosserat rotation satisfy the ODEs:

$$\frac{d\mathbf{F}}{dt} = \mathbf{L}\mathbf{F}, \frac{d\boldsymbol{\Theta}}{dt} = \boldsymbol{\Omega}, \text{ where } \boldsymbol{\Omega} = -\boldsymbol{\epsilon}\cdot\boldsymbol{\omega}, \text{ (i.e. } \Omega_{ij} = -\epsilon_{ijk}w^k) \tag{25}$$

We recall that $\mathbf{K}$ and its conjugate $\mathbf{M}$ are expressible, respectively, in terms of a third-rank wryness or curvature (cf. [29]) and double-stress.

The above relations yield the frame-indifferent description of Cosserat mechanics, with complex asymmetric deformation rate $\mathbb{D}$ and stress $\mathbb{T}$, and stress-power $\dot{w}$ defined, respectively, by

$$\mathbb{D} = \boldsymbol{\Lambda} + \imath\mathbf{K}, \text{ with } \boldsymbol{\Lambda} = \mathbf{D} + \boldsymbol{\Omega}, \mathbb{T} = \boldsymbol{\Sigma} + \imath\mathbf{M},$$
$$\text{and} \tag{26}$$
$$\dot{w} = \mathfrak{R}(\mathbb{T} : \mathbb{D}^*) = \boldsymbol{\Sigma}:\boldsymbol{\Lambda} + \mathbf{M}:\mathbf{K} = \mathbf{T}:\mathbf{D} + \mathbf{S}:\boldsymbol{\Omega} + \mathbf{M}:\mathbf{K} = \mathbf{T}:\mathbf{D} + \mathbf{s}\cdot\boldsymbol{\omega} + \mathbf{M}:\mathbf{K},$$

where

$$\mathbf{T} \stackrel{\text{def}}{=} \text{Sym}\,\boldsymbol{\Sigma}, \mathbf{S} \stackrel{\text{def}}{=} \text{Skw}\,\boldsymbol{\Sigma}, \mathbf{D} \stackrel{\text{def}}{=} \text{Sym}\,\mathbf{L}, \mathbf{s} = -\tfrac{1}{2}\boldsymbol{\epsilon}:\mathbf{S}, \tag{27}$$

with Sym and Skw denoting symmetric and skew-symmetric parts, respectively. Any of the tensors $\mathbf{Z} = \mathbf{D},\ldots,\mathbf{M}$, transforms as $\mathbf{Z}^+ = \mathbf{Q}\mathbf{Z}\mathbf{Q}^T$ under a change of frame $\mathbf{F}^+ = \mathbf{QF}$, where $\mathbf{Q}^{-1} = \mathbf{Q}^T$. Note that one can substitute the spin $\boldsymbol{\Upsilon}$ for $\boldsymbol{\Omega}$ by merely changing the definition of the conjugate stress $\mathbf{S}$.

Although we shall not make use of them in the present work, it is particularly noteworthy that the standard linear and angular momentum balances [28, 29, 30] can now be put into a remarkably compact form:

$$\boldsymbol{\nabla}\cdot\mathbb{T}^T + \mathbb{b} = \rho\mathbb{a}, \text{ with, } \mathbb{b} = \mathbf{b} + \imath(\mathbf{s} + \mathbf{c}), \text{ and } \mathbb{a} = \dot{\overline{\mathbf{v} + \imath\boldsymbol{\Xi}\boldsymbol{\omega}}}, \tag{28}$$

where $\mathbb{a}$ is complex acceleration, $\mathbf{b}$ and $\mathbf{c}$ represent external body-force and body-couple density, respectively, and $\boldsymbol{\Xi}$ is a rotational microinertia tensor, different from but related to that employed in [30]. Hence, insofar as the momemtum balance is concerned, Cosserat mechanics is merely *complex Cauchy* mechanics.

As indicated next, this same complex-variable formulation makes for a rather compact generalization of the preceding hypoplastic constitutive theory.

---

[10] The less conventional form $\varkappa = \varkappa(\varkappa^\circ, t)$, with $\mathbb{F} \stackrel{\text{def}}{=} \partial\varkappa/\partial\varkappa^\circ$, would represent classes of reference states involving different rotations $\boldsymbol{\theta}^\circ$ at each material point.

Elastoplastic forms

With real and imaginary components denoted, respectively, by subscripts $\Re$ and $\Im$, then

$$\mathbf{D}_\Re = \mathbf{\Lambda}, \ \mathbf{D}_\Im = \mathbf{K}, \ \mathbf{T}_\Re = \mathbf{\Sigma}, \ \mathbf{T}_\Im = \mathbf{M}, \tag{29}$$

and complex 4th-rank moduli are defined by:

$$\mathbb{\mu} = \boldsymbol{\mu}_\Re + \imath \boldsymbol{\mu}_\Im, \tag{30}$$

with a similar form for complex compliances, where it is understood that regular boldface denotes real quantities. Then, any *essentially real* linear map $\mathfrak{M}$, i.e., a map restricted to linear combinations with real coefficients, of complex 2nd-rank tensors $\mathbb{X}$ into complex 2nd-rank tensors $\mathbb{Y}$, can be represented by *two* complex moduli $\mathbb{\mu}_i = \boldsymbol{\mu}_{i\Re} + \imath \boldsymbol{\mu}_{i\Im}$, for $i = 1, 2$. Of the several possibilities, we choose

$$\mathbb{Y} = \mathfrak{M} : \mathbb{X} \stackrel{\text{def}}{=} \Re(\mathbb{\mu}_1 : \overline{\mathbb{X}}) + \imath \Re(\mathbb{\mu}_2 : \overline{\mathbb{X}}),$$

with matrix representation

$$\left[\begin{array}{c} \mathbf{Y}_\Re \\ \mathbf{Y}_\Im \end{array}\right] = \left(\begin{array}{cc} \boldsymbol{\mu}_{1\Re} & \boldsymbol{\mu}_{1\Im} \\ \boldsymbol{\mu}_{2\Re} & \boldsymbol{\mu}_{2\Im} \end{array}\right) : \left[\begin{array}{c} \mathbf{X}_\Re \\ \mathbf{X}_\Im \end{array}\right], \tag{31}$$

By an extension of the preceding terminology, we refer to the restricted linear operator $\mathfrak{M}$ as a "modulus", and, by matrix inversion, one then obtains a "compliance" $\mathfrak{R} = \mathfrak{M}^{-1}$. Quotation marks are employed here to emphasize that the terminology should, strictly speaking, be reserved for matrix representations like that in (31), which are necessary because of the well-known limitation on complex-variable representations[11].

With $\mathbb{X} = \mathbb{D} = \mathbf{\Lambda} + \imath \mathbf{K}$, (31) defines a pseudo-linear kinematic form, with associated (pseudo-)quadratic form:

$$w \stackrel{\text{def}}{=} \Re(\mathbb{D}^* : \mathfrak{M} : \mathbb{D}) = \mathbf{\Lambda}^T : \boldsymbol{\mu}_{1\Re} : \mathbf{\Lambda} + \mathbf{\Lambda}^T : (\boldsymbol{\mu}_{1\Im} + \boldsymbol{\mu}_{2\Re}^*) : \mathbf{K} + \mathbf{K}^T : \boldsymbol{\mu}_{2\Im} : \mathbf{K} \tag{32}$$

The requirement that $\mathfrak{M}$ be self-adjoint implies symmetry of the matrix in (31), with the consequence that

$$\boldsymbol{\mu}_{1\Im} = \boldsymbol{\mu}_{2\Re}^*, \ \boldsymbol{\mu}_{1\Re} = \boldsymbol{\mu}_{1\Re}^*, \ \boldsymbol{\mu}_{2\Im} = \boldsymbol{\mu}_{2\Im}^*, \tag{33}$$

and with corresponding reduction of the quadratic form (32). The further condition of positivity obviously requires *inter alia* the positivity of $\boldsymbol{\mu}_{1\Re}$ and $\boldsymbol{\mu}_{2\Im}$.

For the sake of completeness, Appendix B presents the isotropic form of (31)-(32), for which both $\boldsymbol{\mu}_{1\Im}$ and $\boldsymbol{\mu}_{2\Re}$ vanish, leaving only two moduli $\boldsymbol{\mu}_{1\Re} = \boldsymbol{\mu}_\Re$ and $\boldsymbol{\mu}_{2\Im} = \boldsymbol{\mu}_\Im$, say. The resulting block-diagonal form for the matrix in (31) leads to a similar form for the inverse, and the first relation in (31) reduces to the uncoupled form

$$\mathbb{Y} = \mathfrak{M} : \mathbb{X} = \boldsymbol{\mu}_\Re : \mathbf{X}_\Re + \imath \boldsymbol{\mu}_\Im : \mathbf{X}_\Im, \tag{34}$$

with corresponding simplification of (32).

Replacing $\mathbb{D}$ by $\mathbb{T} = \mathbf{\Sigma} + \imath \mathbf{M}$ and $\mathbb{\mu}$ by $\mathbb{\zeta}$ in (32), one obtains an explicit expression for the Cosserat version of the dissipative yield condition (5):

$$Y = |\mathbb{T}|_{\mathfrak{Z}}^2 = \Re(\mathbb{T}^* : \mathfrak{Z} : \mathbb{T}) = 1, \ \text{with} \ \mathfrak{Z} = \mathfrak{R}_P^* : \mathfrak{R}_P, \ \mathfrak{R}_P = \mathfrak{M}_P^{-1}, \tag{35}$$

Special cases of this appear in several past works [6, 44, 9], with some of the prior history given by Tejchman [2, Chapt. 3]. Although most of those special cases are based on specific constitutive models, they appear to represent variants on the general isotropic form inferred from (45) and cited below in Appendix B.

Hypoplasticity

With the above definitions, the evolution of stress in Cosserat hypoplasticity is given simply by

$$\overset{\circ}{\mathbb{T}} = \mathfrak{P}(\mathcal{X}, \hat{\mathbb{D}}) : \mathbb{D}, \ \text{with} \ \mathcal{X} = \{\mathbb{T}, \mathcal{A}\}, \ \hat{\mathbb{D}} = \mathbb{D}/|\mathbb{D}| \ \text{and} \ |\mathbb{D}| = (|\mathbf{\Lambda}|^2 + \xi^2 |\mathbf{K}|^2)^{1/2} \tag{36}$$

---

[11] The hybrid formulation employed here may nevertheless be useful for purposes of numerical computation, given the ability of current computer software to handle complex arithmetic.

where $\mathfrak{P}$ is defined as in (31) in terms of two complex moduli $\varpi_i = \pi_{i\mathfrak{R}} + \iota\pi_{i\mathfrak{J}}$, $i = 1, 2$, depending on $\mathcal{X}, \hat{\mathbb{D}}$. Here, we have chosen the simplest possible norm $|\mathbb{D}|$ involving a material length scale ξ referred to above, which we recall is associated with a representative particle diameter by other workers [7, 8, 21]. With the replacement of **D** by $\mathbb{D}$ in (11), (13) and (14) one obtains the complete parametric Cosserat hypoplasticity, for which some special cases are worth noting.

Whenever $\mathfrak{P}$ in (36) is independent of $\hat{\mathbb{D}}$, we recover Cosserat hypoelasticity, and whenever it is constant, independent of $\{\mathcal{X}, \hat{\mathbb{D}}\}$ we obtain a concise representation of linear Cosserat elasticity, as formulated e.g. by Mindlin [28, 29], simply by replacing stress rate by stress and deformation rate by infinitesimal strain. In this special case, the symmetric positive-definite form of (32) represents strain energy.

With the goal of achieving thermodynamically admissible forms, one can employ the following extended version of the incremental-elastoplastic form (6):

$$\mathfrak{P} = \mathfrak{M}_E - \beta(\mathfrak{M}_E : \mathfrak{K}_P : \mathbb{T}) \otimes \hat{\mathbb{D}} \qquad (37)$$

with positive-definite elastic modulus $\mathfrak{M}_E$ and plastic compliance $\mathfrak{K}_P$, and with yield condition of the form (35). In a similar way, the present formulation makes for a rather straight-forward extension of (8) to account for dilatancy.

For the isotropic case, Appendix B gives the general form of (36), with a more concrete form of (37), the latter of which involves a total of twelve functions representing various elasticities and compliances. At the time of this writing, the author has not attempted a detailed comparison of the latter against various empiricisms [12, 8, 9, 21, 2].

In closing, it should be noted that rate effects can be included in the above models, simply by allowing various moduli to depend on a non-dimensional variable such as $\tau|\mathbf{D}|$, where τ denotes a material time-scale, or by a more general viscoplastic model with viscous stresses of the form (1) added to plastic stresses of the type (5) [1].

CONCLUSIONS

The Abstract and Introduction represent an adequate summary of the present paper. It is hoped that this largely expository work will serve, among other things, to clarify the connections between hypoplasticity, polar as well as non-polar, and more traditional models of elastoplasticity, both of which are currently employed in the mechanics of granular materials. While explicit representations are given here for isotropic models, it would be useful to work out some of the necessary modifications to account for anisotropy based on evolutionary fabric or texture.

ACKNOWLEDGMENTS

I wish to acknowledge the support of the U.S. National Science Foundation and the influential ideas of the late Professor Ioannis Vardoulakis, particularly his visionary applications of structured continuum models to geomechanics.

REFERENCES

1. J. Goddard, *Granular Mat., accepted and to appear* (2009).
2. J. Tejchman, *Shear Localization in Granular Bodies with Micro-Polar Hypoplasticity*, Geomechanics and Geoengineering, Springer, Berlin; Heidelberg, 2008.
3. D. Kolymbas, *Introduction to Hypoplasticity*, A.A. Balkema, Rotterdam; Brookfield, 2000.
4. H. B. Mühlhaus, *Ingenieur Arch.* **56**, 389–99 (1986).
5. H. Mühlhaus, and I. Vardoulakis, *Géotechnique* **37**, 271–83 (1987).
6. J. Tejchman, and W. Wu, *Acta Mech.* **99**, 61–74 (1993).
7. J. Tejchman, and E. Bauer, *Comput. Geotech.* **19**, 221–44 (1996).
8. W. Huang, K. Nubel, and E. Bauer, *Mech. Mat.* **34**, 563–76 (2002).
9. S. Mohan, K. Rao, and P. R. Nott, *J. Fluid Mech.* **457**, 377–409 (2002).
10. F. Froiio, G. Tomassetti, and I. Vardoulakis, *Int. J. Solids Struct.* **43**, 7684–720 (2006).
11. P. Nott, *Acta Mech.* **205**, 151–60 (2009).
12. E. Bauer, *Mech. Cohes.-Frict. Mater.* **5**, 125–48 (2000).
13. C. Truesdell, and R. A. Toupin, "Principles of classical mechanics and field theory," in *Handbuch der Physik*, edited by S. Flügge, Springer, Berlin, 1960, vol. 3/1.
14. C. Truesdell, and W. Noll, *The non-linear field theories of mechanics*, Encyclopedia of Physics,v. III/3, Berlin ; New York : Springer-Verlag, 1965.
15. G. Gudehus, *Soils Found. (J. Jp. Geo. Soc.)* **36**, 1–12 (1996).

16. J. D. Goddard, *Adv. Coll. Interface Sci.* **17**, 241–62 (1982).
17. A. C. Pipkin, and R. S. Rivlin, *ZAMP* **16**, 313–26 (1965).
18. G. Gudehus, *Powder Tech.* **3**, 344–51 (1969).
19. D. Kolymbas, *Arch. Appl. Mech.* **61**, 143–51 (1991).
20. W. Wu, E. Bauer, and D. Kolymbas, *Mech. Mater.* **23**, 45–69 (1996).
21. W. Huang, and E. Bauer, *Int. J. Num. Anal. Meth. Geomech.* **27**, 325–52 (2003).
22. E. Bauer, W. Huang, and W. Wu, *Int. J. Solids Struct.* **41**, 5903–19 (2004).
23. R. Hill, *Proc. Roy. Soc.* **A 193**, 281–97 (1948).
24. V. C. Prantil, J. T. Jenkins, and P. R. Dawson, *J. Mech. Phys. Solids* **41**, 1357–82 (1993).
25. J. Sun, and S. Sundaresan, "A Plasticity model with microstructure evolution for dense granular flows," in *these proceedings*, 2009.
26. R. Hill, *J. Mech. Phys. Solids* **35**, 23–33 (1987).
27. J. Ericksen, and C. Truesdell, *Arch. Rat. Mech. Anal.* **1**, 295–323 (1957).
28. R. D. Mindlin, *Arch. Ratl. Mech. Anal.* **16**, 51–78 (1964).
29. S. C. Cowin, *Int. J. Solids Struct.* **6**, 389–98 (1970).
30. C. B. Kafadar, and A. C. Eringen, *Int. J. Eng. Sci.* **9**, 271–305 (1971).
31. J. D. Goddard, "From Granular Matter to Generalized Continuum," in *Mathematical models of granular matter*, edited by P. Mariano, G. Capriz, and P. Giovine, Springer, Berlin, 2008, vol. 1937 of *Lecture Notes in Mathematics*, pp. 1–20.
32. P. Giovine, "An Extended Continuum Theory for Granular Media," in *Mathematical models of granular matter*, edited by P. Mariano, G. Capriz, and P. Giovine, Springer, Berlin, 2008, vol. 1937 of *Lecture Notes in Mathematics*, pp. 167–92.
33. K.-C. Chen, J.-Y. Lan, and Y.-C. Tai, *Int. J. Solids Struct.* **46** (2009).
34. W. Pietraszkiewicz, and V. A. Eremeyev, *Int. J. Solids Struct.* **46**, 774–87,2477–80 (2009).
35. M. M. Mehrabadi, S. C. Cowin, and J. Jaric, *Int. J. Solids Struct.* **32**, 439–49 (1995).
36. J. Park, and W. K. Chung, IEEE *Trans. Robotics* **21**, 850–63 (2005).
37. A. Iserles, H. Z. Munthe-Kaas, S. P. Nørsett, and A. Zanna, *Acta num.* **9**, 215–365 (2000).
38. W. Spring, K, *Mech. Machine Theo.* **21**, 365–73 (1986).
39. O. A. Bauchau, and J.-Y. Choi, *Nonlin. Dyn.* **33**, 165–88 (2003).
40. J. Gallier, and D. Xu, *Int. J. Robot. Autom.* **17**, 1–11 (2002).
41. H. Xiao, O. T. Bruhns, and A. Meyers, *Acta Mech.* **168**, 21–33 (2004).
42. M. D. Shuster, *J. Aero. Sci.* **41**, 439–517 (1993).
43. J. D. Goddard, "A note on the representation of Cosserat rotation," in *Continuous Media with Microstructure*, edited by B. Albers, Springer, Berlin, 2010, vol. to appear.
44. H. Lippmann, *Appl. Mech. Revs.* **48**, 753–62 (1995).

APPENDICES

A. Plastic potentials

Considering first the case of non-polar plasticity, we note that, if there exists a plastic potential[12] $\psi(\mathbf{T})$, then it must be connected to a plastic compliance $\boldsymbol{\kappa}$ like that in (5) by

$$\hat{\mathbf{D}} = \partial_\mathbf{T}\psi/|\partial_\mathbf{T}\psi| = \boldsymbol{\kappa} : \mathbf{T} \tag{38}$$

The second equality is guaranteed by the existence of a (Legendre) complementary potential φ, provisionally regarded as a function of $\hat{\mathbf{D}}$, such that $\psi = \mathbf{T}:\partial_\mathbf{T}\varphi - \varphi$. However, whenever this relation is invertible, then we may regard φ as a function of $\mathbf{T}$, so that $\partial_\mathbf{T}\psi = \mathbf{T}:\partial_\mathbf{T}\partial_\mathbf{T}\varphi$, and (38) is satisfied by taking

$$\boldsymbol{\kappa} = \frac{\partial_\mathbf{T}\partial_\mathbf{T}\varphi}{|\mathbf{T}:\partial_\mathbf{T}\partial_\mathbf{T}\varphi|} \tag{39}$$

For example, in the case of an anisotropic (von Mises-Hill) potential [23, 26]:

$$\psi = \tfrac{1}{2}\mathbf{T}:\boldsymbol{\psi}:\mathbf{T} = \tfrac{1}{2}T_{ij}\psi^{ijkl}T_{kl}, \text{ with } \boldsymbol{\psi} = \boldsymbol{\psi}^* \text{ constant}, \Rightarrow \varphi \equiv \psi = \tfrac{1}{2}\hat{\mathbf{D}}:\boldsymbol{\phi}:\hat{\mathbf{D}} = \tfrac{1}{2}\mathbf{T}:\boldsymbol{\psi}:\mathbf{T}, \text{ with } \boldsymbol{\phi} = \boldsymbol{\psi}^{-1}, \tag{40}$$

and it is easy to show that $\boldsymbol{\kappa} = \boldsymbol{\psi}$ and, hence, that $\boldsymbol{\mu} = \boldsymbol{\kappa}^{-1} = \boldsymbol{\phi}$. The positivity of $\boldsymbol{\kappa}$ obviously implies the convexity of φ regarded as function of $\mathbf{T}$.

---

[12] The scalar $\psi(\mathbf{T})$ must of course be given as a function of an appropriate set of joint isotropic scalar invariants of $\mathbf{T}$ and a set of tensors sufficient to define the anisotropy. For the Cosserat extension, the joint invariants of $\mathbf{T}$ are replaced by those of $\boldsymbol{\Sigma}, \mathbf{M}$.

The case of Cosserat plasticity is formally covered simply by replacing $\mathbf{D}$ by $\mathbb{D}$ and $\mathbf{T}$ by $\mathbb{T}$ in (38), with

$$\partial_{\mathbb{T}} = \partial_{\Sigma} - \iota\partial_{\mathbf{M}}, \tag{41}$$

and with suitable modifications of (39), which are not recorded here.

B. Isotropic forms

With a slight change of notation, the expression of Cowin [29] for the isotropic form of $\mathbf{Y} = \boldsymbol{\mu}:\mathbf{X}$ given by Mindlin [28] can be written

$$\mathbf{Y} = (\mu + \tau)\mathbf{X} + (\mu - \tau)\mathbf{X}^T + \lambda \mathrm{tr}(\mathbf{X})\mathbf{I}, \text{ i.e. } \mu_{ijkl} = (\mu + \tau)\delta_{ik}\delta_{jl} + (\mu - \tau)\delta_{il}\delta_{jk} + \lambda\delta_{ij}\delta_{kl}, \tag{42}$$

from which it follows that

$$\mathrm{Sym}\,\mathbf{Y}' = 2\mu\,\mathrm{Sym}\,\mathbf{X}', \quad \mathrm{Skw}\,\mathbf{Y} = 2\tau\,\mathrm{Skw}\,\mathbf{X}, \quad \mathrm{tr}(\mathbf{Y}) = (\tfrac{2}{3}\mu + \lambda)\,\mathrm{tr}(\mathbf{X}) \tag{43}$$

representing the orthogonal decomposition into symmetric, skew and spherical parts. The same orthogonality gives the *normal* quadratic form

$$\mathbf{X}^T:\boldsymbol{\mu}:\mathbf{X} = 2\mu|\mathrm{Sym}\,\mathbf{X}'|^2 + 2\tau|\mathrm{Skw}\,\mathbf{X}|^2 + (\tfrac{2}{3}\kappa + \nu)|\mathrm{tr}(\mathbf{X})|^2, \tag{44}$$

and the positivity of $\boldsymbol{\mu}$ obviously requires the positivity of $\mu, \tau, \tfrac{2}{3}\mu + \lambda$. Such positivity is required of strain energy when $\boldsymbol{\mu}$ represents an elastic modulus, or for dissipation when it represents a viscous or plastic modulus.

Also, the inverse form $\mathbf{Y} = \boldsymbol{\kappa}:\mathbf{X}$ follows immediately from (42)-(43) by means of the substitutions:

$$\mathbf{X} \leftrightarrow \mathbf{Y}, \quad \boldsymbol{\mu} \to \boldsymbol{\kappa}, \text{ with } \{\mu, \tau, \lambda\} \to \{\kappa, \sigma, \nu\}$$

and (43) readily gives scalar compliances κ, σ, ν in terms of moduli μ, τ, λ, and vice versa. Without recording these results, we note that self-adjointness, $\boldsymbol{\mu}^* = \boldsymbol{\mu}$ and, hence, $\boldsymbol{\kappa}^* = \boldsymbol{\kappa}$, is also evident from (42).

Hence, another normal quadratic form

$$\mathbf{X}^T:\boldsymbol{\zeta}:\mathbf{X} = 4\kappa^2|\mathrm{Sym}\,\mathbf{X}'|^2 + 4\sigma^2|\mathrm{Skw}\,\mathbf{X}|^2 + (\tfrac{2}{3}\kappa + \nu)^2|\mathrm{tr}(\mathbf{X})|^2, \text{ where } \boldsymbol{\zeta} = \boldsymbol{\kappa}^*:\boldsymbol{\kappa} = \boldsymbol{\kappa}^2 \stackrel{\text{def}}{=} \boldsymbol{\kappa}:\boldsymbol{\kappa}, \tag{45}$$

follows immediately upon two-fold application of the compliance-form of (43) to calculate $\boldsymbol{\kappa}^2$. It is further evident from the now diagonal matrix representation in (31) that

$$\mathfrak{Z} = \mathfrak{K}^*:\mathfrak{K} = \mathfrak{K}^2 = \boldsymbol{\kappa}_{\mathfrak{R}}^2 + \iota\boldsymbol{\kappa}_{\mathfrak{J}}^2, \text{ hence } \mathfrak{R}(\mathbb{X}^*:\mathfrak{Z}:\mathbb{X}) = \mathbf{X}_{\mathfrak{R}}^T:\boldsymbol{\kappa}_{\mathfrak{R}}^2:\mathbf{X}_{\mathfrak{R}} + \mathbf{X}_{\mathfrak{J}}^T:\boldsymbol{\kappa}_{\mathfrak{J}}^2:\mathbf{X}_{\mathfrak{J}} \tag{46}$$

Therefore, the isotropic form of (35) is given simply as the sum of two terms of the form (45), the first obtained by taking $\mathbf{X} = \mathbf{X}_{\mathfrak{R}} = \boldsymbol{\Sigma}$, $\boldsymbol{\kappa} = \boldsymbol{\kappa}_{\mathfrak{R}}$, and the second by taking $\mathbf{X} = \mathbf{X}_{\mathfrak{J}} = \mathbf{M}$, $\boldsymbol{\kappa} = \boldsymbol{\kappa}_{\mathfrak{J}}$ with κ, σ, ν replaced, respectively, by rotational compliances, say, $\check{\kappa}, \check{\sigma}, \check{\nu}$.

Furthermore, with $\boldsymbol{\mu}(\mu, \tau, \lambda)$ defined in (42), we immediately obtain the linear isotropic form of (36) by means of (34) and (36):

$$\overset{\circ}{\boldsymbol{\Sigma}} = \boldsymbol{\pi}_{\mathfrak{R}}:\boldsymbol{\Lambda}, \text{ and } \overset{\circ}{\mathbf{M}} = \boldsymbol{\pi}_{\mathfrak{J}}:\mathbf{K}, \tag{47}$$

with

$$\boldsymbol{\pi}_{\mathfrak{R}} = \boldsymbol{\mu}(\mu, \tau, \lambda) \text{ and } \boldsymbol{\pi}_{\mathfrak{J}} = \boldsymbol{\mu}(\check{\mu}, \check{\tau}, \check{\lambda}) \tag{48}$$

which are uncoupled, linear versions of (9), involving six elasticities[13] $\mu, \lambda, \tau, \check{\mu}, \check{\tau}, \check{\lambda}$.

The nonlinear isotropic version of parametric Cosserat hypoplasticity is also given by (48), with moduli $\boldsymbol{\pi}_{\mathfrak{R}}, \boldsymbol{\pi}_{\mathfrak{J}}$ depending on $\mathbb{T}, \mathcal{A}, \hat{\mathbb{D}}$, where $\mathcal{A}$ represents a set of evolutionary scalar parameters. The resulting equations are now generally coupled, because of their joint dependence on $\mathbb{T} = \boldsymbol{\Sigma} + \iota\mathbf{M}, \mathcal{A}, \hat{\mathbb{D}} = \boldsymbol{\Lambda} + \iota\mathbf{K}$.

If one further adopts the incremental elastoplastic form (6), then each of the moduli $\boldsymbol{\pi}_{\mathfrak{R}}, \boldsymbol{\pi}_{\mathfrak{J}}$ is defined by its own isotropic elastic modulus $\boldsymbol{\mu}_E$ and plastic compliance $\boldsymbol{\kappa}_P$, of the type defined above, which in turn are given by a total of twelve elastic moduli and plastic compliances $\mu_E, \ldots, \nu_P$ and $\check{\mu}_E, \ldots, \check{\nu}_P$. The latter must in general be treated as functions of evolutionary scalar parameters $\mathcal{A}$ and the joint isotropic scalar invariants of $\mathbb{T}, \mathbb{D}$.

---

[13] Which are easily related to Cowin's [29] moduli $\mu, \lambda, \tau, \eta, \eta', \alpha$. The notation employed here leads to appreciably simpler conditions for the positivity of quadratic forms such as (44), with conditions on $\check{\mu}, \check{\tau}, \check{\lambda}$ identical with those on μ, τ, λ.

The stress in a slowly sheared granular column

Vishwajeet Mehandia and Prabhu R. Nott

Department of Chemical Engineering, Indian Institute of Science, Bangalore 560012, India.

Abstract. We report measurements of the wall stress in a granular material sheared in a cylindrical Couette cell, as a function of the distance from the free surface. Our results shows that when the material is static, all components of the stress saturate to constant values within a short distance from the free surface, in conformity with earlier experiments and theoretical predictions. When the material is sheared by rotating the inner cylinder at a constant rate, the stresses are remarkably altered. The radial normal stress does not saturate, and increases even more rapidly with depth than the linear hydrostatic pressure profile. The axial shear stress changes sign on shearing, and its magnitude increases with depth. These results are discussed in the context of the predictions of the classical and Cosserat plasticity theories.

Keywords: granular flow, granular rheology, stress measurement
PACS: 83.80.Fg,83.50.Ax,83.10.Gr

INTRODUCTION

It has been known for several centuries [1] that the normal stress at the base of a column of granular material deviates from the value dictated by the hydrostatic balance. This deviation was explained by Janssen [2] as being due to the shear stress imposed by the confining walls on the granular column, as a result of grain-wall friction. Some of the simplifying assumptions made in Janssen's analysis were dispensed with by Walker [3] and Cowin [4], but they both showed that the principal features of the Janssen solution remain unchanged: the normal and shear stresses at the wall rise with distance from the free surface, but saturate asymptotically to constant values far from the free surface.

Understanding the static stress in a granular column is an important problem in industry, such as in the design of bins and bunkers for the storage of food grains and other granular materials. Equally important is the understanding of the stress field in continuously sheared columns, such as in the emptying of bins, moving bed catalytic reactors etc. Indeed, the dynamic stress during flow is much larger than the static stress, and exhibits large fluctuations; it is therefore critical in the design of equipment. Despite its importance, the stresses developed by flowing granular materials is not well understood, as the rheology of granular materials is complex. To further complicate matters, there are regions of negligible deformation rate, raising the question of whether elastic deformations contribute to the stress.

Historically, granular flow has primarily been studied and analyzed in the two extreme regimes of dense, slow flow and loose rapid flow. For the former, theories derived from soil mechanics and metal plasticity have been applied, and theories for the latter have been derived by extending the kinetic theory of gases to account for grain inelasticity and roughness. This paper is concerned only with the regime of slow, dense flow.

There have been numerous experimental measurements of the forces on immersed bodies and stresses on confining walls in the slow flow regime [5, 6, 7, 8]. Some other studies have determined the stress using used DEM simulations [9, 10]. The common theme in all the observations is that during flow, there are significant fluctuations in the stress. The mean stress during flow is generally higher than the static stress. Most of the studies have measured the stress at a single spatial point in the flow. In an attempt to determine the stress as a function of the height of a sheared column, Tardos et al. [7] determined the torque required to rotate the inner cylinder in a cylindrical Couette device for various heights. They observed a quadratic dependence of the torque with the bed height, and concluded that the shear stress varies linearly with distance from the free surface. Later, Tardos et al. [11] measured the pointwise radial and axial normal stresses, but they used sensors that intruded into the flow. They too reported a linear dependence of normal stresses with depth.

Many studies have conducted theoretical analyses of flow in bins and bunkers using plastic and hypoplastic [12] constitutive relations for the stress. In rectangular bins, the more accurate solution of the equations by the method of characteristics [13] agrees reasonably well with the Janssen solution. Classical plasticity theories predict no deformation in the bulk at steady state, thereby making no distinction between a state of incipient yield and that of finite slip. Hence, the stress field is the same whether the material is static or flowing. However, theories that involve a material length scale, such as some simple Cosserat plasticity and higher-gradient theories [see, e.g., 14], predict a finite strain rate in the material adjacent to the wall; what is more relevant here is that they predict that all the normal stresses increase linearly with distance from the free surface.

In this paper, we address the issue of the difference in the stress field between static and sheared columns of granular materials. We report the results of our experimental study of shear of a granular column in a cylindrical Couette cell, in which all components (one normal and two shear) of the traction on the stationary outer cylinder were measured. The imposed shear rate was small enough that the granular material was well within the dense, slow flow regime. The stress was measured as a function of distance from the free surface. The results of our experiments are intriguing: the stress profile differs fundamentally from that of a sheared fluid column and a static granular column. The data presented here are the preliminary results of an ongoing study.

Before presenting our results, we set the context of our study by discussing the Janssen-Walker solution for a static column, and the predictions of plasticity models for a sheared column, so that we may compare our results with these.

PREDICTIONS OF THE STRESS FIELD IN STATIC AND SHEARED COLUMNS

Statics

Consider a column of granular material at rest in a bin of rectangular cross section $2W \times D$. For ease of exposition, we consider the dimension D in the x direction to be infinite, so that the stress varies only along the y (lateral) and z (vertical) directions. We also assume that the density ρ is a constant. The y and z components of the momentum balance then are

$$\frac{\partial \sigma_{yy}}{\partial y} + \frac{\partial \sigma_{zy}}{\partial z} = 0, \tag{1}$$

$$\frac{\partial \sigma_{yz}}{\partial y} + \frac{\partial \sigma_{zz}}{\partial z} = \rho g, \tag{2}$$

where ρ is the density of the granular material and g is the acceleration due to gravity. Integrating (2) from the centerline ($y = 0$) to the wall ($y = W$), we obtain

$$\frac{d}{d\bar{z}} \langle \overline{\sigma}_{zz} \rangle = 1 - \sigma_{yz}^w \tag{3}$$

where the overline indicates that the variables have been rendered dimensionless, by scaling lengths with W and stresses with $\rho g W$, the angle brackets indicate a lateral average, and the superscript w indicates a property at the wall. Here, we have assumed symmetry about the centerline, thereby imposing the condition of zero shear stress $\overline{\sigma}_{yz} = 0$ at the centerline. Janssen [2] (see [15] for translation) made the assumptions that σ_{zz} is constant across the cross section, and σ_{yz} is constant across the perimeter; Walker [3] and Cowin [4] made less restrictive assumptions, and showed that the qualitative features of the Janssen solution remain. In what follows, we describe the Walker's extension of the Janssen solution.

At the wall, the friction boundary condition

$$\sigma_{yz}^w = \sigma_{yy}^w \tan \delta \tag{4}$$

is assumed to hold, where δ is the angle of wall friction. The problem is closed with the further assumption that

$$\sigma_{yy}^w = K \langle \overline{\sigma}_{zz} \rangle \tag{5}$$

where K is a constant, called the Janssen coefficient. Walker [3] determined the value of K by assuming that the material adjacent to the wall is yielding or in a state of incipient yield, and satisfies the Mohr-Coulomb condition. Two solutions are obtained,

$$K = K_a \equiv \frac{1 - \sin\phi \cos(\psi - \delta)}{1 + \sin\phi \cos(\psi - \delta)}, \quad K = K_p \equiv \frac{1 + \sin\phi \cos(\psi + \delta)}{1 - \sin\phi \cos(\psi + \delta)} \tag{6}$$

where ϕ is the angle of internal friction, $\psi \equiv \sin^{-1}(\sin\delta/\sin\phi)$, and K_a and K_p correspond to the so-called active and passive states of stress [16, p. 210].

Substituting (4) and (5) in (3), and integrating, we obtain

$$\langle \overline{\sigma}_{zz} \rangle = \frac{1}{K \tan \delta} [1 - \exp(-\overline{z} K \tan \delta)] \tag{7}$$

Thus, the shear and normal stresses exerted on the wall increase with the depth, and level off at large depths. The limiting value of the dimensionless shear stress is unity – in other words, the shear force exerted on a horizontal slice of material by the wall balances its weight. Thus, wall roughness is essential for the stress to level off as the depth increases.

If friction is not fully mobilized at the wall, (4) must be replaced with the inequality

$$\sigma_{yz}^w \leq \sigma_{yy}^w \tan \delta, \tag{8}$$

and the yield condition in the bulk is not satisfied. This implies that K is not given by (6), but is an arbitrary constant, and that (7) only provides a lower bound for $\langle \overline{\sigma}_{zz} \rangle$ [4].

Though the Janssen solution is an approximate one, it is in reasonable agreement with the data for wall stress in bins and bunkers [4]. Moreover, the more rigorous solution obtained by using the method of characteristics to solve the balances of momentum, yield condition and flow rule [13], agrees quite well with the Janssen solution [16].

Sheared column

As mentioned in the introductory section, classical plasticity theories make no distinction between a state of incipient yield and that of a sheared material, as they predict no deformation in the bulk at steady state – the material slips at the walls [16]. The thin shear layers adjacent to the walls that are observed in experiments are not captured by these theories. Hence, the dynamic stress field during flow is identical to the static stress field, described above. Tardos et al. [7, 11] used a simple plasticity theory and found that the pressure is linear in z and the shear stress σ_{yz} vanishes; however their solution is incorrect, as it does not satisfy the x momentum balance.

The Cosserat plasticity theory of Mohan et al. [14] captures the thin shears; in what follows, we describe their solution briefly, and refer the reader to the original paper for details. We consider again the semi-infinite column, but with one wall moving in the x direction at constant velocity, and the other kept stationary. Since the geometry is of infinite extent in the x direction, and under the assumption that the velocity profile is independent of z, we have

$$v_x \equiv v_x(y), \quad v_y = v_z = 0, \quad \omega_x = \omega_y = 0, \quad \omega_z \equiv \omega_z(y) \tag{9}$$

where $\mathbf{v}$ is the velocity and ω the mean particle spin. For a classical continuum, ω equals the material spin $\mathbf{w} \equiv \frac{1}{2} \nabla \times \mathbf{v}$, or half the vorticity, but the two can differ in a Cosserat continuum. We assume that the associated flow rule and the Cosserat extension of the Drucker-Prager yield condition [14] hold. The diagonal components of the flow rule yield

$$\sigma_{xx} = \sigma_{yy} = \sigma_{zz} = p. \tag{10}$$

Thus, the normal stresses are isotropic, as in a fluid. The equality of the normal stresses in viscometric flows is a flaw in the model [14], as it is well known that there are normal stress differences – this flaw could be remedied by the use of a more sophisticated yield condition and flow rule.

From the yz and zy components of the flow rule, it follows that $\sigma_{yz} = \sigma_{zy} = 0$. It then follows from (2) that

$$p = \rho g z, \tag{11}$$

i.e. the pressure follows the hydrostatic balance. The shear stresses σ_{xy} and σ_{yx} are, in general unequal, and the couple stress M_{yz} is non-zero. These details are not given here, as we are primarily interested in the normal stress σ_{yy} and the shear stress σ_{yz} at the walls, but they may be found in Mohan et al. [14].

Thus, the profiles of the normal stresses and the vertical shear stress on the wall differ significantly from the respective profiles in a static column. As stated in the introductory section, Tardos et al. [11] found the normal stresses to vary linearly with z, and their indirect measurements [7] of the shear stress indicates that it too varies linearly with z. The former is in agreement with the predictions of the Cosserat plasticity model, but the latter is at variance. We note that the Cosserat model of Mohan et al. [14] is a relatively simple extension of classical plasticity; more complex Cosserat formulations are, no doubt, possible, and may perhaps better represent the observed behaviour.

In an effort to resolve the discrepancies between the various theoretical predictions and the experimental observations, we undertook our experimental study, the details of which follow.

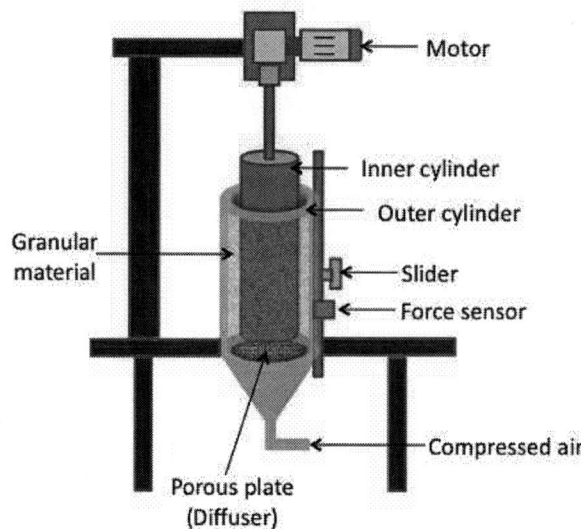

FIGURE 1. The cylindrical Couette apparatus used in the experiment. The diameters of the inner and outer cylinders are 10 cm and 15 cm, respectively. The stress on the outer cylinder was measured using the force sensor.

EXPERIMENTAL MEASUREMENT OF THE STRESS IN CYLINDRICAL COUETTE FLOW

Experiments were conducted in a cylindrical Couette apparatus, shown in Fig. 1a. It consists of two vertical, coaxial cylinders. The outer cylinder (diameter 15 cm) was kept stationary, and the inner cylinder (diameter 10 cm) rotated at a constant rate. The gap between the cylinders was filled with the granular material, here glass beads of 0.8 mm mean diameter, up to a height of $H = 19$ cm. A force transducer capable of measuring three orthogonal components of the force was used to measure the stress on the outer cylinder. The transducer was coupled to a rigid transmitter, which was inserted through a circular aperture on the slider; the transducer-transmitter assembly was mounted firmly on the slider (Fig. 2). The measuring surface of the transmitter and the slider were both flush against the inner surface of the outer cylinder, so that the inner surface of the outer cylinder formed a smooth cylindrical surface. A gap of $\approx 100\,\mu$m was maintained between the transmitter and the aperture in the slider, to ensure that the entire force acting on the measuring surface was transmitted to the transducer. By moving the slider up and down the outer cylinder, the stresses on the outer wall as a function of vertical position was measured. The outer wall of the inner cylinder, the inner wall of outer cylinder, the slider and the measuring surface of the transmitter were all coated with sand paper to increase the roughness of the walls. The granular column could be aerated through a porous diffuser plate at the base of the Couette assembly; this allowed uniform distribution of air from the blower to the granular column. The components of the stress measured are the radial normal stress σ_{rr}, the azimuthal shear stress $\sigma_{r\theta}$, and the axial shear stress σ_{rz}.

Procedure

Each experimental run started with positioning of the slider such that the transducer was at the lowest point in the Couette cell. The Couette gap was then filled with the glass beads, and the material was aerated at minimum fluidization. The purpose of aeration is to get a uniform packing and a reproducible microstructure, and also clear the

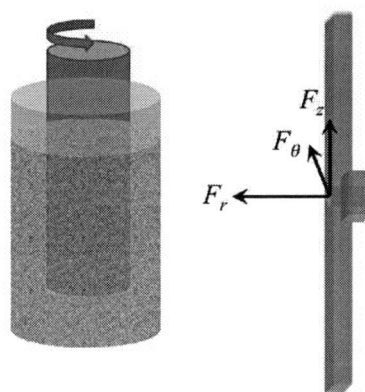

FIGURE 2. An exploded schematic of the slider, transducer and Couette assembly. When assembled, the inner surface of the slider is flush against the inner surface of the outer cylinder. The arrows indicate the coordinate frame of the transducer.

fabric history formed during filling. After aeration, the material was sheared until a steady state was reached, typically about 30 minutes. At the end of this period, compaction of the column by ≈ 0.5 cm was observed. The forces were then recorded for about 5 min. The slider was then moved upward to the next location and the forces recorded for about 5 min. This procedure was repeated until the free surface was reached. Significant fluctuations in the stress were observed (see Fig. 5b); we report here the time-average at each location.

RESULTS

To validate our measurements, we first determined the stress profile for a static column. In the figures below, we have used the cylindrical coordinates (r, θ, z), where z is the vertical axis pointing in the direction of gravity. Its is evident from Fig. 3 that the normal and shear stresses saturate beyond a certain distance from the free surface. The Janssen-Walker solution (equations (6) and (7)) for a narrow-gap cylindrical Couette cell is shown alongside for comparison. It is seen in Fig. 3 that beyond a depth of about 14 cm, σ_{rr} decreases suddenly. This is probably an influence of the base, which supports a part of the weight of the granular column. However, on the whole the agreement between our measurements and the Janssen-Walker solution appears to be satisfactory. Our data for σ_{rr} fall between the active and passive solutions, but the magnitude of σ_{rz} is less than both the solutions.

We now present the measurements of the normal stress during shear. Figure 4 shows the radial normal stress as a function of z alongside the predictions of the classical plasticity theory (the Janssen solution) and the Cosserat plasticity theory (linear hydrostatic profile). It is immediately clear that the profile of σ_{rr} during shear is very different from the Janssen solution; the variation is much more rapid, and there is no indication of a saturation. The stress is also much higher that the hydrostatic pressure profile predicted by the Cosserat plasticity theory.

Figure 5a shows the axial shear stress σ_{rz} as a function of z, alongside the Janssen solution. Recall that the Cosserat plasticity theory predicts $\sigma_{rz} = 0$. A surprising feature of our data is that the shear stress on the wall now acts upwards, which is the opposite of the static stress. The magnitude of the shear stress broadly increases with z, but there are spatial fluctuations, which grow in magnitude with increasing z. It is not clear if the fluctuations are an influence of the base, or due to secondary flows in the Couette gap. As already stated, the stress at a given location exhibits significant temporal fluctuations (Figure 5b). However, the stress always remains positive.

As noted earlier, Tardos et al. [7, 11] report a linear variation of the normal and shear stresses with depth in a sheared column, whereas our data for the normal stress deviate quite dramatically from the linear hydrostatic profile. The reason for the discrepancy is not immediately clear; indeed, the dimensions of the cylindrical Couette device and the size of the particles used in our study and by Tardos *et al.* are quite similar. An important difference is the manner in which the stresses were measured: Tardos et al. [7] inferred linearity of the shear stress with depth from measurements of torque on the inner cylinder, which is questionable – the torque gives only the average shear stress over the entire cylinder. Tardos et al. [11] made local measurements of the normal stresses, but used sensors that intruded into the

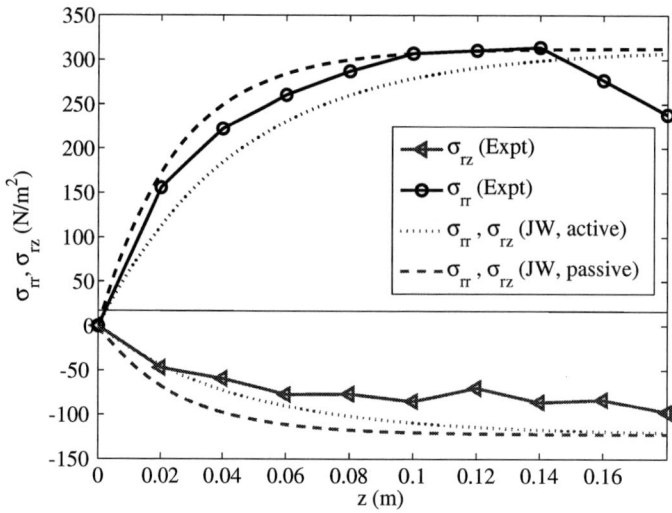

FIGURE 3. Data for the radial normal stress σ_{rr} and the axial shear stress σ_{rz} on the wall in a static column. The Janssen-Walker solution (JW) is shown for comparison. The parameters for the Janssen-Walker solution are $\phi = 23.3°$ and $\delta = 21.4°$. Note that the σ_{rz} is negative, i.e. the material exerts a shear stress on the wall in the direction of gravity.

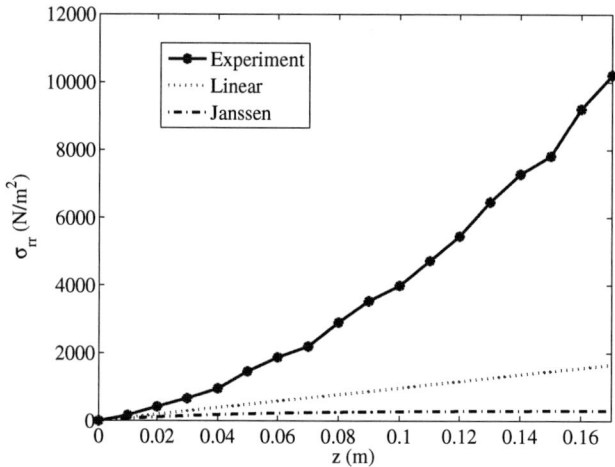

FIGURE 4. The radial normal stress σ_{rr} as a function of z for a column sheared at a shear rate of $0.21\,\text{s}^{-1}$. The Janssen solution for a static column and the linear hydrostatic profile are shown for comparison.

flow, which may have affected their measurements.

CONCLUSIONS

For a fluid sheared in a cylindrical Couette device, the normal stresses follow the hydrostatic balance, and there is no shear stress in the vertical direction on the walls. Our experiments show that the profile of the wall stress in a cylindrical Couette cell differs quite dramatically from the stresses in a static column, from the predictions of the classical and Cosserat plasticity theories for granular materials, and from prior experimental data [7, 11]. We note that these are preliminary measurements; a comprehensive study in which the accuracy, repeatability, effects of the shear rate, Couette gap and column height is underway, the results of which will be reported shortly. However, we

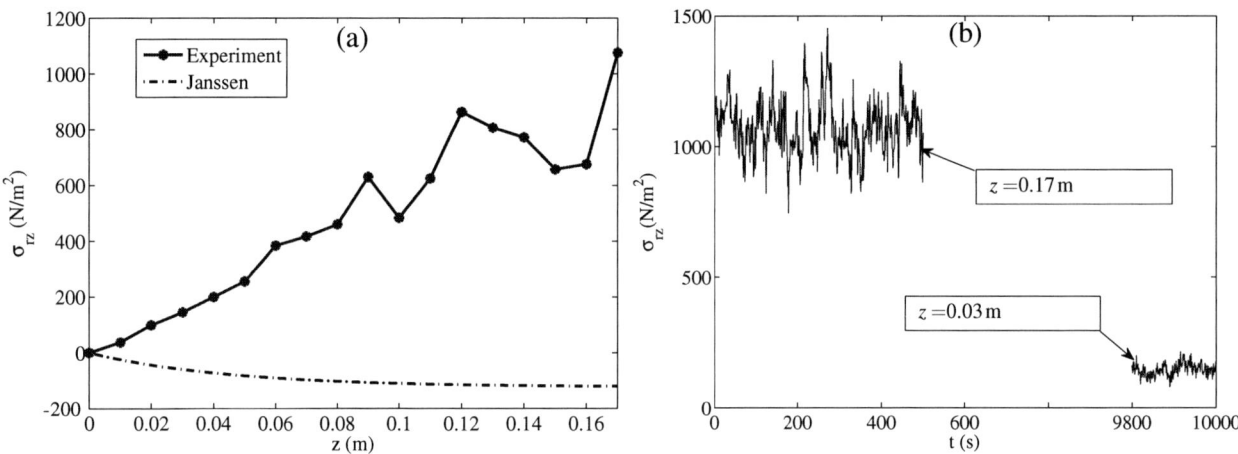

FIGURE 5. (a) The axial shear stress σ_{rz} as a function of z for a column sheared at a shear rate of $0.21\,\text{s}^{-1}$. The Janssen solution is shown for comparison. (b) Time trace of σ_{rz} at two vertical positions on the outer cylinder.

believe that the data presented here is sufficient to convince the reader that the behaviour of a granular column in a simple viscometric flow is remarkably different from that of classical fluids, and hence provoke a re-assessment of the constitutive relations. It is possible that a constitutive relation that incorporates a more sophisticated yield condition and flow may rule represent the data.

ACKNOWLEDGMENTS

This work was funded by the Department of Science and Technology, India, and the Indo-French Centre for the Promotion of Advanced Scientific Research. We have benefitted from discussion with K. Kesava Rao and Olivier Pouliquen. PRN is grateful to the organizers of the IUTAM-ISIMM symposium for partial travel support.

REFERENCES

1. P. Delanges, *Memoare de matematica e fisica della societa Italiana* (1788).
2. H. A. Janssen, *Z. Ver. Deut. Ing.* **39**, 1045–49 (1895).
3. D. M. Walker, *Chem. Engng Sci.* **21**, 975–97 (1966).
4. S. C. Cowin, *J. Appl. Mech.* **44**, 409–12 (1977).
5. K. Wieghardt, *Annu. Rev. Fluid Mech.* **7**, 89–114 (1975).
6. U. Tüzün, and R. M. Nedderman, *Chem. Engng Sci.* **40**, 325–36 (1985).
7. G. Tardos, M. Khan, and D. Schaeffer, *Phys. of Fluids* **10**, 335–41 (1998).
8. D. Howell, R. P. Behringer, and C. Veje, *Phys. Rev. Lett.* **82**, 5241–44 (1999).
9. J. M. Rotter, J. M. F. G. Holst, J. Y. Ooi, and A. M. Sanad, *Phil. Trans. R. Soc. Lond. A* **356**, 2685–712 (1998).
10. C. Thornton, *Géotechnique* **50**, 43–53 (2000).
11. G. Tardos, S. McNamara, and I. Talu, *Powder Tech.* **131**, 23–39 (2003).
12. J. Eibl, and G. Rombach, "Numerical investigations on discharging silos," in *Numerical Methods in Geomechanics*, edited by Swoboda, Balkema, Rotterdam, 1988.
13. R. M. Horne, and R. M. Nedderman, *Powder Technol.* **14**, 93–102 (1976).
14. L. S. Mohan, K. K. Rao, and P. R. Nott, *J. Fluid Mech.* **457**, 377–409 (2002).
15. M. Sperl, *Granular Matter* **8**, 59–65 (2006).
16. K. K. Rao, and P. R. Nott, *An Introduction to Granular Flow*, Cambridge University Press, New York, 2008.

SEGREGATION

Particle size segregation in granular avalanches: A brief review of recent progress

J.M.N.T. Gray

*School of Mathematics and Manchester Centre for Nonlinear Dynamics,
University of Manchester, Manchester, M13 9PL, United Kingdom*

Abstract. Hazardous natural flows such as snow avalanches, debris-flows, lahars and pyroclastic flows are part of a much wider class of granular avalanches, that frequently occur in industrial processes and in our kitchens! Granular avalanches are very efficient at sorting particles by size, with the smaller ones percolating down towards the base and squeezing the larger grains up towards the free-surface, to create inversely-graded layers. This paper provides a short introduction and review of recent theoretical advances in describing segregation and remixing with relatively simple hyperbolic and parabolic models. The derivation from two phase mixture theory is briefly summarized and links are drawn to earlier models of Savage & Lun and Dolgunin & Ukolov. The more complex parabolic version of the theory has a diffusive force that competes against segregation and yields S-shaped steady-state concentration profiles through the avalanche depth, that are able to reproduce results obtained from particle dynamics simulations. Time-dependent exact solutions can be constructed by using the Cole-Hopf transformation to linearize the segregation-remixing equation and the nonlinear surface and basal boundary conditions. In the limit of no diffusion, the theory is hyperbolic and the grains tend to separate out into completely segregated inversely graded layers. A series of elementary problems are used to demonstrate how concentration shocks, expansion fans, breaking waves and the large and small particles paths can be computed exactly using the model. The theory is able to capture the key features of the size distribution observed in stratification experiments, and explains how a large particle rich front is connected to an inversely graded avalanche in the interior. The theory is simple enough to couple it to the bulk flow field to investigate *segregation-mobility* feedback effects that spontaneously generate self-channelizing leveed avalanches, which can significantly enhance the total run-out distance of geophysical mass flows.

Keywords: avalanches, debris-flows, segregation, mixing, shocks, diffusion, recirculation
PACS: 61.43.-j; 62.50.Ef; 64.75.Ef; 64.75.St; 92.40.Gc; 92.40.Ha; 92.40.vw

INTRODUCTION

As a body of grains avalanches downslope it dilates, in order for the particles to shear past one another, and the upper layers of the flow move faster than the lower ones, as shown in the schematic diagram in figure 1. The combination of velocity shear and dilation, acts as a random fluctuating sieve [1], which allows the smaller particles to preferentially percolate down into gaps that open up beneath them under the action of gravity, since they are more likely to fit into the available space than the large grains. Once the smaller particles get underneath they exert a force that squeeze the larger grains upwards. The combination of *kinetic sieving* and *squeeze expulsion* [1] causes the particles to segregate into layers, which have greater concentrations of large particles near the free-surface and higher concentrations of fines near the base of the flow. In geology this is known as *inverse grading* [2] and is often associated with granular flows [3]. The inverse grading of the particle size distribution is not necessarily preserved in avalanche deposits, which can be very complex [4]. Indeed, even a relatively simple two-dimensional inversely graded flow with deposition yields a deposit that is *normally graded* [5] with the fines above the large particles.

Large particles that rise to the upper faster moving layers of the avalanche, tend to be transported to the flow front. Here they are often overrun, but rise to the surface again by particle size segregation. This *recirculation* allows bouldery flow fronts to develop in hazardous geophysical mass flows, such as debris-flows, pyroclastic flows and wet and dry snow avalanches [6–9], which give rise to interesting *segregation-mobility* feedback effects [5]. Larger less mobile particles at the flow front are shouldered to the side by the more mobile interior, to create static coarse grained lateral *levees* [10] that channelize the flow and enhance the total run-out distance. Such segregation mobility feedback effects are also responsible for fingering instabilities on chutes [11–13] and digitate lobate terminations [14].

Granular avalanches frequently occur in much smaller scale processes, such as rotating tumblers, where different modes of deposition create a rich variety of patterns, including Catherine wheels [15], leafs [16] and petals [17, 18]. Similar deposition mechanisms are responsible for the formation of stratification and segregation patterns in heaps

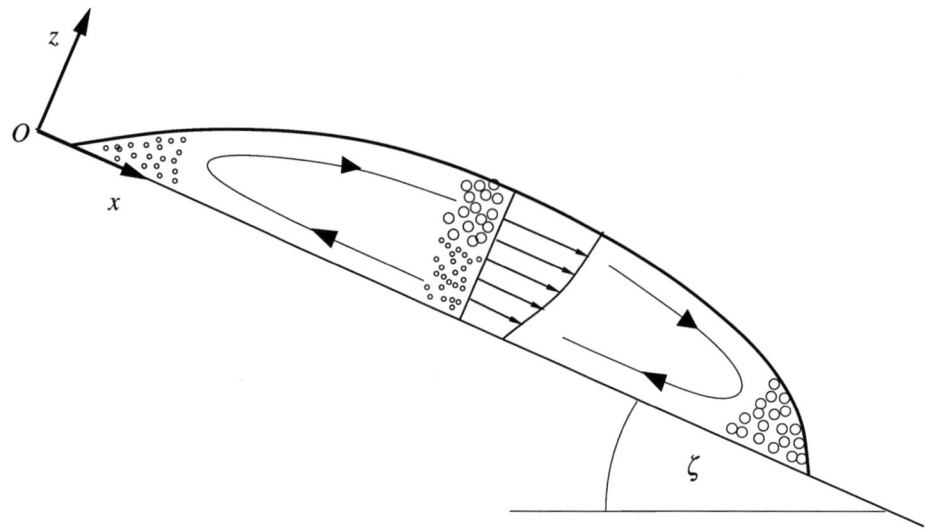

FIGURE 1. A sketch of an avalanche flowing down a plane inclined at an angle ζ to the horizontal. The x axis points down the chute, the z axis is normal to the plane and the y axis goes into the page. The particle size distribution and the downslope velocity profile are illustrated schematically. Large particles tend to rise to the faster moving surface layers, and are transported to the flow front, where they can be recirculated.

and silos [5, 15, 19, 20]. Size segregation is therefore of considerable practical importance to the pharmaceutical, bulk chemical, mining and food industries. Sometimes it is useful, such as in the mineral processing industry, but often it is a source of inconsistency and poor quality and the aim is then to minimize its effect. This paper provides a brief review of recent progress in modelling the particle size segregation process in a bi-disperse avalanche using a relatively simple approach.

DERIVATION OF THE SEGREGATION-REMIXING EQUATION

The segregation-remixing equation can be derived [16, 21, 22] from binary or ternary mixture theory. Here we follow Gray and Chugunov [16]'s derivation, which assumes that the mixture is composed of *large* and *small* particles and that the interstitial pore space is subsumed into the volume fractions, ϕ^l and ϕ^s, of large and small particles per unit mixture volume. This implicity assumes that the solids volume fraction is approximately constant within the avalanche, which is a reasonable first approximation. By definition the volume fractions $\phi^s, \phi^l \in [0,1]$ and they sum to unity

$$\phi^l + \phi^s = 1. \qquad (1)$$

Mixture theory defines overlapping partial densities, ρ^μ, partial velocities, $\mathbf{u}^\mu$, and partial pressures, p^μ, for each of the constituents $\mu = l, s$ per unit mixture volume. Each of the constituents satisfies individual mass and momentum conservation laws [e.g. 23, 24]

$$\frac{\partial \rho^\mu}{\partial t} + \nabla \cdot (\rho^\mu \mathbf{u}^\mu) = 0, \qquad \mu = l, s, \qquad (2)$$

$$\frac{\partial}{\partial t}(\rho^\mu \mathbf{u}^\mu) + \nabla \cdot (\rho^\mu \mathbf{u}^\mu \otimes \mathbf{u}^\mu) = -\nabla p^\mu + \rho^\mu \mathbf{g} + \mathbf{B}^\mu, \qquad \mu = l, s, \qquad (3)$$

where $\otimes$ is the dyadic product, $\rho^\mu \mathbf{g}$ is the gravitational force and $\mathbf{B}^\mu$, is the force exerted on phase μ by the other constituent. The interaction forces in a binary mixture are equal and opposite to one another, $\mathbf{B}^l = -\mathbf{B}^s$, and cancel out in the bulk mass and momentum balances, which are obtained by summing (2) and (3) over all constituents. It is useful to define the bulk density ρ, bulk velocity $\mathbf{u}$ and bulk pressure p as

$$\rho = \rho^l + \rho^s, \quad \rho \mathbf{u} = \rho^l \mathbf{u}^l + \rho^s \mathbf{u}^s, \quad p = p^l + p^s. \qquad (4)$$

The partial and intrinsic densities are related by a linear volume fraction scaling [24], while the partial and intrinsic velocity fields are identical

$$\rho^\mu = \phi^\mu \rho^{\mu*}, \quad \mathbf{u}^\mu = \mathbf{u}^{\mu*}, \tag{5}$$

where the superscript $*$ denotes an intrinsic variable. A coordinate system $Oxyz$ is defined, with the x-axis pointing down a chute inclined at an angle ζ to the horizontal, the y-axis across the chute and the z-axis as the upward pointing normal as shown in figure 1. The constituent velocity $\mathbf{u}^\mu$ and the bulk velocity $\mathbf{u}$ have components (u^μ, v^μ, w^μ) and (u, v, w) in each of these directions, respectively. The large and small particles are assumed to have the same constant intrinsic density, $\rho^{l*} = \rho^{s*}$, and (1), (4) and (5) therefore imply that the bulk density ρ is also equal to the same constant value. It follows from the bulk mass balance that the bulk velocity field is incompressible

$$\frac{\partial u}{\partial x} + \frac{\partial v}{\partial y} + \frac{\partial w}{\partial z} = 0. \tag{6}$$

This is one of the key assumptions that is made in nearly all granular avalanche models [7, 25–37]. The other important assumption for compatibility with existing models, is that the bulk pressure p is lithostatic through the avalanche depth

$$p = \rho g (h - z) \cos \zeta, \tag{7}$$

which is true, provided the acceleration terms are negligible in the normal component of the bulk momentum balance.

During percolation the small grains can not support as much of the overburden pressure and the larger grains have to support proportionately more of the load. The driving force for particle size segregation are therefore perturbations to the lithostatic pressure distribution [16, 21, 22]. Instead of relating the partial pressure to the bulk pressure using a linear volume fraction dependent scaling, as in standard mixture theory, Gray and Thornton [21] introduced a linear scaling

$$p^\mu = f^\mu p, \quad \mu = l, s, \tag{8}$$

with a factor f^μ that could deviate away from ϕ^μ. The functions f^μ satisfy three constraints:-

$$\left. \begin{array}{ll} \text{(i)} & f^l + f^s = 1, \\ \text{(ii)} & f^s = 1 \quad \text{when} \quad \phi^s = 1, \\ \text{(iii)} & f^l = 1 \quad \text{when} \quad \phi^l = 1, \end{array} \right\} \tag{9}$$

which ensure that the partial pressures sum to the bulk pressure (4), and that when either of the constituents are in a pure phase they carry all of the load. Although there are many functions that satisfy these constraints the simplest non-trivial functions are

$$f^l = \phi^l + b\phi^s \phi^l, \quad f^s = \phi^s - b\phi^s \phi^l, \tag{10}$$

where b is the magnitude of the perturbation. To balance the pressure perturbations in the normal momentum balance equations (3), Gray and Thornton [21] and Thornton et al. [22] introduced an interaction drag $\mathbf{B}^\mu$ with a simple linear velocity dependent drag

$$\mathbf{B}^\mu = p \nabla f^\mu - \rho^\mu c (\mathbf{u}^\mu - \mathbf{u}) - \rho d \nabla \phi^\mu, \quad \mu = l, s, \tag{11}$$

with drag coefficient c. Gray and Chugunov [16] introduced a further gradient dependent remixing force that drives the grains of phase μ towards areas of lower concentration. The strength of these diffusive forces is ρd. The interaction drag (11) automatically satisfies the constraint that $\mathbf{B}^l + \mathbf{B}^s = 0$, and when it is substituted into the constituent momentum balances the first term combines with the partial pressure gradient $-\nabla(f^\mu p)$ to leave $-f^\mu \nabla p$. Assuming that the acceleration terms are negligible, the normal component of the constituent momentum balances reduces to

$$\phi^\mu w^\mu = \phi^\mu w + (f^\mu - \phi^\mu)(g/c) \cos \zeta - (d/c) \partial \phi^\mu / \partial z, \quad \mu = l, s. \tag{12}$$

Substituting for the pressure fluctuation functions (10) and dividing through by the volume fraction ϕ^μ implies that the normal velocities of the large and small particles are

$$w^l = w + q\phi^s - D\frac{\partial}{\partial z}(\ln \phi^l), \tag{13}$$

$$w^s = w - q\phi^l - D\frac{\partial}{\partial z}(\ln \phi^s), \tag{14}$$

where the maximum percolation velocity of the grains q and the diffusivity D are

$$q = (b/c)g\cos\zeta, \quad D = d/c. \tag{15}$$

Within the avalanche the segregation velocities $\mathbf{u}^\mu - \mathbf{u}$ are assumed to be of the same order of magnitude as the bulk normal velocity w, which is a lot less than the bulk down and cross slope velocities. It follows that to leading order the down and cross slope constituent velocities are equal to their bulk counterparts

$$u^\mu = u, \quad v^\mu = v, \quad \mu = l, s. \tag{16}$$

Substituting (13), (14) and (16) into the small particle mass balance (2) yields the segregation-remixing equation

$$\frac{\partial \phi^s}{\partial t} + \frac{\partial}{\partial x}(\phi^s u) + \frac{\partial}{\partial y}(\phi^s v) + \frac{\partial}{\partial z}(\phi^s w) - \frac{\partial}{\partial z}(q\phi^s \phi^l) = \frac{\partial}{\partial z}\left(D\frac{\partial \phi^s}{\partial z}\right). \tag{17}$$

The first four terms simply advect the local concentration with the bulk flow, the fifth is responsible for segregation and the sixth diffuses the small particles. The segregation-remixing equation (17) is parabolic for $D > 0$ and reduces to a hyperbolic equation when $D = 0$. The reduced model is termed the *hyperbolic segregation theory* or *segregation theory* for short. The segregation flux has the interesting property that it shuts off when the concentration of small particles equals zero or unity, so ϕ^s automatically stays in the range $[0,1]$.

In the absence of diffusion, equations (13)–(14) show that the large particles rise up until there are no more small particles, while small particles percolate downwards until there are no more coarse grains. This tends to drive the grains into completely separated *inversely graded layers* with all the large particles on top of the fines [1]. The hyperbolic segregation equation is related to the theory of sedimentation [38, 39] and is also very closely related to Savage and Lun [1]'s kinetic sieving and squeeze expulsion model. The link to Savage and Lun [1] is not immediately apparent, as they formulated their theory in terms of the layer number density ratio, η, and the particle diameter ratio, σ, instead of volume fractions. However, by substituting the definitions $\phi^l = 1/(1+\eta\sigma^3)$ and $\phi^s = \eta\sigma^3/(1+\eta\sigma^3)$ into their equations (6.4) and (6.3) we can see that the net percolation velocities have the same leading order concentration dependence as (13) and (14). Savage and Lun [1]'s information entropy approach yields considerably more structure for the segregation rate, but there is no dependence on gravity. The mixture approach does not provide as much structure for q, but it does have an explicit dependence on gravity in (15), which sets a direction for segregation and reflects the gravity driven nature of the kinetic sieving process. Thornton et al. [22] used ternary mixture theory to derive (17) in the presence of an interstitial fluid. This theory had an extra relative density difference factor $\hat{\rho}$ in q, which was able to explain the reduced segregation rates in liquid particle mixtures and the absence of segregation with a density matched fluid, observed in the experiments of Vallance and Savage [40].

The segregation-remixing equation (17) is closely related to Burgers' equation [41], and smoothes out the sharp concentration jumps that develop in the hyperbolic theory. Dolgunin and Ukolov [42] were the first to write down this form of the equation, by ingeniously spotting that the segregation flux must shut off when $\phi^s = 0, 1$, but there was no formal derivation. Khakhar et al. [43] went on to use equation (17) to study the equilibrium segregation of particles of different densities, but the same size, and obtained good agreement with experimental measurements in rotating drums. More recently, Gray and Chugunov [16] showed that steady-state solutions were in good agreement with particle dynamics simulations of chute flows of large and small particles of Khakhar et al. [44], which is reproduced in the bottom panel of figure 2. Both experiments and particle dynamics simulations [45, 46] are likely to be very useful in determining the functional dependence of q and D on other parameters, such as the shear rate, the particle-size ratio and the overburden pressure. An example of this is provided by Hajra and Khakhar [47] who used rotating drum experiments to infer that even small size differences are sufficient to cause segregation, but once the size ratio reaches a critical value the driving force for segregation saturates. It may also be possible to push bi-disperse kinetic theories [48–50] into the dense regime and find a link between the two approaches.

NON-DIMENSIONALIZATION, BOUNDARY AND JUMP CONDITIONS

Avalanches are shallow, with their typical thickness H being much less than their length L. The incompressibility condition (6) implies that, if typical down slope velocities are of magnitude U, then typical normal velocities are of magnitude HU/L. The variables are non-dimensionalized to reflect these scalings

$$(x, y, z) = L(\tilde{x}, \tilde{y}, \varepsilon\tilde{z}), \quad (u, v, w) = U(\tilde{u}, \tilde{v}, \varepsilon\tilde{w}), \quad t = (L/U)\tilde{t}, \tag{18}$$

where the tilde denotes a non-dimensional variable. Substituting these into (17) and dropping the tildes and the superscript s implies that the non-dimensional segregation-remixing equation is

$$\frac{\partial \phi}{\partial t} + \frac{\partial}{\partial x}(\phi u) + \frac{\partial}{\partial y}(\phi v) + \frac{\partial}{\partial z}(\phi w) - \frac{\partial}{\partial z}\left(S_r \phi(1-\phi)\right) = \frac{\partial}{\partial z}\left(D_r \frac{\partial \phi}{\partial z}\right), \qquad (19)$$

where the non-dimensional segregation and diffusive-remixing numbers are

$$S_r = \frac{qL}{HU}, \quad \text{and} \quad D_r = \frac{DL}{H^2 U}, \qquad (20)$$

respectively. Provided there is no erosion or deposition, there is no flux or large or small particles at the surface and basal boundaries of the avalanche. This can be expressed by the nonlinear boundary condition

$$S_r \phi(1-\phi) + D_r \frac{\partial \phi}{\partial z} = 0, \qquad (21)$$

which insulates the avalanche from the exterior. In the hyperbolic theory shocks may also develop on a propagating surface of discontinuity across which a jump condition [51] must be satisfied

$$[\![\phi(\mathbf{u}\cdot\mathbf{n}-v_n)]\!] = [\![S_r \phi(1-\phi)\mathbf{k}\cdot\mathbf{n}]\!], \qquad (22)$$

where the jump bracket $[\![f]\!] = f^+ - f^-$ is the difference of f evaluated on the forward "+" and rearward "-" side of the surface, $\mathbf{k}$ is the unit vector normal to the chute, $\mathbf{n}$ is the unit normal to the surface and v_n is it's normal speed. A Lax entropy condition implies that the shock will be stable if and only if the shock is inversely graded [52].

TIME-DEPENDENT SOLUTIONS OF THE SEGREGATION-REMIXING EQUATION

Gray and Chugunov [16] constructed a general time-dependent solution in a flow of unit depth with no down or cross slope gradients in concentration

$$u = u(z), \quad v = v(z), \quad w = 0, \quad \partial \phi/\partial x = 0, \quad \partial \phi/\partial y = 0, \quad 0 < z < 1. \qquad (23)$$

In this case the segregation-remixing equation (19) reduces to

$$\frac{\partial \phi}{\partial t} - \frac{\partial}{\partial z}\left(S_r \phi(1-\phi)\right) = D_r \frac{\partial^2 \phi}{\partial z^2}, \qquad (24)$$

which is subject to the surface and basal boundary conditions (21) and the initial condition

$$t = 0: \qquad \phi = \phi_i(z), \qquad (25)$$

which must be independent of x and y. The reduced segregation-remixing equation can be mapped directly on to Burgers equation, which can in turn be linearized by using the Cole-Hopf transformation [53, 54]. Gray and Chugunov [16] used this sequence of transformations to linearize both the segregation-remixing equation (24) and the nonlinear boundary conditions (21). The resulting diffusion problem was then solved by using Fourier series. The general solution takes the form

$$\phi = \frac{1}{2}\left(1 - \frac{2z_0}{\omega}\frac{\partial \omega}{\partial z}\right), \qquad (26)$$

where the segregation-remixing length scale $z_0 = D_r/S_r$. Considerable simplification is achieved by splitting the solution ω into a steady-state and time dependent part

$$\omega = \omega_s + \omega_t, \qquad (27)$$

where

$$\omega_s = \chi(1)\frac{\sinh(z/(2z_0))}{\sinh(1/(2z_0))} - \frac{\sinh((z-1)/(2z_0))}{\sinh(1/(2z_0))}, \qquad (28)$$

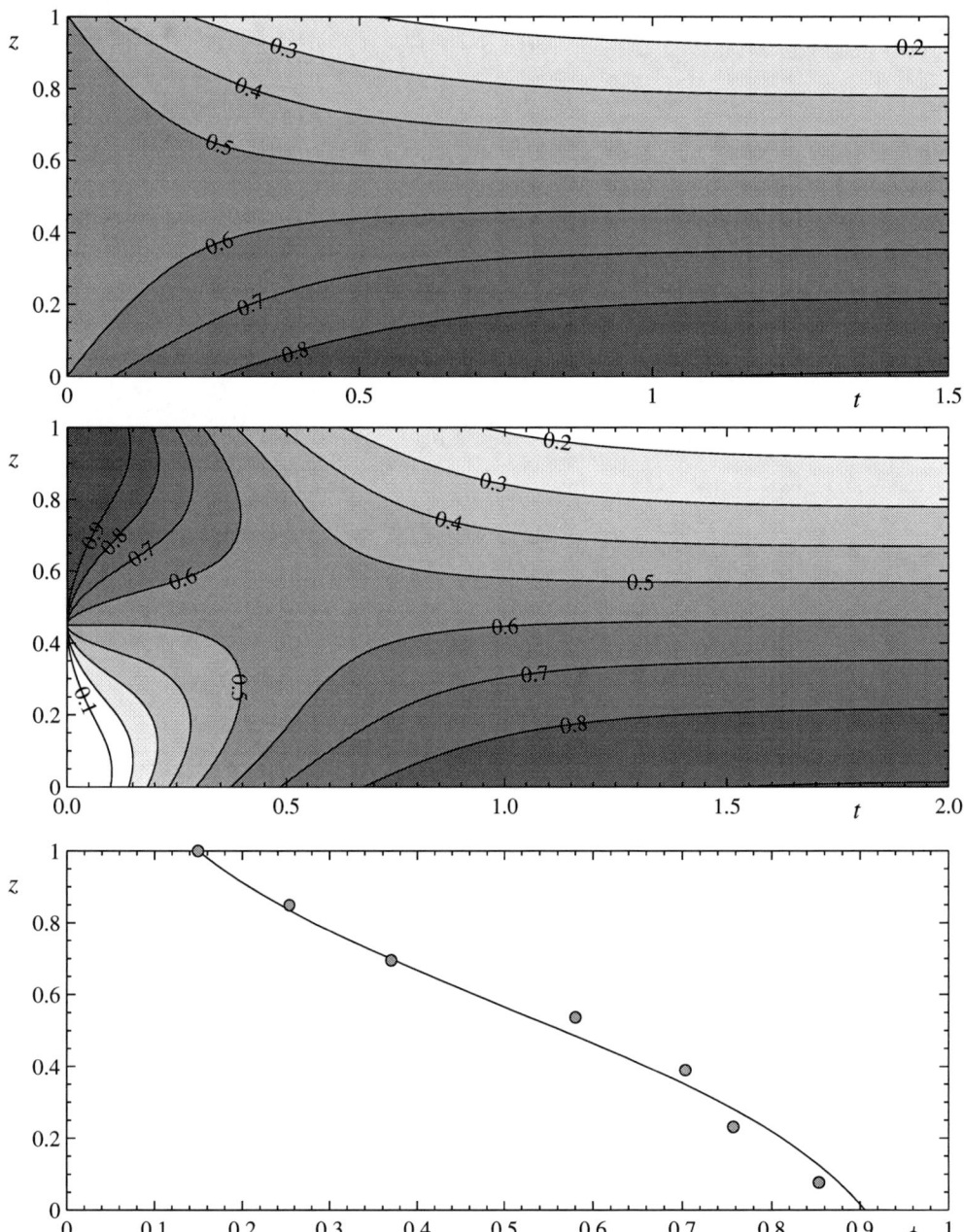

FIGURE 2. The upper and middle panels show two contour plots of the evolution of the the small particle concentration through the depth of the avalanche for $S_r = 1$ and $D_r = 0.25$. In the top panel the avalanche is initially homogeneously mixed with an initial concentration $\phi_0 = 0.55$ and in the middle panel the small grains are initially on top of the large ones and are separated by a sharp interface at $z_r = 0.45$. For these parameters the same steady-state develops, which is shown in the bottom panel as a profile of the small particle concentration ϕ with depth. The dots are the equivalent steady state concentrations derived from the particle dynamics simulations of Khakhar et al. [44] for comparison.

$$\omega_t = \sum_{n=1}^{\infty} A_n \exp\left(-\left(\frac{1}{4} + n^2\pi^2 z_0^2\right)\frac{S_r t}{z_0}\right) \sin(n\pi z). \tag{29}$$

The function χ is defined as

$$\chi(z) = \exp\left(-\int_0^z \frac{2\phi_i - 1}{2z_0} dz'\right), \quad \text{and} \quad \chi(1) = \exp\left(-\int_0^1 \frac{2\phi_i - 1}{2z_0} dz'\right), \qquad (30)$$

which are both dependent on the initial concentration profile $\phi_i(z)$ and the segregation-remixing length z_0. At $t = 0$ equation (29) reduces to a Fourier Sine series for ω_t and the coefficients A_n can be determined by integration in the usual way. The coefficient A_n can be split into two parts, $A_n = B_n - C_n$, where B_n is given by an integral of the transformed initial conditions, $\chi(z)$, and C_n is dependent on an integral of the steady-state solution ω_s. It follows that to construct a specific solution it is necessary to compute

$$B_n = 2\int_0^1 \chi(z)\sin(n\pi z)\, dz, \quad C_n = \frac{8n\pi z_0^2}{1 + 4n^2\pi^2 z_0^2}\left(1 - (-1)^n \chi(1)\right), \qquad (31)$$

where the integral for C_n has already been evaluated by substituting the steady-state solution (28).

Two different initial conditions are shown in figure 2. In the top panel the grains are initially homogeneously mixed with concentration $\phi_0 = 0.55$. As time progresses the small particles percolate down towards the base of the avalanche and the large particles are pushed upwards until a balance establishes itself with the diffusive effects of remixing. For large time the concentration profile therefore approaches an inversely graded steady state, which is shown in the bottom panel of figure 2. The grey circles are points derived from the steady-state particle dynamics simulations of Khakhar et al. [44], which the theory fits well for $S_r = 1$ and $D_r = 0.25$, giving a segregation-remixing length $z_0 = D_r/S_r = 0.25$. The middle panel shows the solution for an unstably stratified initial distribution, with all the small particles above the large ones, separated by a discontinuity at height $z_r = 0.45$. The initial discontinuity is rapidly smoothed out by diffusion, before percolation and squeeze expulsion take over as the dominant means of transport. The small particles percolate down from the top and collect at the base, while the large ones rise from the base to the surface, until the same steady-state distribution as the top and bottom panels is established.

STEADY STATE SOLUTIONS TO THE SEGREGATION EQUATION

In the last section we showed that segregation-remixing theory is able to reproduce steady state one-dimensional profiles produced by particle dynamics simulations, which lends considerable weight to this approach. The chute flow experiments of Savage and Lun [1] suggest that when strongly stratified layers develop the diffusive term in (19) can be neglected. It is therefore of interest to study the hyperbolic segregation equation, which allows complicated physical problems to be treated in a simple way and gives considerable insight into the nature of segregation in granular avalanches. In a steady uniform avalanche of unit depth

$$u = u(z) \geq 0, \quad v = 0, \quad w = 0, \quad \text{in} \quad 0 < z < 1, \quad x > 0, \qquad (32)$$

the steady hyperbolic segregation equation (19) and the no flux condition (21) reduce to

$$\frac{\partial}{\partial x}(\phi u) - \frac{\partial}{\partial z}\left(S_r \phi(1-\phi)\right) = 0, \qquad (33)$$

$$\phi(1-\phi) = 0, \quad \text{at} \quad z = 0,1. \qquad (34)$$

This can be written as a first order quasi-linear equation by expanding out the derivatives

$$u\frac{\partial \phi}{\partial x} + S_r(2\phi - 1)\frac{\partial \phi}{\partial z} = 0, \qquad (35)$$

and solved by the method of characteristics. The small particle concentration is equal to a constant ϕ_λ along the characteristic curve given by

$$u\frac{dz}{dx} = S_r(2\phi_\lambda - 1). \qquad (36)$$

Solutions for general velocity fields can be constructed by defining a depth-integrated velocity coordinate

$$\psi = \int_0^z u(z')\, dz', \qquad (37)$$

that increases monotonically with z. By virtue of the scalings (18), we may assume without loss of generality that at the free surface $\psi(1) = 1$. The mapping (37) transforms (36) into a linear equation

$$\frac{d\psi}{dx} = S_r(2\phi_\lambda - 1), \tag{38}$$

which can be integrated, subject to the initial condition that the characteristic starts at $(x_\lambda, \psi_\lambda)$, to give a straight line

$$\psi = \psi_\lambda + S_r(2\phi_\lambda - 1)(x - x_\lambda). \tag{39}$$

The position in physical space can be calculated by inverting the transformation (37) once $u(z)$ is prescribed. The beauty of the depth-integrated velocity coordinates is that the solutions constructed with it are valid for all velocity fields provided the inverse mapping is well defined. In this paper, we will consider linear velocity profiles

$$u = \alpha + 2(1 - \alpha)z, \quad 0 \leq \alpha \leq 1, \tag{40}$$

which include plug flow ($\alpha = 1$), simple shear ($\alpha = 0$) and shear with basal slip, for intermediate values of α. The integral (37) implies that the depth-integrated velocity coordinate

$$\psi = \alpha z + (1 - \alpha)z^2, \tag{41}$$

which is quadratic and can be inverted to give

$$z = \begin{cases} \psi, & \alpha = 1, \\ \dfrac{-\alpha + \sqrt{\alpha^2 + 4(1 - \alpha)\psi}}{2(1 - \alpha)}, & \alpha \neq 1. \end{cases} \tag{42}$$

Homogeneous inflow

Gray and Thornton [21] considered the case in which there is an inflow at $x = 0$ at which the particles enter in a homogeneously mixed state with concentration

$$\phi = \phi_0, \quad \text{at} \quad x = 0, \quad 0 < z < 1. \tag{43}$$

Through most of the avalanche the inflow concentration ϕ_0 is simply swept into the domain by the characteristics, and the small particles percolate downwards by kinetic sieving and the large grains are squeezed upwards. At the base, however, the no flux condition (34) implies that there are no more large particles to rise up, and instead the small particles separate out across a concentration shock. In the uniform flow field (32) the jump condition (22) reduces to

$$\left[\!\left[\phi u \frac{dz}{dx} + S_r \phi(1 - \phi)\right]\!\right] = 0, \tag{44}$$

provided the shock is stationary, $v_n = 0$. Dividing both sides by $[\![\phi]\!]$ yields

$$u\frac{dz}{dx} = S_r(\phi^+ + \phi^- - 1), \tag{45}$$

where ϕ^+ and ϕ^- are the values of the concentration on the forward and rearward sides of the discontinuity. Using the mapping (37) this reduces to

$$\frac{d\psi}{dx} = S_r(\phi^+ + \phi^- - 1). \tag{46}$$

The position of the bottom shock, which separates the fines from the homogeneous mixture, can be computed from (46) by substituting $\phi^- = \phi_0$ and $\phi^+ = 1$ and integrating, subject to the boundary condition that $\psi = 0$ at $x = 0$, to give

$$\psi_{bottom} = S_r \phi_0 x. \tag{47}$$

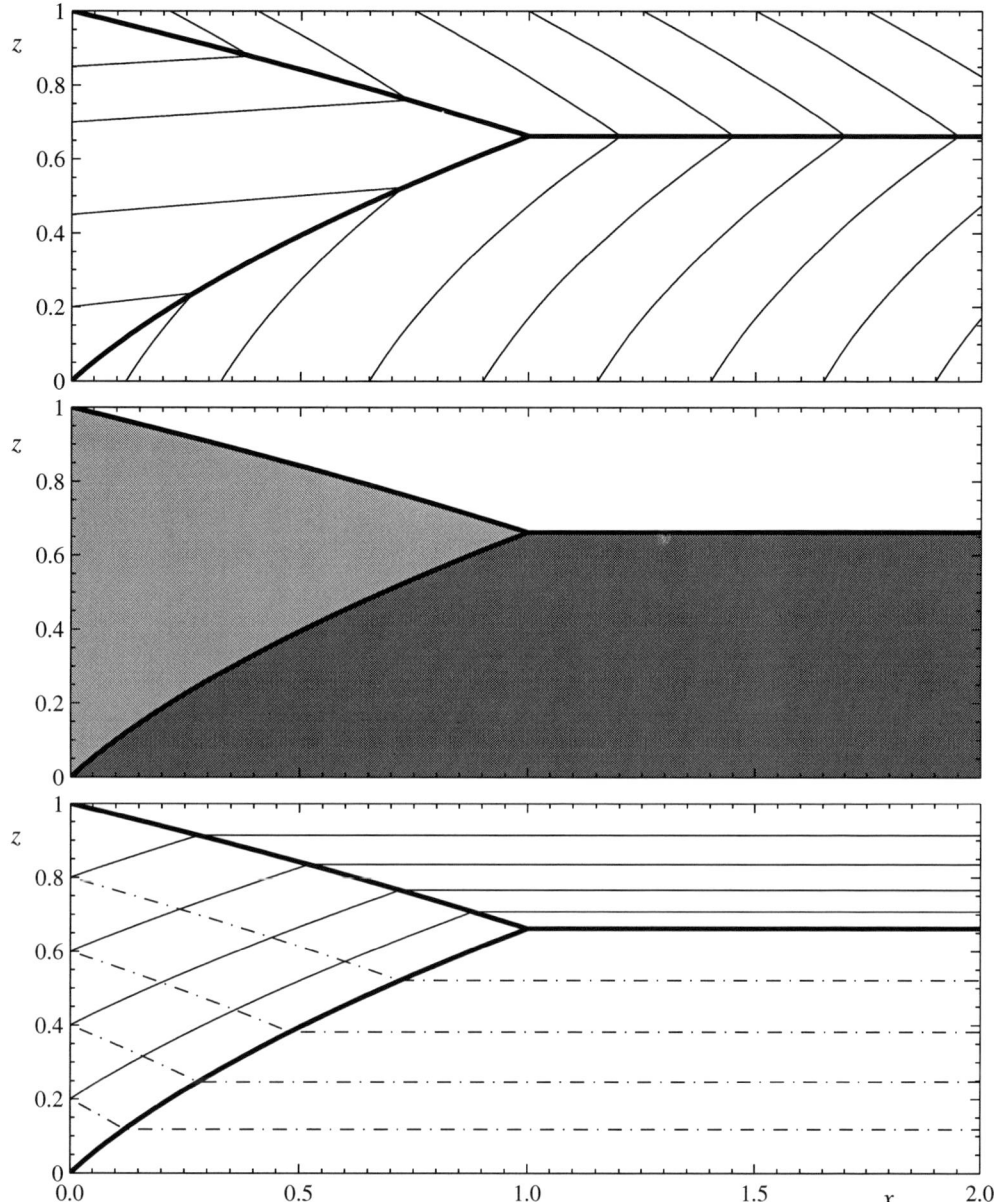

FIGURE 3. The steady-state segregation solution for a homogeneous inflow of concentration $\phi_0 = 0.55$ at $x = 0$ in a unit thickness avalanche with a linear shear profile through its depth ($\alpha = 0.5$) and the non-dimensional segregation number $Sr = 1$. The top panel shows the shocks as thick solid lines and the characteristics as thin lines. The middle panel indicates the concentration, with the white area composed of all large particles, the dark grey area of all fines and the light grey region being at the inflow concentration. The bottom panel shows the large particle paths with thin solid lines and the small particle paths with thin dot-dash lines.

The basal layer of small particles therefore becomes progressively thicker with increasing downstream distance. A similar thing happens at the top of the avalanche, where the boundary condition (34) implies that there are no more small particles to percolate downwards, and the large grains separate out into a pure phase across a shock. Substituting $\phi^+ = 0$ and $\phi^- = \phi_0$ into (46) and integrating subject to the boundary condition $\psi = 1$ at $x = 0$, implies that the top shock

$$\psi_{top} = 1 - S_r(1-\phi_0)x. \tag{48}$$

The top layer of large particles also becomes thicker with increasing x, and the top and bottom shocks meet to form a triple-point at $x_{triple} = 1/S_r$, $\psi_{triple} = \phi_0$. Downstream of x_{triple} a third shock is formed that separates a pure layer of large particles from a pure layer of fines below. Substituting $\phi^+ = 0$ and $\phi^- = 1$ into (46) and integrating implies that the shock separating the inversely graded layers is at constant height

$$\psi_{inverse} = \phi_0, \quad \text{for} \quad x \geq x_{triple}. \tag{49}$$

The solution therefore consists of three shocks that separate regions of homogeneously mixed material (light grey) adjacent to the inflow, from small particles (dark grey) next to the base and large particles (white) at the surface of the avalanche, as illustrated in the middle panel of figure 3. The top panel shows how each shock (thick line) has characteristics (thin lines) intersecting from both sides, which is required by the entropy condition. This problem is the hyperbolic counterpart of the time-dependent segregation remixing problem shown in the top panel of figure 2. The key differences are that (i) x replaces t along the lower axis, (ii) there are sharp shocks instead of a smooth transition to a diffuse steady-state and (iii) the inversely graded layer of large particles at the surface is thinner than would develop in the equivalent time-dependent problem, because there is a higher mass flux near the surface.

Particle paths of the large and small grains

One of the major benefits of using the hyperbolic model is that it is possible to exactly reconstruct the large and small particle paths. They satisfy the equations

$$\frac{dx^\mu}{dt} = u^\mu, \quad \frac{dz^\mu}{dt} = w^\mu, \quad \mu = l, s \tag{50}$$

with the non-dimensional forms of the constituent velocities (13), (14) and (16) are given by

$$u^l = u, \quad w^l = w + S_r\phi, \quad u^s = u, \quad w^s = w - S_r(1-\phi). \tag{51}$$

Using the chain rule and the depth-integrated velocity coordinate transformation

$$\frac{d\psi^l}{dx} = S_r\phi, \quad \frac{d\psi^s}{dx} = -S_r(1-\phi). \tag{52}$$

For the homogeneously mixed inflow problem discussed above, the small particles enter at $x = 0$ at a height ψ^s_{enter} and percolate downwards along the path

$$\psi^s = \psi^s_{enter} - S_r(1-\phi_0)x, \tag{53}$$

until they hit the bottom shock at $x^s_{cross} = \psi^s_{enter}/S_r$ and then move parallel to the base

$$\psi^s = \phi_0 \psi^s_{enter}, \quad x > x^s_{cross}, \tag{54}$$

through a region of pure fines. Conversely large particles entering at height ψ^l_{enter} are pushed upwards, by squeeze expulsion, along the path

$$\psi^l = \psi^l_{enter} + S_r\phi_0 x, \tag{55}$$

until they reach the top shock at $x^l_{cross} = (1 - \psi^l_{enter})/S_r$, after which they move downslope at height

$$\psi^l = 1 - (1-\phi_0)(1-\psi^l_{enter}), \quad x > x^l_{cross}, \tag{56}$$

through a region of purely large grains. The particle paths are illustrated in the bottom panel of figure 3 using solid lines for the large particle paths and dot-dash lines for the fines. From this we see that the top shock is also a small particle path, while the bottom shock is also a large particle path. This can also be proved by observing that the shock condition (46) degenerates to the particle path equations (52) when ϕ^+ is equal to zero or unity.

Unstably stratified inflow

Thornton et al. [22] investigated the case on an unstably stratified inflow in which all the small particles are fed into the chute above the large particles

$$\phi = \begin{cases} 1 & z_r \leq z \leq 1, \\ 0 & 0 \leq z < z_r. \end{cases} \tag{57}$$

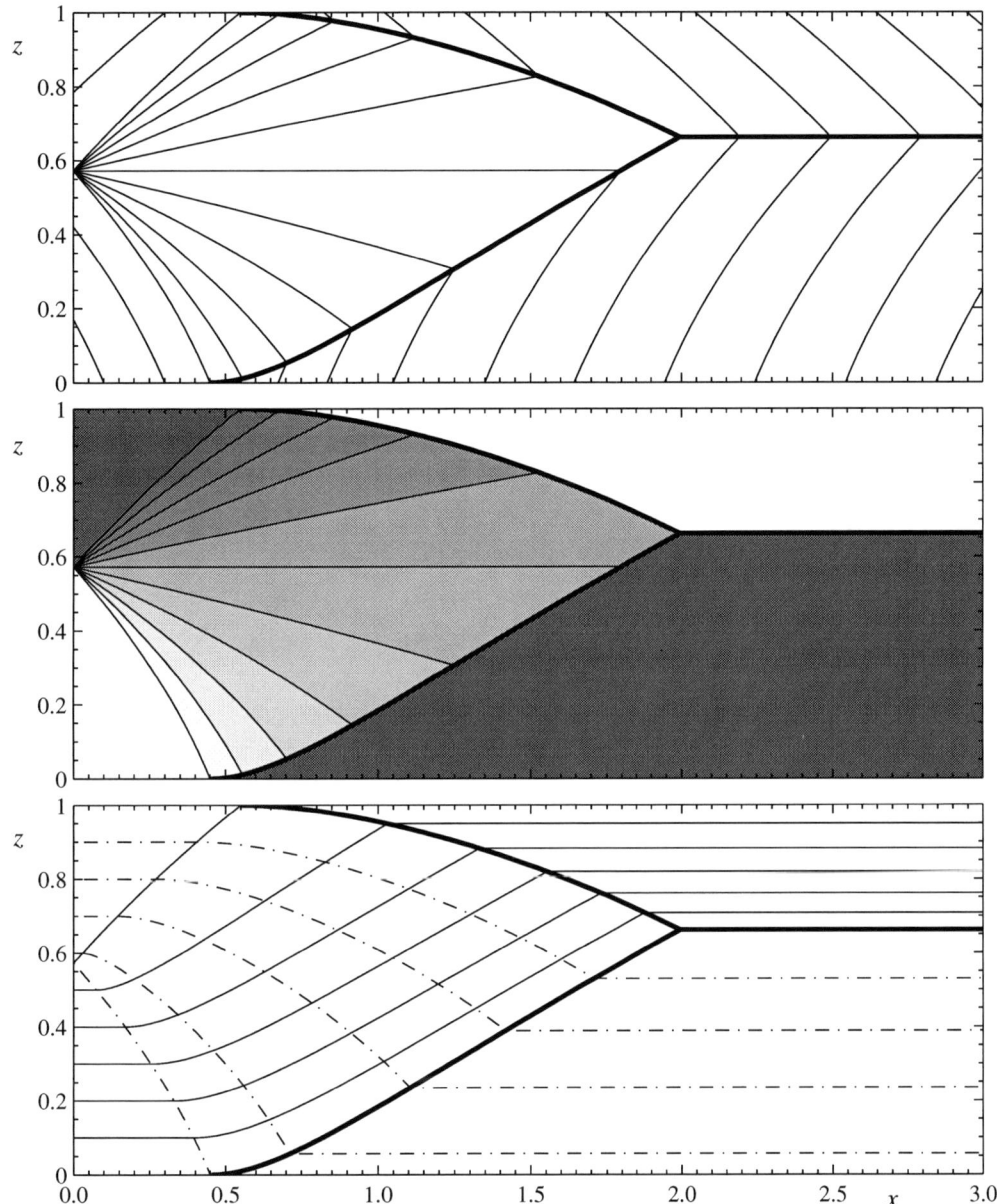

FIGURE 4. The steady-state segregation solution for in flow at $x = 0$, where a layer of small particles enters on top of a layer of coarse grains. The interface between the two layers lies at a height $z_r = 0.5724$. The avalanche is assumed to be of unit thickness with a linear shear profile through its depth ($\alpha = 0.5$) and the non-dimensional segregation number $S_r = 1$. The top panel shows the shocks as thick solid lines and the characteristics as thin lines. The middle panel indicates the concentration, with the white area composed of all large particles, the dark grey area of all fines, and the fan shown with contours of 0.1 unit intervals. The bottom panel shows the large particle paths with thin solid lines and the small particle paths with thin dot-dash lines. The inversely graded steady-state shock height is the same as in figure 3.

One possible solution that satisfies the shock condition (46), is simply to continue the discontinuity at height z_r down the chute. However, this is not admissible, because it does not satisfy the Lax entropy condition [52], which requires that the grains must be inversely graded across the shock. Instead a rarefaction fan is formed, which is centred at $(0, z_r)$. In depth averaged velocity coordinates (37) this corresponds to the position $(0, \psi_r)$, and the characteristic equation (39) then implies that the concentration within the fan is given by

$$\phi = \frac{1}{2}\left(1 + \frac{\psi - \psi_r}{S_r x}\right), \quad |\psi - \psi_r| < S_r x, \tag{58}$$

which is shown in the top panel of figure 4. The lowest characteristic that emanates from the fan has concentration $\phi = 0$. It intersects with the base of the flow at $x_b = \psi_r/S_r$ and represents the first small particle to percolate down to the base of the flow. As in the homogeneous problem, there is no flux of large particles across the basal boundary, and so the small grains separate out into a pure phase across a concentration shock that emanates from $(x_b, 0)$. The shock height can be found by integrating the linear ordinary differential equation that is obtained by substituting the fan concentration (58) and the small particle concentration into the shock condition (46). This implies that the basal shock

$$\psi_{bottom} = \psi_r + S_r x - 2\sqrt{S_r \psi_r x}. \tag{59}$$

Similarly, the first large particle reaches the surface at $x_s = (1 - \psi_r)/S_r$ and the large particles then begin to separate out into a pure phase across a concentration shock that emanates from $(x_s, 1)$. Solving the shock condition (46) with the expansion fan (58) on one side and large particles on the other, yields an equation for the top shock

$$\psi_{top} = \psi_r - S_r x + 2\sqrt{S_r(1-\psi_r)x}. \tag{60}$$

The top and bottom shocks meet at the triple point

$$x_{triple} = \frac{1}{S_r}\left(\sqrt{\psi_r} + \sqrt{1-\psi_r}\right)^2, \quad \psi_{triple} = 1 - \psi_r, \tag{61}$$

and a third shock is formed that separates the inversely graded layers of large and small particles. The shock condition (46) implies that this is parallel to the base of the flow at height

$$\psi_{inverse} = \psi_{triple}, \quad x \geq x_{triple}, \tag{62}$$

which now satisfies the Lax entropy condition [52]. The expansion fan and the three shocks are shown in figure 4. The top panel shows the characteristics, which intersect either side of the shocks, and the centre panel shows the concentration using a grey scale. This solution shows how an unstably stratified inflow readjusts into a stable configuration. Small particles that enter near the surface of the avalanche, move straight downslope until they reach the expansion fan. Here they percolate downwards, until they cross the bottom shock and enter into a pure phase of small particles again implying that they move straight downslope again, but at a much lower level than they started. The small particle paths are shown as dot-dash lines in the bottom panel of figure 4. Conversely large particles entering near the bottom of the flow, move straight downslope until they reach the expansion fan. They are then squeezed upwards until they reach the top shock, after which they move straight downslope again, but at a much higher position than where they started. The detailed formulae of the particle paths can be found in Thornton et al. [22].

BREAKING SIZE SEGREGATION WAVES

Gray et al. [55] and Shearer et al. [52] have gone on to construct fully time and spatially dependent two-dimensional solutions to the segregation equation. These solutions show that inversely graded shocks can develop which have monotonically decreasing sections. As these are transported downstream the velocity shear causes the interface to steepen and it eventually breaks, as small particles are sheared over the top of large grains. Thornton and Gray [56] used shock capturing numerical simulations to show that these breaking waves precess like a spinning rugby ball and move downstream at approximately constant speed. The time-dependent behaviour of these lens-like features is extremely complicated [57], but eventually they settle down towards a steady travelling wave.

An exact solution for the breaking wave [56] can be constructed by transforming equation (19) into a frame

$$\xi = x - u_{lens}t, \tag{63}$$

which moves downslope at speed u_{lens}. The steady-state segregation equation in the moving frame is then

$$\frac{\partial}{\partial \xi}\left(\phi(u - u_{lens})\right) - \frac{\partial}{\partial z}\left(S_r \phi(1-\phi)\right) = 0, \tag{64}$$

which has the same form as (33) with a redefined velocity field. Defining a depth-integrated velocity coordinate

$$\psi(z) = \int_0^z u(z') - u_{lens} dz', \tag{65}$$

the method of characteristics again implies that the concentration is equal to ϕ_λ along the straight line

$$\psi = \psi_\lambda + S_r(2\phi_\lambda - 1)(\xi - \xi_\lambda), \tag{66}$$

emanating from $(\xi_\lambda, \psi_\lambda)$ in mapped coordinates. The shock condition (22) transforms to

$$\frac{d\psi}{d\xi} = S_r(\phi^+ + \phi^- - 1). \tag{67}$$

Equations (66) and (67) are the direct equivalents to equations (39) and (46) in the fixed domain.

Consider a unit depth avalanche that is inversely graded up slope of a breaking wave and has a region of purely large particles downstream of it. The inversely graded shock is assumed to lie at a height $z_{inverse}$ as shown in figure 5. The breaking size segregation wave has the property that there is no net flux of small particles across the wave. This can be expressed by the integral

$$\int_0^{z_{inverse}} \phi(u - u_{lens}) dz = 0. \tag{68}$$

Using the linear velocity field defined in (40) this implies that the speed of the lens is equal to

$$u_{lens} = \alpha + (1 - \alpha) z_{inverse}. \tag{69}$$

The depth-integrated velocity coordinate (65) then becomes

$$\psi = (1 - \alpha) z (z - z_{inverse}), \tag{70}$$

which is zero at $z = 0$ and $z = z_{inverse}$, and attains a minimum, $\psi_{lens} = -\frac{1}{4}(1 - \alpha) z_{inverse}^2$ at height $z_{lens} = z_{inverse}/2$. This height is special, because it is also the height at which there is no net velocity relative to the lens, i.e. $u(z_{lens}) = u_{lens}$. The height z_{lens} is shown as a dotted line in the top panel of figure 5. Particles above z_{lens} move downslope faster than the breaking wave, while grains below z_{lens} move downslope slower then the lens and are overtaken by it. In the moving frame, particles therefore move from left to right above the dotted line and from right to left below it.

The solution starts by assuming that there is an expansion fan, centred on the no mean flow line $\psi = \psi_{lens}$ at an arbitrary downstream position ξ_A, within which the concentration is

$$\phi_{top} = \frac{1}{2}\left(1 + \frac{\psi - \psi_{lens}}{S_r(\xi - \xi_A)}\right). \tag{71}$$

The fan is shown in the top panel of figure 5 emanating from point A. The characteristics for $\phi \in [1/2, 1]$ propagate upwards above the no mean flow line, and the $\phi = 1$ characteristic reaches the inversely graded shock at height $z_{inverse}$ when $\psi = 0$ at point B, which lies at $\xi_B = \xi_A - \psi_{lens}/S_r$. Here there are no more small particles to percolate downwards and the large grains separate out into a pure phase across a concentration shock. Substituting the fan concentration (71) and the large particle concentration into the shock condition (67) and integrating, subject to the condition that the shock starts at $(\xi_B, 0)$ in mapped coordinates, implies that the top shock

$$\psi_{top} = \psi_{lens} - S_r(\xi - \xi_A) + 2\sqrt{-\psi_{lens}}\sqrt{S_r(\xi - \xi_A)}, \tag{72}$$

where the constant $\sqrt{-\psi_{lens}}$ is real. The shock satisfies the Lax entropy condition because the concentration is inversely graded across it, and propagates down until it reaches the no mean flow line $\psi = \psi_{lens}$ at point C. This is the furthest downstream distance of the lens and has position $\xi_C = \xi_A - 4\psi_{lens}/S_r$. The shock (72) could be continued down into the lower domain, but because the flow changes direction this would imply that a pure region of large particles would lie beneath a region of mixed particles. This configuration is not inversely graded and the shock is therefore not admissible by the Lax entropy condition [52]. Instead a lower expansion fan forms, that is centred at (ξ_C, ψ_{lens}), and within which the concentration is

$$\phi_{bottom} = \frac{1}{2}\left(1 - \frac{\psi - \psi_{lens}}{S_r(\xi_C - \xi)}\right). \tag{73}$$

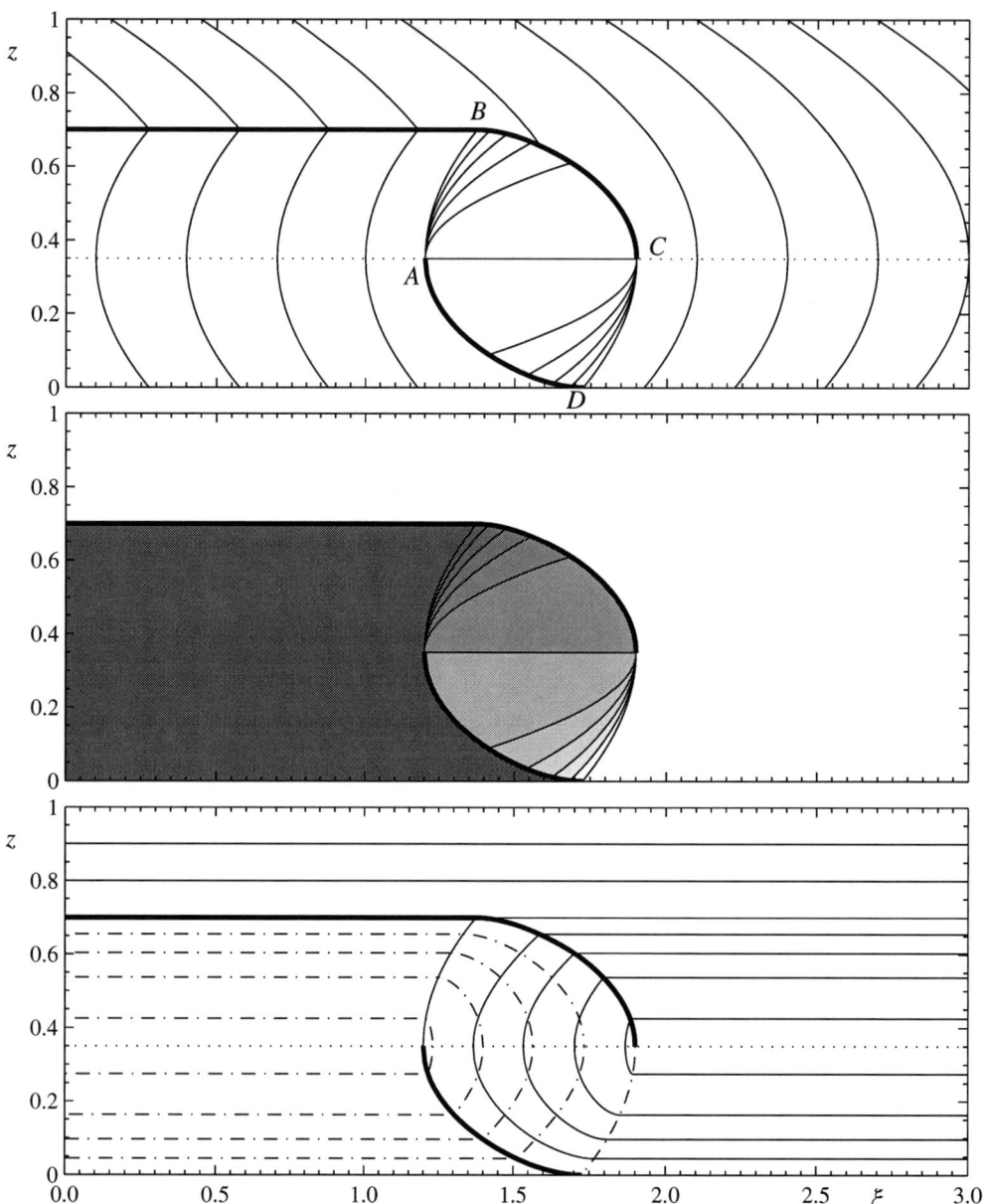

FIGURE 5. A breaking size segregation wave connects an upstream inversely graded avalanche from a downstream region of purely large grains. The height of the inversely graded interface $z_{inverse} = 0.7$, there is a linear velocity profile through the avalanche depth with $\alpha = 0.5$ and $S_r = 0.35$. In the top panel expansion fans are centred at points A and C and the shocks are shown with thick solid lines. The thin lines are characteristics, which change direction at the no mean flow line, which is marked by a dotted line. In the middle panel the concentration is shown using a contour scale with 0.1 unit intervals. Darker regions correspond to greater concentrations of fines and the the white region is composed of large particles. In the bottom panel the particle paths are shown in the moving frame ξ. Large particle paths are shown with solid lines and small particle paths with dot-dash lines.

This matches up with the upper expansion fan (71) centred at point A, since the concentration is equal to 1/2 in both cases on the no mean flow line $\psi = \psi_{lens}$. The characteristics emanating from point C lie in the range $\phi \in [0, 1/2]$ and curve downwards and backwards in the moving frame. The outermost, $\phi = 0$, characteristic hits the base of the avalanche at $\psi = 0$ at point D, which has position $\xi_D = \xi_C + \psi_{lens}/S_r$. Here there are no more large particles to be squeezed upwards and the small particles therefore separate out across a concentration shock. Solving the shock condition (67), with small particles on one side and the expansion fan (73) on the other, and, subject to the boundary condition that the shock starts from $(\xi_D, 0)$, implies that the bottom shock

$$\psi_{bottom} = \psi_{lens} - S_r(\xi_C - \xi) + 2\sqrt{-\psi_{lens}}\sqrt{S_r(\xi_C - \xi)}. \tag{74}$$

This reaches the zero mean flow line, $\psi = \psi_{lens}$, at $\xi = \xi_C + 4\psi_{lens}/S_r = \xi_A$, which is the same as point A. The Lax entropy condition [52] predicts that a continuation of the lower shock is not admissible and it is instead replaced by an expansion fan centred at (ξ_A, ψ_{lens}), justifying the original assumption (71). The solution is now complete and is illustrated using a contour scale in the central panel of figure 5. It consists of two shocks, (72) and (74), and two expansion fans, (71) and (73), that are arranged in a 'lens'-like structure that propagates downstream with speed u_{lens} given by (69). These breaking size segregation waves are a very important feature of granular avalanches, because they allow particles to circulate in the flow. Thornton and Gray [56] have used equations (50) to reconstruct the large and small particle paths, which are illustrated in the bottom panel of figure 5. Large particles below the no mean flow line are caught up by the lens, rise up through it and exit into faster moving regions of the flow. While small particles that are above the no mean flow line catch up with the lens, percolate downwards and exit into a basal layer that is moving slower than the lens. In the example in figure 5 all the small particles are recirculated. While large particles below z_{lens} are recirculated between z_{lens} and $z_{inverse}$, and large particles above $z_{inverse}$ simply move downstream faster than the breaking wave.

PARTICLE SIZE SEGREGATION AT BOULDERY FLOW FRONTS

Gray and Ancey [5] observed that coarse particle rich flow fronts form in small scale stratification experiments [15, 19, 20, 58, 59] performed in a two-dimensional Hele-Shaw cell with a 3mm gap. A particularly interesting feature of these flows was that the coarse rich front remained at almost constant length [5], with those large grains that reached it, being deposited to form a carpet of grains over which the rest of the avalanche flowed. This can be seen in figure 5, where the large particles that form the current avalanche, or have just been deposited by it, are highlighted in white. The large rich flow front is connected to a small particle sandwich in the interior, with a layer of inversely graded large particles on top, and a static carpet of deposited large grains at the base. The stratification experiments of Gray and Ancey [5] are very closely analogous to self-channelizing flows, since large particles that reach the front are removed, by basal deposition in two-dimensions and lateral transport in three-dimensions, allowing the more mobile interior to continue to propagate downslope.

Gray and Ancey [5]'s observations allowed them to construct a travelling wave solution for the particle size distribution at the flow front by switching to a frame (ξ, z) moving downslope at the speed of the front u_F. Introducing the transformation

$$\xi = x - u_F t, \quad \tau = t, \tag{75}$$

the depth-integrated mass balance, the segregation equation (19) and the basal kinematic condition [see e.g. 5, 60], for a steady-state solution in the moving frame, become

$$\frac{\partial}{\partial \xi}\left(h(\bar{u} - u_F)\right) = -d, \tag{76}$$

$$\frac{\partial}{\partial \xi}\left(\phi(u - u_F)\right) + \frac{\partial}{\partial z}\left(\phi w - S_r \phi(1 - \phi)\right) = 0, \tag{77}$$

$$(u_b - u_F)\frac{\partial b}{\partial \xi} - w_b = d, \tag{78}$$

where h is the avalanche thickness, b is the height of the basal deposit and d is the deposition rate. The basal velocity components u_b and w_b are assumed to be zero, which implies that $d = -u_F \partial h/\partial \xi$. Substituting this into (76) allows the depth-averaged mass balance equation to be integrated, subject to the condition that the flow front is located at $\xi = 0$, to show that

$$b = \lambda h, \quad \text{where} \quad \lambda = \frac{\bar{u} - u_F}{u_F}. \tag{79}$$

In order to close the model Gray and Ancey [5] assumed that $\bar{u}$ was equal to a constant throughout the flow. It follows the basal deposit height b is linearly related to the avalanche thickness h by the parameter λ and the constant depth-averaged velocity

$$\bar{u} = (1 + \lambda)u_F. \tag{80}$$

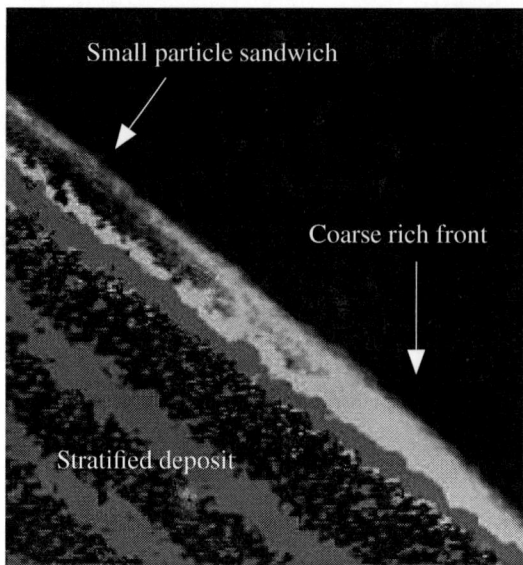

FIGURE 6. A composite image of the stratification pattern experiments of Gray and Ancey [5] that highlights the grains that are part of, or, have just been laid down by, the current avalanche. The flow front is almost entirely composed of large white sugar particles (500–600 μm), but behind it the active region has a layer of small dark iron spheres (210–420 μm) that are sandwiched between recently deposited large grains at the base and inversely graded large grains at the surface. The grains in the static stratified deposit that have been darkened, so that the large particles appear dark grey in colour. A complete sequence of stills can be found in Gray and Ancey [5] together with an animation of the flow in the online version of their paper.

Gray and Ancey [5] used the depth-averaged momentum balance to compute an exact solution for the avalanche thickness using Pouliquen and Forterre [61]'s basal friction law for rough beds. Here $h = h(\xi)$ is given. The downslope velocity is assumed to be linear with no slip at the base

$$u = \begin{cases} 2\bar{u}\left(\dfrac{z-b}{h}\right), & b \leq z \leq b+h, \\ 0, & 0 \leq z < b, \end{cases} \qquad (81)$$

and the incompressibility condition (6) then implies that the normal velocity

$$w = \begin{cases} \dfrac{\bar{u}}{h^2}\dfrac{dh}{d\xi}(z^2 - b^2), & b \leq z \leq b+h, \\ 0, & 0 \leq z < b. \end{cases} \qquad (82)$$

In two-dimensions the depth-integrated velocity coordinate

$$\psi = \int_0^z u(\xi, z') - u_F \, dz', \qquad (83)$$

is equivalent to the stream-function, since $\partial \psi / \partial z = u - u_F$ and $\partial \psi / \partial \xi = -w$. For the downslope velocity field defined in (81) this implies that

$$\psi = \begin{cases} \dfrac{\bar{u}}{h}(z-b)^2 - u_F z, & b \leq z \leq b+h, \\ -u_F z, & 0 \leq z < b, \end{cases} \qquad (84)$$

which is zero along the free-surface $z = b + h$ and the inclined base $z = 0$. Since the flow is steady, bulk particle paths are equal to lines of constant ψ. These are illustrated in the top panel of figure 7 using solid lines for paths that are deposited and dashed lines for those that are recirculated within the avalanche. The dotted line is the height $z_{u_F} = b + u_F h/(2\bar{u})$ where the velocity is equal to the front velocity, i.e. $u(z_{u_F}) = u_F$. Above the no mean flow line the bulk flow is from left to right, and below it is from right to left. The line also marks a local minimum in the stream-function coordinate $\psi_{u_F} = -u_F^2 h/(4\bar{u}) - u_F b$.

The stream-function coordinates are very useful for solving the segregation equation (77). Using incompressibility the segregation equation can be written in the quasi-linear form

$$(u - u_F)\frac{\partial \phi}{\partial \xi} + \left(w + S_r(2\phi - 1)\right)\frac{\partial \phi}{\partial z} = 0, \tag{85}$$

and the method of characteristics then implies that the concentration is equal to a constant value ϕ_λ along the characteristic curve $z = z(\xi)$ given by

$$(u - u_F)\frac{dz}{d\xi} - w = S_r(2\phi_\lambda - 1). \tag{86}$$

When $z = z(\xi)$ differentiating the stream-function ψ with respect to ξ, using Leibniz's rule [62] and the incompressibility condition (6), yields the important identity

$$\frac{d\psi}{d\xi} = (u - u_F)\frac{dz}{d\xi} - w. \tag{87}$$

which linearizes the characteristic equation (86). Solving for the characteristic starting from $(\xi_\lambda, \psi_\lambda)$ therefore yields a straight line

$$\psi = \psi_\lambda + S_r(2\phi_\lambda - 1)(\xi - \xi_\lambda), \tag{88}$$

in streamfunction coordinates. The identity (87) can also be used to show that the large and small particle paths (50) and the shock condition (22) also reduce to the familiar forms

$$\frac{d\psi^l}{d\xi} = S_r\phi, \quad \frac{d\psi^s}{d\xi} = -S_r(1 - \phi), \quad \frac{d\psi}{d\xi} = S_r(\phi^+ + \phi^- - 1), \tag{89}$$

respectively, even in the case of two-dimensional velocity fields. It is useful to note that the equation for the large particle path is identical to the equation for a shock condition when $\phi^+ = 1$. While, the small particle path equation is equivalent to the shock condition with $\phi^+ = 0$. Tracking large particles is therefore equivalent to solving for a shock with small particles on the forward side and tracking small particles is equivalent to solving for a shock with large particles on the other side.

Steady travelling wave solutions in a depositing flow field, only exist if all the large particles that reach the flow front are deposited. This necessarily implies that if the particles are inversely graded in the interior of the avalanche, the interface, ψ_L which lies along a bulk streamline, must lie in the region of those paths that are deposited, which are denoted by solid lines in the top panel of figure 7. Assuming that this is the case, the solution is very similar to the breaking wave solution of the previous section. There is an expansion fan centred at point A, which lies on the no mean flow line $\psi = \psi_{u_F}$. This expands into the upper domain of material moving towards the flow front and the concentration within the fan is

$$\phi_{top} = \frac{1}{2}\left(1 + \frac{\psi - \psi_A}{S_r(\xi - \xi_A)}\right), \tag{90}$$

where (ξ_A, ψ_A) is the position of point A in stream-function coordinates. The leading $\phi = 1$ characteristic AB intersects with the inversely graded layer $\psi = \psi_L$ at point B, which coordinates $\xi_B = \xi_A + (\psi_L - \psi_A)/S_r$. Since there are no more small particles above ψ_L to percolate downwards, a shock BC is generated between the expansion fan (90) on one side and a pure phase of large particles on the other. Solving the jump condition subject to the shock starting at (ξ_B, ψ_L) gives

$$\psi_{top} = \psi_A - S_r(\xi - \xi_A) + 2\sqrt{\psi_L - \psi_A}\sqrt{S_r(\xi - \xi_A)}. \tag{91}$$

The upper shock BC starts at (ξ_B, ψ_L) and propagates downwards, reaching the no-mean-flow line at (ξ_C, ψ_C). Below $\psi = \psi_{u_F}$ the continuation of the shock is unstable by the Lax entropy condition [52] and it breaks into a fan centred at $(\hat{\xi}_C, \hat{\psi}_C)$ in which the concentration

$$\phi_{bottom} = \frac{1}{2}\left(1 - \frac{\psi - \psi_C}{S_r(\xi_C - \xi)}\right). \tag{92}$$

The lead $\phi = 0$ characteristic CD intersects the $\psi = \psi_L$ particle path again at point D, which has coordinates $\xi_D = \xi_C - (\psi_L - \psi_C)/S_r$. For steady states the breaking segregation wave recirculates large particles on the ψ_L particle

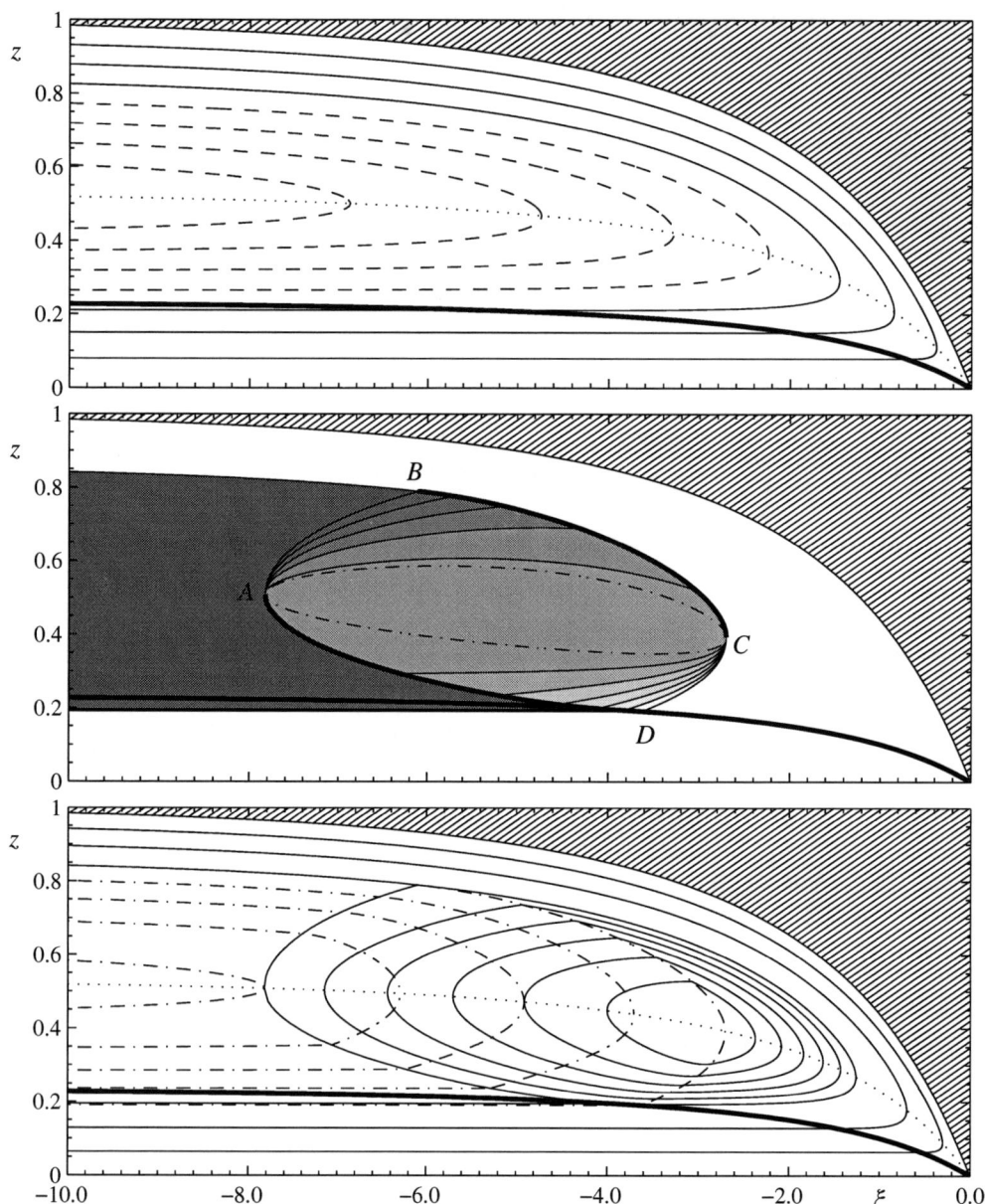

FIGURE 7. The travelling wave solution for the bulk particle paths (top), concentration (middle) and the large and small particle paths (bottom) are shown for an avalanche flow front that is propagating downslope and depositing grains. The deposition surface b is marked by a thick solid line and the dotted line is the no mean flow line z_{u_F}. Thin solid lines are used to indicate bulk flow paths that are deposited, while dashed lines show those paths that recirculate within the avalanche for $\lambda = 0.3$. In the middle panel a breaking size segregation wave ABCD connects an upstream regime that resembles a small particle sandwich, as in figure 6, with a downstream region of purely large grains. There is a "eye" of constant concentration in the centre of the lens, which is marked by a dot-dash line. The concentration is shown using a grey scale with 0.1 unit intervals. Regions of all large particles are white. In the bottom panel the large particle paths are shown with solid lines and small particle paths with dot-dash lines. The inversely graded interface lies along the streamline $\psi_L = -0.1931$ and $S_r = 1$.

path. Part of the path DA also forms a shock across which the small particles separate out, but since the equations for this shock and particle path are identical (89), it is not necessary to determine the location of the transition until later. The large particle path equation adjacent to the fan (92) can be solved subject to the condition that it starts at (ξ_D, ψ_L)

to show that

$$\psi_{bottom} = \psi_C - S_r(\xi_C - \xi) + 2\sqrt{\psi_L - \psi_C}\sqrt{S_r(\xi_C - \xi)}. \tag{93}$$

This intersects with the no-mean-flow line $\psi = \psi_{u_F}$ again at (ξ_A, ψ_A) and breaks into an expansion [52] consistent with our original assumption in equation (90). The small particle sandwich is therefore connected to the large particle front by a breaking size segregation wave as in the middle panel of figure 7. There is an interesting new feature in this case. Since $\psi_A \neq \psi_C$, there is an additional central "eye" of constant concentration

$$\phi_{eye} = \frac{1}{2}\left(1 + \frac{\psi_C - \psi_A}{S_r(\xi_C - \xi_A)}\right), \tag{94}$$

that is bounded above and below by the dot-dash line in the central panel of figure 7. More importantly there is a unique position for the breaking size segregation wave, which connects the upper and lower branches of the ψ_L bulk particle path. If it is too far upstream, part of the lens intersects the basal topography, which violates the assumption that $S_r \neq 0$. While if the lens is too far downstream, not all the large particles above the incoming ψ_L particle path are deposited. The breaking wave must therefore be positioned so that large particles on ψ_L^- are recirculated at the front, while large particles on ψ_L^+ side are deposited. This amounts to the requirement that the ψ_L recirculating particle path is tangent to the basal topography. Gray and Ancey [5] calculated the steady particle paths explicitly, and they are shown in the bottom panel of figure 7. This shows that at steady state all the incoming large particles are deposited, but some of those that had previously reached the front are recirculated, which has been observed in experiment [11, 12]. Most of the small particles are recirculated within the flow, but there are a few that are deposited. Rather intriguingly even though the parent flow is inversely graded, the deposit that is generated by this combination of flow and deposition is normally graded with small particles on top or large. This is diametrically opposite to the inversely graded distribution that is obtained when the flow is brought to rest by a shock wave [15] and raises many questions on the interpretation of deposits that are often made by geologists.

DISCUSSION AND CONCLUSIONS

The segregation-remixing equation (19) provides a simple, but effective way of modelling particle size segregation in granular avalanches. At present little is known about the dependence of the parameters S_r and D_r on the particle size ratio, shear-rate, slope angle and lithostatic pressure, but, both particle dynamics simulations [44] and careful experiments [5], provide important means of calibrating the theory. It may also be possible to find a link with binary kinetic theories [49, 50] as they push towards the dense flow regime. All the results that have been presented in this paper are for a prescribed flow field, but, the theory is sufficiently simple to envisage coupled simulations in which the evolving particle size distribution has a direct feedback on to the bulk flow field, which can be computed with existing avalanche models [7, 25–37]. Such *segregation-mobility* feedback effects are responsible for fingering instabilities in bi-disperse mixtures of dry grains with different frictional properties [11–13], as well as petal formation in rotating drums [18], and the spontaneous formation of leveed channels in both wet and dry geophysical mass flows [6–9, 14].

ACKNOWLEDGMENTS

This research was generously supported by an EPSRC Advanced Research Fellowship GR/S50052/01 & GR/S50069/01 as well as NERC grant NE/E003206/1 and a Royal Society International Travel Grant TG090172.

REFERENCES

1. S. Savage, and C. Lun, *J. Fluid Mech.* **189**, 311–335 (1988).
2. G. Middleton, and M. Hampton, "Subaqueous sediment transport and deposition by sediment gravity flows," in *Marine sediment transport and environmental management*, edited by D. Stanley, and D. Swift, Wiley, 1976, pp. 197–218.
3. R. Bagnold, *Proc. Roy. Soc. Lond. A* **225**, 49–63 (1954).
4. M. Branney, and B. Kokelaar, *Bull Volcanol.* **54**, 504–520 (1992).
5. J. Gray, and C. Ancey, *J. Fluid Mech.* **629**, 387–423 (2009).
6. T. Pierson, "Flow behavior of channelized debris flows, Mount St. Helens, Washington," in *Hillslope Processes*, edited by A. Abrahams, Allen and Unwin, Winchester, Mass., 1986, pp. 269–296.

7. R. Iverson, *Reviews in Geophysics* **35**, 245–296 (1997).
8. V. Jomelli, and P. Bertran, *Geografiska Annaler. Series A, Physical Geography* **83**, 15–28 (2001).
9. P. Bartelt, and B. McArdell, *J. Glaciol.* **55**, 829–833 (2009).
10. R. Iverson, "The debris-flow rheology myth," in *Debris-flow hazards mitigation: Mechanics, prediction and assessment*, edited by D. Rickenmann, and C. Chen, Millpress, Rotterdam, 2003, pp. 303–314.
11. O. Pouliquen, J. Delour, and S. Savage, *Nature* **386**, 816–817 (1997).
12. O. Pouliquen, and J. Vallance, *Chaos* **9**, 621–630 (1999).
13. G. Félix, and N. Thomas, *Earth and Planetary Science Letters* **221**, 197–213 (2004).
14. R. Iverson, and J. Vallance, *Geology* **29**, 115–118 (2001).
15. J. Gray, and K. Hutter, *Continuum Mech. & Thermodyn.* **9**, 341–345 (1997).
16. J. Gray, and V. Chugunov, *J. Fluid Mech.* **569**, 365–398 (2006).
17. K. Hill, G. Gioia, and D. Amaravadi, *Phys. Rev. Lett.* **93**, 224301 (2004).
18. I. Zuriguel, J. Gray, J. Peixinho, and T. Mullin, *Phys. Rev. E* **73**, 061302 (2006).
19. S. Williams, *Powder Technol.* **2**, 13–20 (1968).
20. H. Makse, S. Havlin, P. King, and H. Stanley, *Nature* **386**, 379–382 (1997).
21. J. Gray, and A. Thornton, *Proc. Roy. Soc. A* **461**, 1447–1473 (2005).
22. A. Thornton, J. Gray, and A. Hogg, *J. Fluid Mech.* **550**, 1–25 (2006).
23. C. Truesdell, *Rational Thermodynamics*, Springer, 1984.
24. L. Morland, *Surveys in Geophysics* **13**, 209–268 (1992).
25. S. Grigorian, M. Eglit, and I. Iakimov, *Snow, Avalanches & Glaciers. Tr. Vysokogornogo Geofizich Inst* **12**, 104–113 (1967).
26. S. Savage, and K. Hutter, *J. Fluid Mech.* **199**, 177–215 (1989).
27. J. Gray, M. Wieland, and K. Hutter, *Proc. Roy. Soc. A* **455**, 1841–1874. (1999).
28. M. Wieland, J. Gray, and K. Hutter, *J. Fluid Mech.* **392**, 73–100 (1999).
29. O. Pouliquen, *Phys. Fluids* **11**, 542–548 (1999).
30. R. Iverson, and R. Denlinger, *J. Geophy. Res.* **106**, 553–566 (2001).
31. Y.-C. Tai, S. Noelle, J. Gray, and K. Hutter, *J. Comput. Phys.* **175**, 269–301 (2002).
32. J. Gray, Y. Tai, and S. Noelle, *J. Fluid Mech.* **491**, 161–181 (2003).
33. E. Pitman, C. Nichita, A. Patra, A. Bauer, M. Sheridan, and M. Bursik, *Phys. Fluids* **15**, 3638–3646 (2003).
34. U. Gruber, and P. Bartelt, *Envirn. Model. & Softw.* **22**, 1472–1481 (2007).
35. J. Gray, and X. Cui, *J. Fluid Mech.* **579**, 113–136. (2007).
36. A. Mangeney, F. Bouchut, N. Thomas, J. Vilotte, and M. Bristeau, *J. Geophys. Res.* **112**, F02017 (2007).
37. X. Cui, J. Gray, and T. Johannesson, *J. Geophys. Res.* **112**, F04012 (2007).
38. G. Kynch, *Trans. Faraday Soc.* **48**, 166–176 (1952).
39. H.-K. Rhee, R. Aris, and N. Amundson, *First-order partial differential equations: Volume 1 Theory and applications of single equations*, Prentice-Hall, Englewood Cliffs, New Jersey, 1986.
40. J. Vallance, and S. Savage, "Particle segregation in granular flows down chutes,," in *IUTAM Symposium on segregation in granular materials*, edited by A. Rosato, and D. Blackmore, Kluwer, 2000.
41. G. Whitham, *Linear and nonlinear waves.*, John Wiley, New York, 1974.
42. V. Dolgunin, and A. Ukolov, *Powder Technol.* **83**, 95–103 (1995).
43. D. Khakhar, J. McCarthy, and J. Ottino, *Phys. Fluids* **9**, 3600–3614 (1997).
44. D. Khakhar, J. McCarthy, and J. Ottino, *Chaos* **9**, 594–610 (1999).
45. P. Rognon, J. Roux, M. Naaim, and F. Chevoir, *Phys. Fluids* **19**, 058101 (2007).
46. E. Linares-Guerrero, C. Goujon, and R. Zenit, *J. Fluid Mech.* **593**, 475–504 (2007).
47. S. Hajra, and D. Khakhar, *Phys. Rev. E* **69**, 031304 (2004).
48. J. Jenkins, and F. Mancini, *J. Appl. Mech.* **54**, 27–34 (1987).
49. J. Jenkins, "Particle segregation in collisional flows of inelastic spheres," in *Physics of dry granular media*, edited by H. Herrmann, J.-P. Hovi, and S. Luding, NATO ASI series, Kluwer, 1998.
50. J. Jenkins, and D. . Yoon, *Phys. Rev. Lett.* **88**, 194301–4 (2001).
51. P. Chadwick, *Continuum Mechanics. Concise theory and problems*, George Allen & Unwin, 1976.
52. M. Shearer, J. Gray, and A. Thornton, *European J. Applied Math.* **19**, 61–86 (2008).
53. J. Cole, *Q. Appl. Maths* **9**, 225–236 (1951).
54. E. Hopf, *Comm. Pure Appl. Math.* **3**, 201–230 (1950).
55. J. Gray, M. Shearer, and A. Thornton, *Proc. Roy. Soc. A* **462**, 947–972 (2006).
56. A. Thornton, and J. Gray, *J. Fluid Mech* **596**, 261–284 (2008).
57. M. McIntyre, E. Rowe, M. Shearer, J. Gray, and A. Thornton, *A.M.R.X.* **2007**, abm008 (2007).
58. H. Herrmann, "On the shape of a sandpile," in *Physics of dry granular media*, edited by H. Herrmann, J. Hovi, and S. Luding, NATO ASI series, Kluwer Academic, 1998, pp. 697–702.
59. J. Baxter, U. Tüzün, D. Heyes, I. Hayati, and P. Fredlund, *Nature* **391**, 136 (1998).
60. J. Gray, *J. Fluid Mech.* **441**, 1–29 (2001).
61. O. Pouliquen, and Y. Forterre, *J. Fluid Mech.* **453**, 133–151 (2002).
62. M. Abramowitz, and I. Stegun, *Handbook of Mathematical Functions*, Dover, 1970, p. 3.3.7, 9 edn.

Size Segregation in Dry Granular Flows of Binary Mixtures

Michele Larcher[a] and James T. Jenkins[b]

[a]*Department of Civil and Environmental Engineering and CUDAM, University of Trento, Trento 38050, Italy*
[b]*Department of Theoretical and Applied Mechanics, Cornell University, Ithaca, New York 14853, USA*

Abstract. We phrase and solve a problem of particle segregation in a dry flow of two sizes of spheres down an inclined bed in a wide channel. The flow is assumed to be steady and fully-developed, collisions between particles are dissipative, and the sizes and masses of the particles are not too different. We restrict our analysis to dense flows and use an extension of kinetic theory to predict the profiles of the volume fractions of the two species. For a particular ratio of the radii and of the masses of the two species, segregation is not predicted.

Keywords: size segregation, extended kinetic theory, dense granular flow, binary mixture.
PACS: 45.70.-n, 45.70.Mg, 47.57.Gc

INTRODUCTION

Particle segregation has important implications both for industrial activities, involving the production and transportation of granular media, and for natural phenomena. In some environmental phenomena, such as debris flows, reverse grading takes place, inducing a large concentration of big boulders on the top and in the front of the flow. This has important effects both on the rheology of the mixture, altering crucial parameters like the depth and the velocity of the flow, and on the characteristics and consequences of the impact of the debris flow on infrastructures.

Despite the importance of particle segregation, the theoretical framework developed for its description is still incomplete. We applied in the past [1] the simplest theory for segregation that results from the kinetic theory for binary mixtures, as derived by Arnarson and Jenkins [2], to collisional flows driven by a clear fluid down an inclined erodible bed, in which little energy is dissipated in collisions between two kinds of spheres that do not differ much in size and mass. In this paper we address dry flows of frictionless, inelastic spheres in the dense limit using the kinetic theory developed by Garzo & Dufty [3], simplified by neglecting the small terms introduced by their function $c*$ [4], and a modified expression for energy dissipation to take in to account the formation of particle clusters [5, 6].

The flow model is based on the balance of mass, momentum, and energy for the mixture of fluid and particles. However, in the dense limit, the volume fraction of the granular phase can be roughly considered as constant and the profile of the granular temperature can be derived directly utilizing the normal momentum balance of the mixture. The segregation is described by an equation related to the difference of the momentum balances of the individual species.

We focus on steady, fully developed flows, driven by gravity, of a mixture of spherical particles with two different radii, r_A and r_B, and material mass densities ρ_A and ρ_B. We predict profiles of species volume fractions, or concentrations, c_A and c_B, as well as the particular combinations of radii and density ratios for which there is no segregation.

In what follows, lengths have been made dimensionless by the average particle diameter d, velocities by $(gd)^{1/2}$, where g is the buoyant gravitational acceleration, stresses by $\rho_p g d$, and energy flux by $\rho_p(gd)^{3/2}$, where ρ_p is the ratio of the average particle masses to the average particle volumes.

THE BALANCE EQUATIONS

Segregation

The species momentum balance in normal direction of particle species A and B can be expressed as follows:

$$0 = -\frac{dp_A}{dy} - c_A \cos\phi + \Phi, \qquad 0 = -\frac{dp_B}{dy} - c_B \cos\phi - \Phi, \qquad (1)$$

where p_A and p_B are the partial pressures of the two species, c_A and c_B are the species volume fractions, ϕ is the inclination angle of the bed and Φ is the term representing the interaction between the two types of particles. The coordinate y is oriented upwards, is orthogonal to the bed, and has its origin there.

The sum of the two species' normal momentum balances leads to the global normal momentum balance of the granular mixture, where $c = c_A + c_B$ is the concentration, or volume fraction, of the solid phase and $p = p_A + p_B$ is the total granular pressure:

$$0 = -\frac{dp}{dy} - c\cos\phi. \qquad (2)$$

The weighted difference of the species momentum balances can be used to write an equation suitable for deriving the concentration profile of the two particle species.

$$-c_B \frac{dp_A}{dy} + c_A \frac{dp_B}{dy} + c\Phi = 0. \qquad (3)$$

The partial pressures can be expressed by means of the following expressions [2]:

$$p_A = n_A(1 + K_{AA} + K_{AB})T \quad \text{and} \quad p_B = n_B(1 + K_{BB} + K_{BA})T, \qquad (4)$$

where n_A and n_B are the number of particles per unit volume of type A and B, respectively and the mixture granular temperature T is evaluated as a weighted sum of the temperatures of the two species T_A and T_B:

$$T = (n_A T_A + n_B T_B)/n, \qquad (5)$$

with

$$T_i \equiv \frac{m_i \langle (u'_i)^2 \rangle}{3} \quad i,j = A,B, \qquad (6)$$

where u'_i is the particle velocity fluctuation and $n = n_A + n_B$.

The transport coefficients K_{ij} are given as a function of the radii ratio, of the concentration of species i and of the radial distribution functions g_{ij}:

$$K_{ij} = \frac{1}{2} c_j g_{ij} \left(1 + \frac{r_i}{r_j}\right)^3 \quad i,j = A,B. \qquad (7)$$

In equation (7) we adopted the expression proposed by Mansoori et al. [7] for the radial distribution functions as a function of the radii, of the volume fractions and on the number densities of the two species, in which $\xi = 4\pi(n_A r_A^2 + n_B r_B^2)/3$:

$$g_{ij} = \frac{1}{1-c} + \frac{3r_i r_j}{r_i + r_j} \frac{\xi}{(1-c)^2} + 2\left(\frac{r_i r_j}{r_i + r_j}\right)^2 \frac{\xi^2}{(1-c)^3}. \tag{8}$$

Particle interactions can be described in equation (3) by means of the following expression:

$$\Phi = K_{AB} n_A T \left\{ \frac{m_B - m_A}{m_B + m_A} \frac{d}{dy}(\ln T) + \frac{d}{dy}\left[\ln\left(\frac{n_A}{n_B}\right)\right] \right\}. \tag{9}$$

In what follows, instead of using the granular temperature, we introduce the variable w, with

$$w^2 = 2T/(m_A + m_B). \tag{10}$$

In the case of small differences in radii and masses, equation (3) reduces to a differential equation that can be utilized to predict particle segregation by solving for a parameter x that characterizes the relative presence of the two species:

$$\frac{dx}{dy} = -\left(\frac{1-4x^2}{2}\right)\left[\frac{dw}{dy}\frac{1}{w}(R_1 \delta r + \Gamma_1 \delta m) + \frac{1}{2w^2}(R_2 \delta r + \Gamma_2 \delta m)\cos(\phi)\right], \tag{11}$$

where

$$x \equiv \frac{1}{2}\frac{n_A - n_B}{n}, \tag{12}$$

$$\delta r = r_A/r_B - 1, \tag{13}$$

$$\delta m = (m_A - m_B)/(m_A + m_B). \tag{14}$$

Equation (11), originally proposed by Arnarson & Jenkins [2], was slightly modified in order to introduce the effect of an inclined bed [1]. The parameters R_1, R_2, Γ_1 and Γ_2 are given as functions of the granular concentration c:

$$R_1 = \frac{5}{58}\left[2 + \frac{c(3-c)}{2-c} - \frac{12}{5}G\right] + 2G\left[3 + \frac{c(3-c)}{2-c}\right] - \frac{12cH_2(1+4G)}{1+4G+4cH_2}, \quad R_2 = -\frac{12cH_2}{1+4G+4cH_2}, \tag{15}$$

$$\Gamma_1 = \frac{179}{29}G + \frac{105}{116}, \quad \Gamma_2 = 2, \tag{16}$$

where G/c gives the dependence of the radial distribution function for a pair of particles in contact at the dense limit [9]:

$$G = 5.6916c\frac{0.60 - 0.49}{0.60 - c}. \tag{17}$$

When x is determined, the solid concentration of species A and B can be calculated, if the global solid concentration and the species radii ratio are given:

$$c_A = -\frac{(2x+1)(r_A/r_B)^3 c}{(2x-1)-(2x+1)(r_A/r_B)^3}, \quad c_B = \frac{(2x-1)c}{(2x-1)-(2x+1)(r_A/r_B)^3}. \tag{18}$$

Kinetic theory in the dense limit

Under the hypothesis that the granular concentration c is almost constant throughout the flow depth H, which is reasonable for dense flows [5, 6], the normal momentum balance given by equation (2) can be integrated to a typical height y, obtaining:

$$p = c(H-y)\cos\phi. \tag{19}$$

The global particle pressure can also be given in terms of the global granular concentration and of the granular temperature of the mixture T [1, 2], where $\delta r = r_A/r_B - 1$ and the coefficient of restitution e is assumed to be equal for both species:

$$p = 2(1+e)cGFw^2(1-3x\delta r). \tag{20}$$

Combining equations (19) and (20), we obtain:

$$w^2 = \frac{1}{2(1+e)G}(H-y)\cos(\phi)(1+3x\delta r). \tag{21}$$

Equation (21) can be used to derive the following two expressions:

$$\frac{dw}{dy}\frac{1}{w} = -\frac{1}{2}\frac{1}{(H-y)}, \tag{22}$$

and

$$\frac{\cos(\phi)}{w^2} = 2(1+e)G\frac{1}{(H-y)}(1-3x\delta r). \tag{23}$$

Equations (22) and (23) can then be employed in equation (11), which becomes

$$\frac{dx}{dy} = -\left(\frac{1-4x^2}{4}\right)\left(\frac{1}{H-y}\right)\left[-(R_1\delta r + \Gamma_1\delta m) + 2(R_2\delta r + \Gamma_2\delta m)(1+e)G\right]. \tag{24}$$

In the present form, the equation (24) describing the relative presence of the two species does not exhibit an explicit dependence on the angle of inclination of the bed or the intensity of the gravity acceleration. As a consequence, the concentration profile of the two species in a fully developed flow will be predicted to be independent of gravity and the inclination. Actually, even if it is not evident in equation (24), a dependence on the inclination is still present, because the coefficients R_1, R_2 and Γ_1 depend on the total concentration c that is here assumed to be roughly constant, because we operate in the dense limit. In this limit, the global concentration c can be shown to depend on the inclination angle of the flow, according to the following relation [4], also represented graphically in figure 1:

$$c = 0.60\left\{1 - 0.63\left[\frac{15}{\pi^{3/2}}\left(\frac{4}{5}\right)^3 2E^2\frac{(1-e)}{(1+e)^2}(\tan\phi)^3\right]^{-3}\right\}. \tag{25}$$

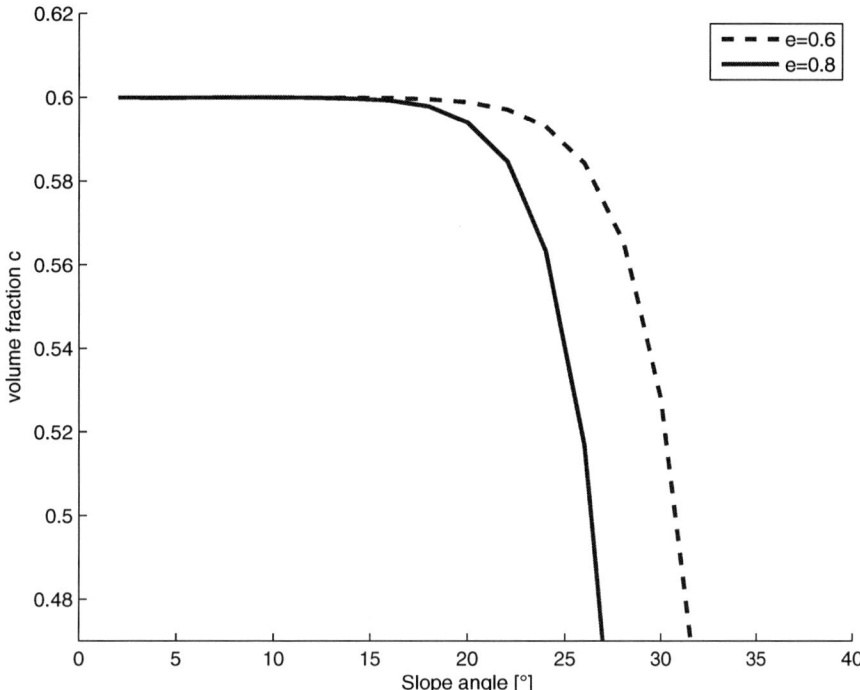

FIGURE 1. Relation between the volume fraction of the granular phase, assumed to be roughly constant in the dense limit, and the inclination angle of the flow.

The absence of the dependence of the steady-state solution on the intensity of the gravity acceleration is substantially in agreement with the original theory based on the kinetic sieving mechanism proposed by Savage and Lun [8]. However, in our case, this is a peculiarity of the steady, fully developed flow only, in which the time dependent terms in the segregation equation are null. On the contrary, the intensity of gravity plays a role in the complete equation that is necessary to solve the unsteady problem. That is, while the steady state is independent of gravity, the process by which it is achieved does depend upon it. This property is also exhibited in the model proposed by Gray & Thornton [10].

The Boundary Value Problem

Based on the considerations above, a very simple boundary value problem can be phrased and solved in terms of the normalized height $z = y/H$. Equation (24) can be associated with a mass balance of the two species that incorporates the knowledge of the relative amount of each in the flow:

$$\begin{cases} \dfrac{dx}{dz} = -\left(\dfrac{1-4x^2}{4}\right)\left(\dfrac{1}{1-z}\right)\left[-(R_1\delta r + \Gamma_1\delta m) + 2(R_2\delta r + \Gamma_2\delta m)(1+e)G\right] \\ \dfrac{d}{dz}I_x = x \end{cases} \qquad (26)$$

where

$$I_x(z) = \int_0^z x\,d\xi. \qquad (27)$$

The boundary conditions are that the integral of x is zero at the bed and equal to $f_A - 1/2$ at the free surface, where $f_A = n_A / n$ is the fraction of the total number of particles of type A, which is assumed to be known:

$$\begin{cases} I_x(0) = 0 \\ I_x(1) = f_A - 1/2 \end{cases}. \qquad (28)$$

Finally, given the inclination of the flow, equation (25) can be used to determine the total concentration c of the granular phase.

The two-point boundary value problem has been integrated using the Matlab solver *bvp4c* [11] in order to obtain the profiles of the concentration c_A and c_B of the two particle species in the case of a flow characterized by $e = 0.8$, $r_A / r_B = 1.8$, $m_A / m_B = 1.8^3$, $c = 0.49$ and $f_A = 0.14$. The results are shown in Figure 2.

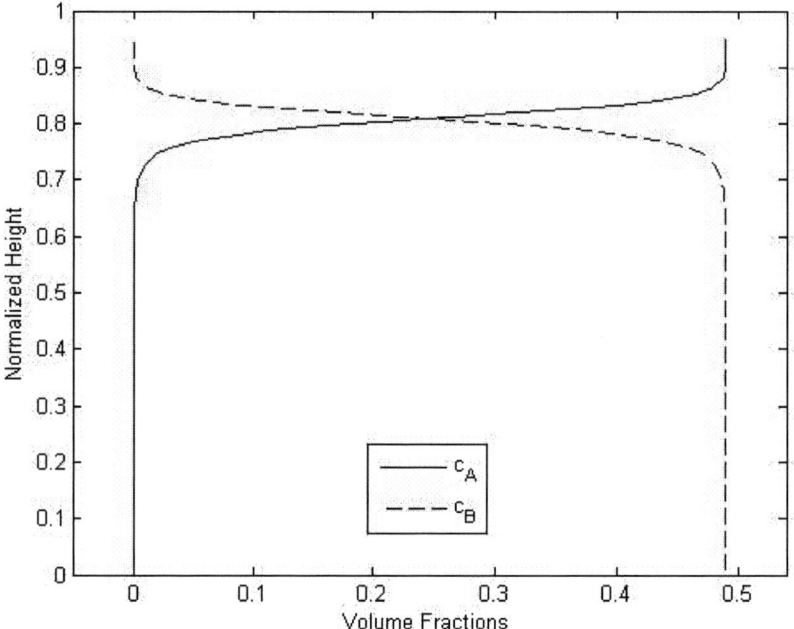

FIGURE 2. Profiles of the concentration c_A and c_B of the two species obtained solving the boundary value problem given by equations (26) and (28), when $e = 0.8$, $r_A / r_B = 1.8$, $m_A / m_B = 1.8^3$, $c = 0.49$ and $f_A = 0.14$.

The concentration profile of the two species, evaluated using equation (18) after obtaining x as a solution of the boundary value problem, presents an inverse-grading structure, in the sense that the larger particles are concentrated at the top of the flow. The two species are mixed only in a rather thin layer of the flow, where a smooth transition from the sole presence of one species to the sole presence of the other takes place.

However, inverse grading is not the only condition that can be predicted by this simple model, as will be pointed out in the next section.

Special case: no segregation

Equation (24) shows that the relative presence of the two species is predicted to be constant throughout the flow depth if the following condition is satisfied:

$$-(R_1 \delta r + \Gamma_1 \delta m) + 2(R_2 \delta r + \Gamma_2 \delta m)(1 + e)G = 0. \qquad (29)$$

As a consequence, no segregation is expected when the mass ratio and the radii ratio of the two species are related by

$$\frac{dm}{dr} = \frac{R_1 - 2R_2(1+e)G}{2\Gamma_2(1+e)G - \Gamma_1}. \tag{30}$$

Moreover equation (30), represented graphically in figure 3, indicates the value of *dm/dr* giving the threshold between inverse grading, with larger particles on top, and normal grading, with larger particles at the bottom.

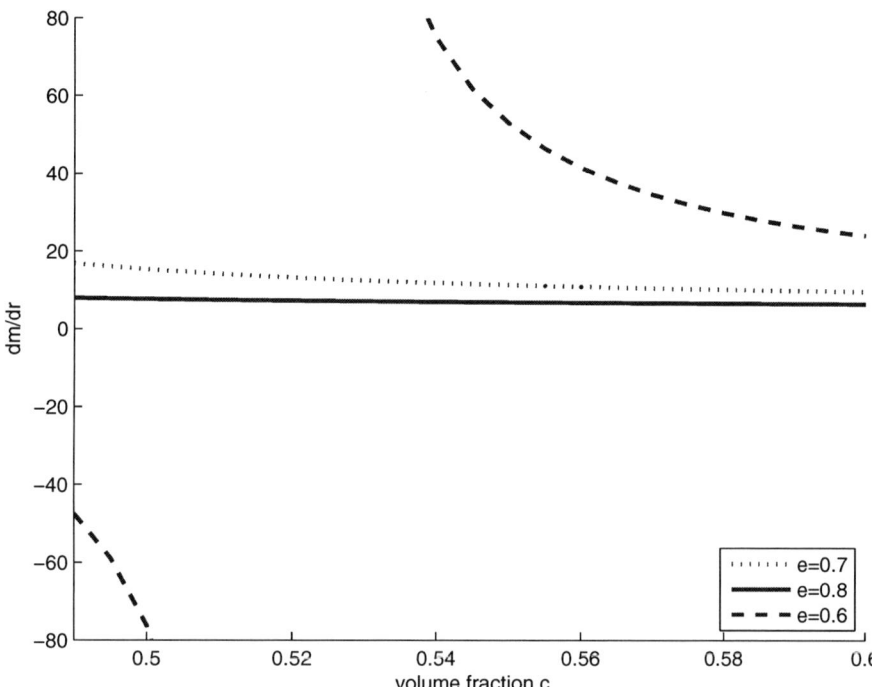

FIGURE 3. Graphic representation of the ratio *dm / dr* corresponding to the absence of segregation, as predicted by equation (30). The curves are given as a function of the granular concentration c and parameterized by the coefficient of restitution e.

CONCLUSION

We proposed a simple model for fully developed, dry, dense granular flows, driven by gravity on an incline, involving two species of spherical particles with small differences in mass and radii and the same coefficient of restitution. When the coefficient of restitution and the granular concentration are given, the larger particles can migrate towards the top or towards the bottom, depending on the relationship between the radii ratio and the mass ratio. For any values of the granular concentration and coefficient of restitution, only one combination of radii and mass ratios exists for which the two species are mixed throughout the flow depth.

The steady state solution is independent on the magnitude of the gravity acceleration, but gravity does influence the transitory process leading to the formation of a fully developed flow.

ACKNOWLEDGMENTS

The authors are grateful to the Departments of Civil and Environmental Engineering at the University of Trento and Theoretical and Applied Mechanics at Cornell University for their financial support and hospitality.

REFERENCES

1. M. Larcher and J.T. Jenkins, *Powders and Grains 2009, American Institute of Physics 978-0-7354-0682-7/09*, 1055-1058.

2. B. Arnarson and J.T. Jenkins, *Phys. Fluids* **16**, 4543-4550 (2004).
3. V. Garzo and J.W. Dufty, *Phys Rev. E* **59**, 5895-5911 (1999).
4. J.T. Jenkins and D. Berzi, *Granular Matter* (In press).
5. J.T. Jenkins, *Phys. Fluids* **18**, 103307 (2006).
6. J.T. Jenkins, *Granular Matter* **10**, 47-52 (2007).
7. G.A. Mansoori, N. F. Carnahan, K. E. Starling, and T. W. Leland, *J. Chem. Phys.* **54**, 1523 (1971).
8. S.B. Savage, C.K.K. Lun, *J. Fluid Mech.* **189**, 311-335 (1988).
9. S. Torquato, *Phys. Rev. E* **51**, 3170-3182 (1995).
10. J.M.N.T. Gray and A.R. Thornton, *Proc. R. Soc. A* **461**, 1447-1473 (2005).
11. L.F. Shampine, J. Kierzenka and M.W. Reichelt, *www.mathworks.com/bvp_tutorial* (2000).

The Gray-Thornton Model of Granular Segregation

Michael Shearer*, Lindsay B. H. May*, Nicholas Giffen* and Karen E. Daniels[†]

*Department of Mathematics, NC State University, Raleigh, NC, USA
[†]Department of Physics, NC State University, Raleigh, NC, USA

Abstract. In this paper, we explore properties of the Gray-Thornton model for particle size segregation in granular avalanches. The model equation is a single conservation law expressing conservation of mass under shear for the concentration of the smaller of two types of particle in a bidisperse mixture. Sharp interfaces across which the concentration jumps are shock wave solutions of the partial differential equation. We show that they can form internally from smooth data, as well as propagate in from boundaries of the domain. We prove a general stability result that expresses the physically reasonable notion that an interface should be stable only if the concentration of small particles is larger below the interface than above. Once shocks form, they are sheared by the flow, leading to loss of stability when an interface becomes vertical. The subsequent evolution of a mixing zone, a two-dimensional rarefaction solution of the equation that replaces the unstable part of the shock can be tracked explicitly for a short time. We conducted experiments to test the continuum model against real flow in a Couette geometry, in which a bidisperse mixture is confined in the annular region between concentric vertical cylinders. Initially, the material is placed in the annulus with a layer of large particles below a layer of small particles. The sample is then sheared by rotating the bottom confining plate, while a heavy top plate is allowed to move vertically to accommodate Reynolds dilatancy. Comparison to predictions of the model show reasonable agreement with the rate at which the sample mixes, and with the rate of the subsequent resegregation. However, the model naturally fails to capture short-time dilatancy, finite size effects, or three-dimensional effects.

Keywords: Segregation,granular materials,shock waves,hyperbolic conservation laws
PACS: 81.05.Rm,83.80.Fg

INTRODUCTION

The mechanisms of kinetic sieving and squeeze-explusion, that induce granular materials of different size but same density to segregate, were incorporated into a mathematical model by Savage and Lun in their classic paper [10] on steady chute flow. More recently, Gray and Thornton [4] gave a different derivation of a time-dependent continuum model using mixture theory. In this paper, we consider the Gray-Thornton model in two space variables (x,z), and time t. The model is a scalar conservation law describing the evolution of the volume fraction $\varphi(x,z,t)$ of small particles as the material is sheared by a given depth-dependent velocity $u(z)$ in the x-direction (for example down a chute). The velocity $u(z)$, assumed to be time-independent, transports particles down the slope, and induces shear through the shear rate $u'(z)$. The Gray-Thornton equation takes the form

$$\varphi_t + u(z)\varphi_x + (S\varphi(\varphi-1))_z = 0, \qquad -\infty < x < \infty,\ -1 < z < 1,\ t > 0, \qquad (1)$$

In this equation, segregation is driven by the normal velocity of small particles, taken to be proportional to the volume fraction $1-\varphi$ of large particles. The constant of proportionality S sets the time scale, the segregation rate. In avalanche flow, the parallel velocity $u(z)$ is roughly linear, and a constant segregation rate is a reasonable assumption. However, for shear induced through a boundary, such as provided by a moving confining plate, the velocity is known to be more closely exponential than linear [9]. In this circumstance, segregation occurs significantly faster in regions of high shear than in those with low shear rate, so that S is properly taken to be a function of z.

Properties of equation (1), and related models, have been explored in a series of papers [3, 8, 12]. The equation is interesting partly because, being a scalar first order equation, it is a simple macroscopic model for segregation, a complicated dynamic process at the grain diameter scale. Moreover, it is a continuum model that seeks to approximate fluid-like flow of a granular material in a context in which flow typically occurs only within a depth of a small number grains. Nonetheless, the model captures the phenomenon of segregation under shear in a reasonably convincing way.

The equation is also interesting mathematically. Although scalar equations are well understood in many ways, equation (1) has some unusual features that create novel solution structures. The structure of the equation is that it transports particles linearly parallel to the x-axis, but with a speed that is depth dependent, giving a non-constant coefficient in the transport term. As a consequence, characteristics are curved. A more subtle effect of the non-constant coefficient is seen on shock waves. These are interfaces in space-time across which the solution has a jump discontinuity. We show in this paper that shocks are stable if there is a greater concentration of large particles above the shock than below. The novel feature that appears in solutions is that, because of the shearing, stable shock waves can become vertical, and then lose stability. The emerging solution structure is still not fully understood, but in this paper we report on some progress in understanding what happens immediately after the shock loses stability.

In [6, 7], we considered equation (1) with exponential function $u(z)$ and $S(z) = s|u'(z)|$. However, we restricted attention to solutions independent of x, mimicking conditions in a Couette cell, in which layers of large and small particles are sheared by rotating a lower confining plate.

In §, we consider a general equation

$$\varphi_t + u(z)\varphi_x + (S(z)f(\varphi))_z = 0, \quad -\infty < x < \infty, \, -1 < z < 1, \, t > 0, \quad (2)$$

Here, $f(\varphi)$ models segregation normal to the x-axis. We assume it is a smooth convex function on the interval $0 \leq \varphi \leq 1$, with $f(0) = f(1) = 0$, consistent with the idea that the normal flux of small particles should be zero if there are either no large particles or no small particles at that location.

CHARACTERISTICS AND SHOCKS

In this section, we present the basic building blocks of solutions of equation (2), and prove a very general stability result.

We write equation (2) as

$$\varphi_t + u(z)\varphi_x + S(z)f'(\varphi)\varphi_z = -S'(z)f(\varphi) \quad (3)$$

If the solution $\varphi(x,z,t) = \varphi_0$ is known at the point $x = x_0, z = z_0$, and time $t = t_0$, then the PDE shows how to continue the solution to $t > t_0$, by tracing the solution along *characteristics,* given by the ODE system

$$\frac{dx}{dt} = u(z); \quad \frac{dz}{dt} = S(z)f'(\varphi); \quad \frac{d\varphi}{dt} = -S'(z)f(\varphi); \quad (4)$$
$$x(t_0) = x_0; \quad z(t_0) = z_0; \quad \varphi(x_0,z_0,t_0) = \varphi_0.$$

Since the final two ODE are independent of x, they form a vector field in the (z,φ)-plane, with a first integral:

$$S(z)f(\varphi) = \text{const.} \quad (5)$$

In particular, when $\varphi = \varphi(z,t)$ is independent of x, this equation is a useful representation of characteristics.

Shock waves are smooth surfaces $z = \hat{z}(x,t)$ across which $\varphi(x,z,t)$ has a jump discontinuity. Let $\varphi_\pm(x,t) = \varphi(x,\hat{z}(x,t)\pm,t)$. Since equation (2) is in divergance (i.e., conservative) form in space-time, the normal component of the divergence-free function $(\varphi, u(z)\varphi, S(z)f(\varphi))$ is continuous across the shock:

$$\hat{z}_t[\varphi] + \hat{z}_x u(\hat{z})[\varphi] - S(\hat{z})[f(\varphi)] = 0. \quad (6)$$

Here, we have used the normal $(\hat{z}_t, \hat{z}_x, -1)$ at the shock; the notation $[g(\varphi)] = g(\varphi_+) - g(\varphi_-)]$ signifies the jump of a function $g(\varphi)$ across the shock. Consequently, the evolution of the shock, coupled to that of the weak solution $\varphi(x,z,t)$ on either side of it, is given by the PDE

$$\hat{z}_t + u(\hat{z})\hat{z}_x = S(\hat{z})G(\varphi_+, \varphi_-), \quad (7)$$

where

$$G(\varphi_+, \varphi_-) = \begin{cases} \dfrac{f(\varphi_+) - f(\varphi_-)}{\varphi_+ - \varphi_-}, & \varphi_+ \neq \varphi_- \\ f'(\varphi_-), & \varphi_+ = \varphi_-. \end{cases} \quad (8)$$

This equation can be solved by the method of characteristics, once $\varphi_\pm(x,t)$ are known. More generally, these functions are found in conjunction with the evolution of the shock wave, as in [1]. To assess stability of the shock, in the sense of hyperbolic conservation laws, we use the Lax entropy condition, which ensures that, for a given initial condition with a shock, the solution can be continued at least for a short time with the same structure, i.e., the solution φ evolves, and the shock evolves with it. The Lax entropy condition simply guarantees that the characteristic surfaces that would emanate from points on the shock would immediately cross. As a consequence, the solution would be double-valued in the overlapping region, but in fact well-posedness is recovered by constructing a shock lying within the region, and satisfying (7). This construction is standard in the hyperbolic equations literature [11] when the solution is constant along characteristics; it corresponds to *structural* stability (i.e., short-time persistence) of the solution rather than the *asymptotic* (i.e., long-time) stability, commonly referred to in dynamical systems.

Stability of shocks

Since φ is not constant along characteristics, the treatment of stability is not completely standard. Nonetheless, for a stable shock, the two characteristic surfaces in space-time overlap, and the single-valuedness of the solutions has to be recovered by continuing the shock into this region.

We suppose there is a shock wave $z = \hat{z}(x,t)$, with well-defined values of φ on either side at time $t = t_0$. Let $\varphi_{\pm}^0(x) = \varphi_{\pm}(x,t_0)$.

Theorem 1 *The interface $z = \hat{z}(x,t)$ is dynamically stable if $\varphi_+^0 < \varphi_-^0$; it is unstable if $\varphi_+^0 > \varphi_-^0$.*

Proof: The idea of stability is that the characteristic surfaces generated by characteristics orginating on the shock at time $t = t_0$, with initial conditions $\varphi = \varphi_{\pm}^0$, should overlap for small $t > t_0$, so that a shock can be fit in between, satisfying the Rankine-Hugoniot condition. To verify this condition, we calculate the speeds of the two characteristic surfaces, and of the shock, normal to the shock at time $t = t_0$. The normal $\hat{N}$ to the shock $z = \hat{z}(x,t)$ is given by

$$\hat{N} = (1, -\hat{z}_x)/(1 + \hat{z}_x^2)^{1/2}. \tag{9}$$

The characteristic speeds $\lambda_{\pm}$ normal to the shock are given by

$$\lambda_{\pm} = (x', z') \cdot \hat{N}, \tag{10}$$

where $x'(t), z'(t)$ are given by the characteristic equations (4). Thus,

$$\lambda_{\pm} = \frac{1}{(1 + \hat{z}_x^2)^{1/2}} \left(-u(\hat{z})\hat{z}_x + S(\hat{z})f'(\varphi_{\pm}) \right). \tag{11}$$

The velocity of the shock at fixed $x = x_0$ is given by $(\dot{x}, \dot{z}) = (0, \hat{z}_t)$. Thus, the normal speed σ is given by

$$\sigma = \hat{N} \cdot (\dot{x}, \dot{z}) = \hat{z}_t / (1 + \hat{z}_x^2)^{1/2}. \tag{12}$$

Now, $\hat{z}(x,t)$ satisfies the PDE (7), so that

$$\hat{z}_t = -u(\hat{z})\hat{z}_x + S(\hat{z})G(\varphi_+, \varphi_-).$$

Substituting into (12), we find

$$\sigma = \frac{1}{(1 + \hat{z}_x^2)^{1/2}} \left(-u(\hat{z})\hat{z}_x + S(\hat{z})G(\varphi_+, \varphi_-) \right). \tag{13}$$

Comparing (11) and (13), we see that, from convexity of $f(\varphi)$, $\lambda_- > \sigma > \lambda_+$ is equivalent to $\varphi_+ < \varphi_-$, as claimed. ∎

Shock formation

In this section, we examine the tendency of shocks to form in the interior of the flow. Shock formation is associated with finite-time blow-up of the gradient of the solution, so that the slope of the graph becomes infinite as a shock forms.

Thus, it makes sense to examine the evolution of the gradient $\nabla\varphi(x,z,t) = (\varphi_x, \varphi_z)$ in a smooth solution $\varphi(x,z,t)$. We do this by differentiating the PDE (2) with respect to x and z:

$$\frac{dv}{dt} = -S(z)f''(\varphi)vw - S'(z)f'(\varphi)v \tag{14a}$$

$$\frac{dw}{dt} = -u'(z)v - S(z)f''(\varphi)w^2 - 2S'(z)f'(\varphi)w - S''(z)f(\varphi). \tag{14b}$$

The derivatives on the left hand side are along characteristics, but note that both z and φ evolve along the characteristics, so that in general, the system of ODE has to include the characteristic equations. In general, this is a complicated system to analyze, and complete results are not available. However, we can treat special cases. For the original Gray-Thornton model, in which $u(z)$ is linear, and $S(z) > 0$ is constant, a complete characterization of shock formation is given in [1]. Here, we consider the case of exponential $S(z)$, which is consistent with the experimental configuration described below. Specifically, we assume

$$u'(z) > 0; \quad S(z) = se^{\beta(z+1)}, \quad -1 \leq z \leq 1, \tag{15}$$

with $s > 0$ constant. We prove the following result.

Theorem 2 *Suppose the conditions (15) hold. If either (a) $\phi_x^0(x_0, z_0) \geq 0$, and $\phi_z^0(x_0, z_0) < 0$, then either a shock forms in finite time, or the characteristic emanating from (x_0, z_0) reaches a boundary $z = \pm 1$ before $\nabla\varphi$ becomes singular.*

Proof: First, observe that the w-axis $v = 0$ is invariant for equation (14a). In case (a), it follows from the assumption $v(0) = \phi_x^0(x_0, z_0) > 0$, that $v(t) > 0$ for all $t > 0$ for which the solution of (14) remains bounded. This is the only information we need concerning v in case (a) in order to analyze finite time blow-up of w in equation (14b).

Differentiating equation (5), we obtain

$$S'(z)f'(\varphi)w = -\frac{S'(z)^2}{S}f(\varphi) \tag{16}$$

Substituting into equation (14b), we get,

$$\frac{dw}{dt} = -S(z)f''(\varphi)w^2 + \left(\frac{S'(z)^2}{S(z)} - S''(z)\right)f(\varphi) - u'(z)v < -S(z)f''(\varphi)w^2, \tag{17}$$

since $v > 0$, $\frac{S'(z)^2}{S(z)} - S''(z) = s\beta^2 e^{\beta(z+1)} > 0$ and $f(\varphi) \leq 0$. Now z and φ are evolving on the characteristic emanating from (x_0, z_0), but both $S(z) > 0$ and $f''(\varphi) > 0$ are bounded from below by positive constants, in the physical domain $-1 < z < 1, 0 \leq \varphi \leq 1$. Thus, there is $k > 0$ such that

$$\frac{dw}{dt} < -kw^2, \tag{18}$$

at least until the characteristic reaches the boundary.

Since $w(0) = \phi_z^0(x_0, z_0) < 0$, then (18) implies that $w(t) \to -\infty$ in finite time, since

$$w(t) \leq \frac{w(0)}{1 + kw(0)t}.$$

■

Remarks: 1. If $w(0) > 0$, then the dynamics are somewhat more complicated, with he characteristics rolling over as $w(t) = \varphi_z$ changes sign. Once this happens, the conditions of the theorem are satisfied, and a shock forms. However, to prove this rigorously, we would need to establish that $w(t^*) < 0$ for some finite time $t = t^*$, in order to show that $w(t) \to -\infty$. The evolution of $w(t)$ is controlled by $v(t)$:

$$\frac{dw}{dt} < -u'(z)v. \tag{19}$$

Now, $u'(z) \geq k_0 = \min_{-1 \leq z \leq 1} u'(z) > 0$. Provided $v(t) > 0$ is bounded away from $v = 0$, then $w(t)$ crosses the v-axis in finite time.

2. In the simpler case of the original Gray-Thornton model, in which $u(z)$ is linear, $S(z)$ is constant, and $f(\varphi) = \varphi(\varphi - 1)$, system (14) simplifies considerably. Solutions φ are constant on characteristics, and the various terms that are now constants rather than variable can be eliminated from the equations by scaling, leaving the system

$$\frac{dv}{dt} = -2vw \tag{20a}$$

$$\frac{dw}{dt} = -v - 2w^2. \tag{20b}$$

Somewhat surprisingly, this system can be solved explicitly:

$$v(t) = \frac{v_0}{q(t)}, \quad w(t) = \frac{w_0 - v_0 t}{q(t)}, \quad q(t) = 1 + 2w_0 t - v_0 t^2, \tag{21}$$

where $v(0) = v_0, w(0) = w_0$. Consequently, conditions for shock formation can be specified precisely, and the time at which the shock forms can be expressed exactly in terms of the initial conditions [1]: shocks form if and only if $q(t)$ has a positive zero.

Shock breaking

Once a shock wave forms, it evolves according to the PDE (7), as discussed above. In this subsection, we show that certain shocks lose stability, due to being sheared by the flow. To avoid the complicated problem of how the solution evolves on either side of the shock, let's simplify the issue. In fact, let's take $\varphi_+ = 0, \varphi_- = 1$, so that the shock is stable. Then $G(\varphi_+, \varphi_-) = 0$ in equation (7). Furthermore, if we take $u(z) = z$, the original form of the Gray-Thornton model, then (7) becomes the inviscid Burgers equation [13] for the shock location $z = \hat{z}(x,t)$:

$$\hat{z}_t + \hat{z}\hat{z}_x = 0. \tag{22}$$

Now solutions of Burgers equation are known to break in finite time, unless they are monotonically increasing. Consequently, any shock wave solution satisfying (22) will become vertical in finite time if $\hat{z}_x < 0$ anywhere, at any time. This makes sense, because the interface is being sheared by the depth-dependent velocity. In the classical theory of Burgers equation, the solution can be continued as a shock wave, but here, the solution itself is a shock, and as it breaks, it loses stability because a middle section is now unstable: it has $\varphi_+ < \varphi_-$, but because the section has turned over, $\varphi = \varphi_+ = 0$ *below* the shock, and $\varphi = \varphi_- = 1$ *above* the shock. (See Theorem 1.)

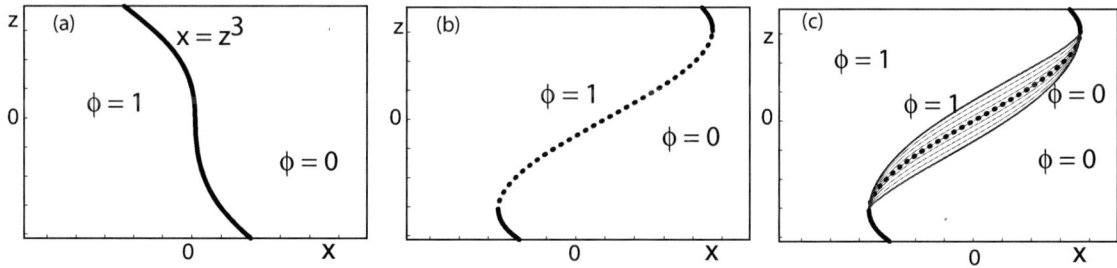

FIGURE 1. Solutions of (1), with $u(z) = z$ and initial condition (23). (a) Initial condition. (b) Evolved shock showing unstable section. (c) Solution with rarefaction wave, $t < 1/12$.

The solution is in fact continued using the method of characteristics to introduce a rarefaction wave, corresponding to a mixing zone. Consider an initial condition

$$\varphi(x,z,0) = \begin{cases} 1, & x < -z^3, \\ 0, & x > -z^3. \end{cases} \tag{23}$$

in which the interface $x = -z^3$ is already vertical at $z = 0$ in the initial condition. Then to see the shock breaking, it is convenient to consider the parameterization of the shock by z rather than x: $x = \hat{x}(z,t) = -z^3 + zt$ satisfies the evolution equation for $\hat{x}$:

$$\hat{x}_t + z\hat{x}_z = 0. \tag{24}$$

Then for $t > 0$, $\hat{x}(z,t)$ is non-monotonic; it is increasing between $z = \pm\sqrt{\frac{t}{3}}$, the unstable section of the evolved shock wave, and decreasing outside this interval. The solution with a rarefaction wave is valid for $t < \frac{1}{12}$, and is shown in Fig. 1. Beyond $t = \frac{1}{12}$, the solution becomes more complicated, but can be calculated with a simple numerical algorithm, as described in a forthcoming paper [1].

EXPERIMENTS WITH A COUETTE CELL

Experiments were conducted in an annular Couette cell, in which a mixture of small and large particles is sheared between concentric vertical cylinders. Here, we summarize results that will appear in a pair of papers [6, 7]. The experimental setup and protocol are described in detail in [2]. A bottom confining plate is rotated at constant frequency $f = 49 \pm 0.5$ mHz, approximately 3 rpm, and a top plate exerts a controlled pressure on the particles $(0.36 \pm 0.008)\,mg$, where mg is the total weight of the particles and the variation in force is due to the stretching of springs partially supporting the plate. The top plate is free to move vertically, thereby accommodating dilation and consolidation. Experiments reported here were carried out with spherical glass beads, of diameters 3 mm and 6 mm. Different size ratios, and variations in other experimental conditions are described in [2].

The purpose of the experiment is to compare predictions of the Gray-Thornton model quantitatively with experimental data. We find that the model captures the gross behavior of mixing and segregation as the material is sheared, even though it cannot reproduce significant features of the flow. To reflect conditions in the experiments, we make several assumptions concerning the model. First, we assume that, although mixing and segregation involve three-dimensional motion of particles, we consider solutions that depend only on the vertical variable z. This assumption is reflected in the initial configurations chosen for the experiments, in which a layer of small particles is placed above a layer of large particles. Second, we assume that the segregation rate is proportional to the shear rate, reflecting the notion that there should be more segregation when the sample is sheared faster. As the beads mix, they occupy significantly less volume, so that the top plate falls slightly. Later, as the mixture resegregates, they begin to occupy more volume, and the top plate rises. We assume that the degree of segregation is quantitatively measured by the height of the top plate, with the connection between model solutions and the height of the top plate being provided by a packing fraction density.

In the experiment, we take two types of measurements. From high speed images we extract particle trajectories, which lead to an average velocity profile $u(z)$ that is roughly independent of time and φ. The second measurement is to record the height $H(t)$ of the top plate as a function of time t.

Figure 2(a) summarizes averaged velocity data, and a fit by an exponential function $u(z)$. The horizontal bars indicate the spread of the data, using the width of a parabolic fit at half height. In Fig. 2(b), we show the corresponding shear rate plot and a fit by an exponential function. The solid line in Fig. 2(a) is derived from the same exponential function. Near the top (resp. bottom) of the cell, a layer of large (resp. small) particles forms quickly, creating an effective boundary. Consequently, the velocity profile and shear rate are determined from data in the middle section shown in the figure, normalized to $0 \leq z \leq 1$.

In Figure 3 we show the time series of the height $H(t)$, together with the result of solving the mathematical initial value problem with shear rate given by the experimental data. The mathematical problem yields a function $\varphi(z,t)$, the volume fraction of small particles on a fixed domain $0 < z < 1$. To extract a physical height from this function, we employ a volume packing fraction $\rho(\varphi)$, based on results from MD simulations of static packings [5]. The packing fraction allows us to translate the local particle fraction into a real volume. Since the cross-sectional area of the apparatus is constant, and the initial volume of particles is known, we then simply have to integrate the effective volume over the fixed domain $0 \leq z \leq 1$ in order to compute a height function $h(t)$, which we refer to as the proxy height. In Fig. 3, we observe that the mixing rate (where $H(t)$ is decreasing) is well captured by the model. Moreover, although the resegregation in the model is delayed, compared to the experimental observation, nonetheless the model and experiment agree remarkably well in the rate of resegregation. In particular, both model and experiment suggest different rates for mixing and resegregation.

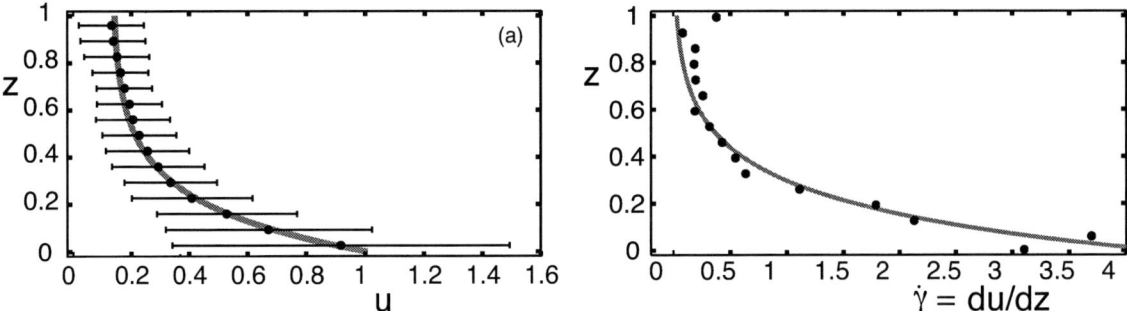

FIGURE 2. (a) Measured velocity profile $u(z)$ (•) in the region $0 \leq z \leq 1$, showing the exponential fit to the shear rate data in (b). Velocities are scaled so that $u(0) = 1$. (b) Shear rate $\dot{\gamma} = |du/dz|$ within the region $0 \leq z \leq 1$. The solid line is the fit to an exponential function.

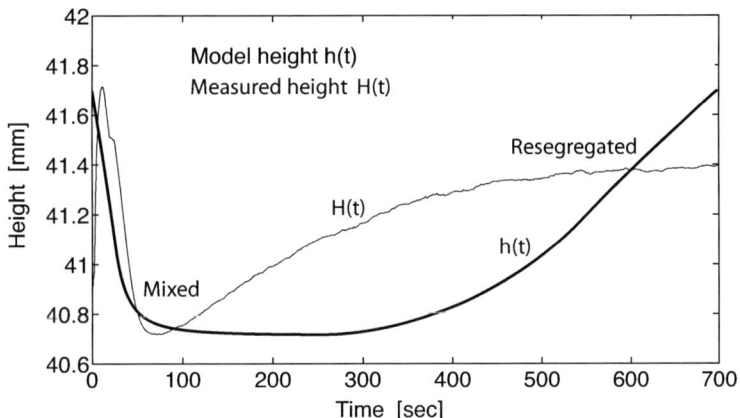

FIGURE 3. The experimentally-measured height $H(t)$ and the calculated height $h(t)$.

ACKNOWLEDGEMENTS

This research was supported by the National Science Foundation under grants DMS-0604047, and the National Aeronautics and Space Agency under grant NNC04GB086.

REFERENCES

1. N. Giffen and M. Shearer. Shock formation and breaking in particle size-segregation. *J. Discrete and Continuous Dynamical Systems*, submitted.
2. L. A. Golick and K. E. Daniels. Mixing and segregation rates in sheared granular materials. *Physical Review E*, 80:042301, 2009.
3. J. M. N. T. Gray, M. Shearer, and A. R. Thornton. Time-dependent solutions for particle-size segregation in shallow granular avalanches. *Proceedings of the Royal Society A*, 462(2067):947–72, 2006.
4. J. M. N. T. Gray and A. R. Thornton. A theory for particle size segregation in shallow granular free-surface flows. *Proceedings of the Royal Society A*, 461:1447–73, 2005.

5. K. D. Kristiansen, A. Wouterse, and A. Philipse. Simulation of random packing of binary sphere mixtures by mechanical contraction. *Physica A*, 358(2-4):249–62, 2005.
6. L. B. H. May, K. C. Phillips, L. A. Golick, M. Shearer, and K. E. Daniels. Shear-driven particle-size segregation of granular materials: comparison of theory, modelling and experiment. *Physical Review E*, submitted.
7. L. B. H. May, M. Shearer, and K. E. Daniels. Scalar conservation laws with nonconstant coefficients with application to particle size segregation in granular flow. 2009. *J. Nonlinear Science*, submitted
8. M. McIntyre, E. L. Rowe, M. Shearer, J. M. N. T. Gray, and A. R. Thornton. Evolution of a mixing zone in granular avalanches. *AMRX*, abm008, 2008.
9. G. D. R. MiDi, *European Physical Journal E*, 14: 341, 2004.
10. S. B. Savage and C. K. K. Lun. Particle-size segregation in inclined chute flow of dry cohesionless granular solids. *Journal Of Fluid Mechanics*, 189:311–35, 1988.
11. D. Serre. *Systems of Conservation Laws: Geometric structures, oscillations, and initial-boundary value problems*. Cambridge Univ. Press, 2000.
12. M. Shearer, J. M. N. T. Gray, and A. R. Thornton. Stable solutions of a scalar conservation law for particle-size segregation in dense granular avalanches. *European Journal of Applied Mathematics*, 19:61–86, 2008.
13. G.B. Whitham. *Linear and Nonlinear Waves*. Wiley, NY, 1974.

Preliminary investigations on the rheology and boundary stresses associated with granular mixtures

B. Yohannes and K. M. Hill

St. Anthony Falls Laboratory, Department of Civil Engineering, University of Minnesota, 55414

Abstract.
Recent advances in rheological models for monodisperse dense granular materials are exciting. However, they do not account for the effect of local particle size distributions on the rheology mixtures of particles. It is well-known that particulate mixtures tend to unmix, and their rheological properties are dependent on species concentration. Typically, expressions for the rheology of dense granular flows are explicitly dependent on particle size. However, there is no indication of what may be used for a representative size in a mixture of particles of different sizes. We find, in the absence of gravity, plane Couette cells present an effective geometry for investigating the rheology of binary mixtures of different sized particles. Unlike the behavior of more sparse systems, we find that the dense systems do not segregate much, indicating the usefulness of the geometry for studying the dependence of the mixture rheology on particle sized distribution systematically. In our preliminary studies we find that the pressure at the boundary has a skewed probability distribution function (pdf). We also find that the pdf of the boundary pressure for a particular mixture scales according to the inertial stress.

Keywords: Boundary stress, granular material, Couette flow, DEM
PACS: 45.70.Mg, 83.10.Rs, 81.05.Rm

INTRODUCTION

The ability to predict and control the movement of particulate mixtures is important for many critical natural and industrial applications. In addition to the transport of the particles themselves, many applications require an understanding of the stresses a particulate system exerts on the boundary. For example, predicting boundary stresses due to bouldery debris flows - massive flows of rocks, boulders and often mud and sand - is important for understanding landscape morphology as well as for maximizing hazard mitigation. The boundary stresses in combination with the bedrock properties determines how much material is taken away from the bedrock, therefore shaping the landscape, as well as adding to the debris flow itself. (See, for example, Refs. [1, 2, 3, 4, 5].)

Effective models exist for relatively slower quasi-static particulate movements and for more sparse, typically more energetic granular systems. In the solid-like state where solids fraction and coordination number are relatively high and there is little movement, plasticity theory models small deformations [6], but not continuous flows [7]. Kinetic theory is arguably the most complete physics-based model for granular flows (see, for example, Ref.[7, 8, 9, 10, 11, 12]) and can account explicitly for mixtures [13, 14, 15, 16, 17]. Given the remarkable versatility of kinetic theory to model granular flows and its firm footing in physics principles, it would be ideal if kinetic theory was applicable to all granular flows. In fact, several tests of kinetic theory at moderate densities suggest it to be valid at higher densities, and some have suggested that kinetic theory may provide a starting point for developing rheology of dense granular flows [18]. However, there is evidence that there is an elasticity-dependent transition at a critical solids fraction above which, while particles flow, kinetic theory is not effective at capturing all of the flow dynamics [18, 19, 20, 21, 22, 23].

In this paper, we briefly review approaches to modeling the behavior of dense granular flows and discuss the need for an extension of these frameworks for granular mixtures. Then, we investigate the effectiveness of using a parallel Couette cell to study the effect of particle size distribution on the rheology and on boundary stresses due to sheared dry cohesionless granular flows. Then, we present preliminary computational results pointing out the suitability of the boundary conditions associated with this geometry for studying granular mixtures. We conclude with discussion of future research that is needed to extend and validate models for dense granular flows to mixtures.

BACKGROUND

To this point, there has been no uniformly accepted method for modeling dense granular flows, though several attempts have been made to model the viscous-like behavior [24, 25, 26, 27, 28, 29, 30, 31, 32, 33, 34, 35, 36, 37, 38]. Some approaches focus on special features of dense granular systems. For example, Mills et al. [25] modeled the flowing granular medium as a network of transient solid chains immersed in an assembly of particles behaving as a viscous fluid. Based on this, they developed a stress tensor that is the sum of three effects: viscous, frictional, and that due to a force contact network - forces that extended throughout the system through interparticle contacts. Kamrin and Bazant [26] used a Mohr-Coulomb plasticity for quasi-two-dimensional granular materials to calculate average stresses and slip planes, but used a stochastic flow rule to account for discreteness of shear.

Other approaches to modeling this flow focus on capturing the transition from liquid-like motion to either solid-like deformation or gas-like flow. For example, Savage [27] developed a model based in part on a model for a frictional plastic continuum with coefficients adjusted to merge smoothly with the kinetic theory for rapid collisional flows. Louge [28] and Aranson and Tsimring [29] developed theories with the shear stresses are composed of two parts: the dynamic part proportional to the shear strain, and the strain-independent or static part independent of strain rate. Savage and Hutter introduced depth-averaged equations to thin flows [39, 40], assuming that the flow is incompressible and the spatial variation of the flow takes place on a scale larger than the flow thickness where the complex rheology of the material is mainly embedded in a Mohr-Coulomb friction with a Coulomb-like basal friction term μ_b for the boundary interaction. This approach was adopted for more complex and erodible flows [41, 42, 43].

While this research represents a wide range of approaches to modeling dense granular materials, it is generally recognized that the rheology should depend in some way on a single non-dimensional number known as the inertial number I, or the square root of the Savage number Sa [24, 44, 45], where:

$$I = \frac{\dot{\gamma} d}{\sqrt{p/\rho_s}}, \text{ and} \tag{1}$$

$$Sa = \frac{(d\dot{\gamma})^2 \rho_s}{(\rho_s - \rho_f) gH\tan(\theta)} \tag{2}$$

Here, $\dot{\gamma}$ is the shear rate, d is the particle diameter, p is local pressure, ρ_s and ρ_f are particle and interstitial fluid density, g is the gravitational acceleration constant, H is the flow depth, and θ is the angle of inclination of the flow. This scaling has been associated with the ratio of collisional stress to total stress [46] and also as a ratio of time scales [38], microscopic time scale (time for particle to fall in a hole of size d under pressure p) and macroscopic time scale $1/\dot{\gamma}$. Notably, Pouliquen and colleagues [33, 34, 35, 36, 37] have suggested new constitutive relationships based on the inertial number and other physical arguments. A number of physical and computational experiments [38] have shown these relationships to be effective in modeling dense flowing granular materials in a wide range of boundary conditions.

There has been somewhat less work done to model pressure, p, and shear stress τ alone. This is perhaps due to the dominance of gravity-driven flows in experiments and, indeed, in many applications. In gravity-driven flow, the average pressure stress is primarily associated with something akin to hydrostatic pressure. However, there are many cases where the effective pressure is not simply hydrostatic. A geological example occurs in the shear band associated with an earthquake fault line. Experiments where stresses associated with side walls rather than gravity alone may be relevant include Couette cells and split bottom cells [8, 47].

Bagnold investigated these stresses in his classic experiments in the 1950's [48]. Based on his experimental observations, Bagnold related stresses to momentum transfer in dynamic inter-particle interactions. Bagnold's experimental apparatus (a concentric-cylinder rheometer) had an inner stationary rubber membrane and an outer wall rotating at a constant speed. The annular space between the inner and outer cylindrical walls was filled with water and neutrally-buoyant particles. When the outer wall was rotated, the resulting radial pressure on the fixed inner membrane and the torque generated were measured. These data supported Bagnold's model for shear and pressure boundary stresses for inertial (collision-dominated) granular flows. According to this model, both normal and shear boundary stresses are proportional to the square of the shear rate and particle size:

$$p \sim \lambda^2 \rho_s d^2 \dot{\gamma}^2 \cos(\alpha) \tag{3}$$
$$\tau \sim \lambda^2 \rho_s d^2 \dot{\gamma}^2 \sin(\alpha) \tag{4}$$

where most parameters are the same as in Equations (1) and (2). α is the average angle of contact between particles relative to the direction of shear. The parameter λ is what Bagnold called the linear concentration of particles, the ratio of particle diameter and the free space between particles in the streamwise direction. Bagnold showed that λ can be expressed by:

$$\lambda = \left[\left(\frac{v_o}{v}\right)^{1/3} - 1\right]^{-1} \tag{5}$$

Here, v is the volume fraction of the granular material, v_o is the maximum volume concentration, approximately 0.74 for a uniform system of spheres.

Equations 1 – 5 contain explicit and implicit dependencies on particle size. For a mixture of two (or more) components, it is not obvious what the representative size should be, nor what one might use for the maximum solids fraction in calculating the linear concentration in equation 5. Indeed, under otherwise similar settings, flow properties have been shown to vary with particle size and mixture concentration [49, 50, 51, 52]. Furthermore, certain macroscopic measures of particle rheology suggest and average particle size does not always capture the behavior of the mixture well. In particular, the dynamic angle of repose of a mixture cannot be simply related to an average of the dynamic angle of repose of the components in the mixture [53, 54].

On the other hand, two sets of results suggest that simply substituting the average particle size into Equations 1–4 is an effective way to predict the behavior of granular mixtures [55],[56]. In the early 1980's, Savage and McKeown [55] studied the rheology of three mixtures, each with a narrow particle size distribution, in a similar experimental set up to Bagnold's classic set-up [48]. They showed that the shear stress scales as the square of the mean particle size for granular flows of high solid volume fraction ($v > 0.53$). However, they did not explicitly investigate the effect of particle size distribution on the stresses. In 2007, Rognon et. al [56] used a 2-D Discrete Element Model (DEM) to study the rheology of a bimodal granular mixtures flowing over an inclined plane. The system segregated. However, they used a local mean grain size in Equation(1), to calculate a local Inertial number I for the mixture. Then, they showed the effective friction coefficient, the ratio of the shear stress to the local pressure was linearly related to I.

A few other studies have investigated boundary stresses due to granular mixtures for natural particulate flows [1], [4], [5]. Takahashi [1] has used the global median particle size for modeling the stresses in granular mixtures that have a narrow particle size distribution. Stock and Dietrich [4] have suggested using the local mean particle size of the boulders at the front (or "snout") of the debris flow to estimate the inertial stresses and their subsequent bedrock wear. Hsu et al. [5] have provided some experimental evidence that the median particle size is directly related to boundary stresses through erosion experiments.

Unfortunately, there is very little data - experimental or computational - on how the rheological behavior of granular mixtures depends on particle size distributions. A limiting factor to a systematic investigation of these effects is the degree to which particulate mixtures segregate or unmix according to particle property. In the next section, we investigate the effectiveness of a parallel Couette cell for studying the rheology of granular mixtures systematically.

COMPUTATIONAL EXPERIMENTS

For our preliminary investigations of the rheology of sheared granular mixtures in parallel Couette cells, we use computational experiments based on a soft sphere Distinct (or Discrete) Element Method (DEM)[57]. Particle-particle and particle-wall contact forces are represented using a non-linear force model in both the normal direction F_n (the direction connecting the centers of two contacting spheres) and the tangential direction F_t (perpendicular to the normal direction and in the direction of relative movement). Both normal and tangential contact force models are nonlinear based on Hertzian and Mindlin contact theories and empirical relationships based on experimental results by Tsuji et al. [58]. The tangential component obeys Coulomb's law.

$$F_n = K_n \delta_n^{3/2} + \eta_n \dot{\delta}_n \delta_n^{1/4} \tag{6}$$

$$F_t = \min \begin{cases} K_t \delta_t \delta_n^{1/2} + \eta_t \dot{\delta}_t \delta_n^{1/4} \\ \mu F_n \end{cases} \tag{7}$$

Here, the subscripts n and t represent the normal and tangential directions, K is the stiffness factor, η is the damping factor, μ is the coefficient of friction and δ is the overlap between the two particles.

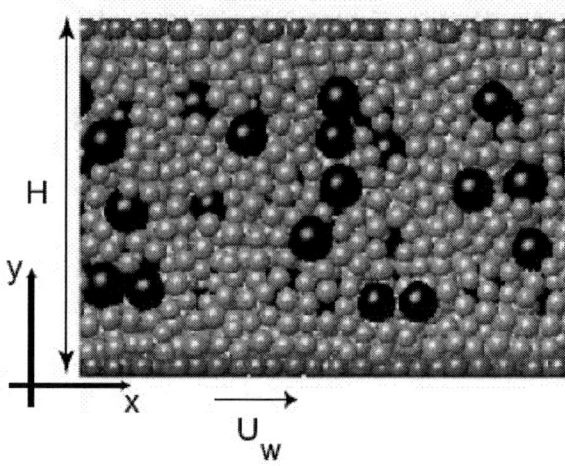

FIGURE 1. Image taken from one time step in a parallel Couette Cell simulation as described in the text. The granular material consists of a mixture of 25% large (d = 20mm +/- 2mm) particles and 75% small (d = 10mm +/- 1mm) particles. The top and bottom walls of the cell are roughened by randomly attaching 10mm spherical particles across the entire boundary.

TABLE 1. Physical properties of the particles used in the computational simulations.

Density, ρ_s	2650 kgm^{-3}
Modulus of elasticity, E	29 GPa
Poisson's ratio	0.15
Coefficient of friction, μ	0.50
Coefficient of restitution, e	0.90

The DEM simulation results described here are performed for two systems. One system is monosized, consisting of particles of diameter d = 10 mm +/- 1mm. The other system is a granular mixture consisting of 75% 10mm +/- 1mm and 25% 20mm +/- 2mm particles. (The polydispersity in the size of each component helps to prevent crystallization.) All computational experiments are performed using a solids fraction of 0.568. of the two sizes. All particles have similar material properties, summarized in Table 1. The material properties of the particles can have a significant effect on the rheology of granular flows, as has been shown for systems of uniform particle size distributions [37, 59]. We discuss this in the last section of this paper.

The particles are sheared in a plane Couette cell (as in Fig. 1) in the absence of gravity. Two parallel walls contain the particles in the y-direction, one of which moves with constant velocity U_w. The boundaries are periodic in the other two directions. To minimize particle slipping at the top and bottom boundaries, we affix 10 mm particles in a random array to the walls as in Ref. [59, 60]. The glued particles have the same material property as the flowing particles. Stresses are obtained by considering all forces on the particles affixed to the boundary. Pressure stresses are found by summing the component of the forces normal to the boundary and dividing the sum by the area of the boundary. Shear stresses are found by summing the component of the forces in the streamwise direction and dividing the sum by the area of the boundary.

We perform both constant volume and constant pressure simulations. In the constant volume simulations, the distance between the two bumpy boundaries H is held constant and the boundary pressure and stress are monitored while the bottom wall is sheared. In the constant pressure simulations, a constant pressure is applied to the upper wall, which is allowed to move in the y-direction in response to the stresses from the flowing particles. Both sets of results yield similar results so, for the sake of space, we report only the results from constant volume simulations here.

Figure 2 shows velocity profiles for the two different granular systems. For both, the wall velocity(U_w) is 550 mm/s, and H = 160 mm. Figure 2(a) depicts the results for the monosized 10 mm particle system, and Figure 2(b) shows

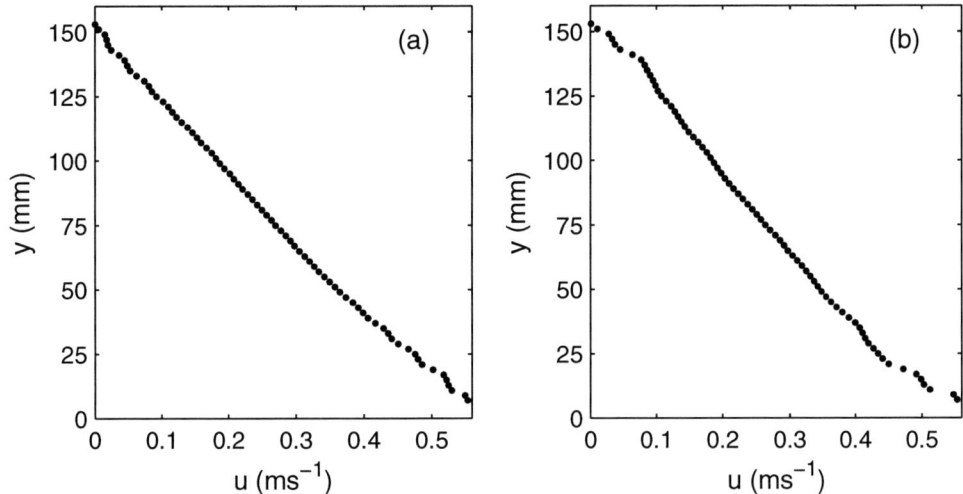

FIGURE 2. Velocity profile (a) for a uniform system of 10 mm +/- 1 mm spherical particles and (b) for a mixture of 75% 10mm +/- 1mm and 25% 20mm +/- 2mm particles in a constant volume Couette flow. For both simulations, the wall velocity is 550 mm/s and the height $H = 160$ mm. All the simulations have a linear velocity profile so we assume a constant shear rate: $(\dot{\gamma} = U_w/H)$. For the results in this figure, then, $\dot{\gamma} = 3.44 s^{-1}$.

the results for the mixture. Because of the wall roughness, the particle slip at the boundary is negligible. The velocity profile is almost linear for both simulations consistent with other results under similar conditions (See for example, Ref. [38, 61]). Therefore, for our calculations we have assumed a constant shear rate: $(\dot{\gamma} = U_w/H)$. For the results in Fig. 2, then, $\dot{\gamma} = 3.44 s^{-1}$.

Figure 3(a) shows the solid fraction profiles $v(y)$ of small and large particles for the mixture in Fig. 2(b) after 18 s of shearing. From these results, it is apparent that the larger particles have segregated slightly away from the walls and small particles have higher concentration near the walls. These results are consistent with similar DEM simulations [61],[62] performed for lower solid volume fractions. Larger particles tend to move away from regions of higher velocity fluctuations [61, 62] and also from walls [63]. Figure 3(b) shows the profile of the streamwise velocity fluctuations (u'^2) for the same mixture. The figure indicates that the velocity fluctuations adjacent to the walls are higher than they are in the middle part of the flow.

While the segregation normal to the direction of flow is not negligible, it appears to be limited in these systems. Figure 4(a) shows a plot comparable to that in Fig. 3(a), the solids fraction for each component in the mixture, though in the case of Fig. 4(a) the results are shown after 200s of run time. The results do not appear qualitatively different after the additional 182s. To investigate the evolution quantitatively, we interrogate the concentration in the central region, between the dashed lines in Fig. 4(a), as a function of time. The concentration of large particles in this central region (F_{LM}) is plotted as a function of time in Fig. 4(b). From this plot it is apparent that, while there is a striking segregation process at the beginning of the simulation, this levels off quickly, and the particles in the center remain relatively well-mixed. This result is completely different from segregation in a shear cell in the presence of gravity [64] presumably because gravity drives most of the segregation observed in these systems. The segregation we observe (Fig. 4) is also limited compared to that reported by others who have performed similar computational experiments to us in the absence of gravity but in sparser systems (i.e., Ref. [61], [62], and references within). The discrepancy might have to do with the size and extent of the temperature gradient, which for us is limited to the region very near the boundary. In any case, unless otherwise noted, we report the results after the initial transients have leveled off.

Figures 5 shows the boundary pressures at a frequency of 450 Hz for the first 18 seconds of the experiments for the monosized system (first column) and the mixture (second column) depicted in Figs. 2(a) and (b), respectively. The plots in the first row of Fig. 5 show the time resolved stresses $p(t)$ for the monosized and mixed systems. Both systems exhibit a transient period that is somewhat different than the subsequent behavior. The monosized system depicted in Fig. 5(a) has a fairly smooth pressure plot for the first second or so before the system becomes more spiky. Fig. 5(b) has a somewhat spikier pressure plot for the first several seconds, presumably while the system segregates somewhat. However, the transient behavior appears to stabilize much more quickly than the segregation is complete, within a

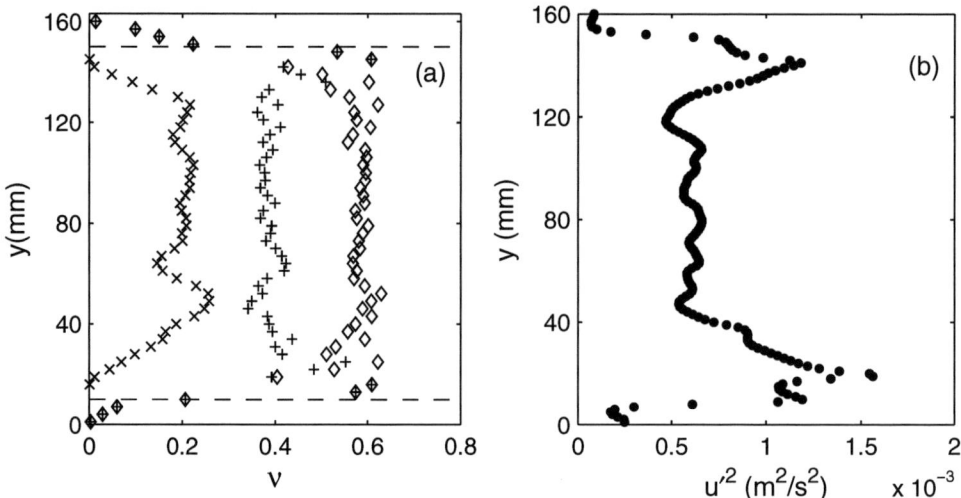

FIGURE 3. (a) Solid volume fraction for constant volume simulation of the mixture described in the text and shown in Fig. 1 sheared at a constant shear rate $\dot{\gamma} = 3.44 s^{-1}$. Data shown include the total solids fraction of the mixture (◊), the solids fraction of large particles (×) and the solids fraction of small particles (+). The dashed lines show the location of the edge of the 10mm particles glued to the wall. (b) The average streamwise velocity fluctuation squared for the sample described in (a). Both the solids fraction of the small particles and the velocity fluctuations are higher adjacent to the walls.

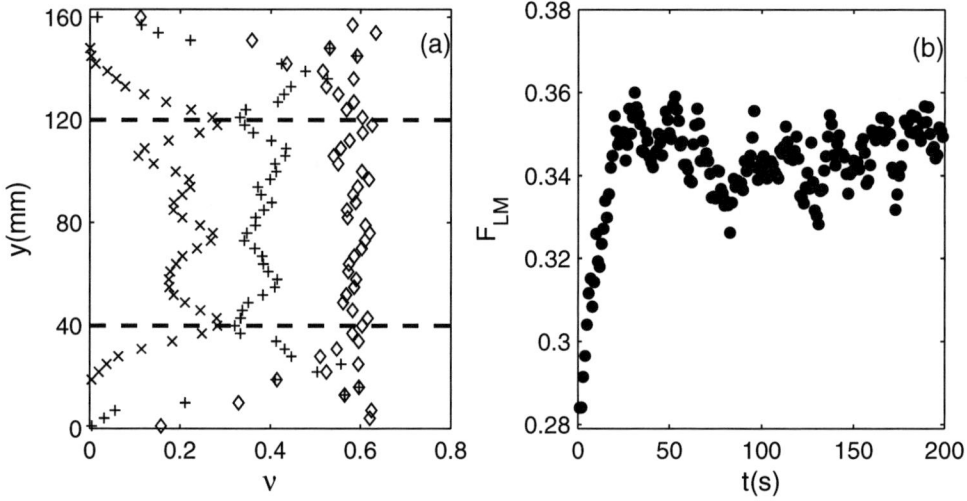

FIGURE 4. (a) Solid volume fraction for constant volume simulation of the mixture described in the text and shown in Fig. 1 after 200 s of shearing. Data shown include the total solids fraction of the mixture (◊), of the large particles (×) and of the small particles (+). (b) The fraction of large particles in the central region (F_{LM}) bounded by the dashed lines in (a).

few seconds. After the transient period, the behavior of the pressure plots for both systems are characterized by spiky behavior throughout the run.

The second row in Fig. 5 shows the probability distribution function (pdf) of the pressure at the boundary for uniformly distributed particles (Fig. 5(c)) and the 25% large particle mixture (Fig. 5(d)) for shear rates of 3.44, 6.13 and 11.44 s^{-1}. The pdf's for both the uniform size particle system and the mixture are characterized by a positively skewed distribution with very long tails. Previous studies have also suggested that the pdf is positively skewed and decays exponentially for large stresses [65]. It is interesting to note that for each shear rate, the peak in the pdf for shear stress is lower for the mixture than for the monosized even though the average particle size is larger for the former.

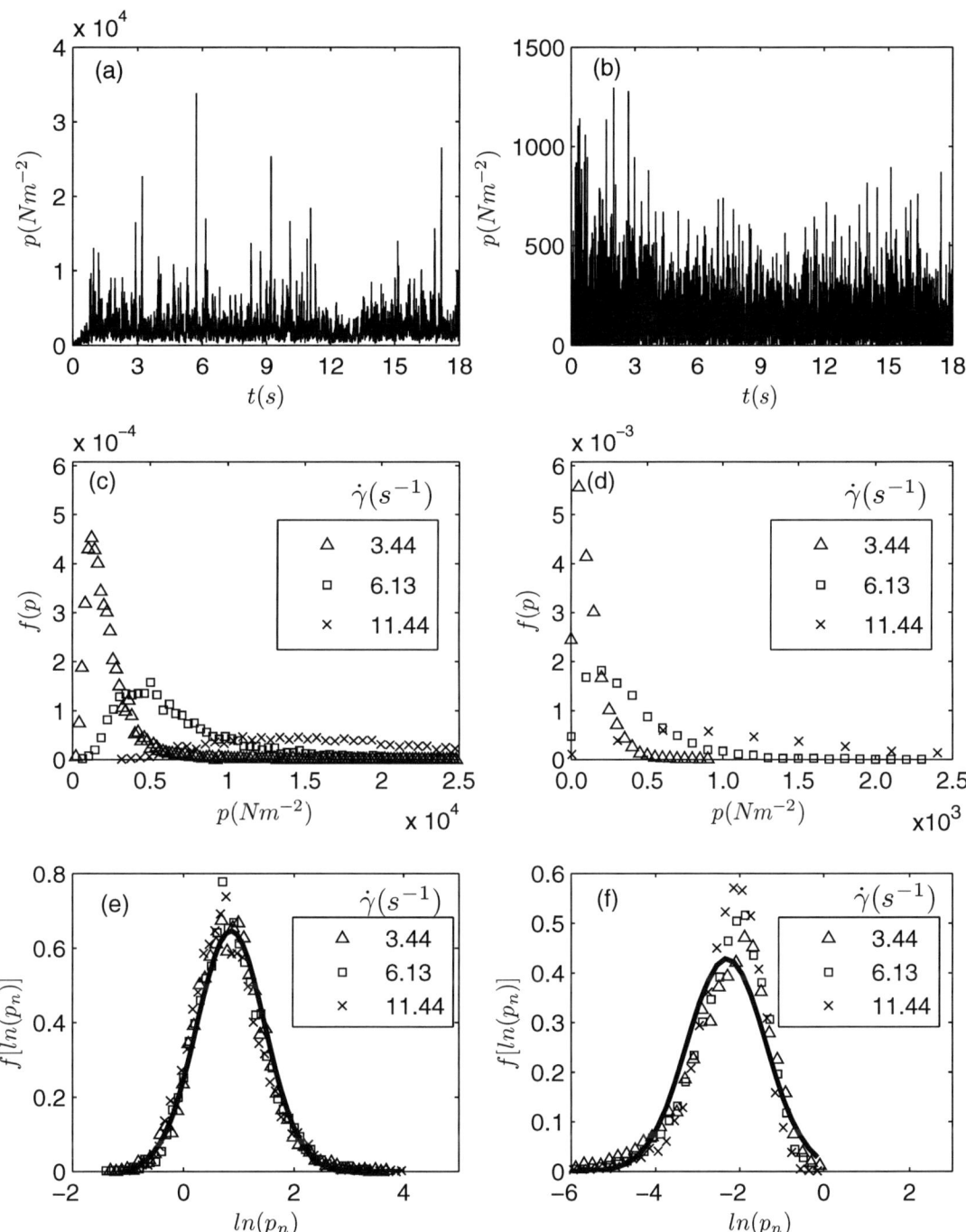

FIGURE 5. Boundary pressures measured at a frequency of 450 Hz for the first 18 seconds of the experiments for the monosized system (first column) and the mixture described in the text (second column). Figures (a) and (b) show pressure as a function of time plots. Figures (c) and (d) show the probability distribution function (pdf) of the boundary pressure. Figures (e) and (f) show the pdf of the natural logarithm of the normalized pressure, $ln(p_n)$, where $(p_n = \rho_s U_w^2)$. The solid line is the best fit curve based on a Gaussian distribution. For the monosized system in Fig. (e), the mean = 0.859 and standard deviation = 0.617. For the mixed system in Fig. (f), the mean = -2.309 and standard deviation = 0.9297.

The third row in Fig. 5 shows the pdf's of the boundary pressures normalized by the inertial stress, $p_n = p/\rho_s U_w^2$. The pdf of the normalized pressure p_n for both the monosized system (Fig. 5(e)) and the mixture (Fig. 5(f)) collapses into a single curve. For these results we have found the lognormal distribution fits the whole range of stresses reasonably well. The solid line in the plot is the best Gaussian fit to the $ln(p_n)$ based on all data points. For the monosized system in Fig. (e), we found the mean to be 0.86, corresponding to a normalized pressure of 2.86. For the mixed system in Fig. (f), we found the mean to be -2.31 corresponding to a normalized pressure of 0.15. There are a few discrepancies between the data and the lognormal distributions. The lognormal distribution predicts slightly lower peak values for the pdf's than the data indicates for all experiments. In addition, the fitted distribution for mixtures is not as good as the fitted distribution for the uniform particle size distribution. However, the lognormal distribution describes the general characteristics of the stress distribution well and it is a common distribution for a measure of a system that is a multiplicative product of many independent random variables. It is interesting to note that the scaling of the pressure pdf is consistent to Bagnold's model indicating the stresses are proportional to the square of the shear rate (3). Here, we see this rule applies to mixtures as well as monosized systems.

SUMMARY

We have found that a parallel Couette cell provides suitable boundary conditions to study the effect of particle size distribution on boundary stresses and other details of the rheology associated with dense granular mixtures. The uniformity of the shear rate and the limited segregation makes the system ideal for a systematic study of the characteristics of the boundary stresses as they depend on particle size distribution. From our preliminary studies, we find the empirical probability distribution function is well-represented by a lognormal distribution function that scales with the inertial stress ($\rho \dot{\gamma}^2 H^2$). This relationship is applicable for systems of single-sized particles and for binary mixtures that we studied.

Our preliminary results for two systems shown in Fig. 5 shows certain details of the rheology may not have a simple dependence on the average particle size. The mixture whose average particle size is larger than the monosized system gives rise to a smaller boundary stress than the monosized system. In contrast, Bagnold's classic model for boundary stress [48] predicts stress should increase with particle size.

There are a number of considerations in interpreting these and future results. The effect of the boundary conditions likely plays an important role in the results. For example, it has been shown that the boundary roughness, the shear rate, and the confining pressure all can have a significant influence on the flow characteristics [66, 67]. They may also play a role in determining the details of the effect of the particle size distribution on the rheology of granular mixtures. In the long run, other factors must be investigated as well. Most natural granular flows are composed not only of a distribution of particle sizes but of a variety of other properties as well, including stiffness, coefficient of restitution and coefficient of friction [37, 48, 59]. These properties have been shown to affect the evolution of the local particle size distribution or segregation properties. They will likely affect the size-dependent rheology as well. In the end, to model these mixtures in the variety of natural and industrial application in which they appear, we as a community need to have both an understanding of how the particles rearrange or segregate according to particle property and then how the local characteristics of the particle properties affect the rheology of the flow.

ACKNOWLEDGMENTS

Funding for this research was provided by the National Center for Earth Surface Dynamics (NCED), a NSF Science and Technology Center funded under agreement EAR-0120914, and by NSF grant CBET-0756480. This work is carried out in part using computing resources at the University of Minnesota Supercomputing Institute.

REFERENCES

1. T. Takahashi, *J. Hydraul. Div. ASCE* **104(8)**, 1153–69 (1978).
2. R. M. Iverson, *Rev. Geophys.* **35(3)**, 245–96 (1997).
3. J. D. Stock, and W. E. Dietrich, *Water Resources Research* **39(4)**, 1089–104 (2003).
4. J. D. Stock, and W. E. Dietrich, *Geol. Soc. Am. Bull.* **118(9/10)**, 1125–48 (2006).
5. L. Hsu, W. E. Dietrich, and L. Sklar, *J. Geophys. Res.* **113**, F02001 (2008).

6. D. C. Drucker, and W. Prager, *Quart. Appl. Math.* **10**, 157–65 (1952).
7. R. M. Nedderman, *Statics and Kinematics of Granular Materials*, Cambridge University Press, Cambridge, 1992.
8. D. Fenistein, J. W. van de Meent, and M. van Hecke, *Phys. Rev. Lett.* **92**, 0940301 (2004).
9. S. B. Savage, and D. J. Jeffrey, *J. Fluid Mech.* **110**, 255–72 (1981).
10. J. T. Jenkins, and S. B. Savage, *J. Fluid Mech.* **130**, 187–202 (1982).
11. V. Garzo, and J. W. Dufty, *Phys. Rev. E* **59**, 5895–911 (1999).
12. C. Campbell, *Powder Tech.* **162**, 208–29 (2006).
13. V. Garzo, and J. W. Dufty, *Phys. Fluids* **14**, 1476–90 (2002).
14. M. L. de Haro, E. G. D. Cohen, and J. M. Kincaid, *J. Chem. Phys.* **78**, 2746–58 (1983).
15. J. M. Kincaid, E. G. D. Cohen, and M. L. de Haro, *J. Chem. Phys.* **86**, 963–75 (1987).
16. J. T. Jenkins, and F. Mancini, *Phys. Fluids* **31**, 2050–7 (1989).
17. J. E. Galvin, S. R. Dahl, and C. M. Hrenya, *J. Fluid Mech.* **528**, 207–32 (2004).
18. N. Mitarai, and H. Nakanishi, *Phys. Rev. Lett.* **94**, 128001 (2005).
19. G. Lois, A. Lemaitre, and J. Carlson, *Phys. Rev. E* **72**, 051303 (2005).
20. J. M. Montanero, V. Garzó, M. Alam, and S. Luding, *Granular Matter* **8**, 103–15 (2006).
21. H. Xu, M. Louge, and A. Reeves, *Continuum Mech. Thermodyn.* **15**, 321–49 (2003).
22. V. Garzo, J. W. Dufty, and C. M. Hrenya, *Phys. Rev. E* **76**, 031304 (2007).
23. V. Garzo, J. W. Dufty, and C. M. Hrenya, *Phys. Rev. E* **76**, 031303 (2007).
24. Y. Forterre, and O. Pouliquen, *Annu. Rev. Fluid Mech.* **40**, 1–24 (2004).
25. P. Mills, D. Loggia, and M. Tixier, *Europhys. Lett. L* **81**, 64005 (2008).
26. K. Kamrin, and M. Z. Bazant, *Phys. Rev. E* **75**, 041301 (2007).
27. S. B. Savage, *J. Fluid Mech.* **377**, 1–26 (1998).
28. M. Y. Louge, *Phys. Rev. E* **67**, 061303 (2003).
29. I. S. Aranson, and L. S. Tsimring, *Phys. Rev. E* **64**, 020301 (2001).
30. L. S. Mohan, K. K. Rao, and P. R. Nott, *J. Fluid Mech.* **457**, 377–409 (2002).
31. R. Brewster, L. E. Silbert, G. S. Grest, and A. J. Levine, *Phys. Rev. E* **77**, 061302 (2008).
32. L. E. Silbert, G. S. Grest, R. Brewster, and A. J. Levine, *Phys. Rev. Lett.* **99**, 068002 (2007).
33. O. Pouliquen, and Y. Forterre, *Adv. Complex Syst.* **4**, 441–50 (2001).
34. O. Pouliquen, C. Cassar, Y. Forterre, P. Jop, and M. Nicolas, *J. of Stat. Mech.: Theory and Experiment* p. P7020 (2006).
35. P. Jop, Y. Forterre, and O. Pouliquen, *Nature* **441**, 727–30 (2006).
36. O. Pouliquen, *Phys. Rev.* **93**, 248001 (2004).
37. F. da Cruz, S. Emam, M. Prochnow, J.-N. Roux, and F. Chevoir, *Phys. Rev. E* **72**, 021309 (2005).
38. G. D. R. MiDi, *Eur. Phys. J. E* **14**, 341–65 (2004).
39. S. B. Savage, and M. Hutter, *J. Fluid Mech.* **199**, 177–215 (1989).
40. S. B. Savage, and M. Hutter, *Acta Mechanica* **86**, 201–23 (1991).
41. J. M. N. T. Gray, M. Wieland, and H. Hutter, *Proc. R. Soc. Lond. A* **455**, 1841–74 (1998).
42. J. M. N. T. Gray, *J. Fluid Mech.* **441**, 1–29 (2001).
43. O. Pouliquen, and Y. Forterre, *J. Fluid Mech.* **453**, 131–51 (2002).
44. C. Ancey, P. Coussot, and P. Evesque, *J. Rheology* **43**, 1673–99 (1999).
45. S. B. Savage, and M. Sayed, *J. Fluid Mech.* **142**, 391–430 (1984).
46. S. B. Savage, *Adv. Appl. Mech.* **24**, 289–366 (1984).
47. K. M. Hill, and Y. Fan, *Phys. Rev. Lett.* **101**, 088001 (2008).
48. R. A. Bagnold, *Proc. R. Soc. Lond. A* **225(1160)**, 49–63 (1954).
49. D. Bonamy, F. Daviaud, and L. Laurent, *Phys. Fluids* **14**, 1666–73 (2002).
50. M. Jain, J. M. Ottino, and R. M. Lueptow, *Phys. Fluids* **14**, 572–82 (2002).
51. N. Taberlet, P. Richard, and E. J. Hinch, *Phys. Rev. E* **73**, 050301R (2006).
52. K. M. Hill, G. Gioia, and V. V. Tota, *Phys. Rev. Lett.* **91**, 064302 (2003).
53. S. D. Gupta, D. V. Khakhar, and S. K. Bhatia, *Chem. Eng. Sci.* **46**, 1513–7 (1991).
54. K. M. Hill, and J. Kakalios, *Phys. Rev. E* **52**, 4393–400 (1995).
55. S. B. Savage, and S. McKeown, *J. Fluid Mech.* **127**, 453–72 (1983).
56. P. G. Rognon, J. N. Roux, M. Naaim, and F. Chevoir, *Phys. Fluids* **19**, 058101 (2007).
57. P. A. Cundall, and O. D. L. Strack, *Geotechnique* **29**, 47–65 (1979).
58. Y. Tsuji, T. Tanaka, and T. Ishida, *Powder Tech.* **71**, 239–50 (1992).
59. C. Campbell, and C. Brennen, *J. Fluid Mech.* **151**, 167–88 (1985).
60. X. Zheng, and J. Hill, *Appl. Math. Modelling* **20**, 82–92 (1996).
61. J. Liu, and A. Rosato, *J. Phys.: Condens. Matter* **17**, 2609–22 (2005).
62. S. L. Conway, A. Lekhal, J. G. Khinast, and B. J. Glasser, *Chem. Eng. Sci.* **60**, 7091–107 (2005).
63. K. M. Hill, and J. Zhang, *Phys. Rev. E* **77**, 061303 (2008).
64. L. A. Golick, and K. E. Daniels, *Phys. Rev. E* **80**, 042301 (2009).
65. O. Baran, and L. Kondic, *Phys. Fluids* **17**, 073304 (2005).
66. O. Pouliquen, *Phys. Fluids* **11(3)**, 542–8 (1999).
67. C. Goujon, N. Thomas, and B. Dalloz-Dubrujeaud, *Eur. Phys. J. E* **11**, 147–57 (2003).

WAVES

Micro-Macro Transition and Linear Wave Propagation in Three-Component Compacted Granular Materials

Bettina Albers

Technische Universität Berlin, Institute for Geotechnical Engineering and Soil Mechanics, Sekr. TIB1-B7, Gustav-Meyer-Allee 25, 13355 Berlin, Germany, albers@grundbau.tu-berlin.de

Abstract. Recently, Albers developed a continuum model for the description of wave propagation in partially saturated, three-component, porous media. Macroscopic parameters have been obtained by a systematic micro-macro transition procedure. Using this model, acoustic properties of sandstone filled by different pore fillings and of several soil types containing a water-air-mixture have been presented. The soil types are classified in the German standard DIN 4220. Originally, both for rocks and soils the shear modulus had been proposed according to the classical elasticity theory. However, it seems that this approach for granular soils yields, at least for the shear wave, wave speeds which are higher than experimentally observed values. Therefore, in this note, phase speeds and attenuations of the four waves appearing in unsaturated compact granular media are calculated using also another approach for the shear modulus, the Mindlin-Duffy approach. The numerical results of both theoretical approaches are compared to experimentally obtained values. It turns out that the latter approach for granular media is much better while for porous media the first approach is concordant with measurements.

Keywords: shear modulus, soil types, wave propagation, unsaturated granular media
PACS: 91.30.Ab, 91.30.Cd, 91.60.Lj

INTRODUCTION

The three-component continuous model introduced by ALBERS e.g. in [2], [5] was originally intended to describe the wave propagation in porous media, especially in rocks, as for example, sandstone [3]. Since rocks have a high compressibility modulus the use of classical elasticity theory was the simplest way to model the rock behavior and it turned out that this approach reflects the characteristics of porous media adequately. Thereupon the model was applied to investigate wave speeds and attenuations of eleven granular soil types [6] which are classified in the German standard DIN 4220 [12]. The wave analysis showed that for these soil types the values of the shear wave speeds are at the upper limit of comparable experimentally obtained results. Therefore another approach for the shear modulus, the MINDLIN-DUFFY approach, which seems to reflect the characteristics of granular media better is checked and the results are compared to these of the elasticity approach. Finally, the wave speeds are compared to the experimentally obtained values of three groups of authors which are arranged in Table 1.

It is not the intention of this work to develop a new theory for the continuum mechanical description of granular media including the detailed investigation of contact mechanisms, rotation of particles and even the flow of particles. The present work is aimed to make the just developed model [5] better applicable to granular media. However, the model is intended to investigate the wave propagation and not flow problems so that the examined deformations are very small. In this case, for porous media, the elasticity approach is the simplest choice to determine material parameters. For granular materials a comparably simple description was thought for. Therefore, one of the first approaches in this direction, the MINDLIN-DUFFY approach has been chosen in this work. Of course, this and other classical approaches have been extended and became more sophisticated in the last 50 years. Especially, there are nowadays several approaches for the computer simulation of flow problems in granular media. Here are some examples of such works: AGNOLIN & ROUX [1], JENKINS et al. [18], MAKSE et al. [23], WANG & MORA [35], DIGBY [11], WALTON [34], KOENDERS [20], HICHER et al. [17], LUDING [22]. For some of these approaches quantities like the number of particles, the number of contacts or the coordination number (i.e. the average number of contacts per particle) play a role but this does not hold true for the present work since a continuum mechanical model for wave propagation in a linear elastic medium with small deformations is considered.

TABLE 1. Left: Extract of a table of SCHULTZE & MUHS [29] specifying shear wave velocity, v_T. Middle: Sound speed in gassy sediment. Extract of Table III, Part II of [8] by ANDERSON & HAMPTON. Right: P-wave velocities of some types of geologic media. From PAVLOVIĆ & VELIČKOVIĆ [26].

Soil type	v_T [m/s]
sands, depending on condition	100-250
sand and gravel	180-550
loam	150-200
clays dep. on water content	120-700
Keuper sandstone, weathered	250
bunter, nonweathered	1100

Sediment type	gas content	sound speed [m/s]
clay	0	1488
	0.001	488
	0.01	359
	0.1	355
silt	0	1552
	0.001	622
	0.01	524
	0.1	528
fine sand	0	1749
	0.001	900
	0.01	826
	0.1	837
coarse sand	0	1907
	0.001	1106
	0.01	1039
	0.1	1054

Medium	velocity [m/s] min	max
air (dep. on temp.)	310	360
weathered soil horizon	100	500
gravel, dry sand	100	600
loam	300	900
wet sand	200	1800
clay	1200	2500
water (dep. on temp.)	1430	1590
friable sandstone	1500	2500
dense sandstone	1800	4000
chalk	1800	3500
limestone	2500	6000
marl	2000	3500
gypsum	4500	6500
ice	3100	4200
granite	4000	5700
metamorphosed rock	4500	6800

CONTINUUM MODEL FOR PARTIALLY SATURATED POROUS AND GRANULAR MEDIA

Soils and rocks are composed of a mixture of grains of different sizes and of voids which may be filled by a gas, a fluid or a mixture of immiscible fluids (which may be liquid or gaseous). In [5] a couple of pore fillings are investigated, e.g. water and air or water and oil. The solid particles are either loose (granular media) or held together by compression and cementing material (porous media). We consider here a three-component medium consisting of a solid (S), a fluid (F) and a gas (G).

The linear field equations for the description of this medium

$$\begin{aligned}
\rho_0^S \frac{\partial \mathbf{v}^S}{\partial t} &= \operatorname{div}\left\{\lambda^S e\mathbf{1} + 2\mu^S \mathbf{e}^S + Q^F \varepsilon^F \mathbf{1} + Q^G \varepsilon^G \mathbf{1}\right\} + \\
&\quad + \pi^{FS}\left(\mathbf{v}^F - \mathbf{v}^S\right) + \pi^{GS}\left(\mathbf{v}^G - \mathbf{v}^S\right), \\
\rho_0^F \frac{\partial \mathbf{v}^F}{\partial t} &= \operatorname{grad}\left\{\rho_0^F \kappa^F \varepsilon^F + Q^F e + Q^{FG} \varepsilon^G\right\} - \pi^{FS}\left(\mathbf{v}^F - \mathbf{v}^S\right), \\
\rho_0^G \frac{\partial \mathbf{v}^G}{\partial t} &= \operatorname{grad}\left\{\rho_0^G \kappa^G \varepsilon^G + Q^G e + Q^{FG} \varepsilon^F\right\} - \pi^{GS}\left(\mathbf{v}^G - \mathbf{v}^S\right), \\
\frac{\partial \mathbf{e}^S}{\partial t} &= \operatorname{sym} \operatorname{grad} \mathbf{v}^S, \quad \frac{\partial \varepsilon^F}{\partial t} = \operatorname{div} \mathbf{v}^F, \quad \frac{\partial \varepsilon^G}{\partial t} = \operatorname{div} \mathbf{v}^G, \quad e \equiv \operatorname{tr} \mathbf{e}^S,
\end{aligned} \quad (1)$$

are satisfied by the six essential fields $\{\mathbf{v}^S, \mathbf{v}^F, \mathbf{v}^G, \mathbf{e}^S, \varepsilon^F, \varepsilon^G\}$. This set of equations (a partial momentum balance and a partial mass balance for each of the three components) coincides with the Biot model if the third component, i.e. the gas, is neglected.

The quantities $\mathbf{v}^S, \mathbf{v}^F, \mathbf{v}^G$ are the velocity fields of the components, $\mathbf{e}^S$ is the deformation tensor. Quantities with subindex zero are initial values of the corresponding current quantity. The relative resistances between fluid and solid π^{FS} and gas and solid π^{GS} are discussed later.

The model is formulated using the volume changes of the components $e, \varepsilon^F, \varepsilon^G$ instead of the partial mass densities of the components, ρ^S, ρ^F, ρ^G. They are related by

$$e = \frac{\rho_0^S - \rho^S}{\rho_0^S}, \quad \varepsilon^F = \frac{\rho_0^F - \rho^F}{\rho_0^F}, \quad \varepsilon^G = \frac{\rho_0^G - \rho^G}{\rho_0^G}. \quad (2)$$

The porosity, n, (i.e. the ratio of the void volume to the volume of a representative elementary volume *REV*) also belongs to the fields and satisfies an own balance equation. However, if we neglect memory effects, the balance equation can be solved and is not further necessary to solve the wave problem. On the other hand the current saturation of the fluid, S, i.e. the fraction of fluid in the voids, is not a field. Instead, a constitutive law will be formulated for this quantity.

MICRO-MACRO TRANSITION

The macroscopic material parameters $\{\lambda^S + \frac{2}{3}\mu^S, \kappa^F, \kappa^G, Q^F, Q^G, Q^{FG}\}$ appearing in the set of equations (1) have to be specified according to the material. This is done by applying a transition from the micro-scale where measurable quantities are available to the macro-scale. To this aim isotropic compression is considered so that the mechanical reactions of the system reduce to pressures

$$\mathbf{e}^S = \frac{1}{3}e\mathbf{1}, \quad p^S = -\frac{1}{3}\operatorname{tr}\mathbf{T}^S, \quad p^F = -\frac{1}{3}\operatorname{tr}\mathbf{T}^F, \quad p^G = -\frac{1}{3}\operatorname{tr}\mathbf{T}^G. \quad (3)$$

p^S, p^F and p^G denote the partial pressures of the components, $\mathbf{T}^S$, $\mathbf{T}^F$ and $\mathbf{T}^G$ are the respective Cauchy stresses. Using the assumption of homogeneous, spherically symmetric deformations we get several macroscopic, microscopic and mixed conditions which have to be combined. First, we have
- *macroscopic constitutive relations*

$$\begin{aligned}
p^S - p_0^S &= -\left(\lambda^S + \frac{2}{3}\mu^S\right)e - Q^F \varepsilon^F - Q^G \varepsilon^G, \\
p^F - p_0^F &= -\rho_0^F \kappa^F \varepsilon^F - Q^F e - Q^{FG} \varepsilon^G, \\
p^G - p_0^G &= -\rho_0^G \kappa^G \varepsilon^G - Q^G e - Q^{FG} \varepsilon^F.
\end{aligned} \quad (4)$$

In the static case the full pressure change must be in equilibrium with a given excess pressure Δp, i.e.

$$\Delta p = \left(p^S - p_0^S\right) + \left(p^F - p_0^F\right) + \left(p^G - p_0^G\right). \quad (5)$$

Secondly,
- *microscopic constitutive relations* are assumed

$$\begin{aligned} p^{SR} - p_0^{SR} &= -K_s e^R, \\ p^{FR} - p_0^{FR} &= -K_f \varepsilon^{FR}, \\ p^{GR} - p_0^{GR} &= -K_g \varepsilon^{GR}, \end{aligned} \tag{6}$$

where e^R, ε^{FR} and ε^{GR} are the volume changes on the microscopic level. This means that e^R describes relative true volume changes of grains, ε^{FR} describes changes of the true mass density of the fluid, $\varepsilon^{FR} = (\rho_0^{FR} - \rho^{FR})/\rho_0^{FR}$, and ε^{GR} describes changes of the true mass density of the gas, $\varepsilon^{GR} = (\rho_0^{GR} - \rho^{GR})/\rho_0^{GR}$. The corresponding pressures are denoted by p^{SR}, p^{FR} and p^{GR}. Furthermore, K_s, K_f and K_g denote the real (true) compressibility moduli of the three components. They can be measured independently of the current morphology of the granular material.

Besides macroscopic and microscopic constitutive relations two sets of compatibility conditions between the quantities of both scales are available. Firstly, the
- *dynamic compatibility relations* are given

$$p^S = (1-n)p^{SR}, \quad p^F = nSp^{FR}, \quad p^G = n(1-S)p^{GR}, \tag{7}$$

where n denotes the current porosity of the medium and S the current fluid saturation. These relations also hold for the initial values of the quantities. We need them in linearized form and arrive at

$$\begin{aligned} p^S - p_0^S &= p_0^S \left(\frac{p^{SR} - p_0^{SR}}{p_0^{SR}} - \frac{n - n_0}{1 - n_0} \right), \\ p^F - p_0^F &= p_0^F \left(\frac{p^{FR} - p_0^{FR}}{p_0^{FR}} + \frac{S - S_0}{S_0} + \frac{n - n_0}{n_0} \right), \\ p^G - p_0^G &= p_0^G \left(\frac{p^{GR} - p_0^{GR}}{p_0^{GR}} - \frac{S - S_0}{1 - S_0} + \frac{n - n_0}{n_0} \right). \end{aligned} \tag{8}$$

Secondly, we have the relations between changes of mass densities and volume changes (2) which also hold for the corresponding realistic (true) quantities e^R, ε^{FR} and ε^{GR} (see above). Combination and linearization of the above relations yields the following
- *geometrical compatibility relations*

$$\begin{aligned} e &= e^R + \frac{n - n_0}{1 - n_0}, \\ \varepsilon^F &= \varepsilon^{FR} - \frac{n - n_0}{n_0} - \frac{S - S_0}{S_0}, \\ \varepsilon^G &= \varepsilon^{GR} - \frac{n - n_0}{n_0} + \frac{S - S_0}{1 - S_0}. \end{aligned} \tag{9}$$

One more equation is needed to solve the problem for the 14 unknown quantities of spherical homogeneous deformations

$$\{e, \varepsilon^F, \varepsilon^G, p^S, p^F, p^G, e^R, \varepsilon^{FR}, \varepsilon^{GR}, p^{SR}, p^{FR}, p^{GR}, n, S\}. \tag{10}$$

The missing equation is the microscopic relation for the capillary pressure. In the present work we use the one proposed by VAN GENUCHTEN [32] in 1980

$$p_c(S) = p^{GR} - p^{FR} = \frac{1}{\alpha} \left[S^{(-1/m)} - 1 \right]^{1/n}, \tag{11}$$

where α, m, n are parameters which depend on the properties of the soil. Also this equation has to be linearized for the purpose of this work. By applying a Taylor expansion around $S = S_0$ we arrive at

$$\begin{aligned} p^{GR} - p^{FR} &= g_0(S_0) - (S - S_0) g_1(S_0), \\ g_0 &:= \frac{1}{\alpha} \left[S_0^{(-1/m)} - 1 \right]^{1/n}, \\ g_1 &:= \frac{1}{\alpha n m} \left[S_0^{(-1/m)} - 1 \right]^{1/n - 1} S_0^{-1/m - 1}. \end{aligned} \tag{12}$$

The combination of the above given 14 equations yields the explicit solution of the problem, i.e. the volume changes e, ε^F and ε^G as well as all other quantities are given in terms of the given excess pressure Δp. However, the excess pressure

$$\Delta p = -\left(\lambda^S + \frac{2}{3}\mu^S + Q^F + Q^G\right)e - \left(\rho_0^F \kappa^F + Q^F + Q^{FG}\right)\varepsilon^F - \left(\rho_0^G \kappa^G + Q^G + Q^{FG}\right)\varepsilon^G, \quad (13)$$

still contains the macroscopic material parameters $\{\lambda^S + \frac{2}{3}\mu^S, \kappa^F, \kappa^G, Q^F, Q^G, Q^{FG}\}$. Due to the presence of six unknown parameters we need six further conditions to solve the problem completely. In the case of static, homogeneous, spherically symmetric problems such conditions can be easily formulated. For the two-component medium, first by BIOT & WILLIS [9] and later by WILMANSKI [37] three simple tests were considered which concern the investigation of the volume changes and the pressures under certain conditions. They are called "jacketed drained", "jacketed undrained" and "unjacketed" tests. However, it became clear that the incorporation of three tests yields too many conditions which are partly equivalent. It turned out that the two-component problem could be solved using only two of the aforementioned tests: either both jacketed tests or, equivalently, the jacketed undrained and the unjacketed test. It was shown by SANTOS, DOUGLAS & CORBERÓ [28] that also in the case of unsaturated porous media only two tests are necessary to set up an adequate number of boundary conditions. They proposed a "generalized jacketed compressibility test" and a "generalized partially jacketed compressibility test" (for details see: [5]). These tests yield the following five conditions:

$$e = \varepsilon^F, \quad \varepsilon^F = \varepsilon^G, \quad p^{FR} - p_0^{FR} = 0 \iff \varepsilon^{FR} = 0, \quad (14)$$
$$p^{GR} - p_0^{GR} = 0 \iff \varepsilon^{GR} = 0, \quad p^{MR} - p_0^{MR} = 0.$$

The latter means that changes of average pore pressure, $p^{MR} = Sp^{FR} + (1-S)p^{GR}$, are zero which follows from the condition $S - S_0 = 0$. Even if the capillary pressure, $p_c = p^{GR} - p^{FR}$, does not enter the formula for the average pressure directly, it still enters the model through Equation (12). Additionally to the five conditions given in (14), the drained compressibility modulus K_d can be measured. It is defined as the negative fraction of the excess pressure Δp to the macroscopic volume change e in the drained jacketed test. Thus

$$K_d := -\frac{\Delta p}{e} = -\frac{\left(p^S - p_0^S\right) + \left(p^F - p_0^F\right) + \left(p^G - p_0^G\right)}{e}. \quad (15)$$

With these six relations at hand we are able to solve the problem for the six unknown material parameters $\{\lambda^S + \frac{2}{3}\mu^S, \kappa^F, \kappa^G, Q^F, Q^G, Q^{FG}\}$. One of the solutions to the six equations which are highly nonlinear in the six unknowns are the relations between macroscopic and microscopic material parameters by SANTOS, CORBERÓ & DOUGLAS [28]. They are not as simple as the GASSMANN relations [14] of the two-component body and, therefore, not quotable here in a compact form. Moreover, for their determination another approach using the total stress tensor of the bulk material and microscopic pressure changes of the two fluid components and another notation are used. However, some conversions yield the dependence of the material parameters on the saturation (for details see: [5]).

SUMMARY OF NECESSARY PARAMETERS

In this note the wave speeds and attenuations of waves appearing in several granular soil types are compared using two different approaches for the shear modulus. However, the shear modulus is not the only parameter which has to be specified. In order to determine the remaining parameters which are listed below we use the classification of the German standard DIN 4220 [12] named "Pedologic site assessment – Designation, classification and deduction of soil parameters (normative and nominal scaling)". Thirty-one soil types are defined in this standard and tables summarizing some of their properties are given. The soil types consist of different fractions of the three main soil types sand, silt and clay. Eleven of them have been chosen for a detailed investigation. They possess the following characteristics:

- **Porosity and mass densities**
  The porosity, n, is interpreted as the ratio of the volume of the voids (whether filled by a single pore fluid or by a mixture of pore fluids) over the entire representative elementary volume REV of a granular medium. The size of REV depends on the considered problem.
  In Table 31 of DIN 4220 air capacity, usable field capacity and field capacity for the 31 soil types for different values of the oven-dry density are given. The oven-dry density is classified in Table 12 of DIN 4220. Further a

medium value of this quantity is assumed, namely the class $pt3$. According to the calculation formula for the total void fraction and thus for the porosity which is given in the standard: the total void fraction is the sum of field capacity and air capacity, the porosity has been calculated (see: Table 2). Using this the mass density of the different soil types can be deduced. The initial true mass density of the skeleton denoted by ρ_0^{SR}, and the initial porosity by n_0 the macroscopic intial mass density is

$$\rho_0^S = (1-n_0)\rho_0^{SR}. \tag{16}$$

The values for different soil types given in Table 2 have been obtained assuming that $\rho_0^{SR} = 2650$ kg/m^3 for each soil type.

- **Compressibility of the skeleton, Poisson's number, shear modulus**
For the compressibility modulus of the skeleton for all granular soil types the value $K_s = 35$ GPa has been chosen. This may be too high if the soil is clayey. Since the particle form of clay is plate-like a model assuming isotropy and spheres can only yield a rough approximation of the soil behavior. However VANORIO, PRASAD & NUR in [33] report in their article of theoretical values for the bulk modulus of clay between 20 and 50 GPa. These values have been widely adopted in the literature because measurements of clay minerals had proven to be difficult. In contrast, the at that time only published experimental measurement of Young's modulus in a clay mineral gave a much lower value of 6.2 GPa. The measurements described in [33] provide a value in between, namely $K_s = 12$ GPa.

The value of Poisson's number has been estimated according to the fractions of the three main soil types. The fractions of clay, silt and sand are given in Table 6 of DIN 4220. The following Poisson numbers for the main soil types are proposed:

$$\nu_{\text{sand}} = 0.31, \qquad \nu_{\text{silt}} = 0.37, \qquad \nu_{\text{clay}} = 0.43. \tag{17}$$

Using these values and approximately the median of the fractions of the main soil types (but taking into account that the fractions do have to sum up to 100%) the granular soil types under consideration possess the values of Poisson's number given in Table 2.

Since in this paper two approaches for the shear modulus will be compared this point will be discussed in an own subsection (see below).

- **Saturation and capillary pressure**
If multicomponent fluid mixtures fill the pores one needs, apart from porosity, at least one additional microstructural variable – a fraction of contributions of these components. If we consider a three-component medium consisting of a solid and two pore fluids (e.g. water and air) the saturation or degree of saturation (S) is defined as the fraction of the volume of the first fluid over the volume of all voids within a representative elementary volume *REV*. The sum of the fractions of the two pore fluids, of course, is equal to one, i.e. that the two pore fluids completely fill out the void volume.

Between two immiscible fluids, of which one may be gaseous, a discontinuity in pressure exists across the interface separating them. The difference is the capillary pressure p_c. It is a measure of the tendency of a porous medium to suck in the wetting fluid or to repel the nonwetting phase. In soil science, the negative of the capillary pressure is called suction. The capillary pressure depends on the geometry of the void space, on the nature of the solids and fluids (e.g. on the contact angle) and on the degree of saturation. Laboratory experiments are probably the only method to derive the relationship $p_c = p_c(S)$.

In DIN 4220 tables of results of such measurements are given. They are plotted in Figure 1 for the 11 chosen soil types. For the presentation of capillary pressure curves by formulas, in the present work the approach of van Genuchten is chosen because in DIN 4220 also the necessary parameters for the various soil types are presented. The normalized form of the van Genuchten formula (compare to (11) where the capillary pressure is given in dependence on the degree of saturation) is

$$S = (1/[1+(\alpha p_c)^n])^m, \tag{18}$$

where the parameters α, n and m are the van Genuchten parameters. Their values for the eleven soil types are presented in Table 3. Parameter m is not given in the tables of DIN 4220 but calculated from n by $m = 1 - 1/n$.

- **Compressibilities of fluid and gas, viscosity, permeability**
The microscopic compressibilities of water and air do not change by consideration of different soil types. Thus, independently of the soil type they have the values

$$K_f = 2.25 \cdot 10^9 \text{Pa}, \qquad K_g = 1.01 \cdot 10^5 \text{Pa}. \tag{19}$$

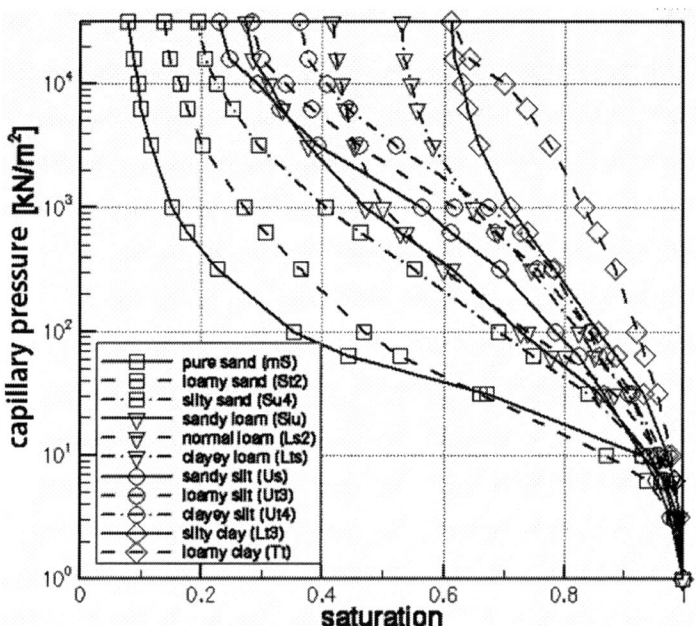

FIGURE 1. Capillary pressure curves for eleven chosen soil types classified in the German standard DIN 4220.

For the wave analysis the incorporation of relative resistances between fluid and solid (FS) and gas and solid (GS) is necessary. They are proposed by

$$\pi^{FS} = \pi^F/k_f, \qquad \pi^{GS} = \pi^G/k_g, \tag{20}$$

where

$$k_f = S^{\frac{1}{2}}\left[1 - \left(1 - S^{\frac{1}{m}}\right)^m\right]^2, \qquad k_g = (1-S)^{\frac{1}{3}}\left(1 - S^{\frac{1}{m}}\right)^{2m}, \tag{21}$$

are relative permeabilities also proposed by VAN GENUCHTEN. The parameter m, is the parameter as appearing in (12) and (18) and is arranged in Table 3. The functions (21) are plotted in Figure 2 for the eleven soil types under consideration. The viscosities of water and air are implicitly incorporated in the parameters π^F and π^G. While the permeability or resistance with respect to the gas is not influenced by the soil type, the permeability with respect to the fluid differs. Thus, the value $\pi^G = 1.82 \cdot 10^5$ Pa is the same for all soil types while π^F is different. This value can be deduced from the hydraulic conductivity in water-saturated soil which is given in Table 3. The conversion of the permeability, κ, into the resistance, π^F, is performed according to

$$\kappa/(\rho g) \sim 1/\pi^F, \qquad \text{i.e.} \qquad \pi^F = (n_0 \cdot 1000 \cdot 9.81)/\kappa \quad [\text{kg}/(\text{m}^3\text{s})]. \tag{22}$$

Two approaches for the shear modulus

- **Elasticity approach**

In the original form of the model the Lamé constants, λ^S and μ^S, have been proposed according to linear elasticity. It had been discussed for a two-component porous medium in ALBERS & WILMANSKI [7] that in this case the shear modulus, μ^S, does not depend on any coupling parameters but only on Poisson's number and on the compressibility modulus of the dry matrix, K_d. Thereby, it depends on the initial porosity n_0 since K_d is used in this work in the form proposed by GEERTSMA [15] $K_d = K_s/(1+50n_0)$. As done in almost all approaches on partially saturated porous media it is assumed that the shear modulus does also not depend on the degree of saturation. Thus, the shear modulus can be expressed by

$$\mu^S = \frac{3}{2}\frac{1-2\nu}{1+\nu}\frac{K_s}{1+50n_0}. \tag{23}$$

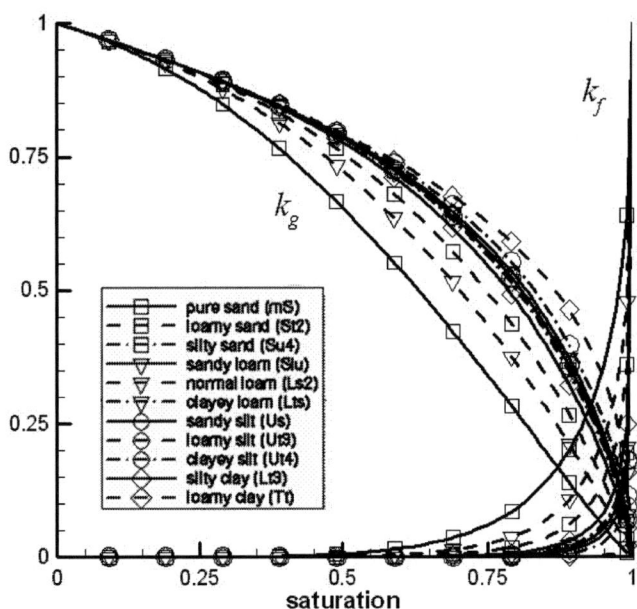

FIGURE 2. Relative permeabilities according to VAN GENUCHTEN (21) for eleven chosen soil types classified in the German standard DIN 4220.

TABLE 2. Porosity, initial mass density, Poisson's number and grain size fraction of eleven chosen soil types.

abbre-viation	porosity	initial mass density ρ_0^S $\left[\frac{kg}{m^3}\right]$	Poisson's number	grain size fraction percent by weight		
				clay	silt	sand
mS	0.42	1537	0.32	3	5	92
St2	0.42	1537	0.33	11	5	84
Su4	0.43	1511	0.34	4	45	51
Slu	0.43	1511	0.35	13	45	42
Ls2	0.43	1511	0.36	21	45	34
Lts	0.43	1511	0.37	35	23	42
Us	0.44	1484	0.35	4	65	31
Ut3	0.43	1511	0.37	15	77	8
Ut4	0.44	1484	0.38	21	74	5
Lt3	0.44	1484	0.38	40	40	20
Tt	0.46	1431	0.41	76	12	12

This form of the shear modulus stems from elasticity. Of course, there are other approaches in predicting it.

- **Mindlin-Duffy approach**

WHITE [36], for example, suggests for granular soils another form of the shear modulus which goes back to a sphere pack model. HARA [16], in attempting to describe the behavior of a carbon-granule microphone, was the first to consider an assemblage of elastic spheres in contact, using a theory due to HERTZ (see: [21], [30]) to compute areas of contact and relative displacements among spheres. DUFFY & MINDLIN [13] combined this theory with calculations of the relative displacement of two spheres subject to tangential forces by CATTANEO [10] and MINDLIN [24] in order to compute the speeds of elastic waves in a cubic array of spheres.

The granular medium under consideration, shown in Figure 3(a), is lined up by identical spheres up in rows parallel to the coordinate axes. A pressure $\bar{p}$ is acting on all faces of the elementary volume. The force supported

TABLE 3. Van-Genuchten-parameters in dependence on soil type (extract of Table 37 of DIN 4220 [12] supplemented by the third van-Genuchten-parameter $m = 1 - \frac{1}{n}$); medium permeability in water-saturated soil for a medium value of the oven-dry density (see Table 35 of DIN 4220).

soil type abbreviation	VAN GENUCHTEN parameters			saturated water permeability κ in m/s
	α [Pa]	n	m	
Slu	0.000425	1.21240	0.17519	$3.24 \cdot 10^{-6}$
St2	0.0009725	1.38158	0.27619	$1.37 \cdot 10^{-5}$
Su4	0.0003996	1.24244	0.195139	$4.40 \cdot 10^{-6}$
Ls2	0.0002664	1.50608	0.33602	$2.66 \cdot 10^{-6}$
Lt3	0.0003005	1.28574	0.22224	$8.10 \cdot 10^{-7}$
Lts	0.0003713	1.25012	0.20008	$1.16 \cdot 10^{-6}$
Us	0.000136	1.23252	0.18865	$2.55 \cdot 10^{-6}$
Ut3	0.0000781	1.23052	0.18734	$1.39 \cdot 10^{-6}$
Ut4	0.0000982	1.17848	0.15145	$1.50 \cdot 10^{-6}$
Tt	0.0000837	1.08011	0.07417	$3.47 \cdot 10^{-7}$
mS	0.0004732	1.75418	0.42993	$5.67 \cdot 10^{-5}$

by one row of spheres is $G = 4a^2 \bar{p}$, where a is the radius of the spheres (see Fig. 3(b)). In order to determine the shear modulus we are interested in the effect of tangential forces. As shown in Figure 3(c) the tangential force $\Delta G'$ is acting at a single contact and yields a relative displacement $\Delta s'$ between the centers of adjacent spheres. As given by Mindlin [24] the following relation between force and displacement occurs

$$\frac{\Delta G'}{\Delta s'} = \frac{\left[6\left(1-v^2\right)aE_S^2 G\right]^{1/3}}{(2-v)(1+v)}, \qquad (24)$$

where E_S, Young's modulus, can be expressed by two other constants of elasticity, e.g. by λ^S and μ^S or by the compressibility modulus, K_s and Poisson's number v by

$$E_S = \mu^S \frac{3\lambda^S + 2\mu^S}{\lambda^S + \mu^S} = 3K_s(1-2v). \qquad (25)$$

One more point must be considered before Relation (24) can yield the desired shear modulus. In Figure 3(d), a typical sphere is shown, subject to horizontal forces $\Delta G'$ in opposite direction from its neighbors above and below. These forces cause the sphere to rotate until a compensating torque is created by vertical forces at the circles of contact with horizontal neighbors. Each of the four forces has the magnitude $\Delta G'$, and a relative displacement of $\Delta s'$ occurs at each contact area by the mechanism just discussed. In addition, each sphere undergoes a clockwise rotation of angle $\Delta s'/2a$. The net vertical displacement of the sphere with respect to its horizontal neighbors is zero, and its net horizontal displacement with respect to a vertical neighbor is $2\Delta s'$. The average strain is therefore $\bar{\varepsilon}_{xy} = 2\Delta s'/2a = \Delta s'/a$. Introduction of average stress and average strain into (24) then yields the elastic modulus governing shear distortion, i.e. the shear modulus

$$\mu_{M-D}^S = \frac{\bar{p}_{xy}}{\bar{\varepsilon}_{xy}} = \frac{\left[3\left(1-v^2\right)E_S^2 \bar{p}\right]^{1/3}}{2(2-v)(1+v)} = \frac{3\left[K_s^2\left(1-v^2\right)(1-2v)^2 \bar{p}\right]^{(1/3)}}{2(2-v)(1+v)}. \qquad (26)$$

I.e. that in this approach the shear modulus depends on the confining pressure $\bar{p}$ and thus on the depth. The relation between the shear modulus obtained by (26) and the depth is illustrated in Figure 4 for one of the soil types, sand mS, for which $\bar{p} = 10^5 + 1537z$ where z is the depth in [m] and $\rho_0^S = 1537 \frac{kg}{m^3}$ is the mass density of the solid. At the surface acts a pressure of 1 atm = 10^5Pa.

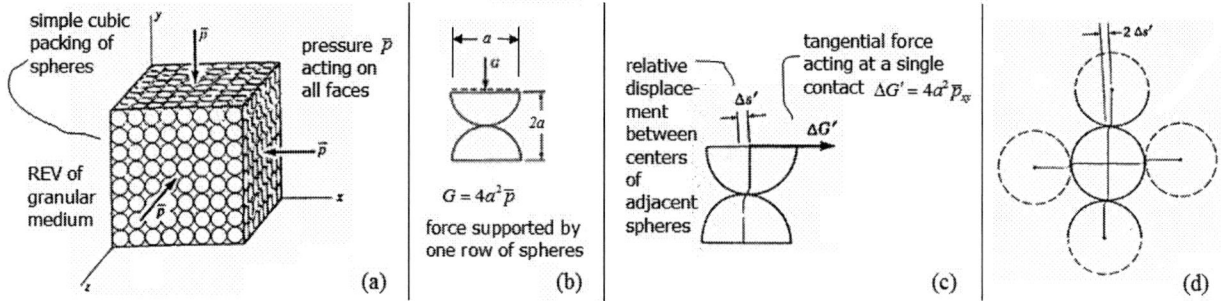

FIGURE 3. Behavior of a cubic sphere pack under incremental loading (modified figures from White [36])

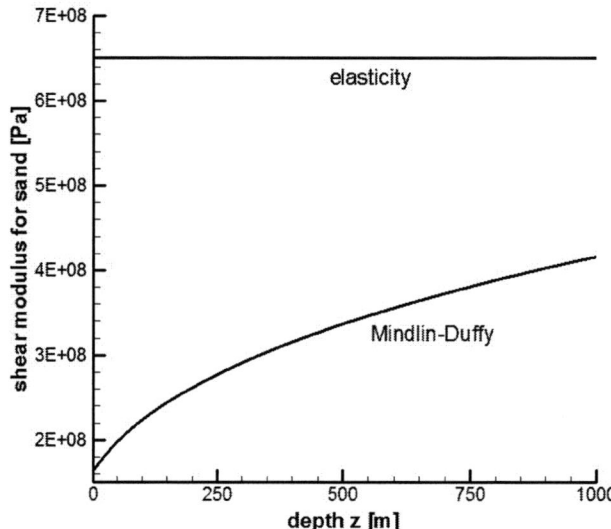

FIGURE 4. Dependence on the depth of the shear modulus obtained by (26) for sand (mS). In comparison the depth-independent shear modulus obtained by the elasticity approach (23) is given.

NUMERICAL RESULTS

In Figures 5 and 6 the results of the linear wave analysis are illustrated (for details see: [3]). The fields satisfy a normal wave ansatz, i.e. for example, $\mathbf{v}^S = \mathbf{V}^S \exp i(\mathbf{k} \cdot \mathbf{x} - \omega t)$, where $\mathbf{V}^S$ is a constant amplitude, ω is a given frequency, $\mathbf{k}$ is the complex wave vector. This means that $\mathbf{k} = k\mathbf{n}$, where k is the complex wave number and $\mathbf{n}$ is a unit vector in the direction of propagation. Such a solution describes the propagation of plane monochromatic waves in an infinite medium whose fronts are perpendicular to $\mathbf{n}$. Separation of the transversal and normal contributions in the resulting equations yields the dispersion relations for the shear wave and the longitudinal waves which are numerically investigated. The dispersion relations have been solved for the complex wave number k. Results for k yield the phase speeds $c = \omega/(\text{Re } k)$ and the attenuations Im k.

It results that besides the transversal wave, S, three longitudinal modes of propagation, $P1$-, $P2$- and $P3$-waves, appear in the unsaturated granular medium.

Figure 5 shows for the eleven chosen soil types the dependence of the shear wave S and the fastest longitudinal wave $P1$ on the initial saturation. The results for the two above described approaches for the shear modulus are compared. Results obtained by use of the elasticity approach are presented by circles, those of the Mindlin-Duffy approach by squares. Figure 5 clearly demonstrates that not all soil types exist for each degree of saturation. For example, clays (e.g. the class Tt) appear in reality in the range of saturations between around 60 and 100%.

S-wave and $P1$-wave are the two waves which propagate mainly through the skeleton and their front speeds are directly dependent on the shear modulus. For the shear wave the front speed ($\omega \to \infty$) independently of the

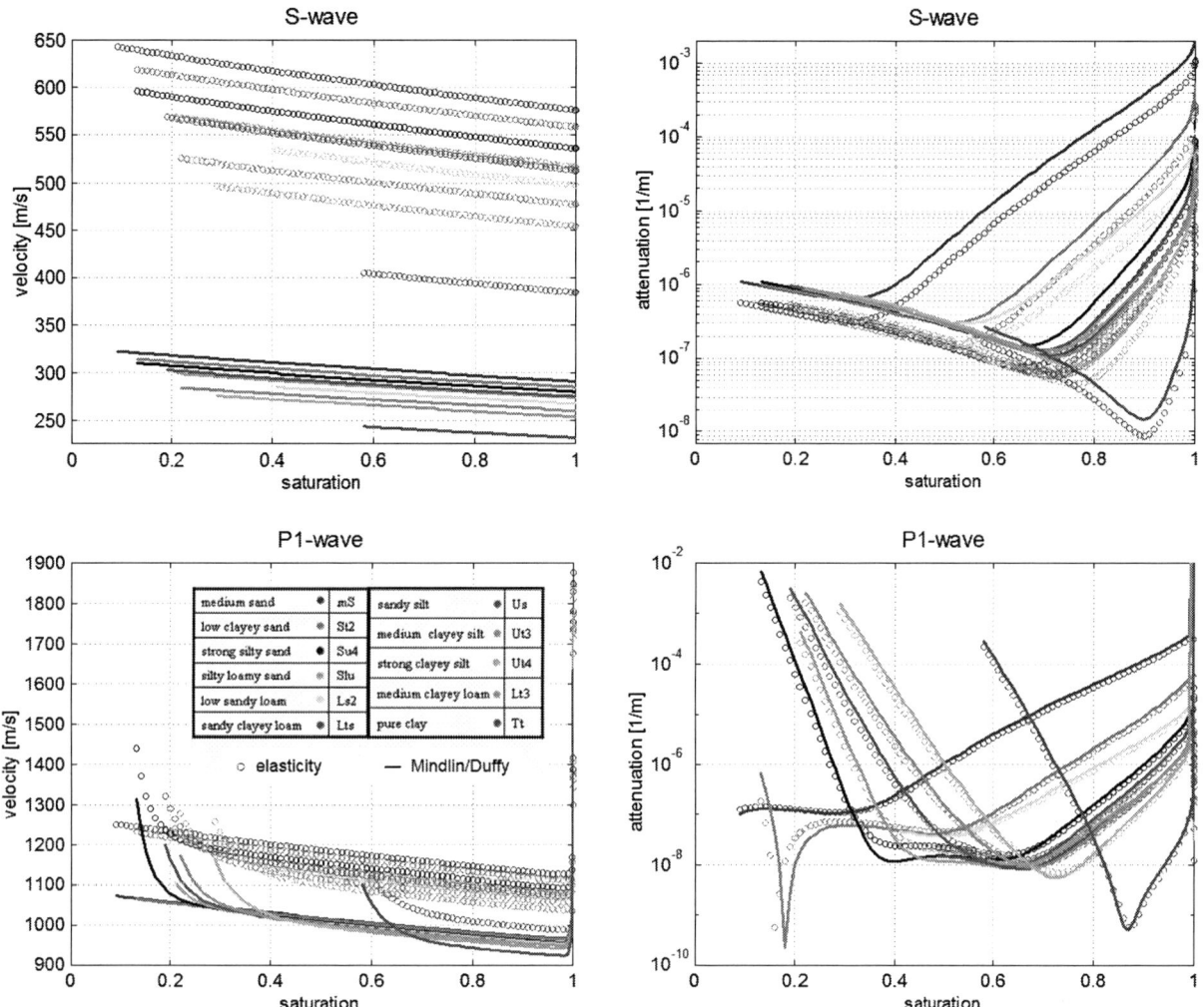

FIGURE 5. Velocities and attenuations of the shear wave S and the fast longitudinal wave $P1$ for several soil types in dependence on the saturation, circles: elasticity approach, squares: Mindlin-Duffy approach.

saturation is $c_{S\infty} = \sqrt{\mu^S/\rho_0^S}$, for the $P1$-wave the expression is more complicated: for the three-component medium it depends on the degree of saturation and on the Lamé parameters, mass densities, compressibilities of the pore fluids and the coupling parameters. Only for two-component media it is comparably simple to the shear wave: $c_{P1\infty} = \sqrt{(\lambda^S + 2\mu^S)/\rho_0^S}$.

By inspection of these relations and of Figure 5 it becomes obvious that the shear wave is mostly affected by change of the shear modulus. Since in both approaches the shear modulus does not depend on the saturation the wave speed in both cases is nearly constant. However, it is obvious that the wave speeds for the elasticity approach are nearly twice as large as for the Mindlin-Duffy approach. Moreover, it is apparent that the curves for different soil types are closer to each other for the Mindlin-Duffy approach. As can be seen in the right panel of Figure 5 the attenuation is not as much influenced by the change of the shear modulus as the speeds are.

As indicated above the speed of the $P1$-wave depends on the shear modulus but also on several other parameters. Therefore it is influenced not as much by the change of this modulus as it is the case for the shear wave. However, it can be seen that also the wave speed of the $P1$-wave is shifted to slightly lower values if the Mindlin-Duffy approach is used instead of the elasticity approach. This does not change the fact that the speed of the $P1$-wave shows an interesting behavior for high degrees of saturation: In this region the speed increases abruptly to nearly the double of its value for other saturations. As shown in [5] this phenomenon is much more distinct for pore fillings for which the

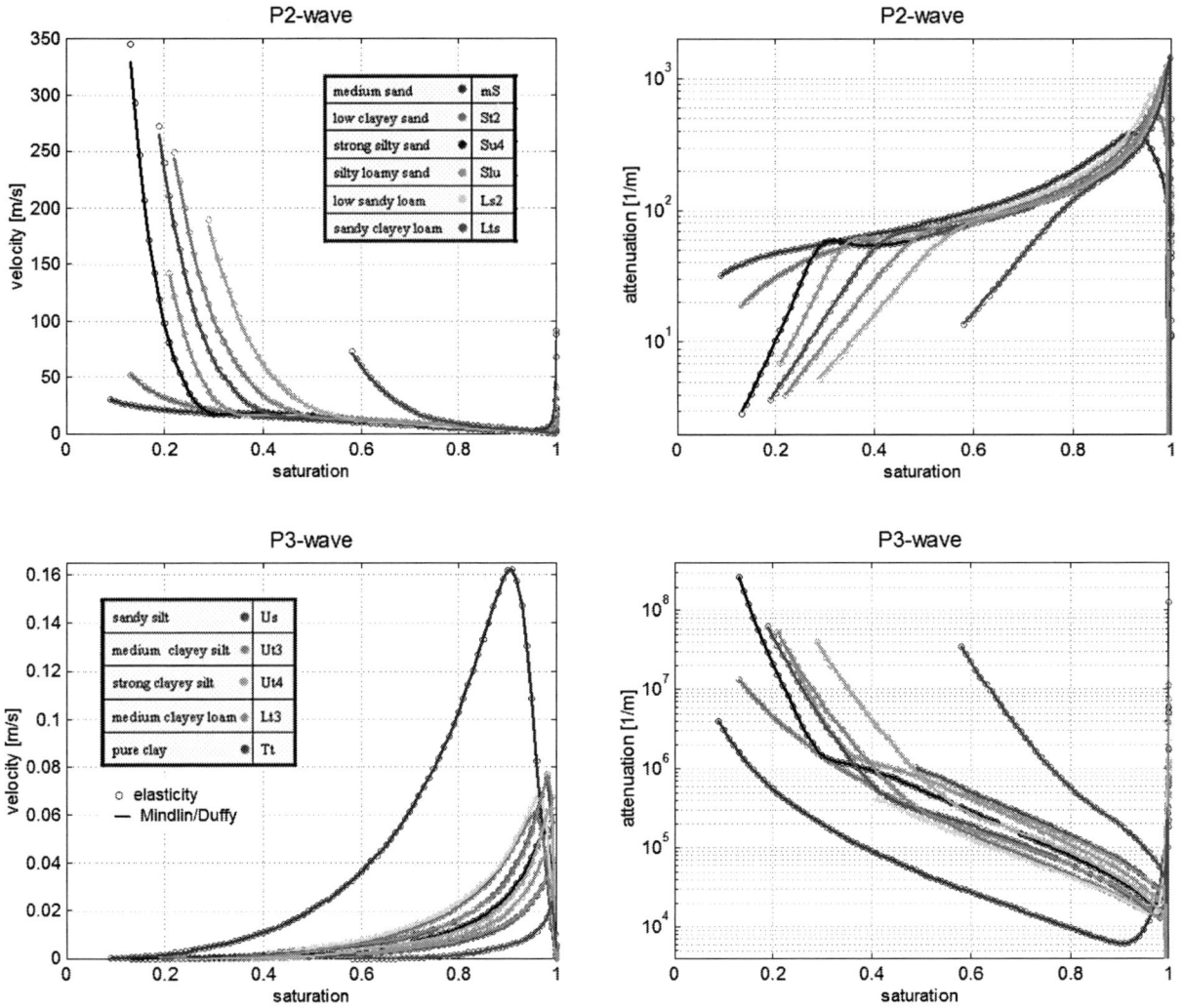

FIGURE 6. Velocities and attenuations of the $P2$-wave and the $P3$-wave for several soil types in dependence on the saturation, circles: elasticity approach, squares: Mindlin-Duffy approach.

orders of magnitude of the microscopic compressibility moduli of the components F and G are more different. This means that already a very small amount of air influences the propagation of waves in water immensely while the same amount of oil in water does not cause an effect of this extent because the differences in the material properties are not that large. Experimental results found in the literature support the occurrence of the sudden increase of the $P1$-wave speed (e.g. MURPHY [25]). It may be an important feature for applications in geotechnics. It provides the hope for the development of a non-destructive testing method to warn against land slides. The latter occur if the degree of saturation exceeds a certain value. It is favourable for such a method that this feature occurs for the $P1$-wave which is the first arrival on an oscillogram. However, it is disadvantageous that it appears in a very narrow range of saturation. The attenuation of the $P1$-wave possesses – as it is the case for the S-wave – very low values. It is barely influenced by the change of the shear modulus but it depends on the degree of saturation. For a certain degree of saturation at least one minimum appears.

Figure 6 shows speeds and attenuations in dependence on the initial saturation for the two slower longitudinal waves which are mainly driven by the properties of the pore fluids. The faster of this two, the $P2$-wave does exist if at least one pore fluid fills the pores of the granular medium, i.e. it does occur also for two-component granular media. It stronger depends on the degree of saturation than S- and $P1$-wave but inspection of Figure 6 points out that the influence of the shear modulus is nearly negligible. The speed of the $P2$-wave decreases as the degree of saturation increases and then

TABLE 4. Comparison of experimental wave speeds of the shear wave S and the fast longitudinal wave $P1$ (see Table 1) and calculated values using the elasticity and Mindlin-Duffy approaches.

wave type	soil/rock type	experimental see Table 1	calculated elasticity	Mindlin-Duffy
S	sandstone/bunter	up to 1110 m/s	$\approx$ 900 m/s	$\approx$ 330 m/s
	sand	up to 550 m/s	$\approx$ 600 m/s	$\approx$ 300 m/s
	clay	120-700 m/s	$\approx$ 400 m/s	$\approx$ 250 m/s
$P1$	loam	up to 900 m/s	$\approx$ 1100 m/s	$\approx$ 920 m/s
	sandstone	1500-2500 m/s	$\approx$ 1700 m/s	$\approx$ 1300 m/s
	sand	up to 1800 m/s	$\approx$ 1250 m/s	$\approx$ 1050 m/s
	clay	1200-2500 m/s or $\approx$ 360 m/s	$\approx$ 1100 m/s	$\approx$ 900 m/s

increases rapidly until saturation is reached. A deep minimum in this curve e.g. for an air-water-mixture in sandstone is experimentally confirmed (see: e.g. [4]). The attenuation of the $P2$-wave is high, nonetheless first PLONA in sintered glass beds [27] and later KELDER & SMEULDERS in sandstone [19] observed this wave in experiments.

The behavior of the speed of the $P3$-wave is opposite to this of the $P2$-wave: it increases with increasing saturation and then decreases. Both for gas saturation and for water saturation this wave disappears. This shows that it is driven by the capillary pressure and thus only exists if a second pore fluid is existent (see: TUNCAY & CORAPCIOGLU [31] whose authors were most likely the first in predicting this form of $P3$-waves in such media). Its speed is much smaller than this of the other waves, however, its attenuation is much higher. Due to the high values of attenuation it is not astonishing that the author is not aware of any attempt to measure this third longitudinal wave, either in artificial media or even less in non-artificial ones. Both speed and attenuation of the $P3$-wave are not influenced by the choice of the theoretical description of the shear modulus.

CONCLUDING REMARKS

Phase velocities and attenuations of some unconsolidated soils have been investigated whereas two approaches for the determination of the shear modulus have been used. These are the classical elasticity approach and an approach proposed by White (the Mindlin-Duffy approach) going back on a sphere pack model. It turned out that the latter approach for unconsolidated soils is much better since the shear wave speed is nearly half of this obtained by the other approach. However, for porous media, like sandstone, the first approach is concordant with measurements.

Comparison of calculated and measured values of S- and $P1$-waves (compare Table 4; the other two types of waves are hard to observe in real media and are barely influenced by the choice of the shear modulus) confirm this statement: the calculated values for sandstone ($c_S \approx$ 900 m/s, elasticity approach), sand ($c_S \approx$ 600 m/s, elasticity; $c_S \approx$ 300 m/s, Mindlin-Duffy) and clay ($c_S \approx$ 400 m/s, elasticity; $c_S \approx$ 250 m/s, Mindlin-Duffy) are qualitatively concordant with the values of Table 1_{left} (bunter: up to 1100 m/s, sand: up to 550 m/s and clay: 120-700 m/s). Also the $P1$-velocities of Table 1_{right} are qualitatively satisfied (loam: up to 900 m/s, calculated: $\approx$ 1100 m/s (elasticity), $\approx$ 920 m/s (Mindlin-Duffy); sandstone: 1500-2500 m/s, calculated: 1700 m/s (elasticity); sand: up to 1800 m/s, calculated: 1250 m/s and 1050 m/s for the elasticity and Mindlin-Duffy approaches). Only the velocity of clay given in the right panel of Table 1 (1200-2500 m/s, calculated: 900 to 1100 m/s) seems to be very high compared to both calculated results. However, in Table 1_{middle} in the limit of small water saturations much lower values of the $P1$-velocity in clay ($\approx$ 360 m/s) are quoted. These differences in the measured speeds by two different groups for the same material mean, that in the measurements there are big variations. But generally, the results of the calculations are anyhow in accordance to experimentally obtained values. For the attenuations a possibility of comparison does not exist.

REFERENCES

1. I. Agnolin, J.-N. Roux, *Phys. Rev. E* **76**, 061302, 061303, 061304 (2007).
2. B. Albers, On a micro-macro transition for a poroelastic three-component model, submitted to *Archives of Mechanics* (2009).
3. B. Albers, *Transport in Porous Media*, **80**(1), 173–92 (2009).

4. B. Albers, On the influence of saturation and frequency on monochromatic plane waves in unsaturated soils, in: *Coupled site and soil-structure interaction effects with application to seismic risk mitigation*, T. Schanz, R. Iankov (Eds.), NATO Science Series, Springer Netherlands, 65–76, 2009.
5. B. Albers, *Modeling and Numerical Analysis of Wave Propagation in Saturated and Partially Saturated Porous Media*, habilitation thesis, Veröffentlichungen des Grundbauinstitutes der Technischen Universität Berlin, Shaker (2010).
6. B. Albers, Linear elastic wave propagation in unsaturated sands, silts, loams and clays, submitted to *Transport in Porous Media* (2009).
7. B. Albers, K. Wilmanski, *J. Engrg. Mech.* **131**, 9 (2005), 974–85.
8. A. L. Anderson, L. D. Hampton, *J. Acoust. Soc. Am.* **67**, 6 (1980), 1865–98, 1890–903.
9. M. A. Biot, D. G. Willis, *J. Appl. Mech.*, **24**, 594–601 (1957).
10. C. Cattaneo, *Acad. Lincei* **27**, 342–8, 434–6, 474–8.
11. P. J. Digby, *J. Appl. Mech.* **48**, 803–8 (1981)
12. DIN 4220: "Pedologic site assessment - Designation, classification and deduction of soil parameters (normative and nominal scaling)" (2005). DIN Deutsches Institut für Normung e.V., Beuth Verlag GmbH (draft, in German), German title: "Bodenkundliche Standortbeurteilung - Kennzeichnung, Klassifizierung und Ableitung von Bodenkennwerten (normative und nominale Skalierungen)"
13. J. Duffy, R. D. Mindlin, *J. Appl. Mech.* **24**, 585–93 (1957).
14. F. Gassmann, *Vierteljahresschrift der Naturforschenden Gesellschaft in Zürich*, **96**(1), 1–23 (1951).
15. J. Geertsma, *Trans AIME* **210** (1957), 331–40.
16. G. Hara, *Elek. Nachr. Tech.* **12** (1935), 191–200.
17. P. Y. Hicher, C. S. Chang, C. Dano, *International Journal of Solids and Structures*, Pergamon Press (2008), Vol. 45, No. 16, pp. 4362Ű-74.
18. J. T. Jenkins, D. Johnson, L. La Ragione, H. Makse, *Journal of the Mechanics and Physics of Solids* **53**, (2005), 197–225.
19. O. Kelder, D. Smeulders, Measurement of ultrasonic bulk properties of water-saturated porous media, in: EAGE, 58th Conference and Technical Exhibition, Amsterdam (1996).
20. M. A. Koenders, *Journal of Physics D* **31**, 15, (1998), 1875–82.
21. A. E. H. Love, *A treatise on the mathematical theory of elasticity*, Dover, New York (1944).
22. S. Luding, *Computer Physics Communications* **147**, 1-2 (2002), 134–40.
23. H. A. Makse, N. Gland, D. L. Johnson, L. Schwartz, *Phys. Rev. E* **70**, 061302 (2004)
24. R. D. Mindlin, *J. Appl. Mech. Trans.*, ASME **71** (1949), 259–68.
25. W. F. Murphy, *J. Acous. Soc. Am.* **71** (1982), 1458–68.
26. V. D. Pavlovic, Z. S. Velickovic, *FACTA UNIVERSITATIS* Series: Physics, Chemistry and Technology **1**, 5 (1998), 63–73.
27. T. Plona, *Appl. phys. Lett.* **36**, 4 (1980), 259–61.
28. J. Santos, J. Douglas, J. Corberó, *J. Acoust. Soc. Am.*, **87**(4), 1428–38 (1990).
29. E. Schultze, H. Muhs, *Bodenuntersuchungen für Ingenieurbauten*, Springer, Berlin, Göttingen, Heidelberg (1950), in German.
30. S. Timoshenko, J. N. Goodier, *Theory of Elasticity*, McGraw-Hill, New York (1951).
31. K. Tuncay, M. Y. Corapcioglu, *Journal of Geophysical Research* **101**, 25, 149Ű59 (1996).
32. M. T. van Genuchten, *Soil Sci. Soc. Am. J.*, **44**, 892–8 (1980).
33. T. Vanorio, M. Prasad, A. Nur, *Geophysical Journal International* **155**, 1 (2003), 319–26.
34. O. R. Walton, *Mech. of Granular Materials*, (eds. J. T. Jenkins and M. Satake), Elsevier Sci. Pub. (1983).
35. Y. Wang, P. Mora, *Journal of the Mechanics and Physics of Solids*, **56**(12), 3459–74 (2008).
36. J. E. White, *Underground sound. Application of seismic waves, Methods in Geochemistry and Geophysics*, vol. 18. Elsevier, Amsterdam, New York (1983).
37. K. Wilmanski, *Geotechnique*, **54**, 9, 593–603 (2004).

Three-dimensional lattice-based dispersion relations for granular materials

N.P. Kruyt

*Department of Mechanical Engineering, University of Twente,
Enschede, The Netherlands; n.p.kruyt@utwente.nl*

Abstract. The propagation of mechanical waves, such as sound, through granular materials is important in many disciplines of engineering. Here wave propagation is studied from the micromechanical viewpoint, in which relationships are investigated between macroscopic wave propagation characteristics and microscopic particle and interparticle characteristics.

The interparticle interaction is described by linear elastic springs in directions normal and tangential to interparticle contacts. Dispersion relations are formulated theoretically, in terms of the micromechanical quantities, using three-dimensional lattice-based approaches. In these approaches the rotational degrees of freedom of the particles are explicitly accounted for. This leads to the presence of 'optical' branches in the dispersion relations.

Dispersion relations for various three-dimensional lattices (simple cubic, body centred cubic, face centred cubic) are given. The resulting dispersion relations are sensitive to the lattice structure and do not show true isotropic behaviour (at larger wave numbers), i.e. they depend on the direction of the wave vector.

Therefore, a statistical formulation of the lattice theory is developed, which does result in isotropic behaviour for all wave numbers. The corresponding dispersion relations show 'optical' branches, just like the lattice theories. The influence on these dispersion relations of the stiffness ratio of tangential over normal stiffness at the contacts is investigated.

Keywords: granular materials, wave propagation, dispersion relation, micromechanics, lattices
PACS: 81.05.Rm, 83.80.Fg, 83.60.Uv

1. INTRODUCTION

The propagation of mechanical waves, such as sound, through granular materials is important in many disciplines of engineering, for example in oil exploration. Here wave propagation is studied from the micromechanical viewpoint, in which relationships are investigated between macroscopic wave propagation characteristics and microscopic particle and interparticle characteristics. In this approach the three-dimensional granular assembly is modelled as a large set of spherical particles that only interact at contacts between particles through linear elastic interparticle springs in directions normal and tangential to contacts. This approach is appropriate for very small deformations.

The dispersion relation gives angular velocities ω that are compatible with plane-wave solutions of the governing equations with a specified spatial variation that is characterised by the wave number k, i.e. relations of the form

$$\omega = \omega(k) . \tag{1}$$

The wave number k is related to the wave length λ by

$$\lambda = \frac{2\pi}{k} . \tag{2}$$

Dispersion relations characterise the internal structure and the behaviour of materials.

Dispersion relations are formulated theoretically here, in terms of particle and interparticle characteristics, using three-dimensional lattice-based approaches. This lattice-based approach forms a step towards the study of dispersion relations for disordered packings. The current study extends the work of Suiker et al. [23, 24], who focussed on two-dimensional lattices, and that of Schwartz et al. [21], who focussed on a Face Centred Cubic lattice (or FCC for short; see also [10]), to general three-dimensional lattices. In these approaches the rotational degrees of freedom of the particles are explicitly accounted for. This leads to the presence of additional, 'optical' branches in the dispersion relations.

In solid state physics (see for example [3, 4, 10, 19]) 'optical' branches in the dispersion relations result from the presence of atoms with varying properties, where the 'optical' branches correspond to movement of the atoms relative to that of the centre of mass of the unit cell. For granular materials consisting of identical particles, the 'optical' branches arise due to the presence of rotational degrees of freedom of the particles.

The wave speeds at small wave numbers (i.e. at large wave lengths) have been studied experimentally (for example [8] and references in [14]), theoretically (for example [6]) as well as numerically (for example [1, 16, 18]), showing a clear dependence on confining pressure.

Here the emphasis is on the dispersion relation over a wide range of wave numbers, so not only at small wave numbers. The pressure dependence is not studied here. This dependence can be incorporated by a proper choice for the dependence of the interparticle stiffnesses and of the coordination number (i.e. the average number of contacts per particle) on confining pressure. These interparticle stiffnesses form important micromechanical parameters in the current approach.

The dispersion relations obtained here from the micromechanical approach incorporates the discrete structure of granular materials. Therefore, suitable continuum theories should result in (qualitatively) similar dispersion relations. Although the study of continuum-based dispersion relations is outside of the scope of this study, the current results can be used for this purpose.

The outline of this study is as follows. Firstly, the basics of micromechanics of granular materials are described. Then dispersion relations are formulated for three-dimensional lattices. This lattice analysis suggests a way to an improvement, the statistical formulation of lattice approach. Finally, findings from this study are discussed.

2. MICROMECHANICS

Three-dimensional assemblies consisting of spherical particles are considered here. The radius of particle p is denoted by R^p. The position of the centre of mass of particle p is given by $\mathbf{X}^p$. For two particles p and q in contact, $\mathbf{r}^{pq}$ is the vector from the centre of particle p to the contact point between particles p and q, with an analogous definition for $\mathbf{r}^{qp}$. For very stiff particles where the deformation at the contact ('overlap') is very small, the unit normal vector $\mathbf{n}^{pq}$ at the contact is obtained from

$$\mathbf{r}^{pq} = R^p \mathbf{n}^{pq} . \tag{3}$$

The direction of the normal vector $\mathbf{n}^{pq}$ determines the orientation, in terms of solid angle Ω, of the contact.

Primary statistical characteristics of the contact geometry are the coordination number C and the contact distribution function [7] $\chi(\Omega)$ that gives the orientational distribution of contacts with solid angle Ω: $\chi(\Omega)$ is defined such that $\chi(\Omega)d\Omega$ gives the fraction of contacts with orientation in $(\Omega, \Omega + d\Omega)$. For the statistically isotropic, three-dimensional assemblies considered here

$$\chi(\Omega) = \frac{1}{4\pi} . \tag{4}$$

The (small) displacement vector of the centre of particle p, relative to the selected reference configuration, is denoted by $\mathbf{u}^p$, while the (small) particle rotation vector is indicated by θ^p. The equations of motion governing the evolution with time t of the displacement $\mathbf{u}^p$ and rotation θ^p of the spherical particle p are

$$m^p \ddot{\mathbf{u}}^p = \sum_q \mathbf{f}^{pq} \qquad I^p \ddot{\theta}^p = \sum_q \mathbf{r}^{pq} \times \mathbf{f}^{pq} \tag{5}$$

where the two superimposed dots denote the second derivative of the quantity involved with respect to time t, m^p and I^p are the mass and moment of inertia of the spherical particle p, respectively, $\mathbf{f}^{pq}$ is the force exerted on particle p by particle q and the sum is over particles q that are in contact with particle p. The moment of inertia I^p of a sphere with radius R^p and mass m^p is given by

$$I^p = Q m^p R^{p2} , \tag{6}$$

with $Q = 2/5$ for a solid sphere and $Q = 2/3$ for a hollow sphere. In the following, solid spheres with $Q = 2/5$ are considered when numerical results are presented. Note that body forces, such as gravitational forces, have been excluded from Eq. (5).

The relative displacement vector Δ^{pq} at the contact point between two particles p and q in contact is defined by

$$\Delta^{pq} = [\mathbf{u}^p + \theta^p \times \mathbf{r}^{pq}] - [\mathbf{u}^q + \theta^q \times \mathbf{r}^{qp}] . \tag{7}$$

Note that it involves contributions from particle translations and from particle rotations.

The contact force $\mathbf{f}^c$ is related to the contact relative displacement Δ^c through the contact constitutive relation. For very small deformations, the contacts can be considered as elastic and bonded (i.e. with fixed contact topology). Thus, interparticle friction and contact creation and disruption are taken into account only indirectly though the set of 'active' contacts. This constitutive relation between force and relative displacement at contacts is determined by the contact stiffness matrix $\mathbf{S}^c$. The employed contact constitutive relation involves two linear elastic interparticle springs in normal and tangential directions at the contact, with spring stiffnesses k_n and k_t, respectively. Hence

$$\mathbf{f}^c = -\mathbf{S}^c \cdot \Delta^c , \qquad (8)$$

where the contact stiffness matrix $\mathbf{S}^c$ is given by

$$\mathbf{S}^c = (k_n - k_t)\mathbf{n}^c\mathbf{n}^c + k_t \mathbf{I} , \qquad (9)$$

with $\mathbf{I}$ the 3 by 3 identity matrix. Note that $\mathbf{f}^{qp} = -\mathbf{f}^{pq}$ due to Newton's third law and $\Delta^{qp} = -\Delta^{pq}$ due to its definition in Eq. (7), and hence $\mathbf{S}^{qp} = \mathbf{S}^{pq}$ (which is satisfied by Eq. (8)). The stiffness ratio Λ of tangential stiffness over normal stiffness at the contacts is defined by

$$\Lambda = \frac{k_t}{k_n} . \qquad (10)$$

For the case of linear contacts considered here, the stiffnesses k_n and k_t are independent of the contact force (and hence independent of confining pressure). Thus contrary to the case of nonlinear Hertzian-type contacts, k_n has the same value at all contacts; the same applies to k_t. Note that Λ may be different from one, however. The assumption of linear contacts is adopted for convenience in the analytical formulation. For numerical approaches this is not a significant advantage.

The equations of motion, Eq. (5), can be expressed concisely in terms of a generalised displacement vector $\mathbf{U}^p$, a generalised force vector $\mathbf{F}^{pq}$ and a generalised mass matrix $\mathbf{M}^p$ as

$$\mathbf{M}^p \cdot \ddot{\mathbf{U}}^p = \sum_q \mathbf{F}^{pq} , \qquad (11)$$

where the generalised displacement and force vectors and the generalised mass matrix are given by

$$\mathbf{U}^p = \begin{bmatrix} \mathbf{u}^p \\ R^p \theta^p \end{bmatrix} \quad \mathbf{F}^{pq} = \begin{bmatrix} \mathbf{f}^{pq} \\ \mathbf{n}^{pq} \times \mathbf{f}^{pq} \end{bmatrix} \quad \mathbf{M}^p = m^p \begin{bmatrix} \mathbf{I} & 0 \\ 0 & Q\mathbf{I} \end{bmatrix} . \qquad (12)$$

Note that the dimension (or unit) of the elements of the vectors $\mathbf{U}^p$ and $\mathbf{F}^{pq}$ and of the matrix $\mathbf{M}^p$ are the same, through the inclusion in $\mathbf{U}^p$ and the exclusion in $\mathbf{F}^{pq}$ of the particle radius R^p. For compactness in notation, the generalised force $\mathbf{F}^{pq}$ is expressed as

$$\mathbf{F}^{pq} = \begin{bmatrix} \mathbf{I} \\ \mathbf{N}^{\times,pq} \end{bmatrix} \cdot \mathbf{f}^{pq} , \qquad (13)$$

where the operator $\mathbf{N}^{\times,pq}$ is defined such that $\mathbf{N}^{\times,pq} \cdot \mathbf{v} = \mathbf{n}^{pq} \times \mathbf{v}$ for all $\mathbf{v}$. Thus

$$\mathbf{N}^{\times,pq} = -\mathbf{E} \cdot \mathbf{n}^{pq} , \qquad (14)$$

with $\mathbf{E}$ the three-dimensional permutation symbol.

3. LATTICE ANALYSIS

The lattice geometry is described as follows, see also Figure 1. A central particle '0' at position $\mathbf{X}^0$ is in contact with C other particles. All particles have the same properties, such as identical particle radius R, mass m and coordination number C. The unit normal vector at the contact c is denoted by $\mathbf{n}^c$. The set of normal vectors $\{\mathbf{n}^c\}$ determines the lattice directions. The position of the other particles in contact with the central particle '0' is given by $\mathbf{X}^c = \mathbf{X}^0 + 2R\mathbf{n}^c$.

For the generalised displacement vector $\mathbf{U}$ a plane-wave solution is assumed

$$\mathbf{U}(\mathbf{x},t) = \begin{bmatrix} \mathbf{u}_a \\ R\theta_a \end{bmatrix} e^{i(\omega t - \mathbf{k} \cdot \mathbf{x})} \equiv \mathbf{V} e^{i(\omega t - \mathbf{k} \cdot \mathbf{x})} , \qquad (15)$$

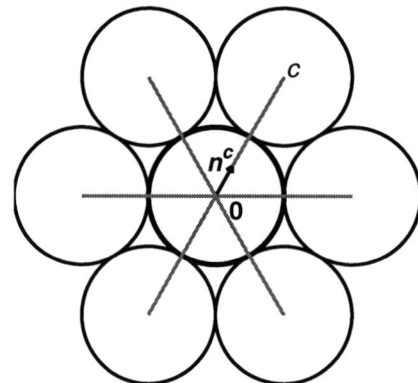

FIGURE 1. Example of a two-dimensional lattice with coordination number $C = 6$.

where $\mathbf{u}_a$ and θ_a are the amplitude vectors for the displacements and rotations, respectively, $\mathbf{x}$ is the position vector, i is the imaginary unit, $\mathbf{k}$ is the wave vector and ω is the angular velocity. The amplitude vector of the generalised displacement vector $\mathbf{U}$ is denoted by $\mathbf{V}$, while the magnitude of the wave vector $\mathbf{k}$ is given by k.

Using the assumed plane-wave solution of Eq. (15), the relative displacement vector Δ^c at contact c becomes, from Eq. (7) with $\mathbf{r}^{0c} = +R\mathbf{n}^c$ and $\mathbf{r}^{c0} = -R\mathbf{n}^c$ for spherical particles with equal radius R

$$\Delta^c = e^{\mathrm{i}(\omega t - \mathbf{k}\cdot\mathbf{X}^0)}[(1-\xi^c)\mathbf{u}_a + (1+\xi^c)R\theta_a \times \mathbf{n}^c] \qquad \xi^c = e^{-\mathrm{i}(2R)\mathbf{k}\cdot\mathbf{n}^c}. \tag{16}$$

Note that the scalar factor ξ^c (also) depends on the wave vector $\mathbf{k}$. The relative displacement vector can be expressed in terms of the amplitude vector $\mathbf{V}$ and the operator $\mathbf{N}^{\times,c}$ (defined in Eq. (14)) by

$$\Delta^c = e^{\mathrm{i}(\omega t - \mathbf{k}\cdot\mathbf{X}^0)} \left[(1-\xi^c)\mathbf{I} \quad -(1+\xi^c)\mathbf{N}^{\times,c} \right] \cdot \mathbf{V}, \tag{17}$$

where the term inside the square brackets is a 3 by 6 matrix, which does not involve a subtraction.

Using Eqs. (8), (11), (13), (15) and (17), it follows that the amplitude vector $\mathbf{V}$ is an eigenvector of the following generalised eigenvalue problem with eigenvalue ω^2 (see also [3])

$$(C\overline{\mathbf{K}} - \omega^2 \mathbf{M}) \cdot \mathbf{V} = \mathbf{0}, \tag{18}$$

where

$$\overline{\mathbf{K}} = \frac{1}{C}\sum_{c=1}^{C} \mathbf{K}^c \qquad \mathbf{K}^c = \begin{bmatrix} \mathbf{I} \\ \mathbf{N}^{\times,c} \end{bmatrix} \cdot \mathbf{S}^c \cdot \left[(1-\xi^c)\mathbf{I} \quad -(1+\xi^c)\mathbf{N}^{\times,c} \right]. \tag{19}$$

The size of the matrix $\overline{\mathbf{K}}$ is 6 by 6. It follows from Eqs. (9), (12), (18) and (19) that the eigenfrequencies (i.e. the square root of the eigenvalues) can be expressed in terms of a dimensionless eigenfrequency $\hat{\omega}$ as

$$\omega = \sqrt{\frac{Ck_n}{m}}\hat{\omega}. \tag{20}$$

3.1. Results

The lattice analysis presented here is applied to three lattices with different coordination numbers C: a simple cubic lattice with $C = 6$, a Body Centred Cubic lattice (BCC for short; see also [10]) with $C = 8$ and an FCC-lattice with $C = 12$. These lattices are shown in Figure 2 (left column).

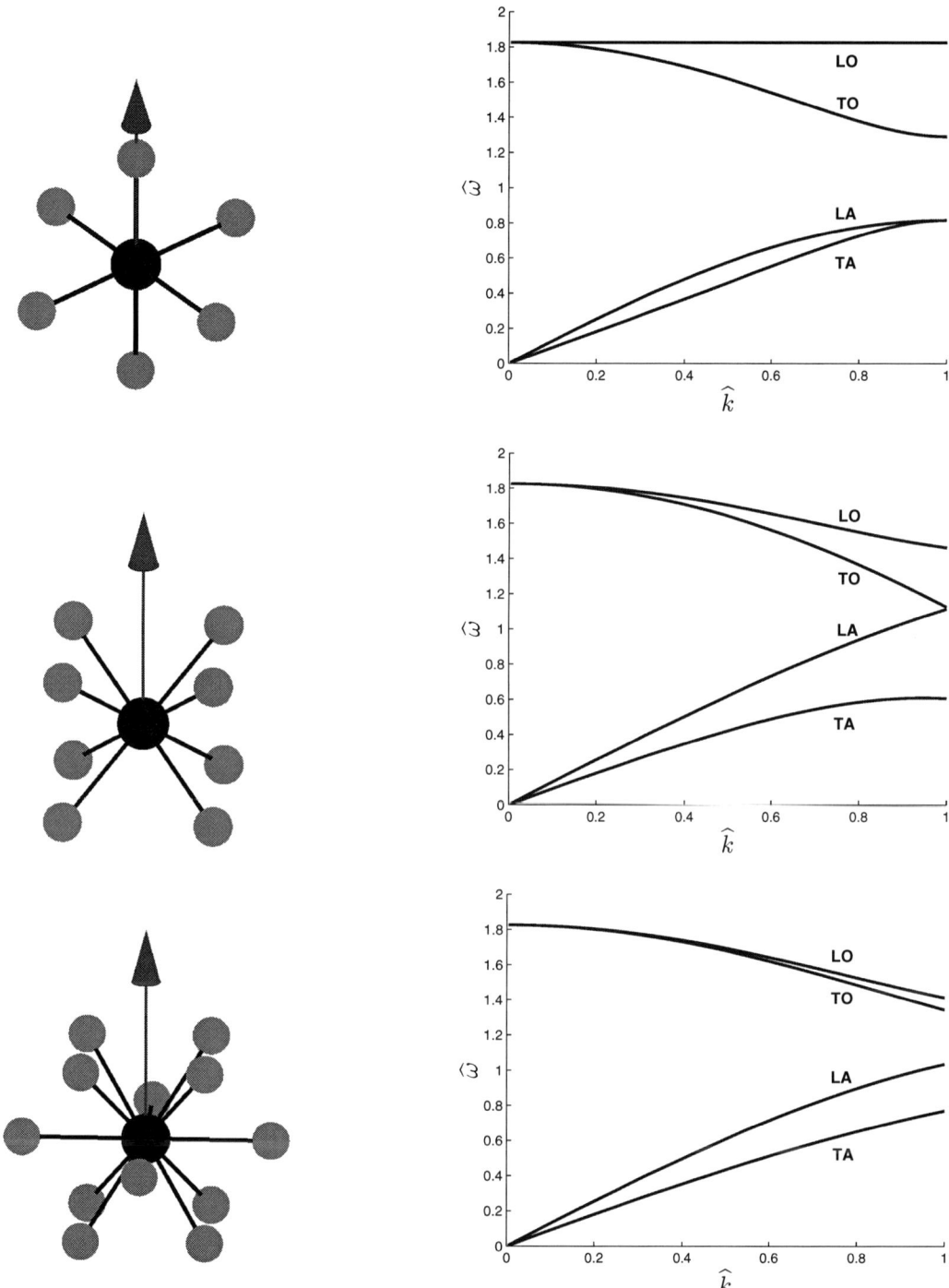

FIGURE 2. Lattice geometry (left column) and dispersion relations (right column). Top row: cubic lattice ($C = 6$); middle row: BCC lattice ($C = 8$); bottom row: FCC lattice ($C = 12$). In the left column, the central sphere of each lattice is shown in black, while the other spheres are shown in grey. For visual clarity the radius of the spheres has been reduced; therefore the spheres in contact appear as if they were not in contact. Contacts between spheres are indicated by black lines connecting particle centres. The direction of the wave vector is indicated by the black arrow. Results for the dispersion relations are for stiffness ratio $k_t/k_n = 1$. The labels **LA**, **TA**, **LO** and **TO** indicate the type of dispersion branch, see main text.

For a single direction of the wave vector **k** (indicated in the left column of Figure 2) and for various values of the wave number k, the six eigenvalues of the generalised eigenvalue problem Eq. (18) have been computed numerically. The corresponding dispersion relations are shown in Figure 2 (right column). The dimensionless wave number $\hat{k}$ along the horizontal axis is defined by

$$\hat{k} = \frac{L_z k}{2\pi} = \frac{L_{\text{cell}}}{\lambda} \qquad L_{\text{cell}} = 4R, \tag{21}$$

where L_{cell} is a cell length that is chosen to be the same for all three lattices for simplicity.

Note that, even though these three lattices have a different coordination number C, the (main) effect of coordination number is removed through the scaling of Eq. (20).

For all three lattices there are four branches in the dispersion relations in Figure 2 (right column). These can be characterised as 'acoustical' branches with $\lim_{k \to 0} \omega(k) = 0$ and 'optical' branches with $\lim_{k \to 0} \omega(k) \neq 0$ [3, 4, 10, 13, 19]. The branches can also be characterised by the dominant type of motion (as reflected in the eigenvectors of the eigenvalue problem Eq. (18)), i.e. as 'longitudinal' or as 'transverse', relative to the direction of the wave vector **k**. This gives the following classification of the branches:

- **LA**: 'longitudinal' and 'acoustical' (also called P-branch);
- **TA**: 'transverse' and 'acoustical' (also called S-branch);
- **LO**: 'longitudinal' and 'optical';
- **TO**: 'transverse' and 'optical'.

Note that the dispersion relations for the three lattices exhibit four branches, and hence two of the six eigenvalues have a multiplicity of two. This is due to the symmetry of these lattices in the plane perpendicular to the selected direction of wave propagation.

The result for the FCC lattice was already reported in [21], employing a (non-documented) method that is, in essence, identical to that reported here in detail [9]. In the study of two-dimensional lattices [23, 24] three branches were observed, two 'acoustical' branches and a single 'optical' branch, since in the two-dimensional case the particles only have a single rotational degree of freedom, besides the two translational degrees of freedom.

Qualitatively, the dispersion relations for the three lattices are roughly the same. The largest difference is the different behaviour of the LO-branch with its constant angular velocity, in comparison with those of the BCC and FCC lattices.

Quantitatively, the angular velocities $\hat{\omega}$ of the 'optical' branches are the same for small wave number $\hat{k}$. The slopes of the 'acoustical' branches at $k = 0$, i.e. the wave speeds, are different for the three lattices. These wave speeds are related to (continuum) bulk and shear modulus of the material. These moduli of lattices have been studied in [25]. At larger values of the wave number, there are significant differences between the dispersion relations of the three lattices. This means that the dispersion relations are sensitive to the lattice structure.

As shown in [23] for two-dimensional lattices, the direction of the wave vector **k** influences the dispersion relations at higher wave numbers k, since the inherent anisotropic structure of the lattices is then apparent. For small wave number k, however, it was noted in [23] that the dispersion relations are independent of the direction of the wave vector **k**. The same is found for the three-dimensional lattices considered here (results not shown).

For cubic packings in periodic space it is possible to determine all eigenvalues of the global stiffness matrix of the system in a nearly analytical way, using results presented here (see Appendix A). In this Appendix the density of state, i.e. the probability density function, for the eigenfrequencies and eigenvalues is also given. The result is obtained that, even for this regular lattice, there is a deviation from Debye scaling [10], due to the presence of rotational degrees of freedom of the particles.

The dispersion relations resulting from the lattice theory have two deficiencies: (i) their anisotropy at higher wave numbers and (ii) their sensitivity to the lattice structure.

In the next section a theory is formulated that gives truly isotropic dispersion relations for all values of the wave number. Its basis is the (regular) lattice formulation of this section.

4. STATISTICAL FORMULATION

To improve upon the lattice theory of the previous section, a statistical formulation is developed here. The basis idea is to 'smear' out the discrete lattice directions $\{\mathbf{n}^c\}$ over all possible directions. The probability of finding a lattice direction $\mathbf{n}(\Omega)$ is described by the *continuous* contact distribution function $\chi(\Omega)$, see also Eq. (4).

The dispersion relation for lattices is obtained from the generalised eigenvalue value problem, Eq. (18), where the matrix $\overline{\mathbf{K}}$ is defined by a sum over the lattice directions, see Eq. (19).

In the statistical formulation the *sum* over lattice directions c is replaced by an *integral* over contact orientations Ω, i.e. $1/C \sum_c \mathbf{K}^c \to \int_\Omega \chi(\Omega)\mathbf{K}(\Omega)d\Omega$. The dispersion relations are (again) obtained from the generalised eigenvalue problem Eq. (18), where the matrix $\overline{\mathbf{K}}$ now is given by (compare Eq. (19))

$$\overline{\mathbf{K}} = \int_\Omega \chi(\Omega) \begin{bmatrix} \mathbf{I} \\ \mathbf{N}^\times(\Omega) \end{bmatrix} \cdot \mathbf{S}(\Omega) \cdot \begin{bmatrix} (1-\xi(\Omega))\mathbf{I} & -(1+\xi(\Omega))\mathbf{N}^\times(\Omega) \end{bmatrix} d\Omega. \tag{22}$$

An important observation is that the matrices $\mathbf{N}^{\times,c}$ and $\mathbf{S}^c$ that are present in the expression for $\mathbf{K}^c$, see Eq. (19), are determined by the normal vector $\mathbf{n}^c$ at the contact (see Eqs. (9) and (14)), i.e. by the contact orientation Ω, so they can also be expressed as $\mathbf{N}^\times(\Omega)$ and $\mathbf{S}(\Omega)$.

To evaluate the integral in Eq. (22), spherical coordinates (ϑ, ϕ) ($\phi \in (0, 2\pi)$ and $\vartheta \in (0, \pi)$), relative to the direction of the wave vector $\mathbf{k}$, are employed. The direction of the wave vector is selected to be the z-direction, without loss of generality. Then the unit normal vector $\mathbf{n}(\vartheta, \phi)$ is given by $\mathbf{n} = (\sin\vartheta\cos\phi, \sin\vartheta\sin\phi, \cos\vartheta)^T$ and $d\Omega = \sin\vartheta d\vartheta d\phi$.

The integration over ϕ can be performed straightforwardly. The subsequent integration over ϑ is simplified by employing the transformation $x = \cos\vartheta$. The resulting matrix $\overline{\mathbf{K}}$ is given in Appendix B. This matrix is independent of (ϑ, ϕ), i.e. of the *direction* of the wave vector. The resulting dispersion relations are therefore truly isotropic at all values of wave number, contrary to those resulting from the lattice analysis of Section 3.

The eigenvalues $\mu \equiv \omega^2$ of the generalised eigenvalue problem Eq. (18) can be determined analytically for the dimensionless eigenfrequencies defined in Eq. (20), after some tedious algebra. The first two dimensionless eigenfrequencies $\hat{\omega}_1$ and $\hat{\omega}_4$ (each with multiplicity of one) are given by

$$\hat{\mu}_1 \equiv \hat{\omega}_1^2 = \frac{2}{3}(1-\Lambda)g - h + \frac{1}{3}(1+2\Lambda) \tag{23}$$

$$\hat{\mu}_4 \equiv \hat{\omega}_2^2 = \frac{2\Lambda}{3Q}(g+1) \tag{24}$$

where the scalar functions $g(k)$ and $h(k)$ are defined by

$$g(k) = 3\frac{\sin k - k\cos k}{k^3} \qquad h(k) = \frac{\sin k}{k}. \tag{25}$$

The functions $g(k)$ and $h(k)$ have been selected such that their limit value at $k = 0$ equals 1.

The remaining four dimensionless eigenvalues $\hat{\mu} \equiv \hat{\omega}^2$ have a multiplicity of two. They are given as solutions of the following quadratic equation:

$$0 = K_2\hat{\mu}^2 + K_1\hat{\mu} + K_0 \tag{26}$$
$$K_2 = Q \tag{27}$$
$$K_1 = \frac{1}{3}(\Lambda + Q - Q\Lambda)g + \Lambda(Q-1)h - \frac{1}{3}(2\Lambda + Q + 2Q\Lambda) \tag{28}$$
$$K_0 = \frac{\Lambda}{9}\left(1 - \Lambda - \Lambda\hat{k}^2\right)g^2 - \frac{\Lambda}{3}(1-2\Lambda)gh + \frac{\Lambda}{3}(h-g) - \Lambda^2 h^2 + \frac{2\Lambda}{8}(1+2\Lambda). \tag{29}$$

Note that the coefficients K_2, K_1 and K_0 depend on the dimensionless wave number

$$\hat{k} = 2Rk \tag{30}$$

through the functions $g(\hat{k})$ and $h(\hat{k})$. Since two of the six eigenvalues have a multiplicity of two, there are four branches in the dispersion relations.

The eigenvalues for large wavelengths (small wave numbers) are obtained by expanding $g(\hat{k})$ and $h(\hat{k})$ in Taylor series around $\hat{k} = 0$. Using the Taylor expansions for $g(k) \cong 1 - 1/10k^2$ and $h(k) \cong 1 - 1/6k^2$, the eigenvalues $\hat{\mu}$ can

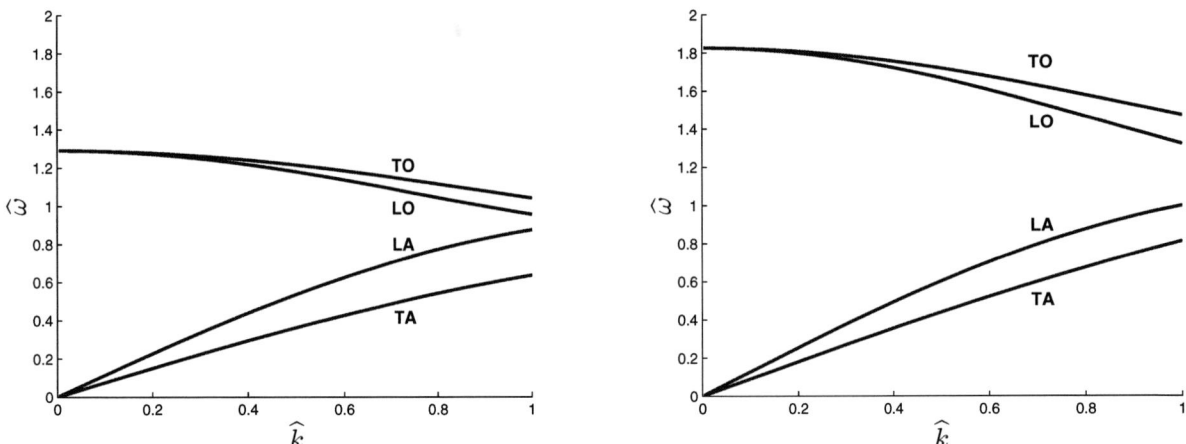

FIGURE 3. Results from the statistical lattice theory for two values of the stiffness ratio. (Left) $k_t/k_n = 0.5$ (Right) $k_t/k_n = 1.0$.

be approximated by

$$\hat{\mu}_1 \cong \left(\frac{1}{10} + \frac{\Lambda}{15}\right)\hat{k}^2 \tag{31}$$

$$\hat{\mu}_{2,3} \cong \left(\frac{1}{30} + \frac{\Lambda}{20}\right)\hat{k}^2 \tag{32}$$

$$\hat{\mu}_{4,5,6} \cong \frac{4\Lambda}{3Q} \tag{33}$$

4.1. Results

For various values of the wave number k, the six eigenvalues of the generalised eigenvalue problem Eq. (18) have been computed from Eqs. (23), (24) and (26). As expected, a direct numerical computation of the eigenvalues from Eq. (18) with the matrix $\overline{\mathbf{K}}$, as given in Appendix B, gave identical results. The corresponding dispersion relations are given in Figure 3 for two values of the stiffness ratio $\Lambda \equiv k_t/k_n$.

The dispersion relations shown in Figure 3 resemble the corresponding results for the three lattices given in Figure 2, especially those for the FCC lattice (compare Figure 3 (right) with Figure 2 (bottom right), which both concern the case $k_t/k_n = 1$).

The 'optical' branches are strongly influenced by the value of the stiffness ratio Λ, with an angular velocity at small wave numbers that increases as the square root of the stiffness ratio, compare also Eq. (33).

The slopes of the 'acoustical' branches at small wave numbers also depend on the stiffness ratio, but not as strongly as the 'optical' branches (compare Eqs. (31) and (32)).

Note that in the statistical lattice formulation, the effect of coordination number C is completely described by Eq. (20), since the dimensionless frequencies $\hat{\omega}$ are independent of C.

For small wave numbers, the analysis of the eigenmodes (eigenvectors) of the branches shows that:

- in the longitudinal modes, LA and LO, the longitudinal degrees of freedom are uncoupled from the other degrees of freedom;
- the transverse modes, TA and TO, do not involve the longitudinal degrees of freedom, but involve coupled transverse displacements and rotations that are out-of-phase.

5. DISCUSSION

The dispersion relations for three-dimensional lattices have been studied from the micromechanical viewpoint. The results show the presence of 'optical' branches, associated with the presence of rotational degrees of freedom of the particles. The analysis of the cubic, BCC and FCC lattices shows that the dispersion relations are sensitive to the lattice structure and that they show anisotropic behaviour (i.e. dependence on the direction of the wave vector) at large wave numbers.

To remedy these deficiencies a statistical formulation of the lattice approach is proposed, in which the discrete lattice directions are 'smeared' out continuously over all directions. The resulting dispersion relations are truly isotropic (i.e. independent of the direction of the wave vector) for all values of wave number. The influence of the primary micromechanical parameter, the stiffness ratio, on the dispersion relations has been studied. An increase in stiffness ratio leads to an increase of the angular velocity for the 'optical' branches at small wave number. A weaker influence is present in the wave speeds for the 'acoustical' branches.

An important characteristic of granular materials is that the particle packings are generally random, contrary to the regular lattice packings considered here. For disordered packings qualitatively different behaviour may be observed for sufficiently large disorder (compare "Anderson localisation" [2]). The study of dispersion relations for disordered packings is being pursued from simulations [12]. Preliminary results show that the statistical lattice formulation of Section 4 gives qualitatively good results. Results for long wave lengths (small wave numbers) are reported in [26] for 'frictionless' (i.e. without tangential stiffness and hence without rotational degrees of freedom of the particles) disordered packings.

Quantitatively, there are differences between dispersion relations from the statistical formulation and for disordered packings, since the geometrical disorder is not accounted for in the statistical formulation. This geometrical disorder leads to significant deviations from mean-field behaviour [11, 14, 15, 20]. This mean-field assumption is also made in the current approach in Eq. (15).

The presented dispersion relations for the lattices, as well as those for three-dimensional disordered packings, reflect the intrinsic discrete character of granular materials. These dispersion relations should therefore be obtained (qualitatively) from appropriate continuum theories. To this end, extended continuum constitutive relations, such as higher-order gradient and micropolar theories (for example [5, 13, 17, 22, 23, 24]), seem suitable. For further research it is therefore recommended to formulate extended continuum constitutive relations that result in dispersion relations with the same shape as those obtained from the three-dimensional lattice-based analyses presented here and from the three-dimensional simulations of disordered packings.

ACKNOWLEDGMENTS

S. Luding and O.J.P. Mouraille (Department of Mechanical Engineering, University of Twente, The Netherlands) and D.L. Johnson and L.M. Schwartz (Schlumberger-Doll Research, Cambridge, MA, USA) are thanked for valuable discussions.

APPENDIX A

A cubic packing in periodic space, with M spherical particles with identical properties along the three perpendicular axes, is considered. The total number of particles then is M^3. For this case it is possible to compute all $6M^3$ eigenvalues of the global stiffness matrix of the full system in a relatively simple way, as is described here.

The particles in the cubic lattice can be identified by three indices (a,b,c) along the axes of the cubic packing, each of which ranges from 1 to M. The position vectors of the particles are $\mathbf{X}_{a,b,c}$. The eigenvectors of the global stiffness matrix of the full system have the following structure

$$\mathbf{U}_{a,b,c} = \mathbf{V}_{\alpha,\beta,\gamma} e^{-i\mathbf{k}_{\alpha,\beta,\gamma} \cdot \mathbf{X}_{a,b,c}}, \tag{34}$$

where $\mathbf{V}_{\alpha,\beta,\gamma}$ is an amplitude vector for the eigenmode (α,β,γ) under consideration. The indices (α,β,γ) each range from 1 to M. This expression gives the values of the generalised displacements for particle (a,b,c). The "wave vectors" being examined are given by

$$\mathbf{k}_{\alpha,\beta,\gamma} = \begin{bmatrix} \frac{2\pi\alpha}{M} & \frac{2\pi\beta}{M} & \frac{2\pi\gamma}{M} \end{bmatrix}^T. \tag{35}$$

When substituting the expression for the eigenvectors, Eq. (34), into the equations of motion, Eq. (11), it follows that these equations are satisfied when the amplitude vector $\mathbf{V}_{\alpha,\beta,\gamma}$ satisfies Eq. (18) with matrix $\overline{\mathbf{K}}$ given by Eq. (19). Each of the M^3 possibilities for $\mathbf{k}_{\alpha,\beta,\gamma}$ yields 6 eigenvalues, resulting in a total of $6M^3$ eigenvalues, i.e. the full spectrum of the periodic cubic lattice.

From this complete spectrum of eigenvalues μ, the resulting density of state (i.e. the probability density function) for the frequencies, p_ω, and for eigenvalues, p_μ, can be determined. The densities of states p_ω and p_μ are related by

$$p_\omega = 2\sqrt{\mu} p_\mu, \tag{36}$$

as follows from the transformation rules for probability density functions, with $\omega = \sqrt{\mu}$.

The results are shown in Figure 4 for $M = 101$ for various numbers of bins. The results for $M = 100$ are practically the same. Two weak singularities are present, implied by the difference in the results obtained with different numbers of bins. An interesting result is the behaviour of the density of state for the eigenvalues, p_μ, at small eigenvalues. Since $p_\mu(\hat{\mu}) \propto \hat{\mu}^0$ at small $\hat{\mu}$, it follows from Eq. (36) that $p_\omega(\hat{\omega}) \propto \hat{\omega}^1$. This differs from Debye scaling in the three-dimensional case, $p_\omega(\hat{\omega}) \propto \hat{\omega}^2$, in solid state physics [10].

APPENDIX B

In this Appendix the Hermitian matrix $\overline{\mathbf{K}}$ resulting from the statistical lattice formulation of Section 4 is given. It can be expressed as

$$\overline{\mathbf{K}} = g(\hat{k})\overline{\mathbf{K}}_g(\hat{k}) + h(\hat{k})\overline{\mathbf{K}}_h(\hat{k}) + \overline{\mathbf{K}}_c(\hat{k}), \tag{37}$$

where the non-dimensional wave number $\hat{k}$ is defined in Eq. (30) and the $g(k)$ and $h(k)$ are defined in Eq. (25).

The matrix $\overline{\mathbf{K}}_g$ is defined by

$$\overline{\mathbf{K}}_g(k) = -\frac{k_n}{3} \begin{bmatrix} 1-\Lambda & 0 & 0 & 0 & i\Lambda k & 0 \\ 0 & 1-\Lambda & 0 & -i\Lambda k & 0 & 0 \\ 0 & 0 & -2(1-\Lambda) & 0 & 0 & 0 \\ 0 & i\Lambda k & 0 & \Lambda & 0 & 0 \\ -i\Lambda k & 0 & 0 & 0 & \Lambda & 0 \\ 0 & 0 & 0 & 0 & 0 & -2\Lambda \end{bmatrix} \tag{38}$$

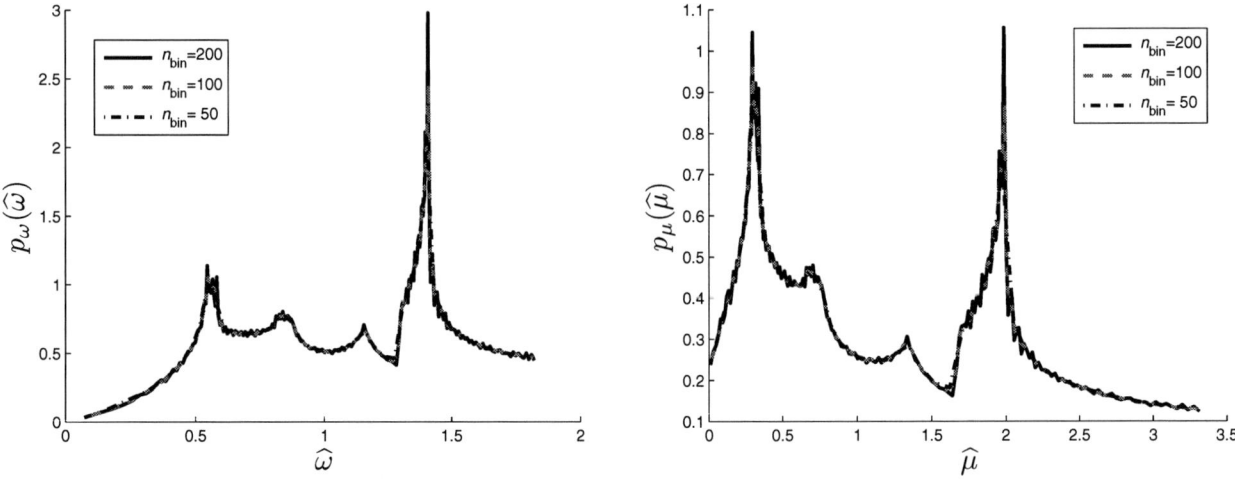

FIGURE 4. Density of state of a cubic packing in periodic space of size $M \times M \times M$, with $M = 101$ for the case $k_t/k_n = 1$. (Left) Density of states p_ω for eigenfrequencies. (Right) Density of states p_μ for eigenvalues. Results are shown for various values of the number of bins, indicating a weak singularity near $\hat{\mu} \approx 0.3$ and 2.0.

and the diagonal matrices $\overline{\mathbf{K}}_h$ and $\overline{\mathbf{K}}_c$ are given by

$$\overline{\mathbf{K}}_h = -k_n \text{diag}(\Lambda, \Lambda, 1, -\Lambda, -\Lambda, 0) \tag{39}$$

$$\overline{\mathbf{K}}_c = \frac{k_n}{3} \text{diag}(1+2\Lambda, 1+2\Lambda, 1+2\Lambda, 2\Lambda, 2\Lambda, 2\Lambda) . \tag{40}$$

REFERENCES

1. Agnolin, I. & Roux, J.N. (2007). Internal states of model isotropic granular packings. III. Elastic properties. *Physical Review E* **76** 061304.
2. Anderson, P.W. (1958). Absence of diffusion in certain random lattices. *Physical Review* **109** 1492–505.
3. Ashcroft, N. W. & Mermin, N.D. (1976). Solid state physics. Saunders College Publishing, Fort Worth, TX, USA.
4. Dekker, A.J. (1962). Solid state physics. Macmillan, London, UK.
5. Eringen, A.C. (1999). Microcontinuum field theories: foundations and solids. Springer-Verlag, New York, NY, USA.
6. Goddard, J.D. (1990). Nonlinear elasticity and pressure-dependent wave speeds in granular media. *Proceedings of the Royal Society of London A* **430** 105–31.
7. Horne, M.R. (1965). The behaviour of an assembly of rotound, rigid, cohesionless particles I and II. *Proceedings of the Royal Society of London A* **286** 62–97.
8. Jia, X. & Caroli, C. & Velicky, B. (1999). Ultrasound propagation in externally stressed granular media. *Physical Review Letters* **82** 1863–6.
9. Johnson, D.L. & Schwartz, L.M. (2009). Personal communication.
10. Kittel, C. (1953). Introduction to solid state physics. John Wiley & Sons, New York, NY, USA.
11. Kruyt, N.P. & Rothenburg, L. (1998). Statistical theories for the elastic moduli of two-dimensional assemblies of granular materials. *International Journal of Engineering Science* **36** 1127–42.
12. Kruyt, N.P. & Luding, S. (2009). Micromechanical study of wave propagation in three-dimensional granular materials. *In preparation*.
13. Kunin, I.A. (1983). Elastic media with microstructure I and II. Springer-Verlag, Berlin, Germany.
14. Magnanimo, V. & La Ragione, L. & Jenkins, J.T. & Wang, P. & Makse, H.A. (2008). Characterizing the shear and bulk moduli of an idealized granular material. *EPL* **81** 34006.
15. Makse, H.A. & Gland, N. & Johnson, D.L. & Schwartz, L.M. (1999). Why effective medium theory fails in granular materials. *Physical Review Letters* **83** 5070–3.
16. Makse, H.A. & Gland, N. & Johnson, D.L. & Schwartz, L. (2004). Granular packings: nonlinear elasticity, sound propagation, and collective relaxation dynamics. *Physical Review E* **70** 061302.
17. Mindlin, R.D. (1972). Elasticity, piezoelectricity and crystal lattice dynamics. *Journal of Elasticity* **2** 217–82.
18. Mouraille, O.J.P. (2009). Sound propagation in dry granular materials: discrete element simulations, theory, and experiments. Ph.D. thesis Department of Mechanical Engineering, University of Twente, Enschede, The Netherlands.
19. Myers, H.P. (1997). Introductory solid state physics. Taylor & Francis, London, UK.
20. Rothenburg, L. & Kruyt, N.P. (2001). On limitations of the uniform strain assumption in micromechanics of granular materials. *Powders & Grains 2001*, pp.191–4, ed. Y. Kishino, Balkema Publishers, Lisse, The Netherlands.
21. Schwartz, L.M. & Johnson, D.L. & Feng, S. (1984). Vibrational modes in granular materials. *Physical Review Letters* **52** 831–4.
22. Song, F. & Huang, G.L. & Varadan, V.K. (2010). Study of wave propagation in nanowires with surface effects by using a high-order continuum theory. *Acta Mechanica* **209** 129–39.
23. Suiker, A.S.J. & Metrikine, A.V. & de Borst, R. (2001). Comparison of wave propagation characteristics of the Cosserat continuum model and corresponding discrete lattice models. *International Journal of Solids and Structures* **38** 1563–83.
24. Suiker, A.S.J. & de Borst, R. (2005). Enhanced continua and discrete lattices for modelling granular assemblies. *Philosophical Transactions of the Royal Society A* **363** 2543–80.
25. Wang, Y. & Mora, P. (2008). Macroscopic elastic properties of regular lattices. *Journal of the Mechanics of Physics of Solids* **56** 3459–74.
26. Xu, N. & Vitelli, V. & Wyart, M. & Liu, A.J. & Nagel, S.R. (2009). Energy transport in jammed sphere packings. *Physical Review Letters* **102** 038001.

The waveguide theory for booming sand dunes

N.M. Vriend* and M.L. Hunt[†]

*Department of Applied Mathematics and Theoretical Physics, University of Cambridge, UK
[†]Department of Mechanical Engineering, California Institute of Technology, USA

Abstract. Sand on the leeward face of a large desert dune may create a persistent, low-frequency sound during a slumping event or natural avalanche. The sound may last for several minutes and be audible at far distances. The current manuscript describes the waveguide model and presents the mathematical derivation for the booming frequency from the waveguide model. The spatial and temporal variations of the waveguide influence the ability of a dune to produce the mysterious sounds of the desert.

Keywords: booming, sand, dunes, waveguide, acoustic
PACS: 45.70.-n, 45.70.Ht, 43.20.+g, 62.30.+d

INTRODUCTION

Booming dunes emit a loud rumbling sound after a man-made or natural sand avalanche is generated on the slip face of a large desert dune. The sound may be created at different locations on the slipface in the hot and dry summer months, but never occurs on shallower windward face or at the base of the dune. The sound emission has a dominant audible frequency (f = 70-105 Hz) that may change by 5-10 Hertz during the slide. Figure 1a presents a microphone recording of a booming slide on Eureka dune in California, USA. The sustained sound has a narrow frequency range and several higher harmonics.

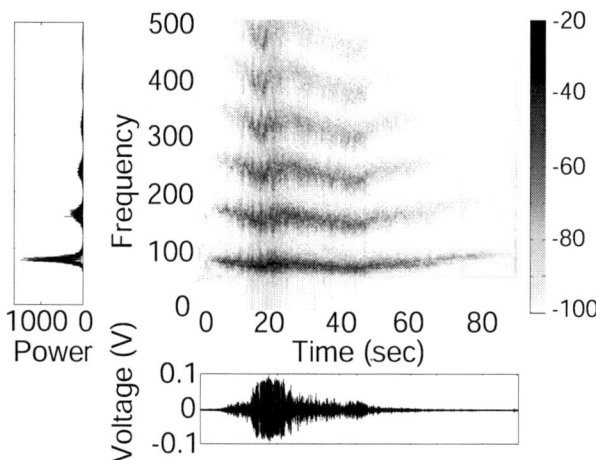

FIGURE 1. Spectrogram, signal and powerspectrum of the booming event measured with a microphone at 12 meters from the crest at Eureka dune on 10/27/2007.

Booming is a rare phenomenon identified at approximately 40 locations in the world and occurs on a variety of dune types, including barchan, linear and star dunes and sand drifts and sheets overlying a rocky outcrop Hunt and Vriend [1]. Booming does not occur on all desert dunes, nor does it occur throughout the entire year, nor everywhere on a booming dune. Experiments on booming dunes in the cold and wet season show that sustained booming cannot be initiated [2]. Avalanches on small dunes (~10 m) within a booming dune field or on the shallower windward face of a booming dune do not generate any acoustic emission. The booming frequency, usually between 70 and 105 Hz [1] for a given location, changes over the course of the season.

Scientists started to explore possible causes of this sound generation at the end of the 19th century. The amount of quantitative data increased significantly in the latter part of the 20th century. Experimental data and observations collected by Criswell et al. [3], Lindsay et al. [4] and Haff [5] incorporated the acoustic activity on booming dunes. Bagnold [6] explained the sound due to a process of dilatation and subsequent free fall of individual grains and provide a mechanical representation of the interaction between sheared grains. Andreotti [7] and Douady et al. [8] related the frequency of oscillations of individual sheared grains in the avalanching layer to the average particle diameter argument. The brief bursts of sound due to local forced shearing of individual grains fundamentally differ from the sustained sound generated by slumping of a large sand avalanche [9]. The booming frequency does correlate directly with the average grain diameter [1] and other length scales are involved. The fluid dynamics approach presented in Patitsas [10] connects stick-slip effects with the creation of regions of failure, similar to the explanation of sound due to local forced compression [11, 12]. A comprehensive theory for the sound generated by slumping should include both the initiation at the grain level and the amplification and frequency selection at the dune scale.

Vriend et al. [2] performed extensive seismic surveys on the dune and suggested that the booming is due to constructive interference of the body P-waves in a waveguide. The dimensions of the waveguide and the speed of sound within the dune set the booming frequency. The sound amplification occurs due to constructive interference of the waves as complete reflection occurs at the interface between a dry sand layer and a denser layer below. The character of the waves responsible for the source of the emission (burping) and the sustained emission from the avalanche itself (booming) have been further investigated in detail by Vriend et al. [9] using in situ measurements. The extent of the internal layering and the origin of the strong stratigraphy in dunes have been investigated by Vriend et al. [13]. The current paper presents the full derivation of the waveguide model and investigates the effects of variations in the subsurface structure on the efficiency of the waveguide.

CONSTRUCTIVE INTERFERENCE IN A WAVEGUIDE

Reflection and Transmission at an Interface

The source of the booming is positioned at a certain location on the surface of the dune. Plane waves exist at some distance from the source in radial direction. The waves also travel in the third direction and are reflected at different angles from subsurface layers. Assume a uniform layer of sand with thickness H, propagation velocity c_1 and density ρ_1. The overlying atmosphere has a velocity c_0 and density ρ_0 and the subsurface half space beneath has velocity c_2 and density ρ_2, as shown in figure 2. A velocity sandwich structure is formed as the dry sand layer has a lower velocity

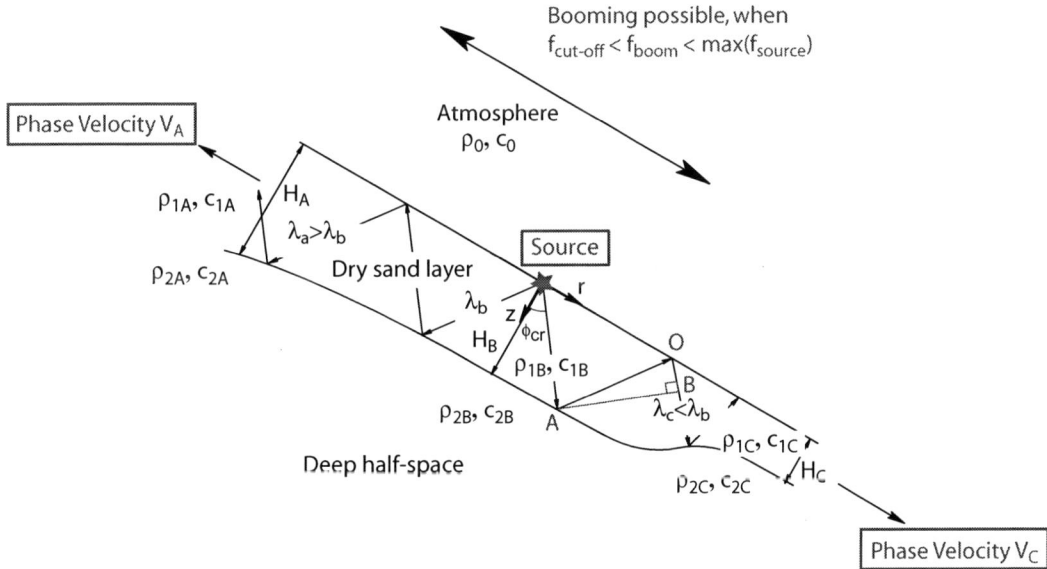

FIGURE 2. The interaction between booming frequency and phase velocity in the waveguide model.

than the surrounding layers: $c_1 < c_0$ and $c_1 < c_2$. The boundary conditions are expressed in terms of the vertical particle velocity and the pressure for acoustic vibrations.

In the case of no partitioning from P- to S-waves at the boundary, the angle of incidence ϕ_1 is equal to the angle of reflection and Snell's law applies for the transmitted wave ϕ_2. When the angle of incidence ϕ_1 is equal to the critical angle $\sin(\phi_{cr}) = \frac{c_1}{c_2}$, the wave is internally refracted ($\phi_1 = \phi_{cr}, \phi_2 = 90°$). For incidence angles equal or larger than the critical angle, no energy propagates into the substrate half space and the phase speed along the interface V_{int} reduces to:

$$V_{int} = \frac{c_1}{\sin(\phi_{cr})} = c_2. \tag{1}$$

For an incidence angle $\phi_1 < \phi_{cr}$, the reflection R and transmission T coefficients have a value between zero and one. For the critical angle and beyond, no energy leaks outside of the waveguide, $T = 0$ and $R = 1$, and the amplitude is preserved for all distances $A = A_0$. Attenuation with distance is expected because of the cylindrical spreading of the waves in the natural waveguide at the desert dunes.

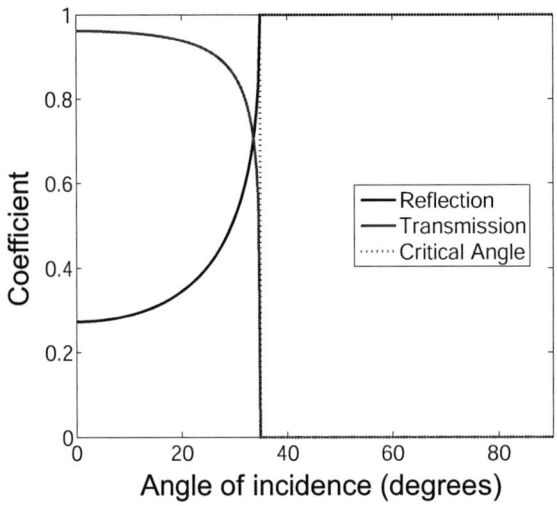

FIGURE 3. Reflection and transmission coefficient [14] as a function of angle. Parameters used are $\rho_1 = \rho_2 = 1500$ kg/m^3, $c_1 = 200$ m/s and $c_2 = 350$ m/s.

Wave Propagation in a Waveguide

Ray tracing

For constructive interference the analysis of the waveguide in terms of ray tracing follows the derivations presented in Ewing et al. [15] and Officer [16]. Assume that the waveguide depth H is constant across a certain length and that the velocities c_0, c_1, c_2 are constant in each layer. The seismic velocity usually increases with depth in a granular material, but Vriend et al. [17] showed that this increase is smaller than the large velocity jump across the interface between medium 1 and 2. The propagating waves are traveling in phase within the waveguide in the case of constructive interference. For a given wave at incident angle ϕ_1, the extra distance traveled by a wave $\bar{AB}$ in figure 2 is:

$$\bar{AB} = \bar{AO} + \bar{OB} = H\left[\frac{1}{\cos(\phi_1)} + \frac{\cos(2\phi_1)}{\cos(\phi_1)}\right] - 2H\cos(\phi_1). \tag{2}$$

Officer [16] defines the condition for constructive interference as the total phase change equal to a factor depending on the mode number $n = 1, 2, 3,...$:

$$k_n\bar{AB} - \varepsilon_{10} - \varepsilon_{12} = 2n\pi. \tag{3}$$

The phase change involved with the ray path from A to B depends on the wave number $k_n = 2\pi/\lambda_n$, the wavelength $\lambda_n = c_1/f_n$ and the distance traveled by the wave AB.

Substituting the wave number and wavelength and taking the tangent on both sides of equation (3):

$$\tan\left(\frac{2\pi H\cos(\phi_1)f_n}{c_1}\right) = \tan\left(\frac{(\varepsilon_{10} + \varepsilon_{12})}{2} + n\pi\right). \tag{4}$$

The phase lag at the surface ε_{10} and the bottom ε_{12} are derived from the reflection coefficient R [16, p. 79] as:

$$\frac{\varepsilon_{12}}{2} = \tan^{-1}\left[\frac{\rho_1\sqrt{\left(\frac{c_1}{V_{int}}\right)^2 - \left(\frac{c_1}{c_2}\right)^2}}{\rho_2\sqrt{1-\left(\frac{c_1}{V_{int}}\right)^2}}\right]. \tag{5}$$

and

$$\frac{\varepsilon_{10}}{2} = \tan^{-1}\left[\frac{\rho_1\sqrt{\left(\frac{c_1}{V_{sur}}\right)^2 - \left(\frac{c_1}{c_0}\right)^2}}{\rho_0\sqrt{1-\left(\frac{c_1}{V_{sur}}\right)^2}}\right]. \tag{6}$$

The current analysis deviates from the treatment by Officer [16], where $\varepsilon_{10} = \pi$ and $\varepsilon_{12} = 0$, as the waveguide in a sand dune has a mirrored velocity structure for which $c_0 = c_2$. In the case of critical refraction, the coupling provides the feedback to the waveguide and the phase speed along the surface interface reduces to $V_{sur} = c_0$, producing zero phase lag $\varepsilon_{10} = 0$ at the top surface [16, p. 228]. Similarly, critical refraction ensures a phase speed of $V_{int} = c_2$ along the interface between medium 1 and 2 and zero phase lag at the bottom surface as well. As a consequence of critical refraction, the phase lag is independent of the density and/or the impedance differences across interfaces. Therefore, equation (4) reduces to:

$$\tan\left(\frac{2\pi f_n H \cos(\phi_{cr})}{c_1}\right) = 0. \tag{7}$$

Solutions are given in terms of the mode number n with $n = 1, 2, 3,...$

$$\frac{2\pi f_n H \cos(\phi_{cr})}{c_1} = n\pi, \tag{8}$$

and

$$f_n = \frac{n}{2}\frac{c_1}{H\sqrt{1-\left(\frac{c_1}{c_2}\right)^2}}, \tag{9}$$

Continuous guided wave

An alternative approach to derive this formula is to analyze the waveguide in the continuous sense following Sleep and Fujita [18]. For wave propagation at long ranges and at moderate to low frequencies, the normal-mode solution combines interference effects from all ray paths. The trial function ϕ of a planar geometry with a standing wave in z-direction and a propagating wave in r-direction with rigid boundaries is:

$$\phi = \cos(k_z z)\exp(i[k_r r - \omega t]). \tag{10}$$

The wave propagates within the waveguide in horizontal direction with a constant phase velocity $V_1 = \omega_1/k_1$, with $k_1 = \sqrt{k_r^2 + k_z^2}$ and $\omega_1 = 2\pi f_1$. The boundary condition prescribes continuity of stress and all displacements at an interface. At the bottom the boundary condition at $z = H$ is satisfied if $k_z H = n\pi$, with integer n. At the interface on the surface (z = 0), traction and particle displacements are matched between the atmosphere and the upper layer of sand. The trial function ϕ including all modes n is:

$$\phi = \sum_1^n \phi_n \cos\left(\frac{n\pi z}{H}\right)\exp(i[k_r r - \omega t]). \tag{11}$$

The wave equation is:

$$\frac{\partial^2 \phi}{\partial t^2} = c_1^2\left[\frac{\partial^2 \phi}{\partial r^2} + \frac{\partial^2 \phi}{\partial z^2}\right], \tag{12}$$

with the p-wave velocity $c_1 = \sqrt{\lambda_1/\rho_1}$.

Substituting the trial function from equation (11) into the wave equation (12) gives:

$$-\omega_n^2 = c_1^2 \left[-k_r^2 - k_z^2\right] = c_1^2 \left[-k_r^2 - \left(\frac{n\pi}{H}\right)^2\right]. \tag{13}$$

The incident angle ϕ_1 is orientated as:

$$\tan(\phi_1) = \left(\frac{k_r}{n\pi/H}\right), \tag{14}$$

such that:

$$\omega_n = c_1 \left(\frac{n\pi}{H}\right)\sqrt{\tan(\phi_1)^2 + 1} = c_1 \left(\frac{n\pi}{H}\right)\sqrt{\frac{1}{\cos(\phi_1)^2}} = c_1 \left(\frac{n\pi}{H}\right)\frac{1}{\cos(\phi_1)}. \tag{15}$$

Restructuring this equation in terms of the frequency and substituting incidence at the critical angle $\phi_1 = \phi_{cr}$ gives:

$$f_n = \frac{\omega_n}{2\pi} = \frac{n}{2}\frac{c_1}{H\sqrt{1-\left(\frac{c_1}{c_2}\right)^2}}, \tag{16}$$

which is the same resonance relation as equation (9) found via ray tracing.

Phase Velocity

The booming waves travel at a phase speed V situated between the seismic speed of the dry layer of sand c_1 and the seismic speed of the denser, deeper layer of sand c_2. As the subsurface structure of the dune changes in the uphill or downhill direction (illustrated in figure 2), the phase velocity also changes independent of the frequency of the source. At a given moment in time a seismic sensor measures the global booming frequency and the local phase velocity. The phase velocity adapts as the waves move into a different velocity or layering structure. A sensor on the desert floor, located roughly 500 meters from the booming dune slope, measured the same booming frequency (82 Hz) as the local recording, as shown in figure 4, but a much higher phase velocity (~500 m/s).

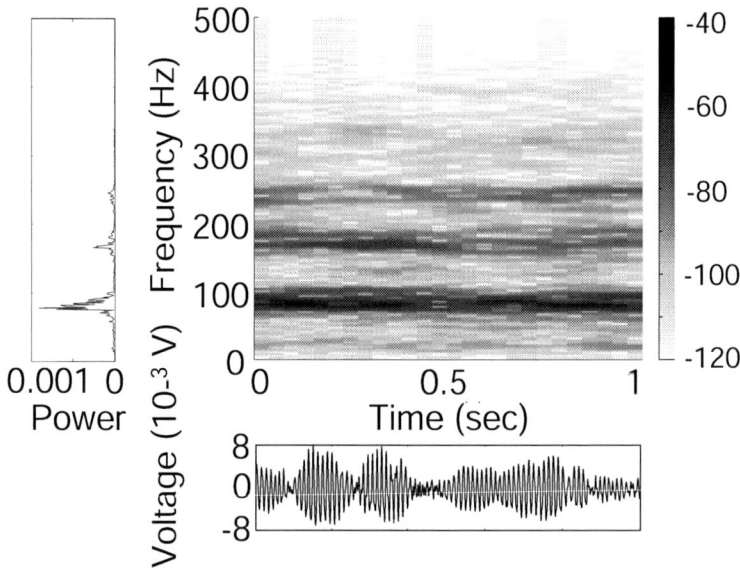

FIGURE 4. Seismic response of a booming emission at a distance of approximately 0.5 km.

The local phase velocity depends on the local depth of the layering H_A, the local velocities c_{1A} and c_{2A} and the global booming frequency $f = f_{\text{cutoff}}$ obtained from equation (16). For known dimensions, velocities and booming frequency, the local phase velocity V_A is obtained by solving the transcendental equation:

$$\tan\left(f_{\text{cutoff}}\frac{2\pi H_A}{c_{1A}}\sqrt{1-\left(\frac{c_{1A}}{V_A}\right)^2}\right) = \frac{\rho_1}{\rho_2}\frac{\sqrt{\left(\frac{c_{1A}}{V_A}\right)^2-\left(\frac{c_{1A}}{c_{2A}}\right)^2}}{\sqrt{1-\left(\frac{c_{1A}}{V_A}\right)^2}}. \tag{17}$$

In the example of figure 2, the depth of the layering increases uphill from the source creating a longer wavelength $\lambda_a > \lambda_b$ and physically signifies the approach of the dune crest [13]. As the subsurface velocity commonly is smaller close to the crest [2], the phase velocity decreases significantly uphill from the source $V_a < V_b$. The example also shows a thinning in downhill direction, where the subsurface velocity increases, occurring in grainfall areas [13]. The local wavelength decreases $\lambda_c < \lambda_b$ and the phase speed increases $V_c > V_b$. An alternative waveguide structure of thickening in downhill direction occurs at the transition of grainfall and grainflow regions. The increase in subsurface velocity and the increase in depth are two competing factors that have an opposite effect on the phase speed. Usually the increase in downhill velocity dominates the increase in depth such that the effective phase speed increases.

THE INTERACTION BETWEEN THE WAVEGUIDE AND BOOMING

A functioning waveguide prohibits energy loss and promotes amplification of the source. There are several consequences for the waveguide theory for booming sand dunes:

1. The source frequency excites the natural resonance frequency of the waveguide and the avalanching of grains provide the energy necessary for the emission.
2. The layering in the region of the source sets the global booming frequency; away from the source the phase speed may change and the amplitude may decrease due to leakage.
3. The dimensions of the waveguide may prevent the excitation to be constructive and limit amplification of the source.

Excitation by a Source

Direct shearing of sand at the surface generates short pulses, defined as the burping emission by Vriend et al. [9]. The burping source provides a continuous energy input to excite a selection of modes in the waveguide and is necessary to initiate the booming emission. The source spectrum of the burp presented in figure 5 involves frequencies in a wide range (50-100 Hz, varies slightly depending on shear rate) at low amplitude.

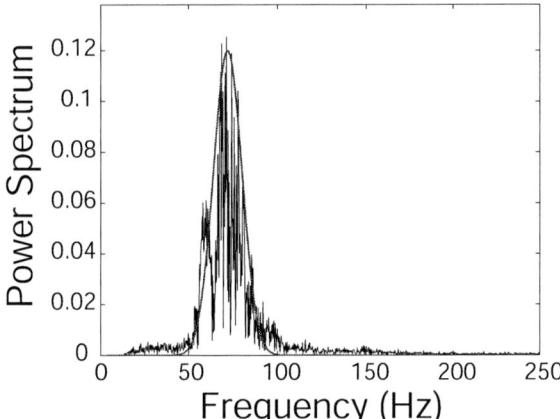

FIGURE 5. Source spectrum of a shearing event on the slip face at Eureka dunes on 07/18/2008. The burping source is fitted with a Gaussian-shaped function with constant $C = 0.12$, center frequency $f_{\text{cent}} = 72$ Hz and width of the pulse $\sigma = 8$.

The total acoustical amplitude of the booming emission is more than an order of magnitude higher than that of a burping emission. The excess energy needs to be supplied by another mechanism as the total energy of the acoustic emission cannot be increased. The release of gravitational energy into kinetic energy due to the avalanching sand provides this additional source of energy. As described in section , the narrow booming frequency is set by external factors such as the waveguide dimensions and the speed of sound. The avalanching of grains supply energy for amplification while the burping emission provides the broad frequency content to excite the booming frequency. When the avalanching of sand stops, the amplitude of the booming stops growing and diminishes slowly with time. However, the sound may continue for up to a minute as the energy continuous to reverberate in the waveguide.

Relation between Frequency and Phase Speed

The dispersion relation given in equation (17) provides a relation between the phase velocity V and the frequency f of a mode. A graphical technique finds the solutions of the transcendental equation for each mode $n = 1, 2, ..$ in figure 6. The intersection of the two functions $g(\omega y)$ and $h(\omega y)$, representing the left- and right-hand side of equation (17)

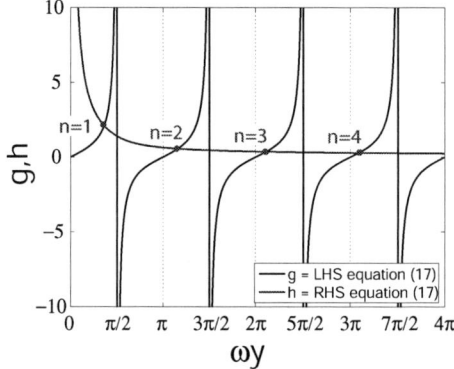

FIGURE 6. Graphical representation of finding the roots of the dispersion relation. Parameters used are $\rho_1 = \rho_2 = 1500$ kg/m^3, $c_1 = 200$ m/s, $c_2 = 350$ m/s and H = 2.0 m. The horizontal axis is defined in terms of the parameter $\omega y = 2\pi f \dfrac{H}{c_1} * \sqrt{[1 - (\frac{c_1}{V})^2]}$.

respectively, are solutions to the dispersion relation. The phase velocity along the interface has a value between c_1 and c_2 and is maximum at the cutoff frequency $\omega_{\text{cutoff},n}$ calculated by equation (9) for a specific mode number. The booming emission propagation speed is between $V_{\text{boom}} = 200$ m/s and $V_{\text{boom}} = 250$ m/s [9]. The mode $n = 1$ needs to be excited by the burping source in order to generate a propagative wave.

Changing Dimensions of the Waveguide

The cutoff frequency determined by the waveguide dimensions needs to overlap with part of the source spectrum of the burping emission. Figure 7a shows the waveguide modes for common parameters found in field experiments. Mode $n = 1$ overlaps slightly with the source spectrum such that energy may be transferred. Figure 7b presents the modes for a smaller, nonbooming dune. Mode $n = 1$ overlaps significantly with the source spectrum, but no sustained booming can be generated. The subsurface velocity $c_2 = 600 m/s$ is very high and the symmetry between the atmosphere and the subsurface deeper layer breaks down as $c_0 \neq c_2$. Furthermore, the length of the waveguide channel in the smaller dune is of the same order of magnitude as the wavelength of booming ≈ 2.5m and its length is insufficient to create an amplification of the sound. A similar situation occurs for a very deep waveguide where the wavelength for constructive interference exceeds the size of the avalanche. The situation for a nonbooming dune in the wintertime is presented in figure 7c. The hard substrate layer is preserved, but the upper layer velocity c_1 increases significantly due to a larger moisture content. The first mode cannot be excited as its cutoff frequency is higher than the source spectrum. A similar mode spectrum occurs if the waveguide depth is very shallow (figure 7d) and the cutoff frequency is beyond the maximum frequency of the source.

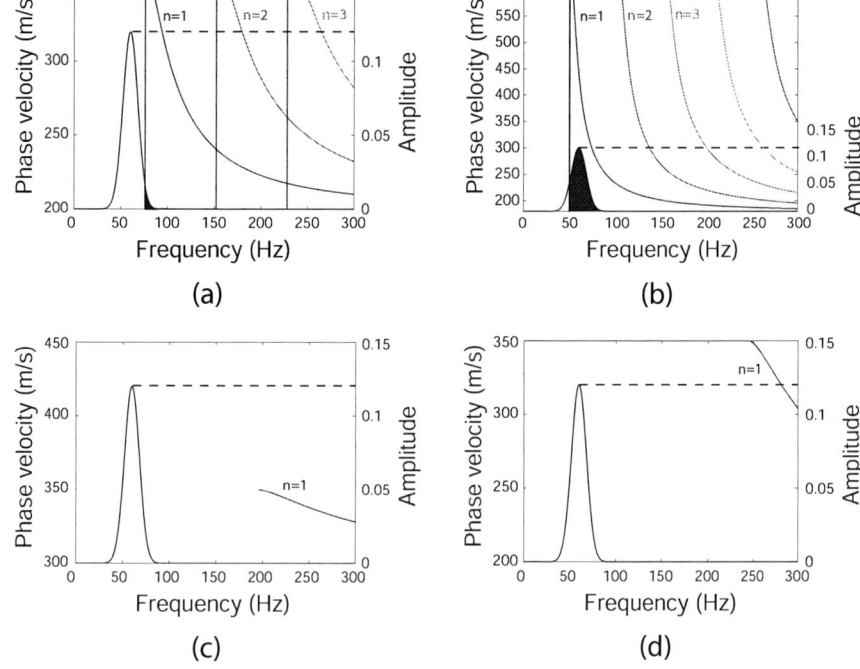

FIGURE 7. Waveguide modes for: (a) booming dune (c_1 = 200 m/s, c_2 = 350 m/s and H = 1.5 m), (b) nonbooming small dune (c_1 = 180 m/s, c_2 = 600 m/s and H = 1.8 m), (c) nonbooming winter dune (c_1 = 300 m/s, c_2 = 350 m/s and H = 1.5 m), (d) nonbooming dune with a shallow waveguide (c_1 = 200 m/s, c_2 = 350 m/s and H = 0.5 m). Booming exist if the excitation frequency is equal or larger than the cutoff frequency.

CONCLUSION

The layered structure in booming dunes is crucial in the explanation of the phenomenon of booming sand dunes. The variation in internal structure and seismic velocities explains the variation in booming frequency with downhill position. The frequency of the waves generated in the waveguide should be larger than the cutoff frequency but smaller than the maximum frequency of the source for constructive interference to be effective.

Booming occurs if three necessary factors are all satisfied. First, the sand grains need to be smoothed and rounded by aeolian processes to create short acoustic pulses when they are rubbed against each other. A natural or man-made avalanche of sand grains on the slipface of the dune create pulses and form the source of booming. Second, the subsurface structure of the dune needs to promote constructive interference by aligning the traveling waves in a regular pattern. An internal layering at a constant depth forms the waveguide and promotes the amplification of the acoustic waves. Third, the velocity structure needs to have structure favoring constructive interference excited by the short-pulsed source. If the layering is too deep or shallow or irregular in structure, the booming frequency cannot be excited by the source. The waveguide is ineffective if the velocity structure is diffuse with depth preventing reflection at a discrete interface. The necessary conditions for the booming emission to develop result in the scarcity in occurrence of booming sand dunes in the world.

AUXILIARY MATERIAL

The Rayleigh reflection coefficient [14] for plane waves is:

$$R = \frac{\dfrac{\rho_2}{\rho_1} - \sqrt{\dfrac{\left(\dfrac{c_1}{c_2}\right)^2 - \left(\dfrac{c_1}{V_{\text{int}}}\right)^2}{1 - \left(\dfrac{c_1}{V_{\text{int}}}\right)^2}}}{\dfrac{\rho_2}{\rho_1} + \sqrt{\dfrac{\left(\dfrac{c_1}{c_2}\right)^2 - \left(\dfrac{c_1}{V_{\text{int}}}\right)^2}{1 - \left(\dfrac{c_1}{V_{\text{int}}}\right)^2}}}, \tag{18}$$

with phase speed along the interface $c_1 < V_{\text{int}} < c_2$ and:

$$V_{\text{int}} = \frac{c_1}{\sin(\phi_1)} = \frac{c_2}{\sin(\phi_2)} \tag{19}$$

The transmission coefficient T is related the reflection coefficient:

$$T = \sqrt{1 - R^2} = \frac{2\sqrt{\dfrac{\rho_2}{\rho_1}}\sqrt{\dfrac{\left(\dfrac{c_1}{c_2}\right)^2 - \left(\dfrac{c_1}{V_{\text{int}}}\right)^2}{1 - \left(\dfrac{c_1}{V_{\text{int}}}\right)^2}}}{\dfrac{\rho_2}{\rho_1} + \sqrt{\dfrac{\left(\dfrac{c_1}{c_2}\right)^2 - \left(\dfrac{c_1}{V_{\text{int}}}\right)^2}{1 - \left(\dfrac{c_1}{V_{\text{int}}}\right)^2}}}, \tag{20}$$

REFERENCES

1. M. Hunt, and N. Vriend, *Annual Review of Earth and Planetary Sciences* **38** (2010).
2. N. Vriend, M. Hunt, R. Clayton, C. Brennen, K. Brantley, and A. Ruiz-Angulo, *Geophysical Research Letters* **34**, L16306 (2007).
3. D. Criswell, J. Lindsay, and D. Reasoner, *Journal of geophysical research* **80**, 4963–4974 (1975).
4. J. Lindsay, D. Criswell, T. Criswell, and B. Criswell, *Geological society of America Bulletin* **87**, 463–473 (1976).
5. P. Haff, *American scientist* **74**, 376–381 (1986).
6. R. Bagnold, *Proceedings of the Royal Society of London, Series A - Mathematical and physical sciences* **295**, 219 – 232 (1966).
7. B. Andreotti, *Physical Review Letters* **93**, 238001/1–238001/4 (2004).
8. S. Douady, A. Manning, P. Hersen, H. Elbelrhiti, S. Protière, A. Daerr, and B. Kabbachi, *Physical Review Letters* **97**, 018002/1–018002/4 (2006).
9. N. Vriend, M. Hunt, and R. Clayton, *In preparation* (2010).
10. A. Patitsas, *Journal of fluid structures* **17**, 287–315 (2003).
11. J. Hidaka, S. Miwa, and K. Makina, *International Chemical Engineering* **28**, 99–107 (1988).
12. S. Miwa, T. Okazaki, and M. Kimura, *The Science and Engineering Review of Doshisha University* **36**, 67–76 (1995).
13. N. Vriend, M. Hunt, and R. Clayton, *In preparation* (2010).
14. T. Lay, and T. Wallace, *Modern global seismology*, Academic Press Limited, 1995, ISBN 0-12-732870-X.
15. W. Ewing, W. Jardetzky, and F. Press, *Elastic waves in layered media*, McGraw-Hill New York, 1957.
16. C. Officer, *Introduction to the theory of sound transmission with application to the ocean*, McGraw-Hill Book company Inc., 1958.
17. N. Vriend, M. Hunt, R. Clayton, C. Brennen, K. Brantley, and A. Ruiz-Angulo, *Geophysical Research Letters* **35**, L08307 (2008).
18. N. Sleep, and K. Fujita, *Principles of geophysics*, Blackwell Science, 1997.

Wave Propagation In Strongly Nonlinear Two-Mass Chains

Si Yin Wang[a], Eric B. Herbold[b], Vitali F. Nesterenko[a,c]

[a]*Materials Science and Engineering Program, University of California, San Diego, CA*
[b]*School of Materials Science and Engineering, Georgia Inst. of Tech., Atlanta, Georgia*
[c]*Department of Mechanical and Aerospace Eng., University of California, San Diego, CA*

Abstract. We developed experimental set up that allowed the investigation of propagation of oscillating waves generated at the entrance of nonlinear and strongly nonlinear two-mass granular chains composed of steel cylinders and steel spheres. The paper represents the first experimental data related to the propagation of these waves in nonlinear and strongly nonlinear chains. The dynamic compressive forces were detected using gauges imbedded inside particles at depths equal to 4 cells and 8 cells from the entrance gauge detecting the input signal. At these relatively short distances we were able to detect practically perfect transparency at low frequencies and cut off effects at higher frequencies for nonlinear and strongly nonlinear signals. We also observed transformation of oscillatory shocks into monotonous shocks. Numerical calculations of signal transformation by non-dissipative granular chains demonstrated transparency of the system at low frequencies and cut off phenomenon at high frequencies in reasonable agreement with experiments. Systems which are able to transform nonlinear and strongly nonlinear waves at small sizes of the system are important for practical applications such as attenuation of high amplitude pulses.

Keywords: granular chains, nonlinear and strongly nonlinear, band gap, oscillating excitations, oscillatory shocks.
PACS: 43.25.Cb, 43.25.Ts, 43.40.Jc, 43.40.Kd, 46.40.Cd

INTRODUCTION

The research in strongly nonlinear wave dynamics of discrete systems, particularly of one-dimensional granular chains has become a growing area of interest [1-4]. They support a linear waves with an amplitude much smaller than static compressive force and nonlinear waves if the amplitude of dynamic force is smaller than the initial compression but of the same order of magnitude. Strongly nonlinear waves can also propagate in these chains with amplitude close to the initial compression or even much higher like in case of "sonic vacuum" [3]. The study of low dimensional granular chains is a logical step forward in wave dynamics with possible "translation" of results to totally different areas, particularly for design of new metamaterials with tunable properties.

Discrete granular chains with a strongly nonlinear interaction law (i.e. the linear part is absent or small in comparison with the nonlinear part of interaction law) are more general systems than traditional linear and weakly nonlinear systems. Strong nonlinearity results in highly tunable systems where relatively small external forces can tune their behavior. It was demonstrated experimentally, numerically and analytically that strongly nonlinear granular chains support a new type of waves – strongly nonlinear solitary waves with qualitatively different properties than well-known solitary waves of the Korteweg-de Vries equation [1-3].

A relatively new area is the investigation of nonlinear and strongly nonlinear waves in two mass granular chains [5,6]. These systems represent a unique opportunity to create tunable band gap materials where the band gap frequencies may be adjusted by external forces [5]. Most of the previous research of wave propagation in two mass chains was related to linear and weakly nonlinear discrete systems [7-14].

This work investigates the propagation of quasiharmonic signals and the influence of band-gaps in nonlinear and strongly nonlinear diatomic periodic chains composed of stainless steel cylinders and stainless steel spheres. The investigated waves were not perfectly harmonic signals. That is why we use the word "quasiharmonic" which means that the oscillating dynamic force contained basic harmonic frequency with relatively high amplitude with additional components having significantly lower amplitudes.

A band gap is a well-known phenomenon for systems with linear interaction law between particles [15]. In the one dimensional elastic granular chain, the force between the contact particles interacting by Hertz law [16] is represented by the following equation:

$$F = A(\delta_0 + \delta_d)^{3/2}, \tag{1}$$

$$A = \frac{4 E_C E_S (1/R_S + 1/R_C)^{-1/2}}{3[E_S(1-\upsilon_C^2) + E_C(1-\upsilon_S^2)]}. \tag{2}$$

The constant A depends on the material properties and the geometry of the contact area [16]. In our case $E_C = E_S$ = 193 GPa and $\upsilon_C = \upsilon_S$ = 0.3 are the elastic modulus and Poisson's ratio for the stainless steel cylinder and stainless steel sphere, and the radii of contact particles are R_S = 1 mm for a sphere and $1/R_C = 0$ for the cylinder with a planar contact surface.

The contact force can be linearized by assuming that the dynamic relative displacement in the wave between neighboring particles is much smaller than their initial displacement caused by static compressive force ($\delta_0 \gg \delta_d$):

$$F \approx A\delta_0^{3/2} + \frac{3}{2} A \delta_0^{1/2} \delta_d. \tag{3}$$

The two force components are the static force $F_0 = A\delta_0^{3/2}$ and dynamic force $F_d = \frac{3}{2} A \delta_0^{1/2} \delta_d = \beta \delta_d$, where

$$\beta = \frac{3}{2} A \delta_0^{1/2} = \frac{3}{2} A^{2/3} F_0^{1/3}. \tag{4}$$

The upper and lower bound of the band gap for this system can be calculated following the classical approach presented, for example, in [15]:

$$f_1 = \frac{1}{2\pi}\left(\frac{2\beta}{M}\right)^{1/2}, f_2 = \frac{1}{2\pi}\left(\frac{2\beta}{m}\right)^{1/2}, \tag{5}$$

where M is the mass of the stainless steel cylinder and m is the mass of the stainless steel sphere. This band-gap can be tuned by changing the value of β which depends on static load (F_0), elastic properties and geometry of the contacting particles. In our experiments a static load equal 7.8 N results in cutoff frequencies of f_1 = 11.3 kHz and f_2 = 109.7 kHz and for a static load 0.8 N the cutoff frequencies are f_1 = 7.7 kHz and f_2 = 75.1 kHz.

We do not know any published data on the propagation of quasiharmonic nonlinear and strongly nonlinear signals in diatomic chains in experiments. Numerical calculations in the paper [5] are related to propagation of harmonic signals which are difficult to realize in experiments. Also system composed from PTFE (Polytetrafluoroethylene) balls and steel cylinders demonstrated a large attenuation masking band gap effects [5]. The response of nonlinear and especially strongly nonlinear discrete diatomic system to periodic boundary data were investigated in this paper. We attempted to address the following questions. Do a granular diatomic chains support strongly nonlinear periodic waves in experiments? Can we develop a mechanical system with a band gap in the audible range of frequencies based on granular chains with a low level of dissipation where the major influence on signal propagation is due to dispersive properties caused by a mesostructure? Is the band gap introduced for a linear elastic interaction relevant to the transmission of nonlinear and strongly nonlinear quasiharmonic waves? What is the length of a discrete chain where some harmonics can be suppressed by the chain structure due to the band gap influence? Can we transform oscillatory shock waves on short distances from the entrance? These questions are related to a basic science and also are important for practical applications to tune a strong amplitude signals.

EXPERIMENTAL PROCEDURES

The one-dimensional granular chains are assembled inside a PTFE cylinder with an inner diameter of 10 mm. The stainless steel cylinders and stainless steel spheres were arranged in alternative order with each other (Fig. 1a). To apply a static force to the chain we used magnetic hollow steel cylinder with a plane bottom at the very top and a small magnet (Fe-Nb-B) interacting with outside magnet (Fe-Nb-B). The static force was tunable by changing the applied weight of the outside magnet. The chain had 55 stainless steel cylinders with height 5mm, diameter 10 mm and mass 3.21g and 57 stainless steel spheres with diameter 2 mm and mass 0.034 g. In experiments with nonlinear systems, we used a magnet with a weight 800 g on the top of the chain. A relatively small weight was used for creating conditions for a strongly nonlinear system.

Three calibrated piezoelectric sensors (RC~$10^3 \mu s$) were embedded in three stainless steel cylinders and were connected to an oscilloscope Tektronix TDS 2014 to detect the force-time curve. The piezoelectric sensors 1, 2 and 3 were placed in the 1st, 5th and 9th cell, so there were 4 cells between them.

A Brüel&Kjær vibration exciter (model 4810) was used to create quasiharmonic signals at the entrance of the diatomic chain (a periodical excitation was generated by a function/arbitrary waveform generator Agilent 33220A 20 MHz). A vibration exciter was driving a stainless steel rod (length of 20 cm and diameter of 5 mm) with prescribed frequency being in contact with top magnetic cylinder of the chain.

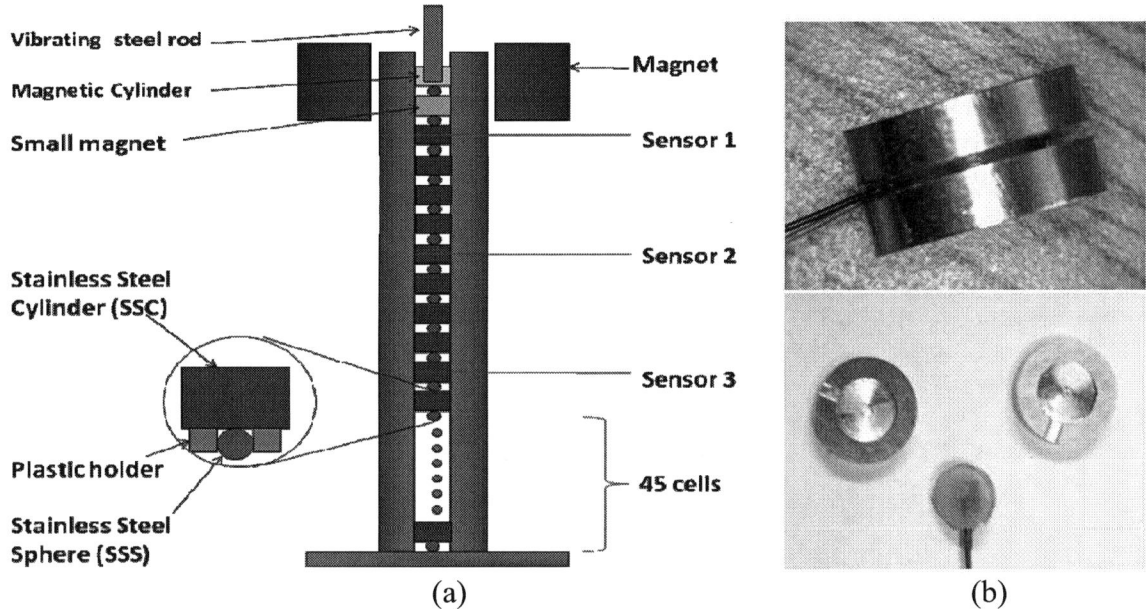

FIGURE 1. Schematic drawing of the experiments to measure strongly nonlinear wave propagation in a chain. (a) The experimental set up, (b) top figure - the SS cylinder with embedded sensor and wires, bottom figure – SS covers with grooves to accommodate piezosensor shown between them.

The contact force between the rod and last particle represented boundary conditions for our system which were controlled by the first gauge close to the end of a system. The generator created an incoming signal with different frequencies and amplitudes that can tune the amplitude of displacement of the rod defining an oscillating contact force at the entrance to the chain. A soft spring supported the vibrator to control the additional static force which may come from the weight of vibrator.

RESULTS AND DISCUSSIONS

Nonlinear Case

To create a nonlinear system we used a relatively large (in comparison with dynamic force in the wave) static compression of 8 N. In this system a quasiharmonic signal with dynamic force amplitude about 2 N was created at the top of the chain. The strength of the signal was smaller than initial precompression but of the same order of magnitude corresponding to the nonlinear regime. A quasiharmonic signal with 5 kHz was first generated. The signal had a frequency lower than the lower bound of band gap corresponding to this static force, which is $f_1 = 11.3$ kHz (upper band frequency $f_2 = 109.73$ kHz). It was observed (Fig. 2) that the signal was transmitted without change of fundamental frequency, but with some attenuation. The high transparency of the system can be understood given that the 5 kHz was outside the band gap calculated in a linear approximation, though we have a nonlinear signal. The attenuation of the amplitude is connected to the unavoidable dissipation in experiments.

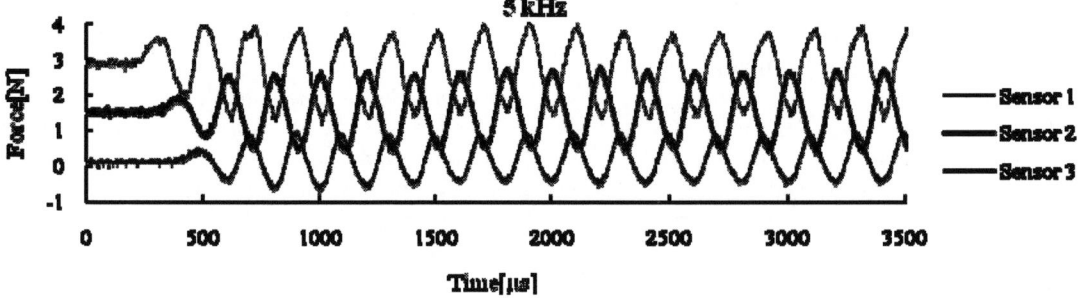

FIGURE 2. Experimental results. The propagation of quasiharmonic nonlinear pulse with main initial frequency of input signal 5 kHz at the entrance of the system. Nonlinear system, static force $F_0 = 7.8$N. The lower range of the forbidden frequency is 11.3 kHz being higher than the frequency of major harmonic in the input signal.

The experimental results were compared with numerical calculation of the chain without dissipation. We tried to create an input signals at the system in numerical calculations similar to input signals in experiments. We were able to match relatively closely the amplitudes of main harmonics though these signals are not identical. It should be mentioned that only elastic properties of particles were used to calculate forces between particles (Eqs. 1-4), no adjustable parameters were used in calculations. As shown in Fig. 3(a) the velocity of the pulse in calculations was in agreement with experiments. The main frequency of 5 kHz was carried on during transmission without attenuation. In the experiments, however, the initial amplitude of the signal was reduced at the 5th cell due to inevitable dissipation (Fig. 2, 3(b)). The shape of transmitted signal remains practically the same with only reduced amplitude. The Fourier coefficients derived from a fast-Fourier-transform similar to [5] are presented for numerical and experimental signals correspondingly at the left and right bottom parts of Fig. 3.

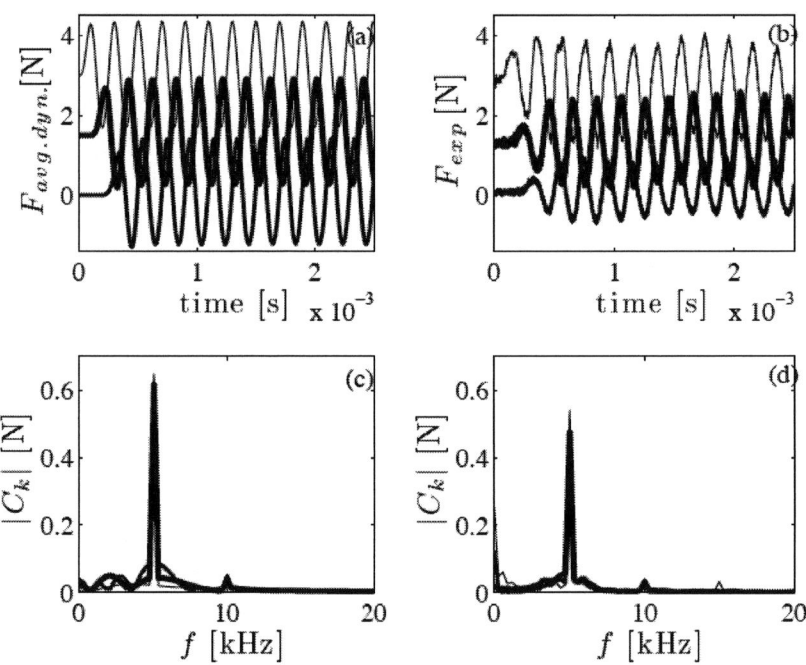

FIGURE 3. Comparison between the numerical calculation and the experimental results for 5 kHz signal.
(a) Numerical calculations: dynamic force inside corresponding particles with embedded sensors 1, 2 and 3 and (c) their Fourier-transforms. Signals from sensor 1, 2 and 3 are overlapping in the vicinity of main frequency with similar amplitudes.
(b) Experimental results: dynamic force from embedded sensors 1, 2 and 3 and (d) their Fourier-transforms. Signals from sensors 1, 2 and 3 are overlapping in the vicinity of main frequency with similar amplitudes for sensors 1 and 2 and amplitude for sensor 3 being two times smaller.

From the comparison of numerical and experimental data we can observe that quasiharmonic nonlinear signal with main frequency 5 kHz below lower band gap corresponding to elastically linear system ($f_1 = 11.3$ kHz, Eq. 5) propagates with little attenuation in calculations and with higher attenuation in experiments.

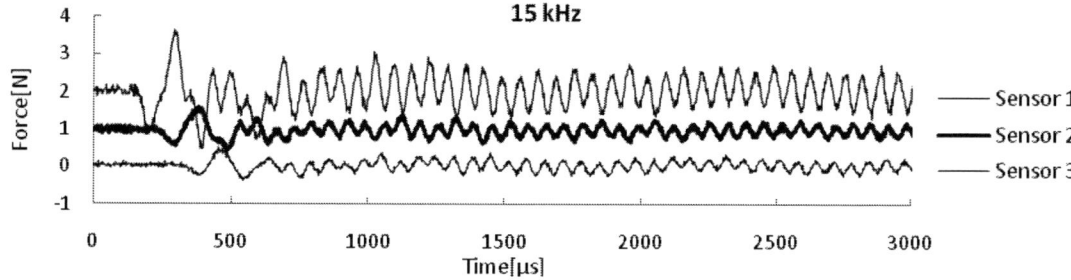

FIGURE 4. Experimental results. The propagation of quasiharmonic nonlinear pulse with main initial frequency of input signal 15 kHz at the entrance of the system. Nonlinear system, static force $F_0 = 7.8$N. The lower range of the forbidden frequency is 11.3 kHz being lower than the frequency of major harmonic in the input signal.

The example of quasiharmonic signal with a frequency 15 kHz, which is inside the band gap is shown in Fig. 4. The comparison between experimental and numerical data is presented in Fig. 5. It is evident that unlike with experiments at 5 kHz (Fig. 2) the nonlinear signal with main frequency 15 kHz propagated with strong attenuation and shape of signal changed dramatically. Still in experiments the fundamental frequency can be clearly picked out after its propagation through 9 cells at sensor 3.

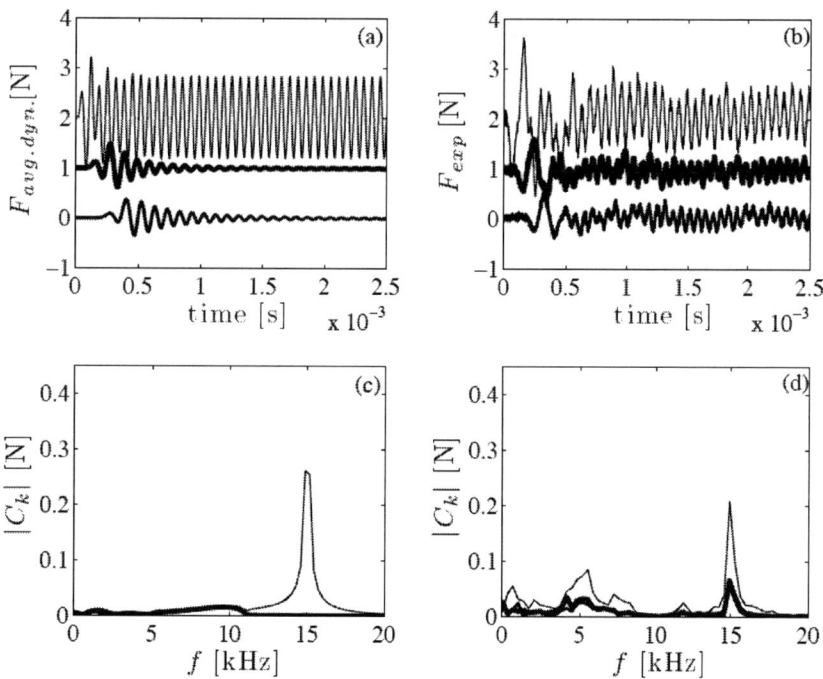

FIGURE 5. Comparison between the numerical calculation and the experimental results for 15 kHz signal.
(a) Numerical calculations: dynamic force inside corresponding particles with embedded sensors 1, 2 and 3 and (c) their Fourier-transforms. Signals from sensors 2 and 3 are overlapping with no harmonics above band gap frequency 11.3 kHz.
(b) Experimental results: dynamic force from embedded sensors 1, 2 and 3 and (d) their Fourier-transforms. At 15 kHz signals from sensors 2 and 3 are overlapping with similar amplitudes being 3 times smaller than amplitude corresponding to sensor 1.

Numerical calculations of nonlinear granular chains were used to clarify a mechanism of attenuation. Fig. 5(c) shows that 15 kHz component was completely wiped out at the second and third gauges in numerical calculations unlike in experiments (Fig. 5(d)). It should be noticed that the cut off frequency at about 11 kHz detected for nonlinear signal in numerical calculations corresponds to lower band gap frequency 11.3 kHz for elastically linear chain. This behavior in numerical calculations is in contradiction with some transparency at the fundamental frequency 15 kHz in experiments (Fig. 5(d)). We can provide a preliminary explanation that this may be due to the nonlinearity of the system where low frequency harmonics present in experiments (input signal in numerical calculations did not have this low frequency components) cause some *in situ* transparency inside band gap in experiments. We plan to investigate this interesting discrepancy between numerical data and experimental results in our future research. This cut off frequency is not clearly reflected in experimental data.

We compared the attenuation of propagating nonlinear quasiharmonic signals with the increase of frequency for interval of 5 – 20 kHz. In the Figure 6 the relative amplitudes of the main harmonics detected at the second gauge (A2) and third gauge (A3) with respect to the amplitude of the same harmonics at the gauge 1 (A1) are presented. The quasiharmonic signal attenuates to a value of about 30% of the input signal strength for the frequencies above 15 kHz at a short distances from the entrance. At this frequency the signal amplitude starts to drop to zero very fast, even at the 5th cell from entrance. We may conclude that at these distances from the entrance the system is partially transparent for frequencies above $f_1 = 11.3$ kHz (Eq.5) and non transparent at frequency about $2 f_1$. We plan more detailed comparison of numerical and experimental results in our future research.

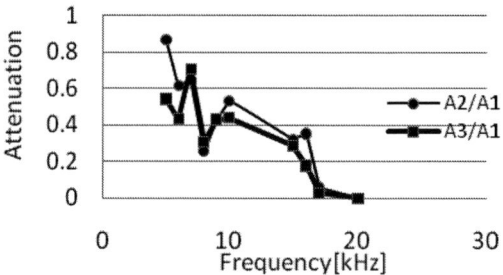

FIGURE 6. The comparison of the attenuation in experiments of the amplitude corresponding to major harmonics with the increase of frequency. A1, A2 and A3 are the amplitudes of the major harmonics in the incoming signal detected by a sensor 1(A1) and in signals measured by sensors 2 (A2) and 3 (A3).

At higher frequencies of the rod vibrations we were able to generate oscillatory shock waves. For example, an oscillatory shock like pulse is presented in Fig. 7.

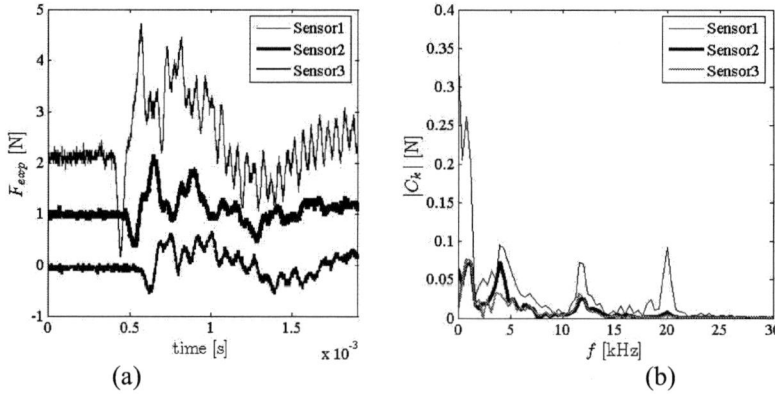

FIGURE 7. The propagation of oscillatory shock through a nonlinear system with static force $F_0 = 7.8$N. The lower range of the forbidden frequency is 11.3 kHz. (a) Experimental profiles of forces detected by three sensors, (b) The fast Fourier transforms of the propagating signals. Signals from sensor 2 and sensor 3 are overlapping for most of frequencies.

We can see dramatic change of the shape of the pulse. From Fourier transforms (Fig. 7(b)) we observe practically complete decay of component at 20 kHz and less attenuation at lower frequencies. As a result a front of the shock wave become less steep, and oscillations on the shock profile were wiped out while lower frequencies in the signal propagated through with some attenuation. This data are in a qualitative agreement with data in [5] for transformation of oscillatory shocks in PTFE-stainless steel system.

Strongly Nonlinear Case

We consider the system as a strongly nonlinear when the amplitude of the signal (~1N) was almost the same as the initial compression (F_0= 0.8 N). We investigated first a quasiharmonic signal at the 3 kHz excited at the entrance to the system. The frequency of this signal was lower than the lower bound of band gap for linear elastic chain for given static compression (f_1=7.7 kHz, f_2 = 75.1 kHz). Experimental data are presented in Fig. 8. We can see that the strongly nonlinear signal was transmitted without change of its shape with some attenuation. However, there was no change in fundamental frequency as in case of low frequency for nonlinear system (Fig.2).

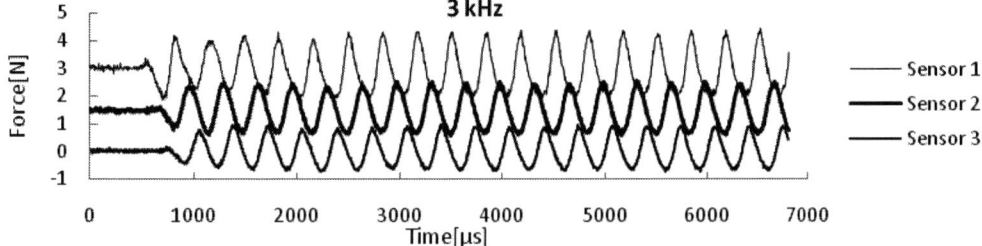

FIGURE 8. Experimental results. The propagation of quasiharmonic strongly nonlinear pulse with main initial frequency of input signal 3 kHz at the entrance of the system. Strongly nonlinear system, static force F_0 = 0.8N. The lower range of the forbidden frequency is 7.7 kHz being higher than the frequency of the major harmonic in the input signal.

Numerical calculations presented in Fig. 9 shows that the main harmonic with frequency of 3 kHz was carried on during transmission in agreement with the experiments. Numerical data also demonstrate transparency of strongly nonlinear system at frequency about 6 kHz and 9 kHz. The former can be due to a strong nonlinearity of the system.

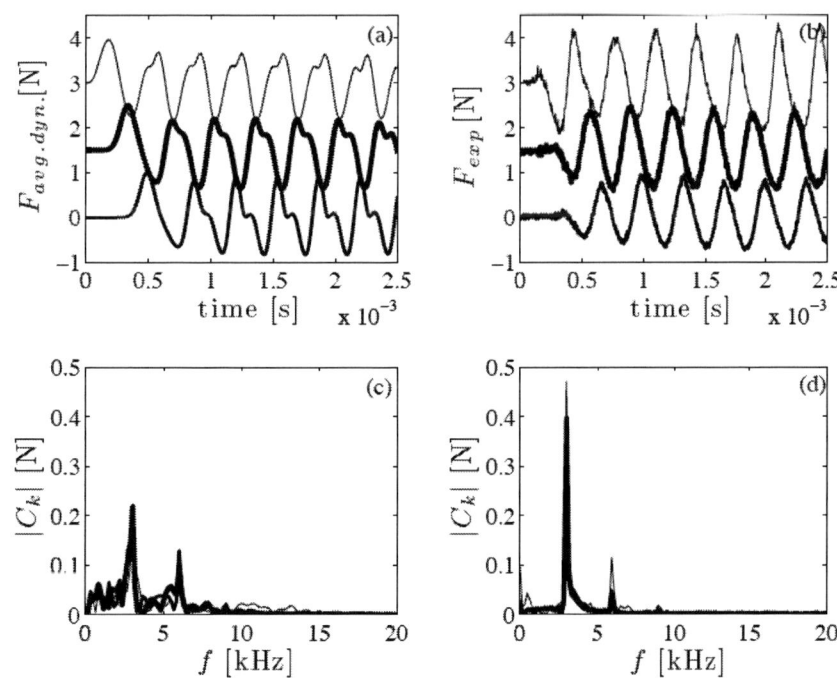

FIGURE 9. Comparison between the numerical calculation and the experimental results for propagating 3 kHz signal. (a) Numerical calculations: dynamic force inside corresponding particles with embedded sensors 1, 2 and 3 and (c) their Fourier-transforms. Signals from sensors 1, 2 and 3 are overlapping in vicinity of major peaks.
(b) Experimental results: dynamic force from embedded sensors 1, 2 and 3 and (d) their Fourier-transforms. Signals from sensors 1, 2 and 3 are overlapping in the vicinity of major peaks with decreased amplitudes 0.48, 0.4, and 0.32 correspondingly.

Experimental results corresponding to the main input frequency 15 kHz are presented in Fig. 10. We can observe that high frequency component of strongly nonlinear signal was practically wiped out within short distances from the entrance. This is demonstrated by the Fourier transforms related to numerical calculations and experiments presented in Fig. 11. Clearly, there was no transmission of signal in either the numerical calculation or the experimental data at 15 kHz. We may also notice a cut off frequency at a vicinity of 8 kHz in numerical calculations which is close to a lower band gap frequency 7.7 kHz for elastically linear chain.

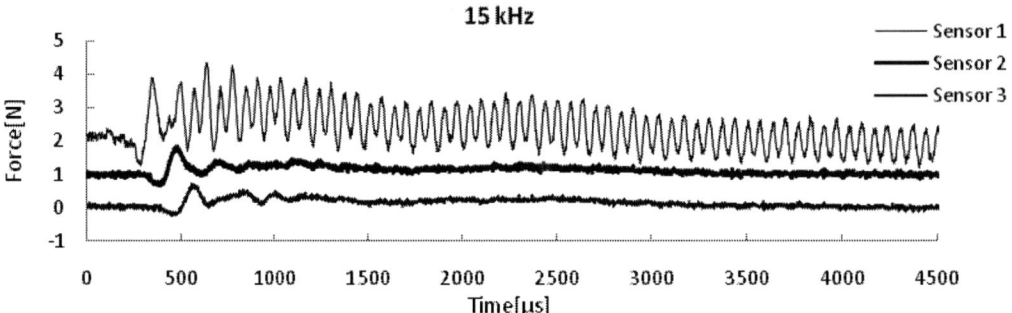

FIGURE 10. Experimental results. The propagation of quasiharmonic strongly nonlinear pulse with main initial frequency of input signal 15 kHz at the entrance of the system. Strongly nonlinear system, static force $F_0 = 0.8$N. The lower range of the forbidden frequency is 7.7 kHz being lower than the frequency of the major harmonic in the input signal.

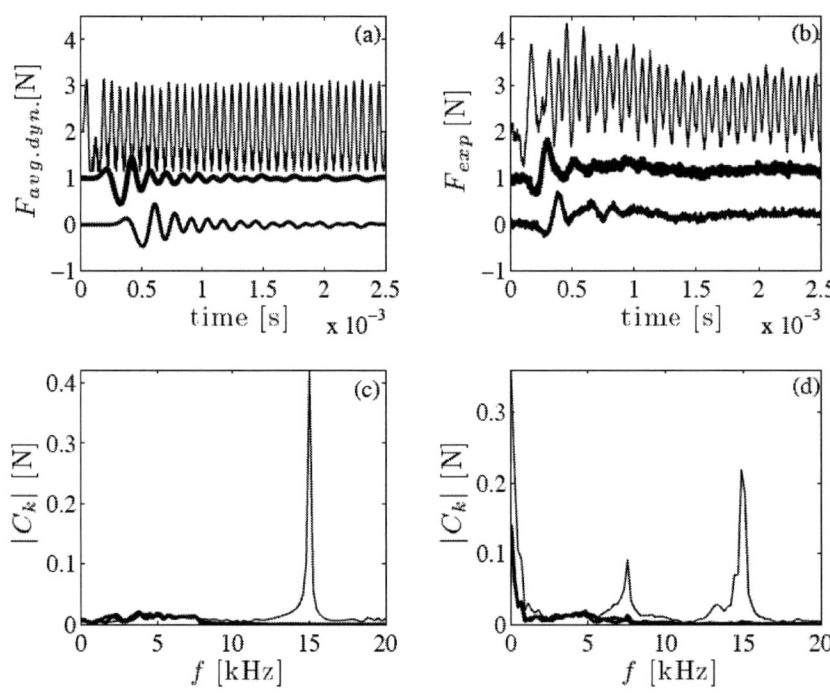

FIGURE 11. Comparison between the numerical calculation and the experimental results for propagating 15 kHz signal.
(a) Numerical calculations: dynamic force inside corresponding particles with embedded sensors 1, 2 and 3 and (c) their Fourier-transforms. Signals from sensors 2 and 3 are overlapping with negligible amplitudes corresponding to frequencies above band gap frequency 7.7 kHz.
(b) Experimental results: dynamic forces from embedded sensors 1, 2 and 3 and (d) their Fourier-transforms. Signals from sensors 2 and 3 are overlapping with negligible amplitudes corresponding to frequencies above band gap frequency 7.7 kHz.

We compared the attenuation of propagating quasiharmonic signals with the increase of frequency for interval of 3 – 18 kHz for strongly nonlinear system. In the Figure 12 the relative amplitudes of the main harmonics detected at the second gauge (A2) and third gauge (A3) with respect to the amplitude of the same harmonic at the gauge 1 (A1) are presented. The cut off of signal transmission at the frequency close to 8 kHz was observed with unexpected transparency at higher frequency 9 kHz followed by second drop in transparency above 10 kHz. The band gap calculated for elastically linear system was 7.7 kHz to 75.1 kHz. This means that strongly nonlinear system also demonstrated transparency inside band gap calculated for the linear system. Again system was practically non transparent at frequency about $2f_1$.

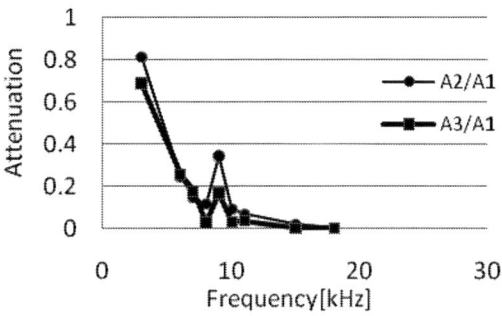

FIGURE 12. The comparison of the attenuation in experiments of the amplitude corresponding to major harmonics with the increase of frequency. A1, A2 and A3 are the amplitudes of the major harmonics in the incoming signal (A1) and in signals measured by sensors 2 (A2) and 3 (A3).

At higher frequencies of the rod vibrations we were able to generate oscillatory shock waves for strongly nonlinear system as it was a case for a nonlinear system (Fig. 13(a)).

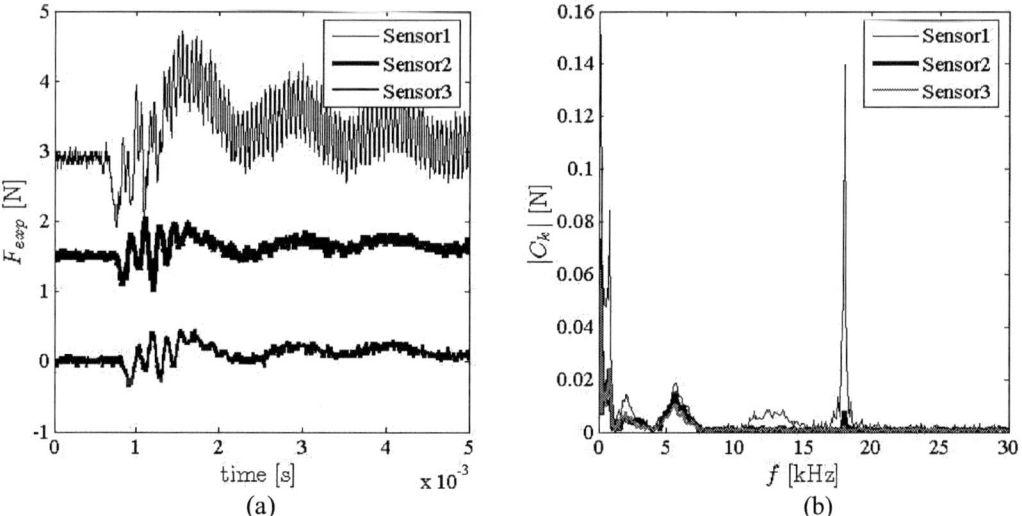

FIGURE 13. The propagation of oscillatory shock through a strongly nonlinear system with static force $F_0 = 0.8$ N. The lower range of the forbidden frequency is 7.7 kHz. (a) Experimental profiles of forces in three sensors, (b) The fast Fourier transforms of the propagating signals. Signals from sensor 2 and sensor 3 are overlapping for most of frequencies.

The propagation of strongly nonlinear oscillatory shock in the system, static force $F_0 = 0.8$ N. The lower range of the forbidden frequency is 7.7 kHz which is lower than the main input signal frequency. (a) The force dependence on time in gauges 1-3, (b) The fast Fourier transforms of the propagating signals. Signals from sensor 2 and sensor 3 are overlapping.

We can observe the transmission of low frequency harmonics and fast decay of high frequency harmonic. As a result the initial front was ramped, and high frequencies were wiped out while the low frequencies remained (Fig. 13(b)). This resulted in less oscillatory shock profile.

CONCLUSIONS

One-dimensional experimental set up was developed to investigate propagation of nonlinear and strongly nonlinear quasiharmonic waves. The quasiharmonic boundary conditions were created by vibrating rod with prescribed frequency contacting with a granular chain. They were monitored by sensor inside particle placed near the corresponding end of the chain. Propagating nonlinear and strongly nonlinear quasiharmonic waves and shock waves were observed in such set ups. A practically perfect transparency of investigated systems was observed at low frequencies and cut off effects at higher frequencies for nonlinear and strongly nonlinear waves. We observed transparency in cases of nonlinear and strongly nonlinear systems within band gap corresponding to linear elastic interaction law between particles. Numerical calculations of signal transformation by non dissipative granular chains demonstrated transparency of the system at low frequencies and cut off phenomenon at high frequencies in reasonable agreement with experiments. Transition from oscillatory to monotonous shock wave was observed for nonlinear system due to the effect related to periodic structure of the system and not due to dissipation as in [17]. Strongly nonlinear systems were less transparent than nonlinear systems at the similar amplitude of the exciting signal and at the same distance from the entrance.

ACKNOWLEDGMENTS

This work was supported by NSF (Grant No. DCMS03013220)

REFERENCES

1. V. F. Nesterenko, *Prikl. Mekh. Tekh. Fiz.* **24**, 136-48 (1983) [*J. Appl. Mech. Tech. Phys.* **24**, 733-43 (1984)].
2. A. N. Lazaridi and V.F. Nesterenko, *Prikl. Mekh. Tekh. Fiz.* **26**, 115-8 (1985) [*J. Appl. Mech. Tech. Phys.* **26**, 405-8 (1985)].
3. V. F. Nesterenko, *Dynamics of Heterogeneous Materials*, New York: Springer-Verlag New York, Inc. 2001, Ch. 1.
4. S. Sen, J. Hong, J. Bang, E. Avalos, and R. Doney, *Physics Reports*, **462**, 21-66 (2008).
5. E. B. Herbold, J. Kim, V. F. Nesterenko, S. Y. Wang, and C.Daraio, *Acta. Mech.* **205**, 85-103 (2009).
6. M. A. Porter, C. Daraio, I. Szelengowicz, E. B. Herbold, and P. G. Kevrekidis, *Physica D* **238**, 666-76 (2009).
7. J. Tasi, *Phys. Rev.* B, **14**, 2358-70 (1976).
8. St. Pnevmatikos, N. Flytzanis, and M. Remoissenet, *Phys. Rev.* B, **33**, 2308-21 (1986).
9. O.A. Chubykalo, A.S. Kovalev, and O.V. Usatenko, *Phys. Rev.* B, **47**, 3153-60 (1993).
10. S. Parmley, T. Zobrist, T. Clough, A. Perez-Miller, M. Makela, and R. Yu, *Appl. Phys. Lett.* **67**, 777-9 (1995).
11. A. Franchini, V. Bortolani and R.F. Wallis, *Phys. Rev.* B, **53**, 5420-9 (1996).
12. A.V. Gorbach and M. Johansson, *Phys. Rev.* E, **67**, Art. 066608 (2003).
13. P. Maniadis, A.V. Zolotaryuk, and G.P. Tsironis, *Phys. Rev.* E, **67**, Art. 046612 (2003).
14. A.C. Hladky-Hennion, G. Allan and M. de Billy, *J. Appl. Phys*, **98**, Art. 054909 (2005).
15. C. Kittel, *Introduction to Solid State Physics,* Atlantic Highlands: John Wiley & Sons, 2005, pp. 95-99.
16. H. Hertz, *J. Reine Angew. Math.* **92**, 156-71 (1881).
17. E.B. Herbold, V.F. Nesterenko, *Phys. Rev E*, **75**, Art. 021304 (2007).

ABSTRACTS

An Advanced Numerical Method to Describe Order Dynamics in Nematics

A. Amoddeo[a], G. Lombardo[b], R. Barberi[b]

[a] *Mechanics and Materials Department, University "Mediterranea" of Reggio Calabria, Via Graziella 1, Feo di Vito, 89122 Reggio Calabria, Italy*
[b] *CNR-INFM LiCryL – Liquid Crystal Lab., Physics Department, University of Calabria, Via P. Bucci, Cubo 32/c, 87036 Arcavacata di Rende (CS), Italy*

Abstract. Nematic liquid crystals are aggregates of calamitic molecules and most related experimental phenomena are well described by their mean molecular orientation, i.e. by the director and by the scalar order parameter[1], considering a perfect uniaxial symmetry. However, there exist situations in which experimental results cannot be fully described by this classic elastic approach. When the nematic distortion is very strong and it occurs over a length scale comparable with the nematic coherence length, the molecular order may be significantly altered, as in the case of the core of a defect [2]. Moreover the standard simplified elastic theory fails also for recent experimental results on phase transitions induced by nano-confinement [3] and for the electric field induced order reconstruction [4,5]. Such systems, where spatial and temporal changes of the nematic order are relevant and biaxial transient nematic configurations arise, require a full Landau-de Gennes **Q**-tensor description [4,6,7]. In this work, we will present the implementation of a **Q**-tensor numerical model, based on a one-dimensional finite element method with a r-type moving mesh technique capable to describe the dynamical electric biaxial transition between two uniaxial different topological states inside a π-cell. The use of the moving grid technique ensures no waste of computational effort in area of low spatial order variability: in fact, the technique concentrates the grid points in regions of large $\nabla \mathbf{Q}$ maintaining constant the total number of the nodes in the domain.

PACS: 64.70.mf, 61.30.Gd, 02.70.Dh

REFERENCES

1. C.W. Oseen, *Trans. Faraday Soc.* **29**, 883-9 (1933).
2. N. Schopohl and T.J. Sluckin, *Phys. Rev. Lett.* **59**, 2582-5 (1987).
3. G. Carbone, G. Lombardo, R. Barberi, I. Musevic, U. Tkalec, *Phys. Rev. Lett.* **103**, 167801-4 (2009).
4. R. Barberi, F. Ciuchi, G. Durand, M. Iovane, D. Sikharulidze, A. Sonnet and E. Virga, *Eur. Phys. J.* E **13**, 61-71 (2004).
5. R. Barberi, F. Ciuchi, G. Lombardo, R. Bartolino, and G. E. Durand, *Phys. Rev. Lett.* **93**, 137801-4 (2004).
6. P. de Gennes, *Phys. Lett.* **30A**, 454-5 (1969).
7. G. Lombardo, H. Ayeb, R. Barberi, *Phys. Rev.* E **77**, 051708-10 (2008).

Mixing equilibrium in two-density fluidized beds by DEM

A. Di Renzo and F. P. Di Maio

Department of Chemical and Materials Engineering
University of Calabria
Via P. Bucci Cubo 44A, 87036 Rende (CS), Italy
alberto.direnzo@unical.it

Abstract. Interaction of fluid and granular flows in dense two-phase systems is responsible for the significantly different behavior of units used in the chemical industry such as fluidized beds. The momentum exchange phenomena involved during gas fluidization of a binary mixture of solids differing in density is such that the continuous mixing action of the fluid flowing upwards counteracts the natural tendency of the two (fluidized) solids to segregate with the heavier component fully settling at the bottom of the bed. In the present work the complex hydrodynamics of two-density gas-fluidized beds is studied by means of a DEM-CFD computational approach, combining the discrete element method (DEM) and a solution of the locally averaged equations of motion (CFD). The model is first validated against experimental data and then used to investigate the role of gas velocity versus density ratio of the two components in determining the distribution of the components in the system. It is shown first that a unique equilibrium composition profile is reached independent of the initial arrangements of the solids. Then, numerical simulations are used to find the equilibrium conditions of mixing/segregation as a function of the gas velocity in excess of the minimum fluidization velocity of the heavier component and as a function of the density ratio of the two solid species. A mixing map on the gas velocity-density ratio plane is finally reconstructed by plotting iso-mixing lines that shows quantitatively how conditions ranging from full mixing to fully segregated components are obtained.

Keywords: Fluidized beds, mixing, segregation, computer simulation, discrete element method
PACS: 47.55.Lm, 45.70.Mg, 83.50.Xa

Plane Waves in Porous Media

Lucia Fiorino

Universita' "Mediterranea" di Reggio Calabria, Italy e-mail: lfiorino@iol.it

Abstract. A porous solid is treated as a continuous medium with ellipsoidal microstructure. The micro-deformation is assumed to consist of a pure straining in which each ellipsoidal cavity or void has a different stretch along its main axes but is assumed to rotate locally with the surrounding matrix. Since the voids in the material are by definition empty, the continuum equations of motion are obtained from the restriction that the microstate is completely determined by macrostrain; thus, the microstructure is only indirectly manifest. Therefore, at the continuum level, a momentum balance having the Cauchy form is sufficient, although some of the usual constitutive principles are no longer applicable. In particular, the stress tensor must depend on temporal and spatial gradients of higher-order order of the map from reference configuration to current configuration. In general, there is an influence of the mechanical energy transmitted through the boundary of a body that results from micro-stress. The resulting interpretation is compatible with Toupin's well-known proposal to treat hyperelastic materials of second grade as a Cosserat media with constrained spins. Finally, the propagation of elastic waves is studied, and solutions involving both transverse and longitudinal waves are found.

Keywords: <plane waves, porous media>
PACS: 46.40.-f

Granular jets and hydraulic jumps on an inclined plane

C. G. Johnson and J. M. N. T. Gray

School of Mathematics and Centre for Nonlinear Dynamics
University of Manchester, Oxford Road
Manchester, M13 9PL, UK

Abstract. We present recent work on a steady jet of granular material impinging on, and subsequently flowing down, an inclined plane [1]. The flow exhibits a wide range of behaviours, depending on the slope angle and fall velocity, ranging from kinetic gas-like flow to intermittent avalanche formation. A steady-state flow regime is described in detail, in which the jet generates a region of fast-moving radial flow on the plane, enclosed by a teardrop-shaped hydrodynamic shock. A depth-integrated hyperbolic flow model [2] is constructed, which predicts with good agreement many of the experimental results, including the location and shape of the shock. A second steady flow regime, in which the teardrop-shape of the shock is 'blunted', is also observed in both experiments and in numerical solutions, and an explanation for the existence of the two regimes given in terms of the flow Froude number. A number of the experimentally observed features cannot be predicted by current shallow-layer models of granular materials, and these are discussed with reference to the constitutive rheology of thin-layer granular flows, and to the assumptions inherent in depth-averaging.

Keywords: hydraulic jump, granular jet, granular avalanche, shallow-water model
PACS: 81.05.Rm, 83.80.Fg

REFERENCES

1. C. G. Johnson, and J. M. N. T. Gray, *Journal of Fluid Mechanics* (sub judice).
2. J. M. N. T. Gray, Y. C. Tai, and S. Noelle, *Journal of Fluid Mechanics* **491**, 161–181 (2003).

Coupled ODEs Model for the Dynamics of Dunes

Hiraku Nishimori[a], Atsunari Katsuki[b], Hiromi Sakamoto[c] and Hirofumi Niiya[a]

[a] *Department of Mathematical and Life Sciences, Hiroshima University, Hiroshima*, Japan
[b] *Department of Physics, College of Science and Technology, Nihon University, Chiba*, Japan
[c] *Department of Mathematics, Hiroshima University, Hiroshima*, Japan

Abstract. A coupled ODEs model is proposed to describe the dynamics of barchans and other types of dunes under a few number of assumptions considering the geometrical characters of dunes and the surface flow over them. Using the model, the transition between coalescence and ejection of two colliding barchans depending on their initial sizes, is numerically and analytically studied. Also, the shape stability of barchans and other types of dunes is discussed through the model. This simple model supplies us with a new tool for a clear-cut understanding of the seemingly complex dynamic of dunes.

Keywords: Dunes, Granular Dynamics, Ordinary Differential Equation.
PACS: 05.65.+b, 05.45.Yv, 05.45.Ac

References

1. Nishimori, H., Katsuki, A. and Sakamoto, H. "Coupled ODEs Model for the Collision Process of Barchan Dunes", Theoretical and Applied Mechanics Japan, Vol.52(2009), pp174-184
2. Endo, N., Taniguchi, K. and Katsuki, A., "Observation of the whole process of interaction between barchans by flume experiments", Geophys. Res. Lett., Vol.31,(2004), ppL12503.
3. Schwmmle, V. and Herrmann, H. J.,"Solitary wave behavior of sand dunes", Nature, Vol.426, (2003), pp619-620.
4. Katsuki, A., Nishimori, H., Endo, N. and Taniguchi, K.,"Collision dynamics of two barchan dunes imulated by a simple model" ,J. Phys. Soc. Jpn., Vol.74,(2005), pp538-541.

Saltating Particles in a Turbulent Boundary Layer

A. Valance[1], A. Ould El Moctar [2], P. Dupont[3], I. Cantat[1] and J.T. Jenkins[4]

[1] *Institut de Physique de Rennes, CNRS UMR 6251, Université de Rennes 1, 35042 Rennes, France*
[2] *Laboratoire Thermocinétique de Nantes, Polytech. Nantes, CNRS UMR 6607, 44306 Nantes, France*
[3] *LGCGM, INSA de Rennes, Campus Beaulieu, 35043 Rennes, France*
[4] *Department of Theoretical and Applied Mechanics, Cornell University, Ithaca, NY 14853, USA*

Abstract. Experiments on aeolian sand transport were carried out in a wind tunnel at the University of Aarhus in Denmark for a wide range of wind speeds. The saltating particles were analyzed using imaging techniques (PIV and PTV). Vertical profiles of particle concentration and velocity were extracted. The particle concentration was found to decrease exponentially with the height above the bed and the characteristic decay height was independent of the wind speed [1]. In contrast with the logarithmic profile of the wind speed, the particle velocity was found to vary linearly with the height. In addition, the particle slip velocity is finite and invariant with the wind speed. These results are shown to be closely related to the features of the splash function that characterizes the impact of the saltating particles onto a sand bed. A numerical simulation was developed that explicitly incorporates low velocity moments of the splash function in a calculation of the boundary conditions that apply at the bed [1]. The overall features of the experimental measurements are well reproduced by the simulation.

Keywords: aeolian transport, saltation, granular flow
PACS: 45.70.Mg, 47.27.-I, 47.55.Kf, 83.10.Gr, 83.80.Hj

REFERENCES

1. M. Creyssels, P. Dupont, A. Ould el Moctar, A. Valance, I. Cantat, J. T. Jenkins, J. M. Pasini and K. R. Rasmussen, J. Fluid Mech. 625, 47 (2009).

AUTHOR INDEX

A

Albers, B., 391
Amar, E. H. B., 151
Amoddeo, A., 437
Aste, T., 157
Azéma, E., 240

B

Barberi, R., 437
Bauer, E., 290
Berzi, D., 31
Blumenfeld, R., 167

C

Cantat, I., 442
Chong, S.-H., 19
Clamond, D., 151
Combe, G., 260
Crassous, J., 79

D

Daniels, K. E., 371
Delaney, G. W., 157
Delannay, R., 79
Delenne, J.-Y., 240
Di Maio, F. P., 438
Di Matteo, T., 157
Di Renzo, A., 438
Dupont, P., 442
Dybenko, O., 89

E

Ehlers, W., 306
El Moctar, A. O., 442
El Youssoufi, S., 240

F

Fiorino, L., 439
Fraysse, N., 151
Froiio, F., 183

G

Giffen, N., 371
Giovine, P., 314
Goddard, J. D., 323
Goldhirsch, I., 198
Gray, J. M. N. T., 343, 440

H

Harris, D., 3
Hayakawa, H., 19
Herbold, E. B., 425
Hill, K. M., 379
Horntrop, D. J., 89
Hunt, M. L., 416

I

Isobe, M., 135

J

Jenkins, J. T., 31, 41, 363, 442
Johnson, C. G., 440

K

Kadau, D., 50
Katsuki, A., 441
Kondic, L., 89
Kruyt, N. P., 405
Kudrolli, A. A., 221
Kumaran, V., 58

L

La Ragione, L., 41
Larcher, M., 363
Lombardo, G., 437
Louge, M., 79
Luding, S., 208

M

Métayer, J.-F., 79
Makse, H. A., 271
May, L. B. H., 371
Mehandia, V., 333
Mineo, C., 72
Mitarai, N., 214

N

Nakanishi, H., 214
Nesterenko, V. F., 425
Niiya, H., 441
Nishimori, H., 441
Nott, P. R., 333

O

Orpe, A. V., 221
Otsuki, M., 19

P

Pignatelli, R., 230

R

Radjai, F., 240
Rajchenbach, J., 151
Ratnaswamy, V., 89
Richard, P., 79

Richefeu, V., 240
Rosato, A. D., 89
Roux, J.-N., 183, 260
Rycroft, C. H., 221

S

Sakamoto, H., 441
Sano, O., 100
Shearer, M., 371
Song, C., 271
Steeb, H., 115
Sun, J., 280
Sundaresan, S., 280

T

Topin, V., 240
Torrisi, M., 72

V

Valance, A., 79, 442
Voivret, C., 240
Vriend, N. M., 416

W

Wakou, J., 135
Wang, P., 271
Wang, S.-Y., 425

Y

Yohannes, B., 379